PRECISION METAL TECHNOLOGY

PRECISION METAL TECHNOLOGY

Mario Rapisarda

Educational Consultant
for
Kelmar Associates

HBJ

Harcourt Brace Jovanovich, Publishers
Technology Publications

San Diego New York Chicago Austin
London Sydney Tokyo Toronto

Cover photo: Ford Motor Company

ISBN: 0-15-571075-3

Library of Congress Catalog Card Number: 90-86244

Printed in the United States of America

PREFACE

There are those who maintain that we live in a Computer Age; others argue that we live in a Video Age. In any case, it's easy to lose sight of the educational values that make engineering advances possible. There is much evidence to indicate that technology "doubles" every two to three years; many technical processes that were once current become obsolete in short order. The result is the development of a "push-button state of mind" wherein the biggest problem facing an industrialized nation is calluses on the forefinger. More than ever trained people are required in engineering if the erosion of our manufacturing and skill base is to stop. No one will be brought in off the street to push that "button."

This book is based on the many years of practical experience working with highly skilled people in the machine tool trade, as well as on industrial research, and on teaching at various educational levels, including adult job training programs. It was in such training programs that I faced my greatest teaching challenge—being understood by individuals speaking different languages and having little or no technical background.

The primary purpose of this book is to help teachers create this background and to help students achieve the hands-on experience necessary for basic machine tool operations. This book also provides a link to more advanced machine tool operations and manufacturing processes. Much of this linkage is achieved through the use of illustrations and by providing "dual dimensioning" throughout the book. The material is organized for both beginners and advanced students, making it suitable for vocational and secondary schools, technical and community colleges, apprenticeship training programs, proprietary career schools, and adult training programs.

ACKNOWLEDGMENTS

While it is impossible to list all those who helped with this project, I cannot fail to thank Al Del Vento, Joe Oriolo, and Fred Lauritano, as well as my sons, Paul and Philip, whose illustrations made my manuscript that much easier to prepare. I also want to thank Steve LaMarine and Steve Krar for providing me with the opportunity to learn something about print and visual media.

Deep appreciation must also be expressed to Ernest Ratterman of GE Superabrasives for reviewing the section of this book that deals with superabrasive technology. Thanks must also be given to Clyde Sluhan and his staff at Master Chemical Corporation for their strong technical support in sections dealing with cutting fluids and coolants. I am also grateful to Susan M. Moehring of Metcut Research Associates, Inc. and to the Latrobe Steel Company for producing special photomicrographs and to Fred Pressman, IBM Senior Engineer, for reviewing the section on metallurgy.

Words of gratitude must be expressed to Emerson Schmidt and William S. Gazdag, who reviewed the sections on metrology, quality control, and quality assurance, and to Tom Farrell of Wilson Instruments, who contributed time, expertise, and photographs for the section on hardness testing. Special thanks go to Jim Washburn and Tom Burmeister for their contributions to the material on numerically controlled machinery and to Gerhard H. Bay and Hans Metternich, who made many of the machine shop photographs possible.

I am also grateful for the help given by colleagues and former colleagues at Norwalk State Technical College, the Platt Vocational Technical School, the J.M. Wright Technical School, the Norwalk Board of Education, and the training classes conducted by the National Tooling and Machining Association. Finally, I would like to thank the following firms who supplied technical information, drawings, and illustrations, or who reviewed sections of the book for completeness, integrity, and accuracy:

Acu-rite Incorporated
Aluminum Company of America
American Iron & Steel Institute
American Society for Metals
American Society of Mechanical Engineers
Anocut, Inc.
Bay State Abrasives, Division of Abrasive Industries Incorporated
Beehive Heat Treating Co.
Bethlehem Steel Corporation
Boyar-Shultz Corporation
Bridgeport Machines, Incorporated
Brown & Sharpe Manufacturing Co.
Bausch & Lomb Inc.
Carpenter Technology Corporation
Cincinnati Milacron Inc.
Clausing Industrial Inc.
Cooper Group
Copper Development Association Inc.
Dillion, W.C. and Co. Inc.
DoAll Company
Electronic Industry Association
Everite Machine Products Co.
Federal Products Corp.
Ford Motor Company
GE Superabrasives
General Motors Corporation
Greenfield Tap and Die Co.
Hardinge Brothers, Inc.
Johnson Gas Co.
Kearney & Trecker Corporation
Kostel Enterprises Ltd.
Laser Applications, Inc.
Latrobe Steel Company
LeBlond Makino Machine Tool Co.
Lincoln Electric Co.
Lindberg, A Unit of General Signal
Litton Industrial Automation, Landis Grinding Machines
Lodge & Shipley
Master Chemical Corporation
Metal Powder Industries Federation
Metcut Research Associates, Inc.
Modern Machine Shop Magazine
MTI Corporation
Nash Engineering Co.
National Machine Tool Builders
National Tooling & Machining Association
Norton Company
Optical Gaging Products
Photocom Productions
Photo Chemical Machining Institute
Pitney Bowes
Producto Machine Co.
Regal-Beloit Corporation
Reynolds Metals Company
Sheffield Measurement Division
Shore Instrument & Mfg. Co. Inc.
L.S. Starrett Co.
South Bend Lathe, Inc.
Suburban Tool, Inc.
Superior Electric Co.
Taft-Pierce Manufacturing Co.
Technet Company, Lasermike Division
Teledyne Post
Tempil Big Three Industries, Inc.
Texaco, Inc.
Union/Butterfield
United States Steel Corporation
View Engineering
Wilson Instruments
Carl Zeiss, Inc.

CONTENTS

Section 1

METALS

UNIT 1 DEVELOPMENT OF THE METALWORKING INDUSTRY

Prehistoric Man

No one knows when tools were first used by humans, but it has been estimated to be about two and one-half million years ago. Crude tools, made of volcanic rock, have been found in remote areas of Africa, some as large as a fist and razor sharp. They were used as knives and choppers to cut skin and crush bones. How important were they? Well, they would mark the beginning of human culture, the start of group behavior. Before that, it was the individual against the environment.

Objectives
After completing this unit you will be familiar with

- the development of tools throughout history.
- the importance of inventions and devices for human progress.
- the advent of metal technology.
- careers in the metalworking industry.

The Stone Age

Down through the ages, tools were continually refined to make them more useful. The Stone Age gave birth to tools, weapons, and implements made of stone, bone, wood, and animal teeth. It is not uncommon to find remains of axes, hammers, hatchets, spears, and arrowheads (Figure 1–1) during excavations for buildings.

The Bronze Age

The Stone Age existed at different times in different parts of the world, but when prehistoric humans found how to smelt copper in crude holes dug in beds of clay (Figure 1–2), the Bronze Age was born. The use of copper and alloying (combining) elements, such as tin and other metals to make weapons, utensils, and ornaments harder and more durable, passed into history with the discovery of iron.

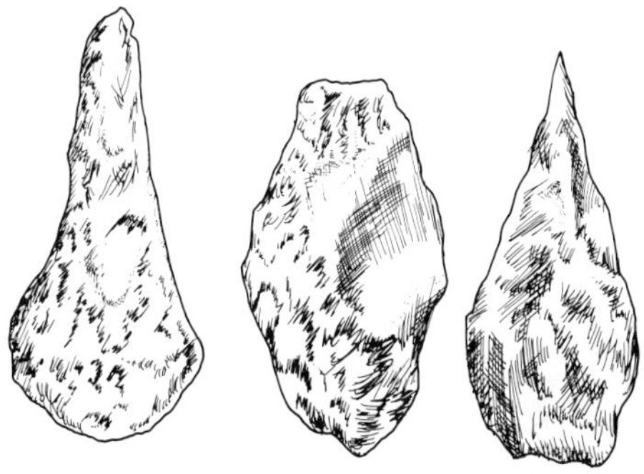

FIGURE 1–1
Stone Age tools were responsible for the start of group behavior. (Credit: Paul Rapisarda)

FIGURE 1–2
Bronze Age smelting: Casting first evolved by gouging grooves in refractory material such as clay. (Credit: Paul Rapisarda)

The Iron Age

The Iron Age began somewhere between 1500 and 1000 B.C. and while weapons and implements were cut of stone in the Stone Age, and cast during the Bronze Age, they were hammered, shaped, and curved during the Iron Age (Figure 1-3).

FIGURE 1-3
Iron Age forging. The Greeks and Romans hammered their weapons and armor. (Credit: Paul Rapisarda)

During this period humans did more with Earth's most abundant metal than their ancestors had with any material in any age before them. But it remained for James Watt to transform society with his invention of the steam engine.

The Industrial Age

Watt, an instrument maker for the University of Glasgow, was asked to make repairs on a steam engine model by Thomas Newcomen. His contributions were so many that the steam engine was patented in his name in 1769 (Figure 1-4).

The steam engine, with its ability to power mechanical equipment, created an industrial revolution. By 1793, Eli Whitney had built the cotton gin (Figure 1-5), a machine that eliminated the need to separate cotton fiber from the seeds by hand, making cotton growing profitable. A few years later, Henry Maudslay came out with the screw cutting lathe (Figure 1-6). This invention led to the development of other power-driven machine tools. Cyrus McCormick and his grain harvesting reaper (Figure 1-7) greatly increased farm production, and so it went. No longer would humans be limited by the strength and endurance of their own bodies or those of their draft animals.

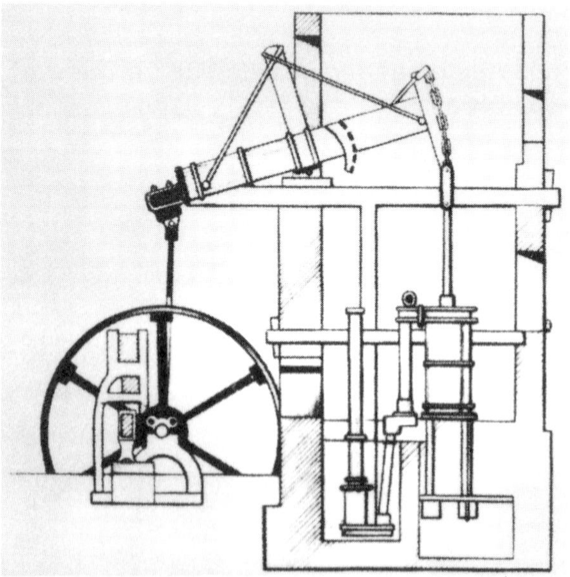

FIGURE 1-4
James Watt successfully developed the first steam engine. (Credit: Sheffield Measurement Division, Warner & Swasey/A Cross & Trecker Company)

FIGURE 1-5
The cotton gin was one of Eli Whitney's many engineering contributions. (Credit: Sheffield Measurement Division, Warner & Swasey/A Cross & Trecker Company)

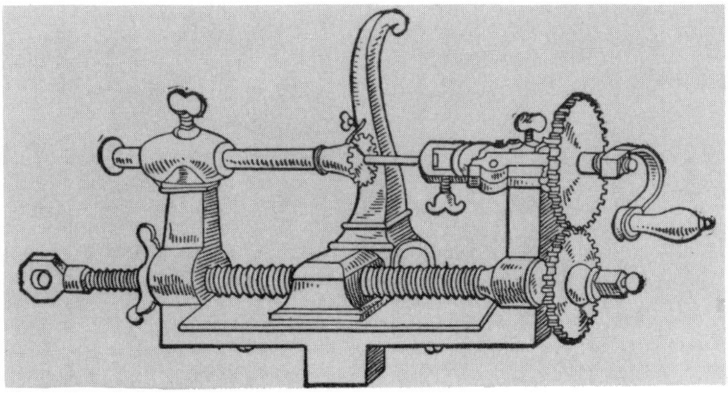

FIGURE 1-6
One of the earliest screw-cutting lathes was built in eighteenth-century France. (Courtesy of South Bend Lathe, Inc.)

FIGURE 1-8
The development of space-age alloys made supersonic flight possible. (Credit: Al Del Vento)

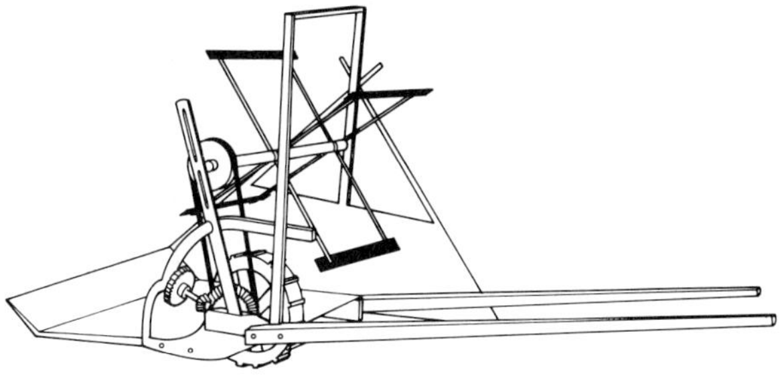

FIGURE 1-7
McCormick's reaper greatly increased farm production.

The Industrial Age exploded with inventions and devices that altered how society would produce goods and services, which was largely agricultural before that. Wind-blown clippers were replaced by steam-driven screw ships. The once grassy prairie was criscrossed with twin ribbons of steel for the Iron Horse; buildings climbed higher and higher to the heavens. A new era of development was at hand.

Space-Age Metals

Through the wide use of common metals, humans have been able to improve their life-style and the surrounding environment. New frontiers, such as outer space and the human body, offer many challenges for the metalworking industry. Some new superalloys have been developed to meet the challenges in these new frontiers; many more will be required in the future.

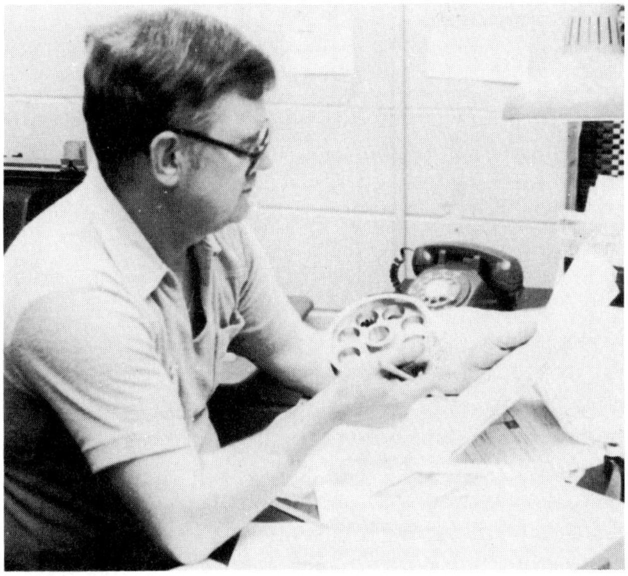

FIGURE 1-9
Due to its lightness and greater tensile strength, more designs are calling for titanium alloys.

To achieve rocket trips to the moon and faster-than-sound aircraft, (Figure 1-8) metals had to be developed that were light, corrosive resistant, performed with high strength, reliability, and withstood high temperatures. Such a metal is titanium. Since heat is one of the biggest factors that can soften or totally destroy the performance of metal, consider this: the cherry glow of a cigarette represents a heat of between 1100 and 1300°F (593 and 704°C). Titanium at 1350°F (732°C) has greater tensile strength (Figure 1-9) than stainless steel at room temperature (68°F or 20°C).

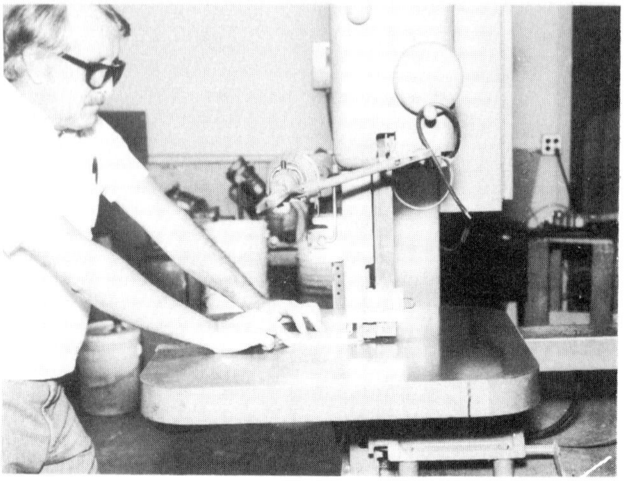

FIGURE 1-10
Machine operators represent limited, semiskilled workers.

FIGURE 1-11
A robot spot welder. (Credit: General Motors Corporation)

Titanium is not restricted to use in spacecraft and space exploration. It is used to make implants in the human body to replace bone sections in the shoulder, leg, and hips that have been damaged or destroyed.

Other metals such as beryllium have come into wide use. Beryllium is characterized by its hardness, high strength, and electrical conductivity. It is frequently alloyed with copper for the aircraft industries where requirements of corrosion resistance, nonmagnetic, and nonsparking properties are most important.

By combining titanium and tantalum (a high-melting element) with base metals of tungsten and vanadium, high-performance superalloys are created. New methods of reinforcement have been made with metal fibers bonded to the alloy base metal. A composite of copper and tungsten can withstand the high temperatures generally found in rocket engines. Another composite of silver and steel can operate in temperatures that range between 122 and 1202°F (50 and 650°C).

Careers

With men and women planning to go on interplanetary travel and colonization in outer space, the need for metals with greater capacities, strength, and heat resistance has never been greater. The metalworking industry requires many skilled workers as well as professional and technically oriented support personnel.

Semiskilled Workers

The metal trades have always had a need for semiskilled people, workers with limited training who hold jobs such as machine operators (Figure 1-10). Their function is to operate drill presses, turret lathes, milling machines, punch presses, and grinding machines that shape metals to exact dimensions. Semiskilled workers are also involved in light assembly, spot welding, and a variety of limited-responsibility jobs of a repetitive nature. Many of these functions may eventually be replaced by automated equipment such as an industrial robot (Figure 1-11).

Skilled Workers

The backbone of the metal trades, the high-precision skilled tradespeople, make up the elite force that turn engineers' dreams into reality. These include tool and die makers, mold makers, model makers, prototype machinists, experimental machinists, instrument makers, etc. The list goes on and on: but while the titles change the requirements and standards are very high, the ability to be creative and work metal and other materials to very close tolerances. Many of these jobs begin with a four-year apprenticeship that includes classroom programs in mathematics, blueprint reading, drafting, metallurgy, and shop theory. Journeymen's wages begin at the end of the apprenticeship program. However, becoming a journeyman is not the end of learning. The apprenticeship has merely scratched the surface of an exciting trade that poses a continual challenge to keep abreast of rapid technological advances.

Even in an age where the "machinist" is being built into the machine tool, there is a need for skilled people.

Technicians

With engineers coming out of colleges with greater training and higher degrees, many of the duties they once performed in industry have now fallen on the shoulders of the technician. Technicians (Figure 1-12) work between the engineer and the shop and help to turn the engineer's ideas into useful products. In production they may be called upon to troubleshoot, follow up on print specifications, and solve a bottleneck that's holding up the finished product. In research and development they may be called upon to assist the engineer by submitting drawings and sketches or perhaps conduct tests. While their training requires college-level technical schooling and their function is to support the engineering staff, in some circles the technician is considered the "nuts and bolts" engineer.

The Professions

Since there are so many different aspects in the metalworking world, it requires highly trained and educated people. The professions offer excellent opportunities for those who have the determination and enjoy the challenge that metalworking offers. People are required to develop new ideas, create new metals, and improve on existing manufacturing processes and develop new ones. The professions in the metalworking industry include teaching and many types of engineering.

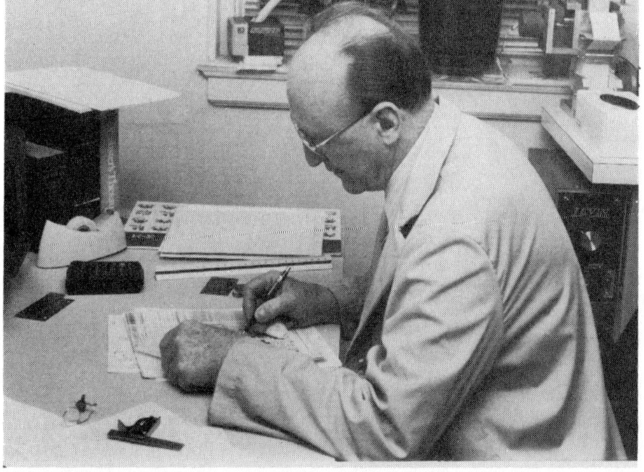

FIGURE 1-12
Technician—the "nuts and bolts" engineer.

Teaching is one of the professions in the metalworking world often overlooked by students. If a person likes to develop the potential of young people, it can be a most satisfying and rewarding profession. Teachers are required for industrial, vocational, and technical metal programs. Most metalworking teaching positions require a college degree and industrial experience can be a great asset.

Mechanical engineers (Figure 1-13) make up a large and ever-growing percentage of the professional ranks. Their prime function is to improve on existing products and develop new products. It has been estimated that the patents developed by engineers since the end of World War II would stack up higher than the Empire State Building. Their degree training in engineering schools include a heavy emphasis on mathematics, drafting, metals, and the principles of physics. It is their job to convert ideas on designs, tools, machines, and things mechanical to sound practices.

Metallurgical engineers (Figure 1-14) generally work in cooperation with chemists and physicists as part of a research team to extract metals from the ore, refine it, and develop new metals. They are responsible for the development of new metal struc-

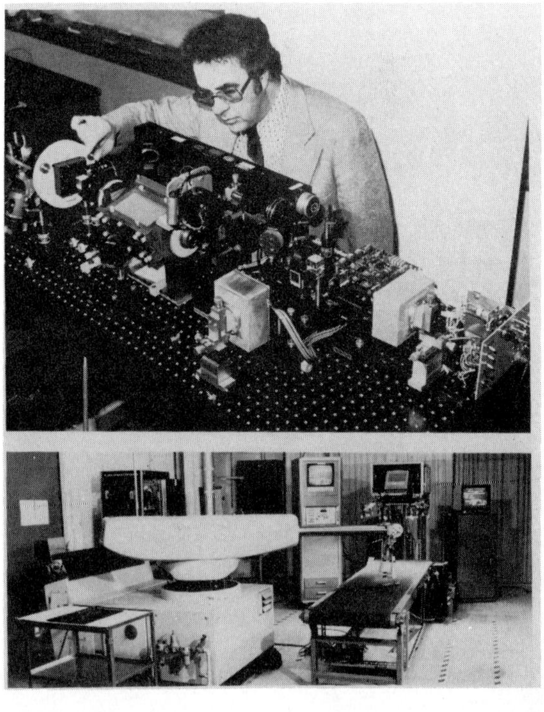

FIGURE 1-13
Mechanical engineers contribute ideas and develop things mechanical to sound practices. (Credit: General Motors Corporation)

FIGURE 1-14
Machining and sonic testing of tool steel slabs. (Credit: American Iron and Steel Institute)

FIGURE 1-16
An industrial engineer oversees a plant's operation. (Credit: Bethlehem Steel Corporation)

FIGURE 1-15
Tool engineers design tools as well as manufacturing processes. (Credit: General Motors Corporation)

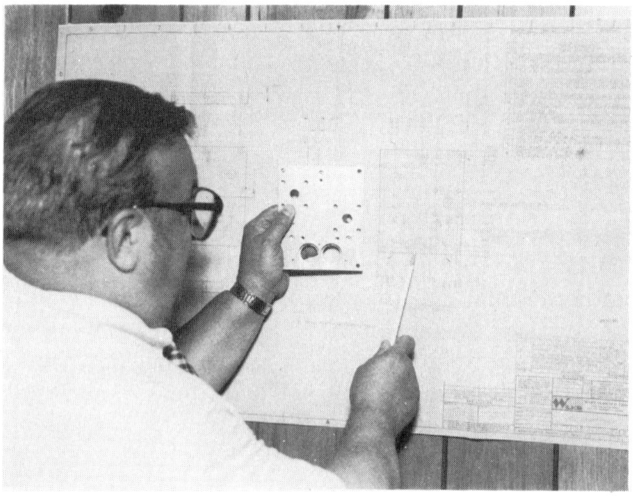

FIGURE 1-17
Quality control engineers have to work closely with manufacturing.

tures by combining existing elements to produce alloys that will improve the performance under various conditions.

The tool and manufacturing engineers (Figure 1-15) design tools and suggest manufacturing processes that help to efficiently produce the product designed by engineers. This could involve designing jigs and fixtures for holding the workpiece, developing cutting tools that shape metal into a product most efficiently, and developing manufacturing systems for mass production of the part.

Industrial engineers (Figure 1-16) are responsible for the overall operation of a plant. They must consider the best use of plant space, production methods, quality control, cost control, as well as the safety of the workers. In short, an industrial engineer oversees and controls a plant's operation.

Quality control engineers (Figure 1-17) are responsible for the quality of products that a plant produces. Sample parts or production items are tested to see that the part meets the necessary standard. The quality control engineer also has to work closely with manufacturing and production heads.

Civil engineers generally transform the landscape by the highways, bridges, power dams, buildings, airports, etc. they help design and construct. They are also involved in surveying because it is important that the things they construct are in the proper place.

Aerospace engineers (Figure 1-18) are involved in designing and constructing aircraft and space vehicles. Their job is as demanding as it is imaginative. Since they work with an environment that is not completely understood, they must develop and use metals that can withstand enormous heat, have high tensile strength, and have high corrosion resistance.

The Future

The technology of the metalworking world is complex and gets more so every day. The ability to improve our standard of living will depend greatly on how we can make our resources and elements work for us. With the technology no longer a male-oriented domain, more and more women are filling positions as technicians, skilled workers, engineers, and teachers. The metalworking trade requires both males and females who can acquire the necessary skills, have the ability to develop new ideas, and enjoy the challenges this industry offers. On every level in the technology, needs in personnel are being created all the time that must be filled by people with the technical know-how and the courage to see them through.

FIGURE 1-18
Aerospace engineers are involved in designing and constructing aircraft and space vehicles. (Credit: General Motors Corporation)

UNIT 2 MANUFACTURE OF IRON AND STEEL

It is hard to imagine a world without iron and steel. Imagine what it would be like to live without refrigerators, automobiles, trains, planes, bridges, or the thousands of things we encounter in our everyday lives. Isn't there some part of our lives that is not affected by the use of iron and steel? Hardly, since these two metals make up some 95% of all the metals used. The Iron Age brought out the use of iron and as society's needs increased, so did the use of iron and steel.

Objectives
After completing this unit you should be able to

1. identify ten properties found in metal.
2. understand the processes by which iron and steel are made.
3. understand the making of steel products.

Properties of Metals

Each metal has a use to which it is best suited. To select the best metal for a specific use, it is important to be familiar with the following terms.

Brittleness (Figure 2-1) in a metal is its tendency to break easily. It allows no permanent distortion before a metal will break. Many types of cast irons will break easily when subjected to impact or shock.

Compression strength (Figure 2-2) is the maximum amount of stress a metal can withstand without deforming or breaking.

Ductility (Figure 2-3) is the ability of a metal that allows it to be permanently deformed without breaking. Wire, automobile bodies, household appliances, and many other products are stamped, formed, or drawn from ductile metals.

Elasticity (Figure 2-4) is the ability of a metal to return to its original shape after it has been bent. Springs are a good example of a product made of elastic material.

Fatigue (Figure 2-5) is the tendency of metal to break under repeated stressing at below the ultimate tensile strength.

Fusibility is the property of metal to turn to liquid quickly and join with other metals. Any metal that welds easily usually has this property.

FIGURE 2–1
Brittle: breaking easily without being able to deform. (Credit: Paul Rapisarda)

FIGURE 2–2
Compression strength: the property to resist forces that tend to crush. (Credit: Paul Rapisarda)

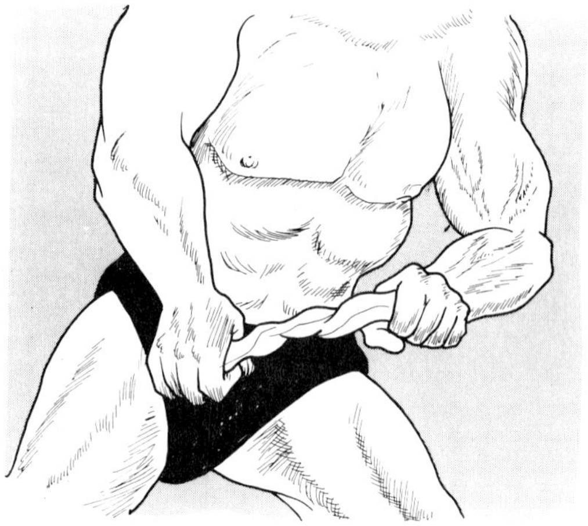

FIGURE 2–3
Ductility: the property that allows permanent change in form without breaking. (Credit: Paul Rapisarda)

FIGURE 2–4
Elasticity: the property that allows metal to return to its original shape after being bent. (Credit: Paul Rapisarda)

Hardness (Figure 2–6) is the resistance of metal to forcible penetration or deformation.

Machinability generally refers to the ability of metal to be machined. The rate and ease of metal removal and the cutting tool wear are the major factors that determine the machinability of a metal.

Tensile strength (Figure 2–7) is the maximum amount of stress or pull that a metal can withstand without breaking.

Toughness (Figure 2–8) is the ability of a metal to absorb or withstand shock or impact. Toughness involves the properties of ductility and strength.

While iron is Earth's most abundant element, it is rarely found in a metallic state except in meteorites. It comes from iron ore that is dug from the ground. The iron becomes steel when small amounts of carbon are added to it during the refining process.

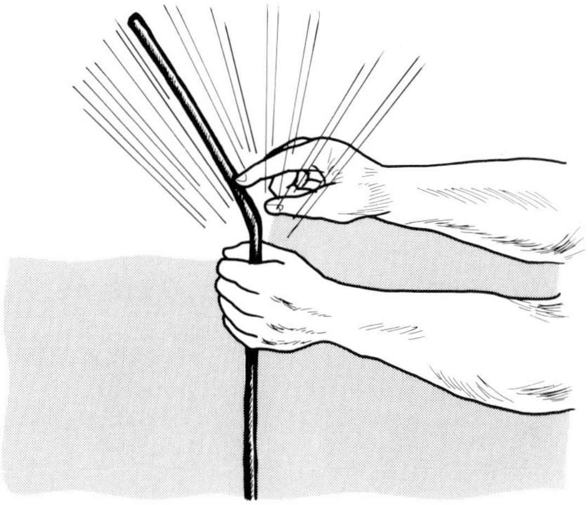

FIGURE 2-5
Fatigue: the collapse or weakening of metal that has been subjected to repeated stressing. (Credit: Paul Rapisarda)

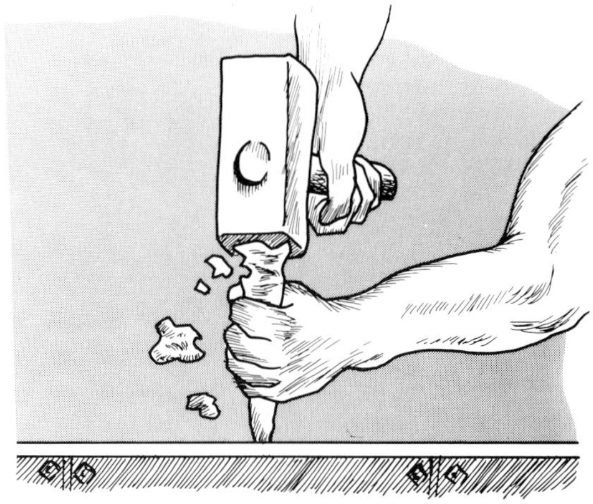

FIGURE 2-6
Hardness: the resistance of metal to be forcibly penetrated or deformed. (Credit: Paul Rapisarda)

FIGURE 2-7
Tensile strength: the amount of direct pull that metal can withstand without rupturing. (Credit: Paul Rapisarda)

FIGURE 2-8
Toughness: the ability of metal to absorb or withstand shock or impact. (Credit: Paul Rapisarda)

For that reason the different processes involved in steelmaking should be understood.

Blast Furnace

While the art of making steel was known for many centuries, it remained for a nineteenth-century Englishman, Henry Bessemer, to make it available in quantity. The Bessemer converter used a system of blown air forced through the bottom of a pear-shaped, heat-resistant lined container to make steel. Several more economical and cleaner methods of steelmaking have been developed since then and the Bessemer converter is no longer used.

Iron is separated from the ore by means of a *blast furnace* (Figure 2-9), which is lined with heat-resistant material. The raw materials of iron ore, coke, and limestone are dropped into the blast furnace chamber by means of a dump wagon called a skip car. A nearby stove heats up the air, which is forced through a bustle pipe that circles the widest

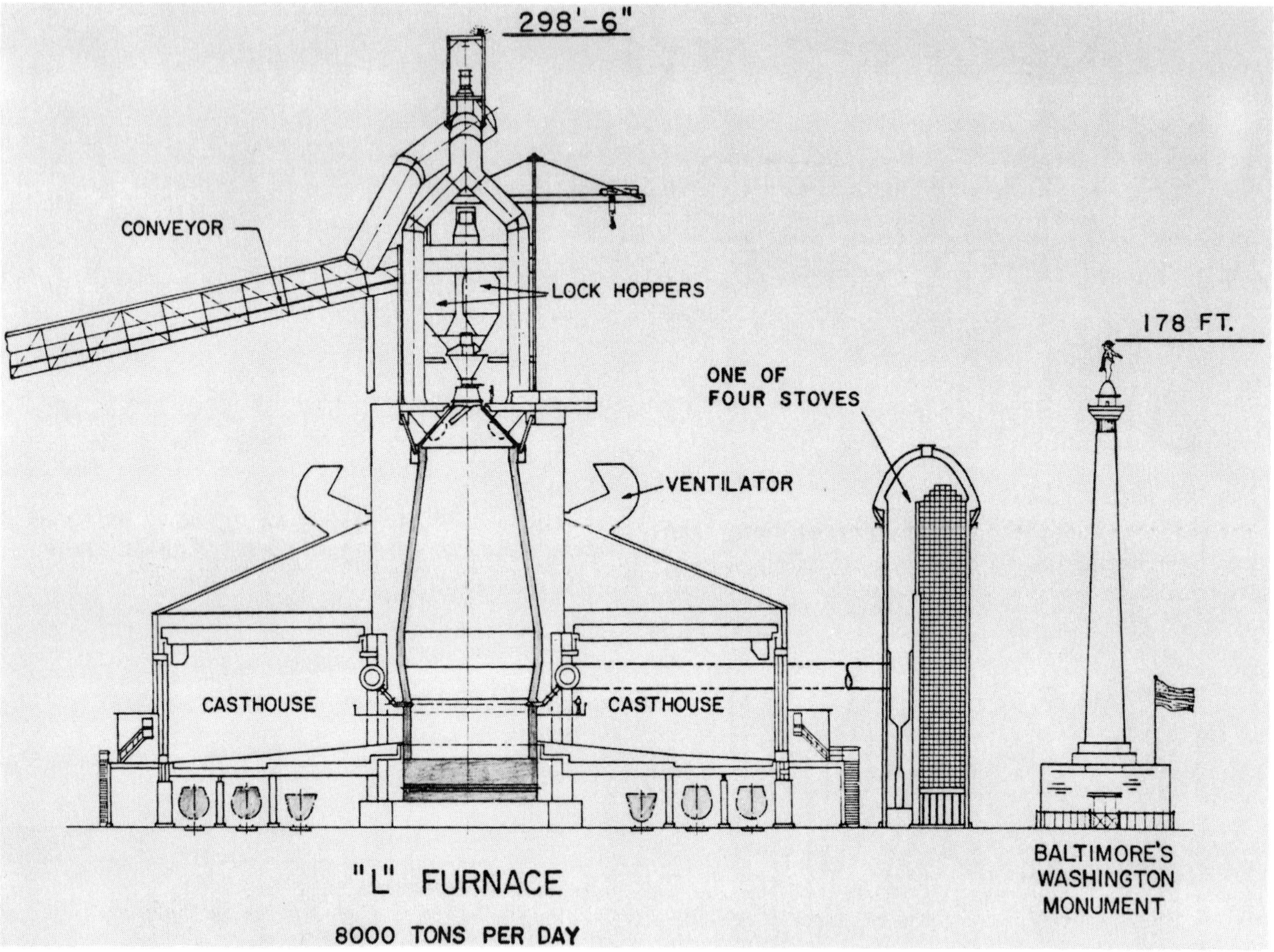

FIGURE 2-9
This L blast furnace is capable of producing 8000 tons per day.
(Credit: Bethlehem Steel Corporation)

part of the furnace. The hot air at 1000°F (538°C) feeds the coke, which burns with intense heat. It is this mounting heat (approximately 3500°F or 1925°C) that helps to separate the iron from the ore. The molten iron drops to the bottom of the furnace where it combines with the limestone. The limestone acts as a cleansing agent and combines with the impurities in the iron to form slag (Figure 2-10).

Because the slag is lighter, it floats to the top as a separate layer. After the blast furnace is tipped and the slag is drawn, the molten iron flows and empties into ladles or a molten iron car (Figure 2-11). The molten pig iron is tapped from the furnace through the lower iron notch, or tapping hole, and is emptied into a ladle car. The slag is reusable for other purposes such as road-building materials, while the molten iron is either processed into steel or cast-iron pigs. Pig iron is continually manufactured and the blast furnaces are only shut down for repair.

Manufacture of Steel

The molten pig iron contains many impurities that must be removed (burned out) before it can be converted into steel. The most common methods of converting pig iron into steel are in the open-hearth furnace, basic oxygen process furnace, and the electric furnace.

Open-Hearth Furnace

The open-hearth furnace resembles a rectangular, round-bottom swimming pool lined with heat-resistant materials (Figure 2-12). A charging machine loads the furnace with limestone, iron ore, and se-

FIGURE 2-10
After a tap hole has been opened, slag overflows into a slag ladle. (Credit: American Iron and Steel Institute)

FIGURE 2-11
Observer reading the temperature of molten iron of a blast furnace cast. (Credit: Bethlehem Steel Corporation)

lected scrap iron (in predetermined amounts) with the scrap iron being equal in weight to the molten iron. A load of limestone is dropped in first, followed by the scrap iron and ore. Burning fuel oil or natural gas combines with oxygen being fed into the furnace to create a fire of intense heat. To this mixture is added preheated air to increase the intensity of heat. The charge is heated for 2 or 3 hr before the molten iron is poured in.

When the oxygen burns out some of the impurities, samples of the steel are taken and analyzed in the laboratory to ensure that the steel contains the proper elements. Impurities are collected by the limestone and float in the boiling mass as slag. It takes from 6 to 10 hr before the furnace is tapped. A jet tapper is fired into the caked mass in the rear of the furnace, releasing a fiery stream of liquid metal. The molten steel is poured into a large ladle, while the slag is collected in a smaller container called a slag pot or thimble.

Many open-hearth furnaces have been updated by introducing oxygen to reduce the time required to make a heat of steel. The oxygen, fed into the furnace at high velocity by an oxygen lance, burns the impurities out of the steel quickly, thereby increasing production.

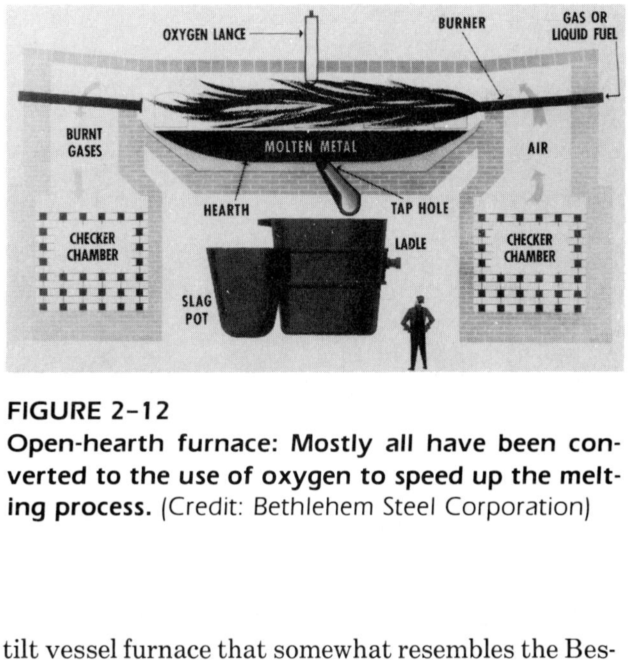

FIGURE 2-12
Open-hearth furnace: Mostly all have been converted to the use of oxygen to speed up the melting process. (Credit: Bethlehem Steel Corporation)

Basic Oxygen Furnace

One of the fastest methods of making steel is with the basic oxygen furnace, which takes about 45 to 50 min to make about 300 tons (272.1t) of steel. A tilt vessel furnace that somewhat resembles the Bessemer converter (Figure 2-13) is capable of making steel of any carbon content. Tilted on an angle, the basic oxygen furnace (Figure 2-14) receives a load of high-temperature molten iron from the blast furnace. The temperature changes with the addition of controlled amounts of scrap. Having had the charge placed within, the vessel is then rotated into an up-

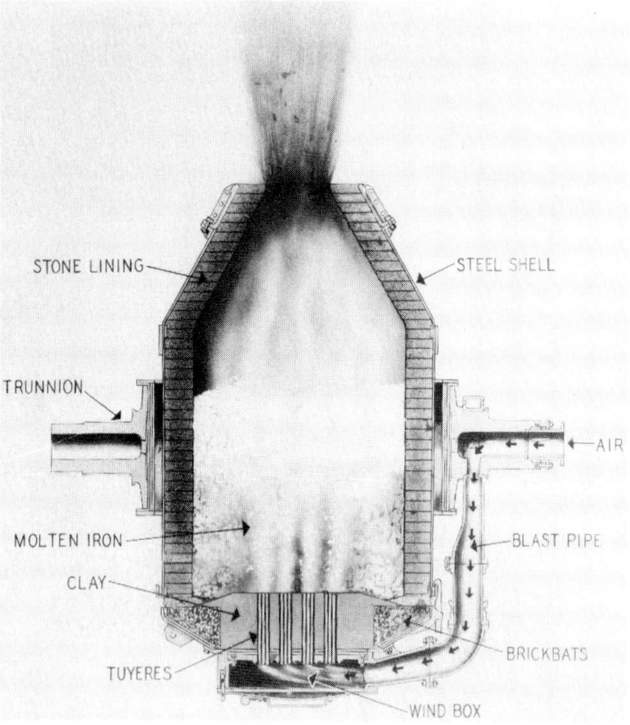

FIGURE 2-13
The Bessemer converter. (Credit: Bethlehem Steel Corporation)

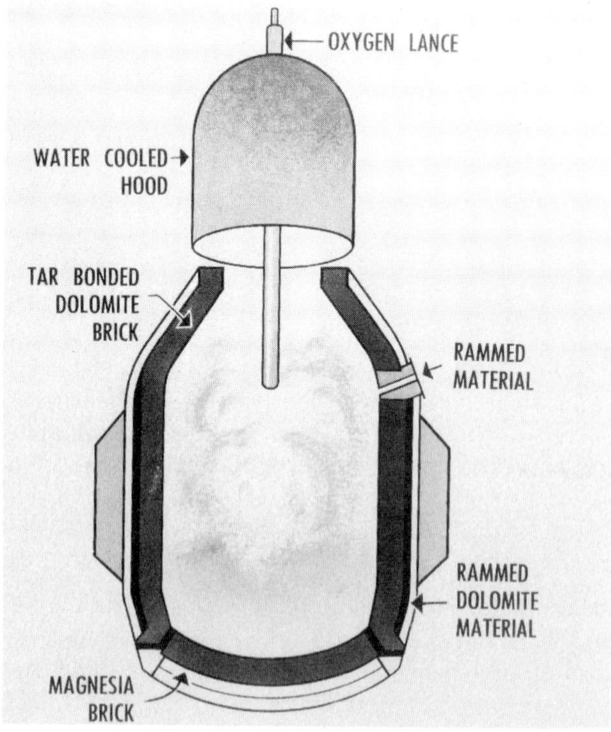

FIGURE 2-14
Basic oxygen furnace. (Credit: Bethlehem Steel Corporation)

right position and a long shaft called a lance is lowered above the bubbling mass, ejecting streams of pure oxygen that intensifies the heat.

As soon as the gas is ignited, creating a steady blow, purifying agents such as limestones and fluxes are added. The oxygen not only intensifies the heat but controls the carbon content of the steel.

During the "cooking process," a sample of the molten steel is tested in the laboratory to make sure that the steel contains the proper elements. Finally, the gas is shut off and the lance is withdrawn. The vessel is again tilted and liquid steel is poured into a ladle. The required alloys are now added to bring the steel to the proper chemical composition.

Electric Furnace

The electric furnace (Figure 2-15) uses selected scrap to produce high-quality alloys such as stainless and tool steel. The tilt vessel is charged with scrap, and large carbon or graphite electrodes are lowered through the top of the furnace to within inches of the metal. A current is sent surging through the electrodes; an arc jumps the gap, generating higher and higher temperatures. These temperatures are controlled by raising or lowering the electrodes.

The close control of the furnace's atmosphere allows alloys to be added without the fear of contam-

FIGURE 2-15
Molten steel is transferred by pouring or "teeming" from the ladle into ingot molds. (Credit: Bethlehem Steel Corporation)

ination, such as oxidation. Fluxes and other purifying agents are also added to help produce high-quality alloy steels. Depending on the size of the furnace, it takes from 4 to 12 hr to make a "heat" of steel.

Making Steel Products

Steel produced in any of the furnaces is generally poured into ingot molds or made into castings. The molten steel is allowed to cool in the ingot molds and then the molds are removed. The steel ingots are reheated to a uniform temperature and then rolled into blooms, slabs, or billets. These rolled products, generally called hot-rolled steel, can then be formed into various shapes of steel by either hot rolling or cold rolling.

Continuous Casting

Continuous casting (Figure 2–16) is a relatively new process that takes molten steel from any of the different types of furnaces and turns it into slabs or blooms. Molten steel is poured into a reservoir or tundish from which it flows into an oscillating mold. The oscillating mold is continually cooled, and as the molten steel flows through, it forms a solid skin on the outside. As the metal flows out of the mold, it is further cooled by a water spray that produces a thicker skin on the metal.

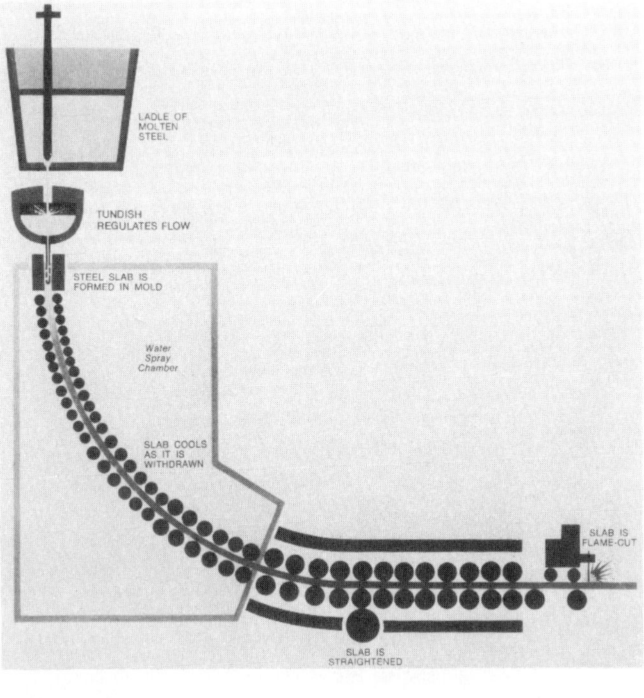

FIGURE 2–16
Continuous casting process. (Credit: Bethlehem Steel Corporation)

As the metal moves beyond the molds, it is formed to the size and shape required by the pinch rolls. After the slabs or blooms, produced by the continuous-casting process, pass through the straightening rolls, they are cut to length (Figure 2–17).

FIGURE 2–17
Many breakdown passes are required to convert a heated bloom into a finished steel beam. (Credit: American Iron and Steel Institute)

UNIT 3 FERROUS METALS

Ferrous metals contain iron. In the pure state (that is, metal in its virgin form), there are gold, platinum, and others that can be mined by washing or panning, as did the adventurers and miners during the gold rush in California, the Yukon, and Australia. Other elements extracted from the earth include chromium, nickel, cobalt, manganese, copper, aluminum, lead, tin, and zinc (all nonferrous metals).

Objectives
After completing this unit you will be familiar with

1. the different classification of carbon steel.
2. the different products made from these steels.
3. the different methods of identifying steel.

Carbon Steels

Ferrous metals fall into three general categories: low-carbon steels, medium-carbon steels, and high-carbon steels. Since the carbon content is a major factor in the grade of steel, it is important to understand the effects of carbon in steel. Carbon content in steel generally varies between a low of 0.10 and a high of 1.65%. In the metal trades, the carbon content of steels is generally referred to as points or hundredths of a percent. For example, a low-carbon steel of 0.10% is called 10-point steel, while a high-carbon steel with 1.65% carbon content is called a 165-point steel. At the low range of carbon, the steel is ductile and can easily be bent or formed to shape, while at the high range the steel can be hardened to resist wear and be sharpened to cut other metals.

Commercially Pure Iron

There is no such thing as pure iron, and even the best specimen of iron used in laboratories contains some impurities. What is known as commercially pure iron has a very small amount of impurities, about 2 points (0.02%) of its total weight in impurities and 3 points (0.03%) of carbon content. It is relatively soft but becomes harder when it is hammered, rolled, or formed by cold working. One of the characteristics of commercially pure iron is its ability to become easily magnetized or demagnetized. Therefore, it is used in magnetic chucks on surface grinders (Figure 3-1).

Low-Carbon Steel

Steels whose carbon content ranges from 15 to 30 points (0.15 to 0.30%) are often called mild steel, machinery steel, or machine steel. Because of its low-carbon content, it is ductile and can be worked, drawn, and shaped easily. Low-carbon steel is used to make wire, screws, washers, nuts, bolts, fences, and automobile bodies. This steel is available in many shapes and forms including sheets, rods, strips, plates, wire, and bars (Figure 3-2).

Many items produced in industry are made of low-carbon steel and must be case hardened to improve their wear-resistant qualities. Case hardening involves heating low-carbon steel to a certain temperature while the steel is in contact with some carburizing (impregnating) material that has a high-carbon content. The heated steel absorbs carbon into its outer surface in a manner similar to what porous chalk would do if it were brushed with a vegetable dye. When the chalk is broken in half, it clearly shows how deep the dye has penetrated. Upon quenching (rapid cooling) the steel, the depth that the carbon has penetrated becomes hardened while the center or core remains soft (Figure 3-3).

Medium-Carbon Steel

The carbon content of medium-carbon steel ranges from 30 to 60 points (0.30 to 0.60%). Because this steel has a higher carbon content than low-carbon

FIGURE 3-1
Magnetic chucks on surface grinders are made from commercially pure iron. (Credit: Philip Rapisarda)

FIGURE 3-2
Low-carbon steel is coiled in many gages for use in industry. (Credit: American Iron and Steel Institute)

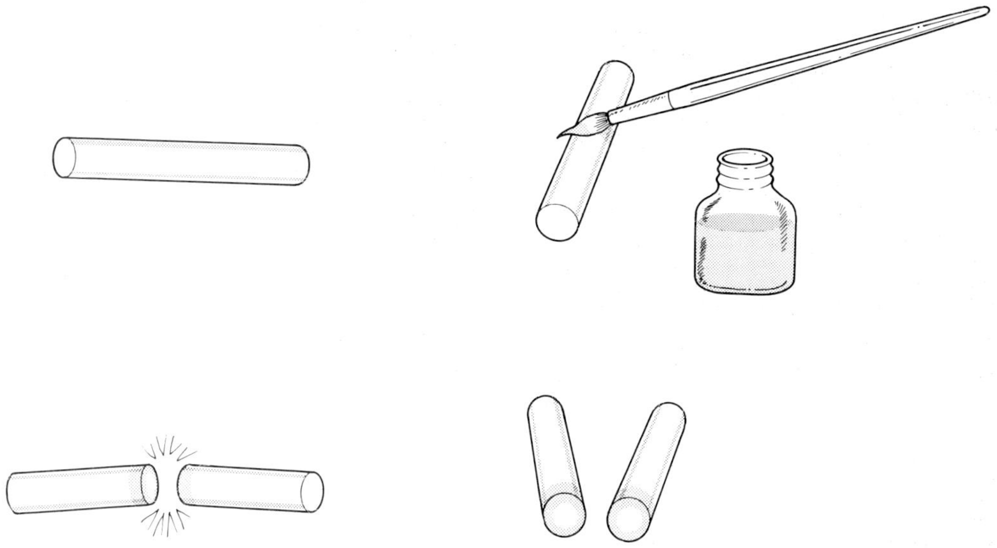

FIGURE 3–3
Illustrating the depth penetration that takes place during case hardening.

steel, it is tougher, does not machine as easily, and has a higher tensile strength. Medium-carbon steels are not as ductile and cannot be formed or flattened as easily as low-carbon steels; however, their carbon content makes them ideal for forgings, which later can be hardened.

Medium-carbon steels are used in parts that require medium strength such as truck axles, gears, connecting rods, shift levers, and many other parts used on automobiles. Medium-carbon steel is available in rods, bars, strips, and structural shapes (Figure 3–4).

FIGURE 3–4
Wide-flange structural shapes are common examples of medium-carbon steel use. (Credit: American Iron and Steel Institute)

High-Carbon Steel

High-carbon steels have a carbon content that ranges from 60 to as high as 165 points (0.60 to 1.65%) and are harder and tougher than steels with lower carbon contents. High-carbon steels can be hardened throughout and therefore are used generally for cutting tools.

There are two basic groups of high-carbon steels: the construction grade and the tool steel.

The *construction grade* (60- to 95-point carbon) is used where greater strength and wear resistance is important. It is used for railroad equipment, automobiles, all types of machinery, and springs where the qualities of hardness and toughness are required. Hammer heads, chisels, and axes are common tools made of construction-grade of high-carbon steel (Figure 3–5).

Tool steels have a carbon content that ranges between 60 and 165 points (0.60 and 1.65%). These steels are used extensively for cutting tools such as rotary lawn mower blades, razors, hacksaw blades, cold chisels, files, reamers, dies, taps, and tools that must cut metal (Figure 3–6).

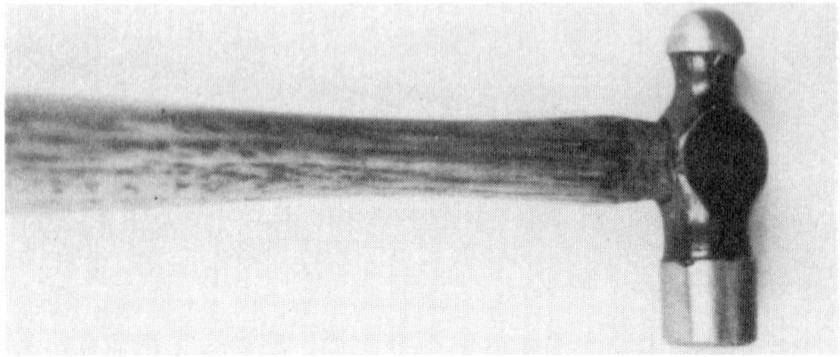

FIGURE 3-5
A hammer head is a good example of a tool made of construction-grade high-carbon steel.

Cast Iron

Cast iron, considered one of the crudest forms of iron, has a carbon content that ranges from 200 to 450 points (2.00 to 4.50%). Cast iron also contains varying amounts of manganese, phosphorous, and sulfur. The gray cast iron contains an excess of graphite, which not only gives it its black color but acts as a lubricant to aid in machining.

Widely used in making pipes and fittings, automobile engine blocks, machine tools, and frames, cast iron has a high compression strength and is used in making punch and draw presses that the tool and die trade uses for metal stampings and drawing shells.

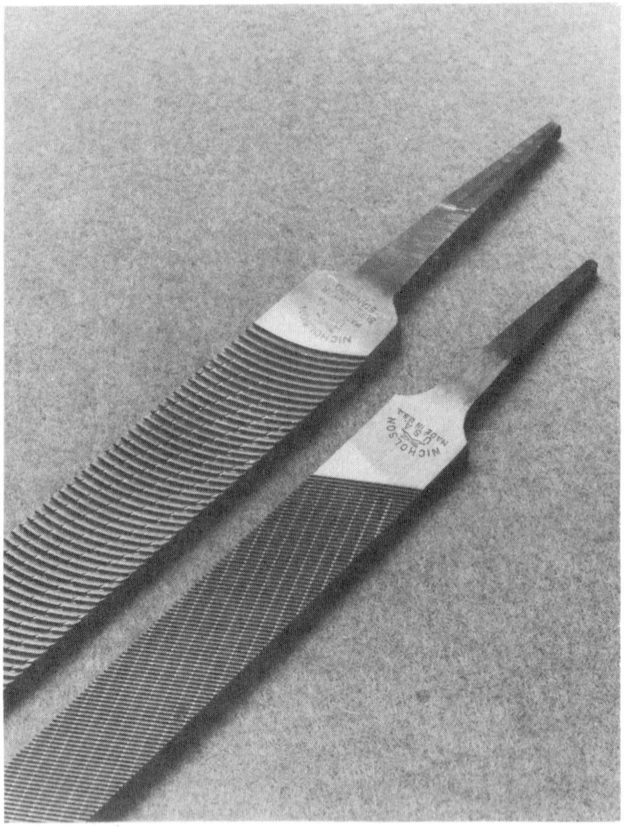

FIGURE 3-6
Files are made of a high-grade tool steel. (Credit: The Cooper Group)

White Cast Iron

White cast iron gets its name from its silvery-white texture. It is very hard and brittle and has little ductility. While it is ideal for use in situations that require a higher wear resistance, care must be taken to avoid designs that will include thin sections and sharp corners. White cast iron is mostly used to produce malleable iron castings.

Malleable Cast Iron

Unlike the cast irons previously mentioned, malleable cast iron is produced so it can be hammered and shaped. It has wide application in industry where resistance to shock is important without sacrificing strength, ductility, and machinability. The key to producing this type of cast iron is the heat treatment of the white cast iron, which is ultimately made malleable.

Annealing is a process by which hardened steel is returned to a workable natural state by heating it to a high temperature and then allowing it to cool off naturally. It is by annealing and graphitization (the conversion of carbon to a graphite condition) that the white cast iron is softened. Batches are placed in a furnace and slowly heated to temperatures that range between 1650 and 1700°F (898 and 926°C) for about 50 hr. It is then allowed to cool off just as slowly.

Different processes also exist to rearrange and soften the carbon forms such as cupola malleable iron, which is produced with an air furnace. Pearlite malleable iron is another process where the heating is interrupted during the transformation to the

FIGURE 3-7
Wrought iron was once widely used for its decorative as well as corrosion-resistant properties.

graphite stage or by reheating above the transformation range.

In addition to pipes and pipe fittings, automobile cams, crankshafts, axles, parts for machinery, and tools have been made from these malleable castings.

Wrought Iron

Wrought iron was once recognizable by the ornamental type of work one saw as fence (Figure 3-7) or as scrolls on doors or window shutters. Made from pig iron where most of the carbon and impurities have been removed, wrought iron uses an amount of slag to make it fibrous and tough but easy to be bent, twisted, hammered, and shaped. It also has other worthwhile characteristics such as resistance to fatigue and corrosion.

In another era, wrought-iron rivets were used extensively as metal fasteners for the Statue of Liberty and the Eiffel Tower (Figure 3-8). With so many technological advances since then, the use of wrought iron has been largely supplanted by steel. Wrought-iron pipes and fittings are still made for the water, oil, and gas industries.

FIGURE 3-8
The Eiffel Tower—a classic example of wrought-iron work.

Steel Identification

There are two basic systems used to denote the chemical composition of steel (Table 3-1). One system is used by the SAE (Society of Automobile Engineers) and one is used by the AISI (American Iron and Steel Institute). Both use a similar number system and both systems are widely used. These numbers identify the steels called out on drawings and prints as well as when ordering steel from distributors.

The first digit in either numbering system usually, but not always, indicates the basic type of steel, the second indicates the amount of alloy, and the last two digits indicate the carbon content. When the carbon content is over 1.00%, it is written with three digits rather than two.

TABLE 3-1
SAE-AISI classifications of steels
(X represents percent of carbon in hundredths)

Carbon steels	1XXX
Plain carbon (not resulfurized)	10XX
Free machining (resulfurized)	11XX
Standard carbon (rephosphurized and resulfurized)	12XX
Manganese steels	13XX
Nickel steels	2XXX
0.50% nickel	20XX
1.50% nickel	21XX
3.50% nickel	23XX
5.00% nickel	25XX
Nickel-chromium steels	30XX
1.25% nickel, 0.65% chromium	31XX
3.50% nickel, 1.55% chromium	33XX
3.00% nickel, 0.80% chromium	34XX
Corrosion and heat-resisting steels	30XXX
Molybdenum steels	4XXX
Chromium-molybdenum	41XX
Chromium-nickel-molybdenum	43XX
Nickel-molybdenum	46XX and 48XX
Chromium steels	5XXX
Low-chromium	51XX
Medium-chromium	51XXX
High-chromium	52XXX
Chromium-vanadium steels	6XXX
Tungsten steels	7XXX
Nickel-chromium-molybdenum steels (triple alloy steels)	8XXX
Silicon-manganese steels	9XXX

The class of steel each number indicates is as follows:

1 Carbon	6 Chromium-vanadium
2 Nickel	7 Tungsten
3 Nickel-chrome	8 Nickel-chromium-molybdenum
4 Molybdenum	
5 Chromium	9 Silicon-manganese

Thus, SAE 1022 steel would be broken down as follows 1, a carbon steel; 0, no alloying agent; and 22, a 0.22% carbon content. SAE 1022 would be a mild, low-carbon steel.

The chief difference between the SAE and AISI systems is that the AISI uses a prefix letter to indicate the steelmaking process. For instance, C1022 would indicate a mild, low-carbon open-hearth steel as identified by the AISI, but it would be identical to the SAE 1022 steel.

Color Coding

Another means of identification is by color coding (Figure 3-9). It should be understood, however, that the system is not universally used by steel producers and can be misleading if one does not understand the color code. It is not uncommon for some steel mills to paint the ends or stripe the length of the steel bars to denote the steel's composition. While the idea is good, the meaning of the color coding for one company may be entirely different for another. Green may signify a form of tool steel for

FIGURE 3-9
Color coding can speed up steel identification.

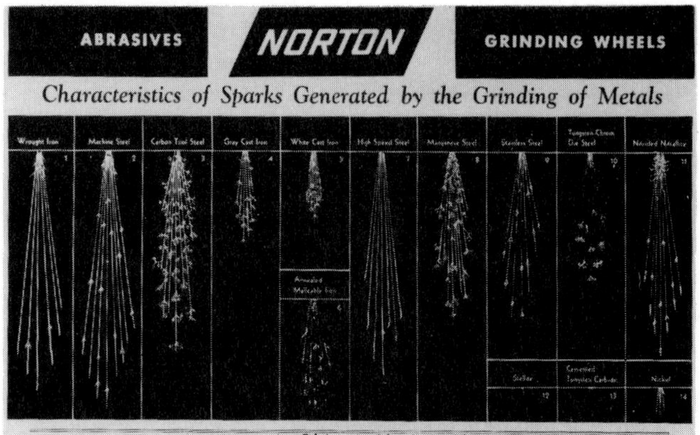

Metal	Volume of Stream	Relative Length of Stream, Inches†	Color of Stream Close to Wheel	Color of Streaks Near End of Stream	Quantity of Spurts	Nature of Spurts
1. Wrought iron	Large	65	Straw	White	Very few	Forked
2. Machine steel (AISI 1020)	Large	70	White	White	Few	Forked
3. Carbon tool steel	Moderately large	55	White	White	Very many	Fine, repeating
4. Gray cast iron	Small	25	Red	Straw	Many	Fine, repeating
5. White cast iron	Very small	20	Red	Straw	Few	Fine, repeating
6. Annealed mall. iron	Moderate	30	Red	Straw	Many	Fine, repeating
7. High speed steel (18-4-1)	Small	60	Red	Straw	Extremely few	Forked
8. Austenitic manganese steel	Moderately large	45	White	White	Many	Fine, repeating
9. Stainless steel (Type 410)	Moderate	50	Straw	White	Moderate	Forked
10. Tungsten-chromium die steel	Small	35	Red	Straw*	Many	Fine, repeating*
11. Nitrided Nitralloy	Large (curved)	55	White	White	Moderate	Forked
12. Stellite	Very small	10	Orange	Orange	None	
13. Cemented tungsten carbide	Extremely small	2	Light Orange	Light Orange	None	
14. Nickel	Very small**	10	Orange	Orange	None	
15. Copper, brass, aluminum	None				None	

†Figures obtained with 12" wheel on bench stand and are relative only. Actual length in each instance will vary with grinding wheel, pressure, etc. *Blue-white spurts. **Some wavy streaks.

FIGURE 3-10
Spark test chart. (Credit: Norton Company)

one mill; it may also imply a low-carbon steel for another.

Spark Test

There is still another way of identifying steel and that is by holding it against a running abrasive wheel to examine the color and formation of the sparks that ensue. As noted in Figure 3-10, each steel gives off a certain color in addition to a particular shape or cluster of sparks. One will notice that the color of the sparks may vary. There may be a difference in color when the stream of sparks is close to the wheel as opposed to the streak near the end of the stream. For example, high-speed steel gives off a forked red stream near the wheel, which turns to straw as it streaks near the end of its stream. On the other hand, machine steel (AISI 1020) gives off a forked white spark from beginning to end. It takes many years of experience to be able to determine by eye the composition of steel using the spark test.

UNIT 4 NONFERROUS METALS

Nonferrous metals contain little or no iron. The more commonly known nonferrous metals include gold, silver, platinum, copper, aluminum, magnesium, lead, tin, and zinc. All these metals exist in a pure state, but they are usually mixed with other metals to form an alloy that has greater properties than the original metal. The precious metal gold is a good example. Kingdoms have been toppled for it; men have become enormously wealthy because of it. It is so ductile that one grain can be drawn into hundreds of feet of gold wire. The gold used in jewelry would be too easily deformed if used as pure gold. It is alloyed with copper (one part copper and three parts gold) to give it some hardness. While over 50% is gold, white gold is an alloy composed of gold, nickel, and silver.

Objectives
After completing this unit, you will be able to

1. identify the more commonly used nonferrous metals.
2. identify their major properties.
3. recognize many of the products made from nonferrous metals.

Copper

Precious metals are used in jewelry, dentistry, and other special cases. Copper remains one of the oldest and most widely used nonferrous metals. It rates behind gold, silver, and platinum in ductility and is next to silver in conductivity of heat and electricity. Historically its use predates the Bronze Age by several centuries. There is evidence to indicate that copper was first discovered around 8000 B.C. in the valleys of the Tigris and Euphrates rivers in what is now Iraq. It is estimated that the Egyptians were hammering copper to make tools and ornaments by

FIGURE 4-1
Copper-nickel alloy hulls make these boats barnacle free for years to come. (Credit: Copper Development Association, Inc.)

FIGURE 4-2
A few of the many products made available through the use of copper and copper alloys (Credit: Copper Development Association, Inc.)

5000 B.C. Copper nails, bands, and battering rams were all features of their shipbuilding and war galleys. Different ages and different people used copper over the centuries. The Renaissance saw great use of the reddish-orange metal.

By 1664, John Winthrop had begun a copper industry in America, establishing a foundry that produced copper and brass kettles at Lynn, Massachusetts. One of the oldest copper rolling mills was developed by Paul Revere after the Revolutionary War. It was used to provide copper sheathing for bottoms of sailing ships to prevent the buildup of barnacles, a serious navigation problem that slows a ship's travel (Figure 4-1).

The properties of copper make it ideal for use in today's technology and life-style. Its high heat conductivity makes it ideal for use in kitchen utensils, electric motor windings, fittings, as well as automobile radiators (Figure 4-2). Copper is so ductile it can be drawn into wire thinner than human hair. It is so malleable it can be shaped, spun, and hammered without cracking; it has superior machinability, can be forged, becomes harder as it becomes cold worked, and can be softened just as easily to be reworked. It is nonmagnetic and in a world that is recycling conscious has a high scrap recovery value.

Copper's resistance to corrosion is well known. Since it does not rust, it is used for roofing materials (Figure 4-3) and fabrication of parts that are exposed to the atmosphere. A protective green film, called patina, forms on copper after long exposure to the atmosphere and further protects it from rusting. Copper can also be highly polished, easily joined, plated, or finished, making it attractive for art metal works in addition to satisfying many needs in industry. Largely dug in open-pit mines, (Figure 4-4), where a ton (1016 kg) of processed ore can yield slightly better than 12 lb (5443 g) of marketable metal, copper is available in cakes, billets, ingots, and extrusions (squeezed forms) in a wide variety of sizes and shapes.

Aluminum

Aluminum was first isolated in 1825 and was produced as a powder. It was used in this state until 1854 when a Frenchman, Henri DeVille, managed to produce some metallic aluminum. The first object to be made with the "new" substance was a rattle for the French prince. Napoleon III was so impressed with its lightness that he asked DeVille to manufacture helmets and breastplates for his army. He thought so much of the metal that honored guests were served with aluminum forks and spoons; lesser guests had to be content with silver and gold. It wasn't until many years later that Charles Hall, an American, and Paul Heroult, a Frenchman, unknown to each other and working in different parts

FIGURE 4–3
All-copper solar collectors are hoisted to the roof of a solar-heated office building. (Credit: Copper Development Association, Inc.)

FIGURE 4–4
Open-pit copper mine. (Credit: Copper Development Association, Inc.)

FIGURE 4–5
Most of the deposits are relatively shallow and close to the surface, making bauxite fairly easy to mine. (Credit: Aluminum Company of America)

of the world, discovered an inexpensive way of producing aluminum.

Light and strong, aluminum is the most abundant metal found in the earth's crust, 50 miles deep. It is not found as a silvery-white metal, but as an ore called bauxite (Figure 4–5). It takes vast amounts of electric power to operate an aluminum smelting plant. It is here that the aluminum is separated from the ore to produce the metal (Figure 4–6). Since hydroelectric power (electricity by water power) is the least expensive source, aluminum plants are usually located near powerful rivers that supply the electrical power required. More efficient methods promoting energy conservation have been introduced by the Aluminum Association to produce aluminum more economically while still conserving natural resources. These would include computerized controls, heat recuperation, use of alternate fuels, combustion and flame management, and consumer recycling. For example, recycling aluminum requires only 5% of the energy needed to produce the molten metal from bauxite (Figure 4–7).

Aluminum is nonmagnetic, nonsparking, nonpoisoning, and reflects light and heat, up to 90% of the sun's rays. It also makes an excellent conductor of electricity (Figure 4–8), greater per pound than any other metal. One of the characteristics of aluminum is that it loses part of its strength at higher temperatures. At low temperatures, however, aluminum increases in strength without loss of ductility. It is this characteristic that makes it a useful metal for low-temperature environments. While only one-third the weight of steel, the resistance to buckling is greater per pound than that of the highest strength steel.

FIGURE 4-6
Hydroelectric plant used for separating the aluminum from bauxite ore. (Credit: Aluminum Company of America)

FIGURE 4-7
Recycling saves up to 95% of the energy needed to make aluminum from bauxite. (Credit: Reynolds Metals Company)

When aluminum is exposed to the atmosphere, its surface is protected by a thin layer of oxide, forming like an invisible skin that protects the metal from further oxidation. It is this protective skin that makes it corrosive resistant to many acids that normally attack other metals. Aluminum will never rust red. It will remain self-protecting unless this invisible barrier is destroyed. Alkalies are one of the few substances that will penetrate its oxide layer and corrode aluminum.

Magnesium

Imagine a metal that is one-third lighter than aluminum, is normally more brittle than silver, can burn with a dazzling light, and is the third most abundant element found in the earth's crust. The element is magnesium, another nonferrous metal that is too soft to be used in its pure state and must be alloyed for extensive use as a structural material.

Magnesium can be found in wide usages that range from leisure-time products to sports to electronic and aerospace components. Magnesium, with an almost unlimited supply in the sea, comes in an alloy form in various shapes such as sheets, bars, and ingots for casting. Widely used in automotive tire mounting rims, this silvery-white element is also used in tools, ladders, television equipment, cameras, bodies of racing cars, and fireworks. Magnesium is also used in underwater flares used by divers, drawing on the oxygen found in water to burn fiercely.

FIGURE 4-8
Aluminum wire is stranded into electrical cable for use in overhead transmission power lines. (Credit: Reynolds Metals Company)

Lead

Lead, which is noted for its dull gray color, is shiny white when freshly cut. A lead ore called galena is mined in many parts of the world where it is smelted and purified through refinement that eliminates elements such as arsenic, copper, and zinc. In its pure state lead is very heavy and malleable but has little tensile strength.

It is widely used as protective shields for people involved in radiation work such as nuclear energy workers or technicians or patients involved in medical X-rays. Protective coatings of lead can be found surrounding electronic cables, as well as lead pipes used in carrying off industrial waters. While ordinary water does not attack lead, rainwater tends to corrode it.

Physical contact with lead may not be dangerous, but swallowing or inhaling compounds containing lead, such as car exhaust, can result in poisoning. Lead is eliminated very slowly from the body, allowing high levels to build up that can result in illness or death. Because of the many accidents that have occurred with lead poisoning, lead is no longer used to manufacture products such as paints, pewter, etc. While it is widely used in the manufacture of storage batteries, lead is used in combination with other alloys to improve the machinability of the metal.

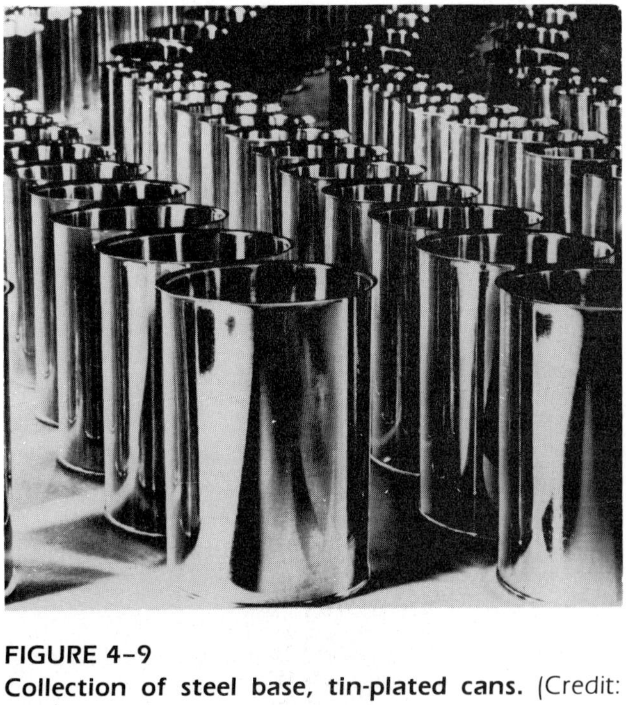

FIGURE 4-9
Collection of steel base, tin-plated cans. (Credit: American Iron and Steel Institute)

Tin

The art of using tin is an ancient one. Mined by the Phoenicians and Romans, tin is processed from ore called cassiternite. The silvery-white metal is soft, nontoxic, doesn't rust, and is too malleable to be used in its natural state. As early as A.D. 1240, it was being hammered as coating on iron.

Today the use of tin is widely recognized in plating products that act as food preservers such as tin cans (Figure 4-9). The cans themselves are made of some 98% steel base, with only a very small amount of tin plating used. Tin increases the hardness of other metals and whitens them. Although industry has modernized its methods of tin plating, there is still some amount of hand dipping. Nonetheless, tin plating can be seen on low-carbon steel sheets that make a variety of products such as cans, roofing materials, metal walls for sheds, and many others. Not only have many items such as wrappers and foils for food, drugs, and beverages been taken over by other materials such as aluminum and plastic, but chrome plating is now replacing some of the heavy demand for tin whose supply is hard to come by.

Zinc

Zinc, a brittle, bluish-white metal, was first discovered by a German scientist in 1746. It is known for the rust protection it gives iron and steel in a plating process called galvanizing. In the galvanizing process cleaned iron and steel are dipped into a molten bath of zinc, which adheres to the surface in a thin layer and crystallizes with a spotted bluish-white coloring (Figure 4-10). Garbage cans, storage tanks, roofing gutters, and downspouts as well as the undersides of automobiles are made of corrosion-resistant galvanized steel.

While zinc is used as an alloying agent, it has other characteristics. It is brittle at room temperature and becomes ductile and malleable at temperatures that range between 212 and 300°F (100 and 149°C). In this range it can be rolled into sheets or drawn into thin wire. It then becomes brittle again at higher temperatures.

UNIT 5 ALLOYS

An alloy is a combination of two or more metals that are melted to form a new metal. The new metal (an alloy) will be different than the original metals and

FIGURE 4-10
Galvanized stains on sheet metal. (Credit: American Iron and Steel Institute)

will contain qualities from each metal. For example, the new metal may become harder, softer, malleable, wear resistant, heat resistant, corrosion resistant, or even change color.

Objectives
After completing this unit, you will

1. understand the difference between ferrous and nonferrous alloys.
2. be familiar with a number of ferrous alloys used in industry.
3. be familiar with a number of nonferrous alloys used in industry.

Ferrous Alloys

There is more to making steel than adding carbon and manganese to the iron. It was found that by adding other elements, steel could be changed to give it properties such as strength, toughness, wear resistance, and corrosion resistance. These alloyed steels are generally named after the major alloying element added to the iron. For example, in nickel steel, nickel is the major alloying element.

Manganese
Manganese is a brittle, hard, gray-white metal that helps to purify steel and give it strength and toughness. It is an essential element used in steelmaking and about 12 lb (5443 g) of it are used to every ton (1016 kg) of steel produced. Manganese in steel gives it strength, wear resistance, and abrasion resistance. Steel with these qualities is very hard and difficult to machine. For this reason manganese steel is used on such things as railway ties, mining equipment, and rock crushers (Figure 5-1).

Chromium
Chromium is used as an alloying element in making stainless steel. Chrome is also added to steels to increase their ability to be hardened to a greater depth, resist shock and impact, and adds toughness while increasing resistance to corrosion and wear. Chrome steels are used in making safes, ball bearings, as well as armor plates for ships and tanks. Chromium is also used as a coating metal to prevent metals from rusting and corroding, and to give metals a bright, shiny silvery gloss. It is also used in candlesticks, automobile parts, table utensils, instruments, etc.

Nickel
Nickel is another element that can be used as a coating metal, known as nickel plating. It is also added to steels to improve impact resistance, increase corrosion resistance, as well as to improve its ductility. Nickel steels are used for a wide variety of products ranging from armor plate to machine parts to aircraft engine parts, as well as heavy-duty structural beams and forgings.

FIGURE 5-1
Railway rail ties are good examples of steel having manganese as an alloying element. (Credit: American Iron and Steel Institute)

Tungsten

Tungsten is a heavy, white metal that adds hardness to steel, gives steel a fine grain structure, and makes it withstand high heat.

Withstanding temperatures as high as 6100 °F (3371 °C) before it melts, tungsten has a wide range of uses as a heat-resistant element alloying with iron and carbon to form carbide properties. Tungsten carbide is one of the hardest alloys known, surpassed only by diamond and cubic boron nitride on the Knoop hardness scale. Tungsten carbide, manufactured by the powder metallurgy process, is used to produce a wide variety of metal-cutting tools where heat and abrasion are the key factors in tool wear and breakdown.

Molybdenum

Molybdenum is a silvery-white metal that gives steel strength and hardness. It also improves and controls the heat treatability of steel rendering it less brittle. Molybdenum, or Molly as it is commonly known in the trade, is used for ball and roller bearings, tool and die steels, automotive parts, construction steels, etc.

Vanadium

Vanadium is a pale, silver-gray metal that gives steel a fine grain structure and imparts lightness, toughness, and strength qualities. It is used extensively for automotive parts that are subject to great shocks such as springs, wheel axles, and parts that vibrate. Vanadium is also combined with chromium to produce chrome–vanadium steels, which are used to make automotive parts such as springs, axles, connecting rods, gears, etc., that must be strong and tough but not brittle.

Silicon

Silicon is used as both a purifying agent and to alloy steel. Silicon steel can contain as much as 60 to 220 points of silicon, which improves the steel's hardenability, tensile strength, and adds to the steel's electrical properties. Steels containing about 3% silicon have excellent magnetic properties and are used in the fabrication of electromagnets and for cores in electrical machinery.

Stainless Steel

Stainless steels are made by adding large proportions of chrome (18%) and nickel (8%) to steel. This makes the steel hard and strong and also imparts rust and corrosion resistance properties (Figure 5–2).

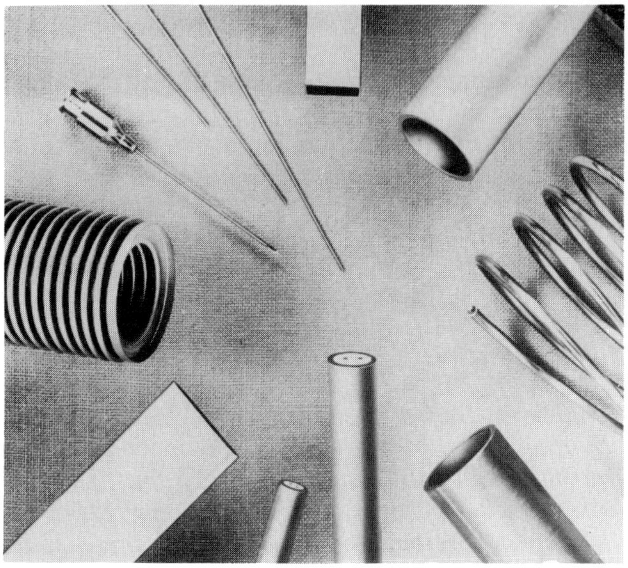

FIGURE 5–2
Specialty steel tubing depends heavily on the use of stainless steel for its quality and corrosion-resistant properties. (Credit: American Iron and Steel Institute)

There are two general grades of stainless steel. The 300 series, which is nonmagnetic, is ductile and cannot be hardened by heat treating. Products made from the 300 series include food handling materials, textile processing, transportation equipment, and many others. The 400 series is magnetic, can be hardened by heat treating, but is not as corrosive resistant as the 300 series.

Nonferrous Alloys

The purpose of using nonferrous metals as alloys is to alter or strengthen the base metal, which might be too soft or subject to rusting or corrosion conditions. The alloying agent may consist of one or more elements with amounts varying in small or large percentages. Consequently, any number of nonferrous alloys is possible.

Alloys are created when the principal or base metal and its alloying agent are heated, mixed together in a molten state, and are allowed to cool. The resultant grain structure dictates the properties the alloy will have. The size of the grain structure is controlled by the rate of cooling undergone by the molten elements. The grain structure is also affected by the manner in which a nonferrous alloy is heat treated or worked by hammering or other cold-working conditions.

Copper-Base Alloys

Copper is used as a base for a great number of alloys, the most common being brass and bronze. Although the chief alloying agents remain tin, zinc, and lead, other elements such as aluminum, phosphorus, manganese, and silver may be added to copper to alter the alloy's melting point. Depending on the amount of alloys, this melting point can vary from 1300 to 1800°F (704 to 982°C).

Brass

Brass is an alloy of copper and zinc. There are many different types of brass, all of which are generally determined by the copper to zinc ratio. "Alpha" brasses contain about 37% zinc and are noted for their cold-working properties. "Beta" brasses, on the other hand, have a higher zinc content (above 37%) and are known for their excellent hot-working properties. Low-zinc brasses range between 5 and 20% zinc content, the most notably in this group being red brass. Red brass, which has an 85 to 15% copper–zinc ratio, is used extensively in making costume jewelry, compacts, medallions, and similar items. High-zinc brasses contain from 30 to 40% zinc and are widely used in the production of automobile radiator tanks and cores, musical instruments, building hardware, valve stems, and many additional applications.

Brass does not discolor as quickly as copper and in cases where corrosion resistance is of great importance one or more elements such as tin is added to produce a series of special brasses known as tin brasses. Inhibited admiralty and naval brass fall in this classification. Containing not more than 1% tin, these special brasses are used for condenser, evaporator, and heat exchanger tubes as well as for propeller shafting and marine hardware (Figure 5-3). Leaded brasses contain from 0.5 to 3% lead in their copper–zinc mixture. The leaded brasses machine freely, breaking into small chips as opposed to producing long spiral curls during turning operations. This type of cutting action tends to increase tool life. The properties of the brass are not diluted by the addition of the lead, however. Leaded brasses are used extensively in the manufacture of clock and watch parts, architectural trim, gears, pinions, and automatic high-speed screw machine parts.

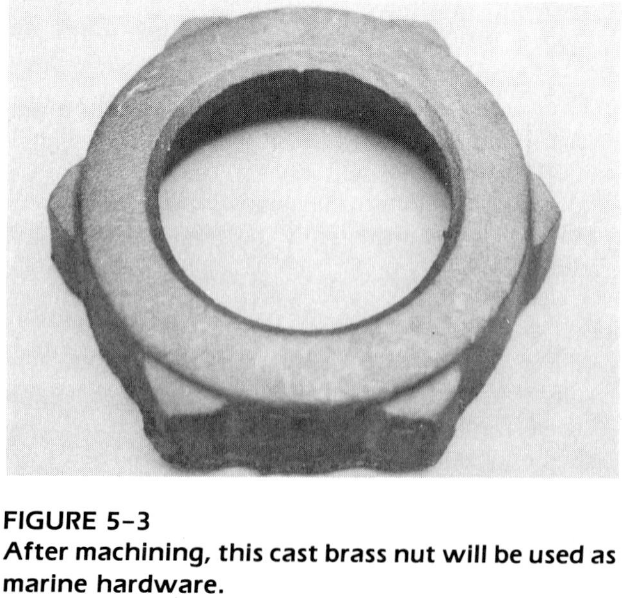

FIGURE 5-3
After machining, this cast brass nut will be used as marine hardware.

Bronze

In ancient times bronze was an alloy of copper and tin. Today there are many different types of bronze, some containing quantities of zinc and other alloys to provide hardness and toughness qualities. Some are copper base alloys that have the appearance of bronze and contain no tin at all.

Phosphor bronze is an alloy that is 95% copper, 4.65% tin, and a minute amount of phosphor (0.35%) to intensify toughness and hardness. The results are a bronze that combines its great corrosion resistance qualities with high resiliency and long wearing properties. Typical applications include springs, clutch discs, shafting, diaphragms, expansion and bridge bearing plates, and fasteners (Figure 5-4).

Silicon bronze is not a true bronze but actually a copper–silicon alloy whose silicon content may vary from 1 to 4%. To this base may be added other elements to engineer different work properties. Commonly added elements include manganese, iron, zinc, tin, nickel, or lead when good machining qualities are desired. Silicon bronzes have the strength and toughness of mild steel and excellent resistance to corrosion. Versatile, they are found in a wide range of engineering and manufacturing applications.

Aluminum bronze is not a true bronze either. Aluminum has been substituted for tin in this alloy that contains up to 13.5% aluminum, small amounts of nickel and manganese, and as much as 4% iron for purposes of hardening. Aluminum bronzes possess good antifrictional characteristics and have high hardness and high strength comparable to medium-carbon steel. They have excellent corrosion resistance and resist scaling and oxidation at high temper-

FIGURE 5-4
Thin-walled bellows are just one of the many applications of phosphor bronze.

FIGURE 5-5
Gears machined in an aluminum-bronze alloy have good corrosion resistance and strength comparable to medium-carbon steel.

atures. Typical applications include bearings, bushings, gears, rollers, and related parts (Figure 5-5).

Manganese bronze is essentially a copper-zinc alloy containing manganese with varying elements added to give it good mechanical properties and high strength. The alloy may consist of 55 to 60% copper, 38 to 42% zinc, 0 to 1.5% tin, 0 to 2% iron, 0 to 1.5% aluminum, and as much as 3.5% manganese. The mechanical properties and corrosion resistance of manganese bronze make it ideal for marine use such as ship propellers, shafting, rudders, and related equipment.

Nickel-Base Alloys

Nickel-base alloys are noted for their strength and ductility and resistance to heat and corrosion. They also have good electrical conductivity. The amount of nickel in nickel-base alloys varies from 62 to 99%, depending on the alloy.

Monel, a nickel-base alloy, is composed of 65% nickel, 30% copper, and small amounts of other elements. Monel takes a silverlike shine and is used in making propeller parts, nonmagnetic aircraft parts, cooking and chemical processing equipment, laundry machines, and the like.

Inconel is another nonmagnetic, hard, tough nickel-base alloy that retains its hardness at high temperatures and resists corrosion and oxidation. These properties make it ideal for use in heat-treating equipment and furnaces as well as jet engine exhaust manifolds. Composed of approximately 76% nickel, 7% iron, 15% chromium, and small amounts of other elements inconel can also be found in use in the food and dairy processing industries.

Zinc-Base Alloys

Zinc is alloyed chiefly with aluminum and varying amounts of copper, iron, lead, magnesium, and tin. The tensile strength of zinc-based alloys will vary with the percentage of elements used, as will the boiling point, ranging in temperatures from about 700 to 787°F (371 to 419°C).

There are more than ten zinc-base alloys in general use. Known for properties that contribute to a good surface finish and electrical conductivity, many of the zinc alloys make their appearances in the form of die castings used in such items as typewriter parts, carburetors, toys, hardware, automotive parts, and other products used by the electrical appliance industries.

Tin-Base Alloys

Tin has its roots in the Bronze Age; its present-day application as a tin-base alloy is in the production of babbitt, solder, and art metal.

Babbitt is a bearing-lining material that contains about 90% tin with varying amounts of copper, lead, and antimony. Babbitt encloses and reacts well with materials that are used to make cranks, axles,

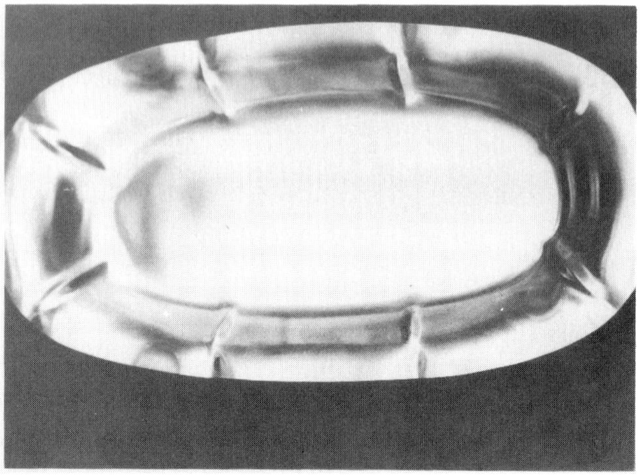

FIGURE 5–6
A pewter serving dish cannot have any lead content.

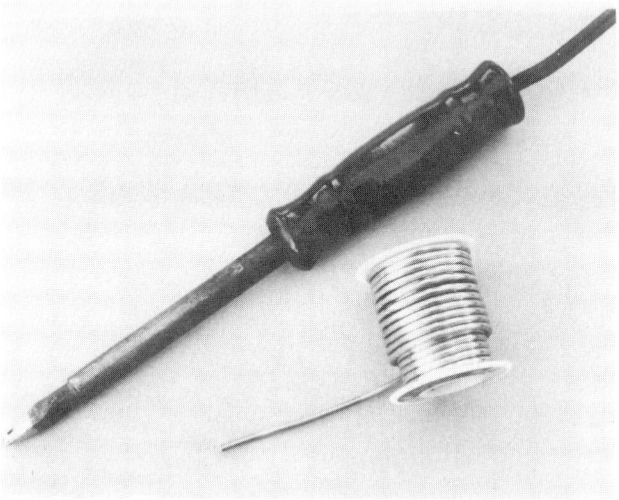

FIGURE 5–7
Some types of soldering wire may be a mixture of tin and a lead-based alloy.

FIGURE 5–8
An aluminum rolling mill, where large ingots are passed back and forth through rollers and gradually reduced in size. (Credit: Reynolds Metals Company)

shafts, and many other parts that move and create friction.

Pewter is also a tin-based alloy composed of 92% tin, 5% antimony, and 3% copper. It takes a beautiful polish and is used for tableware and ornamental purposes (Figure 5-6). Legally, pewter can no longer have any lead content, removing any lead poisoning that can result when lead is mixed with food and water.

Magnesium-Base Alloys

When alloyed with other metals, magnesium gives the new metal moderate strength, weldability, and high mechanical properties. Its physical compositions vary with the alloying agent such as aluminum, manganese, silicon, copper, nickel, tin, and iron. These alloying agents will also affect the tensile strength required for aircraft, tanks, or die-casting requirements.

FIGURE 5-9
A forged automotive wheel undergoing stress analysis using computerized modeling techniques. (Credit: Aluminum Company of America)

Lead-Base Alloys

Lead-base alloys offer protection against corrosion in places where strength is not important. Soft solders carrying varying amounts of tin and lead are good examples, where the composition can be as high as 80% lead (Figure 5-7). Babbitt material have tin and antimony alloyed with as much as 70% lead to make bearing parts that will be subjected to strong mineral acids. Lead and its alloys can be processed in a number of ways, including spinning, casting, extrusion, stamping, or drawing.

Aluminum-Base Alloys

Too soft to be used in the pure state, aluminum is alloyed with magnesium, silver, zinc, manganese, copper, and silicon to take advantage of its naturally high resistance to corrosion, heat and electrical conductivity, as well as structural strength. Thus it becomes the second most used metal (Figure 5-8).

Containing slightly more than 98% aluminum, which is combined with small amounts of silicon, copper, magnesium, and chromium, this aluminum-base alloy (6061) is used to manufacture wheel mounting rims (Figure 5-9). In an industry that is greatly concerned with energy conservation—"It takes energy to move weight"—an automobile body and frame have been produced using an aluminum-base alloy. The product is only slightly heavier than pure aluminum but has the tensile strength of steel (Figure 5-10).

Since a variety of alloying elements are used to give aluminum specific qualities, the Aluminum Association has set up a standard four-digit system

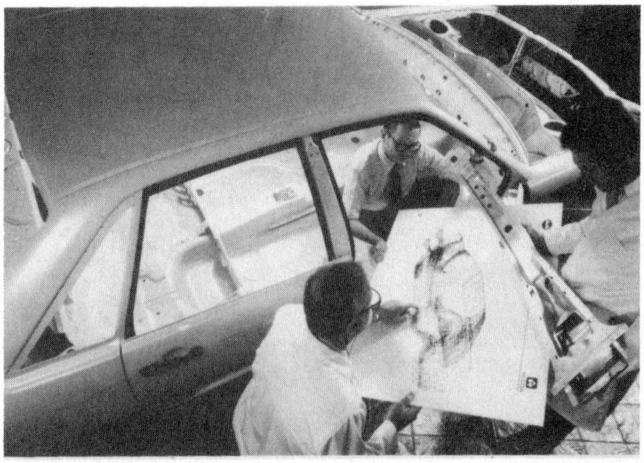

FIGURE 5-10
The prototype aluminum auto body and frame is as strong as steel but nearly 50% lighter. (Credit: Aluminum Company of America)

TABLE 5-1
Aluminum-based alloys identification system

Series	*Alloying agent*
1000	Iron-silicon
2000	Copper
3000	Manganese
4000	Silicon
5000	Magnesium
6000	Silicon-magnesium
7000	Zinc-magnesium
8000	Other elements
9000	Special

(1000 to 9000) to identify each type of aluminum alloy. The 1XXX series contains 99.00% aluminum, while numbers 2 to 9 identify the major alloying element. For example, 7075 identifies the aluminum, zinc-magnesium high-strength alloy that is used in air frame construction (Figure 5-11). See Table 5-1 for the aluminum identification system.

Aluminum Tempers

For all their wondrous properties, aluminum alloys have one common flaw. They tend to lose part of their strength at elevated temperatures. Consequently, temper plays as important a role in aluminum as the alloying agents. The temper will significantly alter or affect the characteristics of the product as to wear, machinability, or mechanical properties.

The symbol used to designate the temper is added to the alloying agent, being separated by a dash, for example, 6061-T6. The numbers that follow indicate the degree of cold work (H) or thermal treatment (T) designations involved, for example, 6063-T81 or 1200-H11.

The temper designation system includes all aluminum alloys except ingot. A partial explanation is listed in Table 5-2. A fuller understanding of the subject can be obtained in materials published by the Aluminum Company of America, the Reynolds Metals Company, or the Aluminum Association.

FIGURE 5-11
The air frame construction, wings, and fuselage of this supersonic aircraft is an aluminum alloy. (Credit: Aluminum Company of America)

TABLE 5-2
Aluminum temper designations

Symbol	*Explanation*	*Use*
F	As fabricated	Applies to alloys that are shaped where the temper is not critical to the process
0 (May be followed by a digit greater than zero)	Annealed	Applies to alloys used in cold working-spinning, drawing, or forming
W	An unstabile temper that follows solution heat treating.	Applies to products whose natural aging is indicated by a time period, i.e., W 1 hr
H (May be followed by one or more digits)	Strain hardened by rolling or drawing. Degree is indicated by the number following the designation.	Applies to wrought products only

TABLE 5-2 *(Continued)*

Symbol	*Explanation*	*Use*
H1	Strain hardened only.	Applies to products whose mechanical properties are obtained without additional thermal treatments
H2	Strain hardened followed by partial annealing.	Applies to products that are equal to H1 and H3 in minimum tensile strength. They also have a higher elongation than H1 tempers
H3	Strain hardening through low-temperature treatment.	Applies to alloys that are stabilized to prevent gradual age softening at room temperature
T (Always followed by one or more digits. Digits 2 to 10 indicate specific sequences of basic treatments)	Thermally treated, quenched, and aged to produce stabile tempers other than F, O, and H.	Applies to products that are thermally treated, with or without supplementary strain hardening, to produce stabile tempers
T2	Annealed	Applies only to cast aluminum to improve ductility, strength, and dimensional stability
T3	Solution heat treated and then cold worked	Applies to wrought products where cold working is recognized as improving strength
T4	Solution heat treated and then stabilized by naturally aging at room temperature	Applies to products that are not cold worked after solution heat treatment
T5	Artificial aging through low thermal cooling	Applies to castings and extrusions to improve their machinability
T6	Solution heat treated and then artificially aged	Applies to alloys that develop high mechanical properties through artificial aging
T7	Solution heat treated and then stabilized by artificial aging	Applies to wrought products that are over aged/stabilized beyond maximum strength to control mechanical properties
T8	Solution heat treated, cold worked, and then artificially aged	Applies to products that are cold worked to provide higher mechanical properties
T9	Solution heat treated, artificially aged, and then cold worked	Applies to products that are cold worked to improve their strength, such as wire or rod that is drawn and used for nails
T10	Artificially aged and then cold worked	Applies to products such as castings or extrusions that are cold worked to improve their strength

SECTION REVIEW

Unit 1

1. Of what importance is the development of tools and toolmaking?
2. Name the three ages that preceded the Industrial Age.
3. What two men were responsible for the development of the steam engine?
4. Who is credited with having invented the screw cutting lathe?
5. Who invented the cotton gin?
6. Name five characteristics required of space-age metals used in the aerospace industries.
7. Why are high-precision skilled tradespeople important to the metalworking industry?
8. Which career personnel are considered the "nuts and bolts" engineers?
9. While working in a team effort with chemists and physicists, what are the principle responsibilities of a metallurgical engineer?
10. What field of engineering is responsible for the overall operation of a factory?

Unit 2

1. Of the many properties of metal, which is the measurement of a direct pull that is exerted without rupturing?
2. Which property characterizes the ability to resist forces that tend to crush?
3. Which property characterizes the ability of metal to return to its original shape after having been bent?
4. Which property characterizes the collapse or weakening of metal that has been subjected to continued stressing?
5. Which property characterizes the ability of metal to absorb or withstand shock or impact?
6. Which property characterizes breaking easily without being able to deform?
7. Which property characterizes the resistance of metal to withstand forceable penetration or deformation?
8. Which property characterizes the metal's ability to permanently change in form without breaking?
9. What property deals with the rate and ease of metal removal and cutting tool wear?
10. Which property deals with a metal's ability to turn to liquid quickly and join other metals?
11. Name two common methods of converting pig iron into steel.
12. What is the purpose of using limestone in making steel and how does it work?
13. Which process of making steel is considered to be the fastest? Describe its output.
14. Which process is used to produce high-quality alloy and tool steels?
15. Which process can take molten steel from any of the furnaces and turn it into slabs or blooms?

Unit 3

1. What are the three general classifications of carbon steels and what is the carbon content of each?
2. How do steel producers term or classify the carbon content of steel and give examples of this classification.
3. What is the range of carbon content of gray cast iron and why is it easily machined?
4. What two systems are used to identify the chemical composition of steel and how do they vary?
5. Name two physical methods of identifying steel.

Unit 4

1. How may nonferrous metals be defined?
2. List eight of the more widely used nonferrous metals.

3. Identify the major properties of six of the more commonly used nonferrous metals used in the metalworking industry.

Unit 5

1. Define the term *alloy*.
2. Name five metals commonly used as base metals in ferrous alloys.
3. What are the principal alloying metals used in stainless steel and what are the two general grades made?
4. What is the purpose of developing nonferrous alloys?
5. Name the two major copper-base alloys.
6. Name one bronze alloy that does not contain tin.
7. Name two nickel-base alloys.
8. Name three uses for tin-base alloys.
9. Explain the difference between lead-base and tin-base babbitt.
10. Name five different metals that are commonly alloyed with aluminum.

Section 2

SEMIPRECISE AND PRECISION MEASURING SYSTEMS

The methods of measurement have evolved over many centuries and in different civilizations. Historical studies indicate that two types of linear measurements were developed in two different parts of the world. Western civilization, which had its development in the valleys along the Nile and Tigris and Euphrates rivers, was rivaled in the Orient by the areas along the Yellow River (China) and the Indus River.

Archaeologists have discovered that people had measuring systems as far back as the very first city known to civilization. The ceramic cylinder shown in Figure 6-1 was used by various agricultural societies even before cities came into existence. Some people believe that the lines on the device represent a counting system to keep track of quantity. Others believe these cylinders could also measure distance: Tied to a string at regular intervals, they became standards of measurement.

Devised about 3000 B.C., the cubit which has many references in the Bible, was an Egyptian linear standard of measurement. A cubit was the length of the arm and was measured from the elbow to the tip of the middle finger of an outstretched hand, roughly 18 to 20 in. (457 to 508 mm) long. A master cubit made of black granite was used as the standard against which other cubits in the realm had to qualify at different intervals. The royal cubit measured 20.62 in. (524 mm) and was subdivided in a complicated system of measurements, all roughly based on the parts of the human hand, such as fingers, digits, palms, and hands (Figure 6-2, p. 36).

Egyptians also devised other systems of measurements. To overcome the flooding of the Nile every spring and restore each landowner's property lines after the floodwaters had receded, an early form of surveying was invented. By tying a long rope to the ends of two sticks, two men would start from a fixed point set in the mud above the flood level and proceed to scribe a series of arcs. These arcs would then enable them to construct parallel and perpendicular lines. The result was an accurate rectangular layout that would form the landowner's field. The Greek mathematician Euclid would later refine and expand upon these principles in the branch of mathematics known today as Euclidean geometry.

The Babylonian Age gave the world a different set of weights and measures, some of which were gradually adopted by other Middle East countries

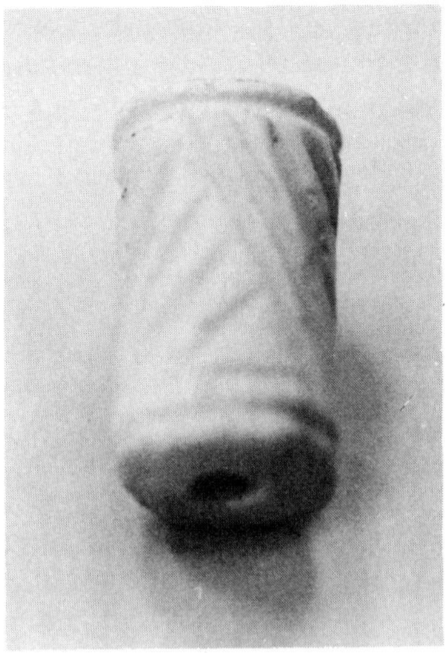

FIGURE 6-1
Discovered in Ur, Mesopotamia, over 5000 years ago, this multipurpose measuring device was made by the Chaldeans. It could measure distance, act as an accounting device, and was used as a standard of measurement.

FIGURE 6–2
The cubit, dating back at least to the Khufu pyramid, was another early standard of measurement. It was used with remarkable accuracy in building all the great pyramids. (Credit: Sheffield Measurement Division, Warner & Swasey/A Cross & Trecker Company)

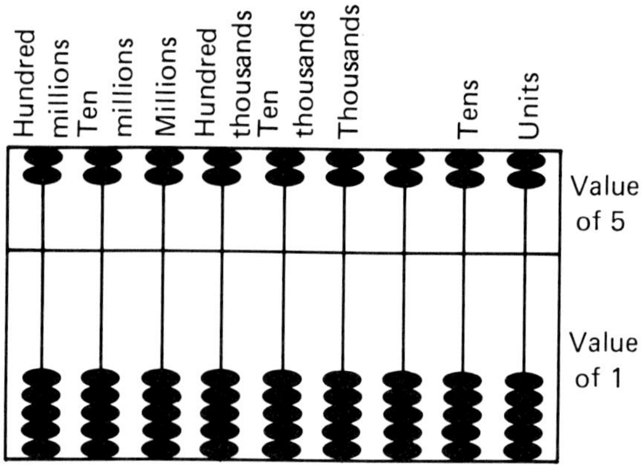

FIGURE 6–3
The abacus is one of the oldest calculating machines known to civilization and is still used in different parts of the world.

due to commercial trading. Another contribution was the study, that is, the measurement of walking distance over a period of time. It was calculated to be the distance a man could travel at a normal walking pace in 2 min time. The study was measured from the time the sun began to rise until it could be seen completely above the horizon. One study was estimated to about 607 to 640 ft (1285 to 1295 m).

Though isolated from the West, ancient China would also devise systems of weights and measures. Some studies indicate that it was the Chinese who originated the abacus as early as 3000 B.C., inventing tablets divided with lines and equipped with markers whose placements represented numerical values. The abacus (Figure 6–3) is still in use in the Middle East and the Orient, and it is claimed that if expertly handled can compete against mechanical calculating machines.

While the length of the flute was used as a standard in the Oshiki era, circa 1000 B.C., the Chinese were responsible for using a system of decimal notation in connection with their foot rulers as early as the sixth century B.C.

Many of the standard and basic units established in ancient times by the Egyptians and Babylonians would be absorbed, modified, and changed by the succeeding eras of Greco-Roman civilization. The Greek unit of length was the finger. In turn, it would influence the Roman system of measuring weights and distances, with both systems gradually becoming intermixed. During the first thousand years B.C., Roman soldiers traveled the length and breadth of the European continent, east into Asia, and south into Africa. It would be the Romans who would rewrite the standards for measuring distances traveled by marching foot soldiers.

The Roman Empire lasted 1500 years. It left behind a legacy of measurements that Western culture accepted and spread, adding contributions and variations chiefly from the Arabs and Scandanavians. When the Roman Empire faded into history, the Western world became enveloped in the Dark Ages, a period where there was a relapse in trade, skills, and industry, cities were overrun by barbaric invaders, and education and learning came to a standstill.

The science of measurement would develop gradually. Doubt exists as to its actual beginning. Some believe that it was not until the signing of the Magna Carta in the thirteen century that the system of using parts of the human arm and leg was replaced with a more durable gage. King Edward I ordered a master gage made of iron to serve as the standard for the yard. The master yardstick was called the "iron ulna," named after the bone, the ulna, in the forearm. While hardly scientific, the reign of Edward II introduced the standard for measuring the inch—"three barleycorns, round and dry." The Renaissance ushered in a new era of academic learning. Men such as da Vinci, perhaps one of the most creative geniuses who ever lived, Michelangelo, and Copernicus would do much to broaden the horizons of civilizations. The Age of Discovery would pave the way for the brilliant minds of men such as Galileo, Newton, Huyhens, Kepler and many others who would do much for advancing the science of metrology.

TABLE 6–1
Common inch units

1 inch	=	1000 thousandths
1 foot	=	12 inches
1 yard	=	36 inches
1 rod	=	198 inches
1 mile	=	63,360 inches

UNIT 6 MEASUREMENT SYSTEMS

The systems of measurement have continually grown, and today we live in a world where products have to be built to very precise tolerances. Two main measuring systems are used throughout the world today: the inch system (U.S. customary, English, or Imperial) and the metric system. Although the United States and Canada are still using the inch system, both countries are in the process of converting to the metric system, which is the standard measuring system used in most countries of the world. In order for the United States and Canada to compete in the world trade with manufactured products, they must be made to the metric standard.

Objectives
After completing this unit you should be able to

1. understand and be able to use the inch system of measurement.
2. understand and be able to use the metric system of measurement.
3. convert measurements from one system to another.

The Inch (Customary) System

The inch standard of measurement has been used in the United States and Canada for many years. The system is derived from early civilizations where parts of the body were used to estimate measurements. The inch, foot, and yard are a few of the measurements derived from the body that are still in use today.

The inch, the base unit of length, evolved from the width of a thumb, until today the *international inch* is based on the wavelength of monochromatic light from krypton 88 gas. The length of a single wave length is 0.0000238 in., and this gives us an absolute standard that never changes. In order for parts to fit and operate properly, measurements to thousandths (0.001) of an inch are very common in industry. One-thousandth of an inch is approximately one-third the thickness of a human hair. It is not uncommon that parts for precision instruments, space vehicles, etc., are made to tolerances of ten-thousandths (0.0001) of an inch or even to a millionth (0.000,001) of an inch. A millionth of an inch is commonly known as a microinch.

The inch is the base unit of length, and other linear units are related to it by odd and unusual factors based on the early system of measurement. The standard inch is usually divided into smaller units such as halves, quarters, eighths, sixteenths, thirty-seconds, sixty-fourths, thousandths, ten-thousandths, and millionths of an inch. Some of the common inch units and their relation to other units are listed in Table 6–1.

The inch system is more complex than the metric system because there is no set relationship between one unit and the next.

Common measurements used in the inch system are given in Table 6–2.

The Metric System

In order to arrive at a common system of measurement throughout Europe in the late eighteenth century, the French Academy of Sciences proposed one

TABLE 6–2
Common measurements in the inch system

Units		*Unit of measurement*
Length	=	Inch
Mass (weight)	=	Pounds
Temperature	=	Degrees fahrenheit (°F)
Volume	=	Fluid ounces or cubic inches
Force	=	Foot-pounds
Pressure	=	Pounds per square inch
Area	=	Square feet and inches

system of measurement in the French Assembly. The system was related to the ignored ideas of measurement suggested by Gabriel Mouton. Mouton, vicar of St. Paul's church in Lyons, had more than a century before advanced the concept of weights and measures based on the physical universe. A line representing the earth's circumference from the North Pole to the equator, drawn at a point where it crossed through Paris, was used as the basis of this new measurement. Since this distance was so great, one ten-millionth of it was used, and the new unit of length was named the meter. The unit of volume would be a cube whose sides would equal this unit of length. Mass would be represented by the amount of water needed to fill this cube. After a dismal trial and repeal by Napoleon, the metric system was passed into law in 1840 and quickly spread throughout Europe and most of the world.

In 1960 the metric system was updated and modified slightly to become the *SI metric* or *Système International d'Unités.* The earth's measurement, long discarded, was replaced by the wavelength of krypton gas, which never changes and remains constant, as the basis of measurement. Used by 90% of the world, the SI metric system consist of seven base units (Table 6-3), two supplementary units, and derived units. A derived unit is one that has been computed from any other type of unit.

Common measurements used in the metric system are shown in Table 6-3.

The metric system is currently in various stages of adoption in the United States and Canada. In the metric system, the base unit of length is the meter, and all other units are related to it by multiples of 10. For example, a meter is divided into 10 *deci*meters, 100 *centi*meters, and 1000 *milli*meters. A *kilo*meter (*kilo* means thousand) is a meter multiplied by 1000. Therefore, in order to convert from a larger unit to a smaller one or vice versa, you must either multiply or divide by 10, 100, 1000, etc. See Table 6-4 for the relationship of all units to the meter.

The metric units most commonly used are the meter, kilogram, liter, cubic meter, and degree Celsius. A brief description of each follows:

Meter (m). The meter is the base metric unit of *length.* It is equal to 39.37 in. and therefore is a little longer than a yard, which is equal to 36 in. For more accurate measurements the meter can be divided into 100 equal parts (*centimeters*) or 1000 equal parts (*millimeters*).

Kilogram (kg). The kilogram is the metric unit of *mass* (weight). It is about 2.2 lb. Anything measured in pounds in the U.S. customary system is measured in kilograms in the metric system.

Liter (L). The liter is the metric unit of *liquid capacity,* and it is a little larger than a quart. Anything normally packaged in quarts and pints is packaged in liters and half liters in the metric system.

Cubic Meter (m^3). The cubic meter is the metric unit of *volume.* Since a meter is slightly larger than a yard, the cubic meter would be about 30% larger than a cubic yard.

Degrees Celsius (C). *Temperature* in the metric system is measured in degrees Celsius. In this system water boils at 100°C and freezes at 0°C. In the Fahrenheit system water boils at 212°F and freezes at 32°F.

TABLE 6-3
Common measurements in the metric system

Unit		*Unit of measurement*	*Symbol*
Length	=	Meter	m
Mass (weight)	=	Kilogram	kg
Temperature	=	Degrees Celsius	°C
Volume	=	Liter (L) or cubic meter	m^3
Force	=	Newton	N
Pressure	=	Pascal	Pa
Speed	=	Meters per minute and meters per second	m/min m/s
Area	=	Square meter	m^2

TABLE 6-4
Relationship of units to the meter

Prefix	*Meaning*	*Multiplier*	*Symbol*
Micro	One millionth	0.000 001	μ
Milli	One thousandth	0.001	mm
Centi	One hundredth	0.01	c
Deci	One tenth	0.1	d
Deca	Ten	10	da
Hecto	One hundred	100	h
Kilo	One thousand	1000	k
Mega	One million	1 000 000	M

Inch and Metric System

Because the change to the metric system in the United States and Canada has progressed very slowly, it may be many more years before the changeover is complete. Therefore, during this changeover period, it is very likely that both systems of measurement may be used, so it would be desirable for anyone associated with the metal trade to be familiar with both systems.

Because SI is an international language, all countries that adopted it follow the approved style. This style is followed throughout this book wherever metric tools or measurements are discussed. However, because the inch system of measurement is still more common in the United States and Canada, wherever measurements are used in this book the inch measurement is given first, followed by its metric equivalent in parentheses, e.g., 1.250 in. (31.75 mm). Wherever possible, both inch and metric tools and information are given. However, in some cases only inch tools or information where available at the time of publication and the soft conversion to metric is given in parentheses.

However, to help a student identify inch and metric dimensions more readily, the following practice is followed. In both the inch and metric systems, a zero (0) is placed to the left of the decimal point for numbers less than 1.

Conversion Factors

Since the student may be required to use both inch and metric system of measurement for some time to come, it is desirable that the conversion factors between both systems be known (Tables 6-5 to 6-7).

UNIT 7 RULES AND CALIPERS

The ability to work in industry or any type of metalworking shop will depend on taking measurements. The accuracy involved will determine the type of

TABLE 6-5
Converting from inch to metric

Inch *to*	*Metric*	*Formula*
Inch	Millimeter (mm)	Multiply by 25.4
Foot	Meter (m)	Multiply by 0.3048
Yard	Meter (m)	Multiply by 0.9144
Mile	Kilometer (km)	Multiply by 0.6213
Ounce	Grams (g)	Multiply by 28.35
Pound	Kilograms (kg)	Multiply by 0.4536
Gallon	Liter (L)	Multiply by 3.7854
Pounds per square inch	Kilopascals (kPa)	Multiply by 6.895
Fahrenheit	Celsius (C)	(°F − 32°) × 5/9

TABLE 6-6
Decimal and millimeter equivalents of fractions

Inches	Inches	Mm	Inches	Inches	Mm
1/64	.01563	0.397	33/64	.51563	13.097
1/32	.03125	0.794	17/32	.53125	13.494
3/64	.04688	1.191	35/64	.54688	13.890
1/16	.0625	1.587	9/16	.5625	14.287
5/64	.07813	1.984	37/64	.57813	14.684
3/32	.09375	2.381	19/32	.59375	15.081
7/64	.10938	2.778	39/64	.60938	15.478
1/8	.125	3.175	5/8	.625	15.875
9/64	.14063	3.572	41/64	.64063	16.272
5/32	.15625	3.969	21/32	.65625	16.669
11/64	.17188	4.366	43/64	.67188	17.065
3/16	.1875	4.762	11/16	.6875	17.462
13/64	.20313	5.159	45/64	.70313	17.859
7/32	.21875	5.556	23/32	.71875	18.256
15/64	.23438	5.953	47/64	.73438	18.653
1/4	.250	6.350	3/4	.750	19.050
17/64	.26563	6.747	49/64	.76563	19.447
9/32	.28125	7.144	25/32	.78125	19.844
19/64	.29688	7.541	51/64	.79688	20.240
5/16	.3125	7.937	13/16	.8125	20.637
21/64	.32813	8.334	53/64	.82813	21.034
11/32	.34375	8.731	27/32	.84375	21.431
23/64	.35938	9.128	55/64	.85938	21.828
3/8	.375	9.525	7/8	.875	22.225
25/64	.39063	9.922	57/64	.89063	22.622
13/32	.40625	10.319	29/32	.90625	23.019
27/64	.42188	10.716	59/64	.92188	23.415
7/16	.4375	11.113	15/16	.9375	23.812
29/64	.45313	11.509	61/64	.95313	24.209
15/32	.46875	11.906	31/32	.96875	24.606
31/64	.48438	12.303	63/64	.98438	25.003
1/2	.500	12.700	1	1.00000	25.400

TABLE 6-7
Decimal inch equivalents of millimeters

mm Inches	mm Inches	mm Inches	mm Inches	mm Inches
0.01 = .00039	0.26 = .01024	0.51 = .02008	0.76 = .02992	2.0 = .07874
0.02 = .00079	0.27 = .01063	0.52 = .02047	0.77 = .03032	3.0 = .11811
0.03 = .00118	0.28 = .01102	0.53 = .02087	0.78 = .03071	4.0 = .15748
0.04 = .00157	0.29 = .01142	0.54 = .02126	0.79 = .03110	5.0 = .19685
0.05 = .00197	0.30 = .01181	0.55 = .02165	0.80 = .03150	6.0 = .23622
0.06 = .00236	0.31 = .01220	0.56 = .02205	0.81 = .03189	7.0 = .27559
0.07 = .00276	0.32 = .01260	0.57 = .02244	0.82 = .03228	8.0 = .31496
0.08 = .00315	0.33 = .01299	0.58 = .02283	0.83 = .03268	9.0 = .35433
0.09 = .00354	0.34 = .01339	0.59 = .02323	0.84 = .03307	10.0 = .39370
0.10 = .00394	0.35 = .01378	0.60 = .02362	0.85 = .03346	11.0 = .43307
0.11 = .00433	0.36 = .01417	0.61 = .02402	0.86 = .03386	12.0 = .47244
0.12 = .00472	0.37 = .01457	0.62 = .02441	0.87 = .03425	13.0 = .51181
0.13 = .00512	0.38 = .01496	0.63 = .02480	0.88 = .03465	14.0 = .55118
0.14 = .00551	0.39 = .01535	0.64 = .0520*	0.89 = .03504	15.0 = .59055
0.15 = .00591	0.40 = .01575	0.65 = .02559	0.90 = .03543	16.0 = .62992
0.16 = .00630	0.41 = .01614	0.66 = .02598	0.91 = .03583	17.0 = .66929
0.17 = .00669	0.42 = .01654	0.67 = .02638	0.92 = .03622	18.0 = .70866
0.18 = .00709	0.43 = .01693	0.68 = .02677	0.93 = .03661	19.0 = .74803
0.19 = .00748	0.44 = .01732	0.69 = .02717	0.94 = .03701	20.0 = .78740
0.20 = .00787	0.45 = .01772	0.70 = .02756	0.95 = .03740	21.0 = .82677
0.21 = .00827	0.46 = .01811	0.71 = .02795	0.96 = .03780	22.0 = .86614
0.22 = .00866	0.47 = .01850	0.72 = .02835	0.97 = .03819	23.0 = .90551
0.23 = .00906	0.48 = .01890	0.73 = .02874	0.98 = .03858	24.0 = .94488
0.24 = .00945	0.49 = .01929	0.74 = .02913	0.99 = .03898	25.0 = .98425
0.25 = .00984	0.50 = .01969	0.75 = .02953	1.0 = .03937	26.0 = 1.02362

*.02520

Credit: The DoAll Company.

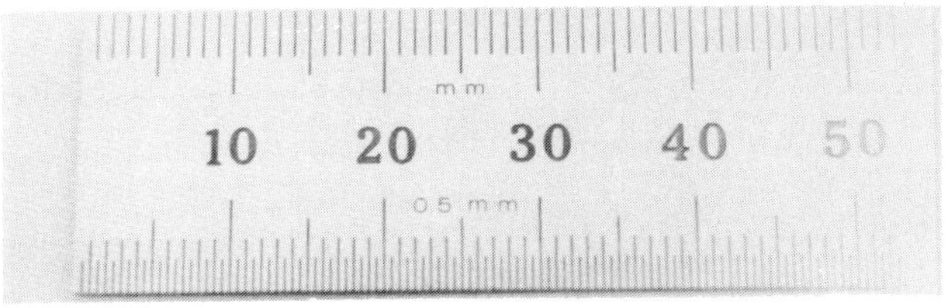

FIGURE 7-1
Metric rule measurement in millimeter and $\frac{1}{2}$ mm.

tool needed. For this reason a variety of precise and semiprecise tools are used.

Objectives
After completing this unit you should be able to

1. identify several types of rules and calipers.
2. measure and use a steel rule properly.
3. measure and use a spring caliper properly.
4. measure and use a precise caliper properly.

Types of Rules and Their Uses

Figure 7-1 reveals a close-up of a metric rule. The divisions or graduations (graduated lines) are subdivisions of the meter. On one edge the graduations represent one millimeter or one-thousandth part of a meter. Subdivisions of one-half millimeter are measured along the opposite edge. Steel rules generally used in shops measure 15 cm long. They also come as long as 1 m.

The 6-in. rule (Figure 7-2) or scale, as it is called in the machine tool trades, is graduated to subdivide the inch into fractional parts. A close-up

FIGURE 7-2
The 6-in. rule.

FIGURE 7-3
The rule should be held on a straight line between measuring points when being used.

reveals that edges may be divided into eighths, sixteenths, thirty-secondths, or sixty-fourths, the last fraction being about the closest you can measure with the naked eye. Some industries use rules that subdivide the inch into one-tenth, one-twentieth, or one-hundredth fractional parts. Additionally, rules vary from the commonly used 6 in.—but can be had even smaller—to many feet in length.

The rule should be held on a straight edge, about 90° to the measured surface (Figure 7-3). If it is not possible to butt against an edge, the rule should be placed so that the 1-in. or 1-cm line is used as a reference point. This eliminates any error caused by wear found on the end of the rule (Figures 7-4A and B).

The combination set (Figure 7-5) can come with the inch or metric measurements and offers semiprecise accuracy in use. Angles can be drawn or measured using the protractor (Figure 7-6). Locating the center of round or octagonal work can be ac-

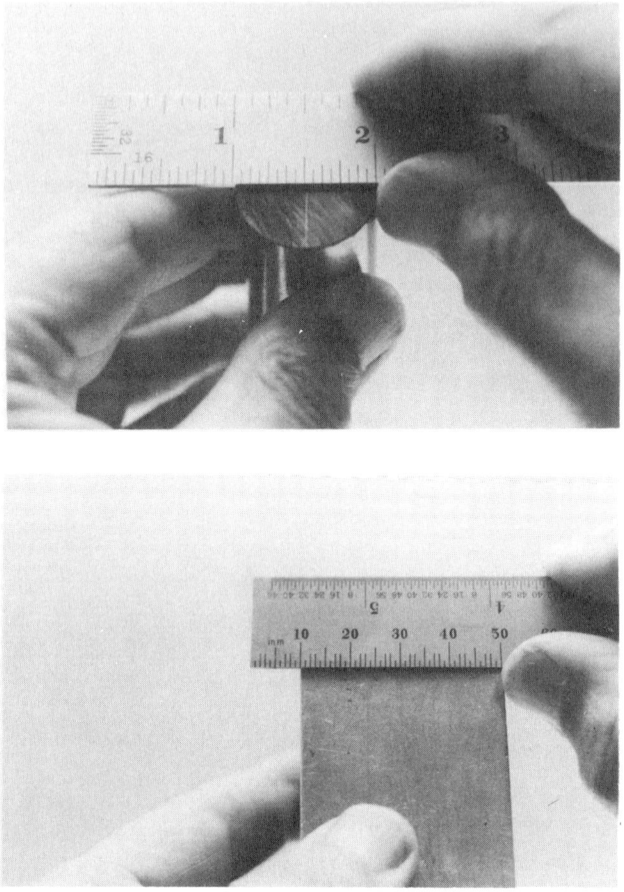

FIGURE 7-4
To eliminate end edge error, use the 1-in. graduation . . . or 10-mm graduation line as a reference point.

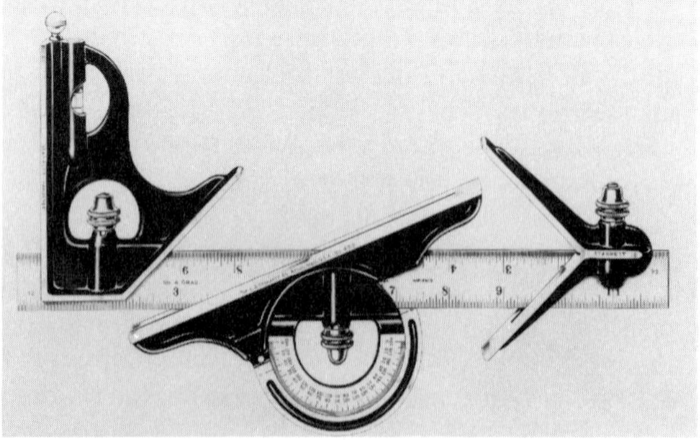

FIGURE 7-5
A combination set. (Credit: L.S. Starrett Company)

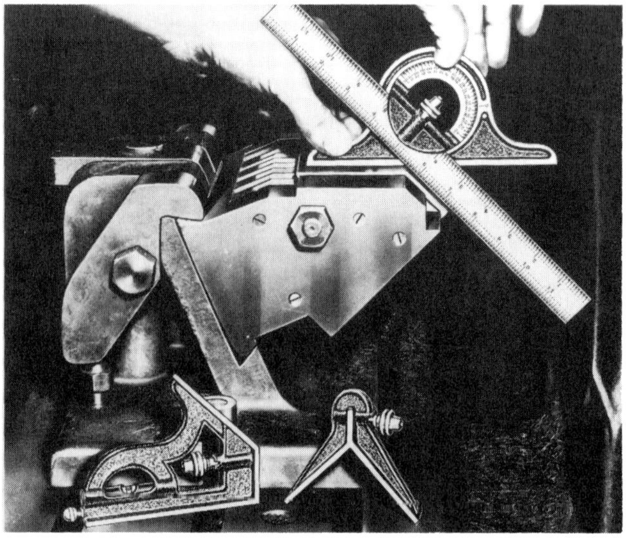

FIGURE 7-6
Measuring angles with a protractor. (Credit: L.S. Starrett Company)

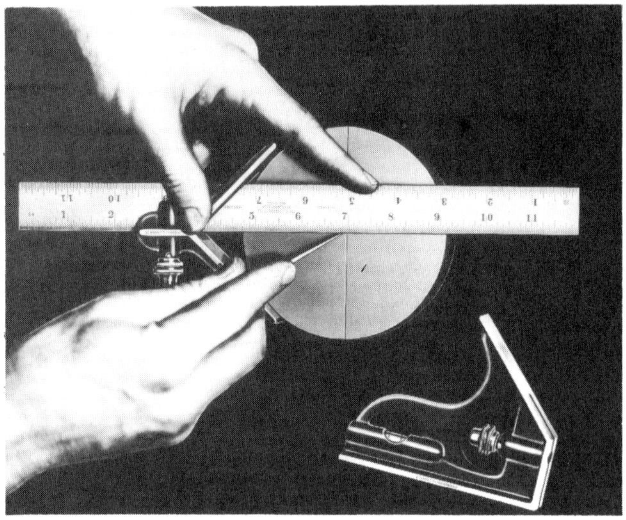

FIGURE 7-7
Using the center head. (Credit: L.S. Starrett Company)

complished using the center head (Figure 7-7). The adjustable square can be used for checking squareness (sides with an included angle of 90°) and 45° angles as well as measuring for depths and locations (Figure 7-8).

Types of Calipers and Their Uses

Outside Calipers

There are two styles of outside calipers (Figure 7-9). The spring joint caliper is controlled by an adjust-

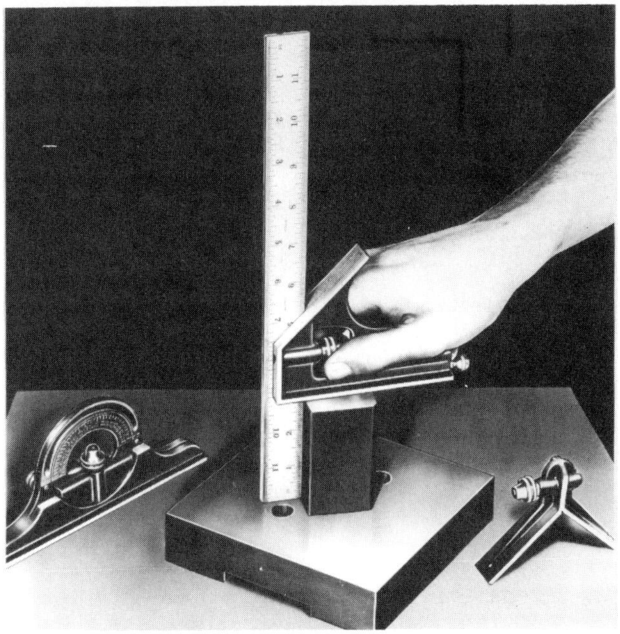

FIGURE 7-8
Using the adjustable square. (Credit: L.S. Starrett Company)

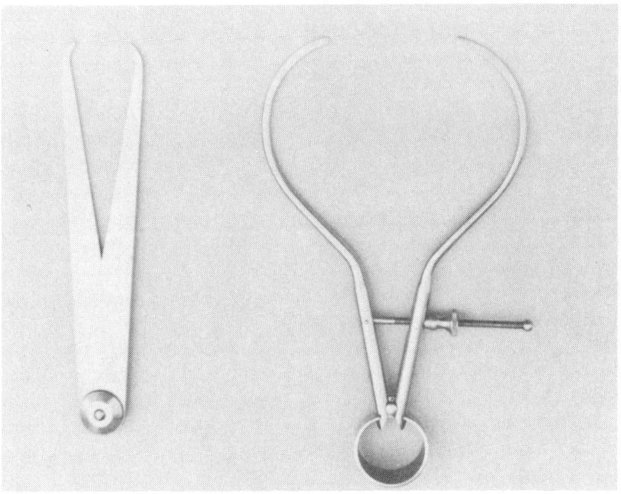

FIGURE 7-9
Spring joint and firm joint outside calipers. (Credit: Photocom Productions)

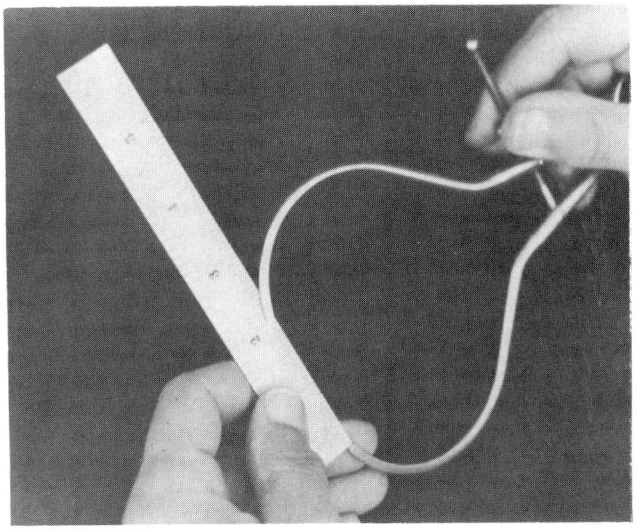

FIGURE 7-10
Setting an outside caliper. (Credit: Paul Rapisarda)

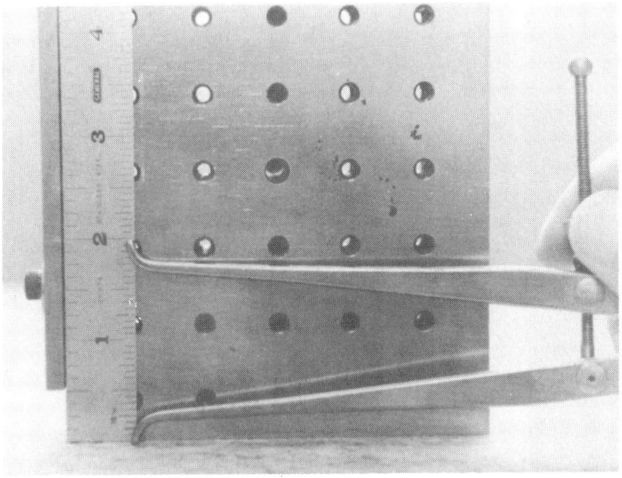

FIGURE 7-11
Setting an inside caliper. (Credit: Photocom Production)

ment nut, which affords greater sensitivity than the hand-controlled firm joint caliper. Outside calipers are semiprecise tools and are usually set against a rule. The trick is to set or read the opening so that the curved legs lie parallel against the straight edge of the graduations. In setting the calipers, one curved leg touches the end of the rule while the other leg is adjusted so that its edge splits the intended measurement graduation (Figure 7-10).

Inside Calipers

Inside calipers also come as a spring or firm joint caliper. They are used to measure slots, holes, and similar openings. In setting the inside caliper for a semiprecise measurement, place the calipers over a rule. One leg should be butted against a common stop, while the other is adjusted on a parallel plane so that its edge splits the graduation line indicated (Figure 7-11). The amount of contact or drag in measuring can be controlled by placing the calipers on

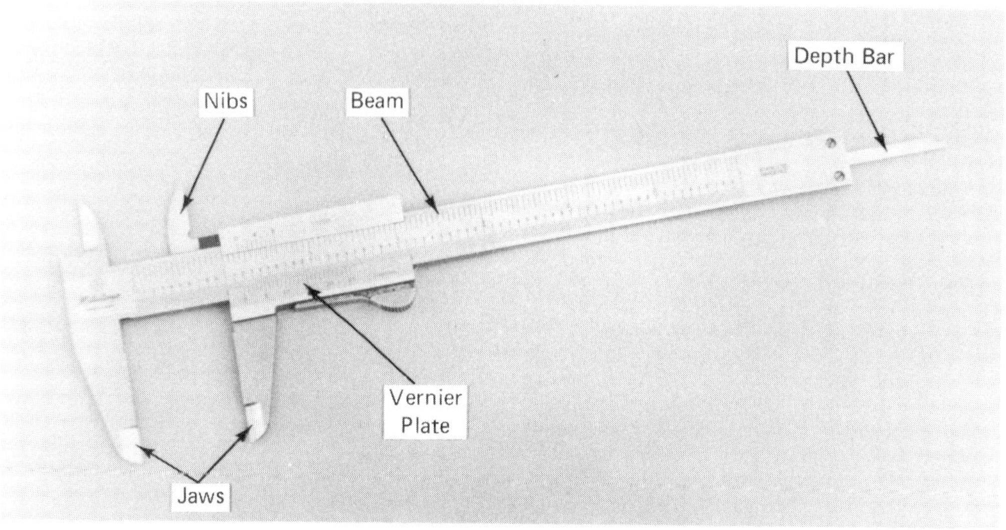

FIGURE 7-12
Principal parts of a vernier caliper. (Credit: Paul Rapisarda)

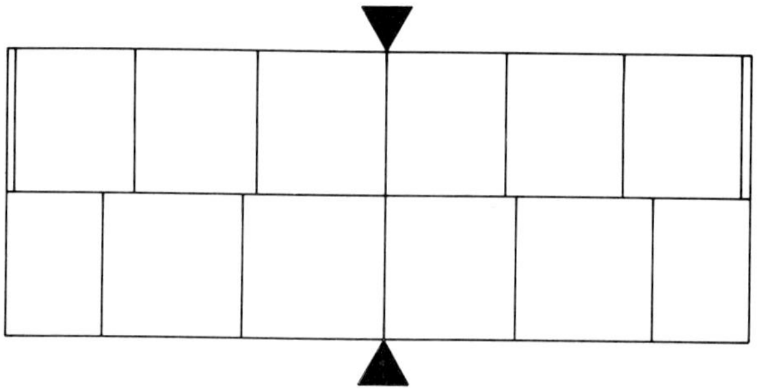

FIGURE 7-13
Matching lines—the key to reading the vernier system. (Credit: Paul Rapisarda)

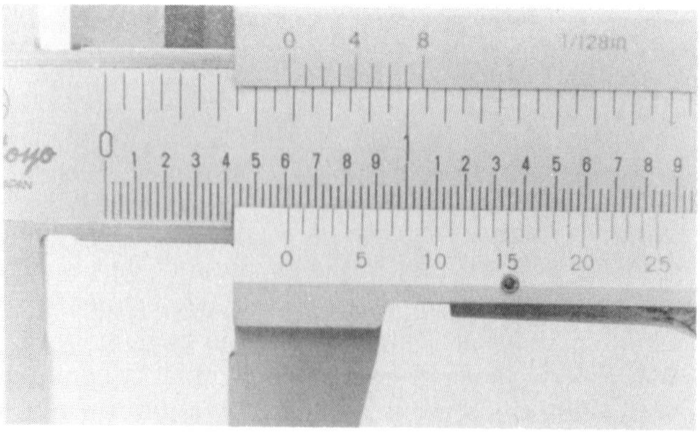

FIGURE 7-14
Setting reads 0.612 in. (Credit: Photocom Productions)

an open palm using the weight of the caliper as the only means of pressure as the caliper is slid through the opening. Never force a caliper inside an opening. It will only give you a false measurement.

Once an opening has been lightly contacted in the manner just described, its measurement can be read by placing the curved legs on a parallel plane against a rule for a semiprecise reading. A more accurate method would be to place the curved legs between an adjusted micrometer to get a reading.

The Vernier Caliper

The vernier caliper is a precision instrument capable of measuring inside, outside, and depth surfaces. Many vernier calipers are equipped with both inch and metric scales. Though their values are different, both are measured and read using the vernier principle of matching lines.

The vernier caliper is an assembly of sliding surfaces with graduated lines on both the beam and the vernier plate (Figure 7-12). Measurement can be taken by sliding the contact points (jaws, nibs, or depth bar) against the surface being tested and then read by combining the values of the beam and vernier plate. The key to reading is based on the merge of graduations on both the beam and vernier plate so as to form one long line (Figure 7-13). Any beam value that falls in front of the zero line on the vernier plate is added to the value indicated on the vernier plate by the merging lines.

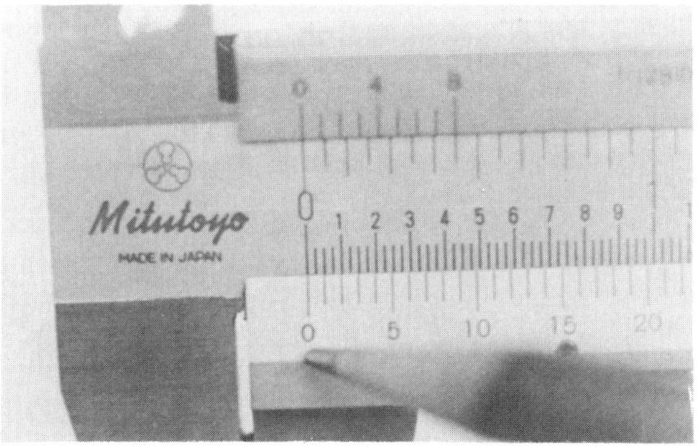

FIGURE 7-15
Zero lines match up when the caliper is closed. (Credit: Photocom Productions)

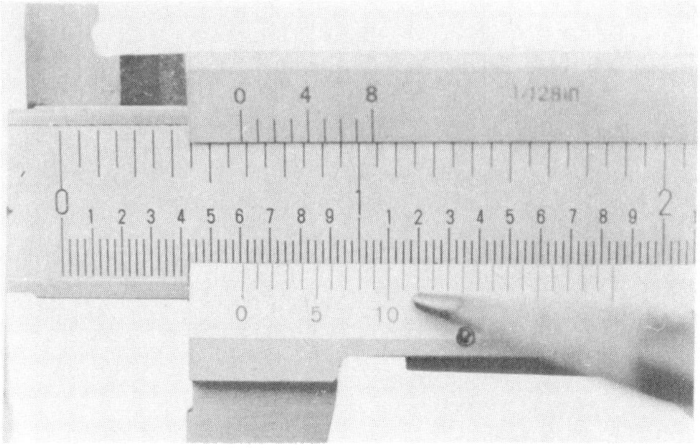

FIGURE 7-17
A match is found on the twelfth graduation. (Credit: Paul Rapisarda)

The Inch System Vernier Caliper

Every inch on the beam of a 25-division caliper is divided into 40 graduations. Since each graduation represents one-fortieth of an inch, converting this fraction will produce a decimal value of 0.025 in. This decimal figure is spoken and read as twenty-five thousandths. The term "thousandths" is the basic mathematical language used when working with a micrometer or vernier caliper in the inch system. Every fourth graduation along each inch length on the beam is slightly longer than the others and is numbered. Since every graduation has a value of 0.025 in., every group of four will be one-tenth of an inch, or 0.1 in. The small 1 on the beam represents 0.100 in. (spoken and read as one hundred thousandths), the small 2 represents 0.200 in. (two hundred thousandths, and the small 3 represents 0.300 in. (three hundred thousandths), etc.

Every graduation on the vernier plate represents 0.001 in. (one thousandth) in face value. Figure 7-14 reveals a setting of 0.612 in. (six hundred twelve thousandths). How this reading was obtained is explained in Figures 7-15 to 7-17.

The Metric Vernier Caliper

The metric vernier caliper (Figure 7-18) is used in the same way as its English counterpart and is read using the following values: each line on the beam represents 1 mm. Since shop drawings are expressed in millimeters rather than larger values, each large number is read in multiples of 10: 10 mm as opposed to 1 cm; 20 mm rather than 2 cm. Each graduation

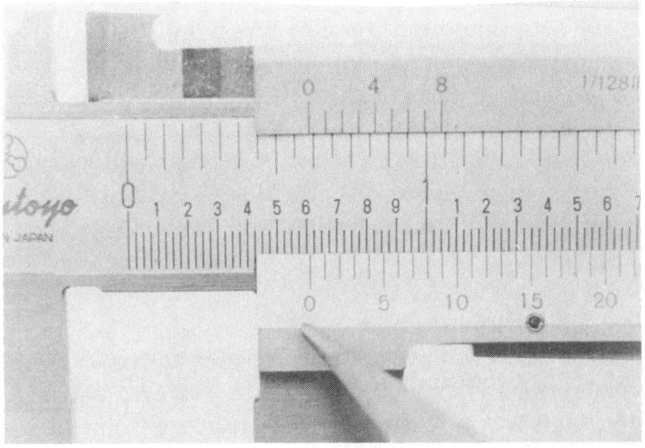

FIGURE 7-16
Zero line on the vernier plate indicates beam reading. (Credit: Paul Rapisarda)

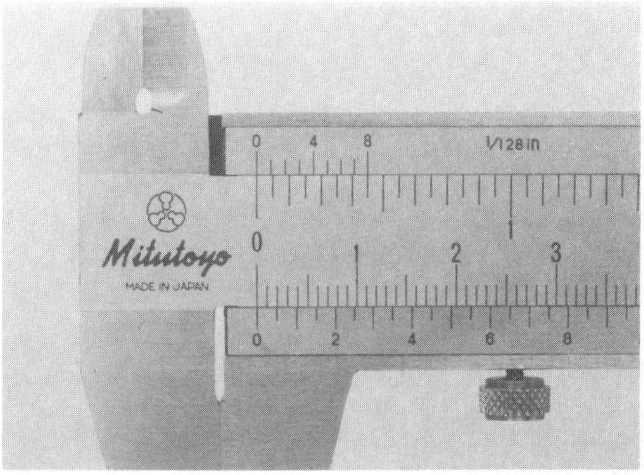

FIGURE 7-18
Metric vernier caliper graduations.

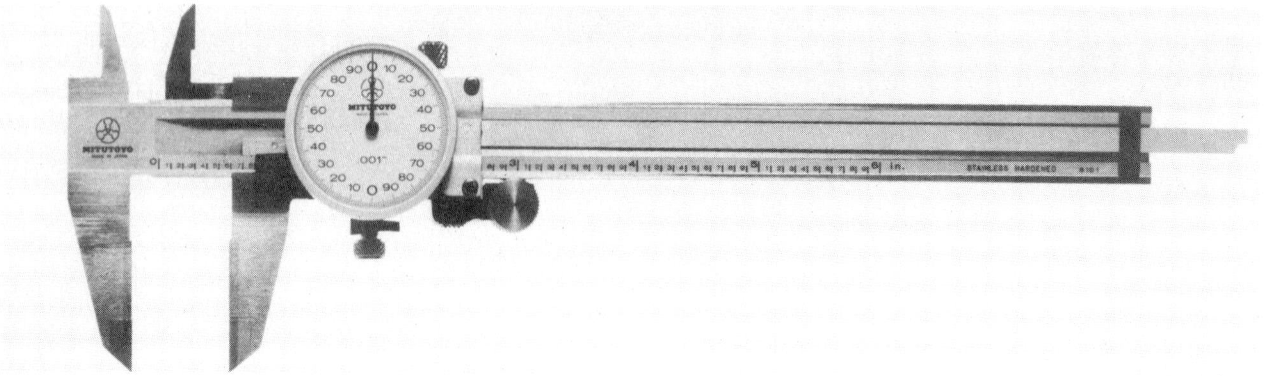

FIGURE 7-19
Dial caliper. (Credit: MTI Corporation)

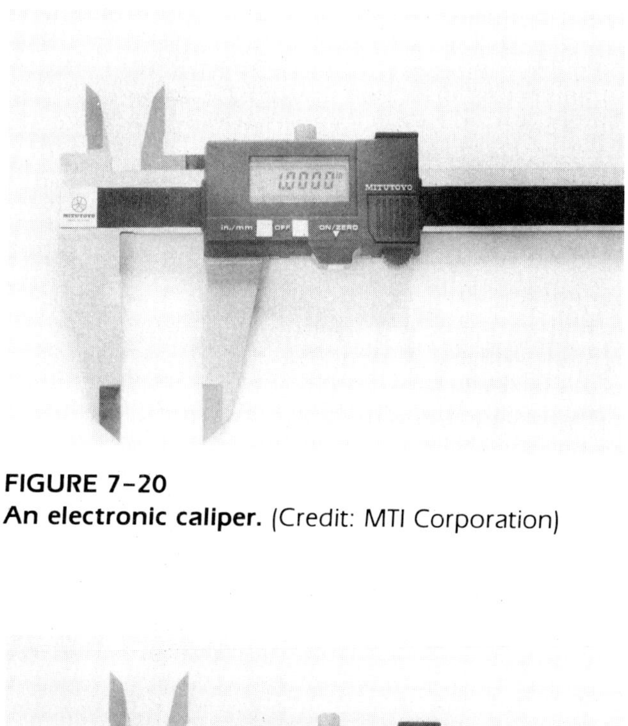

FIGURE 7-20
An electronic caliper. (Credit: MTI Corporation)

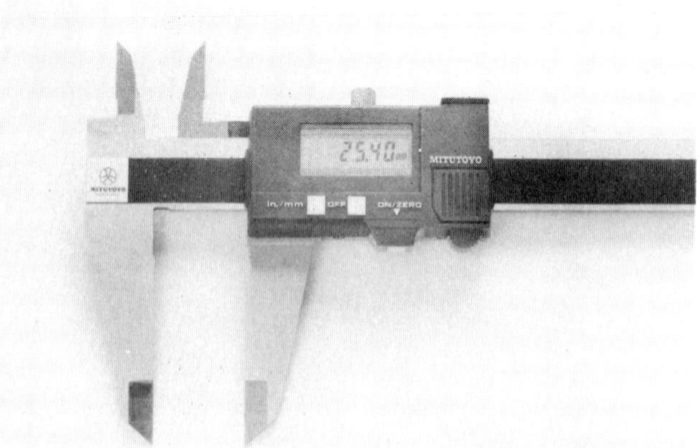

FIGURE 7-21
The inch measurement has been converted to metric at the touch of a button. (Credit: MTI Corporation)

on the vernier plate represents one-twentieth of a millimeter or 0.05 mm (five-hundredths of a millimeter). By combining the beam and the vernier plate values as before, metric readings are obtained.

The Dial Caliper

The dial caliper should not be confused with the vernier caliper. They both have totally different reading systems. While the vernier caliper uses a system of matching lines as its key, the dial caliper is read directly on the face of a dial (Figure 7-19).

Operating on a rack and pinion system, the needle on the face of a dial caliper registers in increments of 0.001 in. (0.02 mm) with each movement of the sliding jaw. One complete revolution of the dial in the English system caliper represents 0.100 in. (2 mm) of travel, with a maximum measuring range of 6 in. (150 mm) for the more widely used instruments. Dial calipers with a maximum reading range of 12 in. (300 mm) and dial revolutions in either 0.100 in. (2 mm) or 0.200 in. (4 mm) are also available. Metric calipers can also be had with graduated face divisions measuring 0.01 mm. Dial calipers can be set and read more quickly than a vernier caliper. Its principal disadvantage is that dirt or a chip can clog the meshing of the rack and pinion gear to give a false reading.

The Electronic Caliper

The electronic caliper (Figures 7-20 and 7-21) features no glass scales, no moving parts to wear or clog, and a switchable inch/metric resolution accurate to a 0.0005-in. (0.01-mm) reading. Its extreme accuracy and repeatability are easily seen on an LCD (liquid crystal display) high-contrast screen and readable in all environments. Not only are absolute measurements made, but at the push of a button the zero mark can be reset for comparative measurements. These battery-powered instruments can also be interfaced with data processing equipment.

UNIT 8 MICROMETERS

In the world of precision machining and measurement, the micrometer is one of the most basic measuring tools (Figure 8–1). Indeed, it is the one tool whose mastery is a requirement to qualify for most jobs in the machine trades. The level that a skilled person will achieve will be largely dependent on the ability to initially measure with a micrometer.

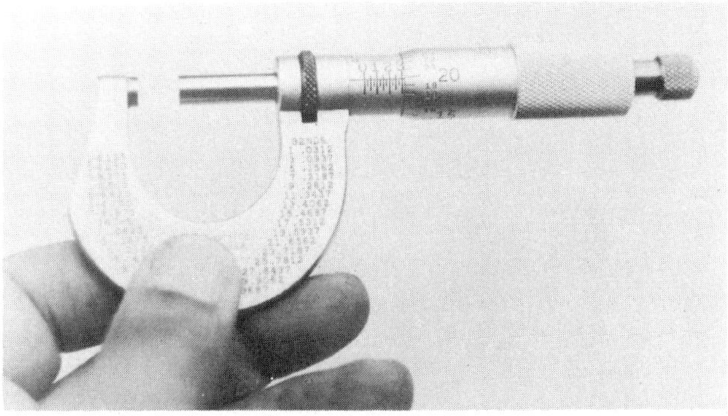

FIGURE 8–1
An outside micrometer caliper. (Credit: Photocom Productions)

Objectives
After completing this unit you should be able to

1. identify the different types of micrometers.
2. use and read the inch outside micrometer to an accuracy of 0.0001 in.
3. use and read the metric outside micrometer to an accuracy of 0.01 mm.

Types and Their Uses

From a humble beginning when its inventor, Jean Palmer, developed a crudelike device whose measuring screw could be seen, the micrometer was redesigned to be a simplified, practical measuring tool. Its measuring screw was also hidden and protected.

Today, the micrometer is used in many instruments to measure depth, threads, inside openings, outside diameters, shoulder thicknesses, thin materials, as well as a variety of special uses, including a nonrotating micrometer to gage relatively soft, compressive material.

The Inch System Micrometer

The inch system micrometer is capable of dividing its 1-in. opening into 1000 equal parts, with its decimal reading called out in thousandths. There are five basic parts to a standard micrometer (Figure 8–2), with calibrated graduations on the sleeve and thimble. A long horizontal line, called the index line, extends the length of the sleeve.

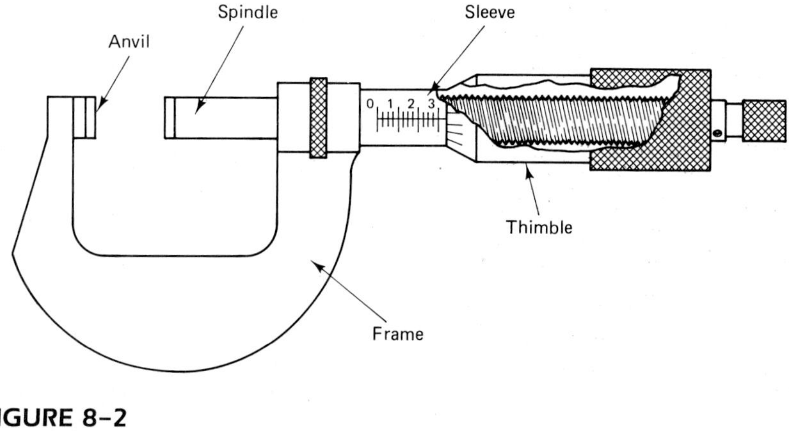

FIGURE 8–2
Principal parts of a standard micrometer. (Credit: Photocom Productions)

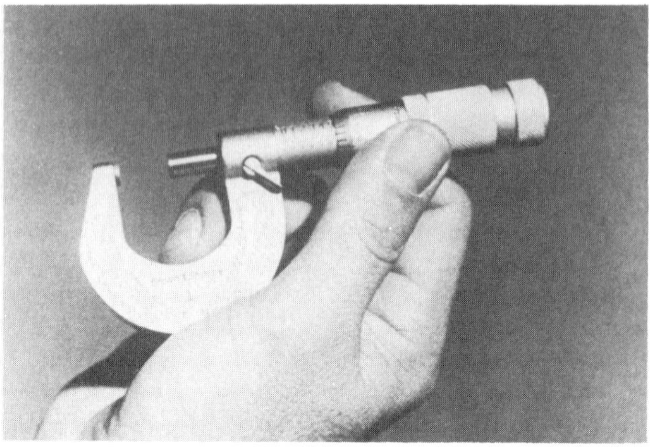

FIGURE 8–3
Hold a micrometer as shown. (Credit: Photocom Productions)

Using the Micrometer

The micrometer works on a very accurate hidden screw designed to measure a 1-in. opening in 40 complete turns. By turning the thimble one complete turn, the measuring surface of the anvil and spindle are opened or closed one-fortieth of an inch or 0.025 in. (twenty-five thousandths). The thimble is marked in 25 equal divisions to register smaller movements.

Procedure

1. Clean the anvil and spindle measuring surfaces by lightly closing on lint-free paper and pulling it through.
2. Roll the knurled thimble across a flat palm or forearm to open or close the micrometer rapidly.
3. The micrometer is generally used with one hand. By anchoring the frame with either of the last two fingers, the thumb and index finger are freed to turn the thimble in either direction (Figure 8-3).
4. Place the clean workpiece so that its full width is on the anvil.
5. Slowly rotate the thimble with the thumb and index finger to make contact with the workpiece. Avoid clamping the object while still retaining light friction.

Reading the Micrometer

Each line on the thimble represents 0.001 in. (one thousandth), while each line on the sleeve represents 0.025 in. (twenty-five thousandths). Every fourth line is numbered and represents one-tenth of an inch or 0.100 (one hundred thousandths). It is by adding the values of the graduations you see on the sleeve to the value of the thimble as indicated by the index line that a reading is provided. See Figures 8-4 and 8-5.

But suppose the index line rests between two graduations on the thimble. How is this measurement read? Readings in ten thousandths of an inch are possible by using a micrometer with a vernier scale located above the sleeve. The vernier scale makes it possible to split each thousandth into ten equal divisions. When a line on the thimble can be found to merge with a line on the vernier scale, this number is added as the fourth decimal place to the regular reading, as shown in Figures 8-6 and 8-7.

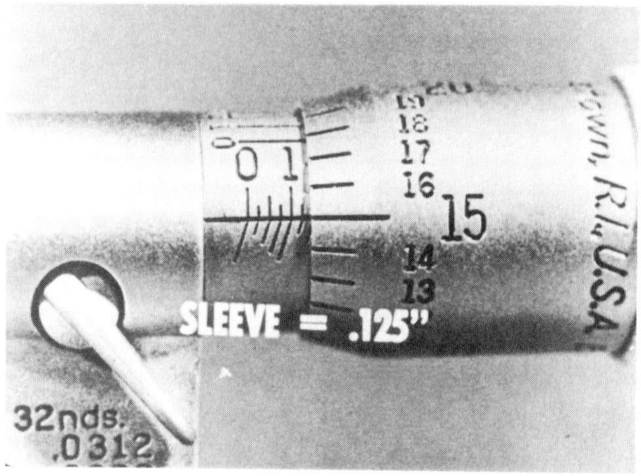

FIGURE 8–4
(Credit: Photocom Productions)

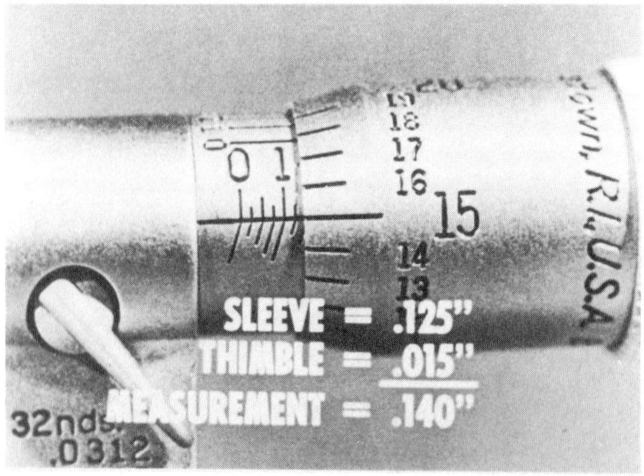

FIGURE 8–5
(Credit: Photocom Productions)

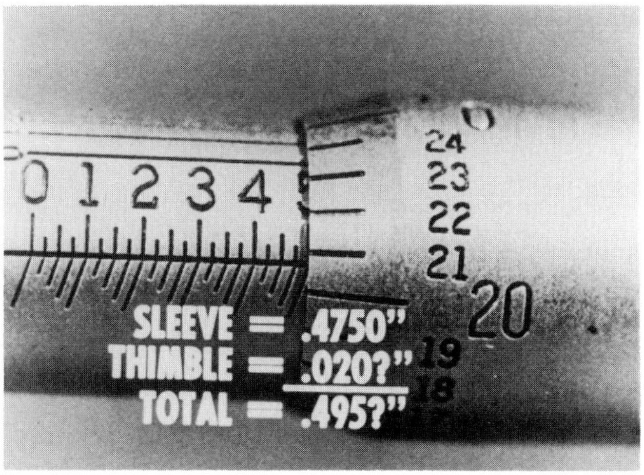

FIGURE 8-6
(Credit: Photocom Productions)

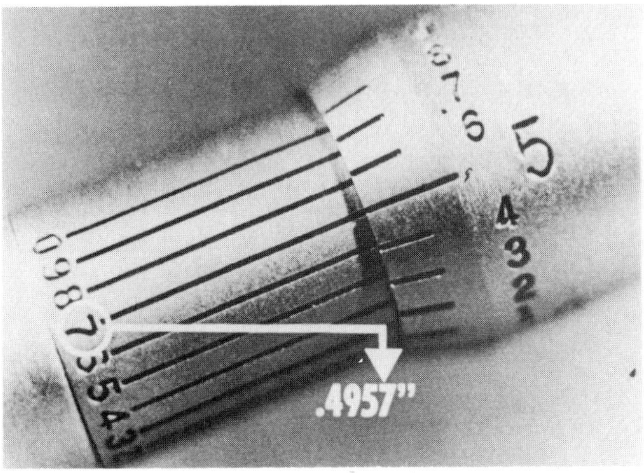

FIGURE 8-7
(Credit: Photocom Productions)

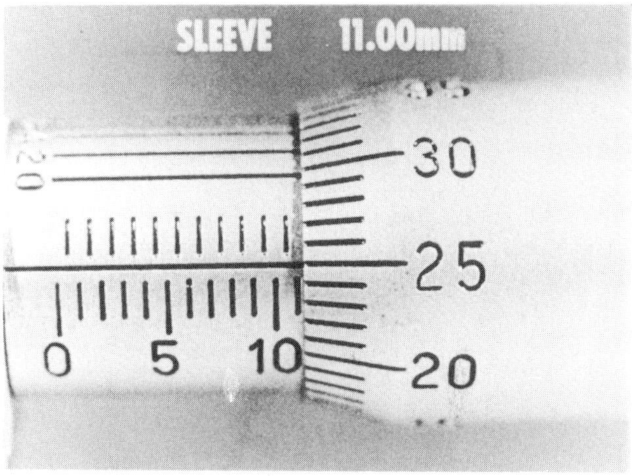

FIGURE 8-8
(Credit: Photocom Productions)

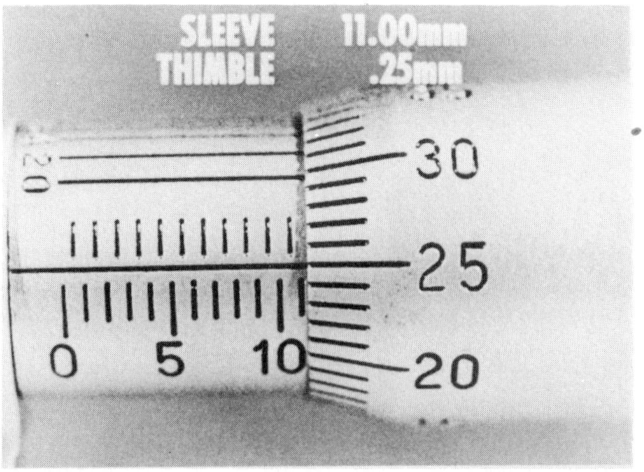

FIGURE 8-9
(Credit: Photocom Productions)

The Metric System Micrometer

The micrometer used for metric measurement is no different in design and basic parts than its inch system counterpart. The difference lies in graduated values on the sleeve and thimble. Whereas the capacity of the inch system micrometer is one inch, the capacity of the metric micrometer is 25 mm. Its hidden measuring screw works on 50 complete turnings to move the spindle 25 mm. Consequently, it will require two complete turnings to move the spindle 1 mm. The thimble is marked with 50 graduations. Each graduation will represent 0.01 mm (one hundredth of a millimeter) in value.

Reading the Metric Micrometer

Reading a metric micrometer will require combining the values of the graduations on the sleeve and thimble. To ease the reading, the sleeve is provided with two rows of graduations. The upper row represents whole millimeters, with every fifth graduation marked for easy identification. Figures 8-8 to 8-11 demonstrate how a metric micrometer is read.

The vernier scale is stamped on top of the sleeve to allow you to read measurements that fall between the graduations on the thimble. As Figures 8-12 and 8-13 illustrate, it is a matter of finding a line on the vernier scale that merges with any line on the thimble. This number is used to add another decimal place to the reading.

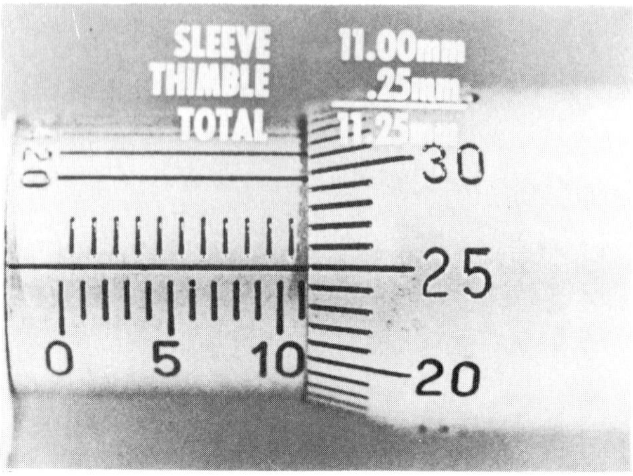

FIGURE 8-10
(Credit: Photocom Productions)

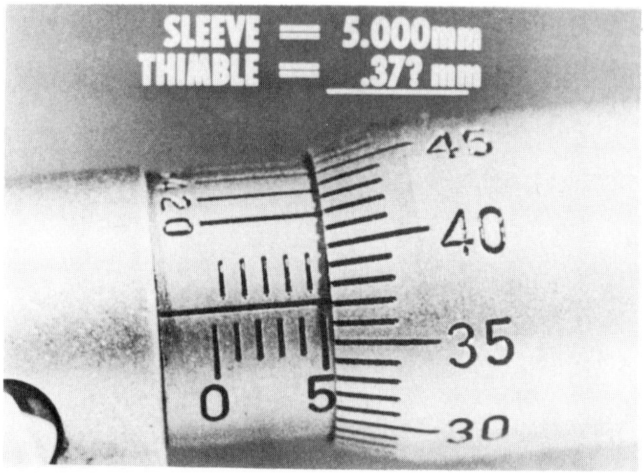

FIGURE 8-11
(Credit: Photocom Productions)

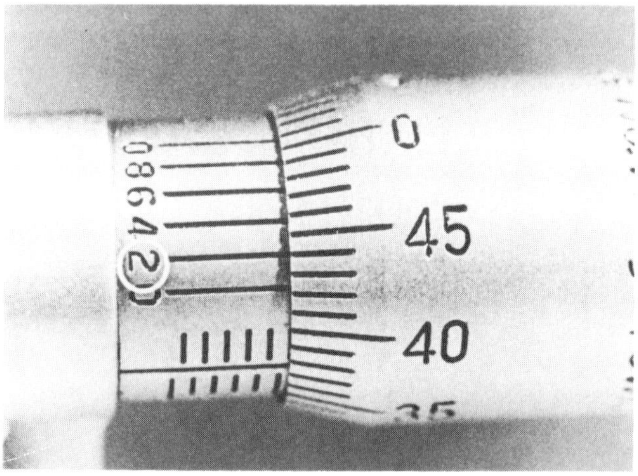

FIGURE 8-12
(Credit: Photocom Productions)

Micrometer Adjustment

Micrometers are quality measuring tools. While they seldom require adjustments, conditions may arise that will require the removal of looseness or play in the spindle threads. These conditions are generally caused by a long time use (wear), or the machinist may simply want to change the measuring "feel" established at the factory. In either case play can be eliminated with a C-spanner wrench (Figure 8-14) and by following the steps outlined here. It is important that any adjustment be carried out while keeping the instrument free of dirt and foreign matter. Do not use an air gun to blow the instrument free of any grit or dust particles. Not only is this practice harmful because it can impact dirt and other substances into hidden crevices to obstruct turning but also moisture from compressed air can be trapped as well.

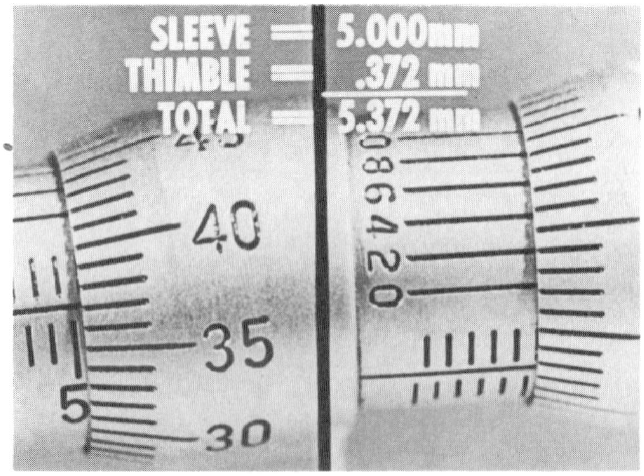

FIGURE 8-13
(Credit: Photocom Productions)

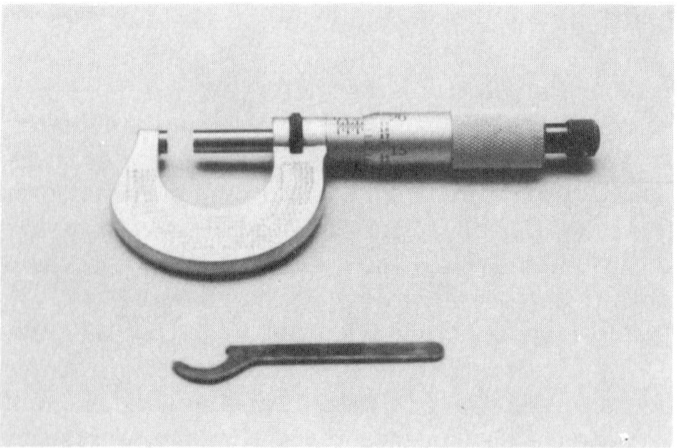

FIGURE 8-14
Micrometer adjustment is accomplished with a C-spanner wrench.

Procedure

1. Back off the thimble to expose the adjustment ring as shown in Figure 8-15.

2. Insert the C-spanner wrench into the notched opening.

3. Turn the adjustment ring clockwise until all play between the threads has been eliminated (Figure 8-16).

4. Close the micrometer to check that the friction thimble or ratchet stop can advance the spindle freely.

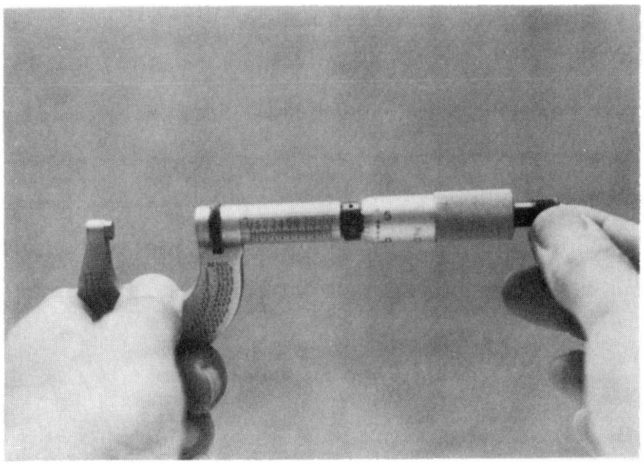

FIGURE 8-15
Back off the thimble to expose the adjustment ring.

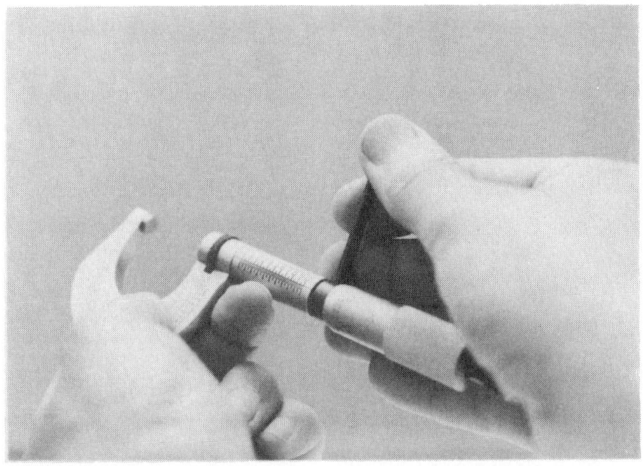

FIGURE 8-16
Tighten adjustment ring with the spanner wrench to eliminate thread play.

English 0-1 in. micrometer				
0.105	0.210	0.315	0.420	0.500
0.605	0.710	0.815	0.920	1.000

Metric 0-25 mm micrometer				
2.5	5.1	7.7	10.3	12.9
15.0	17.6	20.2	22.8	25.0

(According to Federal Specification GGG-C-105b)

FIGURE 8-17
Ten-point gage block check. (Credit: MTI Corporation)

Micrometer Calibration
To assure accuracy, a micrometer should be calibrated periodically using a set of gage blocks as standards. Before attempting any testing, however, all surfaces should be clean. This would include the anvil, spindle, and gage blocks. It is also recommended that testing of a micrometer be done on at least ten different measurements (Figure 8-17), using the friction thimble or ratchet stop, and that the micrometer be held in a clamping device not in one hand. While the ten-point check will accurately measure a micrometer's accuracy along its length, it will not reveal any measuring inaccuracies due to center wear of the anvil and spindle faces. This test would be performed by optical flats (see Unit 15).

Procedure for Calibration

1. Check and clean the anvil and spindle faces for dirt, dust, and damage.

2. Carefully close the anvil and spindle faces by turning the friction thimble or ratchet stop.

3. Examine the micrometer setting. Does the index line split the zero line on the thimble (Figure 8-18)?

4. If adjustment is required, tighten the spindle lock clamp.

5. Insert the C-spanner wrench into the hole provided in the sleeve (Figure 8-19A).

6. Rotate the sleeve until the index line matches the thimble zero line perfectly (Figure 8-19B).

7. Loosen the spindle lock clamp and recheck the setting by using the friction thimble or ratchet stop. Repeat steps 1 thru 6 if not accurately set.

8. Clamp the micrometer to a stand or fixture.

9. Clean the 10 gage blocks.

10. Test the micrometer for wear and variables in the measuring screw by turning the friction thimble or ratchet stop to close the measuring faces against each gage block.

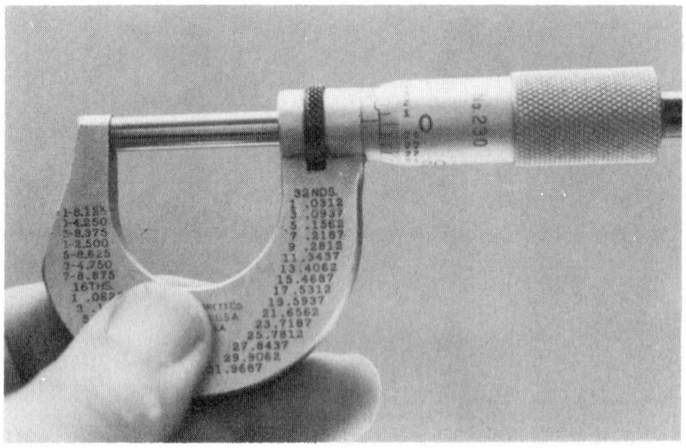

FIGURE 8–18
Inaccurate micrometer closing.

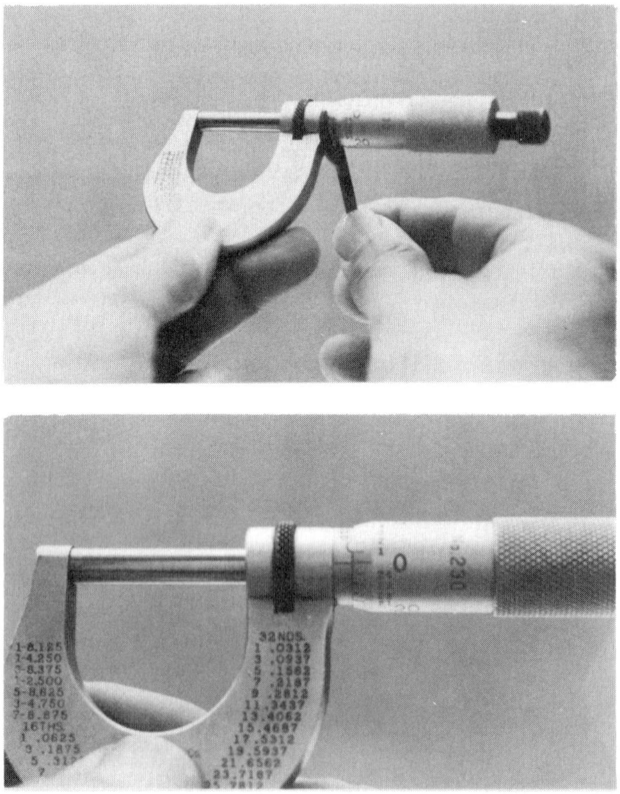

FIGURE 8–19
Use the C-spanner wrench to align the index line with the zero setting.

Micrometer Maintenance
While the micrometer requires little maintenance, it is still a good idea to wipe it clean of moisture and finger marks after use. When storing it, a small gap should be left between the anvil and spindle faces to prevent rust from developing. Finally, if the micrometer is not to be used for a while the measuring surfaces should be covered with a very light film of instrument oil.

Digital Micrometers

Digital readout micrometers, which simplify the reading of a measurement, are also available. These micrometers can be had in combination of inch and metric values (Figure 8–20): The measurement in inch can be had by combining the values on the sleeve and thimble, while its comparable metric value can be read in the digital readout box. Combination micrometers can also reverse the values, with the sleeve and thimble reading in metric and the digital readout in inch. Digital readout micrometers where the digital readout and sleeve and thimble values are read in one system only are also available.

Electronic Micrometers

Advanced technology has made possible the use of the electronic micrometer that boasts a high-contrast LCD screen, instant inch to millimeter conversion at the push of a button, and accuracy to 50-millionths of an inch. The electronic micrometer (Figure 8–21) can be set to any measurement and recalled as well. In addition to absolute and relative zero settings, the battery-powered electronic micrometer can be interfaced with data processing equipment.

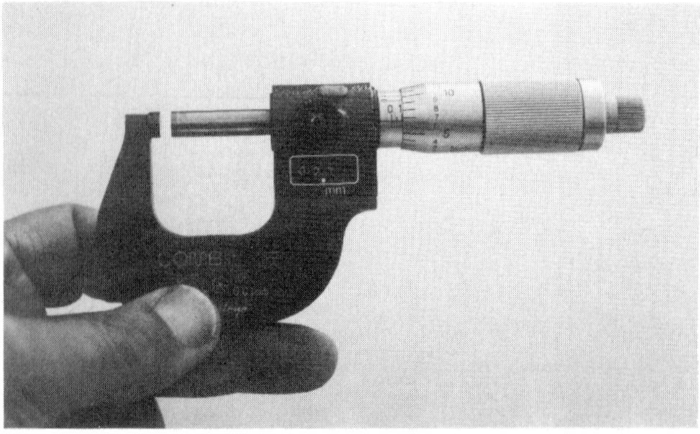

FIGURE 8–20
Digital readout micrometer. (Credit: Paul Rapisarda)

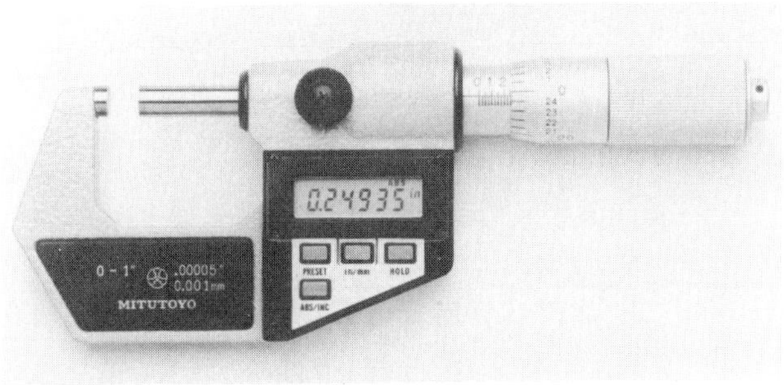

FIGURE 8–21
Electronic micrometer. (Credit: MTI Corporation)

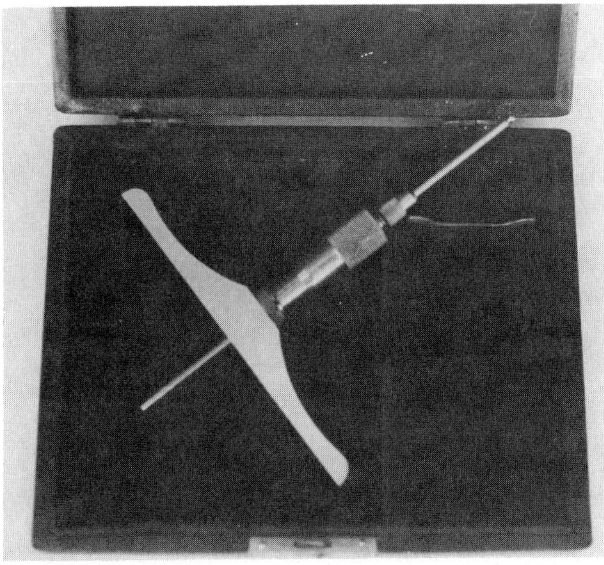

FIGURE 8–22
A wide-base micrometer depth gage.

Micrometer Depth Gage

Constructed with a flat base attached to a micrometer sleeve and thimble, the micrometer depth gage (Figure 8–22) is the most commonly used tool designed to measure the bottom of holes, slots, steps, shoulders, or surface projections. Due to the fact that the instrument has a limited range of 0 to 1 in. (0 to 25 mm) measurement, many depth micrometers come in sets that provide precalibrated depth rods designed to be inserted, properly seated, and locked in place in the end of the instrument to extend through the base. The rods come in varied lengths—as much as 9 in. for inch tools and 225 mm

FIGURE 8–23
Micrometer depth gage sets come with precalibrated depth rods. (Credit: L.S. Starrett Company)

for metric tools—and are hardened, ground, and lapped for long lasting wear and accuracy (Figure 8–23).

Reading the Depth Micrometer
The graduations on a depth micrometer have the same values as the standard outside micrometer only the depth micrometer is read from right to left. Consequently, in taking any reading on the sleeve, count the number that is covered by the edge of the thimble first. Add to this value any number of graduations (0.025 in. or 0.05 mm) also covered by the thimble. Add to this combined reading any value on the thimble indicated by the index line.

1. Clean the surface plate free of dirt, dust, or oil.

2. Remove any burrs from the workpiece edges—hole, slot, shoulder, etc.—and place on the surface plate.

3. Hold the depth micrometer base firmly against the surface of the workpiece with one hand.

4. Carefully turn the friction thimble clockwise with the fingertips of the other hand until the bottom of the extension rod touches the surface depth being measured (Figure 8–24).

5. Carefully read the setting as described above.

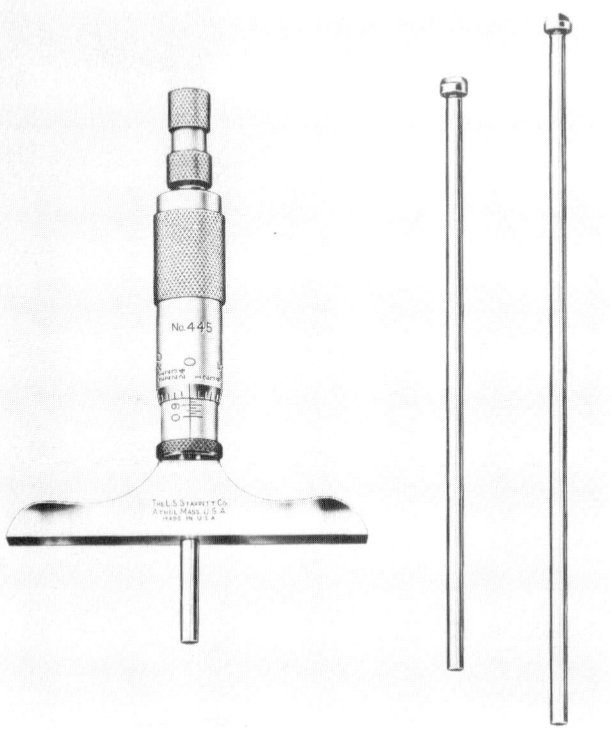

FIGURE 8–24
Measuring the bottom of a shoulder with a micrometer depth gage. (Credit: L.S. Starrett Company)

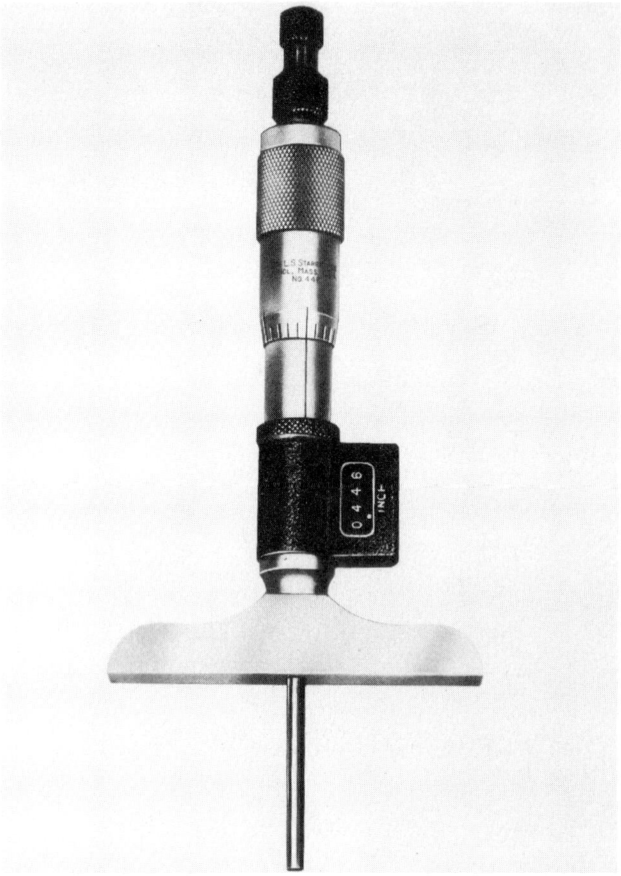

FIGURE 8–25
A digital micrometer depth gage. (Credit: L.S. Starrett Company)

Testing the Depth Micrometer

Like any other precision instrument, depth micrometers should be tested for accuracy, more so when different extension rods have been inserted for use. On some micrometer depth gages, the accuracy of the setting can be controlled by a nut on the end of the extension rod.

Procedure for Testing 0 to 1 in. (0 to 25 mm)

1. Clean the surface plate free of dirt, dust, or oil.
2. Clean off the tip of the measuring rod.
3. Turn the thimble counterclockwise to retract the measuring rod just below the surface of the base.
4. Holding the 0- to 1-in. (0- to 25-mm) instrument flat on the surface plate, turn the friction thimble to test the measuring rod's contact. The reading should be 0.000 in. (0.00 mm).
5. Further check a second point by stacking up two columns of wrung gage blocks of equal height, say 0.500 in. (12.7 mm).
6. Straddle these gage block heights with the base of the depth micrometer.
7. Holding the base firmly with one hand, carefully turn the friction thimble until the measuring rod touches the surface plate. The micrometer reading should be the same as the gage block height.

Digital Micrometer Depth Gage

Held and measured using the same methods previously explained, the digital micrometer depth gage offers accuracy and speed in reading. There are no numbered markings on the sleeve, nor does the instrument have to be studied closely to be read. A digital readout mounted above the base supplies a direct reading. Digital micrometer depth gages are available in either inch or metric models (Figure 8–25).

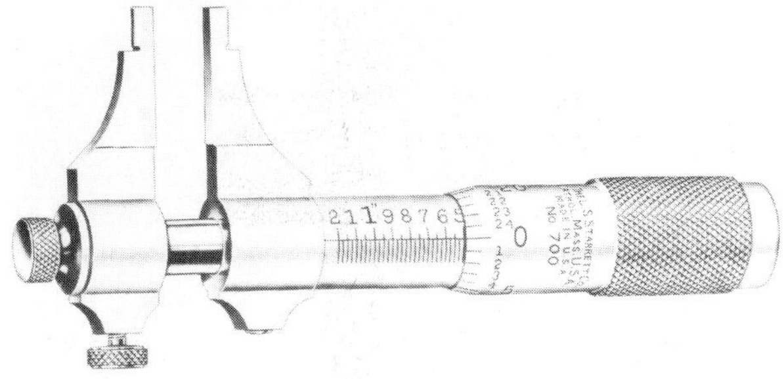

FIGURE 8-26
Inside micrometer caliper. (Credit: L.S. Starrett Company)

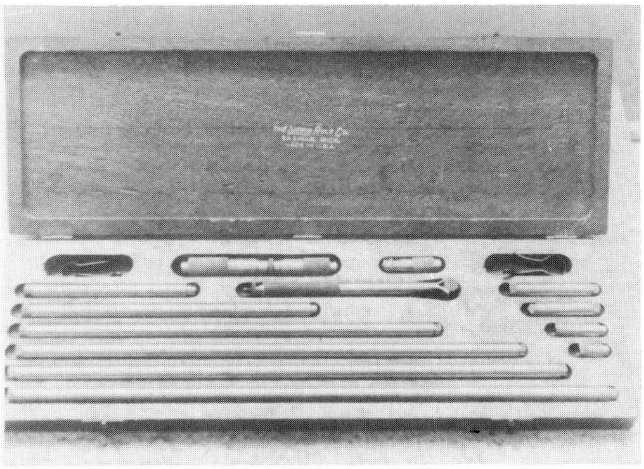

FIGURE 8-27
A set of inside micrometers. (Credit: Photocom Productions)

Inside Micrometer Instruments

The inside micrometer may come as an individual inside micrometer caliper (Figure 8-26) or in a set (Figure 8-27). Either is designed to measure holes, slots, grooves, bores, and similar openings as accurately as a standard micrometer would. Both types are available in inch or metric values.

The nibs on an inside micrometer caliper will not be able to measure openings smaller than 0.200 in. and ranges to 1.220 in.; similar type metric micrometer calipers may range from 5 to 30 mm. While the instruments have the same values as the standard micrometer, the sleeve readings may be reversed, similar to that of a depth micrometer. Consequently, a certain precaution will have to be undertaken in reading this type of instrument.

Using an Inside Micrometer Caliper

1. Close the jaws smaller than the opening to be measured.

2. Hold one nib against the side of the opening and turn the thimble until the movable nib is in light contact with the opposite side.

3. Pivoting the solid nib, adjust the thimble, swinging the movable nib in small arcs, back and forth past the hole's centerline until the proper "feel" of contact is attained.

4. Lock the setting, remove the instrument, and carefully check the readings.

Inside Micrometer Set

A set of inside micrometers is used for openings that are no smaller than 1.5 in. and range as wide as 100 in. or more. The metric counterpart will vary with the manufacturer and may range from 40 to 1000 mm or greater, with the extension rods determining the actual measurement in both cases. The set may consist of a micrometer head (one or more) having a measurement range of 1 in. (25 mm), a handle and contact points on the head, and extension rods made spherical and hardened for longer wear. Extension rod additions will vary with the set being used, with some sets permitting addition on both sides of the body, thereby increasing the span of measurement.

Reading the inside micrometer is no different than reading a standard micrometer. Since there is no locking device to hold the reading on some sets, reading will be greatly affected by the looseness between the mating threads of thimble nut and spindle threads. A tighter union between these threads will prevent accident moving of a setting when removing the inside micrometer from measurement. For closer tolerance work, the inside micrometer with added rod should be tested with a known standard or ring gage prior to using.

Using an Inside Micrometer

1. Measure the hole diameter to be tested with a rule.

2. Subtract 1 in. (25 mm) from the rule dimension and add the proper extension rod to span the measurement.

3. Tighten the extension rod firmly in place.

4. Tighten the handle on the extension rod at about midpoint of the total extension (Figure 8-28).

5. Extend length of the inside micrometer to slightly less than the diameter to be measured.

6. Pivoting the micrometer head in place, adjust the micrometer by moving the rod end in a small arc, back and forth past the centerline.

7. The proper contact "feel" is reached when a light drag is created as the micrometer is pulled out along the centerline of the hole.

8. Carefully remove the inside micrometer by pulling it out by means of the handle (Figure 8-29).

Special Micrometers

Of wide interest is the multianvil micrometer (Figure 8-30). Designed to hold a variety of interchangeable anvils, the multianvil micrometer can be used to measure tube thicknesses, cylinder walls, shoulders, slots, grooves, lands, fins, as well as outside dimensions. Measurements usually done by calipers, such as measuring the distance between a hole and the edge of a workpiece, can also be done with a multianvil micrometer. Removing the clamping device and placing the reference edge against a flat surface, such as a surface plate, allows the mutlianvil micrometer to be used as a miniature height gage. In all cases it is read as any standard micrometer is read.

The V-anvil micrometer (Figure 8-31) is designed to check measurements of three or five fluted end mills, taps, reamers, and cutters with similar sided edges. It is especially suited for checking concentricity of parts ground by the centerless grinder.

The blade micrometer (Figure 8-32) is another measuring instrument that features a nonrotating spindle. It is especially useful for measuring slots, keyways, recessed narrow grooves, and workpieces

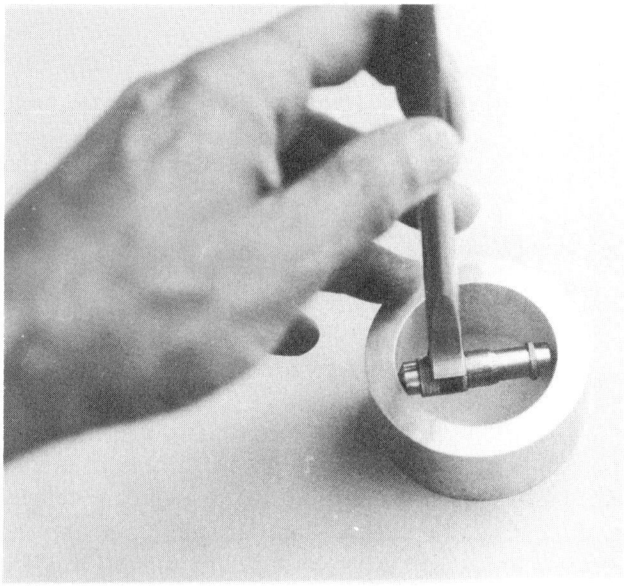

FIGURE 8-29
Carefully withdraw the inside micrometer by the handle.

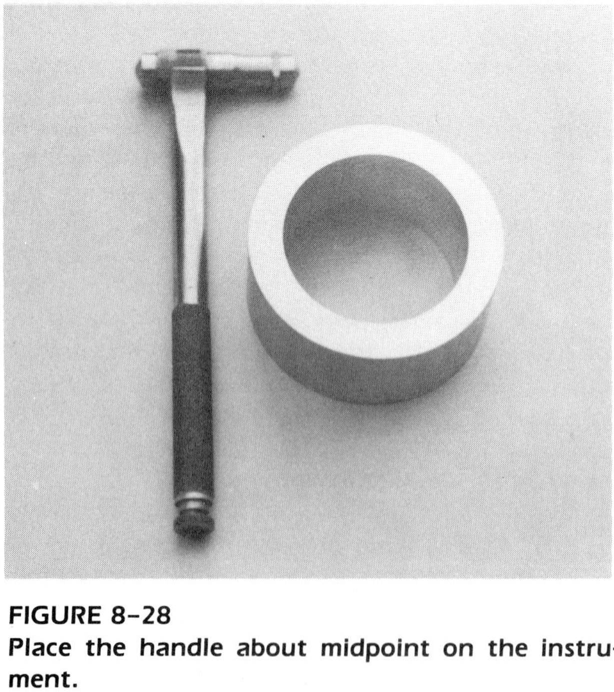

FIGURE 8-28
Place the handle about midpoint on the instrument.

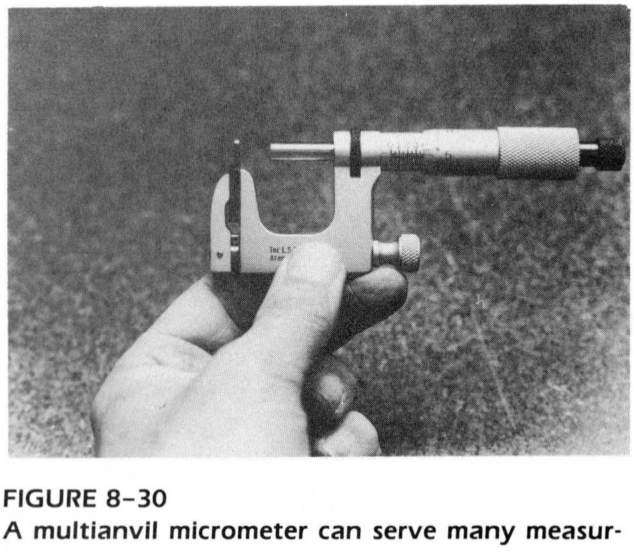

FIGURE 8-30
A multianvil micrometer can serve many measuring purposes.

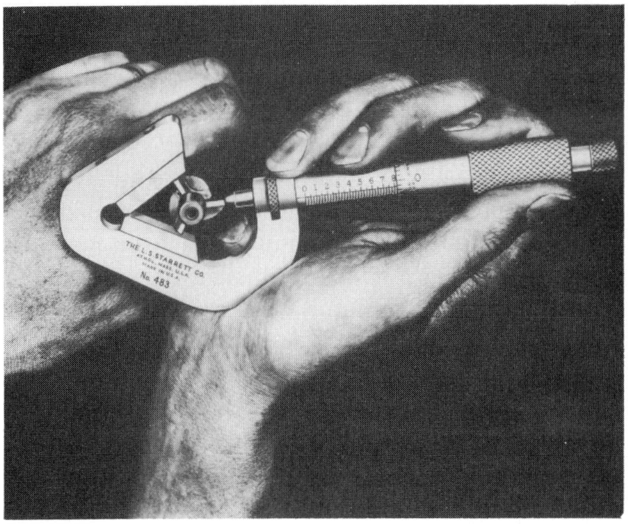

FIGURE 8-31
A V-anvil micrometer can measure out-of-round ground stock as well as 3- or 5-sided cutting tools. (Credit: L.S. Starrett Company)

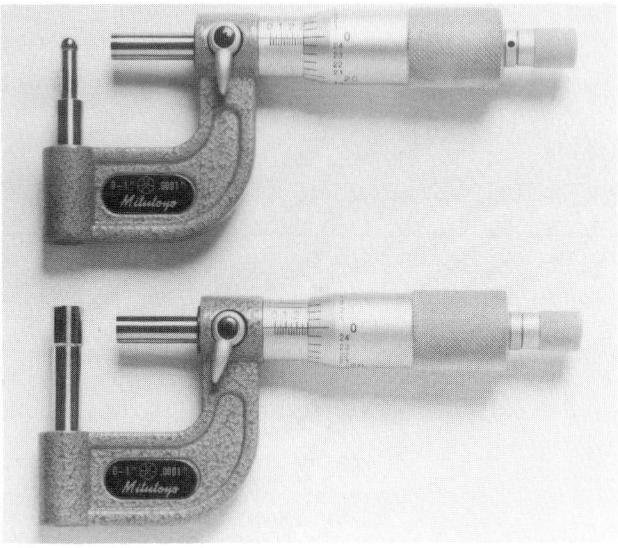

FIGURE 8-33
A ball anvil micrometer is designed to easily measure convex surfaces. (Credit: MTI Corporation)

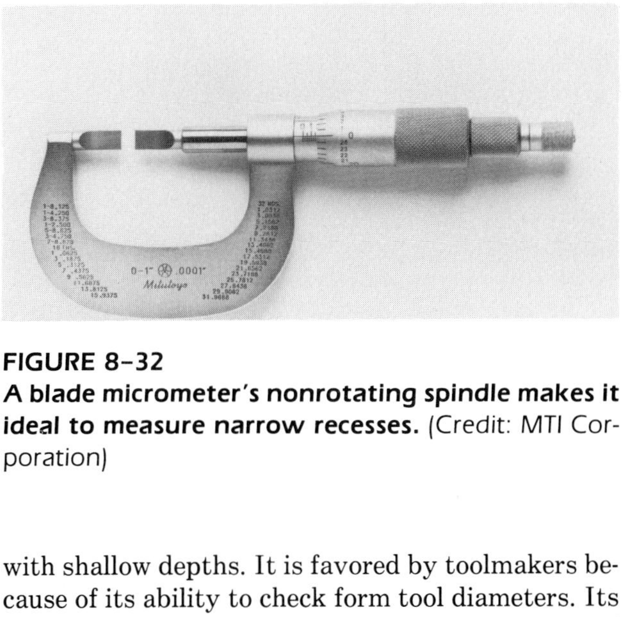

FIGURE 8-32
A blade micrometer's nonrotating spindle makes it ideal to measure narrow recesses. (Credit: MTI Corporation)

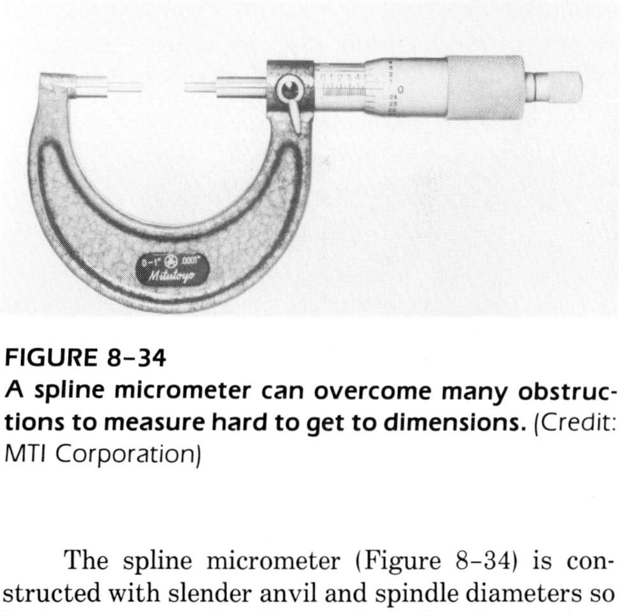

FIGURE 8-34
A spline micrometer can overcome many obstructions to measure hard to get to dimensions. (Credit: MTI Corporation)

with shallow depths. It is favored by toolmakers because of its ability to check form tool diameters. Its straight linear motion is accomplished by means of a keyway located on the spindle axis.

The ball anvil micrometer (Figure 8-33) is basically designed to measure convex surfaces. Wall thicknesses of tubing whose inside diameter can slip over the anvil, which has very good work clearance, are easily measured. Unlike any ball attachment, which can convert a standard outside micrometer and is read by subtracting the attachment ball diameter from the measuring reading, the ball anvil micrometer is read directly.

The spline micrometer (Figure 8-34) is constructed with slender anvil and spindle diameters so it can overcome shoulders and similar obstructions to measure hard to reach diameters.

The disc micrometer (Figure 8-35) is named after the shape of its anvil and spindle measuring faces. These shapes better enable its user to measure narrow grooves, slots, recessed features, as well as take measurements on sections such as ribs, fins, lands, cutting edges on form tools, and chordial thicknesses of gear teeth.

The nonrotating spindle disc type (Figure 8-36) has another advantage. It does not create a

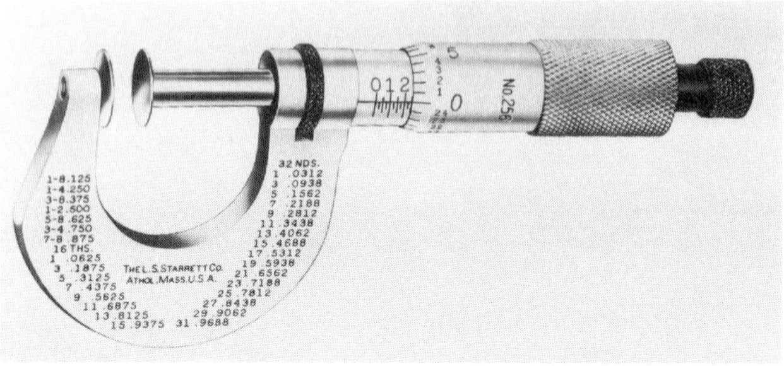

FIGURE 8–35
A disc micrometer can measure flat surfaces as well as cutting edges on form tools. (Credit: L.S. Starrett Company)

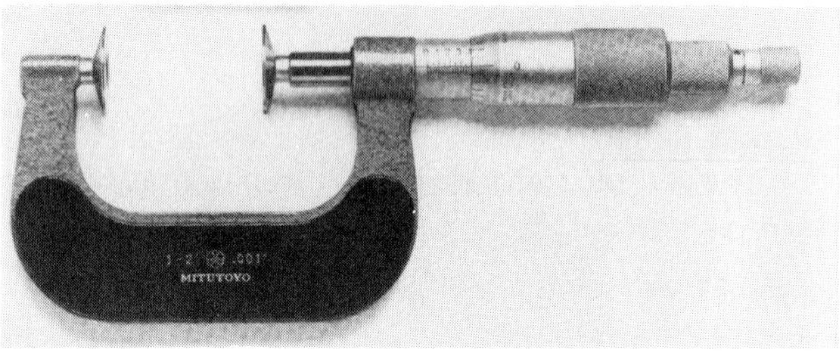

FIGURE 8–36
A nonrotating spindle disc micrometer measures with a uniform linear contact. (Credit: MTI Corporation)

FIGURE 8–37
Micrometer heads are used for precise measurement adjustments on machines and electronic and scientific devices. (Credit: MTI Corporation)

twisting contact (radial torsion) in measuring and is recommended for fragile workpieces such as foil, shim stock, laminated sheets, Mylar-based film, etc.

Micrometers on Machines

Micrometer heads (Figure 8–37) are incorporated in many applications where very precise linear adjustments are required. This may range from machine tools, gages, xyz mounts, and other mechanical devices. A bench micrometer, for example, with a spindle that is adjusted by a wide-diameter thimble construction, uses a micrometer head for repetitive inspection of standard workpieces within very close tolerances (Figure 8–38). In the field of medicine, they are used for surgical procedures. Micrometer heads come in inch and metric measurements as fine as 0.0001 in. or 0.01 mm readings, as well as direct-reading digital models (Figure 8–39).

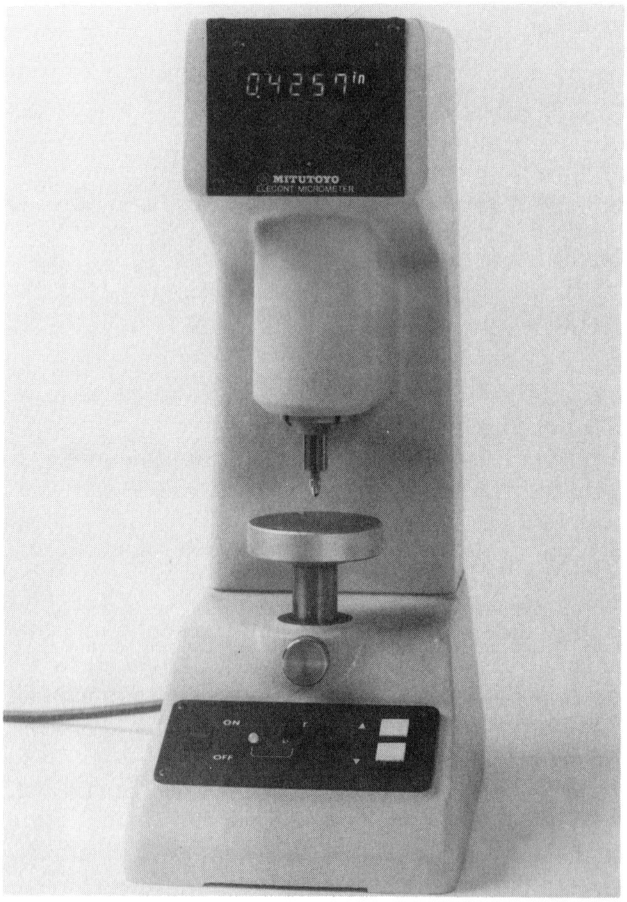

FIGURE 8–38
A bench micrometer with electronic gaging. (Credit: MTI Corporation)

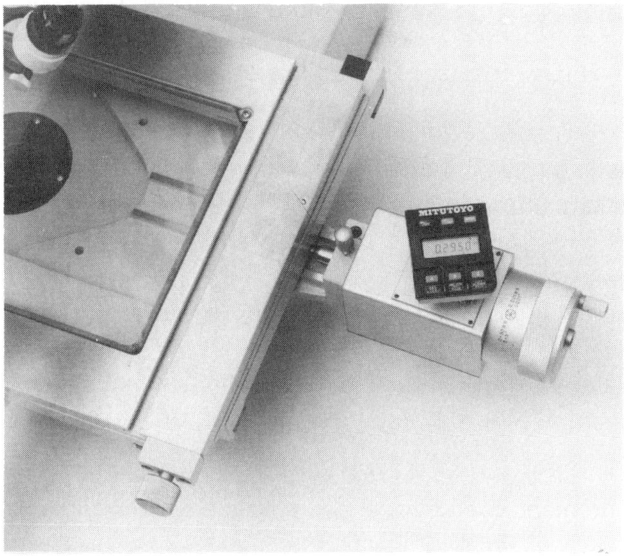

FIGURE 8-39
A direct-reading digital micrometer head. (Credit: MTI Corporation)

UNIT 9 GAGE BLOCKS

The concept of gage block measuring originated with Carl Johansson, an engineer who was working for a Swedish rifle factory around the end of the nineteenth century. At the time that his company received an order to produce magazine rifles, precision measurement was a slow and cumbersome process. The only method available used individual steel blocks that were made for a specific dimension. These rifles would require many precise component parts. Johansson recognized a need for a more practical way to control accuracy. His idea was to produce a set of blocks varying in a wide range of precise measurement. They would have opposite sides lapped precisely flat and parallel to each other, increasing gradually in size in uniform increments so that they could be used individually or collectively in sets. These blocks would meet any production measurement requirement.

Eventually these gage blocks, or "Jo blocks" as they would be called throughout the world, were made and revolutionized the means of practical measurement in the toolroom, the inspection room, as well as in areas where extreme accuracy in metalworking is needed.

Objectives
After completing this unit you will be able to

1. describe how and where gage blocks are used.
2. wring gage blocks together.
3. calculate inch and metric gage block buildup.
4. understand how to take care of gage blocks.

Precision Gage Blocks

Known for their accuracy and reliability, gage blocks come square shaped or rectangular. They are made from alloy steel seasoned in repeated cycles of deep freezing and extreme heat, a process that leaves their crystalline structure free from warping and growing. Hardened and ground, these blocks are precisely lapped and polished to leave their surfaces optically flat and smooth, accurate within a few millionths of an inch (50-millionths of a millimeter). Their surfaces are so fine and parallel that when they are slid against each other with inward pressure (wringing), they will stick together. The exact reason this happens is not clearly understood. What is known is that properly wrung blocks have held as much as 200 lb (90 kg) on direct pull.

Gage Block Classifications

Gage blocks may be purchased in sets varying in combinations of sizes and durability. Chrome-plated sets or solid carbide blocks are available for gages requiring longer wear. They are normally rated in accuracy according to three classifications and calibrated at a temperature of 68°F (20°C). It should be noted, however, that temperature variations will affect the size of the blocks.

The class 1 set, commonly called a laboratory or master set, is used in a temperature-controlled room as a reference in checking the accuracy of measurement and inspection gages.

	Inch	*Metric*
Class 1	0.000002	0.00005 mm
Class 2	0.000004	0.00015 mm
		0.00005 mm
Class 3	0.000008	0.00025 mm
		0.00015 mm

The class 2 set, sometimes called inspection gage blocks, is used for setting and measuring inspection tools and for checking dimensions of manufactured parts.

The class 3 set, referred to as the working set, is used in a variety of machine shop uses, from surface plate layout to machine tool setup to common inspection measurements.

Gage Block Sets

Inch Gage Blocks. Gage blocks can be obtained individually as well as in sets ranging from several to over 100 blocks. An 83-gage block set (Figure 9-1) may consist of the following:

2 Wear blocks, 0.050 in.

9 Blocks, 0.1001 through 0.1009 by increments of 0.0001

49 Blocks, 0.101 through 0.149 by increments of 0.001

19 Blocks, 0.050 through 0.950 by increments of 0.050

4 Blocks, 1.000 through 4.000 by increments of 1.000

Not only are measurements up to 25 in. possible with this set but almost any measurement in four-decimal places can be produced.

Metric Gage Blocks. Manufacturers produce metric gage blocks in sets of 47, 88, and 113 pieces. Comparable to its inch counterpart in common usage, an 88-piece set of metric gage blocks (Figure 9-2) may consist of the following:

3 Blocks, 0.5, 1.00, 1.005 mm

9 Blocks, 1.001 through 1.009 by increments of 0.001 mm

49 Blocks, 1.01 through 1.49 by increments of 0.01 mm

17 Blocks, 1.5 through 9.5 by increments of 0.5 mm

10 Blocks, 10 through 100 by increments of 10 mm

It is estimated that over 120,000 different measurement combination are possible with the use of either of the mentioned sets.

Wringing Blocks Together

The buildup of gage blocks should only be undertaken when the procedure for wringing them together is understood. Care must be taken not to drop the gage blocks or damage them in any way.

Procedure

1. Holding the gage block by squeezing against its side, wipe the contacting surface against a chamois or a clean, soft cotton cloth (Figure 9-3).

FIGURE 9-1
An 83-gage block set.

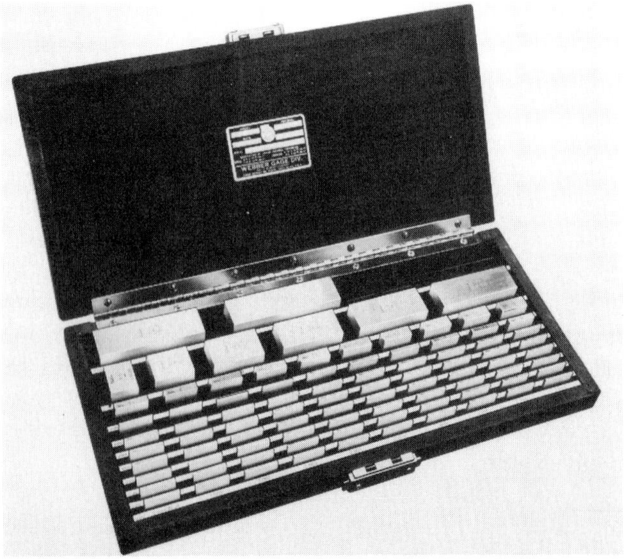

FIGURE 9-2
An 88-gage block metric set. (Credit: L.S. Starrett Company)

FIGURE 9-3
Wipe the gage blocks clean with a chamois cloth.

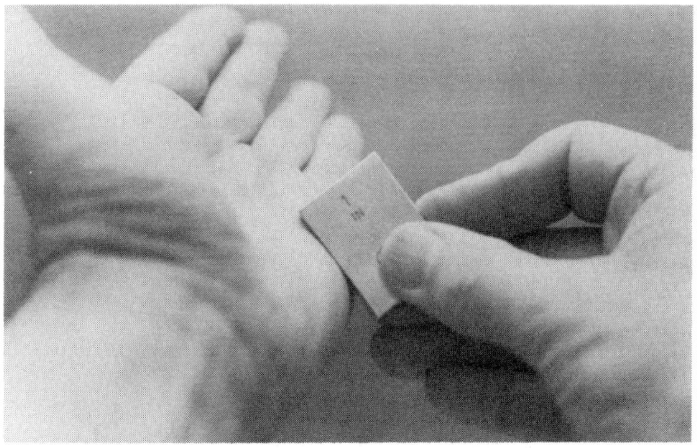

FIGURE 9-4
Remove any remaining dust particles against a dry palm.

FIGURE 9-5
Place the blocks together.

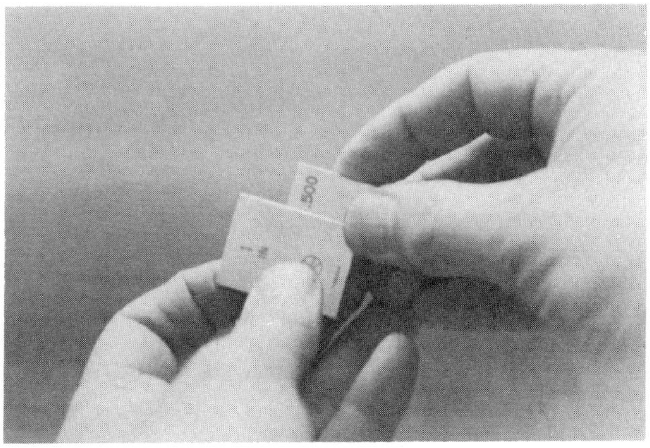

FIGURE 9-6
Slide the top block forward with a slight inward pressure.

2. Clean any remaining dust particles by wiping the surface against a dry palm or wrist (Figure 9-4).

3. Place the building gage blocks as shown in Figure 9-5.

4. Holding the bottom gage block firmly, slide the top block forward with slight inward pressure (Figure 9-6).

5. Blocks will wring together if properly cleaned. If not, clean them again. (Failure to achieve wringing may be due to minute scratches, nicks, or burrs. These areas can be smoothed by moderately using a special gage block stone that will not destroy the gage size.)

6. Continue wringing blocks together until the buildup has been completed.

Building Gage Block Measurements

In the buildup of any gage block combination, there are a few rules to follow. First, acquaint yourself with the size of the blocks of your set, taking into account any wear blocks that will be used. Second, reduce or eliminate the right-hand figure on the decimal measurement you wish to achieve by working from right to left. Continue reducing the measurement by working from right to left and build the combination with as few as possible number of blocks. The following examples illustrate these rules.

Procedure for inch measurement buildup

			Blocks used
	Required measurement	1.8733 in.	
Step 1	Eliminate the size of the two wear blocks (2 × 0.50 in.)	−0.100 in.	0.100 in.
		1.7733 in.	
Step 2	Eliminate the right hand digit	−0.1003 in.	0.1003 in.
		1.673 in.	
Step 3	Select a block that will reduce the hundredths digit to a 0 if the digit is 4 or less; if 5 or greater, reduce it to 5.		
		1.673 in.	
	Eliminate	−0.123 in.	0.123 in.
		1.550 in.	
Step 4	Eliminate	−0.550 in.	0.550 in.
		1.000 in.	
Step 5	Eliminate	−1.000 in.	1.000 in.
	Check	0.000	1.8733 in.

Procedure for metric measurement buildup

			Blocks used
	Required measurement	27.435 mm	
Step 1	Eliminate the size of the two wear blocks (2 × 2 mm)	− 4.000 mm	4.000 mm
		23.435 mm	
Step 2	Eliminate the right-hand digit	− 2.005 mm	2.005 mm
		21.43 mm	
Step 3	Eliminate the right-hand digit	− 2.43 mm	2.43 mm
		19.00 mm	
Step 4	Select the largest size available to eliminate the left-hand digit	−10.00 mm	10.00 mm
		9.00 mm	
Step 5	Eliminate the left-hand digit	− 9.00 mm	9.00 mm
	Check	0.00	27.435 mm

Care of the Gage Block

As you would respect any precise measuring tool, attention should be given to the handling of gage blocks.

1. When not in use, gage blocks should be kept in a closed box to protect them against dirt, dust, or moisture.

2. Gage blocks should not be stored while still wrung together. They should be slid apart, not pulled apart. Any moisture that collects between the blocks can cause them to rust.

3. Care should be taken not to drop the gage block in handling. The polished surface may be nicked or damaged that way (Figure 9–7).

4. Do not allow fingerprints or dust to remain on the polished surfaces. They should be wiped clean with chamois cloth that is free of sand or abrasive powder to prevent rusting, tarnishing, or scratching.

5. Before storing, gage blocks should be wiped clean with a solvent to remove finger marks.

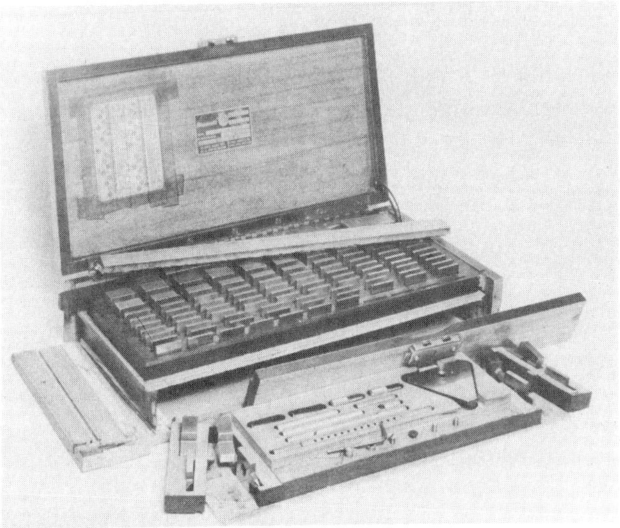

FIGURE 9–7
Care should be used handling gage blocks. (Credit: L.S. Starrett Company)

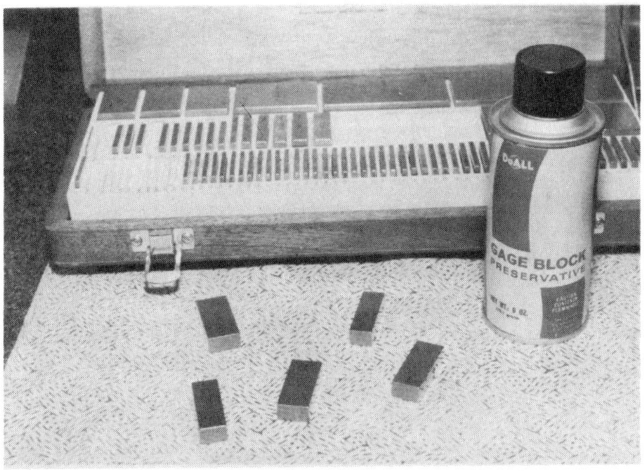

FIGURE 9–8
Gage blocks should be protected with a film of oil before storing.

Note: Do not use alcohol or ether for this purpose as they may contain water. Upon cleaning, gage blocks should be protected with a film of antirust oil (Figure 9–8).

6. If gage blocks are not to be used for long periods of time, they should be stored in a room free of humidity and direct sunlight.

7. Gage blocks that are not frequently used should be wiped free of oil and checked for rust at least three times a year.

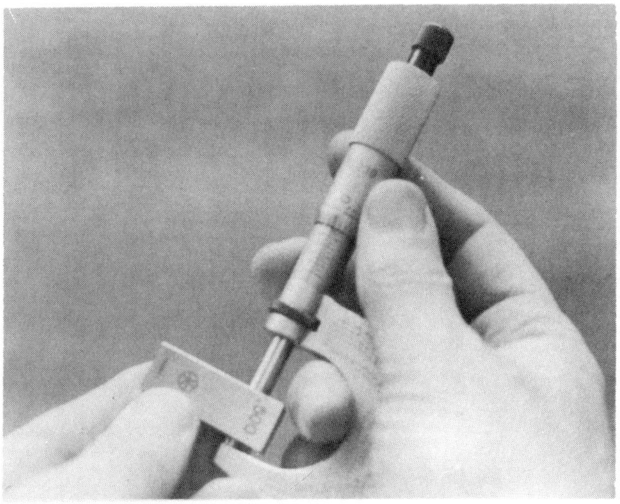

FIGURE 9–9
Testing a micrometer's accuracy with a gage block.

8. Due to the fact that the accuracy of the gage block is measured at 68°F (20°C), excessive handling should be avoided. The heat of the hand can change the size of the block. If gage blocks are handled for any length of time, say for several minutes, they should be rested at room temperature before using.

Common Gage Block Applications

Accepted as a precise standard, gage blocks serve as an invaluable aid in policing the accuracy of many measuring tools and industrial gages. Gage blocks serve the following functions:

1. Test and calibrate movable measuring tools such as micrometers and vernier calipers (Figure 9–9).
2. Set the accuracy of height for tools such as dial indicators, electronic measuring instruments, and height gages.
3. Inspect fixed gages to determine wear, warpage, shrinkage, or other factors causing dimensional inaccuracies (Figure 9–10).
4. Establish machine tool setups.
5. Aid in precision layout work.
6. Aid in inspection of fixed workpieces such as measuring fixed heights or fixed locations.
7. Set the accuracy of height required for the use of a sine bar or sine plate measuring, checking, or laying out angular work (Figure 9–11).

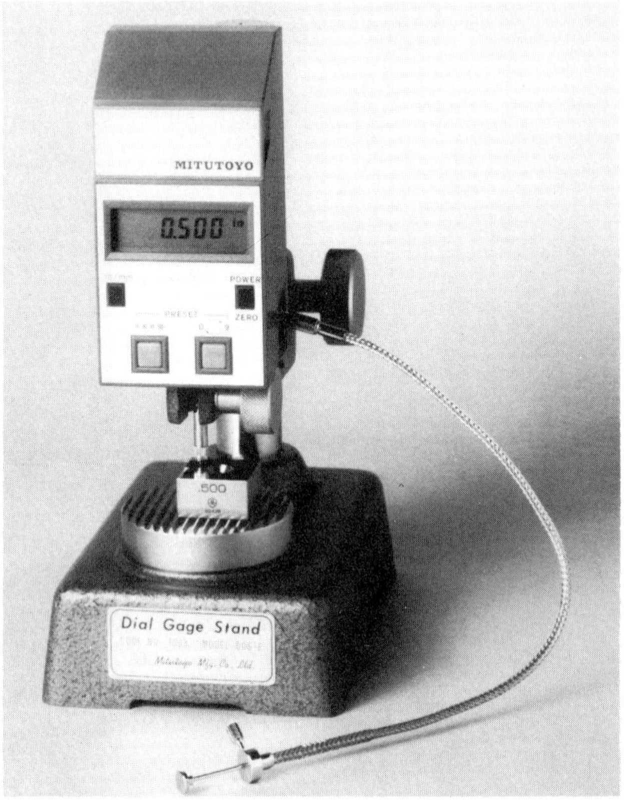

FIGURE 9-10
Checking the calibration of a digital indicator with a gage block. (Credit: MTI Corporation)

Ceramic Gage Blocks

Zirconia is a material never dreamed of by Johansson for making gage blocks. Zirconia ceramic gage blocks (Figure 9-12) are highly chip resistant and possess a thermal expansion coefficient very close to that of steel. They offer ten times the wear resistance of the traditional steel gage blocks and are every bit as accurate. Zirconia ceramic gage blocks are made in sets to accommodate both inch and metric measurements (Figure 9-13). Their durability and flatness is directly attributable to their dense grain structure. Their wringing quality is very high, made possible by the fact that these surfaces are not easily burred. Any burrs that might appear due to surface scratches are easily smoothed with the use of an Arkansas stone. Furthermore, their accuracy

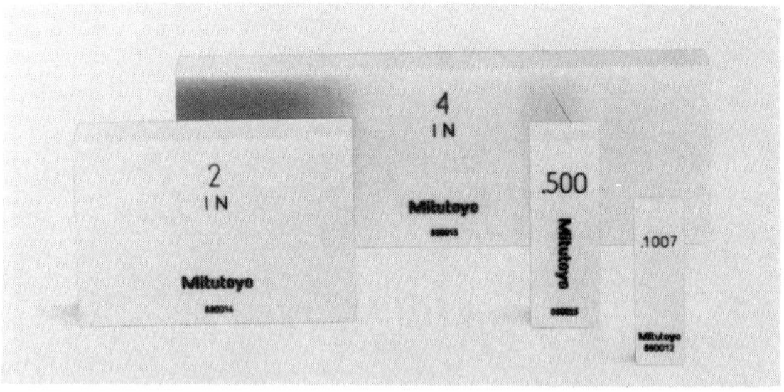

FIGURE 9-12
Ceramic gage blocks are highly chip resistant and do not require special maintenance. (Credit: MTI Corporation)

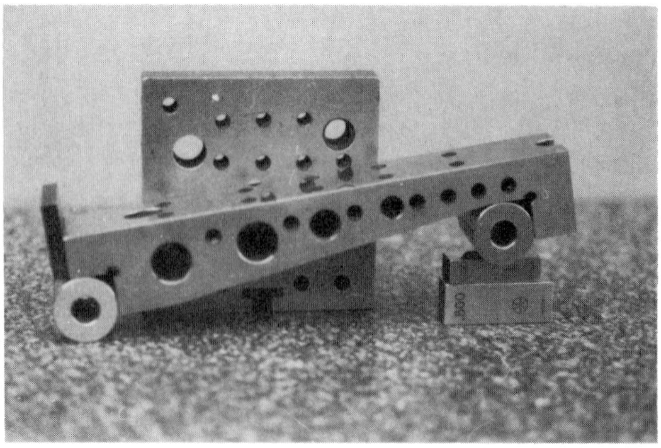

FIGURE 9-11
Sine bar setups.

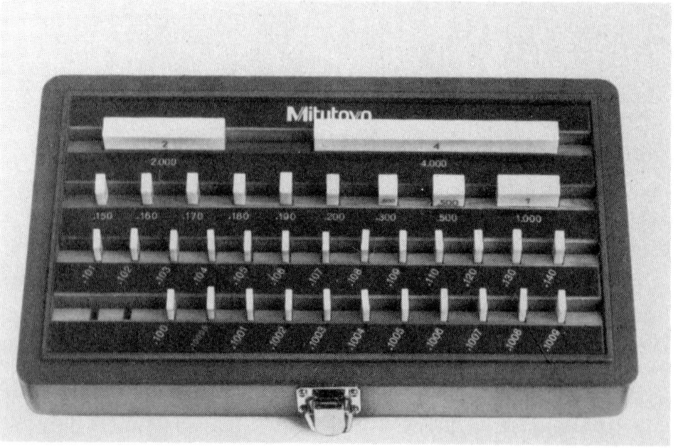

FIGURE 9-13
A set of ceramic gage blocks. (Credit: MTI Corporation)

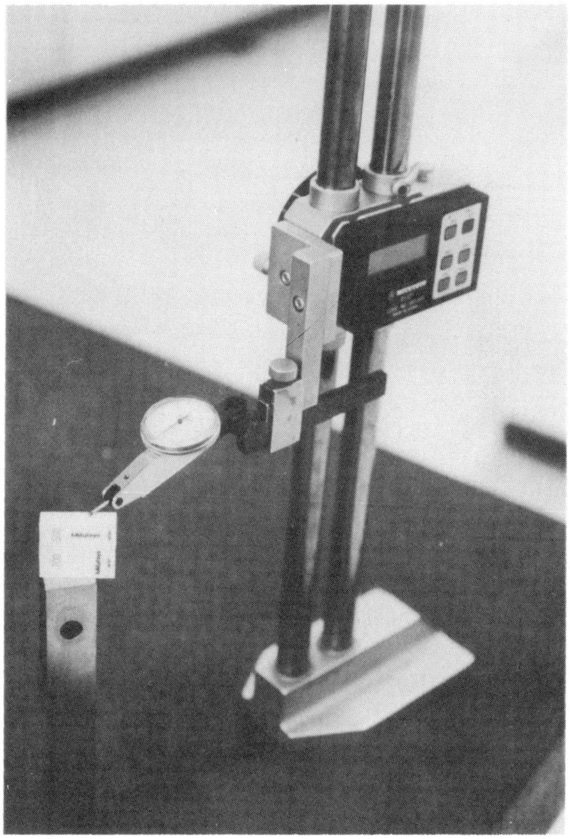

FIGURE 9-14
Establishing height measurement with the use of ceramic gage blocks. (Credit: MTI Corporation)

is not affected by the everyday handling and normal use to which gage blocks are subjected (Figure 9-14).

Advantages of Ceramic Gage Blocks

- Do not rust or stain due to environmental or finger marks.
- Are not affected by temperature changes.
- Are nonmagnetic.
- Do not have internal stresses; consequently, they will not warp or distort.
- Do not require special maintenance.
- Do not require special storage consideration.
- Surfaces can be wiped clean using a chamois cloth.
- Will not absorb moisture, dust, or abrasive particles, making them easier to use near wet or dry grinders.

UNIT 10 FIXED GAGES

For industry to function smoothly, it must use methods and instruments that are simplified and capable of precision. The need for quick, accurate, and easy-to-understand devices in machine shops, inspection rooms, and on the manufacturing floor is partly answered by the use of fixed gages (Figure 10-1).

Objectives
After completing this unit you should be able to

1. identify various standard fixed gages.
2. understand how a fixed gage size is determined.
3. use standard fixed gages properly.

Fixed Gages

Fixed gages are used widely in the inspection of manufactured parts. They are designed to measure within the specific limits of tolerance—the permissible difference between maximum and minimum dimensions of a product. Made with two ends to gage the limits of acceptability, referred to as "go" and "not-go," these gages can quickly accept or reject a part. Due to the fact that temperature changes affect the length and size of a gage, the temperature environment adopted for use for precision gages in industry is 68°F (20°C). The adopted industrial temperatures for metric gages are 0°C (32°F) and 20°C (68°F).

Tolerances for gages vary. In practice, gages are made to an accuracy of 10% of the product's total tolerance applied in the proper direction to the part tolerance, see Table 10-1. Tolerances for plain cylindrical plug and ring gages have been standardized in four classes by gage makers: XX, X, Y, and Z. XX represents the highest quality and, depending on the size involved, is precise to 0.00008 in. (0.00203 mm). In some cases higher accuracy is classified XXX and XXXX. The Z class is not as demanding and might be as accurate to 0.0001 in. (0.002 mm). Thus, a manufacturer will consider the part function and added cost for accuracy and wear longevity in choosing a particular type gage.

The many types of fixed gages have their holders clearly marked to identify the sizes of the go/not-go end. The inch dimensions are always marked to at least three decimal places, four places when called for. Millimeter markings are expressed to at least two decimal places. For example, 5.7 mm would be marked 5.70 mm; three decimal places are stamped as is.

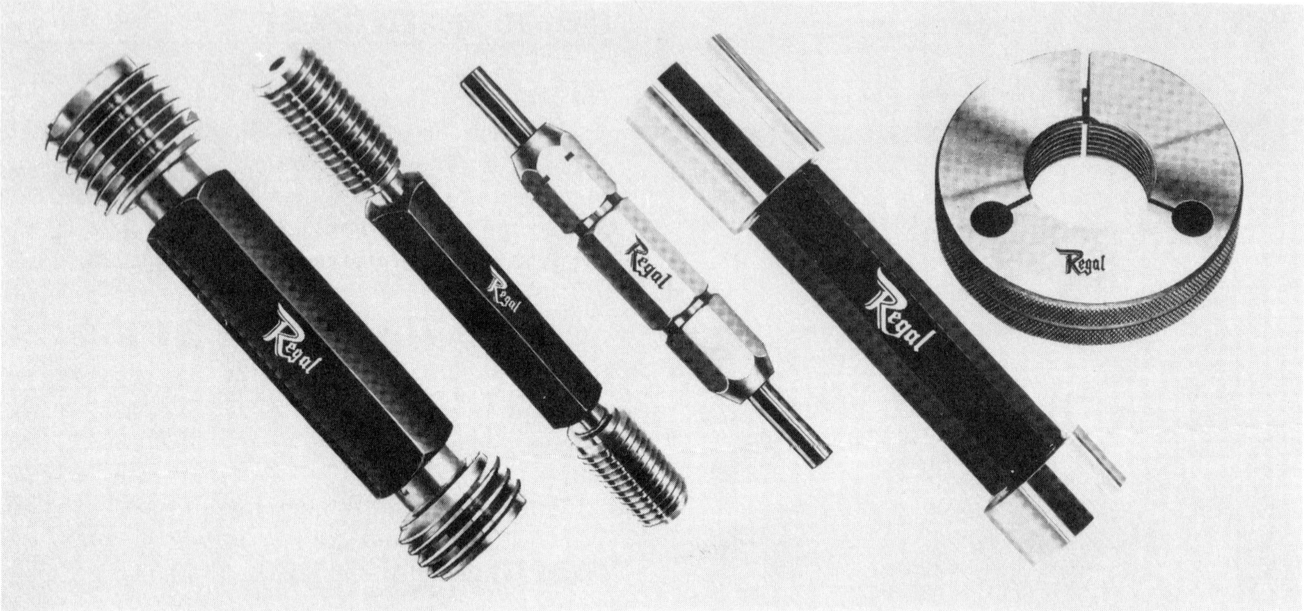

FIGURE 10-1.
A variety of fixed gages commonly used in industry. (Credit: Regal-Beloit Corporation)

Cylindrical Plug Gage

Made to test the uniformity of holes for out of roundness, wear, and conformance to engineering specifications, the cylindrical plug gage is made to a predetermined size to test the maximum and minimum limits of requirements (Figure 10-2). If, for example, engineering specifications call for a hole to measure 0.500 in. ± 0.002 (12.7 mm ± 0.5) in diameter, the go gage size would be made by applying the 10% tolerance rule in the following manner:

Hole size	0.498 in.	(12.64 mm)
Part tolerance (0.10 × 0.002)	+0.0002	(0.005 mm)
Male go end	0.4982 in.	(12.645 mm)

The not-go end (larger size) would be made so that it would not enter into a hole smaller than 0.5018 in. (12.735 mm) in diameter. This would be calculated in the following manner:

Hole size	0.502 in.	(12.74 mm)
Part tolerance (0.10 × 0.002)	−0.0002	(0.005 mm)
Male not-go end	0.5018 in.	(12.735 mm)

As an added visual feature, the go gage is made longer than the not-go gage. Ends of the holders are often color coded to specify usage. A green color is used for the go size, while red is used for the not-go diameter. On jobs requiring frequent inspection, cemented carbide has proven successful not only for wear purposes, but also in corrosion resistance.

TABLE 10-1
10% Tolerance rule

Wear allowance: Add (+) 10% of total tolerance	Wear allowance: subtract (−) 10% of total tolerance
Male go	Male not-go*
Female not-go*	Female go

*Cannot be applied to outside and root diameters of thread gages.

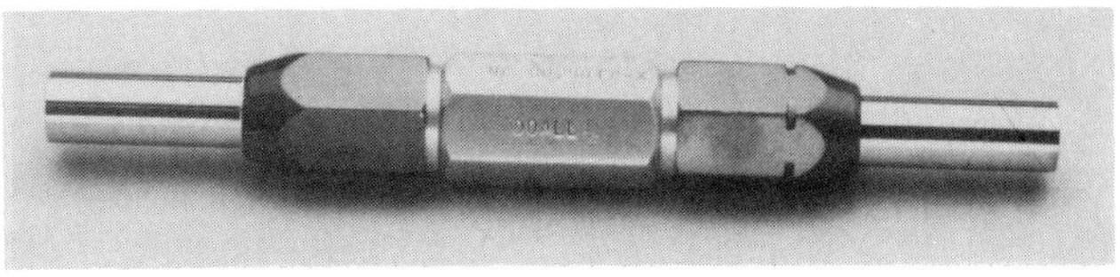

FIGURE 10-2
A cylindrical plug gage that features go and not-go ends. (Credit: DoAll Company)

Procedure for Using a Cylindrical Plug Gage

1. Inspect both ends of the lintended gage as well as the workpiece hole for nicks and burrs.

2. Clean both ends of the gage with a clean, dry cloth.

3. Wipe clean the workpiece hole.

4. Coat the gage surfaces with a thin film of oil by wiping them with an oily rag.

5. Carefully enter the go end evenly into the hole but do not twist or force it in place. Any play or movement between members will indicate the hole is too large. If entry is denied, the hole is undersize.

6. Complete the inspection by attempting to enter the not-go gage. If there is entry, the hole is too large. A partial entry will indicate a tapered opening or a bell mouth condition.

FIGURE 10-3
Diameters of round work are measured for accuracy with a plain ring gage. (Credit: DoAll Company)

Plain Ring Gage

The counterpart to the cylindrical plug gage is the plain ring gage. Made to inspect external diameters such as rods and shafts, inside dimensions of plain ring gages are ground and lapped to engineering specifications (Figure 10-3). When made in sets, both rings are identified to the upper and lower size limits involved. The not-go ring gage has an added annular ring grooved into its knurled surface for visual identification.

When a part is to a specified size, the go gage will effectively slip over it, being stopped from entering a not-go gage. If both gages are able to pass over them, the part is undersize. If both gages are prevented from slipping over it, the part is oversize.

Tapered Plug Gage

Resembling the frustum of a cone (Figure 10-4), the tapered plug gage is an external gage that is used to test the diameter, taper, and length of internal tapered dimensions according to engineering specifications. The go/not-go system on tapered plug gages varies. On some gages the limits of acceptability are denoted by two grooved lines. If the edge of the tapered hole being tested rests between the grooved lines of the inserted gage, the hole is considered within tolerance. Other gages substitute ground steps on the large end for the grooved lines. These, too, measure the limits of the hole size. Any mismatch between hole and plug gage is denoted by freedom of movement between mating parts.

Procedure for Testing

1. Examine the gage and work hole for nicks and burrs.

2. Wipe clean the surfaces of both the hole and tapered plug gage with a dry, clean cloth.

3. Chalk three lines equidistant along the length of the plug gage. (A very thin coating of Prussian blue can also be used for highly finished inspection work.)

FIGURE 10-4
A tapered plug gage is designed to measure a taper ring size of equal dimension.

4. Insert the marked gage into the hole as far as possible.

5. Check to see if the edge of the hole falls between the steps or rings. If not, the hole is not deep enough.

6. Holding the gage evenly with light pressure, rotate it partially in a counterclockwise direction. Any freedom of movement at the large end indicates the taper is too large; movement at the small end reveals the taper is too small.

7. Remove the gage to examine the chalk marks for rubbings. A matching taper will show uniform rub marks along its entire length; an improper match will show up as a rub mark on one end or the other.

Taper Ring Gage

Externally sloping diameters such as lathe centers, twist drill shanks, and reamers are checked for dimensional and taper accuracy by means of a taper ring gage. Similar to the taper plug gage in its go/not-go application, taper ring gages will also use grooved lines or ground steps to indicate limits of acceptance.

Testing with a taper ring gage is done in much the same manner as using the taper plug gage. Upon cleaning the surface, three equally spaced chalk lines are drawn the length of the external taper. Light Prussian blue can be applied to ground surfaces. The ring is then slipped over this surface and, bearing down lightly for good surface contact, it is rotated clockwise and counterclockwise in small movements. Upon removing the ring gage, the external tapered surface is examined for rub marks. Uniform rub marks that are distributed along its length indicates the taper is correct. If the rub marks are more prominent on one end or the other, the taper has to be corrected or the part is no good. The correct diameter is indicated by the end position of the external taper in relation to the grooved lines or ground steps.

Thread Plug Gage

The inspection of screws and threads is of paramount importance in industry as well as in the toolroom. The major diameter, minor diameter, thread angle, lead, and the measurement that controls the class of fit (pitch diameter) must be all checked.

Internal threads, such as tapped holes and nuts are inspected with thread plug gages (Figure 10-5). Thread plug gages are provided with a chip groove for ease in collecting and cleaning inspected thread material. The longer end represents the go gage and is made to the minimum pitch diameter. The shorter end, the not-go gage represents the maximum pitch diameter limit (Figure 10-6).

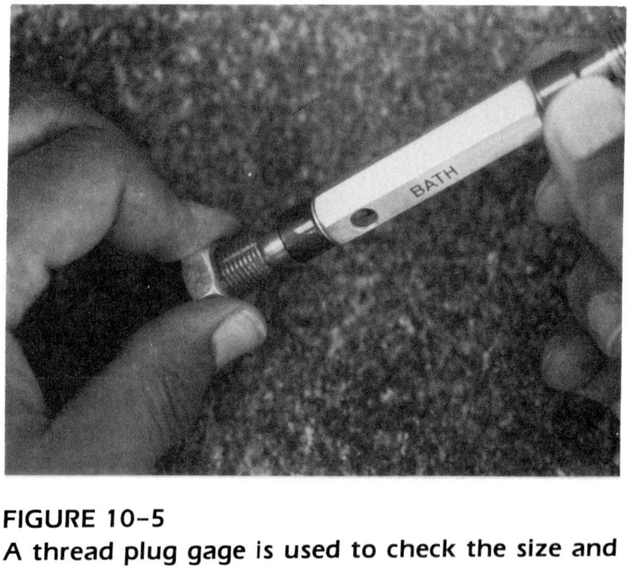

FIGURE 10-5
A thread plug gage is used to check the size and accuracy of an internal thread such as a nut or tapped hole.

FIGURE 10-6
The not-go gage represents the maximum pitch diameter limit.

FIGURE 10-7
A go and not-go set of thread ring gages that may be used for checking the accuracy of screws.

Procedure for Use

1. Check tapped hole to see that it is clear of chips or other debris. If not, chase out all foreign matter with an old tap.
2. Apply a drop of light oil on both gages before using.
3. Screw the go gage into the tapped hole its full thread length but do not force it. The gage should be square with the work and display no movement.
4. Attempt the not-go gage. It may or may not enter. If it does, it will become snug on or before the third thread. Movement beyond the third thread will indicate the thread is oversize.

FIGURE 10-8
The not-go gage should not accept the thread more than $1\frac{1}{2}$ turns.

Thread Ring Gage

Designed to inspect the accuracy of externally threaded work, thread ring gages generally come in a set clamped in a holder. They could have a solid design or may be slotted and provided with a small set screw that will permit adjustment (Figure 10-7). Like its threaded gage counterpart, each gage is designed to inspect the principal size and measurements of a thread and should be used with equal precaution.

The go gage is made to inspect the maximum pitch diameter of the external thread in question. If the thread is acceptable, it should fully enter the go opening. The not-go gage, which bears an annular ring grooved in its outer knurling, is designed to inspect the minimum pitch diameter and should not accept the thread more than $1\frac{1}{2}$ turns (Figure 10-8).

Pin Gages

Pin gages (Figure 10-9) are designed to test the preciseness of small holes as to size and out of roundness. Coming in sets of various sizes, they are made to several classes of accuracy ranging from XXXX to Z. These gages are also available in sets that are stepped in increments of 0.0001 in. (0.002 mm) or larger. Pin gages are found to be very practical and

FIGURE 10-9
Pin gages are designed to check the accuracy of small holes for size and out of roundness. (Credit: DoAll Company)

provide a more positive means of checking small-diameter holes rather than relying on the use of a pair of sharp pointed nibs on a dial caliper, for instance.

UNIT 11 ANGLE MEASUREMENTS

Precise angle measurements are as necessary to inspection as they are to layouts. In order to measure, however, the tools and means for using them must be known. The methods for angular measurements can range from the practical to the very sophisticated system of measuring with light waves (interferometry). In this unit we will concentrate on the precise but practical means.

Objectives
After completing this unit you will be able to

1. identify various precise instruments of angle measurement.
2. identify and correctly use angle gage blocks.
3. identify and correctly use a sine bar.

The Vernier Bevel Protractor

The combination set protractor does not measure up to the accuracy of the universal bevel protractor

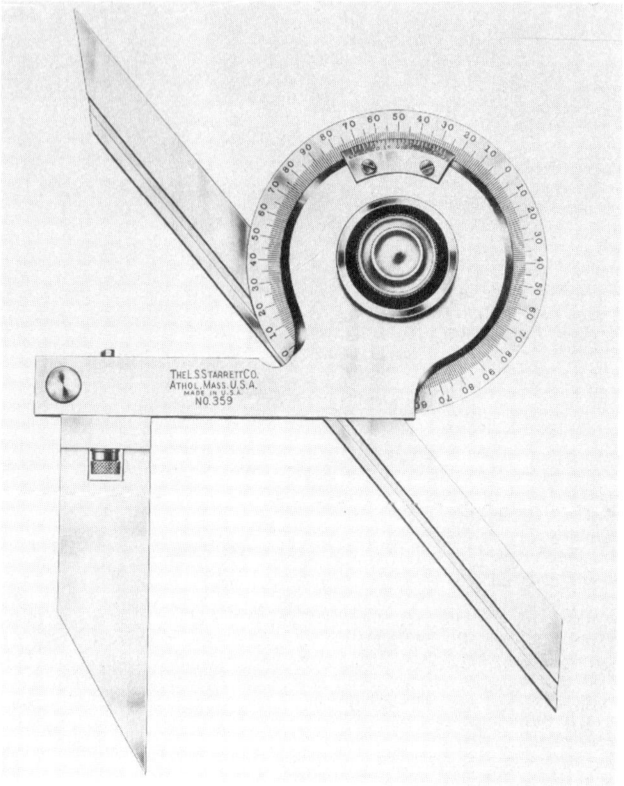

FIGURE 11-1
A universal bevel protractor can measure angles precisely. (Credit: L.S. Starrett Company)

(Figure 11-1). This instrument is precise enough to measure an angle to 5 min of an arc (one-twelfth of one degree). A sliding blade, protractor dial, and base is part of the basic construction. At the turn of a dial clamp nut, the sliding blade may be positioned and locked to any degree desired. A fine angle adjustment of the blade is also possible by the turning of a small pinion located on the instrument. Its versatility can be seen when used in connection with other measuring devices for checking angles (Figure 11-2), for instance. Measuring an obtuse angle (Figure 11-3) is no problem, and with an acute angle attachment the laying out or measurement of small angles is also possible.

The protractor dial is circled 360° in increments of one degree and sectioned off in quadrants marked 0° to 90°, 90° to 0°, 0° to 90°, and 90° to 0°. Graduations, which are numbered for easier readings, are divided finer with the addition of a vernier plate marked in 5-min increments, a total of 60 min to the right and left of the zero line.

FIGURE 11-2
Using the vernier bevel protractor for measuring and laying out varied angles. (Credit: L.S. Starrett Company)

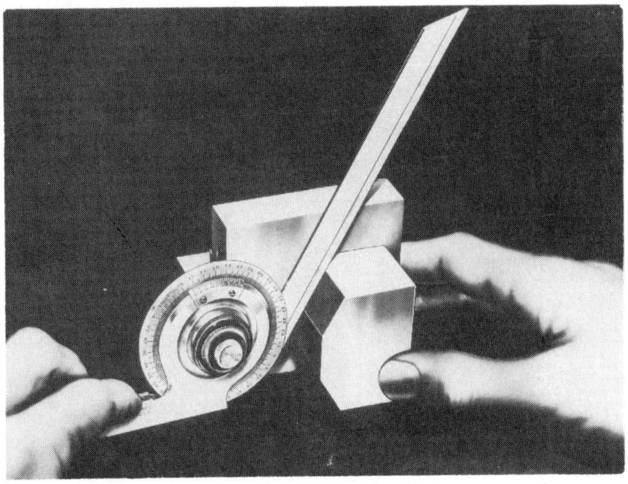

FIGURE 11-3
Measuring an obtuse angle. (Credit: L.S. Starrett Company)

Reading the Vernier Protractor

1. Look for the vernier plate zero line.
2. Examine the reading between the zero graduation on the protractor dial and the zero graduation on the vernier plate and record it.
3. Going in the same direction beyond the vernier plate zero line, look for a matchup of lines on adjoining surfaces.
4. Use the number graduation as is or multiply the total vernier plate graduations beyond the zero by 5 and add this value to the number of whole degrees registered on the protractor dial.

As Figure 11-4 illustrates:

Protractor dial reading	=	50°
Vernier plate reading (direct)	=	20′
Total reading	=	50° 20′

FIGURE 11-4
A vernier protractor reading of 50° 20′. (Credit: L.S. Starrett Company)

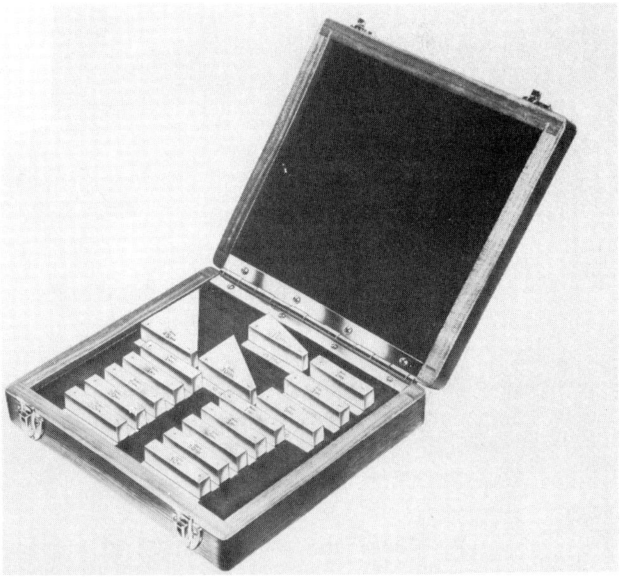

FIGURE 11-5
A set of angle gage blocks. (Credit: L.S. Starrett Company)

Angle Gage Blocks

Excellent for machine tool setups or measurements, angle gage blocks are precisely made and can accurately measure any angle from 0° to 99°. Made very much like their precision gage block brothers, angle gage blocks are manufactured from high-grade steels that are heat treated and artificially stabilized. Consisting of as many as 18 angular blocks that can be wrung together (Figure 11-5), up to 356,400 angle measurements are possible in steps of

TABLE 11-1
Angle gage block specifications

	Toolroom	*Inspection*	*Laboratory or ref. grade*
Gaging surface base	5/8 × 4 in.	1 × 2 in.	1 × 2 in.
Accuracy	±1°	±1/2°	±1/4°
Flatness across width	0.000004 in.	0.000002 in.	0.000001 in.
Flatness along width	0.000006 in.	0.000002 in.	0.000001 in.
Flatness of sides	0.000010 in.	0.000004 in.	0.000004 in.
Blocks	6 in. parallel 6 in. knife edge 1°, 3°, 5°, 15°, 30°, 45° 1, 3, 5, 20, and 30 min 1, 3, 5, 20 and 30 s	1°, 3°, 5°, 15°, 30°, and 45° 1, 3, 5, 20, and 30 min 1, 3, 5, 20, and 30 s	

1 s, 1 min, or 1°, making them superior to sine bar setups.

Three grades of classifications are obtainable: Laboratory master, inspection, and toolroom accuracy. The laboratory master angle gage blocks have a $\pm\frac{1}{4}$-s accuracy, the inspection grade boasts a $\pm\frac{1}{2}$-s accuracy, while the toolroom grade has a ±1-s accuracy (Table 11-1).

The gage blocks are manufactured so that the angular dimensions may be assembled by combining the plus or minus values marked on each end of the gage blocks, enabling the user to generate greater or lesser angles. To illustrate, the plus end of a 10°-angle block may be wrung together with a plus end

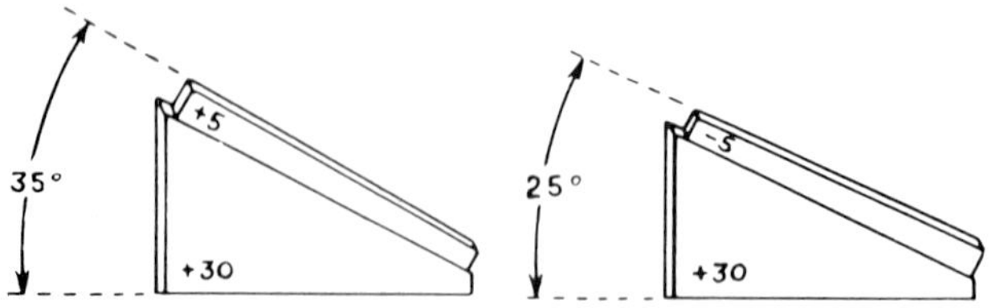

FIGURE 11-6
It is possible to increase or decrease an angle by wringing angle gage blocks together. (Credit: L.S. Starrett Company)

of the 15°-angle block to achieve a 25°-angle block; however, wringing the plus end of a 20°-angle block with the minus end of a 15°-block will generate a 5°-angle block. In this manner various combinations may be wrung together to obtain angles in degrees, minutes, or seconds. As will be seen, this method is more direct and less complicated than using the sine bar (Figure 11-6).

The Sine Bar

Layout or measurement of angles closer than 5 min in accuracy is carried out by toolmakers on a sine bar. Consisting of a stabilized bar of high-grade steel that has been hardened and highly finished, a sine bar rests on two accurately ground and lapped steel rollers of exact diameters (Figure 11-7). The center distance of the rollers are in precise alignment and parallel with the top surface of the sine bar within very close tolerance. Their center distance may be 5 or 10 in. or multiples of 5 in.

The right triangle and the sine trignometric function of one of its angles is the basis of accuracy when using the sine bar. In a right triangle the sine of an angle is defined as the ratio of the length of the side opposite this angle over the hypotenuse. This is illustrated in the chart in Figure 11-8, which states that the sine of angle C is equal to side c over side a, the hypotenuse. Since B can also be obtained by dividing side b by side a.

When one side of the sine bar is raised, a right triangle is formed with the bottom of the rollers acting as the hypotenuse. To achieve the included angle, the sine bar is elevated and rested on stacked gage blocks, which represents side c. Since the hypotenuse represents the standard distance between rollers center, all that remains is to look up the sine of the given angle in the tables of natural trignometric functions (Table 11-2) and complete the mathematics. (The same function can be obtained using an engineering-type calculator.) The resultant answer will be the stacked height of gage blocks to be used.

FIGURE 11-7
A 5-in. sine bar.

Example:
Determine the gage block height to set a 5-in. sine bar to an angle of 15°.

Solution:

$$\text{sine } C = \frac{c}{a}$$

$$\text{sine } 15° = \frac{c \text{ (gage block height)}}{a \text{ (sine bar length)}}$$

$$0.25882 = \frac{c}{5 \text{ in.}}$$

$$c = 0.25882 \times 5$$

$$\text{Gage block height} = 1.2941 \text{ in.}$$

Example:
Determine the gage block height to set a 5-in. sine bar to an angle of 27°30′.

Solution:

$$\text{sine } C = \frac{c}{a}$$

$$\text{sine } 27°30' = \frac{c \text{ (gage block height)}}{a \text{ (sine bar length)}}$$

$$0.46175 = \frac{c}{5 \text{ in.}}$$

$$c = 0.46175 \times 5$$

$$\text{Gage block height} = 2.30875$$

So it can be seen that the formula for gage block buildup can be stated as

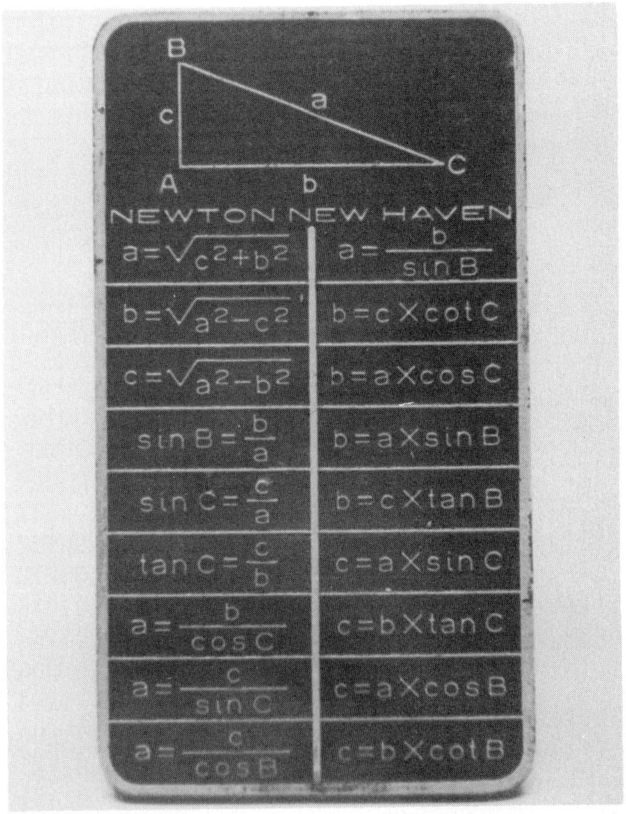

FIGURE 11-8
A right-angle triangle chart.

$$\text{sine of given angle} = \frac{\text{side opposite}}{\text{hypotenuse}}$$

or

$$= \frac{\text{gage block height}}{\text{length of sine bar}}$$

Applying this formula to a 5-in. sine bar, it would read

$$\text{sine of given angle} = \frac{\text{gage block height}}{5 \text{ in.}}$$

Transposing this formula, we have

$$\text{Gage block height} = \text{sine of given angle} \times 5 \text{ in.}$$

A table of constants for setting the inch sine bars can be found in technical handbooks or are sometimes provided by manufacturers of sine bars. Using Table 11-3, the height of gage block stacking can be determined simply by referring to the angle and minute involved.

The process can be reversed to determine the included angle using Table 11-3. If you know the gage block height, all you have to do is read the table until you find a dimension to the nearest whole minute that corresponds to the height and read across the table directly to the angle and minute involved.

When using the trignometric method, the angle is determined by inverse function, calculating its sine, and then finding a value in the sine column that corresponds to the nearest whole minute.

Example: Using the table of natural trignometric functions (Table 11-2), determine the include angle when a 5-in. sine bar is elevated to a height of 0.98685 in.

Solution:

$$\text{sine } C = \frac{c}{a}$$

$$= \frac{c \text{ (gage block height)}}{a \text{ (sine bar length)}}$$

$$= \frac{0.98685}{5 \text{ in.}}$$

$$= 0.19737$$

$$\text{Included angle} = 11°23'.$$

Example:
Determine the included angle when a 5-in. sine bar is elevated 1.9268 in.

$$\text{sine } C = \frac{c}{a}$$

$$= \frac{c \text{ (gage block height)}}{a \text{ (sine bar length)}}$$

$$= \frac{1.92685}{5 \text{ in.}}$$

$$= 0.38537$$

$$\text{Included angle} = 22°40'$$

TABLE 11-2
Trigonometric functions

Angle	*Sine*	*Cosine*	*Tan.*	*Cotan.*	*Secant*	*Cosec.*	
0°00′	.00000	1.0000	.00000	Infinite	1.0000	Infinite	90°00′
5	.00145	1.0000	.00145	687.55	1.0000	687.55	55
10	.00291	.99999	.00291	343.77	1.0000	343.77	50
15	.00436	.99999	.00436	229.18	1.0000	229.18	45
20	.00582	.99998	.00582	171.88	1.0000	171.89	40
25	.00727	.99997	.00727	137.51	1.0000	137.51	35
30	.00873	.99996	.00873	114.59	1.0000	114.59	30
35	.01018	.99995	.01018	98.218	1.0000	98.223	25
40	.01163	.99993	.01164	85.940	1.0001	85.946	20
45	.01309	.99991	.01309	76.390	1.0001	76.396	15
50	.01454	.99989	.01454	68.750	1.0001	68.757	10
55	.01600	.99987	.01600	62.499	1.0001	62.507	5
1°00′	.01745	.99985	.01745	57.290	1.0001	57.299	89°00′
5	.01891	.99982	.01891	52.882	1.0002	52.891	55
	Cosine	Sine	Cotan.	Tan.	Cosec.	Secant	Angle

TABLE 11-2
Continued

Angle	*Sine*	*Cosine*	*Tan.*	*Cotan.*	*Secant*	*Cosec.*	
10	.02036	.99979	.02036	49.104	1.0002	49.114	50
15	.02181	.99976	.02182	45.829	1.0002	45.840	45
20	.02326	.99973	.02327	42.964	1.0003	42.976	40
25	.02472	.99969	.02473	40.436	1.0003	40.418	35
30	.02618	.99966	.02618	38.188	1.0003	38.201	30
35	.02763	.99962	.02764	36.177	1.0004	36.191	25
40	.02908	.99958	.02910	34.368	1.0004	34.382	20
45	.03054	.99953	.03055	32.730	1.0005	32.745	15
50	.03199	.99949	.03201	31.241	1.0005	31.257	10
55	.03344	.99944	.03346	29.882	1.0005	29.899	5
2°00′	.03490	.99939	.03492	28.636	1.0006	28.654	88°00′
5	.03635	.99934	.03638	27.490	1.0007	27.508	55
10	.03781	.99928	.03783	26.432	1.0007	26.450	50
15	.03926	.99923	.03929	25.452	1.0008	25.471	45
20	.04071	.99917	.04075	24.542	1.0008	24.562	40
25	.04217	.99911	.04220	23.694	1.0009	23.716	35
30	.04362	.99905	.04366	22.904	1.0009	22.925	30
35	.04507	.99898	.04512	22.164	1.0010	22.186	25
40	.04652	.99892	.04657	21.470	1.0011	21.494	20
45	.04798	.99885	.04803	20.819	1.0011	20.843	15
50	.04943	.99878	.04949	20.205	1.0012	20.230	10
55	.05088	.99870	.05095	19.627	1.0013	19.653	5
3°00′	.05234	.99863	.05241	19.081	1.0014	19.107	87°00′
5	.05379	.99855	.05387	18.564	1.0014	18.591	55
10	.05524	.99847	.05532	18.075	1.0015	18.103	50
15	.05669	.99839	.05678	17.610	1.0016	17.639	45
20	.05814	.99831	.05824	17.169	1.0017	17.198	40
25	.05960	.99822	.05970	16.750	1.0018	16.779	35
30	.06105	.99813	.06116	16.350	1.0019	16.380	30
35	.06250	.99804	.06262	15.969	1.0019	16.000	25
40	.06395	.99795	.06408	15.605	1.0020	15.637	20
45	.06540	.99786	.06554	15.257	1.0021	15.290	15
50	.06685	.99776	.06700	14.924	1.0022	14.958	10
55	.06830	.99766	.06846	14.606	1.0023	14.640	5
4°00′	.06976	.99756	.06993	14.301	1.0024	14.335	86°00′
5	.07121	.99746	.07139	14.008	1.0025	14.043	55
10	.07266	.99736	.07285	13.727	1.0026	13.763	50
15	.07411	.99725	.07431	13.457	1.0027	13.494	45
20	.07556	.99714	.07573	13.197	1.0029	13.235	40
25	.07701	.99703	.07724	12.947	1.0030	12.985	35
30	.07846	.99692	.07870	12.706	1.0031	12.745	30
35	.07991	. 99680	.08016	12.474	1.0032	12.514	25
40	.08136	.99668	.08163	12.250	1.0033	12.291	20
45	.08281	.99656	.08309	12.035	1.0034	12.076	15
50	.08426	.99644	.08456	11.826	1.0036	11.868	10
55	.08571	.99632	.08602	11.625	1.0037	11.668	5
5°00′	.08715	.99619	.08749	11.430	1.0038	11.474	85°00′
	Cosine	Sine	Cotan.	Tan.	Cosec.	Secant	Angle

TABLE 11-2
Continued

Angle	*Sine*	*Cosine*	*Tan.*	*Cotan.*	*Secant*	*Cosec.*	
5	.08860	.99607	.08895	11.242	1.0039	11.286	55
10	.09005	.99594	.09042	11.059	1.0041	11.104	50
15	.09150	.99580	.09189	10.883	1.0042	19.929	45
20	.09295	.99567	.99335	10.712	1.0043	10.758	40
25	.09440	.99553	.09482	10.546	1.0045	10.593	35
30	.09584	.99540	.09629	10.385	1.0046	10.433	30
35	.09729	.99525	.09776	10.229	1.0048	10.278	25
40	.09874	.99511	.09922	10.078	1.0049	10.127	20
45	.10019	.99497	.11069	9.9310	1.0050	9.9812	15
50	.10163	.99482	.10216	9.7882	1.0052	9.8391	10
55	.10308	.99467	.10363	9.6493	1.0053	9.7010	5
6°00′	.10453	.99452	.10510	9.5144	1.0055	9.5668	84°00′
5	.10597	.99437	.10657	9.3831	1.0057	9.4362	55
10	.10742	.99421	.10805	9.2553	1.0058	9.3092	50
15	.10887	.99406	.10952	9.1309	1.0060	9.1855	45
20	.11031	.99390	.11099	9.0098	1.0061	9.0651	40
25	.11176	.99373	.11246	8.8918	1.0063	8.9479	35
30	.11320	.99357	.11393	8.7769	1.0065	8.8337	30
35	.11465	.99341	.11541	8.6648	1.0066	8.7223	25
40	.11609	.99324	.11688	8.5555	1.0068	8.6138	20
45	.11754	.99307	.11836	8.4489	1.0070	8.5079	15
50	.11898	.99290	.11983	8.3449	1.0071	8.4046	10
55	.12042	.99272	.12131	8.2434	1.0073	8.3039	5
7°00′	.12187	.99255	.12278	8.1443	1.0075	8.2055	83°00′
5	.12331	.99237	.12426	8.0476	1.0077	8.1094	55
10	.12476	.99219	.12574	7.9530	1.0079	8.0156	50
15	.12620	.99200	.12772	7.8606	1.0080	7.9240	45
20	.12764	.99182	.12869	7.7703	1.0082	7.8344	40
25	.12908	.99163	.13017	7.6821	1.0084	7.7469	35
30	.13053	.99144	.13165	7.5957	1.0086	7.6613	30
35	.13197	.99125	.13313	7.5113	1.0088	7.5776	25
40	.13341	.99106	.13461	7.4287	1.0090	7.4957	20
45	.13485	.99086	.13609	7.3479	1.0092	7.4156	15
50	.13629	.99067	.13757	7.2687	1.0094	7.3372	10
55	.13773	.99047	.13906	7.1912	1.0096	7.2604	5
8°00′	.13917	.99027	.14054	7.1154	1.0098	7.1853	82°00′
5	.14061	.99006	.14202	7.0410	1.0100	7.1117	55
10	.14205	.98986	.14351	6.9682	1.0102	7.0396	50
15	.14349	.98965	.14499	6.8969	1.0104	6.9690	45
20	.14193	.99844	.15648	6.8269	1.0107	6.8998	40
25	.14637	.98923	.14796	6.7584	1.0109	6.8320	35
30	.14781	.98901	.14945	6.6911	1.0111	6.7655	30
35	.14925	.98880	.15094	6.6252	1.0113	6.7003	25
40	.15068	.98858	.15243	6.5605	1.0115	6.6363	20
45	.15212	.98836	.15391	6.4971	1.0118	6.5736	15
50	.15356	.98814	.15540	6.4348	1.0120	6.5121	10
55	.15500	.98791	.15689	6.3737	1.0122	6.4517	5
	Cosine	Sine	Cotan.	Tan.	Cosec.	Secant	Angle

TABLE 11-3
Constants for 5″ sine bars

Figures are in Inches

Min.	40°	41°	42°	43°	44°	45°	46°	47°	48°	49°	50°	51°	52°	53°	54°	55°	56°	57°	58°	59°	Min.
0	3.2139	3.2803	3.3456	3.4100	3.4733	3.5355	3.5967	3.6567	3.7157	3.7735	3.8302	3.8857	3.9400	3.9932	4.0451	4.0957	4.1452	4.1933	4.2402	4.2858	0
2	.2161	.2825	.3478	.4121	.4754	.5376	.5987	.6587	.7176	.7754	.8321	.8875	.9418	.9949	.0468	.0974	.1468	.1949	.2418	.2873	2
4	.2184	.2847	.3499	.4142	.4774	.5396	.6007	.6607	.7196	.7773	.8339	.8894	.9436	.9967	.0485	.0991	.1484	.1965	.2433	.2888	4
6	.2206	.2869	.3521	.4163	.4795	.5417	.6027	.6627	.7215	.7792	.8358	.8912	.9454	.9984	.0502	.1007	.1500	.1981	.2448	.2903	6
8	.2228	.2890	.3543	.4185	.4816	.5437	.6047	.6647	.7235	.7811	.8377	.8930	.9472	4.0001	.0519	.1024	.1517	.1997	.2464	.2918	8
10	3.2250	3.2912	3.3564	3.4206	3.4837	3.5458	3.6068	3.6666	3.7254	3.7830	3.8395	3.8948	3.9490	4.0019	4.0536	4.1041	4.1533	4.2012	4.2479	4.2933	10
12	.2273	.2934	.3586	.4227	.4858	.5478	.6088	.6686	.7274	.7850	.8414	.8967	.9508	.0036	.0553	.1057	.1549	.2028	.2494	.2948	12
14	.2295	.2956	.3607	.4248	.4879	.5499	.6108	.6706	.7293	.7869	.8433	.8985	.9525	.0054	.0570	.1074	.1565	.2044	.2510	.2963	14
16	.2317	.2978	.3629	.4269	.4900	.5519	.6128	.6726	.7312	.7887	.8451	.9003	.9543	.0071	.0587	.1090	.1581	.2060	.2525	.2978	16
18	.2339	.3000	.3650	.4291	.4921	.5540	.6148	.6745	.7332	.7906	.8470	.9021	.9561	.0089	.0604	.1107	.1597	.2075	.2540	.2992	18
20	3.2361	3.3022	3.3672	3.4312	3.4941	3.5560	3.6168	3.6765	3.7351	3.7925	3.8488	3.9039	3.9579	4.0106	4.0621	4.1124	4.1614	4.2091	4.2556	4.3007	20
22	.2384	.3044	.3693	.4333	.4962	.5581	.6188	.6785	.7370	.7944	.8507	.9058	.9596	.0123	.0638	.1140	.1630	.2107	.2571	.3022	22
24	.2406	.3065	.3715	.4354	.4983	.5601	.6208	.6805	.7390	.7963	.8525	.9076	.9614	.0141	.0655	.1157	.1646	.2122	.2586	.3037	24
26	.2428	.3087	.3736	.4375	.5004	.5621	.6228	.6824	.7409	.7982	.8544	.9094	.9632	.0158	.0672	.1173	.1662	.2138	.2601	.3052	26
28	.2450	.3109	.3758	.4396	.5024	.5642	.6248	.6844	.7428	.8001	.8562	.9112	.9650	.0175	.0689	.1190	.1678	.2154	.2617	.3066	28
30	3.2472	3.3131	3.3779	3.4417	3.5045	3.5662	3.6268	3.6864	3.7448	3.8020	3.8581	3.9130	3.9667	4.0193	4.0706	4.1206	4.1694	4.2169	4.2632	4.3081	30
32	.2494	.3153	.3801	.4439	.5066	.5683	.6288	.6883	.7467	.8039	.8599	.9148	.9685	.0210	.0722	.1223	.1710	.2185	.2647	.3096	32
34	.2516	.3174	.3822	.4460	.5087	.5703	.6308	.6903	.7486	.8058	.8618	.9166	.9703	.0227	.0739	.1239	.1726	.2201	.2662	.3111	34
36	.2538	.3196	.3844	.4481	.5107	.5723	.6328	.6923	.7505	.8077	.8636	.9184	.9720	.0244	.0756	.1255	.1742	.2216	.2677	.3125	36
38	.2561	.3218	.3865	.4502	.5128	.5744	.6348	.6942	.7525	.8096	.8655	.9202	.9738	.0262	.0773	.1272	.1758	.2232	.2692	.3140	38
40	3.2583	3.3240	3.3886	3.4523	3.5149	3.5764	3.6368	3.6962	3.7544	3.8114	3.8673	3.9221	3.9756	4.0279	4.0790	4.1288	4.1774	4.2247	4.2708	4.3155	40
42	.2605	.3261	.3908	.4544	.5169	.5784	.6388	.6981	.7563	.8133	.8692	.9239	.9773	.0296	.0807	.1305	.1790	.2263	.2723	.3170	42
44	.2627	.3283	.3929	.4565	.5190	.5805	.6408	.7001	.7582	.8152	.8710	.9257	.9791	.0313	.0823	.1321	.1806	.2278	.2738	.3184	44
46	.2649	.3305	.3950	.4586	.5211	.5825	.6428	.7020	.7601	.8171	.8729	.9275	.9809	.0331	.0840	.1337	.1822	.2294	.2753	.3199	46
48	.2671	.3326	.3972	.4607	.5231	.5845	.6448	.7040	.7620	.8190	.8747	.9293	.9826	.0348	.0857	.1354	.1838	.2309	.2768	.3213	48
50	3.2693	3.3348	3.3993	3.4628	3.5252	3.5866	3.6468	3.7060	3.7640	3.8208	3.8765	3.9311	3.9844	4.0365	4.0874	4.1370	4.1854	4.2325	4.2783	4.3228	50
52	.2715	.3370	.4014	.4649	.5273	.5886	.6488	.7079	.7659	.8227	.8784	.9329	.9861	.0382	.0891	.1386	.1870	.2340	.2798	.3243	52
54	.2737	.3391	.4036	.4670	.5293	.5906	.6508	.7099	.7678	.8246	.8802	.9347	.9879	.0399	.0907	.1403	.1886	.2356	.2813	.3257	54
56	.2759	.3413	.4057	.4691	.5314	.5926	.6528	.7118	.7697	.8265	.8820	.9364	.9896	.0416	.0924	.1419	.1902	.2371	.2828	.3272	56
58	.2781	.3435	.4078	.4712	.5335	.5947	.6548	.7138	.7716	.8283	.8839	.9382	.9914	.0433	.0941	.1435	.1917	.2387	.2843	.3286	58
60	3.2803	3.3456	3.4100	3.4733	3.5355	3.5967	3.6567	3.7157	3.7735	3.8302	3.8857	3.9400	3.9932	4.0451	4.0957	4.1452	4.1933	4.2402	4.2858	4.3301	60

Constants in table are for five inch Sine Bars or Plates—when using ten inch Sine Bars or Plates, multiply Constants by two.

Metric and English Conversion

Measures of Length

1 millimeter (mm) = 0.03937 inch
1 centimeter (cm) = 0.39370 inch
1 meter (m) = 39.37008 inches
= 3.2808 feet
= 1.0936 yards
1 kilometer (km) = 0.6214 mile
1 inch = 25.4 millimeters (mm)
= 2.54 centimeters (cm)
1 foot = 304.8 millimeters (mm)
= 0.3048 meter (m)
1 yard = 0.9144 meter (m)
1 mile = 1.609 kilometers (km)

Measures of Area

1 square millimeter = 0.00155 square inch
1 square centimeter = 0.155 square inch
1 square meter = 10.764 square feet
= 1.196 square yards
1 square kilometer = 0.3861 square mile
1 square inch = 645.2 square millimeters
= 6.452 square centimeters
1 square foot = 929 square centimeters
= 0.0929 square meter
1 square yard = 0.836 square meter
1 square mile = 2.5899 square kilometers

Measures of Capacity (Dry)

1 cubic centimeter (cm^3) = 0.061 cubic inch
1 liter = 0.0353 cubic foot
= 61.023 cubic inches
1 cubic meter (m^3) = 35.315 cubic feet
= 1.308 cubic yards
1 cubic inch = 16.38706 cubic centimeters (cm^3)
1 cubic foot = 0.02832 cubic meter (m^3)
= 28.317 liters
1 cubic yard = 0.7646 cubic meter (m^3)

Measures of Capacity (Liquid)

1 liter = 1.0567 U.S. quarts
= 0.2642 U.S. gallon
= 0.2200 Imperial gallon
1 cubic meter (m^3) = 264.2 U.S. gallons
= 219.969 Imperial gallons
1 U.S. quart = 0.946 liter
1 Imperial quart = 1.136 liters
1 U.S. gallon = 3.785 liters
1 Imperial gallon = 4.546 liters

Measures of Weight

1 gram (g) = 15.432 grains
= 0.03215 ounce troy
= 0.03527 ounce avoirdupois
1 kilogram (kg) = 35.274 ounces avoirdupois
= 2.2046 pounds
1000 kilograms (kg) = 1 metric ton (t)
= 1.1023 tons of 2000 pounds
= 0.9842 ton of 2240 pounds
1 ounce avoirdupois = 28.35 grams (g)
1 ounce troy = 31.103 grams (g)
1 pound = 453.6 grams
= 0.4536 kilogram (kg)
1 ton 2240 pounds = 1016 kilograms (kg)
= 1.016 metric tons
1 grain = 0.0648 gram (g)
1 metric ton = 0.9842 ton of 2240 pounds
= 2204.6 pounds

Credit: The DoAll Company.

TABLE 11-3
Continued

Figures are in Inches

Min.	0°	1°	2°	3°	4°	5°	6°	7°	8°	9°	10°	11°	12°	13°	14°	15°	16°	17°	18°	19°	Min.
0	0.00000	0.08725	0.17450	0.26170	0.34880	0.43580	0.52265	0.60935	0.69585	0.78215	0.86825	0.95405	1.0395	1.1247	1.2096	1.2941	1.3782	1.4618	1.5451	1.6278	0
2	.00290	.09015	.17740	.26460	.35170	.43870	.52555	.61225	.69875	.78505	.87110	.95690	.0424	.1276	.2124	.2969	.3810	.4646	.5478	.6306	2
4	.00580	.09310	.18030	.26750	.35460	.44155	.52845	.61510	.70165	.78790	.87395	.95975	.0452	.1304	.2152	.2997	.3838	.4674	.5506	.6333	4
6	.00875	.09600	.18320	.27040	.35750	.44445	.53130	.61800	.70450	.79080	.87685	.96260	.0481	.1332	.2181	.3025	.3865	.4702	.5534	.6361	6
8	.01165	.09890	.18615	.27330	.36040	.44735	.53420	.62090	.70740	.79365	.87970	.96545	.0509	.1361	.2209	.3053	.3893	.4730	.5561	.6388	8
10	0.01455	0.10180	0.18905	0.27620	0.36330	0.45025	0.53710	0.62380	0.71025	0.79655	0.88255	0.96830	1.0538	1.1389	1.2237	1.3081	1.3921	1.4757	1.5589	1.6416	10
12	.01745	.10470	.19195	.27910	.36620	.45315	.54000	.62665	.71315	.79940	.88540	.97115	.0566	.1417	.2265	.3109	.3949	.4785	.5616	.6443	12
14	.02035	.10760	.19485	.28200	.36910	.45605	.54290	.62955	.71600	.80230	.88830	.97405	.0594	.1446	.2293	.3137	.3977	.4813	.5644	.6471	14
16	.02325	.11055	.19775	.28490	.37200	.45895	.54580	.63245	.71890	.80515	.89115	.97690	.0623	.1474	.2322	.3165	.4005	.4841	.5672	.6498	16
18	.02620	.11345	.20065	.28780	.37490	.46185	.54865	.63530	.72180	.80800	.89400	.97975	.0651	.1502	.2350	.3193	.4033	.4868	.5699	.6525	18
20	0.02910	0.11635	0.20355	0.29070	0.37780	0.46475	0.55155	0.63820	0.72465	0.81090	0.89685	0.98260	1.0680	1.1531	1.2378	1.3221	1.4061	1.4896	1.5727	1.6553	20
22	.03200	.11925	.20645	.29365	.38070	.46765	.55445	.64110	.72755	.81375	.89975	.98545	.0708	.1559	.2406	.3250	.4089	.4924	.5755	.6580	22
24	.03490	.12215	.20940	.29655	.38360	.47055	.55735	.64400	.73040	.81665	.90260	.98830	.0737	.1587	.2434	.3278	.4117	.4952	.5782	.6608	24
26	.03780	.12505	.21230	.29945	.38650	.47345	.56025	.64685	.73330	.81950	.90545	.99115	.0765	.1615	.2462	.3306	.4145	.4980	.5810	.6635	26
28	.04070	.12800	.21520	.30235	.38940	.47635	.56315	.64975	.73615	.82235	.90830	.99400	.0793	.1644	.2491	.3334	.4173	.5007	.5837	.6663	28
30	0.04365	0.13090	0.21810	0.30525	0.39230	0.47925	0.56600	0.65265	0.73905	0.82525	0.91120	0.99685	1.0822	1.1672	1.2519	1.3362	1.4201	1.5035	1.5865	1.6690	30
32	.04655	.13380	.22100	.30815	.39520	.48210	.56890	.65550	.74190	.82810	.91405	.99970	.0850	.1700	.2547	.3390	.4228	.5063	.5893	.6718	32
34	.04945	.13670	.22390	.31105	.39810	.48500	.57180	.65840	.74480	.83100	.91690	1.0016	.0879	.1729	.2575	.3418	.4256	.5091	.5920	.6745	34
36	.05235	.13960	.22680	.31395	.40100	.48790	.57470	.66130	.74770	.83385	.91975	.0054	.0907	.1757	.2603	.3446	.4284	.5118	.5948	.6772	36
38	.05525	.14250	.22970	.31685	.40390	.49080	.57760	.66415	.75055	.83670	.92260	.0082	.0935	.1785	.2631	.3474	.4312	.5146	.5975	.6800	38
40	0.05820	0.14540	0.23265	0.31975	0.40680	0.49370	0.58045	0.66705	0.75345	0.83960	0.92545	1.0110	1.0964	1.1813	1.2660	1.3502	1.4340	1.5174	1.6003	1.6827	40
42	.06110	.14835	.23555	.32265	.40970	.49660	.58335	.66995	.75630	.84245	.92835	.0139	.0992	.1842	.2688	.3530	.4368	.5201	.6030	.6855	42
44	.06400	.15125	.23845	.32555	.41260	.49950	.58625	.67280	.75920	.84530	.93120	.0168	.1020	.1870	.2716	.3558	.4396	.5229	.6058	.6882	44
46	.06690	.15415	.24135	.32845	.41550	.50240	.58915	.67570	.76205	.84820	.93405	.0196	.1049	.1898	.2744	.3586	.4423	.5257	.6085	.6909	46
48	.06980	.15705	.24425	.33135	.41840	.50530	.59200	.67860	.76495	.85105	.93690	.0225	.1077	.1926	.2772	.3614	.4451	.5285	.6113	.6937	48
50	0.07270	0.15995	0.24715	0.33425	0.42130	0.50820	0.59490	0.68145	0.76780	0.85390	0.93975	1.0253	1.1106	1.1955	1.2800	1.3642	1.4479	1.5312	1.6141	1.6964	50
52	.07565	.16285	.25005	.33715	.42420	.51105	.59780	.68435	.77070	.85680	.94260	.0281	.1134	.1983	.2828	.3670	.4507	.5340	.6168	.6991	52
54	.07855	.16580	.25295	.34010	.42710	.51395	.60070	.68720	.77355	.85965	.94550	.0310	.1162	.2011	.2856	.3698	.4535	.5368	.6196	.7019	54
56	.08145	.16870	.25585	.34300	.43000	.51685	.60355	.69010	.77645	.86250	.94835	.0338	.1191	.2039	.2884	.3726	.4563	.5395	.6223	.7046	56
58	.08435	.17160	.25875	.34590	.43290	.51975	.60645	.69300	.77930	.86540	.95120	.0367	.1219	.2068	.2913	.3754	.4591	.5423	.6251	.7073	58
60	0.08725	0.17450	0.26170	0.34880	0.43580	0.52265	0.60935	0.69585	0.78215	0.86825	.95405	1.0395	1.1247	1.2096	1.2941	1.3782	1.4618	1.5451	1.6278	1.7101	60

Figures are in Inches

Min.	20°	21°	22°	23°	24°	25°	26°	27°	28°	29°	30°	31°	32°	33°	34°	35°	36°	37°	38°	39°	Min.
0	1.7101	1.7918	1.8730	1.9536	2.0337	2.1131	2.1918	2.2699	2.3473	2.4240	2.5000	2.5752	2.6496	2.7232	2.7959	2.8679	2.9389	3.0091	3.0783	3.1466	0
2	.7128	.7945	.8757	.9563	.0363	.1157	.1944	.2725	.3499	.4266	.5025	.5777	.6520	.7256	.7984	.8702	.9413	.0114	.0806	.1488	2
4	.7155	.7972	.8784	.9590	.0390	.1183	.1971	.2751	.3525	.4291	.5050	.5802	.6545	.7280	.8008	.8726	.9436	.0137	.0829	.1511	4
6	.7183	.8000	.8811	.9617	.0416	.1210	.1997	.2777	.3550	.4317	.5075	.5826	.6570	.7305	.8032	.8750	.9460	.0160	.0852	.1534	6
8	.7210	.8027	.8838	.9643	.0443	.1236	.2023	.2803	.3576	.4342	.5100	.5851	.6594	.7329	.8056	.8774	.9483	.0183	.0874	.1556	8
10	1.7237	1.8054	1.8865	1.9670	2.0469	2.1262	2.2049	2.2829	2.3602	2.4367	2.5126	2.5876	2.6619	2.7354	2.8080	2.8798	2.9507	3.0207	3.0897	3.1579	10
12	.7265	.8081	.8892	.9697	.0496	.1289	.2075	.2855	.3627	.4393	.5151	.5901	.6644	.7378	.8104	.8821	.9530	.0230	.0920	.1601	12
14	.7292	.8108	.8919	.9724	.0522	.1315	.2101	.2881	.3653	.4418	.5176	.5926	.6668	.7402	.8128	.8845	.9554	.0253	.0943	.1624	14
16	.7319	.8135	.8946	.9750	.0549	.1341	.2127	.2906	.3679	.4444	.5201	.5951	.6693	.7427	.8152	.8869	.9577	.0276	.0966	.1646	16
18	.7347	.8162	.8973	.9777	.0575	.1368	.2153	.2932	.3704	.4469	.5226	.5976	.6717	.7451	.8176	.8893	.9600	.0299	.0989	.1669	18
20	1.7374	1.8189	1.899	1.9804	2.0602	2.1394	2.2179	2.2958	2.3730	2.4494	2.5251	2.6001	2.6742	2.7475	2.8200	2.8916	2.9624	3.0322	3.1012	3.1691	20
22	.7401	.8217	.9026	.9830	.0628	.1420	.2205	.2984	.3755	.4520	.5276	.6025	.6767	.7499	.8224	.8940	.9647	.0345	.1034	.1714	22
24	.7428	.8244	.9053	.9857	.0655	.1447	.2232	.3010	.3781	.4545	.5301	.6050	.6791	.7524	.8248	.8964	.9671	.0369	.1057	.1736	24
26	.7456	.8271	.9080	.9884	.0681	.1473	.2258	.3036	.3807	.4570	.5327	.6075	.6816	.7548	.8272	.8988	.9694	.0392	.1080	.1759	26
28	.7483	.8298	.9107	.9911	.0708	.1499	.2284	.3061	.3832	.4596	.5352	.6100	.6840	.7572	.8296	.9011	.9718	.0415	.1103	.1781	28
30	1.7510	1.8325	1.9134	1.9937	2.0734	2.1525	2.2310	2.3087	2.3858	2.4621	2.5377	2.6125	2.6865	2.7597	2.8320	2.9035	2.9741	3.0438	3.1125	3.1804	30
32	.7537	.8352	.9161	.9964	.0761	.1552	.2336	.3113	.3883	.4646	.5402	.6149	.6889	.7621	.8344	.9059	.9764	.0461	.1148	.1826	32
34	.7565	.8379	.9188	.9991	.0787	.1578	.2362	.3139	.3909	.4672	.5427	.6174	.6914	.7645	.8368	.9082	.9788	.0484	.1171	.1849	34
36	.7592	.8406	.9215	2.0017	.0814	.1604	.2388	.3165	.3934	.4697	.5452	.6199	.6938	.7669	.8392	.9106	.9811	.0507	.1194	.1871	36
38	.7619	.8433	.9241	.0044	.0840	.1630	.2414	.3190	.3960	.4722	.5477	.6224	.6963	.7694	.8416	.9130	.9834	.0530	.1216	.1893	38
40	1.7646	1.8460	1.9268	2.0070	2.0867	2.1656	2.2440	2.3216	2.3985	2.4747	2.5502	2.6249	2.6987	2.7718	2.8440	2.9153	2.9858	3.0553	3.1239	3.1916	40
42	.7673	.8487	.9295	.0097	.0893	.1683	.2466	.3242	.4011	.4773	.5527	.6273	.7012	.7742	.8464	.9177	.9881	.0576	.1262	.1938	42
44	.7701	.8514	.9322	.0124	.0920	.1709	.2492	.3268	.4036	.4798	.5552	.6298	.7036	.7766	.8488	.9200	.9904	.0599	.1285	.1961	44
46	.7728	.8541	.9349	.0150	.0946	.1735	.2518	.3293	.4062	.4823	.5577	.6323	.7061	.7790	.8512	.9224	.9928	.0622	.1307	.1983	46
48	.7755	.8568	.9376	.0177	.0972	.1761	.2544	.3319	.4087	.4848	.5602	.6348	.7085	.7815	.8535	.9248	.9951	.0645	.1330	.2005	48
50	1.7782	1.8595	1.9402	2.0204	2.0999	2.1787	2.2570	2.3345	2.4113	2.4874	2.5627	2.6372	2.7110	2.7839	2.8559	2.9271	2.9974	3.0668	3.1353	3.2028	50
52	.7809	.8622	.9429	.0230	.1025	.1814	.2596	.3371	.4138	.4899	.5652	.6397	.7134	.7863	.8583	.9295	.9997	.0691	.1375	.2050	52
54	.7837	.8649	.9456	.0257	.1052	.1840	.2621	.3396	.4164	.4924	5677	.6422	.7158	.7887	.8607	.9318	3.0021	.0714	.1398	.2072	54
56	.7864	.8676	.9483	.0283	.1078	.1866	.2647	.3422	.4189	.4949	.5702	.6446	.7183	.7911	.8631	.9342	.0044	.0737	.1421	.2095	56
58	.7891	.8703	.9510	.0310	.1104	.1892	.2673	.3448	.4215	.4975	.5727	.6471	.7207	.7935	.8655	.9365	.0067	.0760	.1443	.2117	58
60	1.7918	1.8730	1.9536	2.0337	2.1131	2.1918	2.2699	2.3473	2.4240	2.5000	2.5752	2.6496	2.7232	2.7959	2.8679	2.9389	3.0091	3.0783	3.1466	3.2139	60

Constants in table are for five inch Sine Bars or Plates—when using ten inch Sine Bars or Plates, multiply Constants by two.

FIGURE 11-9
A simple one axis sine plate.

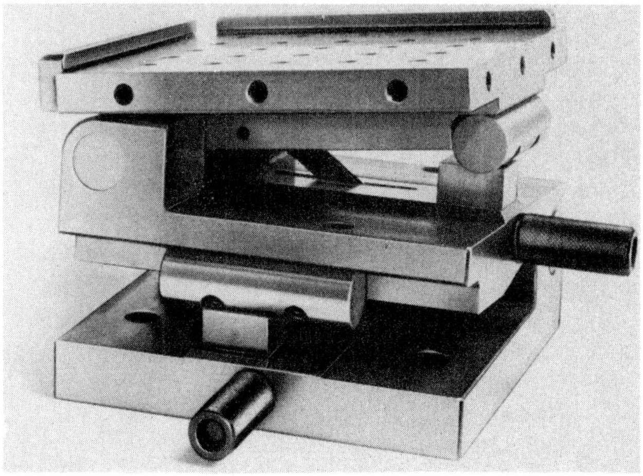

FIGURE 11-10
A compound sine plate. (Credit: MTI Corporation)

The Sine Plate

Much wider than the sine bar, the sine plate (Figure 11-9) will accept work generally 2 in. (50 mm) or wider. Working on the same principle as the sine bar, that is, the wringing and stacking of gage blocks under one end to obtain an angle, some sine plates can be used for simple inspection work, angular measurement, or layout.

Like its sine bar counterpart, the sine plate functions on two hardened rollers, only it is hinged for elevation. It also contains tapped holes that will permit work to be strapped to its surface and may also contain an end stop to prevent the workpiece from slipping during machining. The end stop will also act as a reference base. Tapped holes on its sides may be found that will allow the fastening of side plates or clamping devices.

Sine plates are generally found in 5- and 10-in. lengths and may be used on a surface plate as well as be clamped to the table of a machine tool.

The Compound Sine Plate

Consisting of two sine plates, one above the other, the compound sine plate is designed to create angles in two directions at the same time (Figure 11-10). The lower sine plate is gaged to create an angle anywhere between 0° and 60°, merely by stacking gage blocks under the free roll. The upper table is hinged to move at right angles to its lower base. This, too, is swivable to create an angle as large as 60°.

Like the sine plate, the compound sine plate's surface is pock marked with tapped holes, featuring end plates and side rails that are fastened to the tables by means of tapped holes. Small angles can also be set up due to grooved steps of 0.100 or 0.200 in. deep in the lower table base, which permits gage blocks to be placed in these slots. It should also be mentioned that compound angles require special calculations, formulas that can be found in technical handbooks such as the *Machinery's Handbook*.

The Magnetic Sine

The magnetic sine (Figure 11-11) is similar to the sine plate in setup and measurement of angular measurement with an added feature: It contains a magnetic chuck of its own. By the turn of a handle, the magnetic sine is energized and can either hold or release a workpiece from the chuck, making it ideal for surface grinding work. Like its sine plate cousin, the magnetic sine can also be clamped to a machine table. Similarly, compound sine plates are also available.

FIGURE 11-11
Magnetic one and two axis sine plates. (Credit: MTI Corporation)

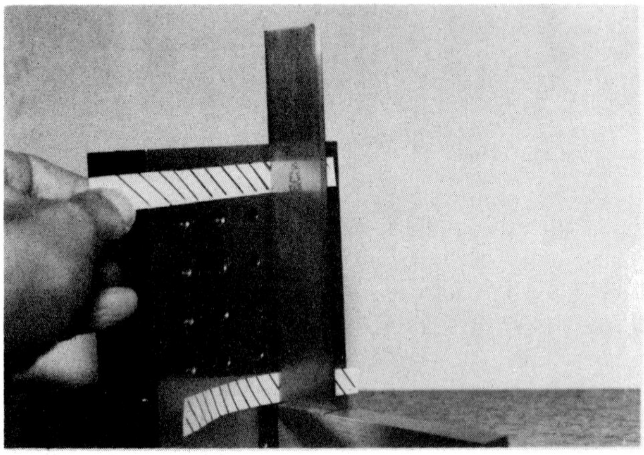

FIGURE 11-12
Checking for squareness using paper feelers between the blade and the work surface.

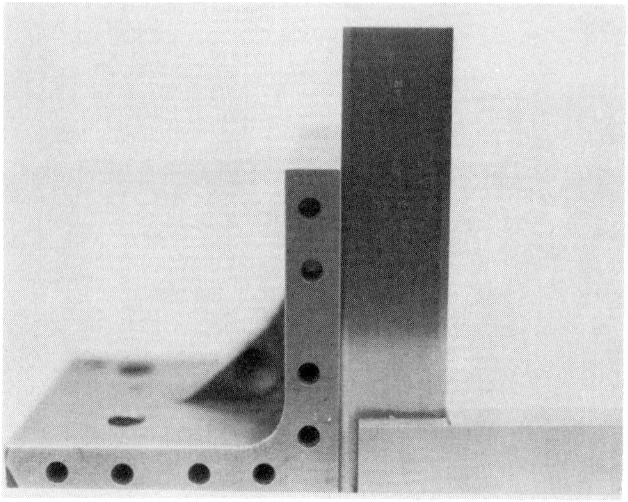

FIGURE 11-13
A bevel edge square.

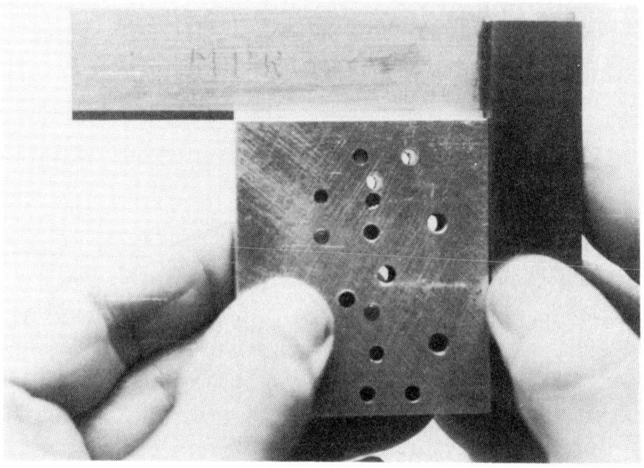

FIGURE 11-14
A bevel edge provides a thin line contact with the work surface.

The Precision Square

The square is one of the oldest tools used for checking the trueness of right angles. It is used as a basic tool by skilled tradespeople when measuring adjacent sides of flat work to see that its included angle is "square" (at 90°, as in Figure 11-12). Precision squares are quality tools and come in a variety of shapes, none of which should be abused if their accuracy is to be maintained.

The bevel edge square (Figure 11-13) is similar to the solid steel square in that the beam and blade is hardened and tempered as well as ground and lapped to ensure straightness and parallelism. The beveled surface provides another feature. Its knife-like edge provides a thin line contact with the work surface (Figure 11-14). A ground notch at the inner corner is provided for burr and dirt clearance on both types.

The cylindrical square can be used as a master square against which other squares are checked for out of squareness. Hardened and ground, one end of the instrument is lapped perfectly square with the axis. The other end is purposely lapped out of square to a predetermined angle. The cylindrical shape provides a straight line of sight when work is placed

FIGURE 11-15
A direct-reading cylindrical square.

against it. All testing by either end is conducted on the work area of a surface plate.

The direct-reading cylindrical square (Figure 11-15) is provided with several elliptical curved lines composed of a series of black dots calibrated in ten-thousandths of an inch (0.0001 in.). Reading is taken at the top of the cylinder from the dotted curve that is in contact with the test part.

Procedure for Testing

1. Place the test part on a clean surface plate.
2. Place the cylindrical out-of-square end on the surface plate next to it.
3. Lightly butt the upright surfaces together.
4. Carefully rotate the cylindrical square until no light can be seen between it and the test part.
5. Note the uppermost curved line in contact against the test part and follow it to the top of the cylinder. Reading indicates out of squareness of the test part.

The Die Makers' Square

The die makers' square (Figure 11-16) is widely used by tool and die makers to examine and measure the degree of die clearance. Pattern makers also have use for this tool to check the angles and drafts of patterns (Figure 11-17). Made in a few styles, one

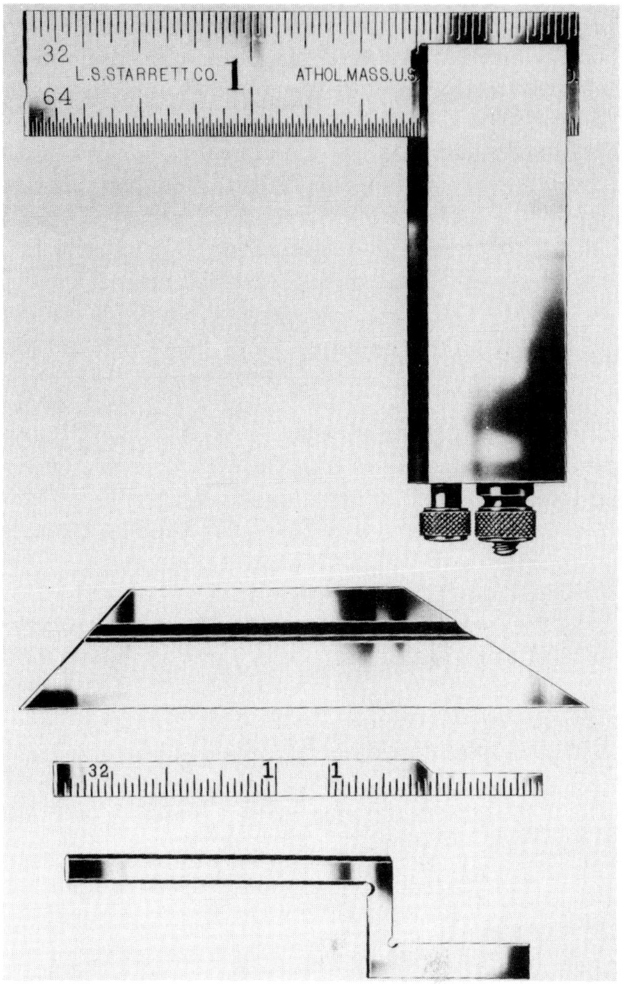

FIGURE 11-16
A die makers' square with sliding blades adjustment. (Credit: L.S. Starrett Company)

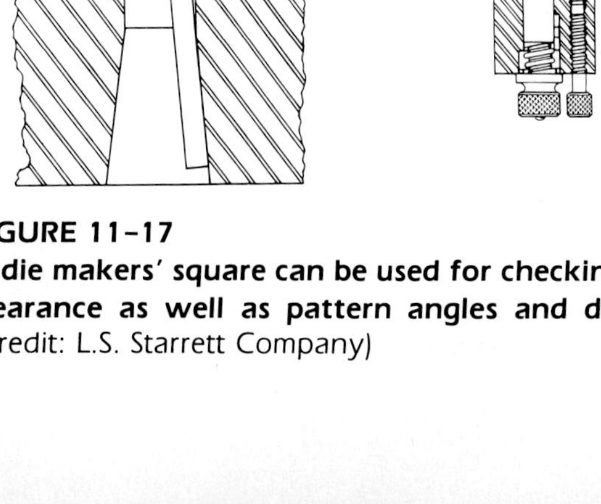

FIGURE 11-17
A die makers' square can be used for checking die clearance as well as pattern angles and drafts. (Credit: L.S. Starrett Company)

die makers' square comes with hardened and ground beam and sliding blade. The blade is adjusted for length by the turn of a clamping nut. A smaller knurled screw works in conjunction with the clamping screw and tilts the blade to any desired angle. This model's angle measurement must then be read against a protractor or compared against a known angle for accuracy. A second style is made with a direct-reading scale on its hardened beam.

UNIT 12 COMPARISON GAGES

Comparison gaging entails the measurement, transference, and comparison of the measurement with a known standard or graduated measuring device. The use of the surface gage (Figure 16–3), the center gage (Figure 45–13), and the drill plate gage (Figure 30–8) are described elsewhere in the book. What follows are a few of the more commonly used comparison gages in inspection and in the machine shop.

Objectives
After completing this unit you should be able to

1. understand and define comparison measurement.
2. identify the commonly used comparison gages.
3. select the proper comparison gage for the measurement involved.

Snap Gage

Providing one of the quickest means of comparison measurement, snap gages are capable of inspecting diameters as well as flat surfaces. Snap gages are available in different styles and sizes, generally having a U-shaped frame with adjustable anvils.

Adjustable Snap Gage

An adjustable snap gage (Figure 12–1) contains one stationary anvil and two adjustable pin or button anvils. The limits of the gage may be preset and secured with sealing wax by the manufacturer, their size limits stamped on the frame. These gages may also be set in the inspection room using gage blocks as standards, locking in the adjustment screw with sealing wax. Snap gages are also available with dial indicator reading (Figure 12–2).

Procedure for Use

1. Clean the surface of the workpiece.
2. Clean the surfaces of the anvils.
3. Hold the gage so that the measuring surfaces may enter evenly with the work.
4. Slide the go opening past the work surface but do not force it through.
5. Continue advancing the gage until contact has been made with the not-go opening.

FIGURE 12–1
A square button-type anvil snap gage. (Credit: DoAll Company)

FIGURE 12–2
A dial indicator snap gage. (Credit: L.S. Starrett Company)

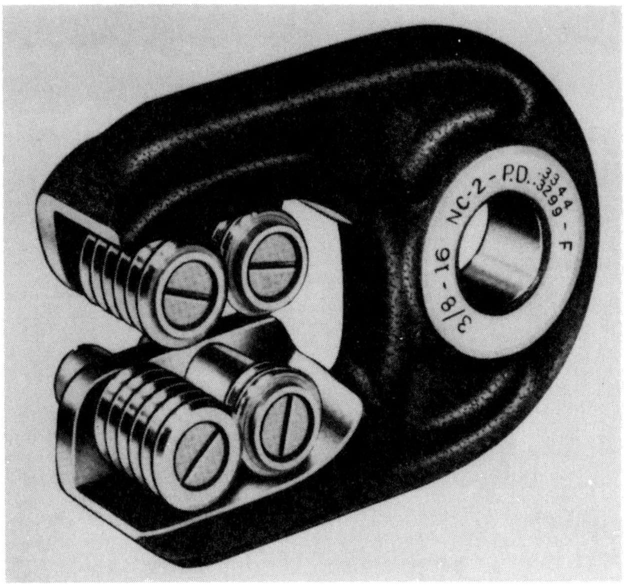

FIGURE 12-3
A roll thread snap gage can rapidly check external threads for profile, size, straightness, or roundness in one pass. (Credit: DoAll Company)

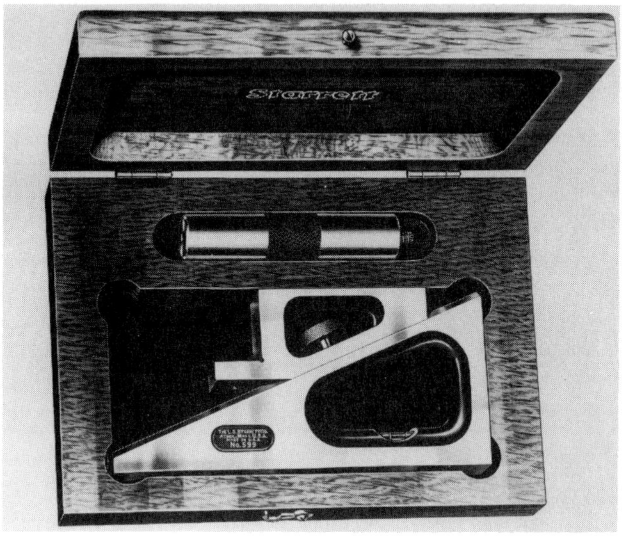

FIGURE 12-4
A planer gage. (Credit: L.S. Starrett Company)

FIGURE 12-5
Setting a planer gage. (Credit: Philip Rapisarda)

If the workpiece is acceptable, the gage will slip by the go opening, stopping only when the work contacts the not-go anvils. If the gage slips by entirely, the work is too large. If the workpiece cannot enter at all, it is considered too large and the workpiece is rejected.

The Roll Thread Snap Gage

External threads are quickly inspected with the use of a roll thread snap gage (Figure 12-3). The gage contains a set of go/not-go rolls and is used to measure the size limits for the pitch diameter of screw threads as previously described. It can also check profile, roundness, and straightness at the same time.

Planer Gage

Originally intended for use as a comparison device with which to establish the depth of cutting tools used on planer and shaper machine tools, the planer gage can be used to measure widths and slots of workpieces. The planer gage (Figure 12-4) works by sliding two inclined planes that permit its measuring surfaces to increase or decrease the distance from the base. Accuracy can be established by setting the distance with a micrometer (Figure 12-5). One use would allow the planer gage to elevate the sine bar for inspection or layout work. Gage blocks and the use of a dial test indicator fastened to a surface gage or height gage can also be used to set the planer gage. Once so set, the planer gage can then act as a comparison inspection gage (Figure 12-6).

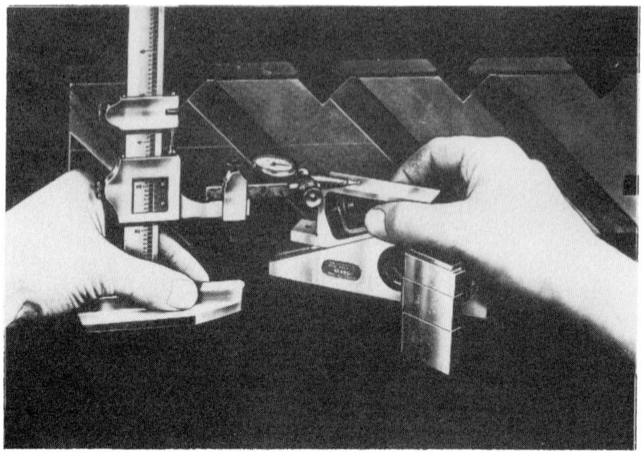

FIGURE 12-6
Using a planer gage as a comparison gage. (Credit: L.S. Starrett Company)

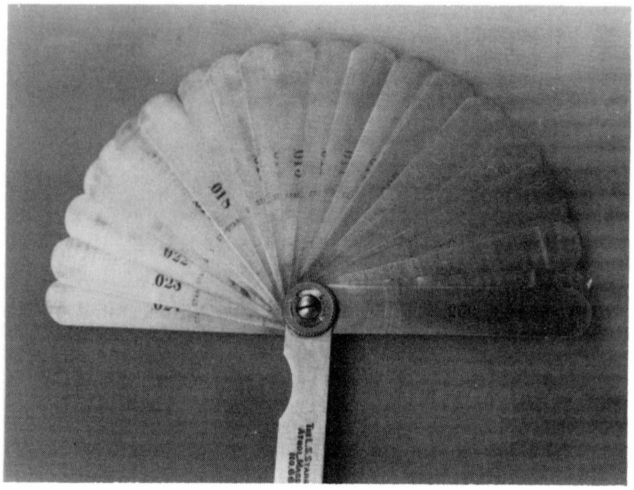

FIGURE 12-8
A thickness gage is also called a feeler gage.

FIGURE 12-7
Checking a contour with a radius gage.

Radius Gage

Made to measure directly, radius gages are thin blades whose profiles are contoured to check the size and radii of convex and concave surfaces. Available in both inch and metric values, these blades come with the radius measurement marked on them (Figure 12-7).

Thickness Gage

Often referred to as a feeler gage (Figure 12-8), a series of tempered steel blades are used to measure spaces, gaps, slots, and small openings. The blades are marked and range in size from 0.0015 to 0.025 in. (0.03 to 0.64 mm) with each blade varying in thickness by 0.001 in. (0.02 mm). Because the blades are combined to obtain various dimensions and are inserted, not forced, to make contact with the divided surfaces, the thinner foil-like blades should be sandwiched between the heavier ones to protect them against distortion, tearing, or wrinkling.

Telescoping Gages

Made to measure slots, holes, and recesses ranging in size from $\frac{5}{16}$ to 6 in. (7.9 to 15 mm), telescoping gages come in a set of five or six T-shaped instruments (Figure 12-9). Connected to each knurled handle are two dome-shaped, spring-loaded plungers that telescope at right angles and make contact with the surfaces to be measured. Another type of gage has one fixed contact from which a spring-loaded plunger is set in motion. By proper rotation of the knurled screw at the end of the handle, the plungers can be set free or locked into position. Dimensional distances are transferred by means of a micrometer measuring over the dome-shaped contact surfaces.

Measuring with Telescoping Gages

1. Select the proper gage for the hole involved.

2. Clean the hole and the gage's contact surfaces.

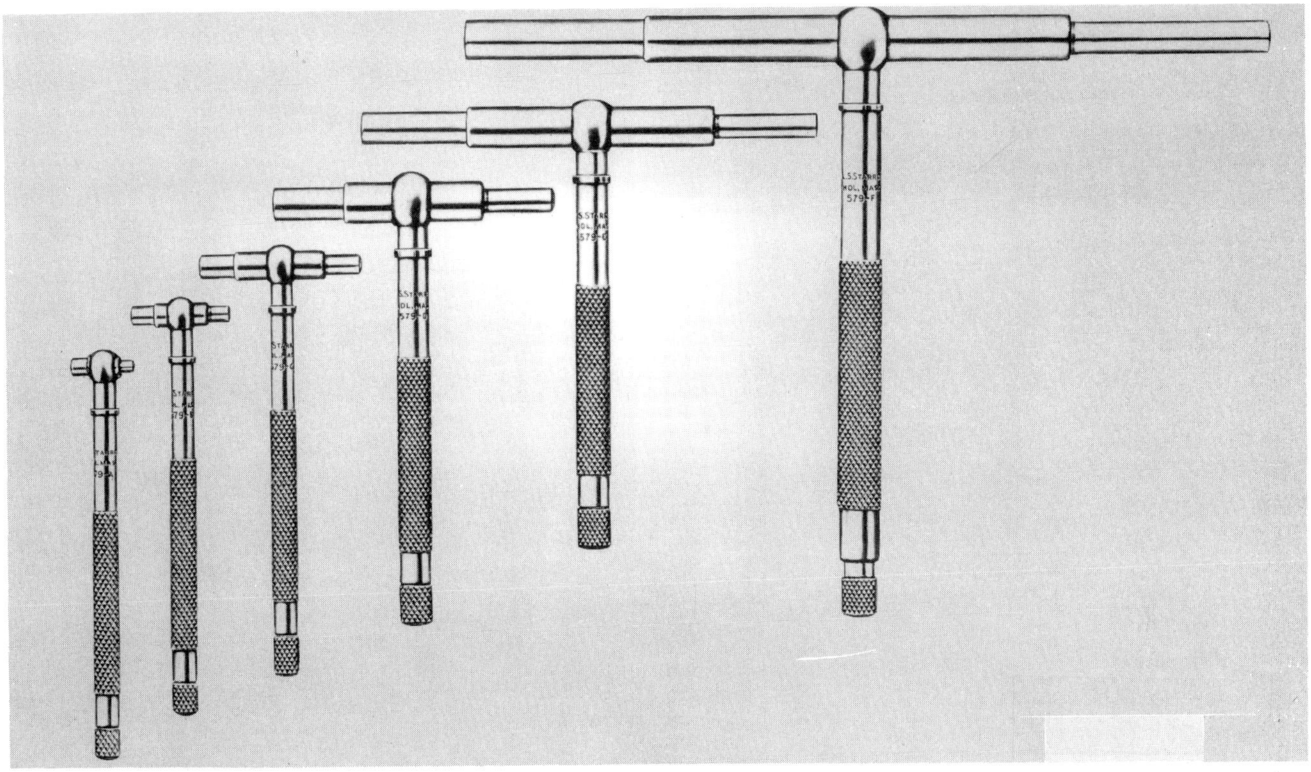

FIGURE 12–9
A set of telescopic gages. (Credit: L.S. Starrett Company)

3. Compress the plungers and lock them by turning the knurled screw.

4. Insert the gage into the hole at a slight angle and release the plungers.

5. Holding one contact firm against the hole's diameter, rock the handle in small movements so that the other plunger can contact the opposite wall.

6. Holding the handle evenly with the hole, tighten the knurled screw (Figure 12–10).

7. Move the gage straight up and down to check the drag of the contact surfaces.

8. Once satisfied the feel is correct, measure over these surfaces with a micrometer.

Small-Hole Gages

Using a measuring technique similar to that of telescoping gages, small-hole gages range in size from 0.125 to 0.500 in. (3.18 to 12.7 mm) and are available in two styles: ball contact and a ball contact with a flat bottom. Both styles are used for measuring

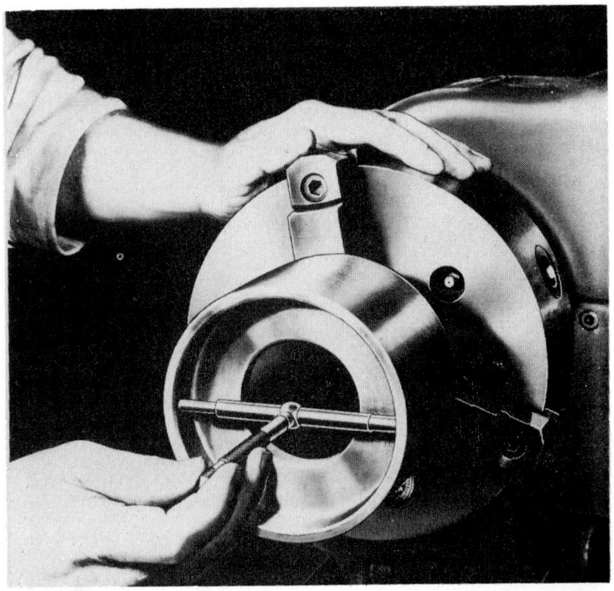

FIGURE 12–10
Accurately measuring a diameter with a telescopic gage. (Credit: L.S. Starrett Company)

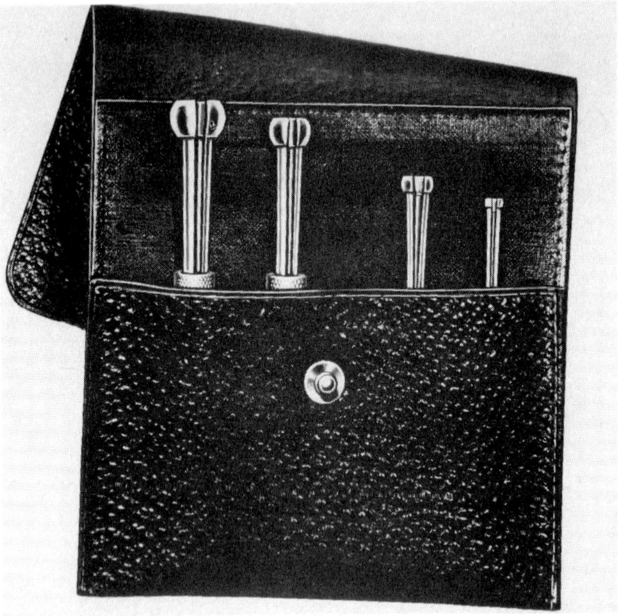

FIGURE 12–11
A set of small-hole gages. (Credit: L.S. Starrett Company)

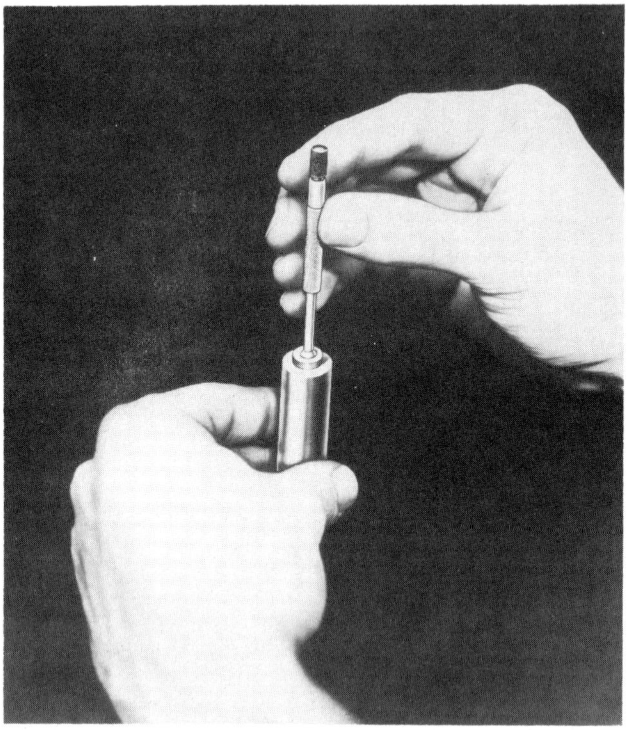

FIGURE 12–12
Measuring a narrow diameter with a small-hole gage. (Credit: L.S. Starrett Company)

slots, holes, grooves, or recesses and generally come in sets of four or five knurled handled instruments whose tightening screw at one end controls the expansion of a hardened split ball shape that is used for gaging. The flattened ball style permits measuring of shallow slots, holes, or recesses. Once distance has been obtained, its accuracy is read by a micrometer measuring over its ball surface (Figure 12–11).

Measuring with Ball Gages

1. Select the proper gage for the hole involved.
2. Clean the hole and gage surfaces.
3. Close the ball width to allow the gage to enter the hole freely.
4. Insert the gage into the hole at a slight angle.
5. Tighten the adjustment screw until you can feel the ball contacting both sides of the surfaces to be measured.
6. Hold the handle evenly with the hole or recessed sides and lightly adjust the screw until you have created the proper drag.

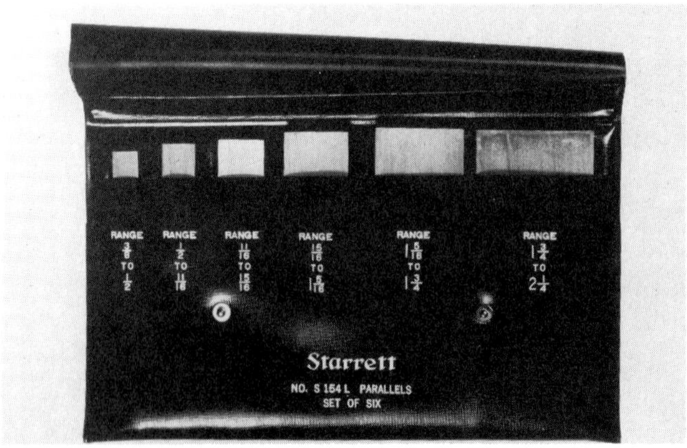

FIGURE 12–13
A set of adjustable parallels. (Credit: L.S. Starrett Company)

7. Move the gage in short up and down movements to check the feel or drag (Figure 12–12).
8. Remove the instrument once satisfied with the feel and measure over the contact surfaces with a micrometer.

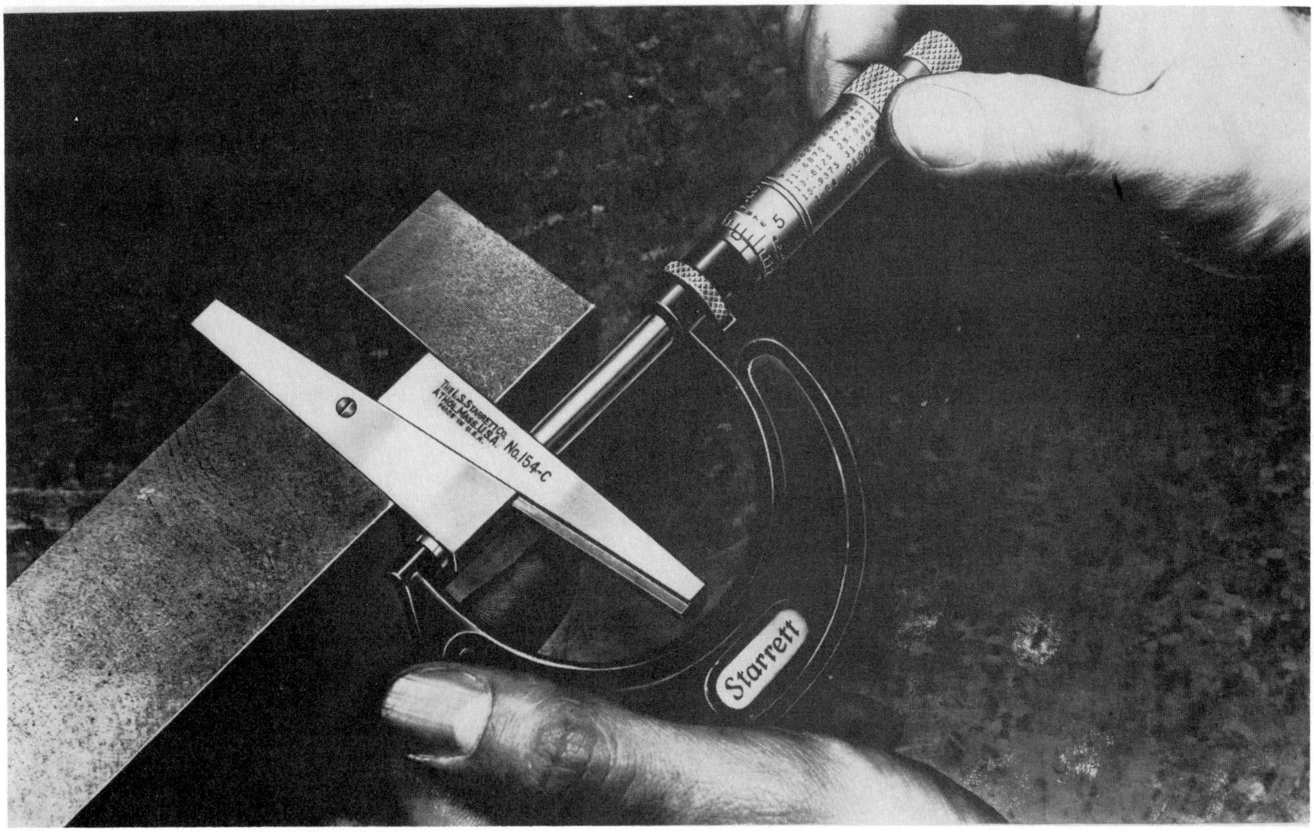

FIGURE 12-14
Measuring the overall width of an adjustable parallel (Credit: L.S. Starrett Company)

Adjustable Parallels

Adjustable parallels (Figure 12-13) comes in sets and range in size as small as $\frac{3}{8}$ in. (9.5 mm) and as wide as $2\frac{1}{4}$ in. (56 mm). Their adjustability make them useful for measuring purposes as well as being used as parallels in machine tool applications. They are also useful for layout purposes.

Made of two sliding members dovetailed and locked in place by one or two screws, adjustable parallels can be set to micrometer measurements (Figure 12-14). Once set, the adjustable parallel can be used as an inspection tool for gaging slots, grooves, and openings of this size. The adjustable parallel can also be used in reverse order. That is, slide the adjustable parallel until its outer edges touch the sides of the opening, lock this setting with a small screwdriver and then measure over the width of the parallel with an outside micrometer to determine its size.

Precision Height Gage

Precision height gages are manufactured by a number of quality toolmakers. They provide the accurate and quick means of precision inspection by means of comparison. All work is carried out on a clean surface plate in connection with a dial test indicator or sensitive electron gage held by a height transfer gage. These instruments can also be fastened to a surface gage or a height gage when heights to be inspected are of a short measurement.

The precision height gage (Figure 12-15) may measure either with hardened and lapped round steps or staggered gage blocks, with each step spaced 1 in. or, in the metric system, 20 mm apart. A large knurled micrometer thimble atop the structure houses the measuring step column and raises or lowers the steps in increments of 0.001 in. (0.02 mm), with the smallest graduation possible being 0.0001

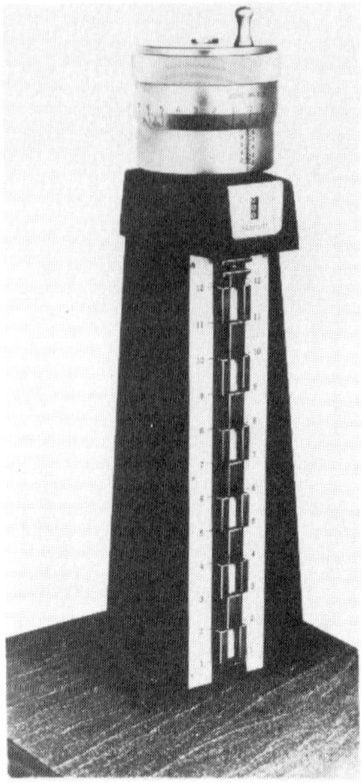

FIGURE 12-15
A digital height gage. (Credit: L.S. Starrett Company)

in. (0.002 mm). In addition to digital readouts, some models are equipped with an adjustable vernier scale that permits readings to 10 millionths.

Precision height gages come as small as 6 in. Standard 12-in. models can be elevated with riser blocks to measure precisely to heights of 18, 24, 36, and 40 in. Metric precision height gages also come as small as 150 mm, with the 300-mm models being elevated to heights of 450 and 600 mm.

Measuring Step Heights

1. Thoroughly clean the surface plate.
2. Thoroughly clean the workpiece base of grime or oil.
3. Wipe clean the base of the precision height gage and transfer gage.
4. Mount a dial test indicator on a suitable holder such as a surface gage, a height gage, or a height transfer gage.
5. Make contact with the lower surface (step 1) so that the dial test indicator's face registers with an approximate 0.005-in. (0.12-mm) reading.
6. Turn the bezel until the indicator reads zero.
7. Move the transfer gage close to the precision height gage, establishing the indicator's contact point near or slightly above an appropriate measuring step.
8. Turn the micrometer thimble until the measuring step surface will zero out the dial indicator.
9. Record the micrometer reading.
10. Withdraw the transfer gage and prepare to measure step 2 by adjusting the dial test indicator to its approximate height.
11. Make contact with the surface of step 2 until the dial test indicator's face registers with an approximate 0.005-in. (0.12-mm) reading.
12. Repeat step 6.
13. Repeat step 7.
14. Repeat step 8.
15. Record the micrometer reading.
16. Subtract reading 1 from reading 2 to obtain the step height difference.

UNIT 13 ADJUSTABLE GAGING

Another form of comparison gaging registers the change or variation from a set standard (deviation). This type of measuring tool is called an adjustable gage and is highly visible and relatively easy to use. Display devices in this category may include indicating instruments, comparators, air gages, or the more sophisticated electronic gages, all of them rendering rapid measuring.

Objectives
After completing this unit you will be able to

1. understand the principles of adjustable gaging.
2. identify four different categories of adjustable gaging devices.
3. be able to use any of the four groups to measure within 0.0005 in. (0.01 mm) accuracy.

Dial Indicator Specifications

By 1945, specifications had been established and published by the American Gage Design (AGD) committee, which governs the minimum requirements for accuracy and performance as well as design dimensions that permit mounting and interchangeability of precision dial indicators. A brief summary of these specifications included:

1. Accuracy: one graduation at any point.
2. Repeatability: one-fifth graduation at any point.
3. Spindle range: minimum of $2\frac{1}{2}$ revolutions.
4. Hand position: 9 o'clock position or $\frac{1}{4}$ revolution to the left of the zero mark when the spindle is in its extended position.
5. Smallest graduation: Inch: 0.00005, 0.0001, 0.0005, and 0.001 in. Metric: 0.001, 0.002, 0.005, and 0.01 mm.
6. Dial numbering: Inch: regardless of magnification, dial numbers always indicate thousandths of an inch. Metric: dial numbers always indicate hundredths of a millimeter.

These standards were later superseded by ANSI standards. While many of the AGD standards were kept, the following changes were made:

1. Metric dial faces will be yellow.
2. Smallest graduation: Inch: 0.00005, 0.0001, 0.0005, and 0.001 in. Metric: 0.001, 0.002, 0.01, and 0.02 mm.

While not all changes are reflected in this listing, it will be seen that the 0.005-mm graduation observed by the AGD was eliminated by ANSI. For further information concerning dial indicators, please consult the *ANSI Standards for Dial Indicators,* ANSI B89.1.10-1978.

Dial Indicators

Referred to by some as dial gages, dial indicators are commonplace measuring devices. Instruments such as the balanced-type dial indicator (Figure 13-1) operates by using a rack and pinion system. The main mechanism consists of a contact point located at the end of a stainless steel spindle, a length of which is

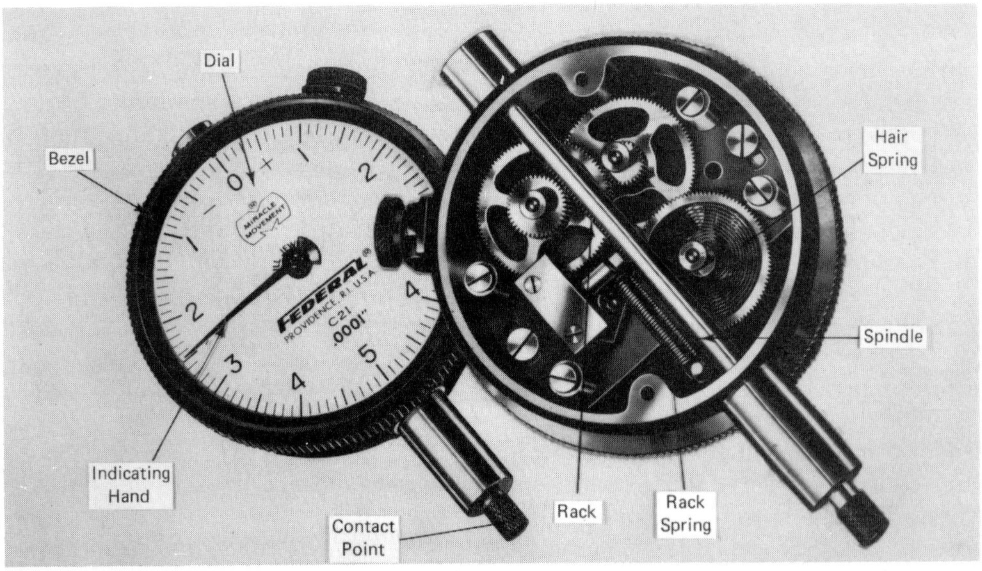

FIGURE 13-1
The principal parts of a dial indicator. (Credit: Federal Products Corporation)

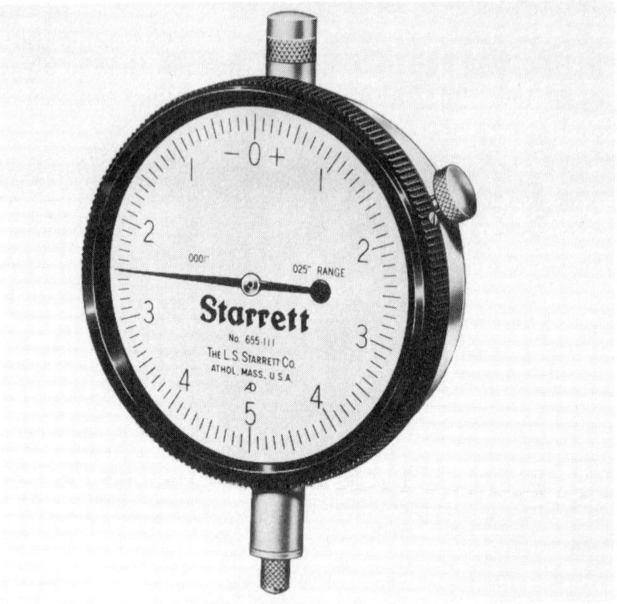

FIGURE 13-2
Somewhat resembling a clock, dial indicators can come with a variety of numbering faces. (Credit: L.S. Starrett Company)

FIGURE 13-3
Metric dial indicators may have graduation values as small as 0.002 mm. (Credit: MTI Corporation)

accurately cut with teeth, forming a rack to precisely mesh with a pinion connected to a gear train inside the instrument's housing. A hairspring and take-up gear eliminate backlash, while a heavy-duty pin and slide assembly maintains the accuracy between the rack and pinion. Uniform spindle contact against the workpiece surface is kept by a pull-back spring that never wears out. Any pressure against the contact point is transferred visually as graduated movements by the hand (needle) on the dial face.

The dial face is circled with graduated lines to somewhat resemble the face of a watch (Figure 13-2). A dial indicator will visually show the variation or amount of error measured in size, location, or alignment of a part. Due to the different working environments, these faces are available with graduations that are etched rather than painted, glass crystal as opposed to plastic, plated not painted housings, and can be operated in extreme temperatures from -40 (-40°C) to 600°F (315°C).

Dial indicators generally come in two styles. The continuous-reading style has an upward reading scale that starts at zero and circles its 100 graduated lines in a clockwise direction. Within the style may be found two types of gages. The most common indicator used (ADG group 2 has a 1-in. travel). Metric indicators (Figure 13-3) come in various ranges from 0 to 0.5, 1, 2.5, 10, and 25 mm. The regular range indicator has a small capacity of measurement, being circled in hand movement about $2\frac{1}{2}$ revolutions. Thus, if the value between each of the 100 graduations represents 0.001 in., the total range of measurement would only be 0.250 in. A comparable metric dial indicator having a value of 0.01 mm between adjacent lines would have a total measurement of 2.5 mm.

The long-range indicator can be found on lathes and other machine tools to gage table or tool travel and has a greater measuring capacity, so much so that an inner counting dial is set within the primary dial face (Figure 13-4).

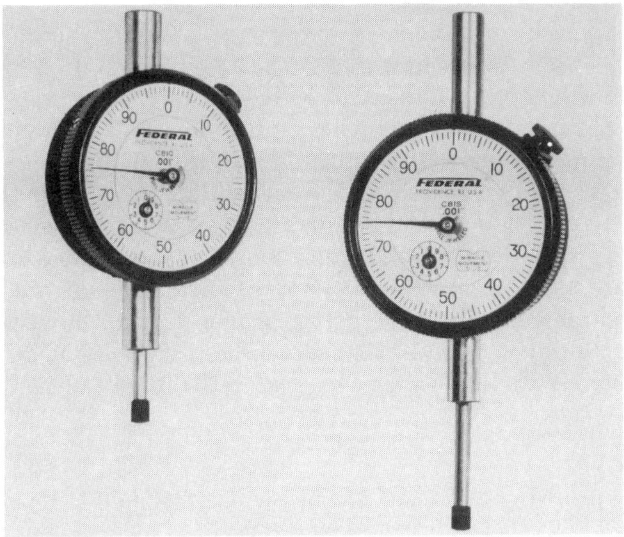

FIGURE 13-4
A long-range dial indicator. Notice its inner revolution hand. (Credit: Federal Products Corporation)

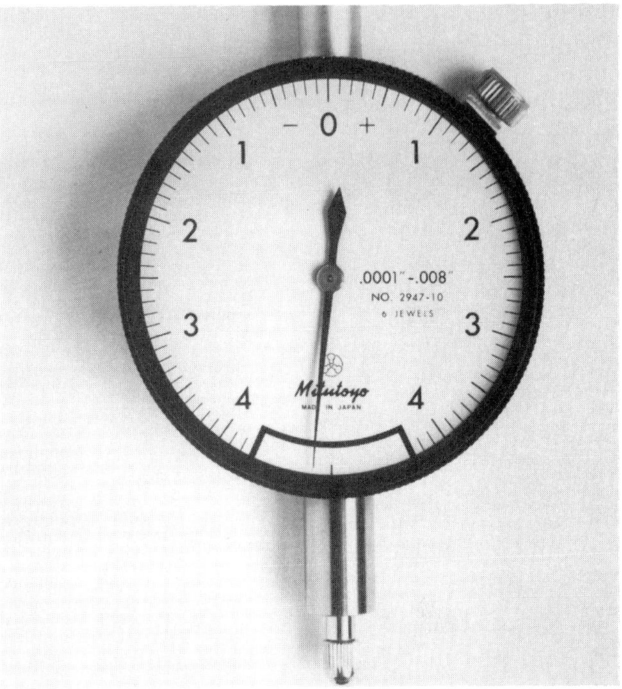

FIGURE 13-5
A balance-type dial indicator can be read to the right (plus value) or to the left (minus value) from the zero graduation. (Credit: MTI Corporation)

The balance-type indicator (Figure 13-5) is more widely used in inspection work than is the continuous-reading type. In addition to being able to set the dial face on zero by rotating the outer dial ring (bezel) at any time, a plus or minus reading from the zero mark will be shown. Many of the smaller indicators will only travel one revolution while wider faced indicators will normally travel $2\frac{1}{2}$ revolutions. An additional feature may be the inclusion of tolerance hands on the dial indicator. They are especially useful for visually establishing the tolerance or limits of acceptability that dimensions may have (Figure 13-6).

Universal Dial Indicator Set

The universal dial indicator set (Figure 13-7) can be used for many applications such as measurement, inspection, and setup work. Its easily read dial, many contact points, and clamping features make it simple and flexible to use. Mounted on a surface gage, it can be used for inspection work on a surface plate or a machine tool table. Set in the toolpost of a lathe, it can inspect out of roundness, inaccuracies of mechanical parts, as well as surfaces with interrupted cuts, face plate locations, or four-jaw chuck setups. Held on a machine tool by means of its clamping feature or magnetic base holder, the universal dial indicator can be made to check runouts and alignments of vises, milling arbors, spindle turnings, etc.

FIGURE 13-6
Tolerance hands can mark the limits of acceptability. (Credit: Federal Products Corporation)

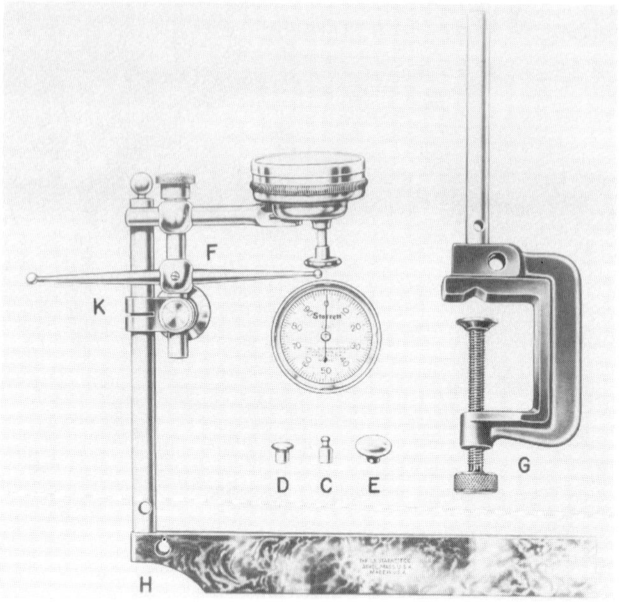

FIGURE 13-7
A universal dial indicator offers a great deal of flexibility in general machine shop work. (Credit: L.S. Starrett Company)

FIGURE 13-8
A dial test indicator. (Credit: Paul Rapisarda)

Dial Test Indicator

Made by a number of quality toolmakers, the dial test indicator (Figures 13-8 to 13-10) is favored by machinists, toolmakers, model makers, inspectors, and other personnel who require precision usage in universal shop operations (Figure 13-11). The sensitivity and repetitive accuracy of these indicators is transmitted and magnified by a gear train mechanism. Not only can the dial test indicator come with different size contact points, which may be hard chrome-plated or tungsten carbide tipped, but the points are replaceable and can be positioned at different angles so they can reach into difficult or hard

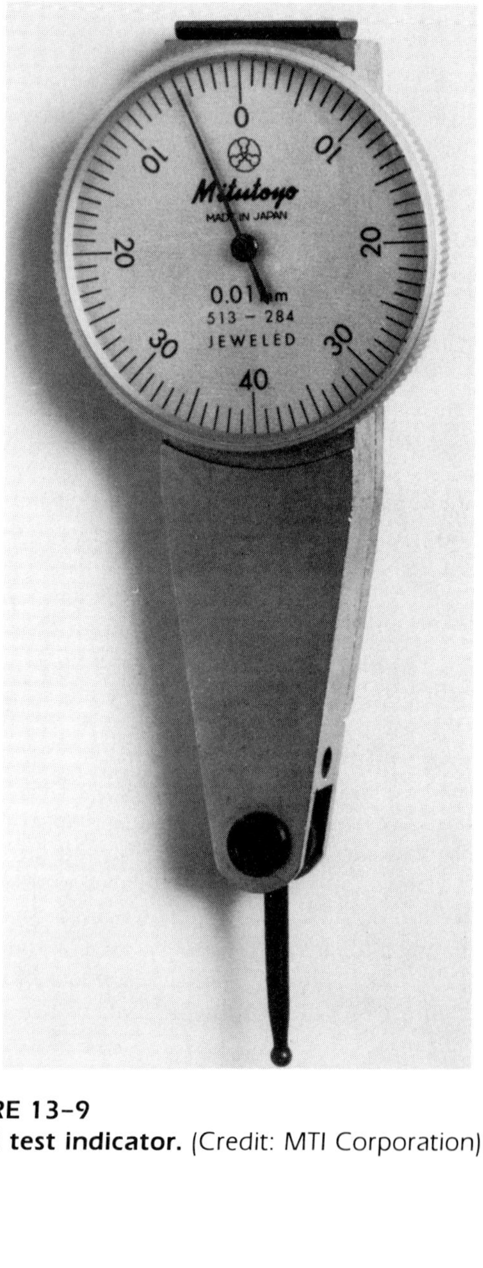

FIGURE 13-9
A dial test indicator. (Credit: MTI Corporation)

to get areas (Figures 13-12 and 13-13). The dial faces vary from 1 to $1\frac{1}{2}$ in. and range from 0.030 to 0.040 in., graduated in 0.001-in. increments or as small as 0.0005 in. Metric dial faces may range to 0.8 mm, with each graduation reading 0.01 mm. There are some dial indicators that are used for very fine measurements, with each graduation marked 0.0001 in. or 0.002 mm for their respective systems. The bezel on both types of indicators can be rotated to position the dial to be read on any reference graduation. Another feature includes a shaded area dial or a series of rise and fall numbers. These dial features allow for rapid readings as the hand of the indicator can be made to measure in a clockwise or counterclockwise movement.

What follows is the procedure used for determining the centerline distance of exact diameter holes using a height gage and a universal dial test indicator:

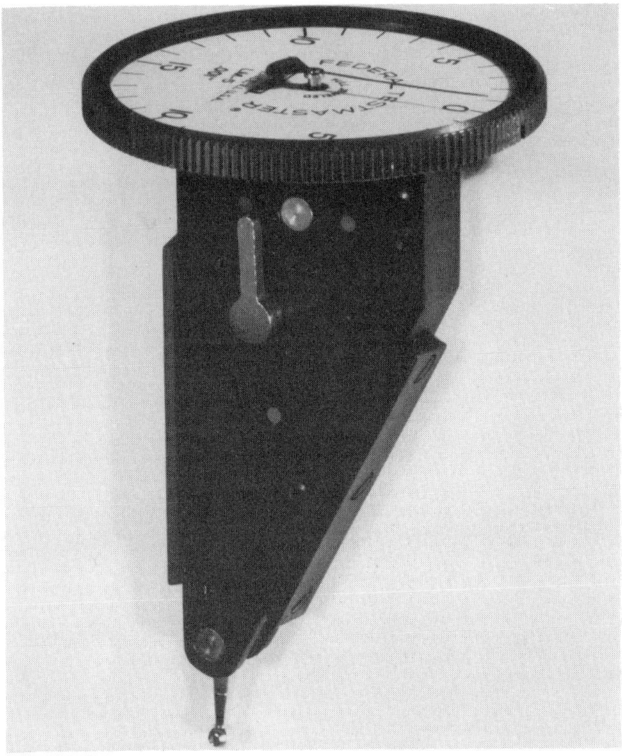

FIGURE 13-10
A dial test indicator. (Credit: Federal Products Corporation)

FIGURE 13-11
Preparing to test a fixturing device. (Credit: Paul Rapisarda)

FIGURE 13-12
Checking concentricity of a setup with a dial test indicator. (Credit: Federal Products Corporation)

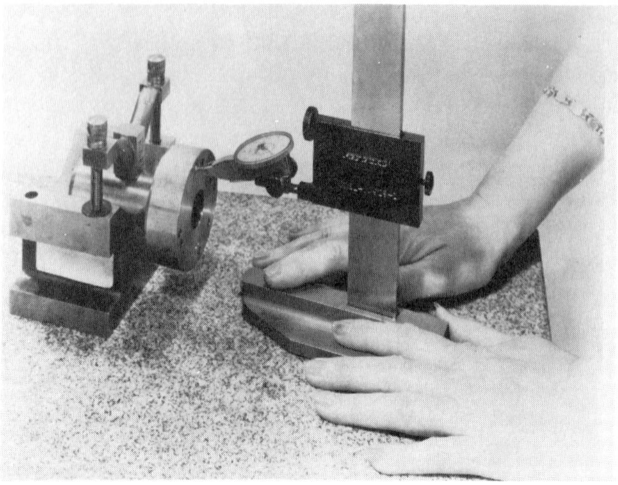

FIGURE 13-13
Checking hole dimensions with a dial test indicator. (Credit: Federal Products Corporation)

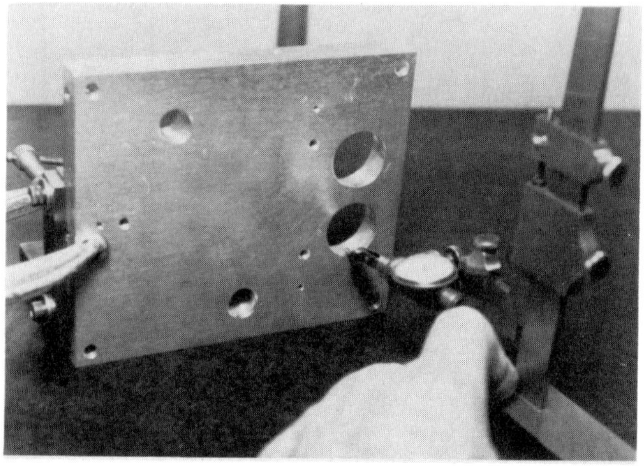

FIGURE 13-14
Barely touch the bottom of the first hole . . . (Credit: Paul Rapisarda)

1. Thoroughly clean the face of the surface plate and base of the height gage.

2. Clamp the workpiece upright to an angle plate, resting its base on the face of the surface plate.

3. Clamp the dial test indicator on the movable jaw of the height gage.

4. Loosen the clamp screws and lower the movable jaw until the indicator's contact point barely touches the bottom of hole 1 (Figure 13-14).

5. Lightly snug the clamp screw controlling the movable jaw.

6. Tighten the clamp screw controlling the adjustment nut carrier.

7. Slowly turn the adjustment nut to increase the dial face reading by five graduations.

8. Carefully move the height gage base so that the indicator's contact point slowly sweeps by either side of the hole's centerline until you can pick out the lowest reading on the dial face.

9. Lock the setting by tightening the clamp screw controlling the movable jaw.

10. Rotate the bezel to set the dial face to zero (Figure 13-15).

11. Note the height gage's reading and record it on a pad of paper.

12. Loosen the height gage's clamps and raise the indicator to the height of the second hole.

FIGURE 13-15
. . . set the dial face to zero . . . (Credit: Paul Rapisarda)

13. Lower the movable jaw until the indicator's contact point barely touches the bottom of hole 2 (Figure 13-16).

14. Repeat step 5.

15. Repeat step 6.

16. Slowly turn the adjustment nut to obtain a zero reading on the dial face.

17. Carefully move the height gage base until the lowest reading can be established as in step 8. If

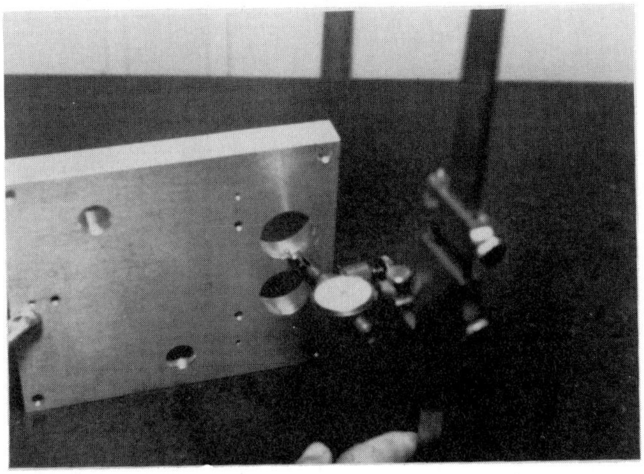

FIGURE 13-16
. . . and establish the reading at the bottom of the second hole. (Credit: Paul Rapisarda)

it is above or below the zero mark, turn the adjustment nut to achieve the zero reading.

18. Repeat step 10.

19. Repeat step 11.

20. Subtract reading 1 from reading 2 to obtain the centerline distance between the holes.

The Indicator Micrometer

Employing a movable anvil and a fixed spindle, the indicator micrometer (Figure 13-17) is widely used for inspection work. Unlike the standard micrometer, which uses the vernier principle to read its ten-thousandths of an inch (0.002 mm) accuracy, the indicator micrometer can be used as a micrometer or a comparator instrument designed to read a preset measurement on an indicating dial that is set into the frame of the instrument to a plus or minus reading of 0.001 in. (0.02 mm). Each 0.001 in. (0.02 mm) is subdivided into ten equal graduations. Thus, each graduation on the dial is read as ten-thousandths of

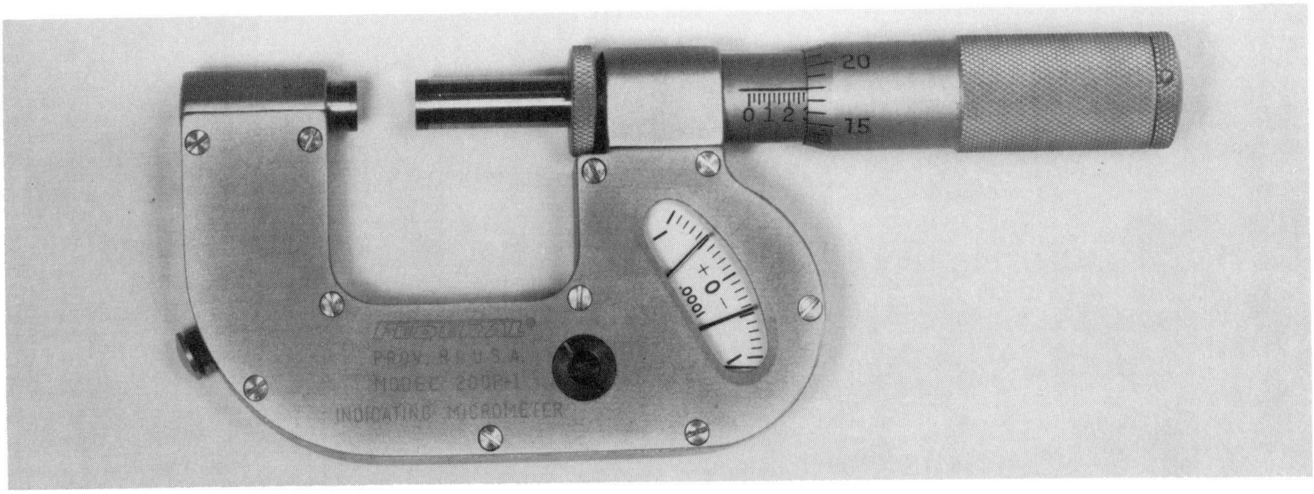

FIGURE 13-17
An indicating micrometer can serve as a micrometer or a comparator instrument. (Credit: Federal Products Corporation)

an inch (0.002 mm). The range of measurement can also be set by the placement of red-colored tolerance hands on the window of the dial (Figure 13–18).

Procedure for Setting Indicator Micrometer as a Snap Gage

1. Clean the faces of the spindle and anvil.
2. Select and clean the surfaces of the standard or gage block of measurement.
3. Place the measuring piece against the anvil, which should be in a fixed position.
4. Turn the friction thimble so that the spindle can make contact against the measuring piece, establishing the indicator hand on the zero mark.
5. Lock the spindle in place by turning the clamping ring.
6. Push the retracting button on the bottom of the frame to move the anvil, releasing the measuring piece.

To Test Setting

1. Push the retracting button.
2. Place the measuring piece against the anvil.
3. Release the button.

The indicating hand should again be on the zero mark. If repeatability is not there after several attempts, reset the indicator micrometer following the steps just outlined.

Procedure for Using as a Micrometer

1. Turn the friction thimble so that the spindle can make contact against the measuring piece.
2. Adjust the thimble reading to the nearest thousandth mark on the dial.
3. Read the micrometer to the nearest ten-thousandth of an inch (0.002 mm).

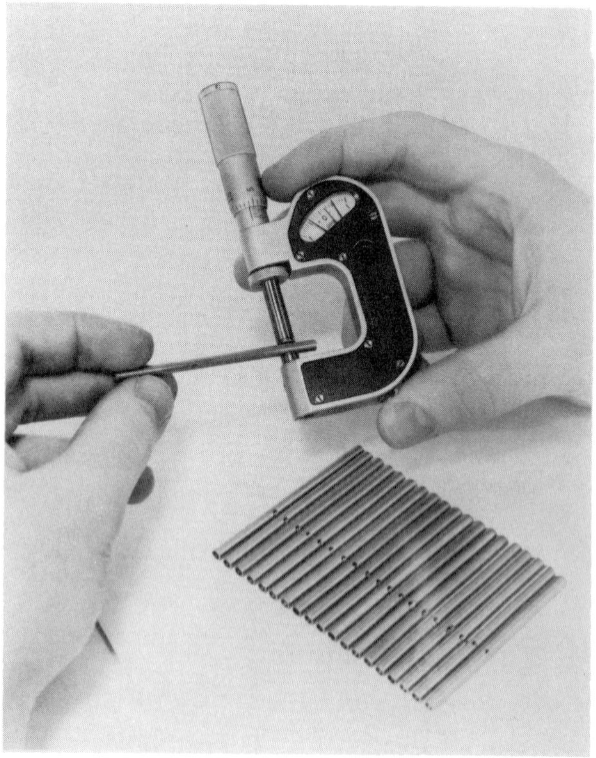

FIGURE 13–18
Using the dial indicating micrometer as a snap gage. (Credit: Federal Products Corporation)

FIGURE 13–19
Comparator stand and dial indicator. (Credit: L.S. Starrett Company)

Bench-Type Comparator

Bench-type comparators are designed for quick, dependable measurements and are adaptable for handling thicknesses, hole depths, and outside dimensions on a variety of production and general shop uses. An upright column that is mounted in a granite or precision ground cast-iron base acts as a pivoting post for a mounted dial indicator that can be swung on an arc above the base (Figure 13-19). The pivoting base also permits the dial indicator support (beam) to be raised or lowered vertically. An adjustment screw located on the beam permits fine movements of the indicator spindle. A safety ring is positioned below the beam and prevents any accidental dropping of the beam onto the base. Located above the indicator's housing is a hand lever, which can be positioned to the right or to the left. By depressing the lever, the indicator's spindle is raised; releasing it, the spindle is extended (Figure 13-20).

Balanced-type indicators are generally used and set against a standard or master gage. Part measurements are then registered on this dial face, with all variations bearing a plus or minus (inch or metric) value.

Procedure for Using a Bench-Type Comparator

1. Clean the surface of the indicator's base.
2. Clean the tip of the indicator's spindle.
3. Clean the standard or master gage.
4. Place the standard or master gage on the base.
5. Carefully lower the indicator beam until the spindle touches the standard or master gage to register needle movement.
6. Lock the beam to the column.
7. Fine turn the beam's adjustment screw to set a zero reading.
8. Test setting by depressing the hand lever and removing the standard or master gage.
9. Replace the standard or master gage and release the hand lever.
10. Set the tolerance hands to the upper and lower limits of the workpiece being tested.

Optical Comparator

The optical comparator (Figure 13-21) is produced by quality optical gage makers. Frequently called a shadowgraph, it is also called a profile projector (Figure 13-22) or a contour projector (Figure 13-23). The system's optical projector throws off a concentrated light beam that illuminates any object placed

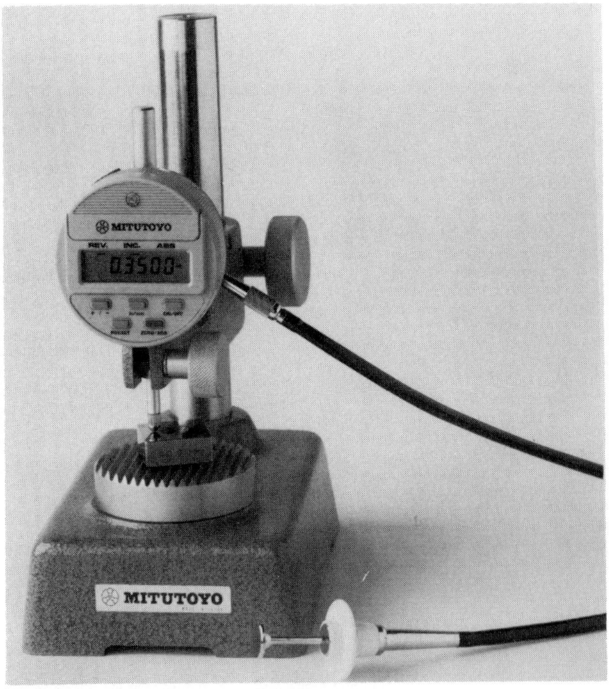

FIGURE 13-20
Digital comparator with cable release. (Credit: MTI Corporation)

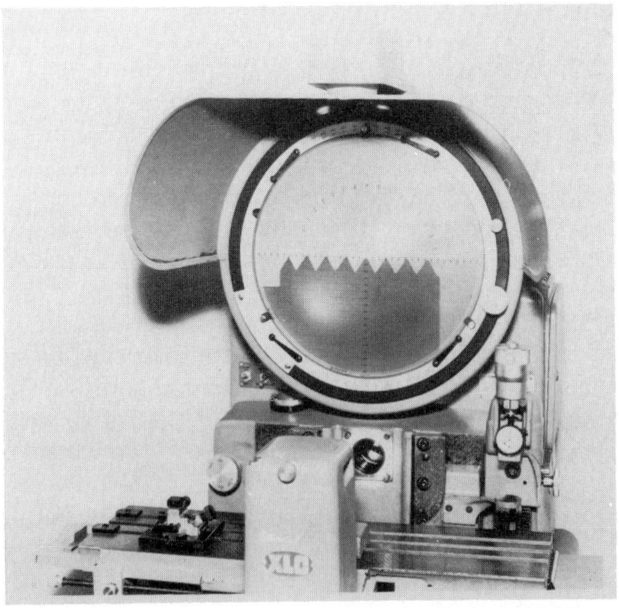

FIGURE 13-21
The optical comparator . . .

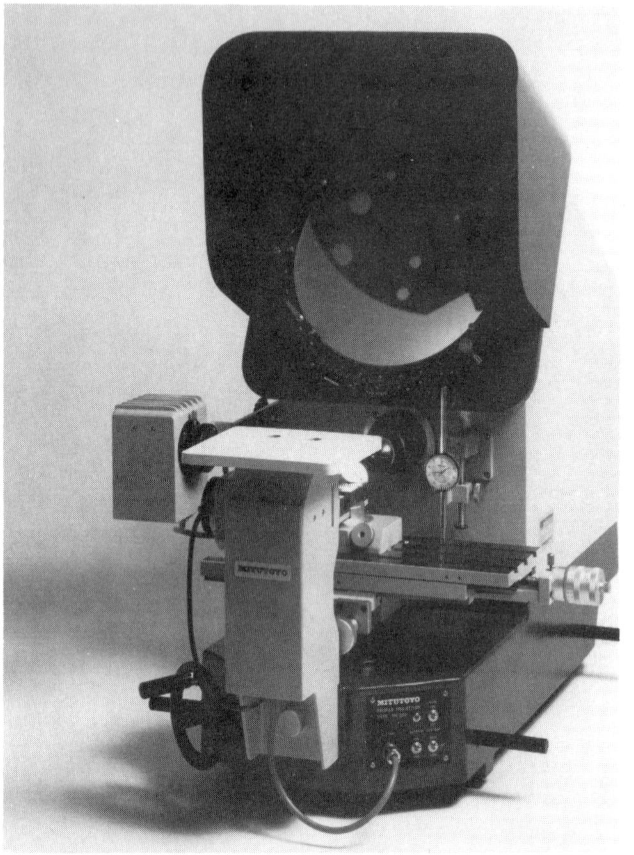

FIGURE 13-22
Is also called the profile projector. (Credit: MTI Corporation)

FIGURE 13-23
And the contour projector. (Credit: Optical Gaging Products)

before it. As with any close-up instrument such as a camera or a microscope, an optical projector only "sees" a thin line of sharpness and will focus this shadow sharply. The projector lens then transmits and enlarges the image onto a projector screen, having first reflected this shadow on one or several mirrors. The result is a large, highly accurate, dark silhouette of the object being checked. An optical comparator can also magnify an illuminated surface.

There are several types of measurement that can be performed on an optical comparator. One measurement relies on the projection of a workpiece's profile, which is compared to a form, a set of lines, or a master template that is fastened to the projector screen (Figure 13-24). It is this outline that dictates the limits of acceptability or configuration to which a workpiece must conform. Due to the vertical and horizontal projection at the same time, a user is able to read a two-dimensional check at a glance. Essentially, the projector screen acts as a

FIGURE 13-24
An optical comparator overlay. (Credit: Optical Gaging Products)

large layout, putting inspectors/users on equal footing with draftspersons, model makers, engineers, and the like who depend on engineering/work layouts to tie dimensions together.

Measurement by movement is also possible and is used on single-piece inspection where nonrepetitive measurements are needed. Once a workpiece's shadow is accurately positioned against a reference line on the screen that is created by a beam of light from the projector, measuring attachments along the front edge of the movable support table make it possible to identify any distance or length traveled past the reference line by reading out the position of the table.

Another feature found on an optical comparator is a protractor ring. Located along the periphery of the projector screen, the rotational device is marked off in degrees, which are subdivided into minutes with the aid of a vernier reading. The protractor ring is used to measure angles by aligning the screen's reference line with each side of the angle. In addition to reading flat angles, protractors are capable of "seeing" and reading compound edge angles, compound surface angles, and triangular calculations.

Angles can also be read using an angular template. A transparent overlay chart is placed on the screen to offer a direct comparison between the workpiece shadow and the graduations on the chart, which are ruled in degrees. The chart is laminated on both sides and may be made from a frosted plastic material (Figure 13-25) or glass.

Optical comparators are available in bench and floor models, ranging in screen sizes from 10 to 30 in. (254 to 762 mm). The magnification on the screen will depend on the array of interchangeable lenses available. Most optical comparators have standard magnifications of 5×, 10×, 20×, 31.25×, 50×, 62.5×, and 100×. To eliminate any error, it is important to match the magnification of the screen with the enlargement of any template or overlay used. These magnifications make optical comparators ideal for testing alignment, squareness, flatness, as well as measuring small irregularly shaped workpieces that would otherwise be difficult to check. Objects that tend to deform or distort under pressure such as springs, plastic, or rubber are also easily checked. The optical comparator can be found in engineering development, toolrooms, assembly departments, production areas, and final inspection, to name but a few precision areas in industry (Figure 13-26).

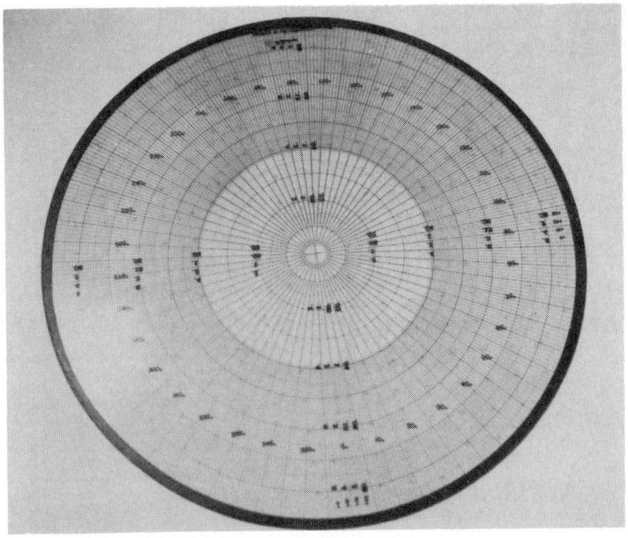

FIGURE 13-25
Angles may be read using a transparent overlay chart.

FIGURE 13-26
Floor model optical comparator. (Credit: Optical Gaging Products)

AIR GAGES

Air gaging, or pneumatic gaging, plays an important role in inspection and quality control. It is another form of comparison gaging whose industrial application began with the measurement of carburetor jets. Sensitivity to measurement is based on the behavior of a fluid or gas (air). Dimensional measurement is indicated by the escape of air through a series of holes in the periphery of a gage that registers against the workpiece surface of the movement of air through the workpiece itself. In all cases this air movement is compared to a setting or reading as indicated by a master or known standard gage.

Air gages are classified by the principles used, such as free flow, back pressure, differential, or a combination of these principles. In this unit free-flow and differential gaging will be examined.

Free-Flow Air Gaging

Free-flowing air gaging (Figure 13–27) is measured by a gaging element (rotometer), which consists of a graduated tapered glass tube, a lightweight float, and inlet and outlet connections. Measurement is conducted by air pressure that passes between the gaging element and the workpiece standard. This acceptable pressure is then fixed by a regulator, with the setting identified on the glass scale by the position of the float. The lightweight float moves up and down in direct ratio to the restriction of air caused by the tightness or looseness of the workpiece being tested. Amplification ranging from 1000 : 1 to 40,000 : 1 is possible with this type of instrumentation.

Differential Air Gaging

The differential system uses parallel circuits. Filtered air is passed through a regulator and master pressure gage that is then divided into two opposite legs or channels (Figure 13–28). The air in one leg (the reference channel) exhausts into the atmosphere through an adjustable zero setting valve (restrictor). The air in the opposite leg (measuring channel) escapes to the atmosphere through the jets of the gage head. The two channels are connected by an extremely precise indicating meter that reacts immediately to any differential in air pressure between the channels.

In setting up the gage, a master is placed over the gaging plug. Air is then let in and the zero setting valve is adjusted until the gage needle registers zero. Any test piece deviation from the standard will change the reading in the following manner.

If the test piece is too large, air is restricted in the measuring channel, causing the gage needle to move counterclockwise. The reading will indicate how much larger the test piece is.

If the test piece is too small, air escaping in the measuring channel will cause the gage needle to move clockwise. The reading will indicate how much smaller the test piece is. Amplification may range between 2500 and 20,000 : 1 with this type of air gage.

Differential air gages can be used with a variety of gaging heads, such as plug gages, ring gages, snap gages, height gages, depth gages, and for a variety of applications.

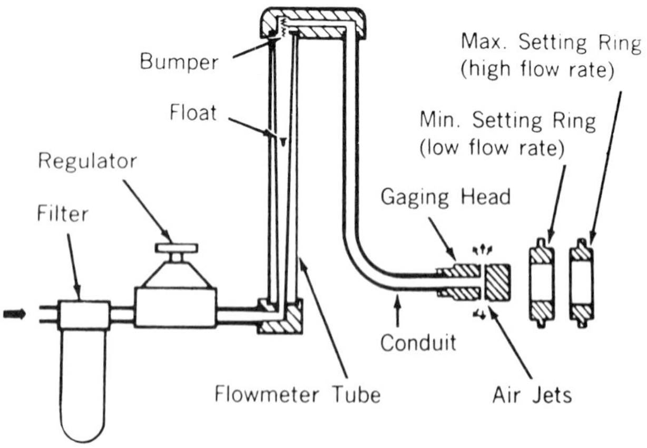

FIGURE 13–27
Rotometer—a free-flowing air gaging instrument. (Credit: Sheffield Measurement Division, Warner & Swasey/A Cross & Trecker Company)

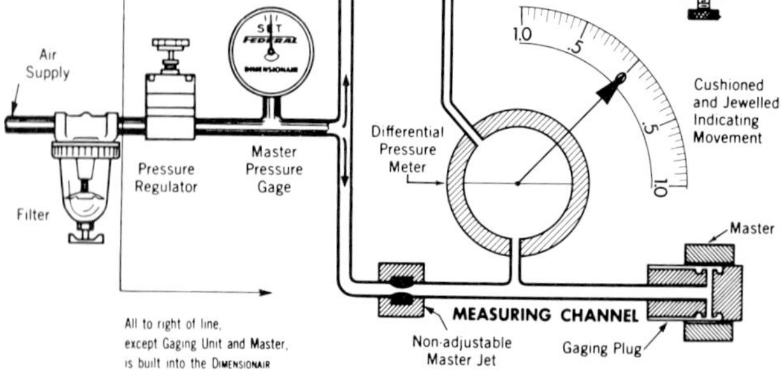

FIGURE 13–28
Differential air gaging system. (Credit: Federal Products Corporation)

Advantages of Air Gaging

1. Air gages are adaptable to tool positioning while eliminating operator error.

2. An air gage can be passed through an internal metal bore without scratching the insides.

3. Air gages can be made with multiple jet tooling to check a number of dimensions at the same time.

4. These measuring devices are adaptable for machine or bench setup (Figure 13–29).

5. They are not dependent on operator skill to measure accurately.

6. They outwear their fixed gage counterpart since friction is reduced between the gaging head and the workpiece.

7. A flexible extension can be connected to air plug sensors, making it possible to machine gage from a distance.

8. Bores can be checked more easily for tapers, out of roundness, concentricity, and irregularity than with mechanical hardware.

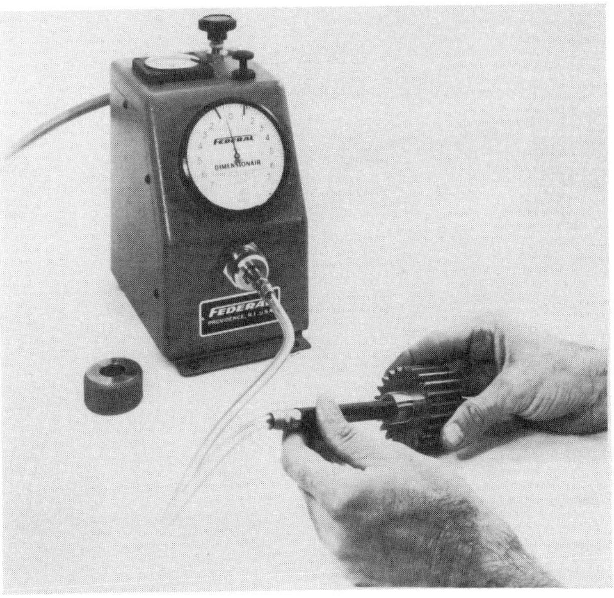

FIGURE 13–29
Using a bench model air gaging system. (Credit: Federal Products Corporation)

Electronic Comparator

The electronic comparator employs an electronic gage amplifier with a readout that measures or compares dimensions in English (inch) or metric values at the push of a button. The instrument produces a continuous output (analog) that is amplified and shown on a meter or a numbered display board or a recorder or a combination of these systems. The high magnification is produced by switching points that are controlled electronically rather than mechanically. The range of magnification will depend on the scale used, but this can be from 0.0001 to 0.00001 in. (0.002 to 0.0002 mm) (Figure 13–30).

There are three basic types of electronic gage heads used with electric gage amplifier. (1) The lever type (Figure 13–31) has a movable contact point suspension and is frequently used for calibration and surface plate work. (2) The cartridge-type gaging head (Figure 3–32) is designed with gaging pressure measured in a vertical position with its contact point in a down position. It is protected against outside magnetic fields and may be activated with a camera-style cable release, achieving magnification of 15,000× or more. (3) The reed-type electronic gage (Figure 13–33) uses the principle of the reed mechanism. The reed mechanism consists of flat steel springs or reeds attached to two blocks, one fixed, the other movable to "float" with any reed movement. Attached to the sensitive reeds is a spindle mounted at one end of the movable block. Above it is extended a pointer or indicator hand that is moved along a vertical arc at the slightest reed deflection and registers on a meter display.

Depending on the length of the reed, amplification of 200:1 can be expected; however, when coupled with a lens system where light is projected on a graduated scale and the pointer casts a shadow on the dial, amplification varying from 500 to 20,000 : 1 is possible. This electronic gage head is friction free, with no working parts, coils, or magnets. Repeatability is 100% and accuracy can be maintained indefinitely.

UNIT 14 SURFACE FINISH

The purpose of surface measurement is to assure that parts will be made to work better as well as last longer. From a manufacturing and design point of view, this is an important consideration. In addition to controlling wear in parts such as pistons, bearings, and gears, friction is another problem to contend with. Other factors to be considered might deal with corrosion, lubricant distribution, seating or sealing capacity, electrical conductivity, as well as other functional demands. To a great extent, these factors will depend on the surface texture produced

FIGURE 13-30
Electronic gage amplifier. (Credit: L.S. Starrett Company)

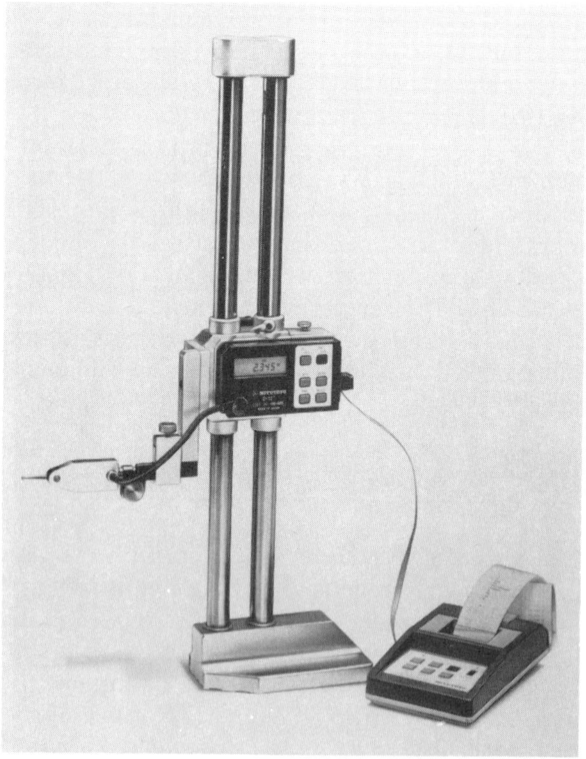

FIGURE 13-31
Lever-type electronic height gage. (Credit: MTI Corporation)

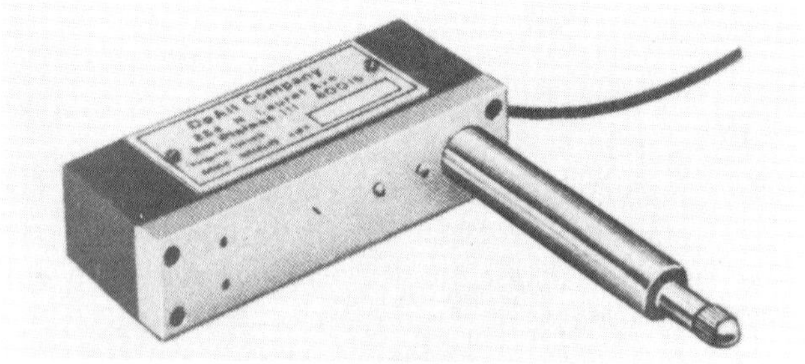

FIGURE 13-32
Cartridge-type gaging head. (Credit: DoAll Company)

FIGURE 13-33
Reed-type electronic gage head has 100% repeatability and can be maintained indefinitely. (Credit: DoAll Company)

and will require economic considerations. In simple terms manufacturing and design talents will have to deal with part function and production costs. How much will it cost to solve the surface finish problem? There is no economic advantage to having a part machine to a high-quality finish when it is not necessary.

In the past it was common practice for a designer to indicate surface finish by lettering drawings with various designations such as a "f" or "g" for "finish" or "grind." While simple, this practice was found to be vague and inaccurate and was consequently supplanted by surface texture standards, ANSI-B 46.1-1978, published by the American Society of Mechanical Engineers (ASME). Not only have these standards replaced all domestic standards, they also conform to most ISO international standards.

Objectives
After completing this unit you will be able to

1. use a microfinish comparator to test the finish of a part.
2. understand the lay surface finish symbols on engineering drawings.
3. understand four average roughness reading systems.
4. use a surface texture instrument to measure the surface finish of a machined part.

The Surface Finish Comparator

Each machining operation leaves distinctive surface characteristics. These irregularities make it possible to identify the different surfaces produced by milling, turning, grinding, lapping, honing, and polishing. These same inherent characteristics also make it possible to compare the surface texture produced by a given process to a set of surface texture standards (Figure 14–1). These figures have been translated in the form of a practical gage (Figure 14–2), which conforms in accuracy of detail to 22 specimens of machined surfaces. The surface finish comparator is made of pure nickel, will not corrode, and is electroformed with great fidelity to produce deposited shapes and contours ranging from 2 to 500 μin. (0.05 to 12.70 μm) in finish. It is the accuracy of detail that makes the gage easy to use, both visually as well as to the touch. It should be noted, however, that the results are general. There is more to surface measurement than meets the eye.

How to Use a Surface Finish Comparator

1. Place the surface finish comparator alongside the workpiece (Figure 14–3).
2. Position the rectangular blank with the desired finish next to the workpiece surface (Figure 14–4).
3. Draw the tip of the fingernail across each surface at right angles to the tool marks (Figure 14–5).

In order to be identical, the nail scratching along the surfaces must feel the same.

Basic Surface Finish Definitions

In order to understand how a profilometer measures surface finish, it is important to understand the definitions of surface texture limits, measurements, and definitions. Some of the definitions that follow have been condensed from previously mentioned surface texture standards.

Surface: The boundry that separates one object from another object, substance, or space.

Surface texture: The characteristic "fingerprints" or pattern caused by the interaction of the cutting tool and material during the machining process. These major features would include roughness, waviness, lay, and flaw (Figure 14–6).

Nominal surface: The theoretical accurate design that is shown and dimensioned on a drawing of parts.

Profile: A line that is created by the crossing of a surface with a plane that is, unless otherwise specified, vertical to the nominal surface.

Measured profile: The reproduction of the actual profile obtained by the use of a measuring instrument. This line is formed by the intersection of the measured surface and a plane vertical to the lay of the measured surface.

Nominal profile: A curved or straight line that is developed by the crossing of the nominal surface with a plane that is, unless otherwise specified, vertical to the nominal surface.

Modified profile: The minimizing of certain texture characteristics while maximizing others when examining a measured profile by the use of filtering mechanical or electromechanical instrumentation.

Type of surface	Roughness height (μin.)	Roughness height (μm)	Lay	Feed (in.)	Feed (mm)
Honed Lapped or Polished	2 4 8	0.05 0.10 0.20	Parallel to long dimension of specimen		
Ground with periphery of wheel	8 16 32 63	0.20 0.40 0.81 1.60	Parallel to long dimension of specimen		
Ground with flat side of wheel (Blanchard)	16 32	0.40 0.81	Angular in both directions		
Shaped or turned	32 63 125 250 500	0.81 1.60 3.18 6.35 12.70	Parallel to long dimension of specimen	0.002 0.005 0.010 0.020 0.030	0.050 0.127 0.254 0.508 0.762
Side Milled End Milled or Profiled	63 125 250 500	1.60 3.18 6.35 12.70	Circular Angular in both directions	0.010 0.020 0.100 0.100	0.254 0.508 2.54 2.54
Milled with Periphery of Cutter	63 125 250 500	1.60 3.18 6.35 12.70	Parallel to short dimension of specimen	0.050 0.075 0.125 0.250	1.27 1.90 3.18 6.35

FIGURE 14–1
Physical specimens of surface roughness and lay.

FIGURE 14–2
A microfinish comparator. (Credit: Philip Rapisarda)

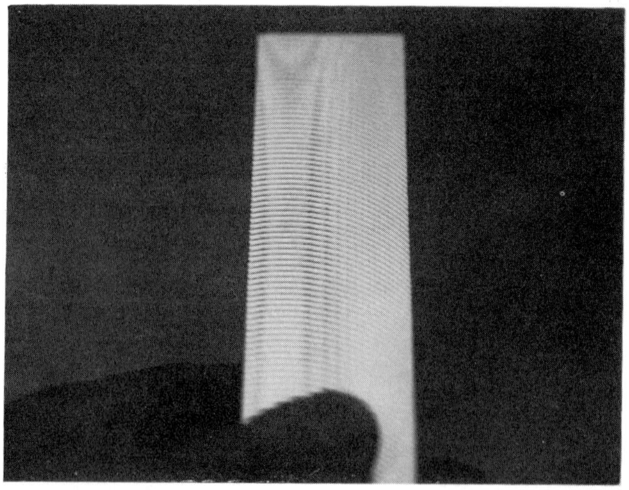

FIGURE 14–3
Prepare to examine the work surface finish . . . (Credit: Philip Rapisarda)

FIGURE 14-4
. . . by positioning the desired finish blank next to the workpiece surface . . . (Credit: Philip Rapisarda)

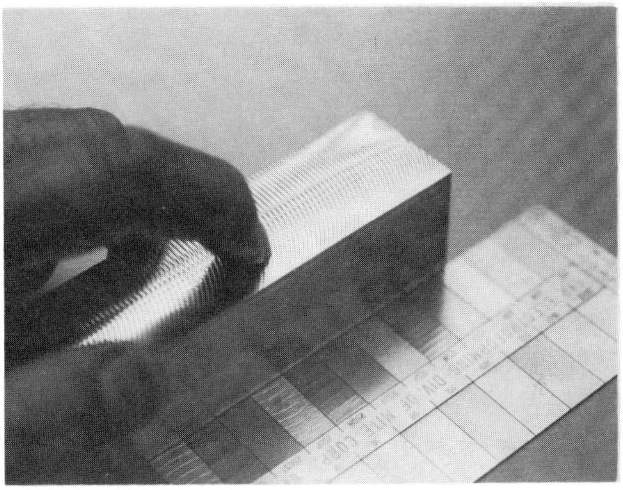

FIGURE 14-5
. . . and drawing the tip of the fingernail across each surface. (Credit: Philip Rapisarda)

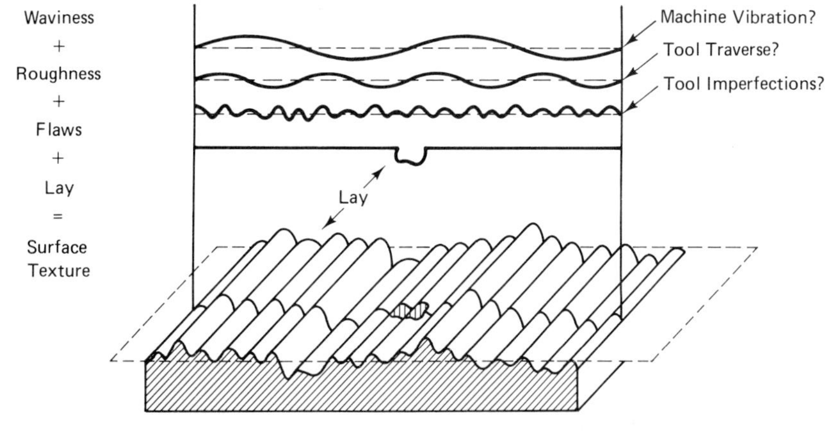

FIGURE 14-6
Surface texture. (Credit: Sheffield Measurement Division, Warner & Swasey/A Cross & Trecker Company)

Centerline: A line that runs parallel to the general direction of the profile so as to divide the sums of the areas between the centerline and the parts above and below the line to be equal (Figure 14–7). The centerline is also called the electrical mean line.

Actual profile: A section through the actual surface measured on a plane vertical to that surface.

Peak: The highest point of a profile between two intersections with the centerline.

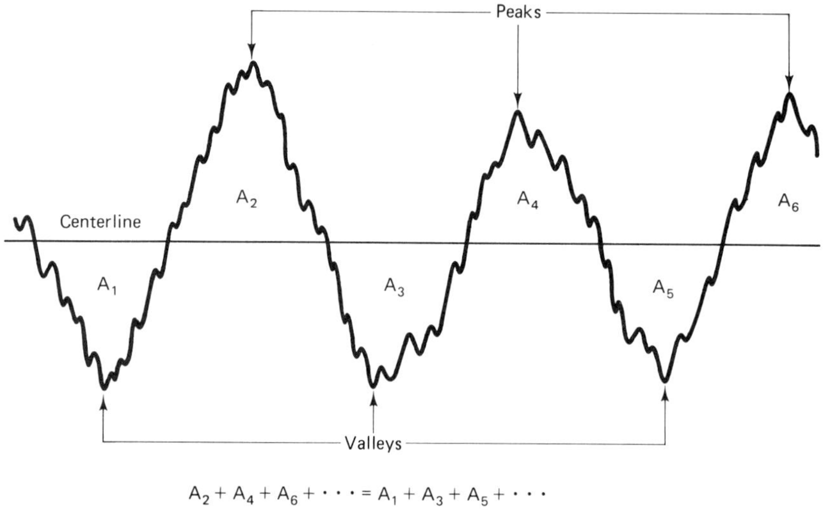

FIGURE 14–7
Centerline, peaks, and valleys. (Credit: Sheffield Measurement Division, Warner & Swasey/A Cross & Trecker Company)

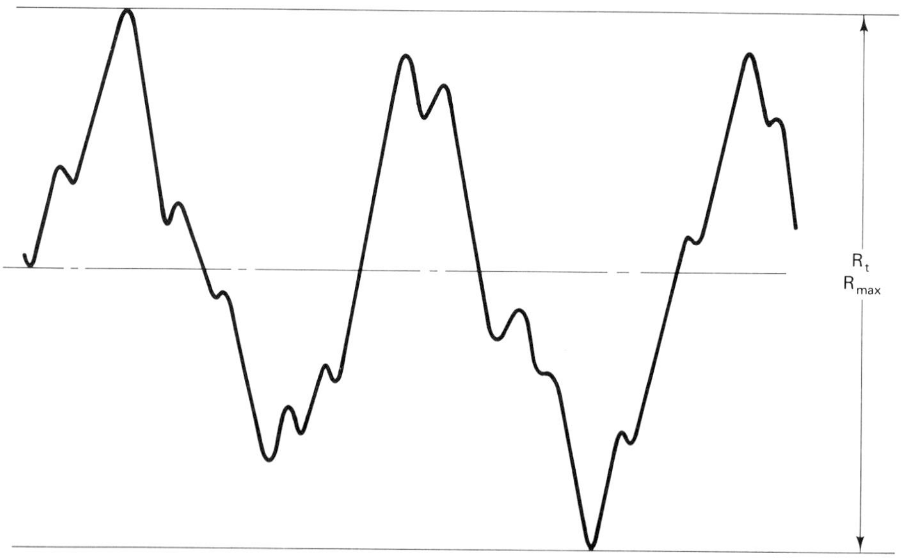

FIGURE 14–8
Peak-to-valley measurement. (Credit: Sheffield Measurement Division, Warner & Swasey/A Cross & Trecker Company)

Valley: The lowest point of a profile between two intersections with the centerline.

Height: The elevation distance of any point of a profile measured vertically from the centerline to the nominal profile.

Peak to valley height: The length of the rise and fall of the measured profile as it extends above and below the centerline within the roughness sampling length (Figure 14-8).

Lay: The direction of the predominant surface pattern usually determined by the machining process used such as milling, turning, or grinding. Symbols indicating direction of lay can be seen in Figure 14-9. Casting, peening, or sand blasting leave no such trail and are considered nondirectional, particulate, or protuberant lay.

Flaws: Irregularities such as cracks, blow holes, scratches, ridges, and similar defects that do not follow a regular pattern or occur infrequently in widely varying places on the surface. Unless otherwise specified, flaws are not included in the roughness average measurement.

Waviness: Widely spaced irregularities found on the mean surface in the form of waves. This condition may be caused by vibration, chatter, or deflection on the machine or workpiece.

Waviness height: The peak to valley distance that may be measured in inches or millimeters.

Waviness width: The distance between successive spacings of waviness peaks or waviness valleys, measured in inches or millimeters.

Roughness: The relatively finely spaced irregularities of the surface texture that are narrower than the waviness pattern. Roughness may be caused by the cutting tool, feed, or other inherent actions of the production process. The range of surface roughness made by various machining processes can be seen in Figure 14-10.

Roughness spacing: The average distance between specified points on a profile measured parallel to the nominal profile.

Roughness height: The deviation from the centerline measured in microinches or micrometers computed by arithmetical averaging within the roughness sampling length (Figure 14-11).

Roughness width: The distance between roughness peaks measured parallel to the nominal surface in inches or millimeters.

Roughness width cutoff: Also called *cutoff,* it represents the longest wavelength to be included in the measurement of roughness

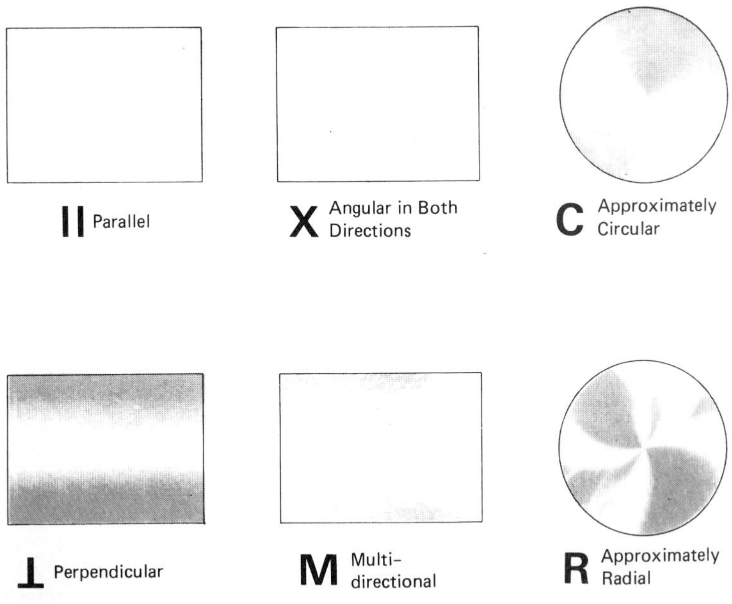

FIGURE 14-9
Lay symbols. (Credit: Sheffield Measurement Division, Warner & Swasey/A Cross & Trecker Company)

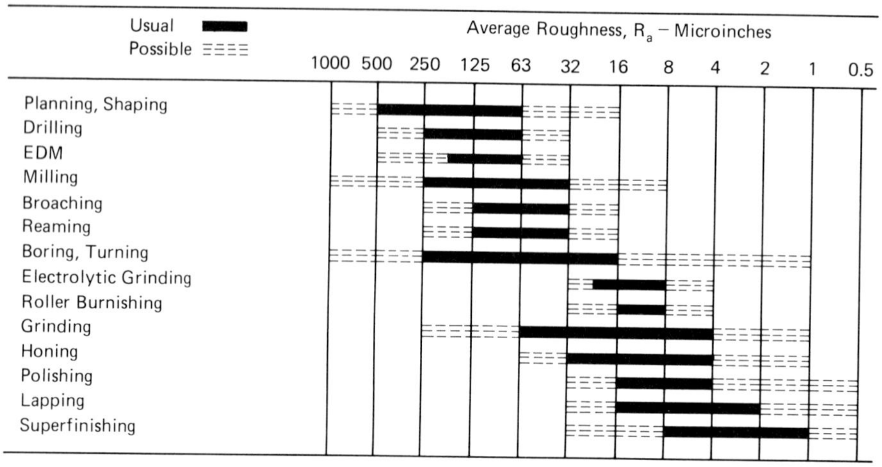

FIGURE 14–10
Surface roughness produced by common production methods. (Credit: Sheffield Measurement Division, Warner & Swasey/A Cross & Trecker Company)

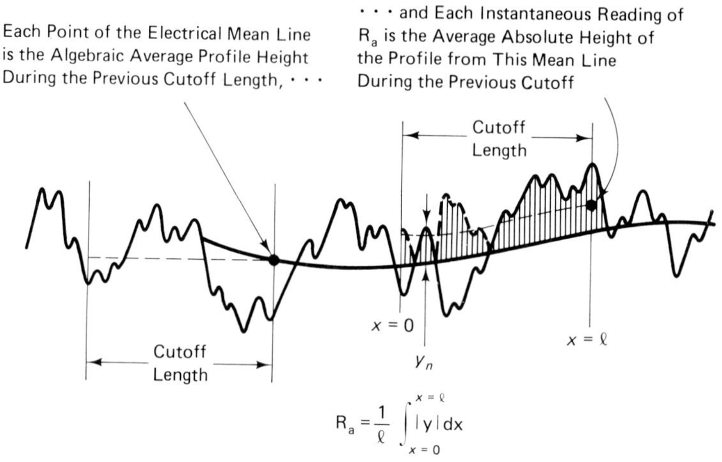

FIGURE 14–11
Roughness sampling length. (Credit: Sheffield Measurement Division, Warner & Swasey/A Cross & Trecker Company)

height, generally in inches or millimeters per cycle. A predetermined cutoff length of 0.003, 0.010, 0.030, 0.100, 0.300, and 1 in. (0.075, 0.25, 0.76, 2.54, 7.62, and 25.4 mm) is provided by the measurement instrument. Unless otherwise specified, a cutoff of 0.030 in. (0.76 mm) is assumed. In any case these lengths will always be greater than the roughness width.

Roughness sampling length: The sampling length within the roughness average is determined. The length is chosen to separate the irregularities, distinguishing the roughness from the waviness (Figure 14–12).

Microinch: One-millionth of an inch (0.000,001 in.) and may be abbreviated or identified by the symbol μin. or mμ inch.

Micrometer: One-millionth of a meter (0.000, 001 m) and may be identified by the symbol μm.

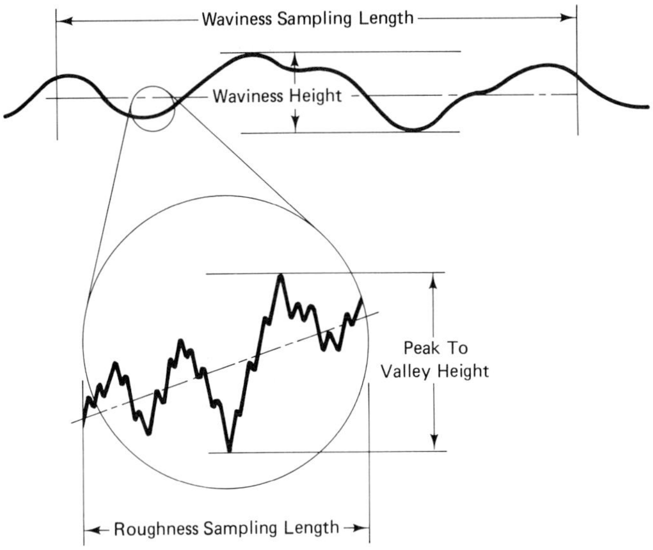

FIGURE 14-12
Peak-to-valley height; waviness height. (Credit: Sheffield Measurement Division, Warner & Swasey/A Cross & Trecker Company)

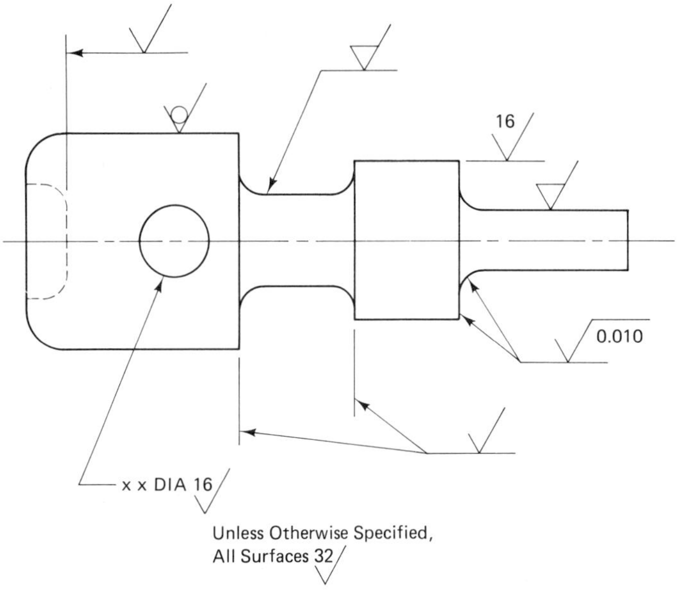

FIGURE 14-14
Application of surface finish symbols. (Credit: Sheffield Measurement Division, Warner & Swasey/A Cross & Trecker Company)

The surface symbol:

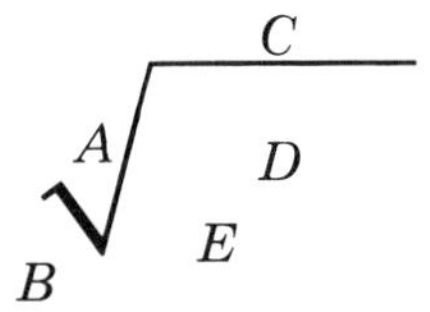

A = roughness height average or maximum-minimum roughness height average
B = machining allowance
C = waviness and waviness spacing
D = roughness sampling length
E = lay

FIGURE 14-13
Surface texture symbol and designation of dimension placements.

Surface Texture Symbols

The basic symbol designation for this system is a checkmark with a horizontal extension line (Figure 14-13). The extension line may be omitted, however, if the only designation shown is that of roughness height. It is by filling the checkmark with dimensions and other symbols that surface finish specifications are established. For example, drawing a triangle at the base of the checkmark indicates a machining allowance requirement. The insertion of a small circle in the base of another prohibits the removal of any material. These symbols may also be used with leader lines pointing to the surface indicated. Where the symbol is used with a dimension, it will affect all the surface identified by the symbol. Application of these symbols and specifications for surface finishes on engineering drawings can be seen in Figure 14-14.

The Surface Profiler

Due to the composite characteristics of surface texture, surface profiling is a complex operation. Profiling instruments can be tied in with computers to obtain the most sophisticated data analysis. The process is slow and requires a high degree of operator skill to use the laboratory equipment involved (Figure 14-15). What is generally used in a shop environment to analyze surface finish is a tracer-type instrument that renders any of several mathematical averaging methods. This type of electronic de-

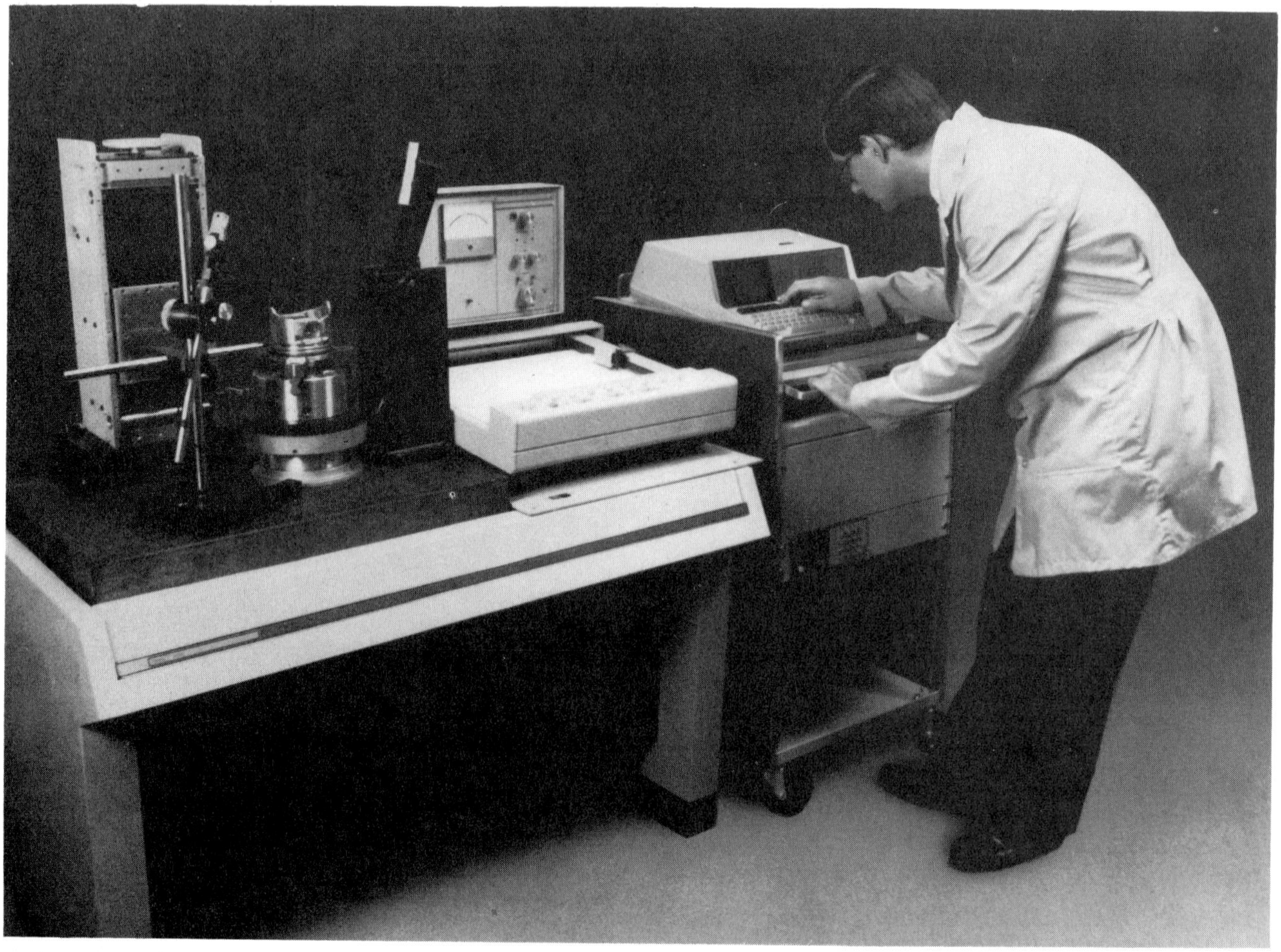

FIGURE 14–15
Laboratory-type profiling system. (Credit: Sheffield Measurement Division, Warner & Swasey/A Cross & Trecker Company)

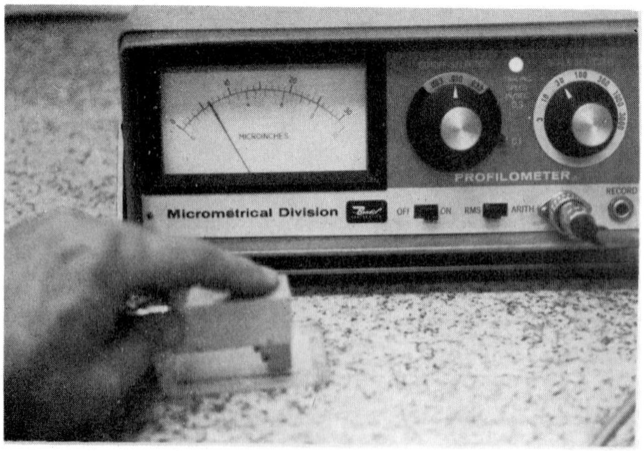

FIGURE 14–16
Portable-type surface texture measurement instrument.

vice employs a fine tracer point or diamond stylus, which can be motor driven or hand drawn across the surface to be measured. The movement of the tracer is amplified as it rises and falls monitoring and successive peaks and valleys of the irregularities being probed. On portable instruments these measurements are displayed on a clear digital readout screen (Figure 14–16). On mounted profiling instruments these fluctuations are plotted to graphically record and define a profile or are fed into an averaging meter.

Numerical Averaging

While there have been many numerical averaging systems developed since the stylus-type surface finish instrument was first developed in the early

1930s, only four of the more widely used systems will be discussed here. They are symbolized as follows:

Ra: Roughness averaging, formerly known and used as AA (arithmetic averaging), as well as CLA (centerline averaging). It is used to average the profile height deviation from a mean line and is officially recognized as the measuring standard used with the checkmark symbol.

RMS: Root-mean-square averaging squares the height deviation from a mean line and renders an 11% higher reading than the Ra system. RMS is considered obsolete and is no longer used.

Rmax: Also referred to as Rt, Rmax measures the distance between the single highest peak to the lowest valley parallel to the mean line within the roughness sampling length.

Rz: Also referred to as the "ten-point roughness height," the Rz system measures the five highest profile peaks to the five lowest valleys distance, which is averaged out and divided by the number of whole measurements to obtain the mean diameter.

Measuring Ra with a Portable Surface Texture Instrument

1. Place selector in the Ra mode of operation.
2. Turn the switch on and wait for the LCD screen to light up.
3. Check the machine's calibration if need be, using a clean standard test block traceable to the Bureau of Standards.
4. Use the instrument's recommendation concerning the standard cutoff range for surface roughness. Unless otherwise specified, the value of 0.030 in. (0.76 mm) is assumed.
5. Thoroughly clean the surface of the texture to be measured.
6. Push the on button to allow the detector arm to be driven over the traverse length. [The traverse length will include the evaluation length plus the start up length, a distance that can range between 0.090 and 0.540 in. (2.25 and 13.5 mm).]
7. Note readings on screen, but allow several readings to confirm measurement taken.

Note: While Ra is standardized over one cutoff width, measurement is not concluded by taking just one reading over one sampling length. Many readings are taken.

UNIT 15 MEASURING WITH LIGHT WAVES

Interferometry, the measurement of distance using the interference of two light waves as standards, represented a search for a scientific means of precise measurement. The journey was long, thought provoking, and arduous. Did the journey begin with Galileo who attempted to measure the speed of light with a hand lantern? Or did it originate with Christian Huygens, the Dutch mathematician who developed the wave theory of light? (A century later, this body of work would play an important role in the Thomas Young explanation of the remarkable occurrence of light wave interference.) Perhaps it all began with Sir Isaac Newton, who noted the formation of rainbowlike color bands one day when two very flat panes of glass were pressed together. These Newton rings, as they are called, became the basis to his spectrum theory.

It remained for Albert Abraham Michelson, an American physicist, to determine the speed of light. It was while working on this project that Michelson devised an interferometer (Figure 15–1). Using the

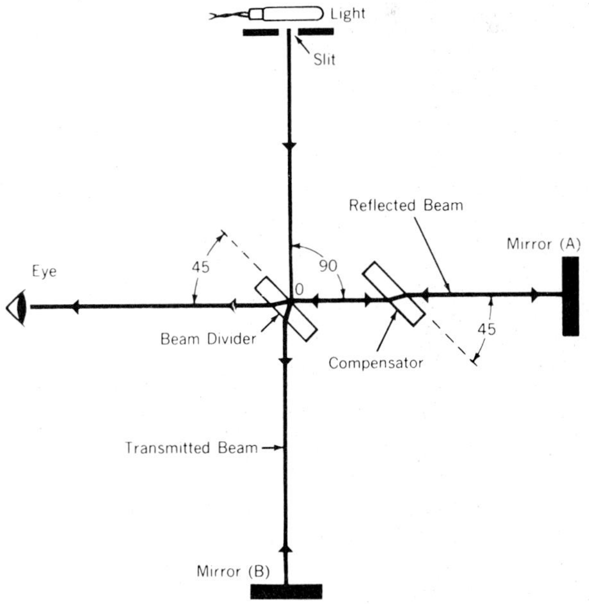

FIGURE 15–1
Michelson's interferometer. (Credit: Sheffield Measurement Division, Warner & Swasey/A Cross & Trecker Company)

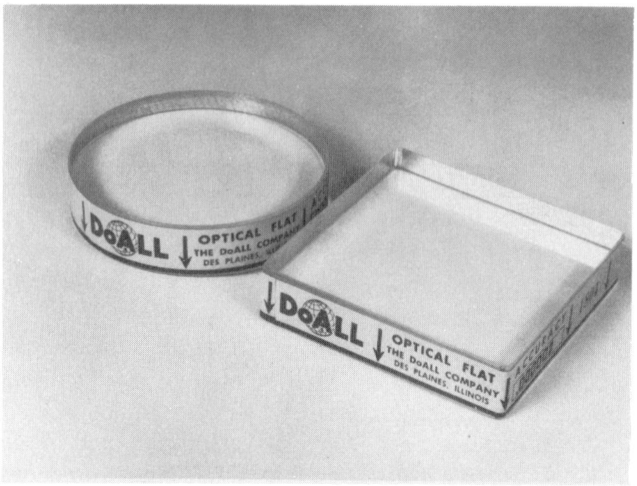

FIGURE 15-2
Optical flats. (Credit: DoAll Company)

FIGURE 15-3
Working optical flat. (Credit: DoAll Company)

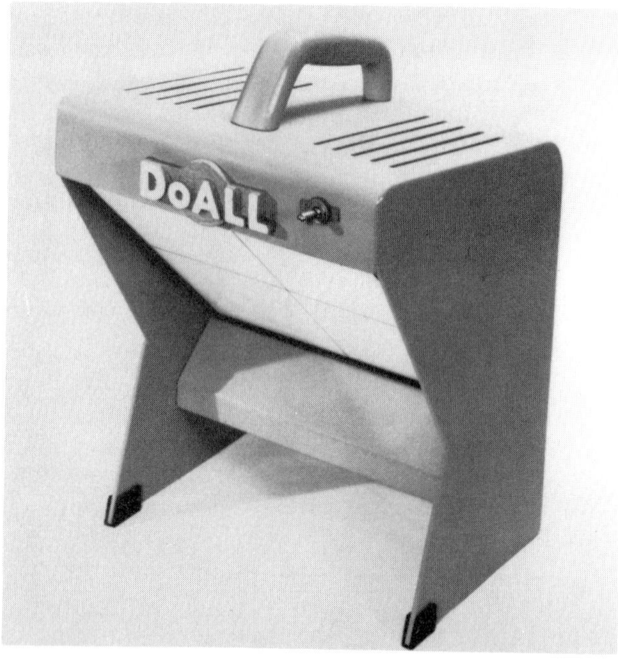

FIGURE 15-4
A monochromatic light source. (Credit: DoAll Company)

wavelength of a red band of cadmium light, which measured slightly more than 25-millionths of an inch (625-millionths of a millimeter), Michelson proceeded to measure the international meter bar for the Paris Bureau International des Poids et Mesures in 1892.

Objectives
After completing this unit you will be able to

1. understand the principles of interferometry.
2. check test objects for flatness, parallelism, or size using optical flats.

Optical Flats

The simplest method of applying the science of interferometry is by means of optical flats (Figure 15-2). Optical flats are made from a variety of materials, sizes, and shapes. While made as large as 10 in. (254 mm) or greater, they commonly range in size between 1 and 3 in. (25 and 75 mm) in materials varying from inexpensive glass to very expensive saphire to high-quality optical quartz. Measuring surfaces, which may have one or both sides finished, also differ in flatness. Working flats (Figure 15-3) are accurate to 4-millionths of an inch (100-millionths of a millimeter), master flats are accurate to 2-millionths of an inch (50-millionths of a millimeter), while reference flats are precise to 1-millionth of an inch (25-millionths of a millimeter). The accuracy involved concerns the amount of deviation in height from one point to another; so that if working flats are used, for example, the difference in height between two points would not be greater than 4-millionths of an inch (100-millionths of a millimeter). Not only are optical flats used to detect errors in flatness of precise parts, they are also used as comparators to measure their sizes.

Let There Be Light

The light used with optical flats originates from a monochromatic light source (Figure 15-4). A pale yellow light is given off from a helium gas tube and is projected through the optical flat, which is placed with its finished lapped surface against the test piece. As the light wave passes through the flat, it is split in two. Since the optical flat is transparent, it can also reflect light. While one light wave strikes the test object and is reflected in the upper surface of the flat, the other is reflected back through the lower surface or the flat that causes an interference as the two light waves interact. The interaction of light rays becomes visible as dark interference bands or fringe lines, forming patterns of surface characteristics of size, flatness, or parallelism. This happens due to an air wedge between the bottom surface of the optical flat and the top surface of the test piece being measured. The distance between them measures one-half of a wavelength or multiples thereof (Figure 15-5).

The wavelength of helium light measures 23.2 millionths of an inch (589 millionths of a millimeter). Each half wavelength, therefore, is equal to 11.6-millionths of an inch (293-millionths of a millimeter) above the point of contact between the optical flat and the test piece. Consequently, when a height is checked, the number of bands shown between two points on a surface is multiplied by 11.6-millionths of an inch (293-millionths of a millimeter) to represent the height difference between the two surfaces.

Measuring with Optical Flats

Checking a gage block for wear would be a good example of how optical flats can be used to measure flatness or can be used as a comparator to check size. Initially, an optical flat is placed on a slight angle with the gage block surface to form an air gap between it and the gage block surface. Incidentally, the angle, which is often exaggerated in illustrations, is generally caused by dust particles and measures approximately 1 to 10 s of an arc.

If the interference fringe lines are straight and parallel (Figure 15-6), the gage block surface is considered flat. If the lines are parallel but curved (Figure 15-7), the surface is either convex or concave. Any irregularities are shown by irregular fringe patterns.

By optically measuring a known standard and the gage block at the same time, the gage block size can be determined. Observe the difference in fringe pattern as two fingers are depressed, one finger above each block. If the pressure above the standard

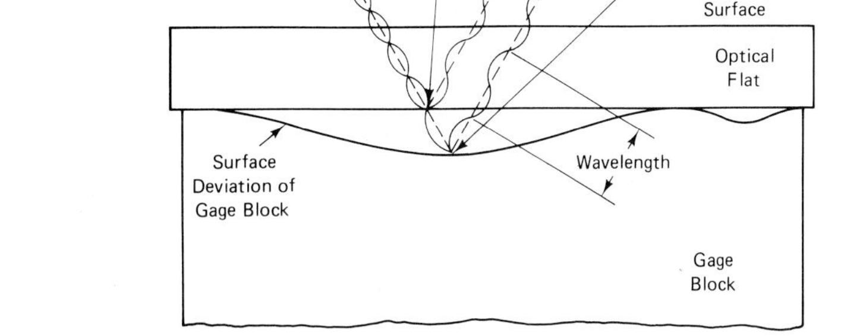

FIGURE 15-5
The principal of an optical flat.

FIGURE 15-6
Fringe lines indicating flatness.

FIGURE 15-7
Fringe lines indicating convex or concave surfaces.
(Credit: DoAll Company)

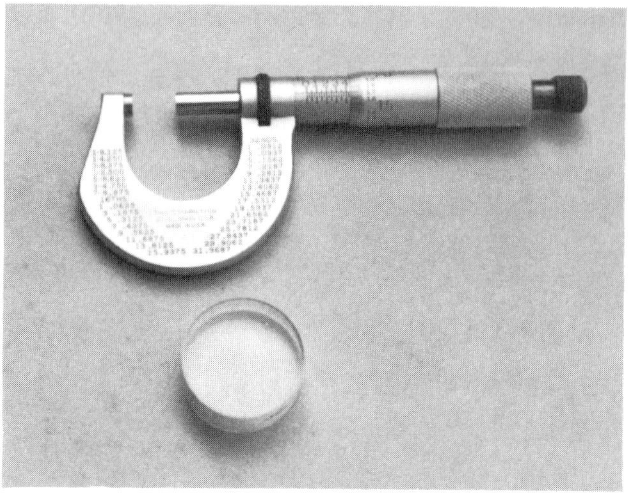

FIGURE 15-8
An optical flat for checking micrometer measuring surfaces.

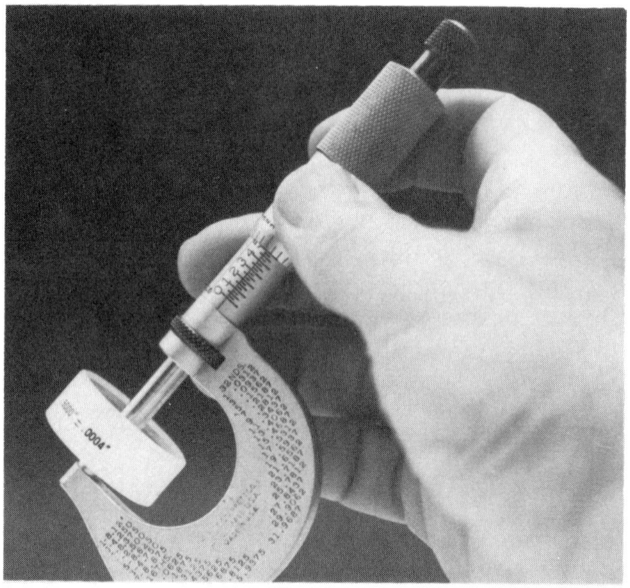

FIGURE 15-9
Make light contact against the optical flat surface.

appears to cause the bands to separate, the standard is larger. If this reaction happens above the gage block, then it is the larger of the two.

How to Measure a Micrometer Measuring Face
An anvil or spindle measuring face can be checked for wear using an optical flat (Figure 15-8) within the monochromatic light source.

1. Lay an optical flat on the anvil measuring face and lightly depress the top surface by advancing the spindle face against it (Figure 15–9).

2. Examine the fringe patterns within the monochromatic light. One of three conditions will be revealed:

(a) If the fringe pattern is straight and equally spaced (Figure 15–10), the measuring face is perfectly flat.

(b) If concentric rings are visible and appear to move outwardly (Figure 15–11), the face is convex.

(c) If the concentric rings appear to move inwardly (Figure 15–12), the surface is concave.

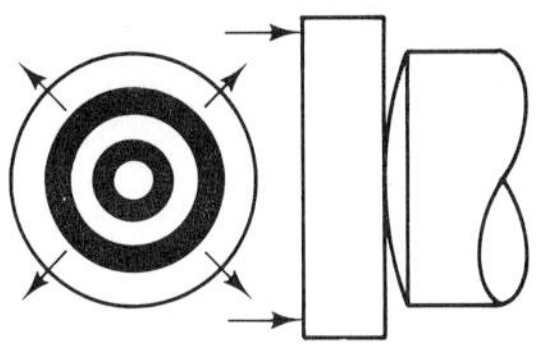

FIGURE 15–11
Concentric rings that appear to move outwardly indicate a convex surface. (Credit: MTI Corporation)

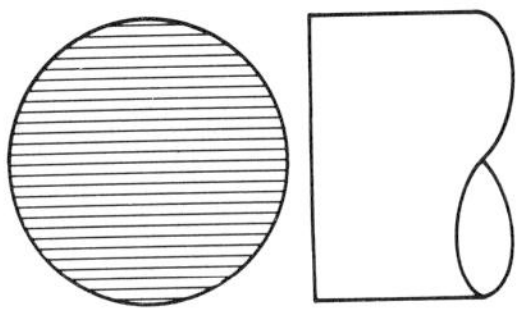

FIGURE 15–10
A straight and equally spaced fringe pattern indicates perfect flatness. (Credit: MTI Corporation)

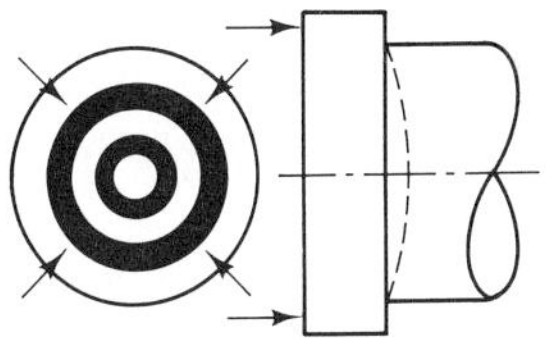

FIGURE 15–12
Concentric rings that appear to move inwardly indicate a concave surface. (Credit: MTI Corporation)

SECTION REVIEW

Unit 6

1. What are the two major measuring systems used throughout the world?
2. On what is the standard of length for the international inch based?
3. What do the letters SI represent?
4. What is the common unit of length in SI?
5. What was the original standard for the unit of length in the metric system?
6. What is the relationship between one unit and the next one in the inch system? The metric system?
7. How are the units of temperature measured in the inch system? The metric system?
8. What is the value of a microinch?
9. What are the meanings of the following prefixes: Milli? Centi? Deci? Mega?
10. The quart is one unit of liquid capacity in the inch system. What is the closest equivalent in the metric system?

Unit 7

1. How is each inch on a 6-in. rule used in the machine shop commonly graduated?
2. How are metric rules commonly graduated?
3. What is the smallest graduation that can be clearly seen on an inch rule?
4. Briefly describe how to use a rule.

5. Briefly explain how accurate measurements can be made if the end of a rule is worn.

6. Name two types of outside calipers.

7. Briefly explain how to set an outside caliper to size.

8. For what purpose are inside calipers used?

9. Briefly explain how to set an inside caliper to size.

10. Name three different types of precision calipers.

Unit 8

1. How many threads per inch are there on standard inch micrometer?

2. What is the value of each line on the sleeve?

3. What is the value of each numbered line on the sleeve or barrel?

4. What is the value of each line on the thimble?

5. Briefly describe how to read the vernier scale on a micrometer.

6. How many threads are there on the measuring screw of a standard metric micrometer?

7. What is the value of one numbered division on the sleeve above the index line?

8. What is the value of one unnumbered division on the sleeve below the index line?

9. What is the value of one division on the thimble?

10. Briefly explain how to adjust a micrometer.

Unit 9

1. Who originated the concept of gage block measuring?

2. How are steel gage blocks seasoned and why is it necessary?

3. What are the three classes of gage blocks and for what purposes are they used?

4. Briefly outline how gage blocks should be wrung together.

5. How should gage blocks be used to minimize the effect of body heat?

6. List five common gage block applications.

7. Outline five precautions to take for the proper care and storage of steel gage blocks.

8. List five advantages of using ceramic gage blocks.

Unit 10

1. What are fixed gages and what purpose do they have in industry?

2. To what accuracy of tolerance are fixed gages made?

3. How are size dimensions identified on holders of go/not-go gages for accuracy in the inch system? The metric system?

4. What two methods may be used to identify the go or not-go ends of a cylindrical plug gage?

5. What type of fixed gage would be used for inspecting external diameters of rods and shafts?

6. What parts of a thread are checked with a thread gage.

7. What type of fixed gages are used for testing the size and out of roundness of small holes?

Unit 11

1. How accurately can a universal bevel protractor measure an angle?

2. Name the three grades of classifications identifying angle gage blocks and the accuracy involved for each.

3. Describe the construction of a sine bar.

4. What is the basis of accuracy when using a sine bar?

5. Determine the gage block height to set a 5-in. sine bar to an angle of 25°.

6. Determine the included angle when a 5-in. sine bar is elevated to a height of 2.4240 in.

7. Describe the construction of the compound sine plate and state its purpose.

8. Name three different types of precision squares used in machine shop and inspection work.

Unit 12

1. Define comparison gaging.

2. Describe an adjustable snap gage.

3. What is another name for a thickness gage?

4. Name three uses for adjustable parallels.

5. In the absence of gage blocks, what advantage is there to using a precision height gage?

Unit 13

1. Define adjustable gaging.

2. List four different types of adjustable gages.

3. Describe an indicator micrometer and how it differs from a standard micrometer.

4. Describe the operation of an optical comparator.

5. List some of the advantages to be found in using an optical comparator.

6. What is another term for air gage?

7. Describe how an air gage operates.

8. Name three principles used in air gaging.

9. List six advantages of air gages.

10. List three basic types of electronic gage heads used with electric gage amplifiers.

Unit 14

1. What is the purpose of surface measurement and why is it important to industry to have standards of surface finish established?

2. What is a surface finish comparator and how does it operate?

3. Illustrate the basic symbol designation for the surface texture system.

4. Define the following surface finish terms: lay, flaw, waviness, nominal surface, and roughness.

5. What do the following lay symbols represent: X, M, C, R, P.

6. Briefly explain the difference between the Ra and RMS numerical averaging systems.

Unit 15

1. What is interferometry?

2. What is the simplest method of applying the science of interferometry?

3. What is the wavelength of helium light?

4. Name three materials from which optical flats are made.

5. What do the following fringe patterns shown under optical glass within the monochromatic light indicate of the object being measured:
   (a) Fringe pattern is straight and equally spaced?
   (b) Fringe pattern of concentric rings appear to move outwardly?
   (c) Fringe pattern of concentric rings appear to move inwardly?

Section 3

LAYOUT PRACTICES

UNIT 16 SEMIPRECISION LAYOUTS

The marking or drawing of lines, angles, circles, or patterns on metal and other materials for the purpose of guidance for material removal is called laying out. The model or prototype that results is called a layout. The accuracy of the layout is dependent on the accuracy of the reference plane upon which varying holding devices and layout tools are used. Many tools will be used, both precise and semiprecise, with many of the instruments serving as measuring as well as layout tools. Throughout the layout procedure, careful and proper use of the tools, instruments, and working surfaces should be maintained in order to maintain accuracy.

Objectives
After completing this unit you will be able to

1. properly prepare a work surface for laying out.
2. use and maintain the various types of surface plates and tables.
3. identify and properly use the semiprecise layout tools and holding devices.

The Surface Plate

Laying out is generally performed on surface plates or tables. Joseph Whitworth, credited with having developed a thread form that is still in use in England and bears his name—the Whitworth thread—was also instrumental in introducing the method for developing a true surface plate. Utilizing a rigid cast iron form, which had been used years earlier by the man under whom he had served his apprenticeship, Henry Maudslay, Whitworth took three cast-iron plates and proceeded to scrape the surface flat and true.

Using the flatest of the three plates as a prototype, Whitworth smeared the surface with a contact dye. He then took the second plate, rubbed the two surfaces together to mark the high spots on plate 2 and using a sharpened metal bar scraped these flat, repeating the process until both plates conformed with each other. He then repeated the process, using the prototype with plate 3 until both surfaces conformed with one another. Whitworth then rubbed and scraped the second and third surfaces until both surfaces matched each other in flatness. The plates were then interchanged until all three plates were brought into conformity with one another. Such was the origin of producing a true reference plane—hand scraping.

The two principal types of surface plates and tables used in industry are made of cast iron and granite (Figures 16–1 and 16–2). Both surfaces are hard and extremely accurate in flatness. Granite

FIGURE 16–1
Granite surface plate.

FIGURE 16-2
A cast iron surface plate.

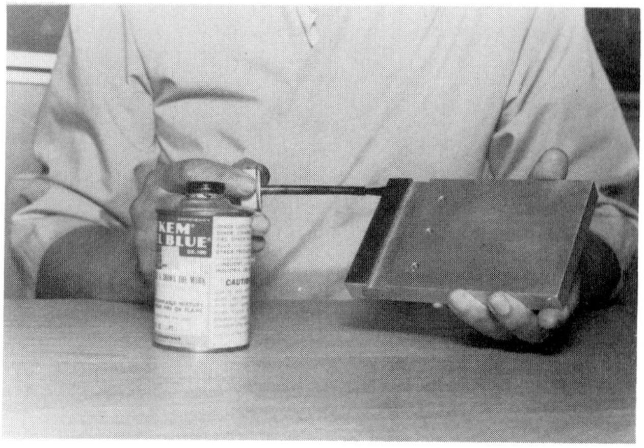

FIGURE 16-3
Applying layout dye.

surface plates are not affected by temperature changes, are nonmagnetic, and will not rust or become nicked or burred as will the cast iron types. Additionally, they will not warp or distort due to internal stresses.

Granite surface plates and tables are used for inspection and layout work and come in three colors: black, pink, or gray. Their accuracy is protected by a hard, highly polished surface. The most important element to be found in any granite plate that assures its quality of hardness and wear resistance is quartz. It is this high-quartz content that prevents penetration of moisture or abrasive particles, presenting a more carefree work surface.

Using and Maintaining Surface Plates and Tables

- Cast iron surface plates should be cleaned of any protective oil prior to use.
- Upon cleaning, examine the surface for burrs and/or dents; remove by oilstone if necessary.
- Granite tables or surface plates should be wiped clean of dirt, dust, or grime using a surface plate cleaner.
- Carefully lay work on the surface plate or table, supporting it on parallels whenever possible.
- Never use the surface to hammer against or for center punching at any time.
- Cast-iron surface plates should be wiped clean of finger marks and foreign matter and protected with a thin film of oil prior to being covered when not in use.
- Cover granite plates or tables when not in use.

Preparation for Laying Out

The workpiece should just as carefully be prepared for the layout. In addition to being free of burrs, it should be kept clean of dirt and grease so that its surface can be coated to produce a background that will eliminate eyestrain. The most widely used method is to apply a blue layout dye (Figure 16-3), which dries quickly and is commercially available. The coating may be applied by brush, dauber, cloth, or sprayed on by aerosol can. Caution should be used to avoid a heavy, thick application. This will result in peeling, scratches, or indistinct lines. Heavy coatings of the dye may be removed by a rag moistened with denatured alcohol.

Bluing can also be accomplished by heating and quenching the workpiece to preserve the color. Care must be taken not to overheat so as to distort or warp the workpiece. Another method of bluing is to immerse the workpiece in a molten bath of sodium nitrate and potassium nitrate. When the color has been achieved, the workpiece is quenched in boiling water.

The rough surfaces of hot-rolled steel and castings may be coated with white chalk to prepare them for laying out. Another source of coating material is a mixture of powdered chalk and alcohol. Water should not be used for this purpose as it tends to rust the surface. The mixture, which will stick to rough surfaces, can then be applied by brush.

Layout Accessories

The process of laying out will require more than just a clean surface plate and layout tools. Quite often it will require the use of an angle plate, against which work can be fastened in an upright position. The ideal type of angle plate can be turned 90°, allowing lines to be scribed at right angles to each other. Another feature to look for on an angle plate is an adjustable side bar that can be position beyond the reference plane of the angle plate to form a butting edge. This will allow its user to quickly position a straight-edged workpiece against it for scribing lines or inspecting holes at 90° to this surface (Figure 16–4).

Parallels are also important to layout work. Often workpieces will have to be elevated while at the same time have a work edge rest parallel with the top of the surface plate. Parallels of different thicknesses and heights should be on hand for laying out (Figure 16–5).

V-blocks are as important in laying out and inspecting round work as an angle plate is to flat work. They may be used individually or in precisely ground matched pairs. The more versatile are designed to allow the user to turn the V-block 90° so that lines may be scribed at right angles if need be without disturbing the workpiece (Figure 16–6).

Rules are fastened with keyseat clamps to form a keyseat assembly that is placed along the length of a diameter, which enables the user to easily layout keyways or for scribing parallel lines on round work or shafting.

No layout of any kind should be undertaken without a collection of holding devices. Depending on the size and type of workpiece involved, small C-clamps and toolmaker's parallel clamps are generally used to fasten the workpiece to the angle plate (Figure 16–7). When using a C-clamp, however, it is advised to keep the solid jaw against the work surface and not the movable swivel portion. In tightening against the work surface, there is a tendency to dig in and scratch the layout blue with the swivel.

A ball-peen hammer and a set of punches, a prick punch and a center punch, are instrumental in indenting center hole locations. The prick punch, with its very sharp point, is able to pick out intersecting lines, making a neater indent (Figure 16–8), while the 90° center punch provides a larger mark for a center drill to pick up during the drilling operation.

FIGURE 16–4
Angle plate.

FIGURE 16–5
Parallels are useful in layout work.

FIGURE 16–6
Laying out lines at right angles on round work using a V-block.

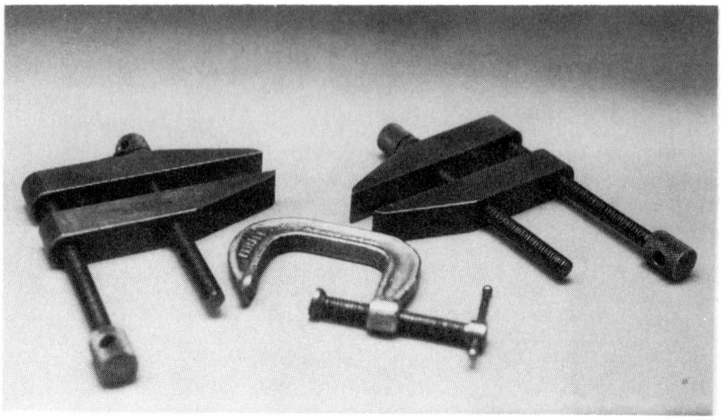

FIGURE 16-7
C-clamp and toolmaker's parallel clamps.

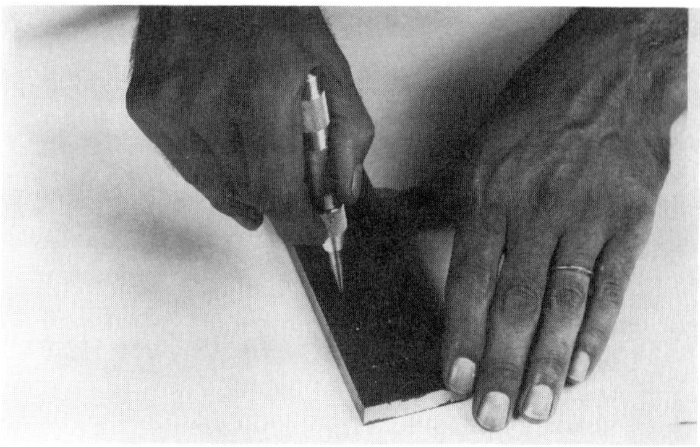

FIGURE 16-9
An automatic center punch can speed up the indenting of layout locations. (Credit: L.S. Starrett Company)

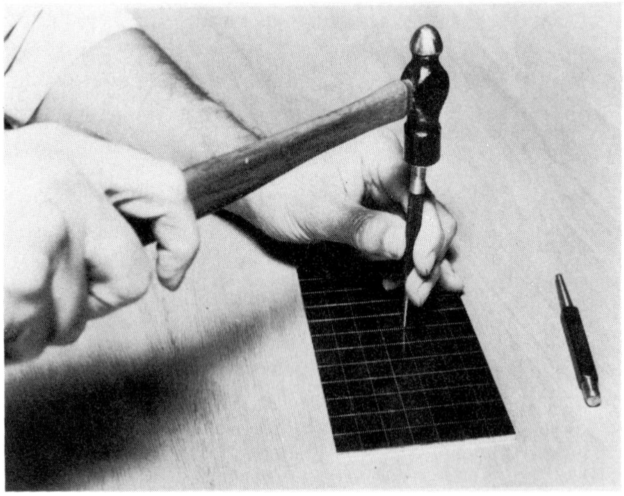

FIGURE 16-8
Indenting intersecting lines with a prick punch.

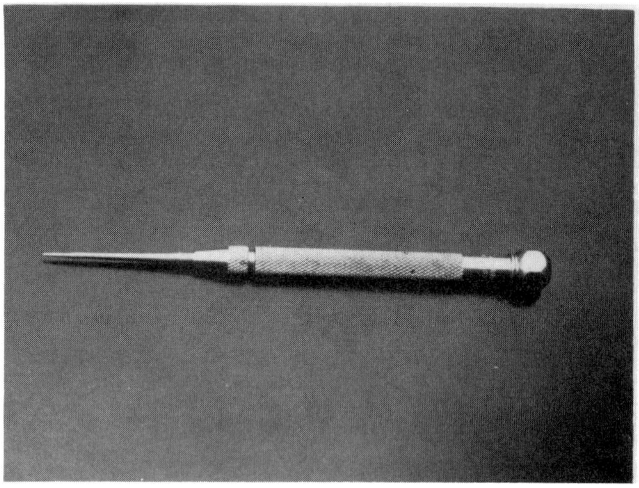

FIGURE 16-10
A pocket scriber.

An automatic center punch (Figure 16-9) requires no hammering at all to indent the workpiece location. All that is required is to hold the punch in an upright position with one hand and press down on the handle. The built-in, spring-loaded mechanism releases a blow against the center punch that strikes a perfect mark every time. Heavy or light indenting can be controlled by turning the upper knurled cap in a clockwise or counterclockwise direction. Once the desired depth has been found, the center punch will repeat the indenting without further adjustment. Some automatic center punches are also provided with spacing attachments to regulate the center to center locations of layout punch marks.

Layout Tools

One of the basic tools for drawing lines in laying out is the scriber (Figure 16-10). Equipped with a sharp steel point on one end, it can also come with a point bent at a right angle on the other end for drawing in awkward places. The scriber is used against a series of edges such as the machinist's scale or the multipurpose combination set (Figure 16-11) to draw a straight line, very much like running a pencil point against a ruler. It is recommended that the scriber point be honed against an oilstone prior to using it to achieve sharp clean lines against the layout dye (Figure 16-12).

The square becomes a handy measuring tool in semiprecise layouts though it is the adjustable square and not the solid square referred to in this case. As discussed in Unit 7, the adjustable square with its sliding blade lends itself to depth measurements as well as for laying out lines parallel or at right angles to an edge or for laying out angles at 45°.

A semiprecise method of drawing lines parallel to a base is accomplished by means of a surface gage (Figure 16-13). By setting and locking the scriber point against an upright rule on the surface plate (Figure 16-14), the height of the line is measured. The line can then be drawn against the workpiece by holding the scriber point on an angle as the base of the surface gage is slid along the surface plate (Figure 16-15).

The other features of the surface gage permit it to be used in situations other than sliding it against the top of the surface plate. Two little pins located in the base can be pushed out to allow the base to be slid against the surface edge of a work-

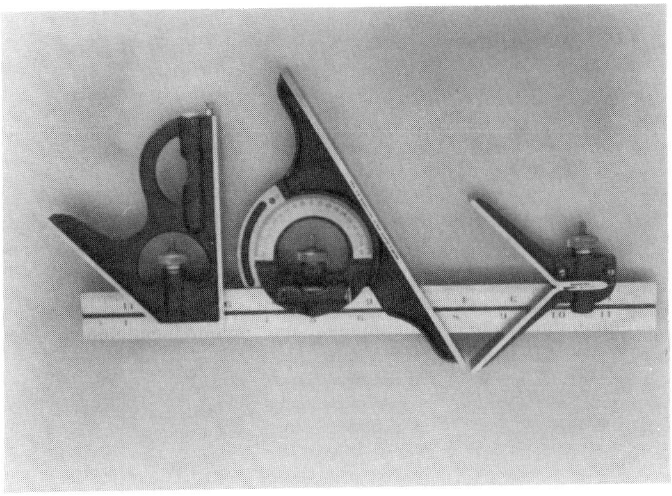

FIGURE 16-11
A combination set.

FIGURE 16-13
A surface gage.

FIGURE 16-12
Honing a scriber point.

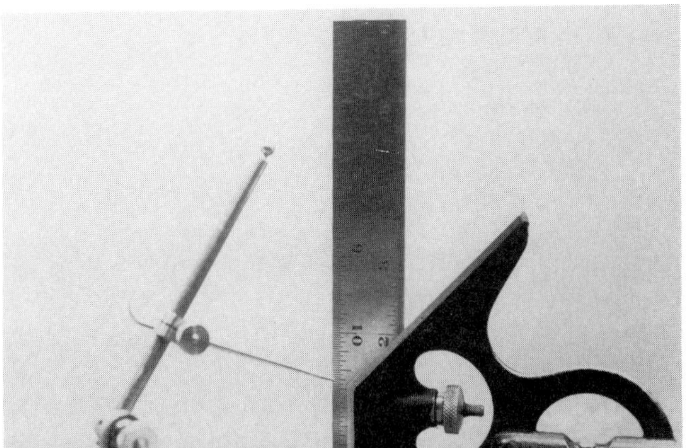

FIGURE 16-14
Setting the surface gage scriber height.

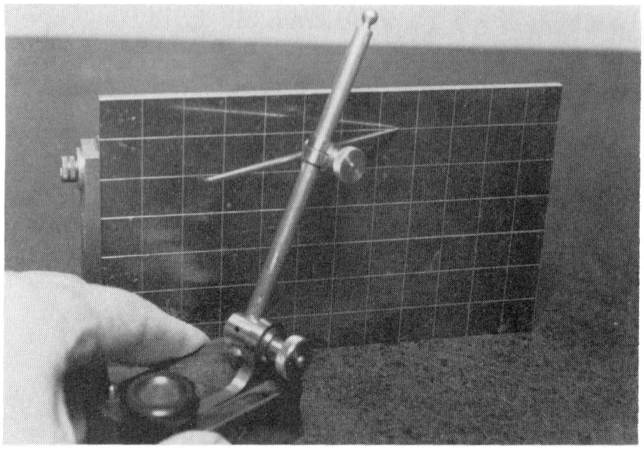

FIGURE 16-15
Slide the surface gage base slanting the scriber point on an angle.

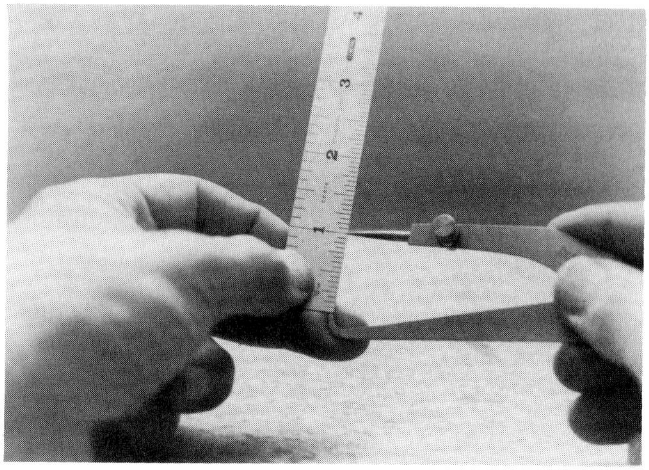

FIGURE 16-18
Hermaphrodite calipers can be set against a rule to draw a line that follows the edge of an irregular contour.

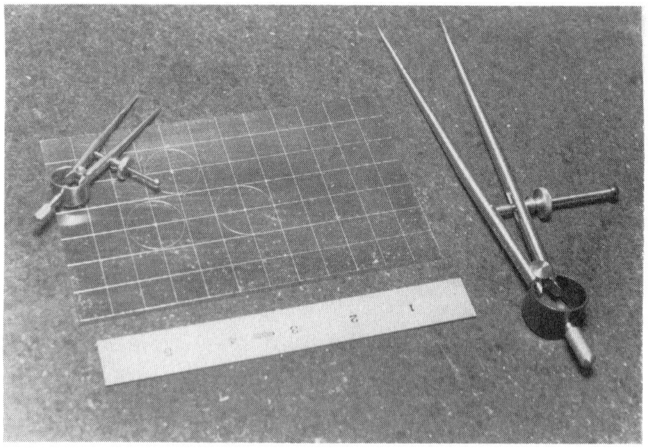

FIGURE 16-16
Dividers are used to draw circles.

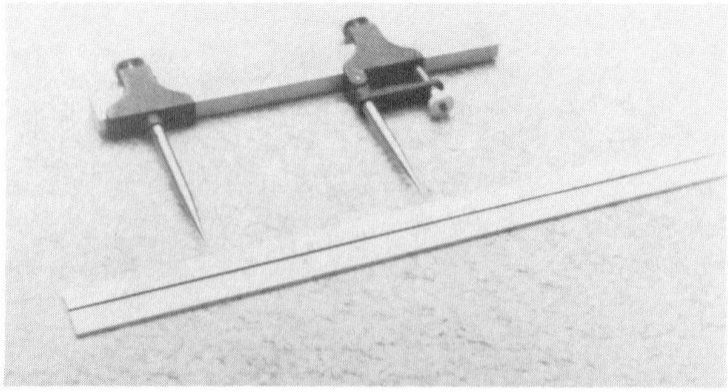

FIGURE 16-17
Trammels are adjustable to draw large arcs and circles.

piece. This will permit a line to be drawn parallel to that surface. The V-shaped groove in the bottom of the base will also permit it to be slid on round surfaces.

Circular Layouts

Holes and circles are scribed with dividers (Figure 16-16) while larger arcs and radii can be drawn by use of a set of trammels (Figure 16-17). Both are set by means of a rule by establishing the radius of the intended circle. For irregular contours a hermaphrodite caliper (Figure 16-18) is used. With a scriber point on one end and a curved leg on the other, the hermaphrodite caliper can be used to draw straight lines or contours as the curved leg is made to follow the edge of a workpiece.

UNIT 17 PRECISION LAYOUT PRACTICES

The effectiveness of a finished part will greatly reflect the thinking and accuracy with which it was laid out. While semiprecise tools may be adequate to scribe lines to fractional accuracy, they cannot compare to the layouts made by a precision height gage (Figure 17-1). These instruments can scribe lines as close as 0.001 in. (0.02 mm). A thorough understanding and transference of drawing dimensions is one of the requirements in precision layouts. In addition to an angle plate (Figure 17-2), parallels (Figure 17-3),

FIGURE 17–1
This precise layout was made with the use of a vernier height gage.

FIGURE 17–2
An angle plate is instrumental in using a height gage accurately.

FIGURE 17–3
Parallels are often used to raise work, permitting scribing of difficult-to-reach areas.

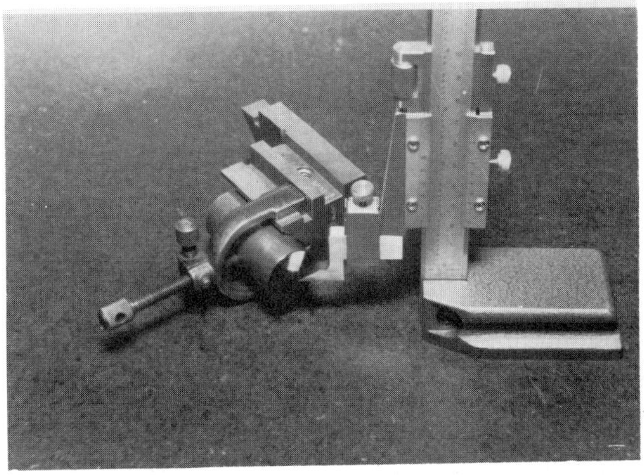

FIGURE 17–4
Round work may also be scribed by a height gage.

gage blocks, and clamping devices whose gripping surfaces should be smooth or taped to prevent scratching the layout surface in use, V-blocks may also be used for round work (Figure 17–4). Laying out holes and angular dimensions may also require the use of trigonometric calculations.

Objectives
After completing this unit you will be able to

1. identify the principal parts of a vernier height gage.

2. use a vernier height gage to layout precise lines.

The Vernier Height Gage

Precision height gages in the machine shop are available in three styles: one using the vernier principle (Figure 17–5), the others using a simplified dial system (Figure 17–6), or a digital readout system (Figure 17–7).

Lines drawn using a height gage are done with the use of a scriber, either straight or offset as the

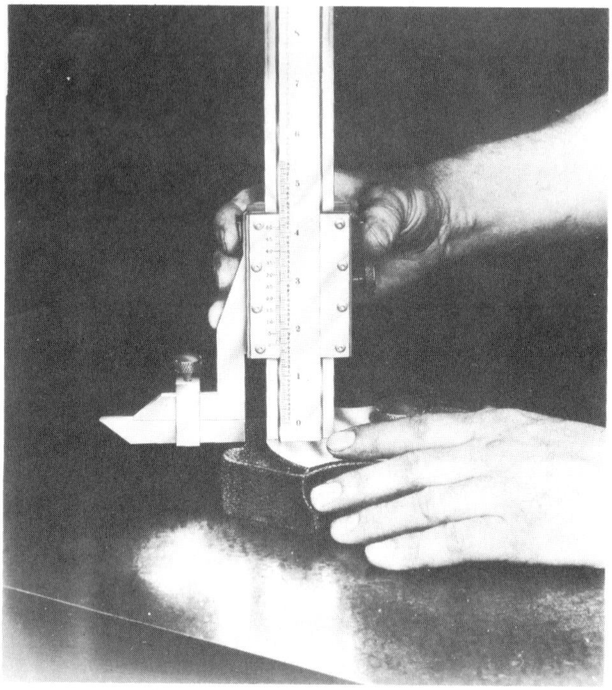

FIGURE 17–5
A 50-division inch vernier height gage. (Credit: L.S. Starrett Company)

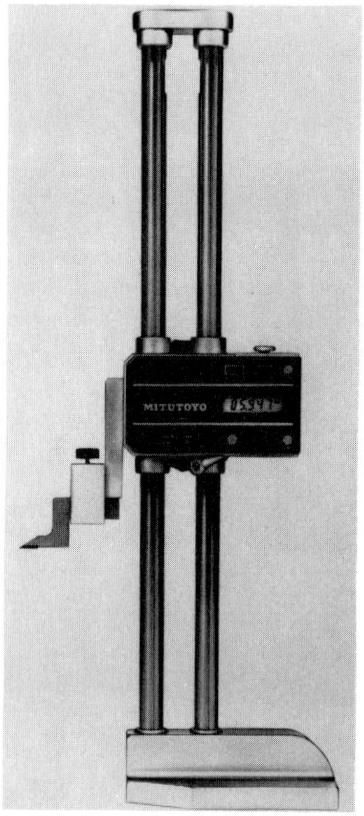

FIGURE 17–7
An error-free LCD readout height gage. (Credit: MTI Corporation)

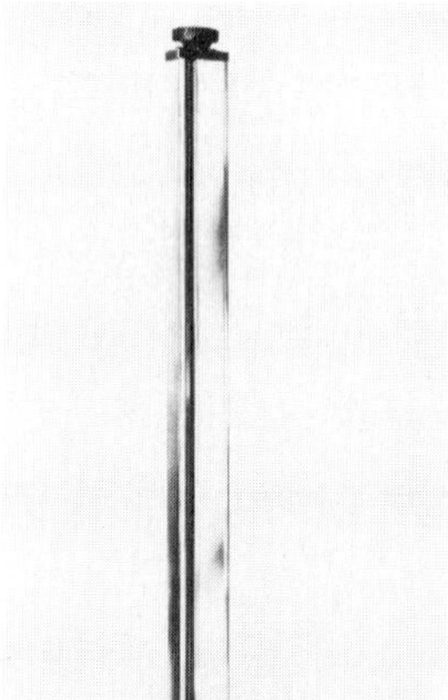

FIGURE 17–6
A direct-reading dial height gage. (Credit: L.S. Starrett Company)

one shown in Figure 17–8. These scribers are generally made from tool steel but are also available carbide tipped for longer life and sharper cutting edges. Care must be taken not to abuse these very hard scribers as carbide may chip easily.

The use of an offset scriber permits a direct reading on the height gage graduations. The height gage will have no reading when the bottom of the offset scriber touches the surface plate. There is less convenience when using a straight scriber. The work must be elevated on parallels or the reading will have to be adjusted to compensate for the mathematical difference between the bottom of the straight scriber and the surface plate.

The vernier height gage is set and read in a manner similar to a vernier caliper except that its graduated surfaces, the beam and vernier plate, are looked on and read in an upright or vertical position. Every graduation on the beam of a 25-division vernier has a value of 0.025 in. and totals 40 graduations to the inch. Every fourth graduation ($\frac{1}{10}$ in.) is

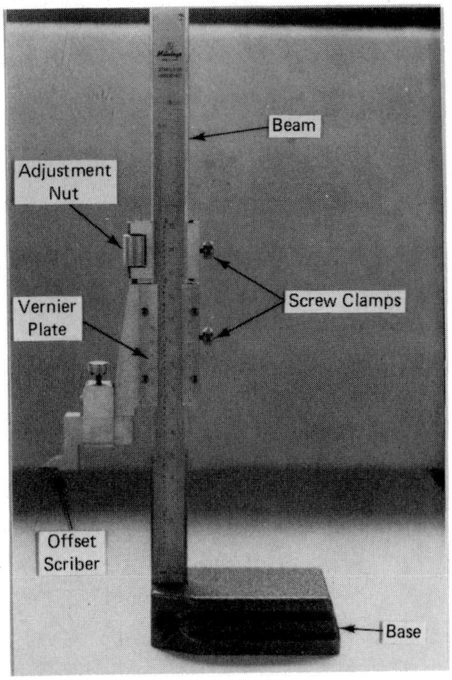

FIGURE 17-8
The main parts of a vernier height gage.

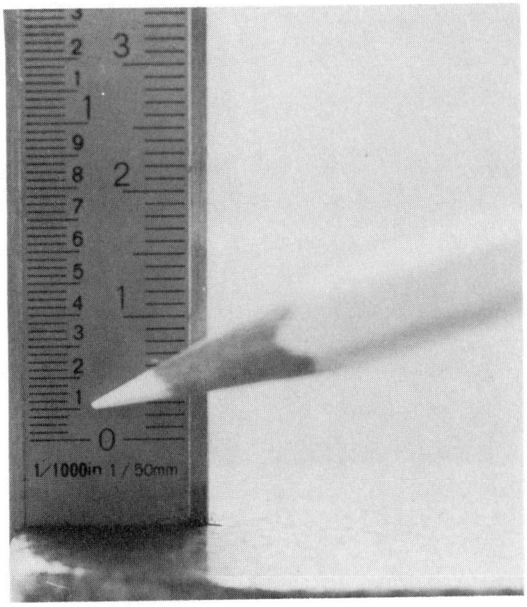

FIGURE 17-9
Every fourth graduation on the inch beam is numbered and represents 0.100 in. in face value.

numbered and represents 0.100 in. (Figure 17-9). There are 25 graduations on the vernier plate, with each graduation representing 0.001 in.

The vernier principle of measurement uses calibrated lines on adjoining surfaces that can be read when two lines match or merge to form one long line. To read the setting correctly, first examine the zero line on the vernier plate, record the beam value below it, and add to it the value of the vernier plate, which is determined by matching lines. This is illustrated in setting the vernier height gage to an example reading of 1.315 in.

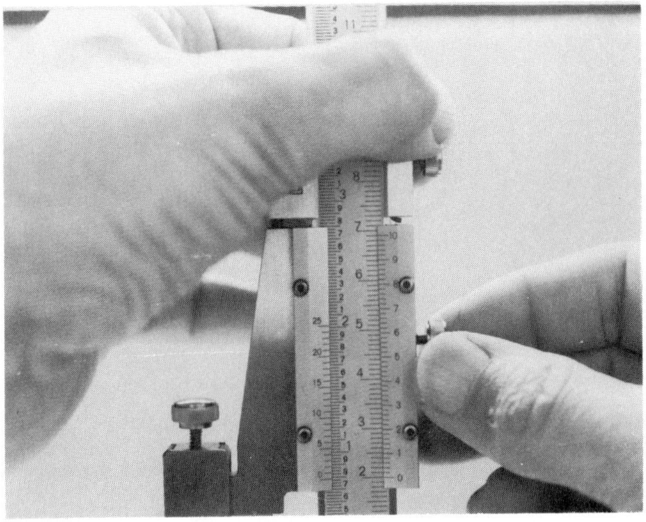

FIGURE 17-10
Loossening the clamp screws of the vernier plate assembly.

Procedure

1. Prepare to read the setting through a magnifying glass if necessary.
2. Loosen the clamp screws of the vernier plate assembly (Figure 17-10).
3. Slide the assembly so that the zero line of the vernier plate rests close to the 1.300 in. mark on the beam (Figure 17-11).
4. Tighten the clamp screw of the fine adjustment nut carrier.

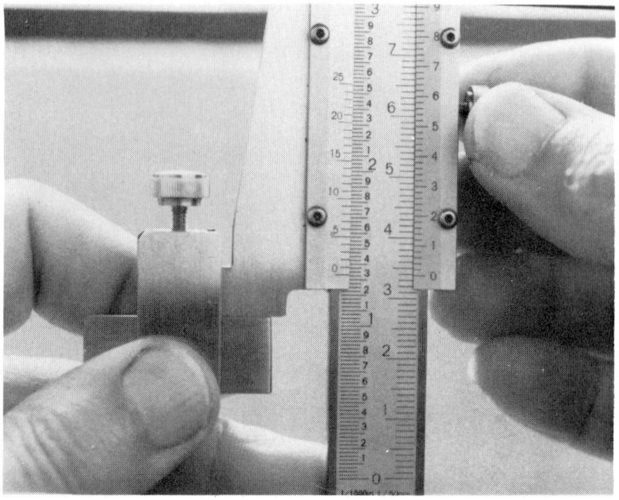

FIGURE 17-11
Slide the assembly so that the zero line on the vernier plate is close to the 1.330 reading on the beam.

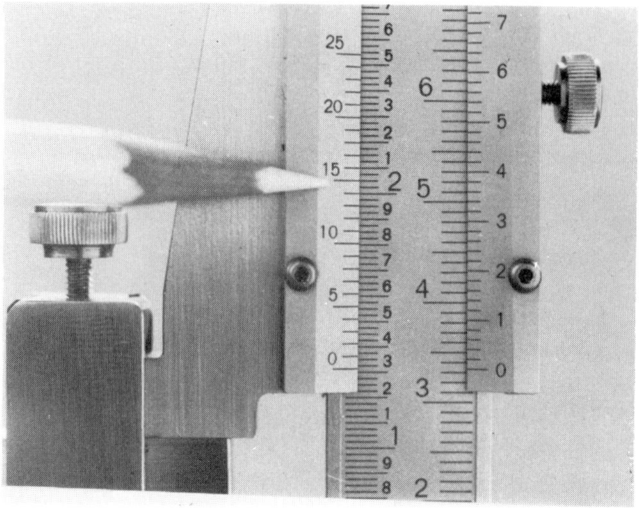

FIGURE 17-13
The height gage is now set to a 1.315-in. reading.

FIGURE 17-12
Slowly turn the adjustment nut to align the vernier plate and beam graduation line.

5. Lightly snug but do not tighten the clamp screw of the vernier plate slide.

6. Looking through the magnifying glass, turn the adjustment nut so that the zero line on the vernier plate lies between the 1.300 and 1.325 in. graduations (Figure 17-12).

7. Keep turning the adjustment nut until the fifteenth graduation on the vernier plate merges with a graduated line on the beam (Figure 17-13).

8. Once the reading has been established, tighten the clamp screw on the vernier plate assembly to lock the measurement.

The Metric Vernier Height Gage

The metric vernier height gage is used in the same manner as its inch counterpart and is read in millimeters using the following values:

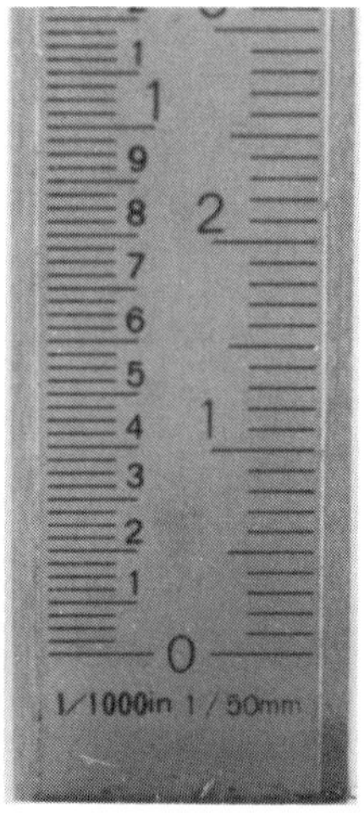

FIGURE 17-14
The number 1 on the metric beam scale represents 10 mm.

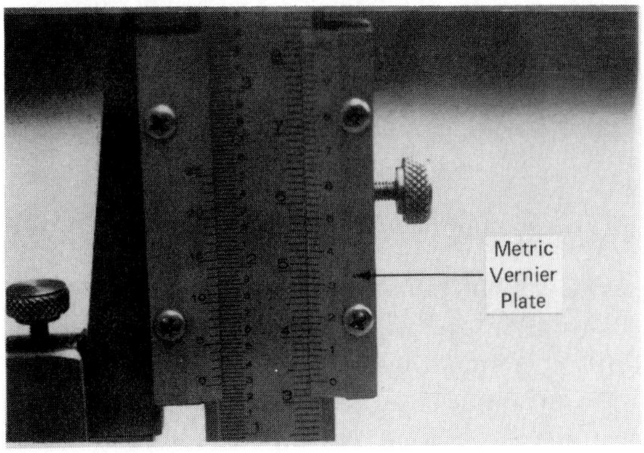

FIGURE 17-15
The metric vernier plate contains 50 graduations.

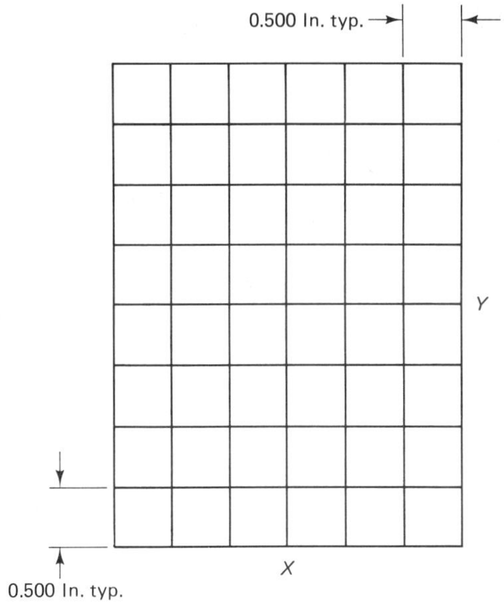

FIGURE 17-16
An illustration of a grid pattern layout identifying the X and Y reference planes.

- Each line on the beam represent 1 mm.
- The number 1 on the beam represents 10 graduations or 10 mm (Figure 17-14).
- The number 2 represent 20 graduations or 20 mm, right through 10 or 100 mm.
- The vernier plate contains 50 graduations, identified as 1, 2, 3, and so on up to 10 (Figure 17-15).
- Every graduation on the vernier plate has a value of $\frac{1}{50}$ of one millimeter or 0.02 mm.

Coordinate Layout Device

A height gage can be used as a coordinate layout device. In order to accurately scribe outlines, angles, or hole locations, the workpiece must have two adjacent sides that are finished and at 90° to each other. If these sides do not exist, they will have to be machined. These finished surfaces become reference surfaces from which the *X* and *Y* coordinates will be drawn to form the intersection of lines. The *X* represents the horizontal plane on a two-dimensional drawing, while the *Y* represents the vertical plane set at 90° to the *X* plane. It is the same method used for referencing when using numerical control, vertical milling machines, jig borers, or coordinating measuring devices.

The measurement of these locations may require different setups and different mathematical calculations, for example, the layout of a grid pattern on a rectangular workpiece (Figure 17-16).

Procedure

1. Examine the workpiece for any burrs and sharp edges and remove them if necessary.
2. Check two adjacent sides for squareness.
3. Apply an even coating of layout dye on the workpiece surface.
4. Clean the surface plate and height gage base. Sticky surfaces will not allow accurate lines to be drawn.
5. Select an angle plate with adjustable side rail to butt a reference surface against.
6. Position the *X* reference surface on the surface plate and clamp the workpiece against the angle plate in this setup.
7. Loosen the height gage clamp screws and slide the assembly down so that the offset scriber touches the surface plate.
8. Check the height gage reading to see that the zero lines on the beam and vernier plate coincide to indicate no value reading (Figure 17-17). If the lines do not coincide, recheck the slide assembly and offset scriber.
9. Set the height gage reading to 0.500 in. and lock the setting.

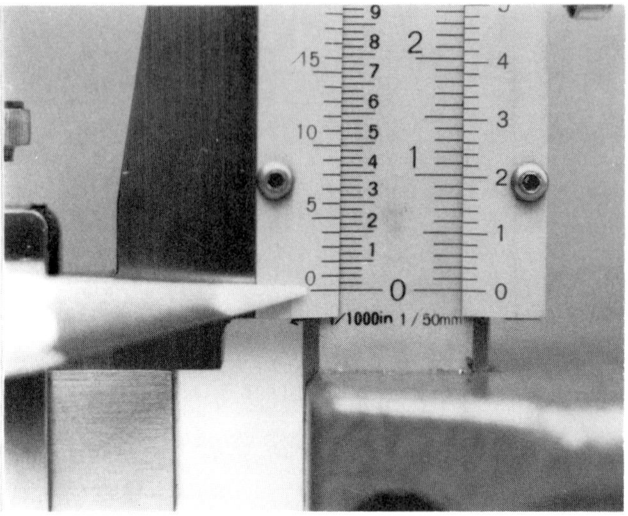

FIGURE 17-17
When the zero lines on the vernier plate and beam match up, there is no reading.

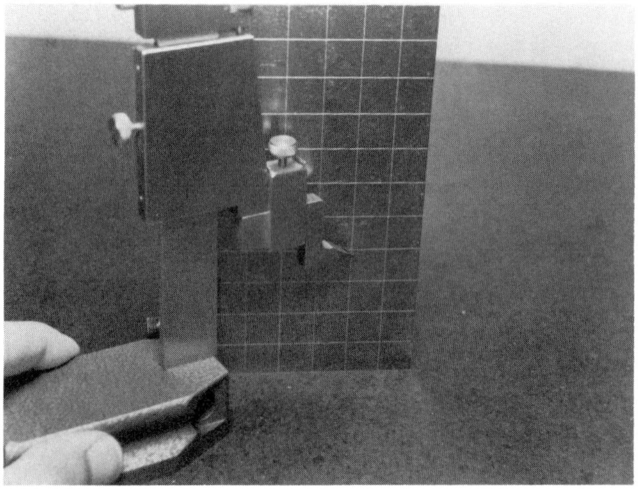

FIGURE 17-18
Slide the height gage base so that the scriber is being pulled against the work surface.

10. Position the scriber on a slight angle to the workpiece so as to draw with a point. *Note: Do not attempt to draw lines using the full width of the flat scriber.* It will leave an uneven, chattering mark.

11. Slide the height gage base so that the corner of the scriber is being pulled against the workpiece (Figure 17-18).

12. Set the height gage for each successive line another 0.500 in. After each setting is locked, scribe the line, and continue this method until the complete surface has been drawn.

13. Remove the clamping device and turn the workpiece so that the *Y* reference side rests on the surface plate and the *X* reference surface butts against the side rail.

14. Clamp the workpiece against the angle plate in this position.

15. Repeat steps 9 through 12 until the complete surface has been drawn.

Circular Hole Spacing

The coordinates for locating equally spaced holes on an established circle can be calculated using trigonometry. They can also be determined using tables of coordinate measures, which can be found in technical handbooks such as the *Machinery's Handbook*. For example, calculate the position of three equally spaced holes on a 4-in.-diameter circle of a 6-in. square plate (Figure 17-19). The basis of calculation appear in Figure 17-20.

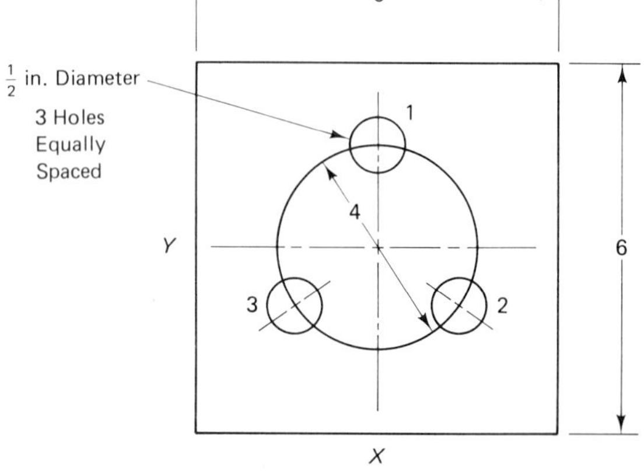

FIGURE 17-19
Three equally spaced holes are required on a 4-in.-diameter circle.

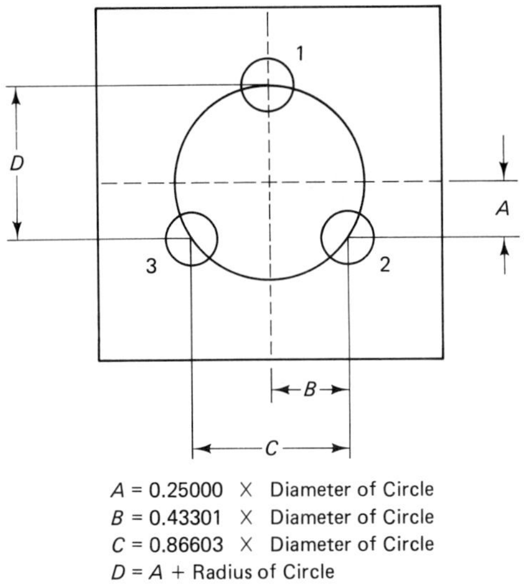

FIGURE 17-20
The basis of mathematical dimensioning.

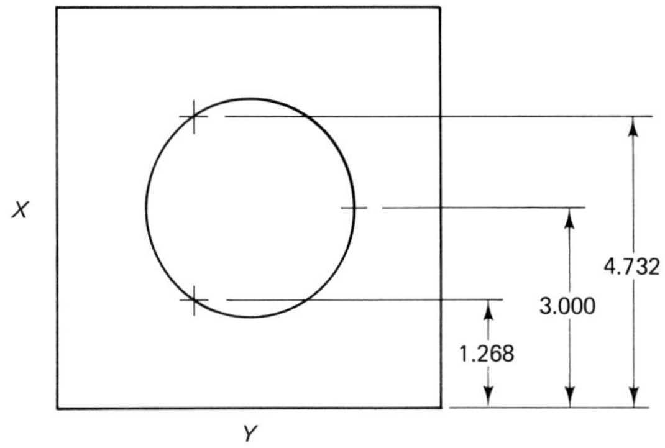

FIGURE 17-22
The horizontal location of each hole.

1. Determine the vertical centerlines of holes 2 and 3 from the X edge. This is shown in Figures 17-21 and 17-22 calculated as follows:

Vertical centerline of holes 2 and 3

= center location of circle − distance A

= 3.000 − (0.25000 × diameter of circle)

= 3.000 − (0.25000 × 4)

= 3.000 − 1.000

= 2.000 in.

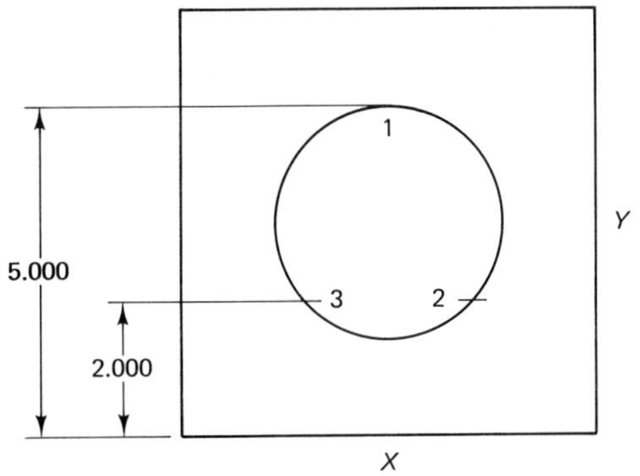

FIGURE 17-21
The vertical distance of each hole location.

2. Mount the circle-scribed plate against the angle plate and clamp it in place with the X edge resting on the surface plate.

3. Set the height gage to 2.000 in. reading and lock it in place.

4. Scribe two short intersecting lines with the circle at hole locations 3 and 2 by drawing the scriber across the locations at a small angle.

5. Calculate the vertical centerline of hole 1 (tangent to the circle) from the X edge.

Vertical height of hole 1

= center location of circle + radius of circle

= 3.000 + 2.000

= 5.000 in.

6. Set the height gage to a 5.000-in. reading and lock it in place.

7. Scribe a short tangent line with the circle.

8. Remove the workpiece to rotate it to a new position.

9. Place the Y reference surface against the surface plate with the X reference surface butting against the side rail and clamp it in place.

10. Determine the height distance from the Y surface to locate center of hole 2 (Figure 17-22) as follows:

Vertical centerline of hole 2

= center location of circle − distance B

= 3.000 − (0.43301 × diameter of circle)

= 3.000 − (0.43301 × 4)

= 3.000 − 1.732

= 1.268 in.

11. Set the height gage to a 1.268-in. reading and lock it.

12. Scribe a short intersecting line at hole location 2.

13. Set the height gage to a 3.000-in. reading and lock it.

14. Scribe a short intersecting line at hole location 1. (This measurement will also be the centerline measurement of the circle.)

15. Calculate the height distance from the Y surface to locate centerline of hole 3:

Vertical centerline of hole 3

= 2 hole location + distance C

= 1.268 + (0.86603 × diameter of circle)

= 1.268 + (0.86603 × 4)

= 1.268 + 3.46412

= 4.732 in.

16. Set the height gage to a 4.732-in. reading and lock it.

17. Scribe a short intersecting line at hole location 3.

18. Remove the workpiece and prepare to mark the intersecting lines with a sharp prick punch and a light ball-peen hammer.

19. Place the workpiece on the workbench with its scribed surface up.

20. Looking through a loupe or a magnifying glass, lightly tap the prick punch to indent the center hole locations.

21. Check the accuracy of the punch marks at two locations 90° apart; correct if necessary.

22. Set the dividers to a $\frac{1}{4}$-in. opening and scribe the three $\frac{1}{2}$-in circles.

23. Carefully prick punch the circumference of the circles at four equally spaced points to preserve the layout.

FIGURE 17–23
A sine bar, gage blocks, and an angle plate will be required to layout angular dimensions.

Using the Height Gage to Lay Out Angles

Angles may be laid out with a height gage by using a sine bar and gage blocks (Figure 17–23) and trigonometry to calculate the dimension of the sine bar elevation. An example of this type of layout is shown in Figure 17–24.

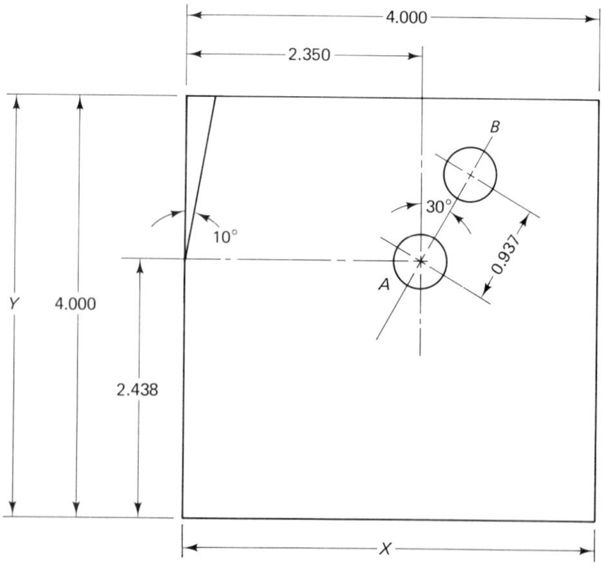

FIGURE 17–24
An example of an accurate angular layout.

Procedure

1. Examine the drawing dimensions required.
2. Machine a workpiece square to its 4.000 × 4.000 in. dimension.
3. Clean the work surface and cover it with an even coating of layout dye.
4. Position the workpiece so that the X surface rests on the surface plate and clamp it to an angle plate.
5. Set the height gage to a 2.438-in. reading to scribe the start of the 10° taper and the centerline of hole A.
6. The position for hole B is calculated by forming a right triangle with line AB. The length of these sides, $X - 1$ and $Y - 1$, must be determined. (See Figure 17–24(a).)

Length of side X–1: $= 0.937 \times \cos 30°$

$= 0.937 \times 0.86603$

$= 0.8114$

Length of side Y–1:

$$\frac{Y\text{–}1}{\sin 30°} = 0.937$$

$$Y\text{–}1 = 0.937 \times \sin 30°$$

$$= 0.937 \times 0.500$$

$$= 0.4685$$

The centerline of holes A and B can now be laid out using the coordinate method. The vertical centerline for hole B would be located 0.811 above hole A. To scribe the centerline, the height gage would have to be set 2.438 + 0.811 or 3.249 in. and drawn above the X reference surface (Figure 17–25).

7. Remove the workpiece from the angle plate and prepare to scribe the horizontal centerline for hole B by positioning the Y reference surface on the surface plate and clamping it to an angle plate.
8. Set the height gage to a reading of 2.350 + 0.469 or 2.819 in. and scribe the horizontal centerline for hole B.
9. Remove the workpiece from the angle plate.
10. Looking through a loupe or magnifying glass, lightly tap a sharp-pointed prick punch to indent the center hole locations.
11. Check the punch mark locations.
12. Set the dividers to a spacing equal to the hole radius and scribe the two holes.

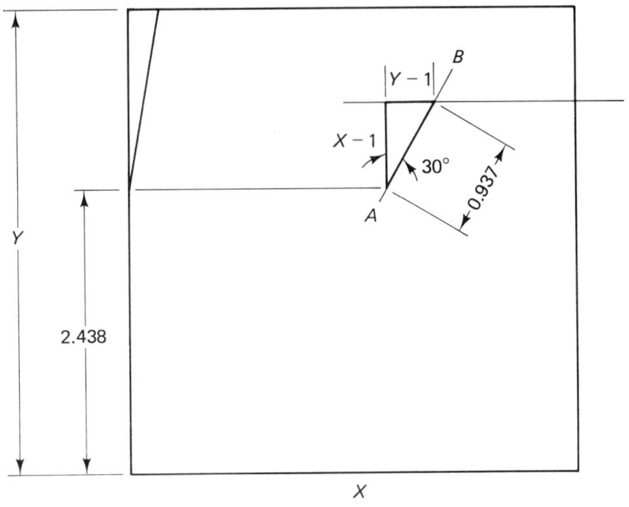

FIGURE 17–24a
Lines *X*-1 and *Y*-1 form a right angle triangle with line *AB*.

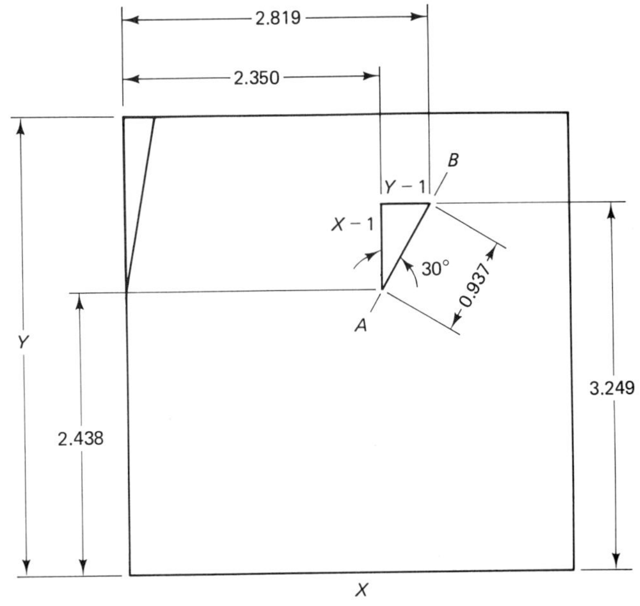

FIGURE 17–25
The position of the hole is done by trigonometry.

13. To lay out the 10° angle at the corner of the plate, determine the gage block height for 10° using a 5-in. sine bar.

$$\begin{aligned}\text{Gage block height} &= 5 \times \sin 10^\circ \\ &= 5 \times 0.17365 \\ &= 0.86825 \text{ in.}\end{aligned}$$

14. Using a surface plate as a base, place one end of the sine bar on stacked gage blocks measuring 0.8683 in.

15. Place the workpiece on the sine bar, angled edge up, and clamp it to an angle plate.

16. Determine the length of MN in the triangle formed in the corner of the workpiece by the 10° line in Figure 17-26 as follows:

$$\begin{aligned}\frac{MN}{4.000 - 2.438} &= \tan 10^\circ \\ \frac{MN}{1.562} &= 0.17633 \\ MN &= 0.17633 \times 1.562 \\ &= 0.27542 \text{ in.}\end{aligned}$$

17. With line MN known, determine the vertical distance below the corner of the plate (MO) to the line (PON), which will form the 10° angle as follows:

$$\begin{aligned}\frac{MO}{MN} &= \cos 10^\circ \\ MO &= \cos 10^\circ \times MN \\ &= 0.98481 \times 0.27542 \\ &= 0.2712 \text{ in.}\end{aligned}$$

18. Raise the vernier height gage assembly so that the scriber point is above the corner of the plate.

19. Carefully lower the offset scriber so that it barely touches the corner of the plate.

20. Examine the reading.

21. Lower the scriber point 0.2712 in. from this reading, lock the height gage, and scribe line PON.

22. Remove the workpiece from the angle plate.

23. Lightly indent the layout line with a sharp-pointed prick punch.

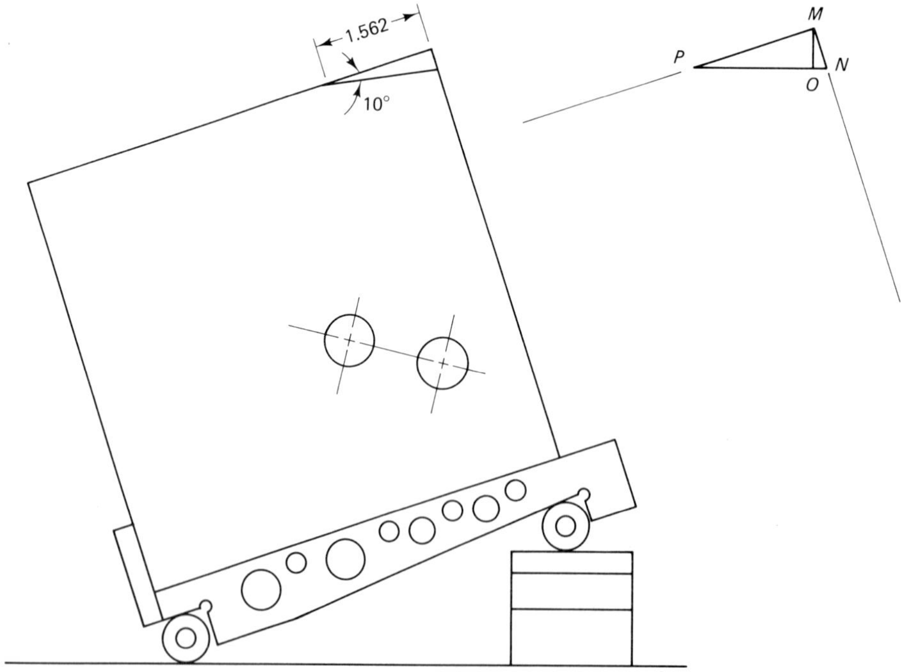

FIGURE 17-26
The 10° angle cannot be scribed unless dimension of line *MO* is calculated.

SECTION REVIEW

Unit 16

1. What is the purpose of a layout?
2. Explain the function of each of the following in the layout process:
   (a) surface plate and
   (b) layout dye.
3. Explain the function of each of the following accessories used during the layout process:
   (a) angle plate,
   (b) parallels,
   (c) V-block,
   (d) C-clamp,
   (e) ball-peen hammer, and
   (f) surface gage.
4. What tools are used during the semiprecise layout for holes, circles, and irregular contours?

Unit 17

1. List three background requirements any machinist should have in order to make a precision layout.
2. List three different styles of precision height gages.
3. List four principle parts of a vernier height gage construction.
4. How can a vernier height gage be used as a direct-reading instrument when scribing lines from a zero reference?
5. Describe the procedure and principle for reading a vernier height gage setting.
6. Describe how a height gage may be used as a coordinate layout device for form intersecting lines.
7. What tools are used in conjunction with a height gage and angle plate to layout angular lines?

Section 4

ENGINEERING DRAWINGS

Mechanical parts and components that are produced by machine tools must meet the standards established by the ASME (American Society of Mechanical Engineers). The ASME represents a raft of professional societies in the engineering community that is geared toward what is generally described as mechanical engineering. This society is composed of 32 technical divisions that are involved in very specific fields such as strength of materials, design engineering, production engineering, safety, lubrication, gas turbine, aerospace, petroleum engineering, applied mechanics, and a host of other specialized topics dealing with the mechanical technology.

ASME also has a very large codes and standards operations. When approved, these data are published by ANSI (American National Standards Institute). ANSI manages and coordinates the development of voluntary standards by a membership of about 900 companies and hundreds of technical, trade, professional, labor, and consumer organizations, all of which contribute expertise and financial support. ANSI represents U.S. interests in voluntary international standardization and serves as a clearinghouse for some 9000 approved standards.

UNIT 18 READING ENGINEERING DRAWINGS

There is a visual system of communication that binds the people involved in a technology. In the machine trades, the visual system is represented in engineering drawings. Ideas printed on engineering drawings are given to impart information, dimensions, sizes, positional relationship of the part, assembly, and other data necessary to make a part, machine, check, or transfer measurement.

Objectives
After completing this unit you should be able to

1. understand the role that ANSI plays in establishing standards.
2. understand the different types of orthographic drawings.
3. recognize the role played by the "alphabet of lines."
4. understand the different dimensioning and tolerancing terms used.

Prints

Reproduction of drawings mentioned in the introduction are referred to as prints and while they often appear as black lines on a white background, the lines may also appear blue, brown, or purple, depending on the paper used (Figure 18–1). Initially, these reproductions were called blueprints. The term "blueprint" is traditional, dating back to the days when drawings were printed in white lines on a blue background. The system and symbol representations remain fairly the same, however. Special lines, symbols, constructions, and dimensions are put on paper to render a mechanical drawing. This drawing generally acts as the skeletal representation of a finish part or construction. It is important to understand the symbols and construction, for without them the interpretation or reading of a print or drawing would not be possible. There would be no guidance or mutual understanding of the job required.

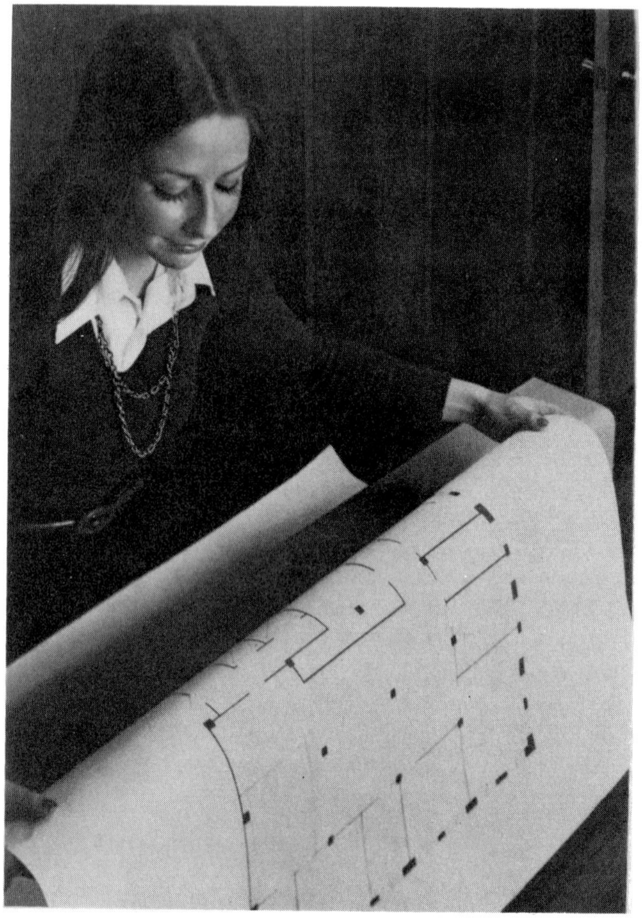

FIGURE 18-1
Reproduction of a print. (Credit Teledyne Post)

Types of Drawings

Draftpersons are largely responsible for inking or penciling the mechanical drawing. They use a variety of tools such as scales, triangles, protractors, French curves (plastic forms for connecting or drawing irregular contours), in addition to a set of drawing instruments to pursue the job of drafting. Special drawing paper is taped down on drafting boards or tables upon which layouts, details, and assemblies are penciled, erased, and redrawn if need be. So that lines can be drawn with a degree of accuracy, a T square is used. The T square is actually a two-part construction. It consists of a fixed head with a flat blade extending so as to form an included angle of 90°. The T square is used as the reference or working base for all construction that takes place. A drafting machine or a cable-guided straight edge replaces or eliminates the necessity of using a T square, protractor, triangle, or scale and saves a great deal of time in completing a drawing (Figure 18-2). Any number of types of drawings are possible in creating an engineering drawing. Only a few are described here.

Orthographic Drawing. Lines that extend at 90° from a surface in two or more planes so as to represent the exact shape of the object are said to be orthographic projections. Up to six views of a rectangular solid, for example, can be projected in this manner: front, top, right side, left side, bottom, and back. Orthographic drawings are the basis of engineering drawings, for they show a surface of an object in form and dimension as it really is.

One-View Drawing. Prints on which multiviews are not needed are represented in one view. The drawing of a uniform cylinder, ball shape, and very thin stock is often done in one view.

FIGURE 18-2
Using a drafting machine. (Credit: Teledyne Post)

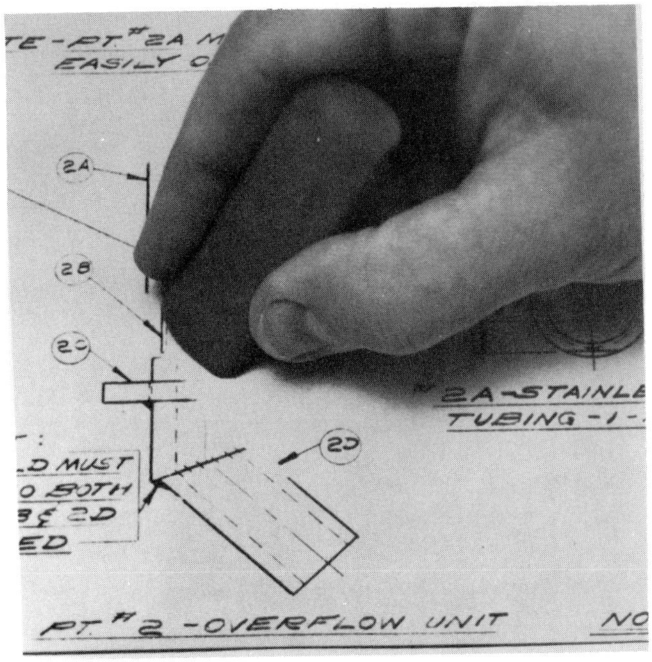

FIGURE 18-3
Correcting a multiview drawing. (Credit: Teledyne Post)

Two-View Drawing. Simple drawings depicting heavy flat objects or one-stepped cylinders, rods, pins, or bolts are completed in two views. Generally, a front view and a side view will suffice in cylindrical work, while a front view showing most of the features can be drawn with a top view to better understand a flat object.

Multiview Drawing. The three-view drawing is commonly referred to as a multiview drawing. The three most commonly used views are the top, front, and right-side view. They are evolved through orthographic projections. Sometimes the multiview might include three views and a little section view or it might be two views and an enlarged detail. A multiview drawing can contain four or five views too (Figure 18-3).

Scale Drawing. Drawings in which the object is drawn, larger or smaller, in proportion are said to be scale drawings. While the shape relationship between the drawings and the actual part varies, the dimensions listed have to be accurate (Figure 18-4).

Auxiliary Views. On prints where parts are difficult to show or describe, auxiliary views are added. This would include inclined or slanted surfaces that could not be shown or understood any other way. The auxiliary view is projected to show the true shape, size, and location of features on the surface (Figure 18-5).

Working Drawing. An engineering drawing that contains the correct shape, exact dimensions, and related information such as heat treatment, finish, and material specification necessary for the purpose of fabricating a part or an object is considered to be a working drawing (Figure 18-6). It may be a pictorial drawing or it may be an orthographic projection consisting of one or multiviews.

Assembly Drawing. A print that represents the coordinated position and how the parts are put together to form a machine or structure is said to be an assembly drawing.

Pictorial Drawing. Whether shaded or not, a pictorial drawing describes an object as it appears in one view while revealing two or three of its surfaces. Due to the nonexacting details, it is easily understood and is readily accepted by those who have no technical background (Figure 18-7).

Exploded View. A type of assembly drawing, the exploded view is a projection of a part that is sepa-

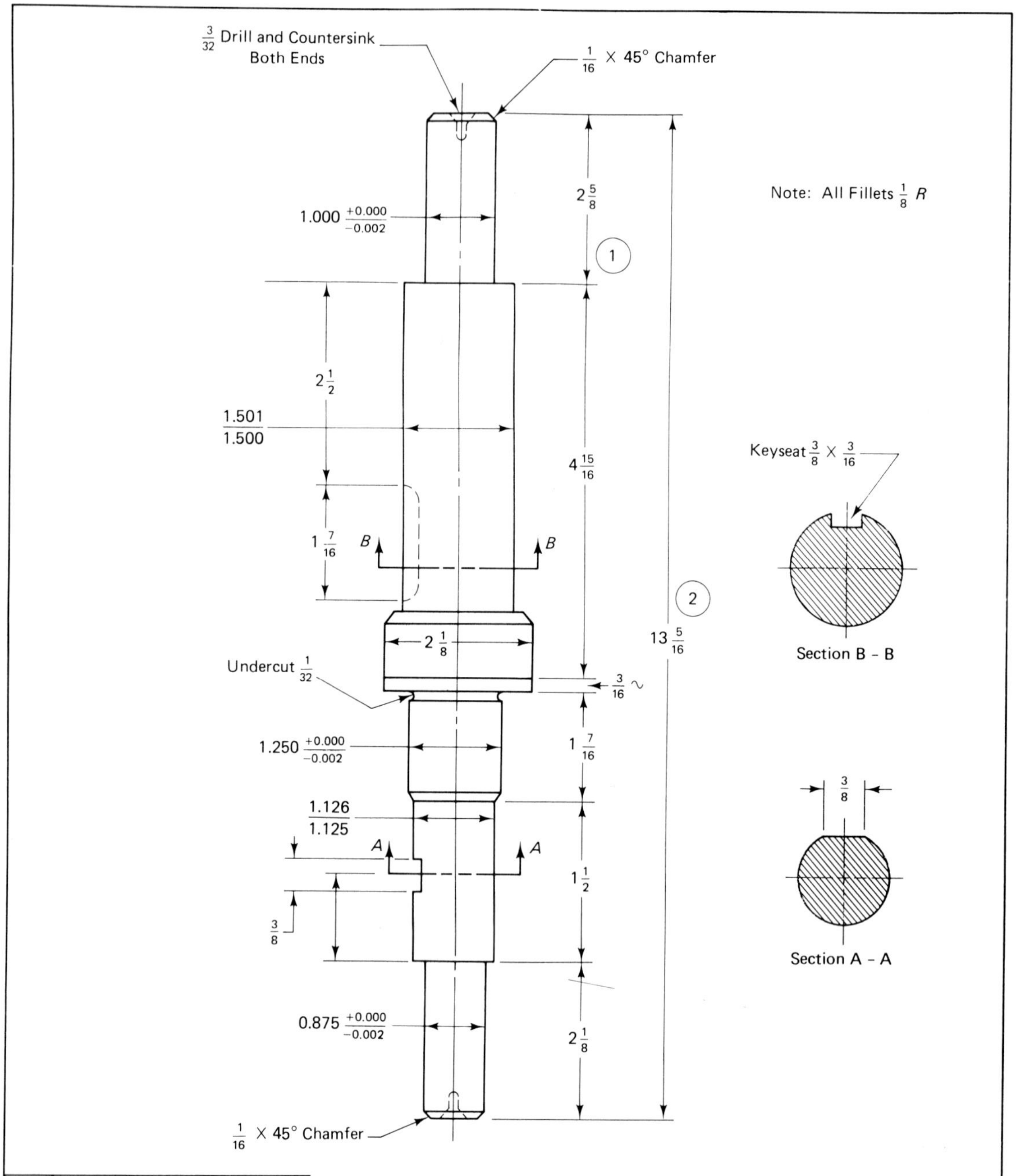

2	Was 13 1/16	9/22/80	
1	Was 1 5/8	9/22/80	
	Change	Date	By

Qty.	Order No.	Date	Dwg. By	Ck. By	App. By	
	16-172	8/27/80	Scale 6 = 12	Matl.	SAE 3135	H.T.

Unless otherwise specified: Limits on decimal dimension within ± 0.005″ Limits on fractional dimensions within ± 1/64″ Limits on angular dimensions within ± 1/2°	Spindle Shaft	
	Finish: Break All Sharp Edges	D-20

FIGURE 18–4
Scale drawing.

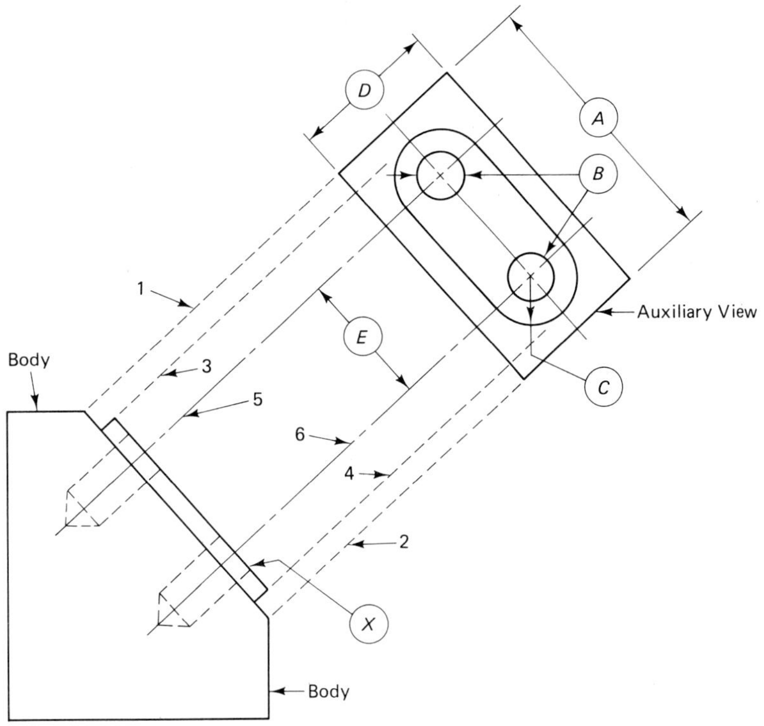

FIGURE 18–5
Auxiliary view.

rated or pulled away from the position it normally occupies in the pictorial drawing. It illustrates how devices, machines, and other appliances are assembled (Figure 18–8).

Section View. The representation of the internal shape and construction of a part that would be too difficult to show by conventional methods of hidden lines is shown by a section view. It is made by cutting through the part with an imaginary cutting plane and removing all in front of it to expose the internal details. The cut surface is identified by section lining (Figure 18–9).

Isometric View. Another type of pictorial drawing is the isometric view. In this view the object is drawn around three major lines called isometric base lines or axis. The base lines that extend right or left form an angle of 30° with the horizontal, while the vertical axis forms a 90° angle. A cube such as a die presented in this manner would appear tilted at the corner, exposing two adjoining numbered sides as well as the top numbered surface (Figure 18–10).

Types of Lines

The graphic language of visual communication depends heavily on the use of lines. Lines vary in thickness, length, and shape. In order to properly define a shape or an object, special lines, which have been adopted by the ANSI, are referred to as the "alphabet of lines" (Figure 18–11). These lines are used for specific reasons and should be understood to both interpret as well as complete a drawing. The major lines follow:

Visible line Sometimes called the object line, the visible line is used to form the outline of the part and is generally dark and thick.

Hidden line Sometimes referred to as an invisible edge line, a series of small dashes are used to denote an edge that is hidden or blocked from vision outside the object.

Centerline Used to form the center location of an object, the centerline is drawn as a thin line

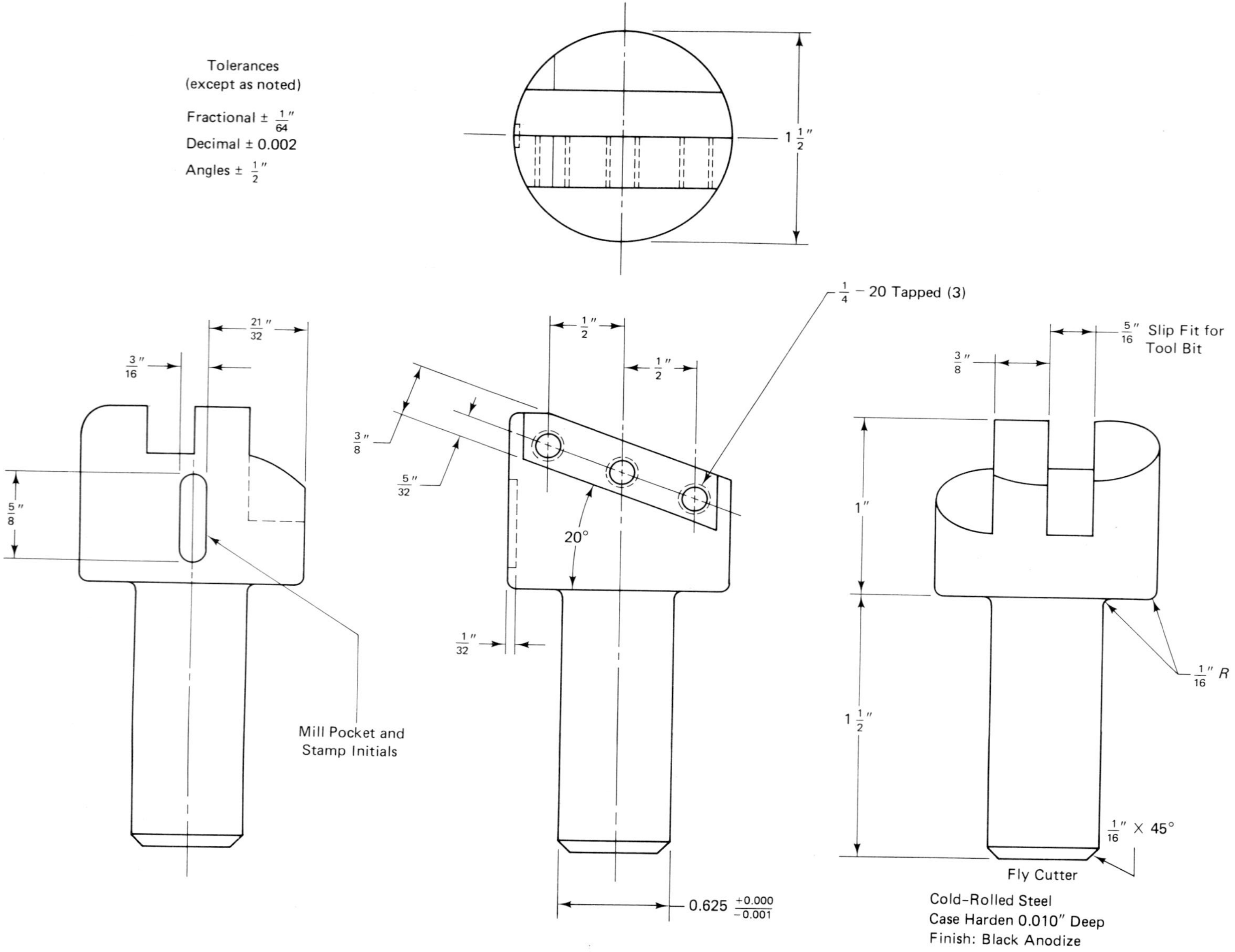

FIGURE 18–6
A working drawing. (Credit: Sight, Sound & Story)

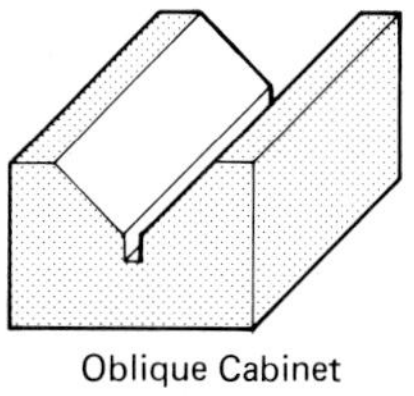

FIGURE 18-7
Pictorial drawing.

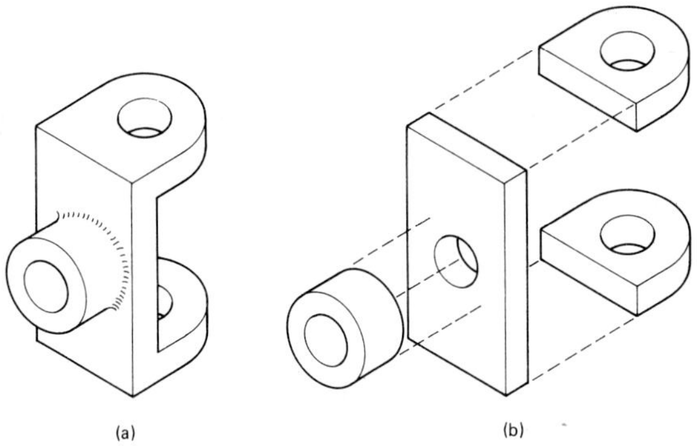

FIGURE 18-8
An exploded view.

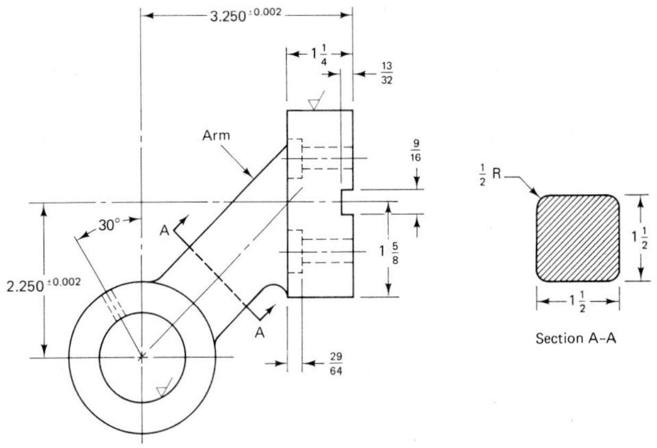

FIGURE 18-9
A section view.

and is composed of a repetitive long, short, long, short series of dashes.

Extension line Lines that are projected from the end of a surface but do not touch the surface are called extension lines. Between the extension lines is placed the exact dimension of the measurement involved.

Dimension line Lines that extend from the dimension toward the extension lines are called dimension lines. These thin, solid lines are capped with arrowheads and are normally drawn at right angles to touch the extension lines (Figures 18-12 and 18-13).

Leader line A thin, solid inclined line with a note or dimension on one end and an arrow or dot touching the part on the other is called a leader line (Figure 18-14).

Cutting plane line When the inside section of an object has to be explored and presented as if the section view is split open, a cutting plane line is drawn through the object. The cutting plane line, also called viewing plane line, may also change direction so as to favor a surface that would otherwise be neglected.

Section line A series of slanted lines across the surface exposed by a cutting plane line. It is also referred to as a cross-hatch line.

Phantom line Sometimes called repeat lines, phantom lines are drawn to indicate alternate positions or adjacent parts of an object or the location of parts not drawn or where material is to be removed. These lines are made by drawing two evenly spaced short dashes that are straddled by evenly drawn long dashes on either side.

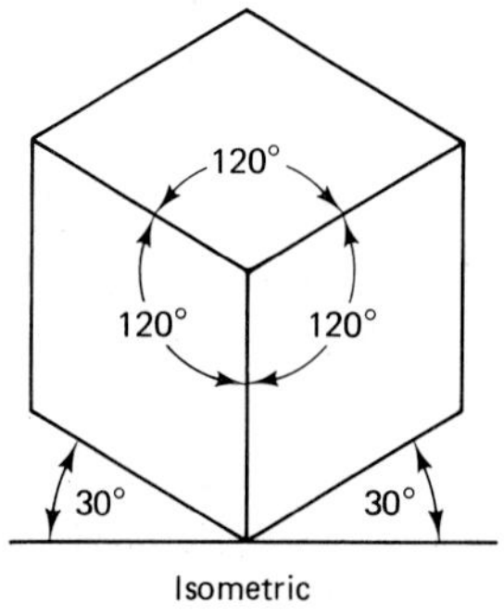

FIGURE 18-10
An isometric view.

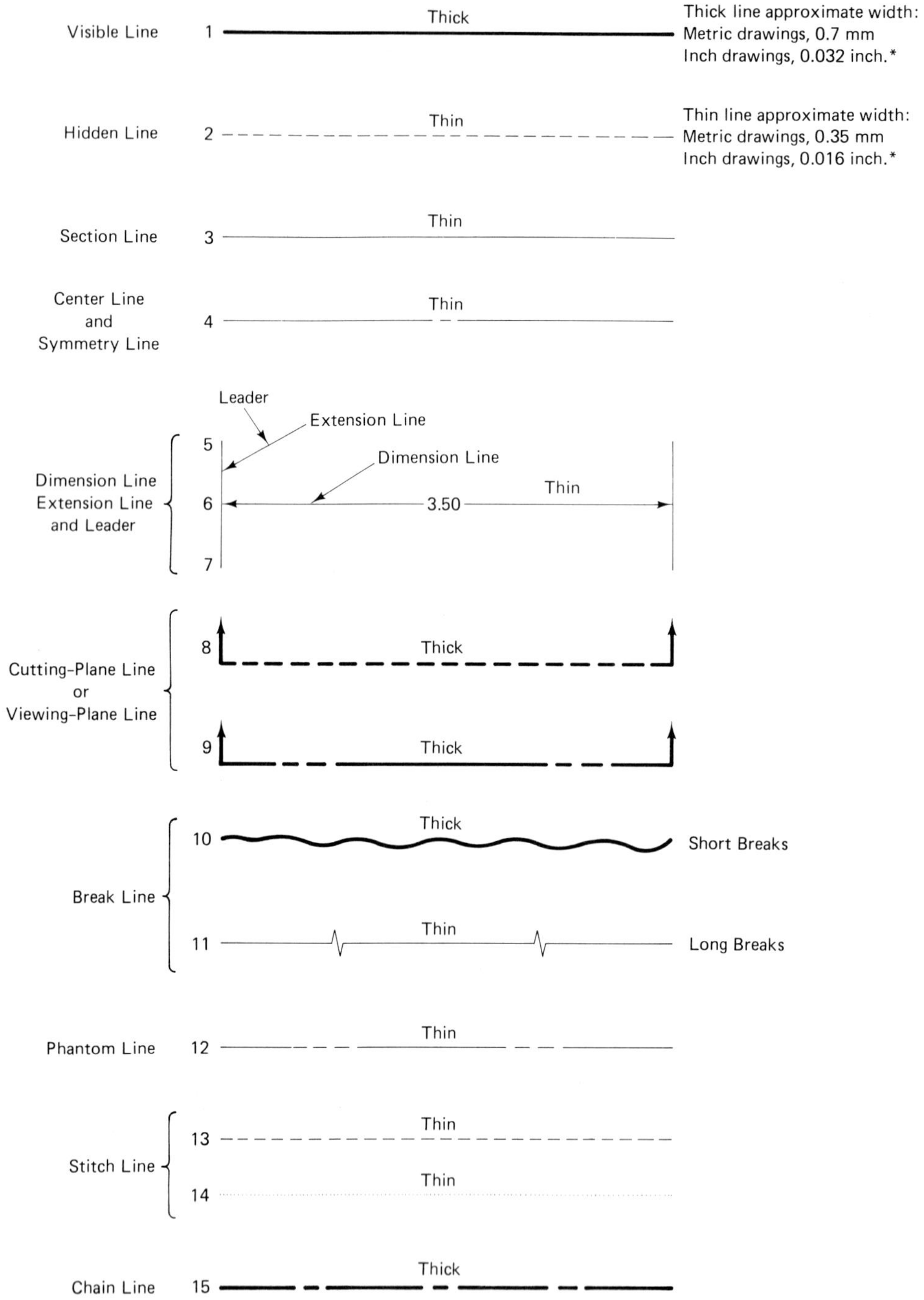

*The metric line widths agree with ISO/DIS/128 (June 1977) and are not a soft metric conversion of the inch value.

These approximate line widths are intended to differentiate between THICK and THIN lines and are not values for control of acceptance or rejection of the drawings.

FIGURE 18–11
The alphabet of lines. (Credit: ANSI Y14.2M-1979)

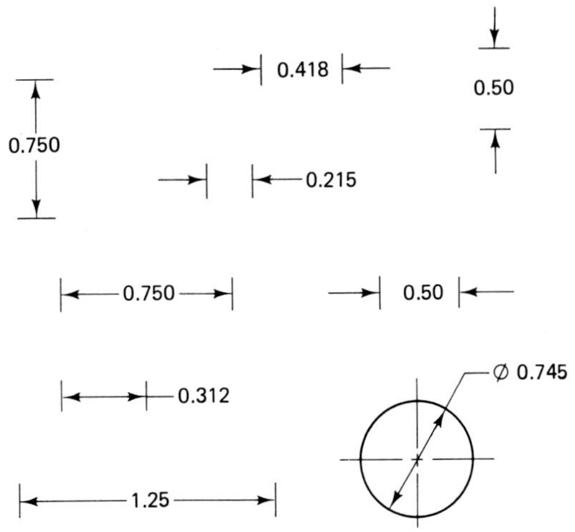

FIGURE 18–12
Examples of decimal inch dimensioning. (Credit: ANSI Y14.5M-1982)

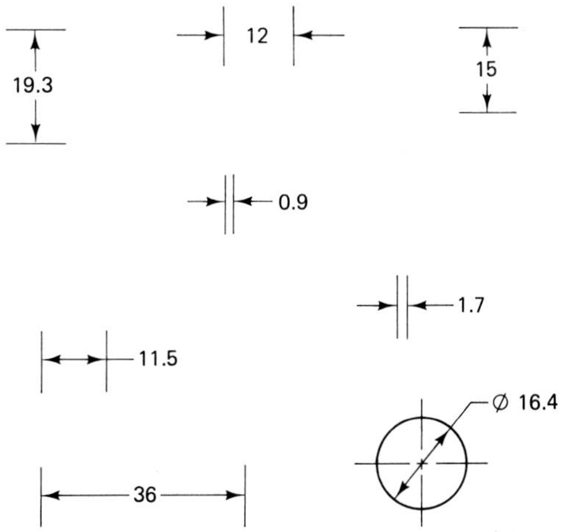

FIGURE 18–13
Millimeter dimension lines and values. (Credit: ANSI Y14.5M-1982)

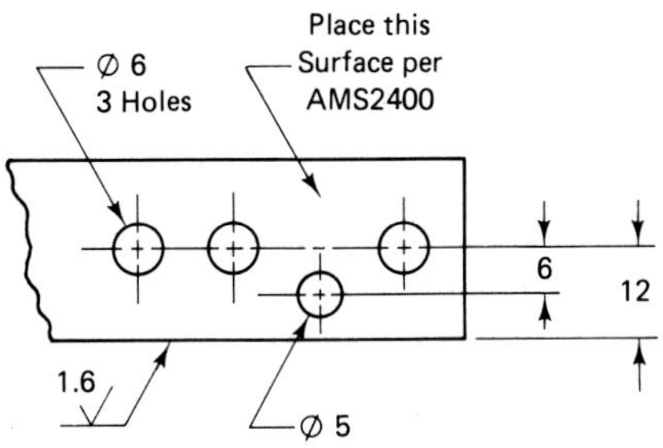

FIGURE 18–14
Leaders. (Credit: ANSI Y14.5M-1982)

Break lines The representation of long objects such as rectangular or round sections, as well as rolled shapes that would not fit on paper if it were drawn full size, is effectively shortened with the use of long and short break lines.

Engineering Drawing Definitions

Since it would be very costly and time consuming to manufacture parts to exact dimensions, interchangeability demands that parts be produced within specific limitations. There are terms connected with dimensioning that should be understood if these goals are to be met. The following manufacturing variations are based on terms used by ANSI:

Dimension A measurement of size and/or geometric design, or both, based on an appropriate numerical value.

Basic dimension The theoretical size from which variations are made.

Nominal size A term used for general identification, e.g., $\frac{1}{4}$-in. tubing.

True position The theoretical exact location of a surface or object created by basic dimensions.

Reference dimension A dimension used as auxiliary information. Since it may be a repeat dimension or derived from other values on direct or related drawings, it is not intended to influence production or inspection processes.

Datum A theoretical exact point, axis, or plane of reference forming the origin from which the location and dimensions are taken and geometric calculations are made.

Feature The representation of a part, such as a surface, groove, slot, hole, or similar physical characteristics.

Actual size The measured size.

Limits of size The permissible maximum and minimum sizes for producing a part. Often referred to simply as "limits," the maximum size is called the upper limit. The minimum size is called the lower limit (Figure 18–15).

Allowance The intentional difference in dimensions between mating parts with regards to intended minimum clearance or maximum interference (Figures 18–16 and 18–17).

Tolerance The amount of variation allowed between the upper and lower limits (Figure 18–18).

Unilateral tolerance The total tolerance in one direction, as related to a basic dimension, being either a plus or a minus value.

Bilateral tolerance The total tolerance in two directions as composed of plus and minus dimensions.

Geometric tolerances A general term applied to the listing of tolerances used to control shape, size, form, profile, positioning, location, and runout.

Fit The relationship of tightness between mating parts when assembled with regards to existing interference or clearance.

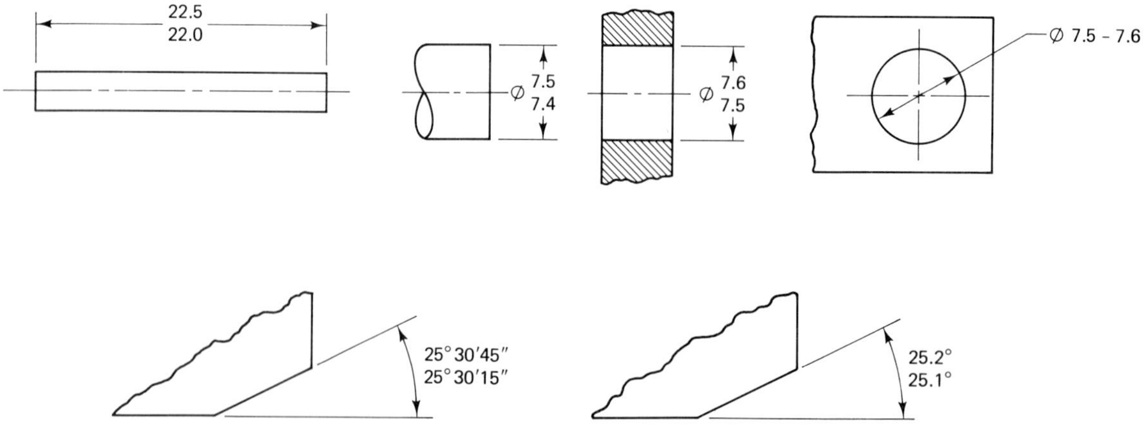

FIGURE 18–15
Limit dimensioning. (Credit: ANSI Y14.5M-1982)

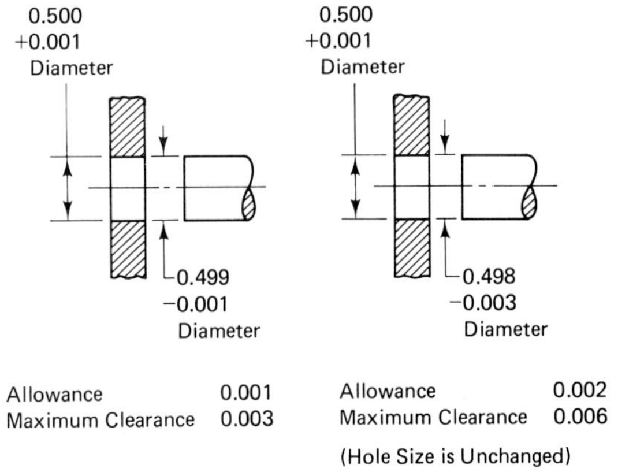

FIGURE 18–16
Basic hole allowance. (Credit: ANSI Y14.5M-1982)

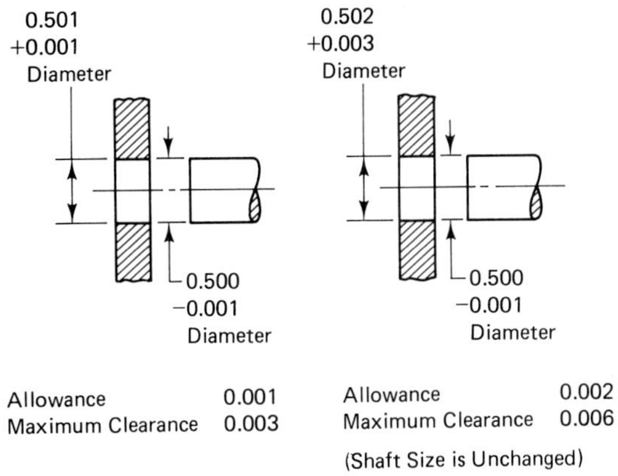

FIGURE 18–17
Basic shaft allowance. (Credit: ANSI Y14.5M-1982)

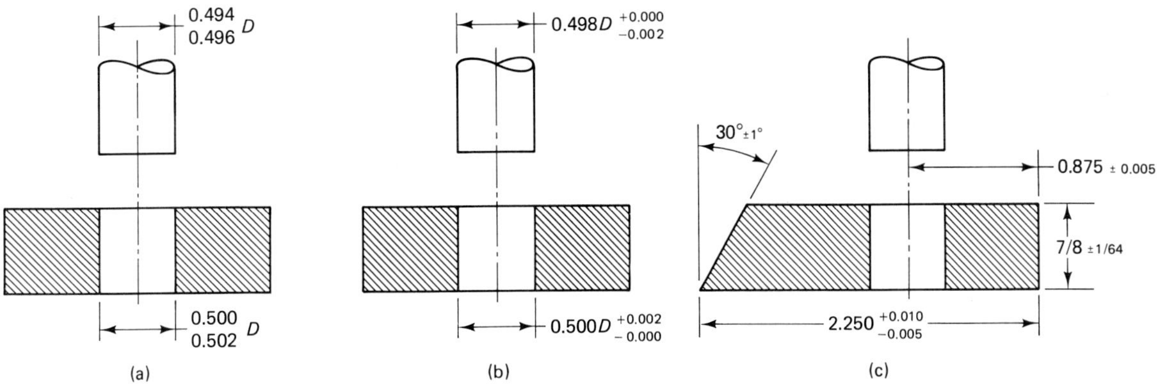

FIGURE 18–18
Methods of general tolerance dimensioning.

Clearance fit Also known as positive allowance, clearance fit is the free space, movement, or "play" between mechanisms such as a shaft and hole.

Interference fit Negative allowance that allows parts to be pressed fit together. It is also known by the term "press fit."

Maximum material condition (MMC) A reference that exists when the feature contains the maximum amount of material within the limits of size, e.g., high limit dimension on a shaft diameter, a low limit on the hole dimension (Figure 18–19).

Least material condition (LMC) A reference that exists when the feature contains the minimum amount of material within the limits of size, e.g., high limit dimension on a hole diameter, the low limit dimension on a shaft diameter.

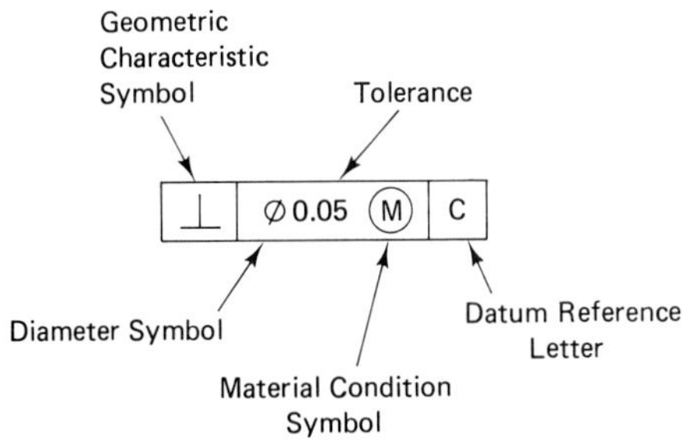

FIGURE 18–19
MMC on a print is designated by the symbol M. (Credit: ANSI Y14.5M-1982)

Regardless of feature size (RFS) A term that indicates that a geometric tolerance or form must be met regardless of where the feature lies within its size tolerance.

Full indicator movement (FIM) At one time known as TIR (total indicator reading) and FIR (full indicator reading), FIM is an indication of measurement variation that is read on the dial face of an indicator when the measuring device is put in contact against the full length or diameter of a work surface.

Dual dimensioning A dimensioning system whereby both the inch and the metric units of measurement are shown on the same engineering drawing. Two methods are widely used in practice, the position method (Figure 18–20) and the bracket method (Figure 18–21).

Graphic Symbols

Equally as important are the many symbols and technical terms used as visual shortcuts. Many industries have their own set of symbols — the electrical, welding, building, and plumbing, to name but a few. One of the symbols used on engineering drawings to denote material composition, section lining, is listed in Figure 18–22.

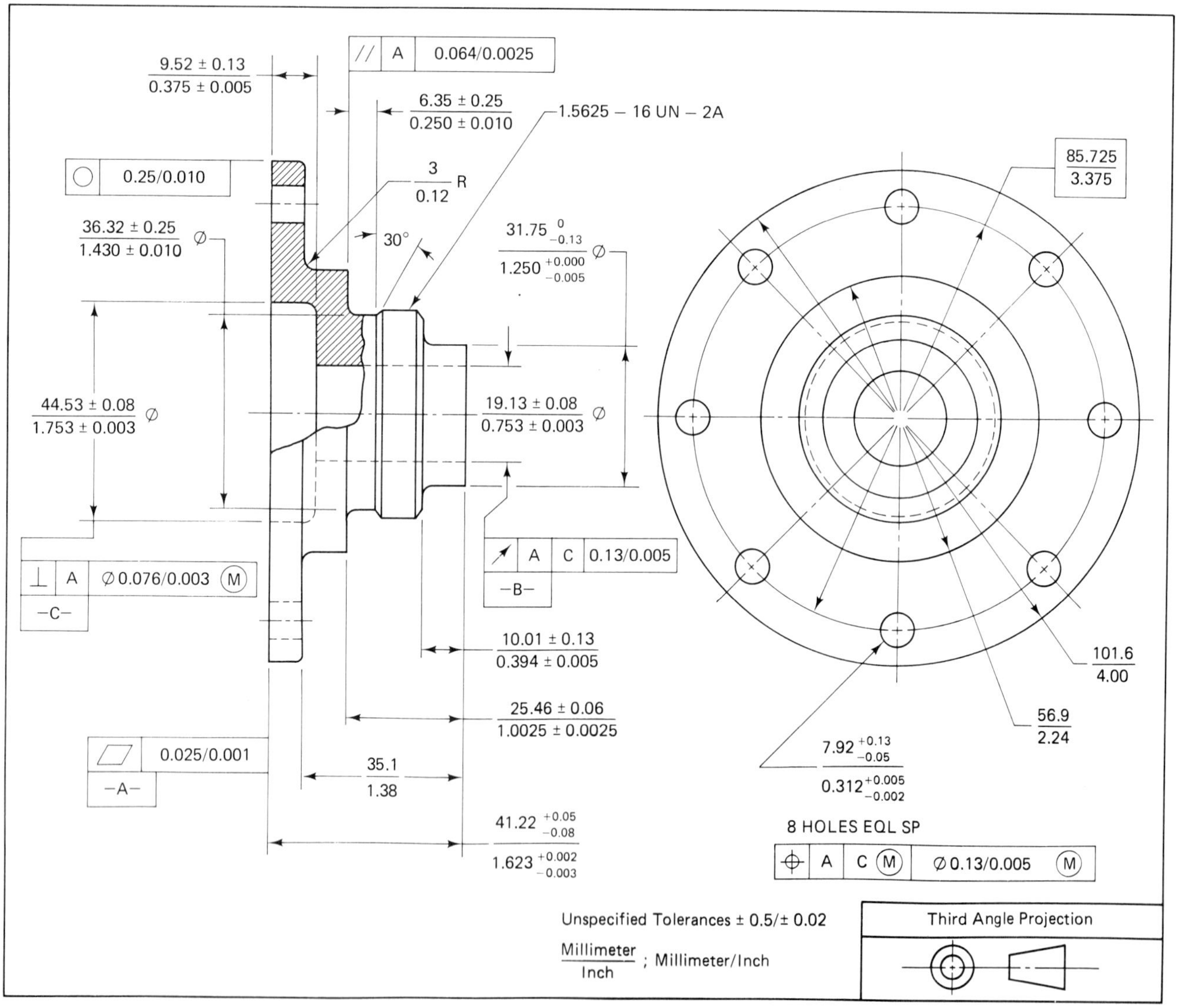

FIGURE 18–20
Dual dimensioning using the position method. (Credit: ANSI Y14.5M-1982)

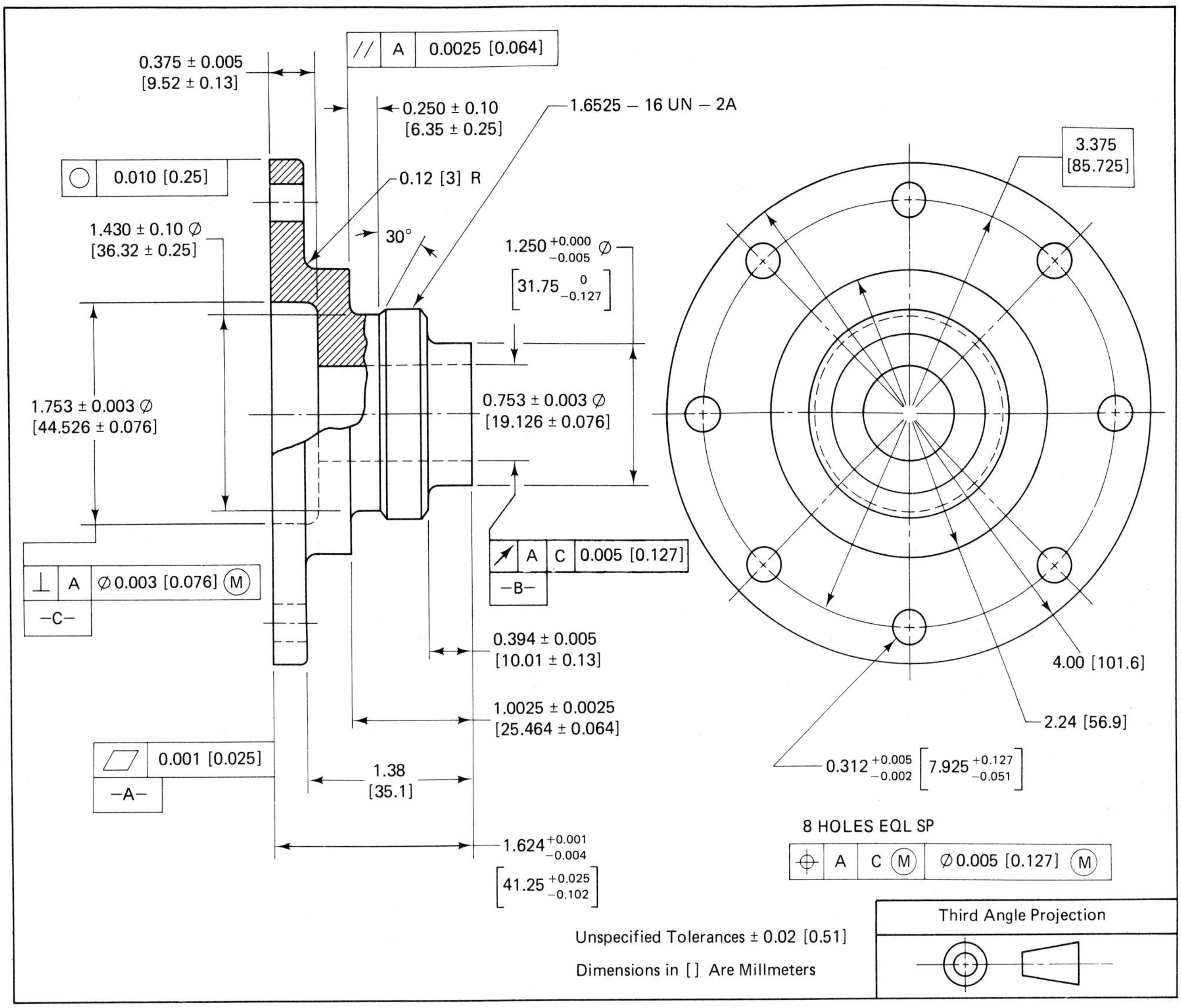

FIGURE 18–21
Dual dimensioning using the bracket method. (Credit: ANSI Y14.5M-1982)

Another set of symbols is used to indicate geometric characteristics (Figure 18–23). These symbols may be used singularly or in connection with other symbols and notes to

1. detail form, position parts, or align assemblies.
2. control and identify information concerning tolerancing and dimensions.
3. establish part features for interchangeability.
4. establish standard tolerances if not given.

What sets this system apart is the fact that its geometric dimensions are tied to the basic dimensions and datum features. As previously explained, the basic dimension is the theoretical size from which variations are made. Datum is the theoretical exact point, axis, or plane of reference forming the origin. In other words the system is based on theoretical exactness, not general tolerancing.

Datum features are referenced by a letter of the alphabet (Figure 18–24) or in combination of letters except for the letters *I*, *O*, and *Q*, which are easily misread. A set of modifying symbols (Figure 18–25)

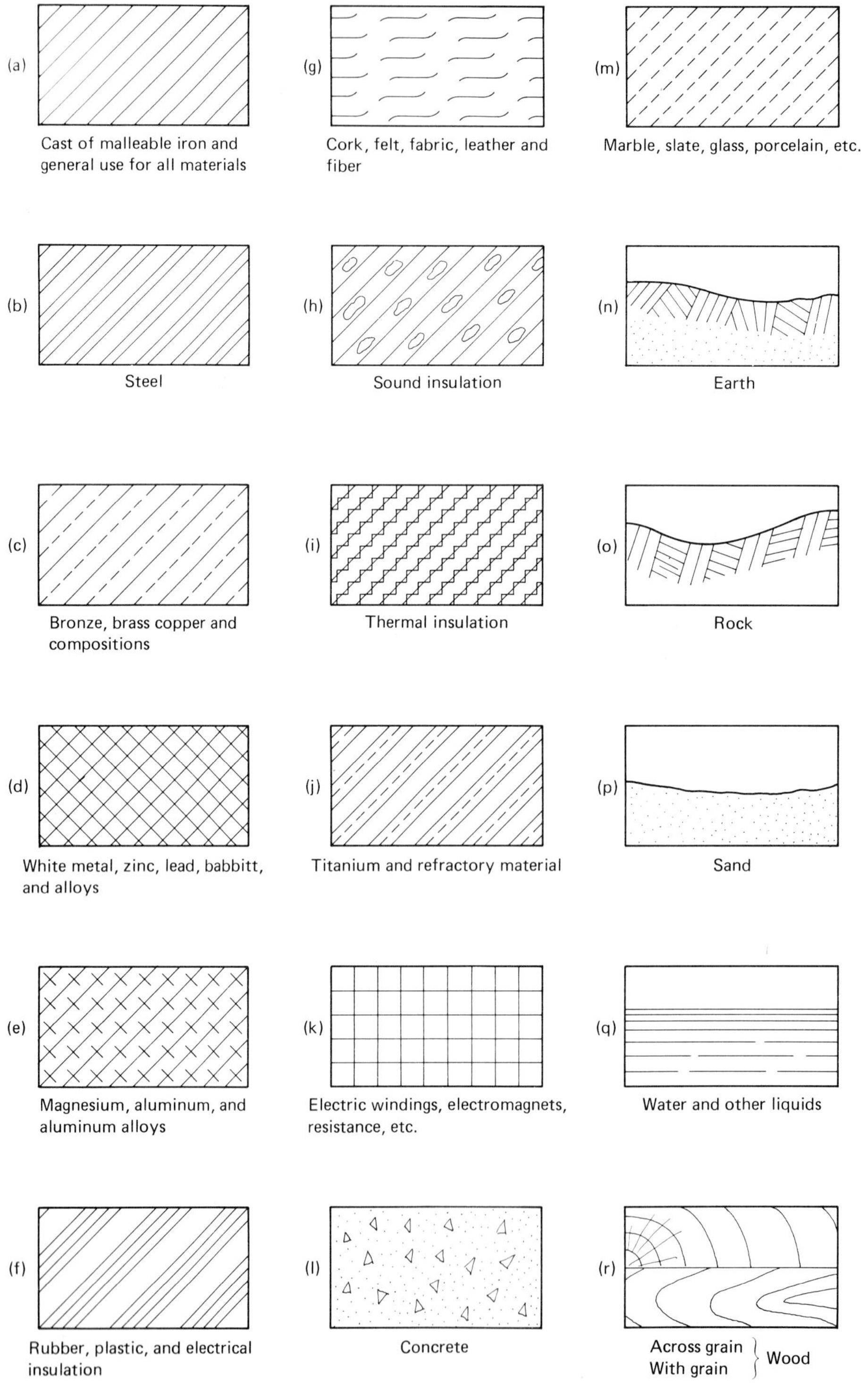

FIGURE 18–22
Section lining symbols. (Credit: ANSI Y14.2-1979)

	Type of Tolerance	Characteristic	Symbol	See:
For Individual Features	Form	Straightness	—	6.4.1
		Flatness	▱	6.4.2
		Circularity (Roundness)	○	6.4.3
		Cylindricity	⌭	6.4.4
For Individual or Related Features	Profile	Profile of a Line	⌒	6.5.2 (b)
		Profile of a Surface	⌓	6.5.2 (a)
For Related Features	Orientation	Angularity	∠	6.6.2
		Perpendicularity	⊥	6.6.4
		Parallelism	//	6.6.3
	Location	Position	⌖	5.2
		Concentricity	◎	5.11.3
	Runout	Circular Runout	↗ *	6.7.2.1
		Total Runout	⌰ *	6.7.2.2

*Arrowhead(s) may be filled in.

FIGURE 18-23
Geometric characteristics symbols. (Credit: ANSI Y14.5M-1982)

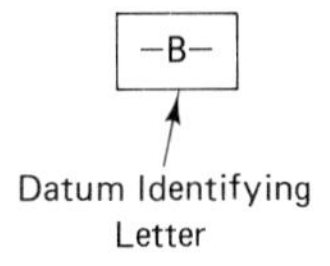

FIGURE 18-24
Datum feature symbol. (Credit: ANSI Y14.5M-1982)

Term	Symbol
At Maximum Material Condition	Ⓜ
Regardless of Feature Size	Ⓢ
At Least Material Condition	Ⓛ
Projected Tolerance Zone	Ⓟ
Diameter	∅
Spherical Diameter	S∅
Radius	R
Spherical Radius	SR
Reference	( )
Arc Length	⌒

FIGURE 18-25
Modifying symbols. (Credit: ANSI Y14.5M-1982)

can also be incorporated to convey additional information such as radius, diameter, arc length, etc. A feature control frame (Figure 18-26), consisting of frames of information where the geometric characteristic, permissible tolerance, modifying symbol, and datum reference letters are stored to make the system highly visible. Examples of this type of tolerancing can be seen in Figures 18-27 through 18-29. Further information is available in *Dimensioning and Tolerancing,* ANSI Y14.5M-1982.

Technical Terms

Technical terms, such as the standard abbreviations listed in Figure 18-30, are found on the print as well as in the title box. The drawing's title box generally appears in the lower right-hand corner and varying amounts of space are blocked off (Figure 18-31). Within each block different bits of information are provided. Besides the title of the part, date drawn, and by whom, other pieces of information recorded might include the choice of material, stock size, heat treatment, and scale of drawing.

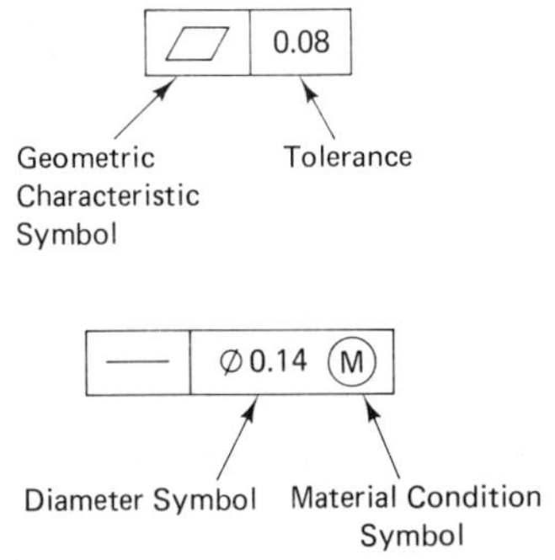

FIGURE 18-26
Feature control frame. (Credit: ANSI Y14.5M-1982)

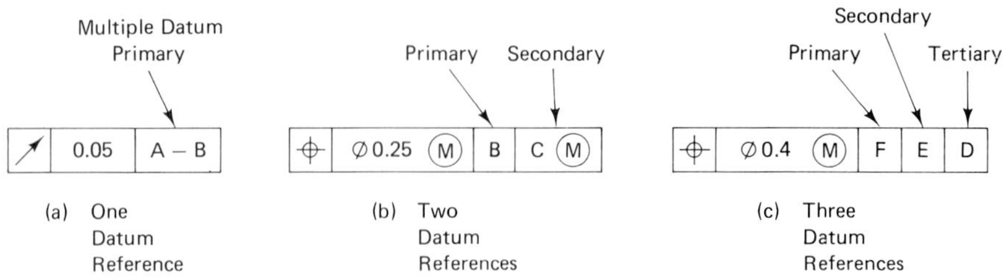

FIGURE 18–27
Order of precedence of datum references. (Credit: ANSI Y14.5M-1982)

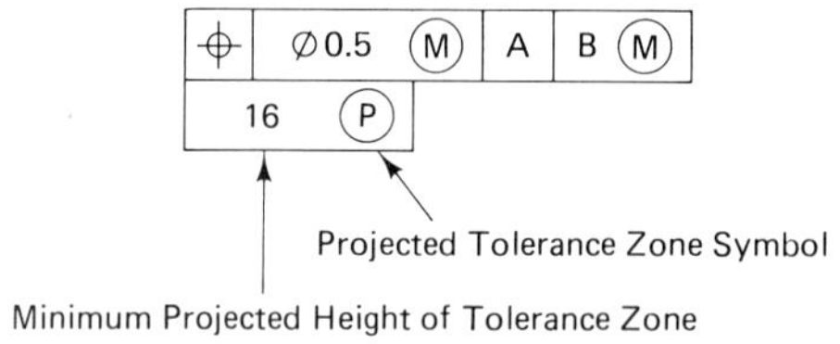

FIGURE 18–28
Feature control frame with a projected tolerance zone. (Credit: ANSI Y14.5M-1982)

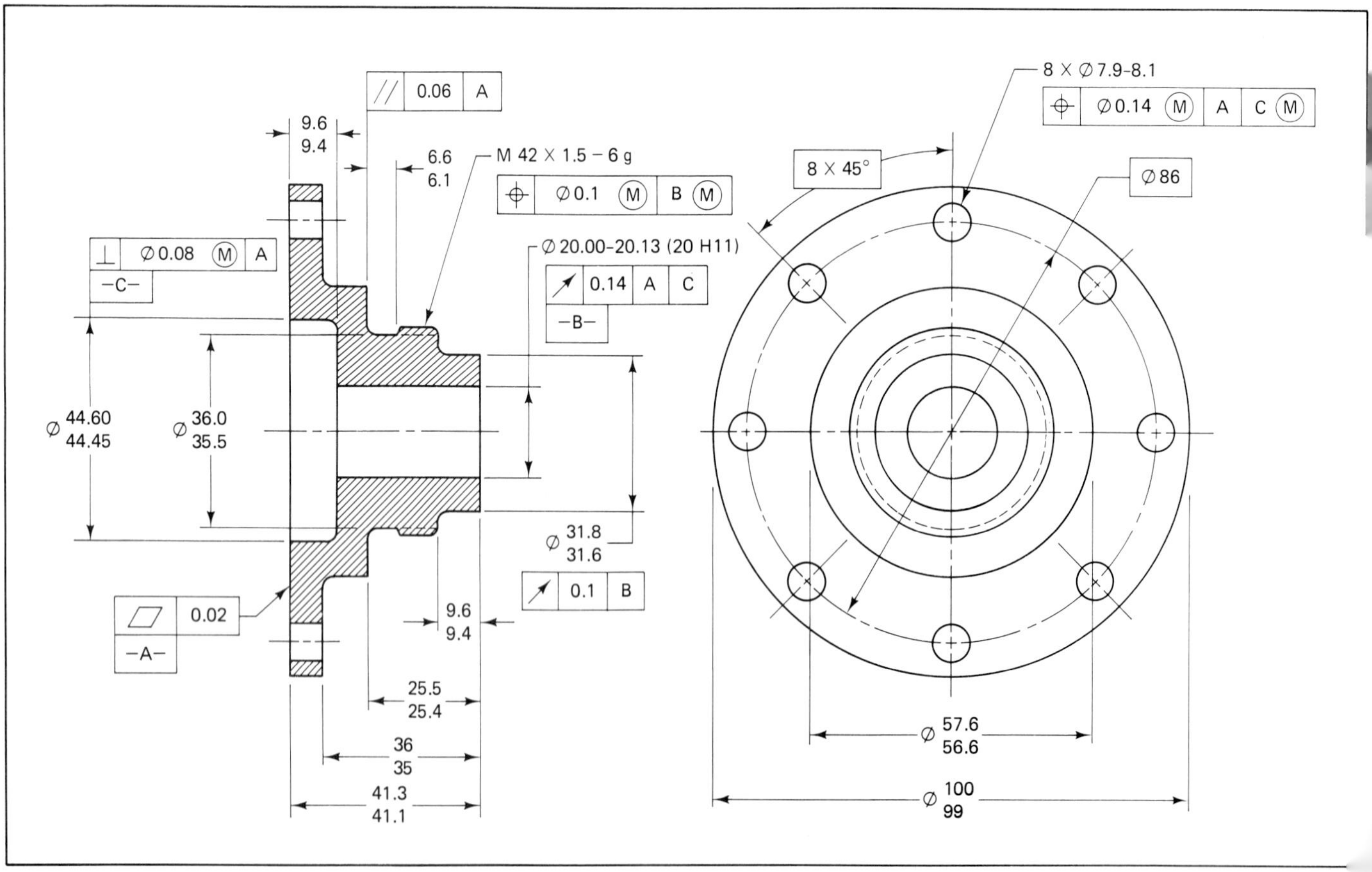

FIGURE 18–29
Feature control frame placement. (Credit: ANSI Y14.5M-1982)

Acme screw thread	ACME	Length	LG
Allowance	ALLOW	Lubricate	LUB
Alloy steel	ALY STL	Machine steel	MST
Aluminum	AL	Magnesium	MAG
Anneal	ANL	Malleable	MAL
Approved	APVD	Malleable iron	MI
As required	AR	Material	MATL
Bevel	BEV	Maximum	MAX
Bracket	BRKT	Millimeter	MM
Brass	BRS	Minimum	MIN
Brazing	BRZG	National coarse thread	NC
Break	BRK	National extra fine thread	NEF
Brinell hardness number	BHN	National fine thread	NF
Bronze	BRZ	Nickel steel	NS
Burnish	BNSH	Nominal	NOM
Bushing	BSHG	Not to scale	NTS
Case harden	CH	Number	NO
Cast iron	CI	Outside diameter	OD
Center	CTR	Plastic	PLSTC
Chamfer	CHAM	Quantity	QTY
Chrome vanadium	CR VAN	Radius	R or RAD
Clearance	CL	Ream	RM
Cold-drawn steel	CDS	Reference	REF
Cold-rolled steel	CRS	Required	REQD
Copper	COP	Right hand	RH
Counterbore	CBORE	Round	RD
Countersink	CSK	Screw	SCR
Cylinder	CYL	Society for Automotive Engineers	SAE
Degree	DEG	Spotface	SF
Diagonal	DIAG	Square	SQ
Diameter	DIA	Standard	STD
Dimension	DIM	Stock	STK
Drawing	DWG	Straight	STR
Drill	DR	Surface	SUR
Fabricate	FAB	Symbol	SYM
Fillet	FIL	Tangent	TAN
Finish	FIN	Tapping	TAP
Finish all over	FAO	Tensile strength	TS
Flange	FLG	Thick	THK
Forged steel	FST	Thread	THD
Gauge	GA	Threads per inch	TPI
Galvanized	GALV	Tolerance	TOL
Grind	GRD	Tool steel	TS
Harden	HDN	Tungsten	TU
Heat treatment	HT	Typical	TYP
High-carbon steel	HCS	Unified national coarse	UNC
Hot-rolled steel	HRS	Unified national extra fine	UNEF
Inch	IN	Unified national fine	UNF
Inside diameter	ID	Volume	VOL
Keyway	KWY	Width	W
Laminate	LAM	Wrought iron	WI
Left hand	LH		

FIGURE 18–30
ANSI standard abbreviations.

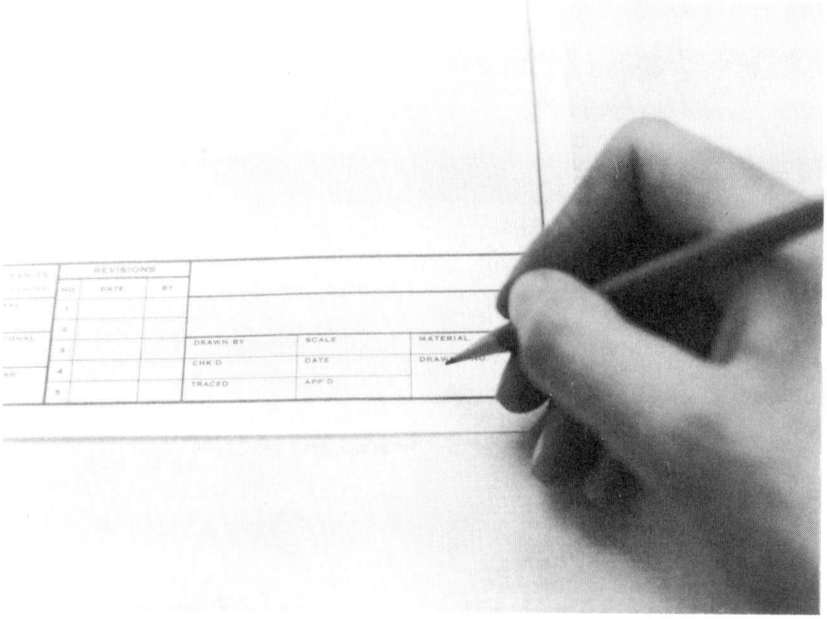

FIGURE 18-31
Title box.

SECTION REVIEW

Unit 18

1. What does ASME represent and what is its function?
2. What does ANSI represent and what is its function?
3. Briefly describe what an orthographic drawing is.
4. How is a working drawing defined?
5. What type of drawing would be used to represent the internal section and construction of a part that would be too difficult to draw?
6. How does an (a) isometric view differ from a (b) pictorial view?
7. What is another term for the following: (a) visible line, (b) hidden line, (c) cutting plane line, and (d) section line?
8. Define the following terms: (a) basic dimension, (b) limits of size, (c) tolerance, (d) dual dimensioning, (e) bilateral tolerance, and (f) datum.
9. Describe the information that might be found in the title box of drawing.
10. Identify the following ANSI standard abbreviations: BHN, CDS, CBORE, CSK, HDN, IN, QTY, RH, THD, and TYP.

Section 5

SAFETY

UNIT 19 WORKING SAFELY

Every year thousands of lives are lost, people are maimed, or injured due to accidents caused by carelessness, thoughtlessness, or lack of concern. Safety is a combination of common sense, good judgment, and mutual respect. Any shop has all sorts of situations that can lead to people being hurt. Yet people work in metal shops everyday and never get seriously hurt. Why? Because they are able to work with positive work habits that free them from potential dangers. Essentially, safety is a matter of mutual concern and respect for yourself, your fellow worker, and the equipment you will be using. Without this foundation, a suit of armour would not be sufficient to protect you (Figure 19–1).

Objectives
After completing this unit you will be able to

1. recognize potential work hazards.
2. identify good work habits.
3. understand how to work more safely.

Eye Protection

When a person goes to work in a shop, all of the person has to be protected. It goes without saying that the eyes need every bit of protection they can get. Imagine how long a welder would last without a protective shield. Not only is the shield provided with a special glass to screen out harmful rays, but it also prevents the welder from being hit in the eyes with the molten metal. The same thing would hold true for any person involved in brazing. Special goggles have to be worn in brazing.

FIGURE 19–1
Without positive work habits of safety, a suit of armour could not protect you in a metal shop.
Credit: Paul Rapisarda)

FIGURE 19-2
Safety glasses with side shields offer good eye protection in any shop.

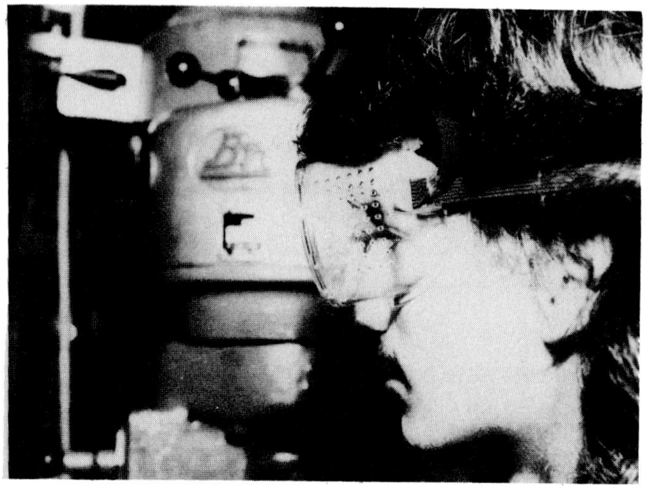

FIGURE 19-4
Goggles offer eye protection and can be worn over prescription glasses.

FIGURE 19-3
In addition to eye protection a face shield can also protect the wearer against fumes and splashes.

The eyes should also be protected from blown waste or other types of metal fragments, broken tools, hot sparks, and other minute particles of solid matter that not only can irritate or injure eyes but can also lead to visual impairment and/or blindness. The cardinal rule in any metal shop: Always wear eye protection while working or visiting the area. Your eyes are too valuable to be wasted.

Eye protection in any shop will fall in one of three basic types:

1. Safety glasses with side shields (Figure 19-2) are the type most widely recommended. Not only should the lens be able to withstand shock if dropped on the floor, the lens must be able to resist the penetration of a steel ball shot if it hits with force without shattering the lens. Regular prescription glasses may be shockproof but they are not shatterproof and do not qualify as safety glasses. Safety glasses are designated as such by the manufacturer or bear the ANSI safety code requirement label somewhere on the lens or frame.

2. A face shield (Figure 19-3) can also be used to protect the face as well as the eyes. This type is also highly recommended for jobs that deal with fumes or splashes of any nature.

3. Goggles (Figure 19-4) are a third recommended type that fit over prescription glasses if need be and still give the wearer the protection required.

Personal Safeguards

Safety hazards will fall into a number of categories. High on the list would include the individual's mode of dress. It requires working in something besides jeans.

FIGURE 19-5
All jewelry should be removed by anyone working in a machine shop.

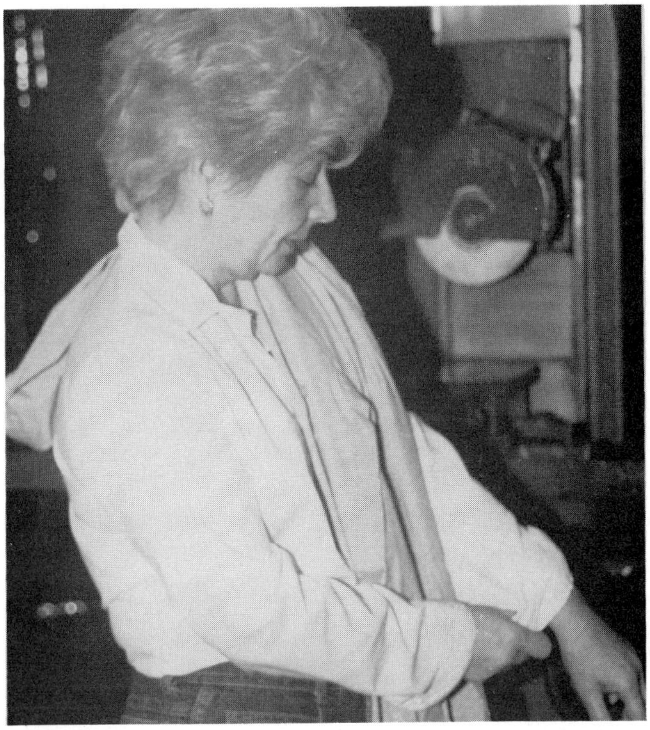

FIGURE 19-6
Long sleeves should be rolled up above the elbow.

1. Remove all jewelry, rings, and wrist watches (Figure 19-5). They can get easily caught, tearing or mangling arms and fingers in the process.

2. Long hair should be neatly tied or tucked under a cap or hair net. Long hair is an invitation to having it painfully pulled out of the scalp by machinery.

3. Beards and other facial hairs should be neatly trimmed and shortened to avoid having them caught in rotating spindles and workpieces.

4. Loosely woven or loose fitting clothing such as sweaters, exposed shirt tails, or neck ties should not be worn at all. Long sleeves should be avoided; but if worn, should be rolled up above the elbow (Figure 19-6).

5. Shop aprons must be properly strung and neatly tied in back with no loose ends.

6. Shop coats should not be baggy and must not have buttons or sleeves that can get easily caught (Figure 19-7).

7. While gloves may be permitted transporting sharp-edged materials, they should never be worn operating machinery. Gloves can easily get caught or be drawn into moving parts or mechanisms, compressing or mangling hands and fingers in the process.

8. Safety shoes are recommended when working in a metal shop. The steel toe insert (Figure 19-8) provides a great deal of protection against seen and unseen falling objects that can damage, maim, or in-

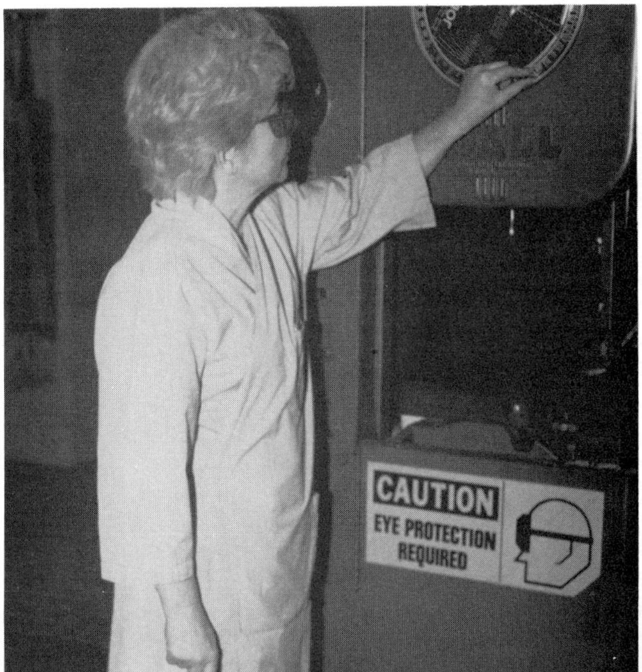

FIGURE 19-7
Shop coats or aprons must not be baggy or have buttons or loose ends that can get caught.

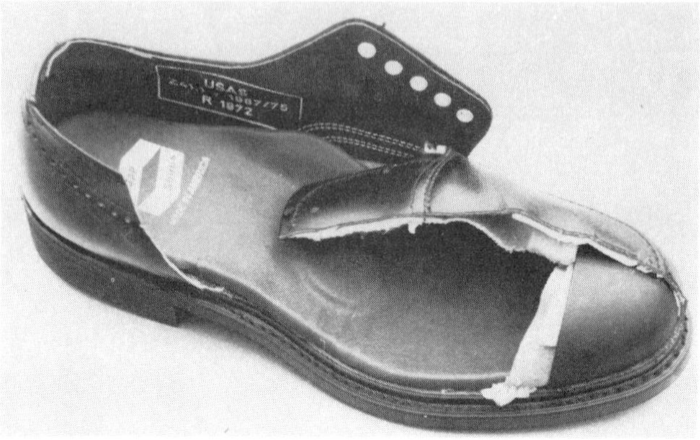

FIGURE 19-8
Safety shoes provide protection against hard or falling objects.

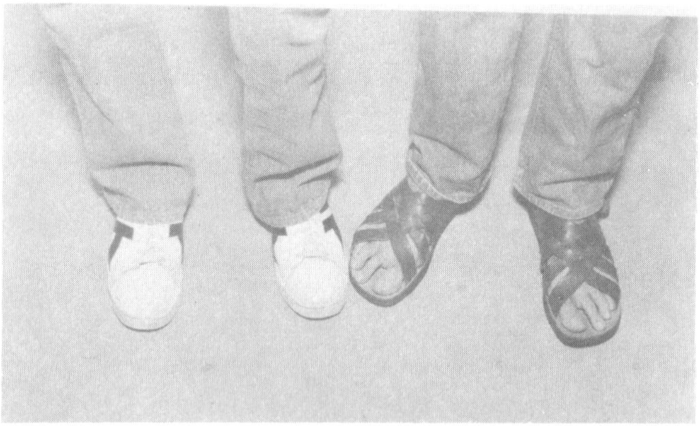

FIGURE 19-9
Sneakers or sandals should never be worn in a metal shop. They offer no protection against sharp chips or falling objects.

jure the foot. They are just as important as any safety apparel worn. Sneakers or sandals (Figure 19-9) should never be worn as chips can easily penetrate this type of footwear.

9. Protective clothing must be worn when working in welding, forging, or foundry areas (Figure 19-10).

10. Hearing should be protected against high-volume or high-intensity noise levels. In areas of this nature, ear protectors should be worn (Figure 19-11).

Electrical Safeguards

1. All electrical cords should be inspected for damage. Frayed or loose wiring (Figure 19-12) should be reported to your supervisor.

2. Know where all safety power switches are and how to use them if need be.

3. Whenever using any electrical switches, always handle them by the insulated portion, avoiding contact with any surrounding metal surface.

4. Electrical tools are a hazard around flammable gasses and vapors. Never use such tools under these conditions as it may cause an explosion.

5. Never stand in water when using electrical tools.

Fire Safeguards

1. Always store oily rags or combustible waste materials in the proper metal container after use.

2. Always close containers of paints, oils, and other flammable substances and return them to the proper metal storage area.

3. Learn the proper method of lighting a furnace or a gas torch.

4. Know where every fire extinguisher is located and how to use it in an emergency.

5. Know the location of the nearest fire alarm box and how to use it.

6. Know the location of the nearest fire exit and the procedure to follow in case of fire.

Machinery Safeguards

1. Do not attempt to run any machine unless supervised or allowed to by your instructor.

2. Never attempt to operate or machine work without using guards or shields to deflect against waste or collect chips.

3. Always make sure that safety guards, such as pulley covers, are in place before attempting to run a machine.

4. No machine should be run by two people. Only one person should control the operating switch.

5. Avoid touching surfaces of moving parts or moving machinery.

6. Do not lean on or against a machine someone else is operating.

FIGURE 19–10
Protective clothing must be worn when welding. (Credit: Lincoln Electric Company)

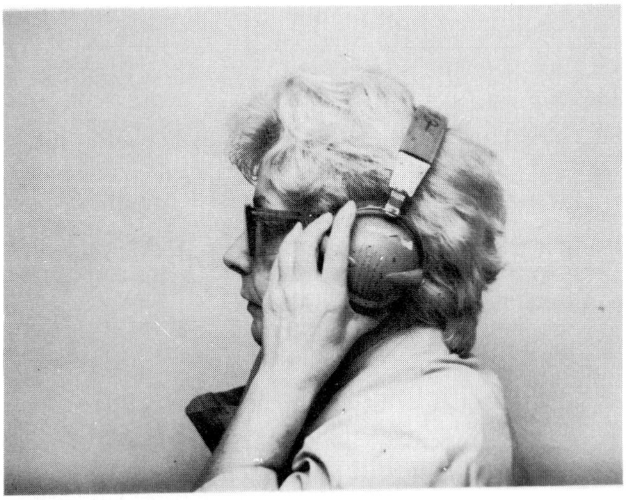

FIGURE 19–11
Hearing should be protected against noise pollution.

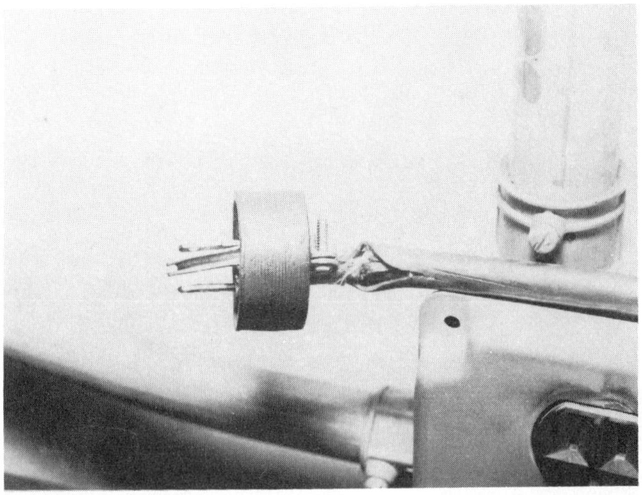

FIGURE 19–12
Frayed or loose wiring should be reported to your shop supervisor. Never use wiring in this condition.

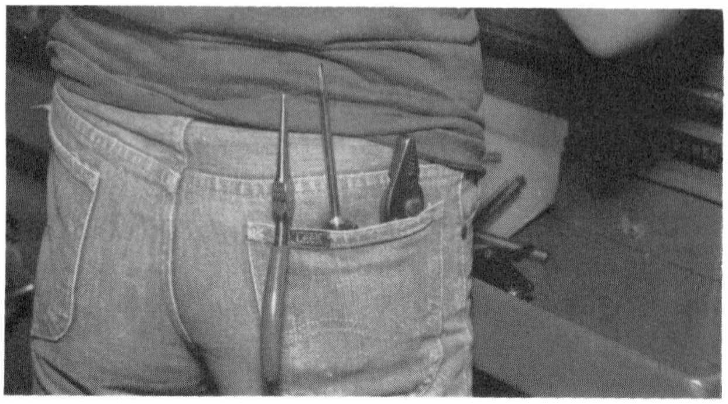

FIGURE 19-13
Tools should never be carried in your hip pocket. You can be crippled or severely injured if you slip and fall.

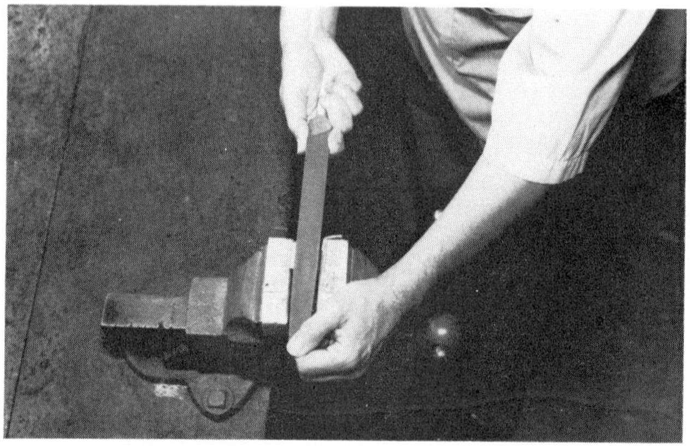

FIGURE 19-14
No file should ever be used without a snug-fitting handle. The tang can be driven into the hand or body, crippling or injuring the individual.

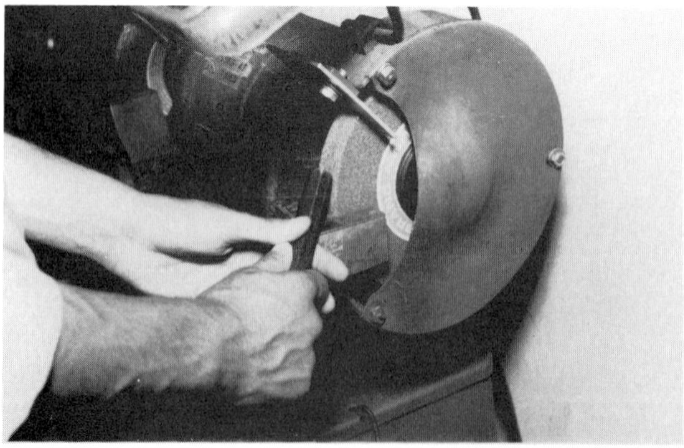

FIGURE 19-15
Always grind away any tool's striking surface that is cracked or mushroomed.

7. Always check to see that cutting tools and pieces are fastened securely before attempting to operate a machine.

8. Make sure a machine is shut off before leaving it. Never leave a machine running unattended.

9. Do not get in the habit of attempting to stop the rotation of a moving part such as a drill press or lathe spindle by hand. Let it coast to a stop.

10. Never attempt to change the belt on any machine unless it is stopped.

11. Do not attempt to measure machined surfaces unless the machine has been stopped.

12. Never attempt to use a rag near moving parts or mechanisms. It can easily get caught, possibly drawing the hand in with it.

13. Be thoroughly familiar with the operation of a machine. Knowing how to disengage a machine safely will be one of your best safeguards.

14. Keep fingers and hands away from moving parts and cutters.

15. Know the location of every emergency power shutoff switch (panic button) and be prepared to use it if necessary.

Tool Safeguards

1. Never carry screwdrivers, pliers, or other common hand tools in your hip pocket (Figure 19-13). Being speared with a sharp tool in falling can leave one crippled.

2. Pointed or sharp-edged tools should never be carried in your pocket. Carry these tools with a protective cover.

3. Never use a file without a snug-fitting handle (Figure 19-14).

4. Cutting tools such as chisels should never be used with dull cutting edges. A dull cutting edge requires more force than would be normally used.

5. If involved in chiseling, screen off the area in front of the chipping to prevent flying chips from striking others like shrapnel.

6. The striking edge of a chisel or punch should not be allowed to mushroom or crack. They should be resurfaced to eliminate this condition (Figure 19-15).

7. Make sure that the tools as well as your hands are free of oil, grease, or dirt before working with them. Slippery tools are dangerous to use.

8. Never use a tool that is too large for the job.

9. Carefully select the wrench to fit the nut involved. Knuckles or hands can be cut or bruised due to turning a loose-fitting wrench as well as result in rounded corners.

10. Report all broken or unsafe tools to your shop instructor.

Cleaning Safeguards

1. Work areas should be cleaned daily, with brooms and dust pans properly stored after use.

2. Floors should be kept clean of grease, oil, and other liquids that can become slipping hazards.

3. Stop the machine and shut off the power before attempting to clean, oil, or adjust.

4. Always use a brush to clean the machine of chips (Figure 19-16). Never use your hands or a rag.

5. Always stop the machine to clear it of chips. Avoid pulling metal chips free of parts being machined. These chips are hot, sharp, and oily. They can cut, burn, and infect the careless operator.

6. Always store long metal bars in proper storage bins (Figure 19-17). Do not leave them out for someone to trip over.

7. Only cut the unpainted end of a metal bar. Bars that are cut should have their ends cleaned and deburred before being placed back in storage.

8. Aisles should be kept free of metal bars, waste, or other remnants.

9. Never attempt to use compressed air to clean a work area. It will scatter the waste and pollute the surrounding air. If compressed air is to be used for any purpose, an air hose should have an OSHA- (Occupational Safety and Health Administration) approved nozzle with 30 psi (pounds per square inch) pressure or less.

10. Never use compressed air to clean off a machine. It will drive and wedge particles between moving parts that can ruin or damage its operating function.

11. Never use compressed air to blow chips off a machine. It will scatter the chips and pollute the air with oil, lubricants, or other contaminants (Figure 19-18).

12. Always leave your work area clean, removing all waste and chips when the job has been completed (Figure 19-19).

13. Clean your workbench of unnecessary tools and accessories when not using them.

14. Always clean all oil spills.

FIGURE 19-16
Always use a brush, never your hands, to clean any waste and metal chips.

FIGURE 19-17
Metal bars should be neatly stored in racks or bins when not being used to avoid blocking work areas or tripping over them.

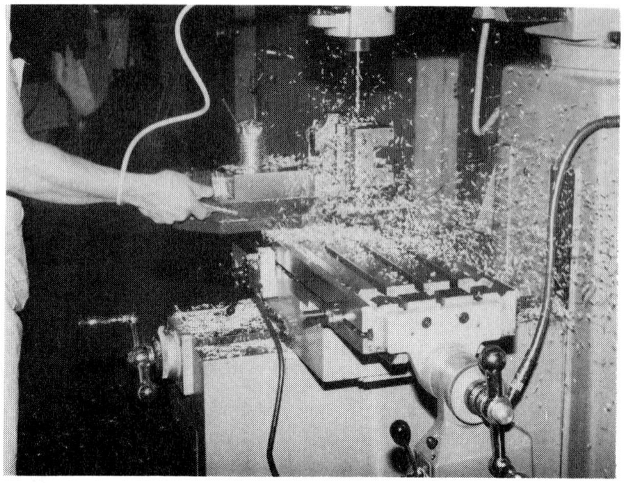

FIGURE 19-18
In addition to polluting the air, blown chips can cause serious eye injuries.

FIGURE 19-19
Always leave your work area clean after completing your job.

General Safeguards

1. When lifting, keep the back straight, bend the knees, and lift with the legs. When an object is too heavy, get someone to help you or use a hoist.

2. Observe and respect any safety zone that has been marked off for your protection.

3. Never attempt to get the attention of someone who is busy at work in a rude or alarming manner. Breaking a train of thought can lead to getting hurt.

4. There is no room for horseplay or practical jokes in a workshop. Accidents can easily happen when distracting others or being distracted by others in this manner.

5. Report every injury to your shop instructor no matter how slight it is. Unattended cuts and bruises can lead to serious infections or other problems.

Learn how to do a job using positive work habits. Do not attempt crude, unsafe short cuts. To walk away from any job unhurt will require that you and others around you think, act, and work safely.

SECTION REVIEW

Unit 19

1. What type of safety attitude should any individual have before working in a shop?
2. Name three types of eye protection that may be found in a shop.
3. List four clothing precautions to be observed in a shop.
4. When are the use of gloves permitted? Never permitted?
5. How should the following personal safeguards be treated? (a) jewelry, (b) long hair, (c) beards, (d) sneakers or sandals, and (e) hearing.
6. How should frayed or loose wiring be treated?
7. Where should oily rags or combustible waste materials be stored? Paints, oils, or other flammables?
8. What three precautions should an individual become familiar with in regards to fire prevention in any shop?
9. List four precautions to observe before operating machinery.
10. What safety precaution should be followed in the use of a metal cutting file?
11. Why should compressed air not be used to clean off a machine of chips and lubricants?
12. Briefly describe how to lift a heavy object.
13. What should you do immediately after receiving an injury?

Section 6

BENCH METALWORKING

UNIT 20 CUTTING METALS

The art of cutting metal is as old as the process of making metals. While one of the most basic and primitive means of shearing metal was by means of a crude hammer and chisel, today's technology has advanced to a point where laser beams are used as cutting tools. Whether basic or advanced, both methods have one thing in common—safety in execution. There is a proper way and an improper way of cutting metal. The improper way is wasteful, time consuming, and dangerous. The proper way is to have a solid setup and take advantage of the way a tool is designed to cut without getting hurt. Maximum achievement should be pursued with minimum risk.

Objectives
After completing this unit you should be able to

1. identify the different methods of cutting metal.
2. select and use the proper hacksaw blade for the material involved.
3. set up and use the horizontal bandsaw.
4. use the contour bandsaw properly.

Types of Metal Saws

There are five common methods of cutting material in the metalworking industry. These would include the hacksaw, the abrasive saw, the cold saw, the bandsaw, and the friction saw. In order to work efficiently, all will require that both the workpiece and the cutting tool be held securely. This is true for a machine cutter as it is for a hand cutter. All five methods will be discussed at varying lengths.

The Hacksaw

The hacksaw (Figure 20-1) is one of the most commonly used hand tools for cutting metal. The hacksaw frame is simple in construction. A blade is hung on two pins and tightened. Adjustability allows different blades to be used from 8 to 16 in. (203 to 406 mm) in length.

There are two types of hand hacksaw blades, both made of a high-grade tool steel. One type is flexible, while the other is hardened and tempered throughout. The flexible blade has only its teeth hardened and tempered, leaving the rest of it soft and springy, making it ideal for general-purpose cutting. In contrast, the all hard blade is hardened and tempered throughout. The blade is rigid and will break easier. It should not be strained when using.

FIGURE 20-1
The hand hacksaw is commonly used for cutting metals.

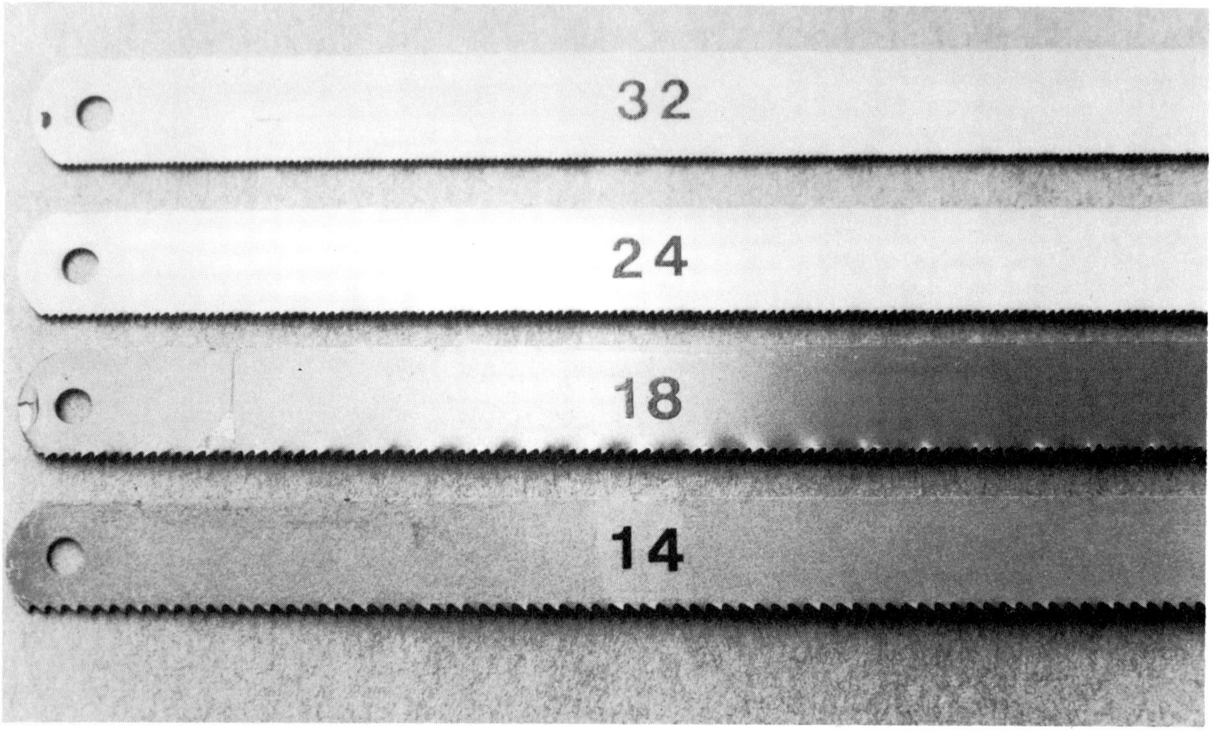

FIGURE 20-2
Hacksaw blades are made with four standard pitches: 14, 18, 24, and 32 teeth per inch.

FIGURE 20-3
The bending of teeth on either side of the blade's length is designed to prevent binding during cutting.

Besides hardness, the two factors that go into the makeup of a hacksaw blade are the number of teeth to the inch it contains, known as pitch (Figure 20-2), and the free cutting that is created by bending the teeth at an angle, known as set (Figure 20-3). The width of the cut is called the kerf (Figure 20-4). The pitch and set will dictate selection of a blade, and while lubrication need not be used, there should be at least two full teeth on the cutting section when using it.

Blade Selection

14-Pitch blade: used for cutting large sections 1 in. (25 mm) or greater.

18-Pitch blade: used for general-purpose work ranging in thickness between $\frac{1}{4}$ and 1 in. (6 and 25 mm).

24-Pitch blade: used on tubing, medium sheet metal, and drill rod.

32-Pitch blade: used on thin-walled sections (Figure 20-5). For extra thin material, sandwich work between hardwood (Figure 20-6) and cut through.

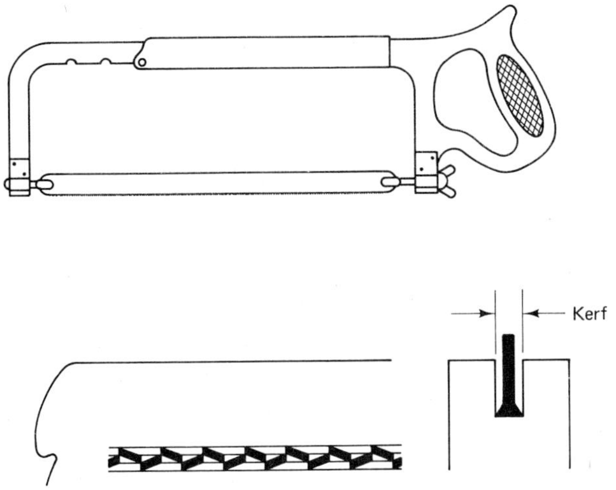

FIGURE 20-4
The width of a saw cut is called kerf.

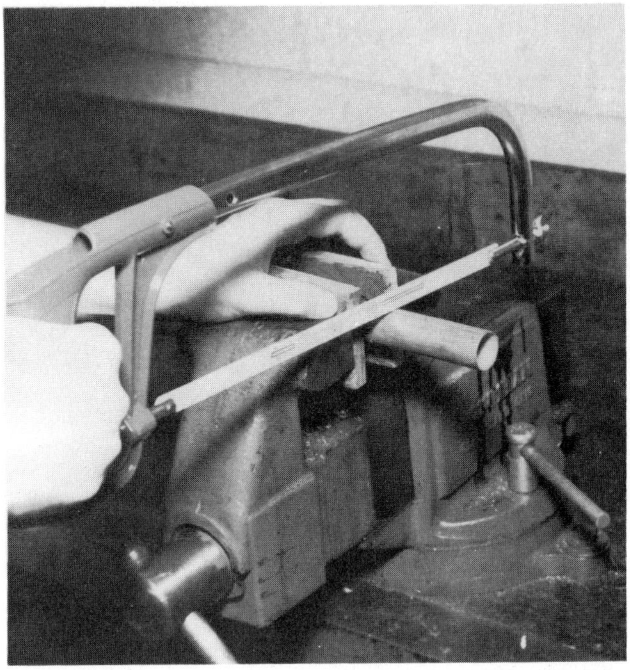

FIGURE 20-5
Thin-walled sections require a 32-pitch blade.

General-Purpose Cutting

1. Insert an 18-pitch blade, teeth pointing down and away from the handle.
2. Holding the blade in position, turn the wing nut as tight as possible with the other hand.
3. Mark off the held workpiece $\frac{1}{4}$ in. (6 mm) beyond the bench vise jaws.
4. Placing a thumb on the scribed line (Figure 20-7), stroke the blade until a groove $\frac{1}{16}$ in. (1.6 mm) deep has been cut.
5. With the cut started, place your feet in a comfortable stance to distribute the body weight to reduce fatigue.
6. Begin cutting by applying an even downward pressure going forward, easing on the return stroke, using about 50 strokes per minute.
7. Ease up on the cutting pressure as the blade nears the end of its cut.
8. If the cut is interrupted due to a broken or worn blade, restart the cut in another section with a new blade.

The Power Hacksaw

The power hacksaw (Figure 20-8) is designed to cut lengths of metal of various sizes, shapes, and materials. While there are a number of types of power hacksaws ranging from a single speed to a four-speed selection, most school shops use a single-speed

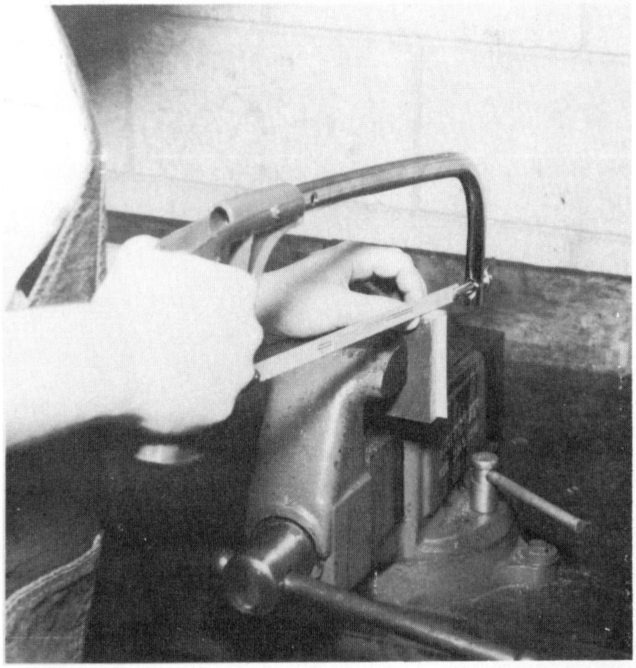

FIGURE 20-6
Extra thin material should be sandwiched between hardwood for cutting.

FIGURE 20-7
The thumb acts as a guide in starting the saw cut.

FIGURE 20-8
A dry-cutting power hacksaw.

FIGURE 20-9
A wet-type power hacksaw blade in use. (Credit: DoAll Company)

machine that draws a blade containing 10 or 14 teeth per inch for general-purpose cutting. Blades with 3, 4, and 6 teeth per inch are available for work that increases in thickness. For 1 to 2 in. (25 to 50 mm) in thickness, a 6-tooth blade is recommended; greater than 2 in. (50 mm) in thickness would use a 3- or 4-tooth blade.

There are two basic types of power hacksaws: wet (Figure 20-9), where a stream of coolant is used, and dry cutting. Both work on a reciprocating principle, that is, a forward and return motion is used. The blade is fed and cuts during the forward stroke and does not cut as it is repositioned for the next forward stroke.

Power hacksaws cut hard materials as well as soft brass and are therefore considered utility machines. Their cutting speeds vary from 35 to 150 strokes per minute but generally fall in a range between 60 and 120 strokes per minute.

The Abrasive Saw

The abrasive cutoff saw (Figure 20-10) uses an abrasive cutoff wheel as well as carbide-tipped or high-speed steel blades for its operation. The abrasive wheels are thin and are designed to cut a variety of materials including ferrous, nonferrous, plastics, fiber, glass, rubber, and ceramic. These wheels can be used wet or dry, depending on the nature of the material and machine. Wet cutting offers several advantages not to be found in dry cutting. Due to the fact that there is less heat buildup, the wheels remain sharper longer. In turn, there is less discoloration and burring of the workpiece.

FIGURE 20-10
A chop action abrasive cutoff machine can cut solids up to 8 in. (203 mm) thick. (Credit: DoAll Company)

FIGURE 20-11
A cold circular saw can accurately cut off material to length.

FIGURE 20-12
For reasons of safety, a cold circular saw blade is carefully surrounded by a guard.

The size of the wheel should not exceed the size recommendation made by the cutoff machine's manufacturer. The wheel speed is also critical to its performance and safety. Turning it at a faster rpm can result in the wheel shattering. As a rule of thumb, the horsepower needed should be approximately one horsepower per inch (25 mm) of wheel.

The Cold Saw

Circular in structure, cold-saw blades are designed to be mounted on hand-operated bench-top machines (Figure 20-11) as well as automated machines. These machines are fast cutting and can accurately cut off materials to length. Generally equipped with blades made of chrome-vanadium steel or bearing carbide-tipped cutting edges, a guard is placed around the outer edge to protect the operator as well as the machine (Figure 20-12). Cold saws can be used on copper, brass, stainless steel, machine steel, and aluminum. Their selection of more than one speed enables them to efficiently cut angles, channels, and I-beams.

The Horizontal Bandsaw

Like the horizontal power hacksaw, the horizontal bandsaw can cut either wet or dry. They range from the manually operated machine generally found in the toolroom (Figure 20-13) to the larger model used in production that can come with more automated

FIGURE 20–13
A toolroom size horizontal bandsaw.

FIGURE 20–14
A production-type horizontal bandsaw has a greater capacity and can come equipped with automated features. (Credit: DoAll Company)

features (Figure 20–14). In either case it is equipped with a looped blade that passes over the outside edge of two large pulley wheels and is made to drive through saw guides with lead-in rollers in the continuous direction of cutting. It is the positioning of these saw guides that supports the saw blade and prevents it from cutting through material on a slant. The versatility of the larger machine will also feature a band tension handwheel, a support table, a supply of cutting fluid as well as variable feeds and speeds.

Bandsaw blades are made in a selection of alloy steels. They are available made of high-speed steel, alloy tool steel, or a bimetal composition where high-speed steel or tungsten carbide teeth are welded to a carbon steel back. These blades are made thin and vary in pitch from 4 to 14 teeth and in width from $\frac{1}{16}$ to $1\frac{1}{4}$ in. (1.6 to 31 mm). As a general rule, a 4-pitch blade should be used on large, heavy sections for the greatest penetration and chip clearance. A 14-pitch blade is recommended for thin material, while a 10-pitch blade is recommended for general-purpose cutting. Metric blades are also available, but their pitch is based on the number of teeth per 25-mm length. A blade bearing 5 teeth per 25 mm length would be classified in the following manner:

$$5 \div 25 \text{ mm} \quad \text{or} \quad 0.2\text{-mm pitch}$$

Because bandsaws cut in one continuous direction, there is no wasted time in cutting as with a power hacksaw. Another advantage is the little waste of material in the width of the cut. Bandsaws cut a kerf measuring about $\frac{1}{16}$ in. (1.6 mm) as opposed to an $\frac{1}{8}$ in. (3 mm) for abrasive saws and power hacksaws. All things being equal, it can be seen that by reducing the waste cut, more pieces will be obtainable in any length of stock that is cut.

General Cutting Procedures

1. Check to see that the vise is set at the proper angle to the blade.

2. Open the vise wide enough to receive the workpiece.

3. Position the workpiece shape as illustrated (Figure 20–15).

4. Support any long rod or bar that extends beyond the machine (Figure 20–16).

5. Lower the cutting blade and position it just above the workpiece.

6. Slide the workpiece so that any cutoff line is under the blade or position the stock by measuring the length to the blade by rule, allowing $\frac{1}{16}$ in. (1.6 mm) extra.

7. Now tighten the vise securely.

8. With the machine running, carefully lower the frame to allow the blade to enter smoothly to make a shallow slot.

9. Shut off the machine and raise the frame to check the cut. If correct, carefully lower the frame to engage the running cut smoothly.

10. When through cutting, most machines will automatically shut themselves off.

11. Deburr sharp edges on cutoff workpieces and stock and return the stock to the storage bin.

The Vertical Bandsaw

There are two types of vertical bandsaws used in the metalworking industry. The cutoff model (Figure 20–17) is generally used in production on larger stock than the horizontal bandsaw. These machines tend to have a greater capacity in many areas, which may include automated controls, a tilt-frame head for making angular cuts, a recirculating coolant, variable speeds and feeds, as well as a hydraulic vise for more positive gripping.

Friction Cutting

Friction cutting is another method used in band sawing. Faster than the conventional teeth method, cutting is accomplished through heat. With the blade moving at a rate of better than 15,000 ft (4572 m) per minute, the contact between the workpiece and blade produces a friction to the point of softening the metal just ahead of the blade. Since friction is the cutting device, a dull edge blade will work more effectively as it will produce more heat. For this reason this method is not used on soft materials such as copper, aluminum, brass, or bronze, which tend to melt or become sticky. This method is very efficient and will cut armour plate $\frac{1}{2}$ in. (13 mm) thick. The depth of the heat penetration is very shallow and will not affect the characteristics of the metal (Figure 20–18).

Heavier metals can be cut using a rocker technique. By slightly lifting the back end of the workpiece, the front edge of the material comes in contact with the blade. As the cutting action begins, the workpiece is lowered, constantly presenting a new

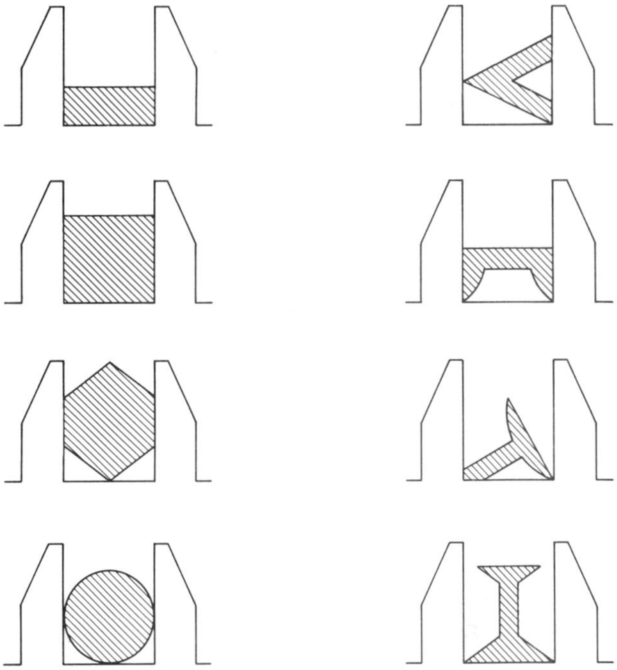

FIGURE 20–15
Correct methods of holding metal shapes in a power saw vise.

FIGURE 20–16
Work extending beyond the vise jaws of any power cutoff machine should be supported.

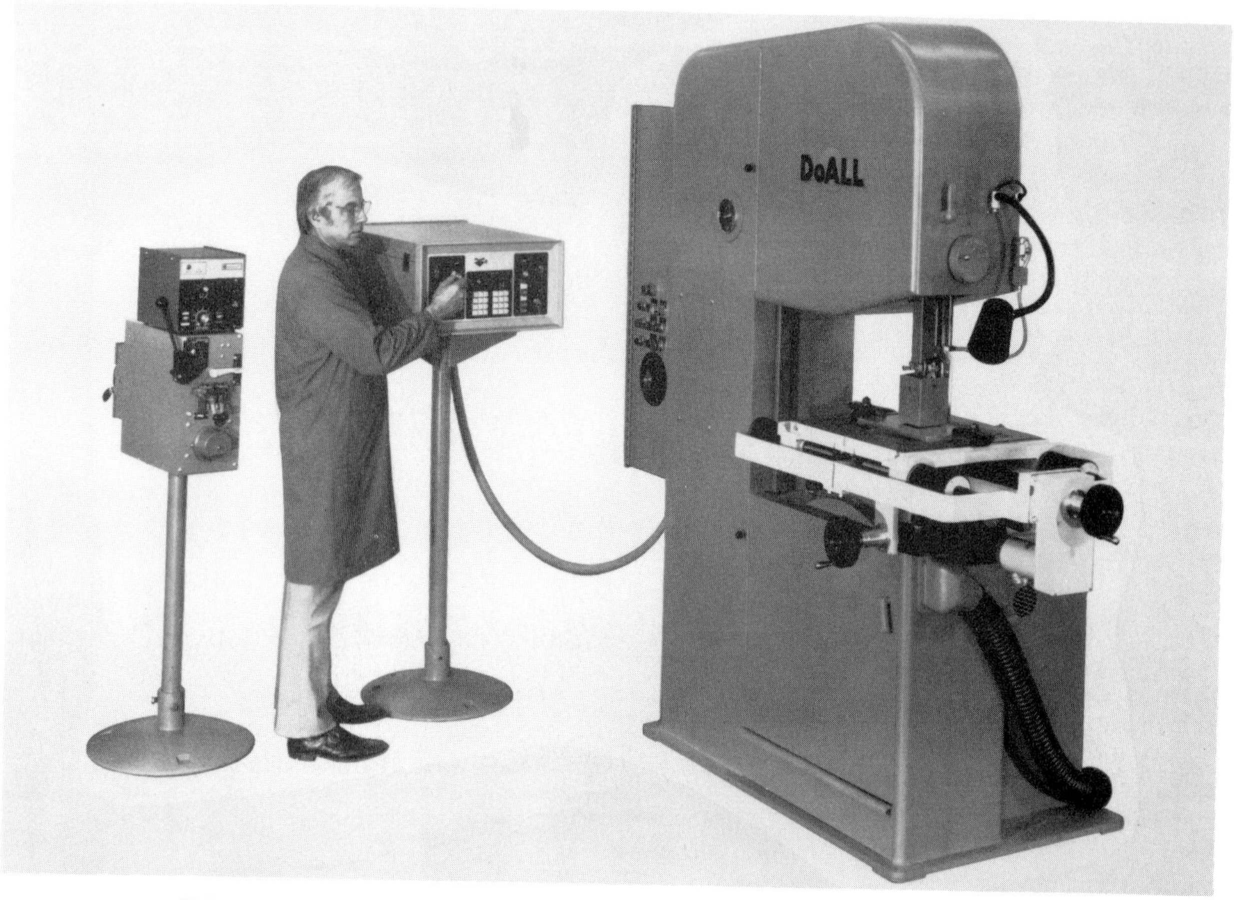

FIGURE 20–17
A numerically controlled vertical bandsaw that cuts intricate shapes faster and at lower costs than conventional methods. (Credit: DoAll Company)

FIGURE 20–18
Friction sawing cuts at tool velocities up to 30 times faster than conventional sawing will allow. (Credit: DoAll Company)

FIGURE 20–19
Trimming casting sprues and risers by friction sawing. (Credit: DoAll Company)

FIGURE 20–20
This friction sawing will require safety face shield, leather gloves, and appropriate coveralls. (Credit: DoAll Company)

edge to cut until, by alternating the raising and lowering of the workpiece, the cut is complete (Figure 20–19).

In addition to the use of approved safety glasses, friction sawing will also require leather gloves and appropriate coveralls (Figure 20–20).

The Contour Bandsaw

The contour bandsaw represents one of the most versatile and economically used methods of cutting metal and other materials used in the metalworking industry. This highly productive machine tool is solidly fabricated and consists of three main parts: base, column, and head, illustrated in Figures 20–21 and 20–22.

Base

The base of the contour bandsaw supports the column and head and is constructed with a heavy base plate to provide extra structural rigidity.

- The *lower pulley* is a cast-aluminum wheel with replaceable crowned rubber tire and is driven by a variable-speed upper pulley wheel.
- The *table* on a contour bandsaw may be mounted directly to the base in a fixed position or may be mounted on a trunnion (a movable support) connected to the base (Figure 20–23). The trunnion allows the table to be tilted 10° to the left and as much as 45° to the right.
- The *lower saw guide* is located on the underside of the table and prevents the saw blade from twisting.

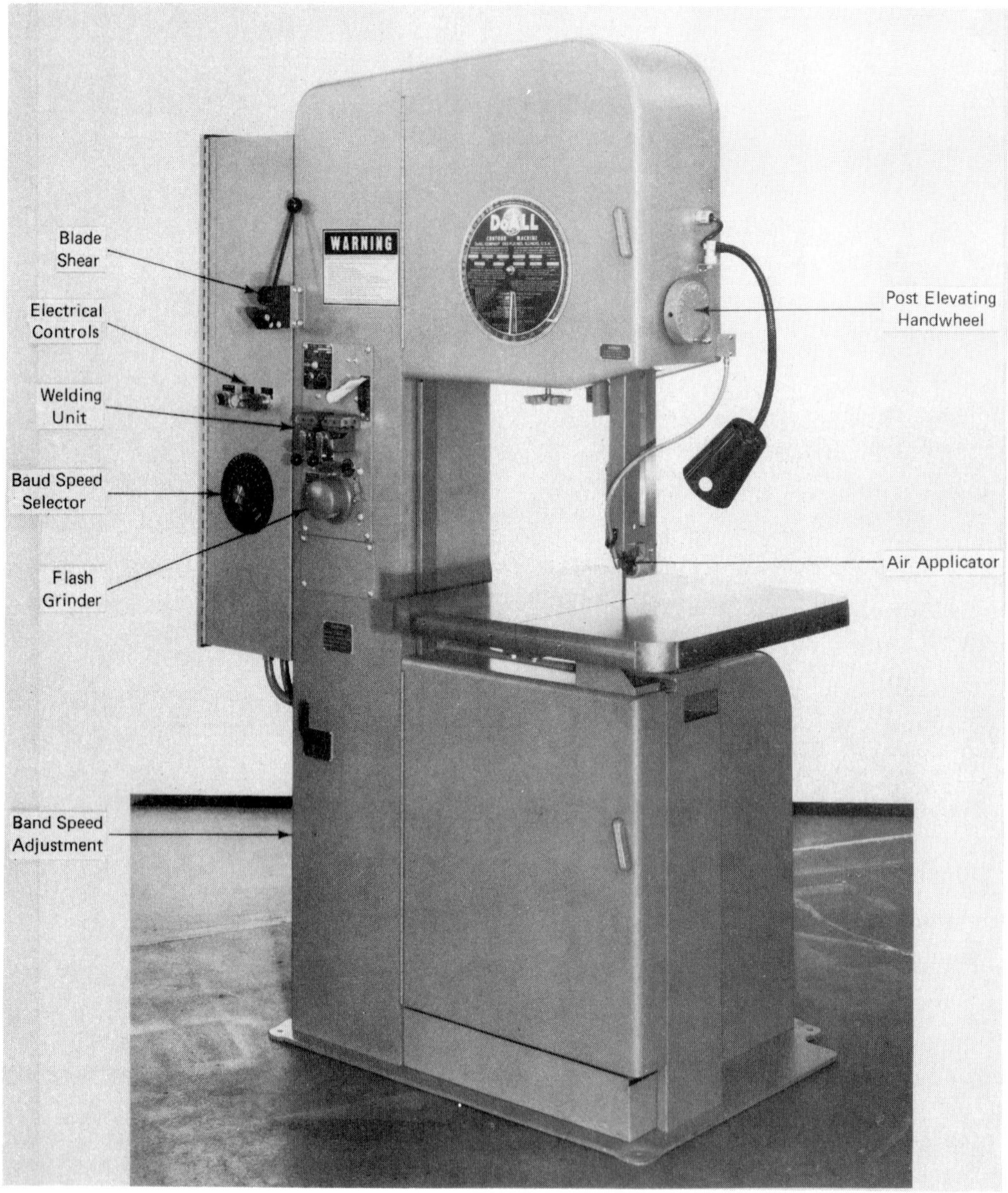

FIGURE 20-21
The main parts of a contour bandsaw. (Credit: DoAll Company)

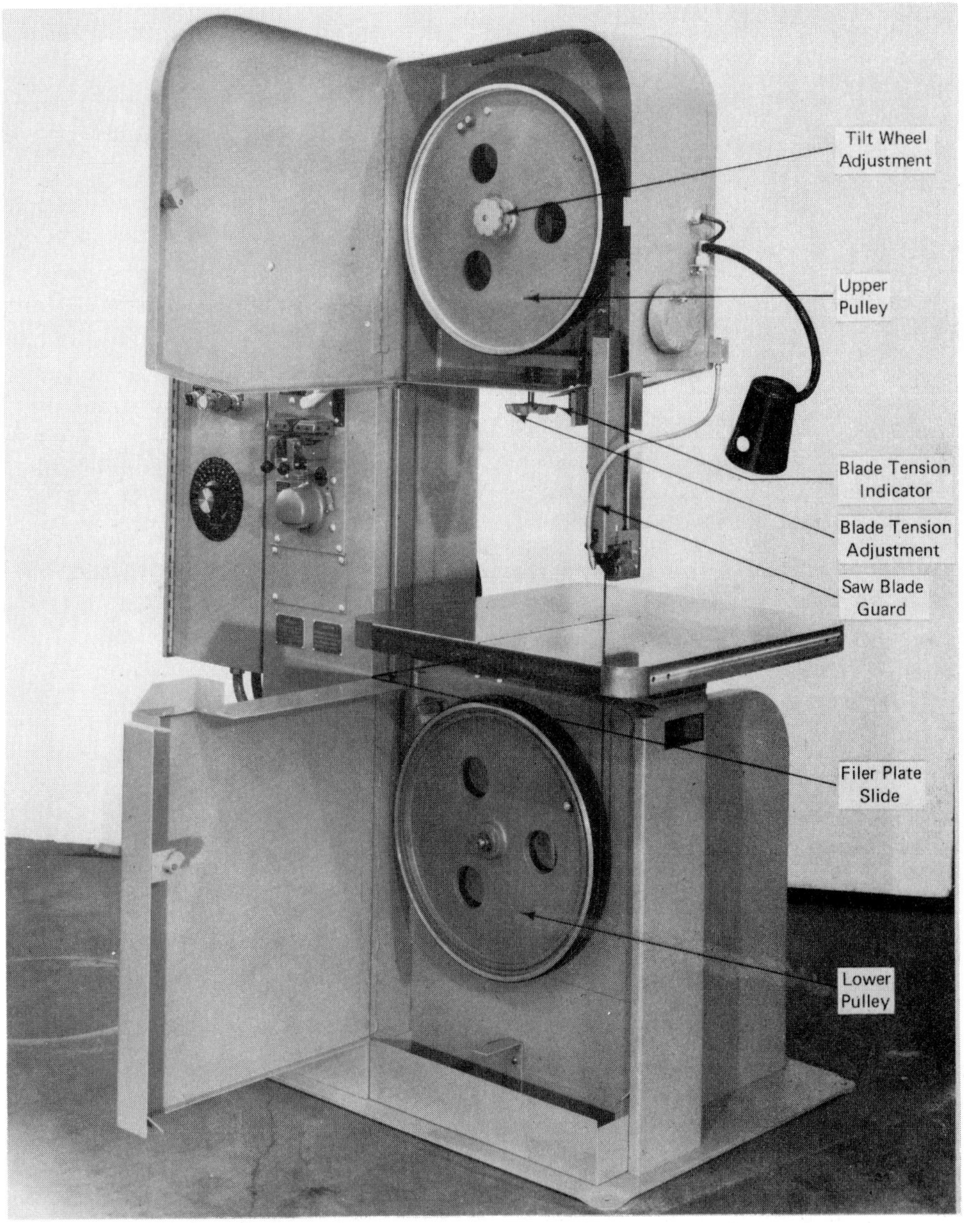

FIGURE 20–22
A two-pulley wheel contour bandsaw. (Credit: DoAll Company)

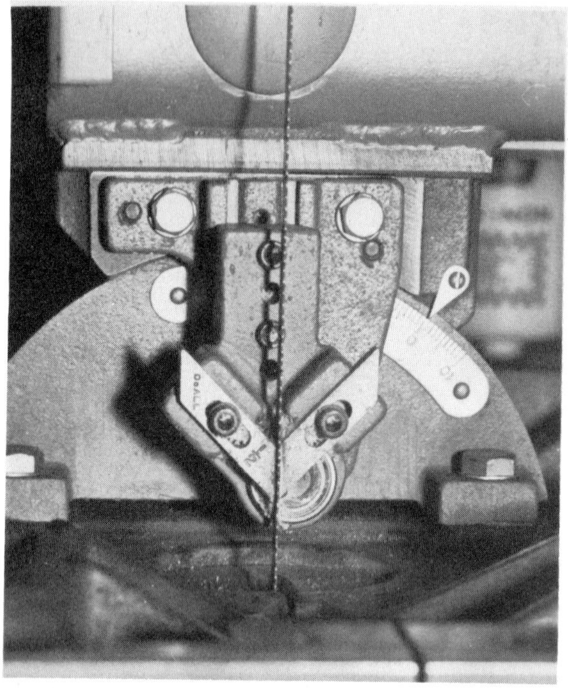

FIGURE 20–23
The trunnion support will allow the table to tilt 10° to the left and as much as 45° to the right.

Head
The construction of the head consists of parts designed to support or guide the saw blade.

- The *upper pulley* is a cast-aluminum wheel that is crowned with a rubber tire. It turns at variable speeds and can be adjusted for tracking (saw blade runout).
- The *blade tension adjustment* and the *blade tension indicator* are located on the head underneath the upper pulley.
- The *upper saw guide* supports and directs the saw blade through a set of adjustable V-guides.
- The *post elevating handwheel* positions the upper saw guide to clear work and automatically locks it in place.
- The *saw guard* is designed as a safety device against exposure to the running saw blade, while the *air applicator* can be used to blow away chips from the workpiece while prolonging blade life.
- The *job selector* (Figure 20–24) is generally found on the door that encloses the upper pulley wheel and enables an operator to select the correct saw and saw speed for the material being cut.
- The *filler plate* and *center plate* are located in the table and are removable to allow saw blade changing, saw guide adjustment, filing band insertion, etc.

Column
The column is a welded, unitized box frame that supports the head as well as houses the safety electrical controls, the welding unit, the blade shear, and band speed selector.

- The *band speed selector* is dial adjusted to set the bandsaw to a range of variable speeds while running.
- The *band speed adjustment,* with its high or low range, is located on the lower back side of the column.
- The *blade shear* is used for cutting blades clean and square, enabling them to be welded to length with a minimum of end waste.
- The *welding unit* is a self-contained butt welder that is also equipped with a flash grinder and thickness gage to check the finish weld area.

FIGURE 20–24
Job selector wheel.

The advantages of contour bandsawing are many and are illustrated in Figure 20–25. They are capable of straight, angular, or irregular contour cutting (Figure 20–26), as well as notching, slotting, and internal and external radius cutting. Additionally, the vertical contour bandsaw can be equipped to polish or file metal to a finish (Figure 20–27), a feature that is widely used by die makers to complete the shaping of dies and punches (Figure 20–28). It can also be equipped with a diamond tooth blade to cut refractory materials such as glass, china, porcelain, granite, silicon carbide, and marble (Figure 20–29).

FIGURE 20–26
Intricate saw cutting from a single disc. (Credit: DoAll Company)

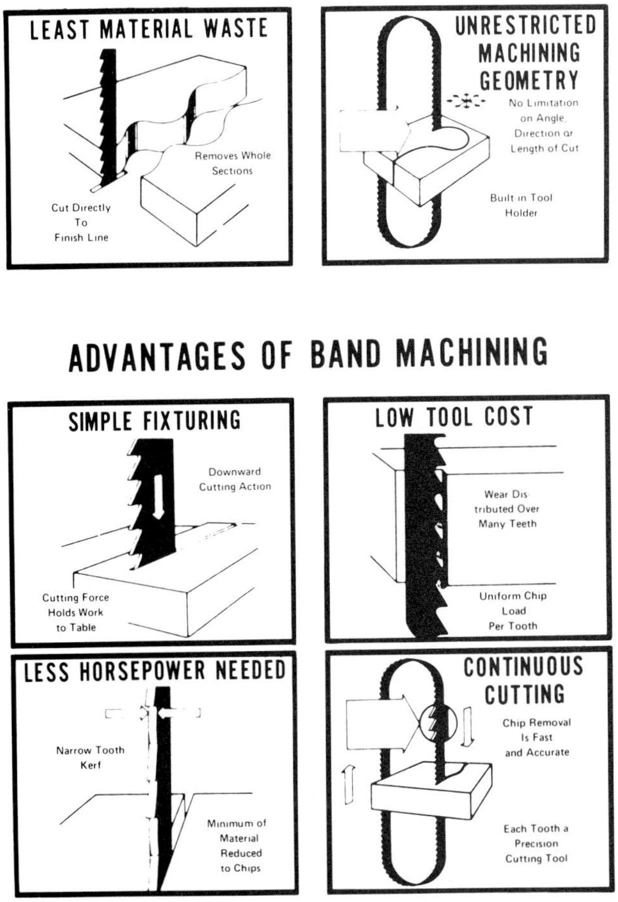

FIGURE 20–25
The advantages of band machining. (Credit: DoAll Company)

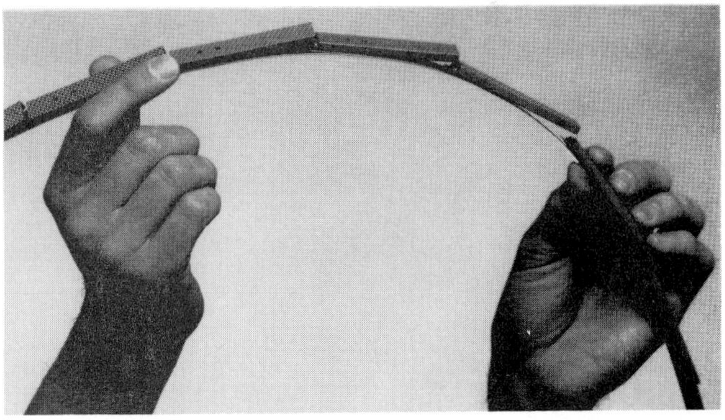

FIGURE 20–27
A band file coupled for use. (Credit: DoAll Company)

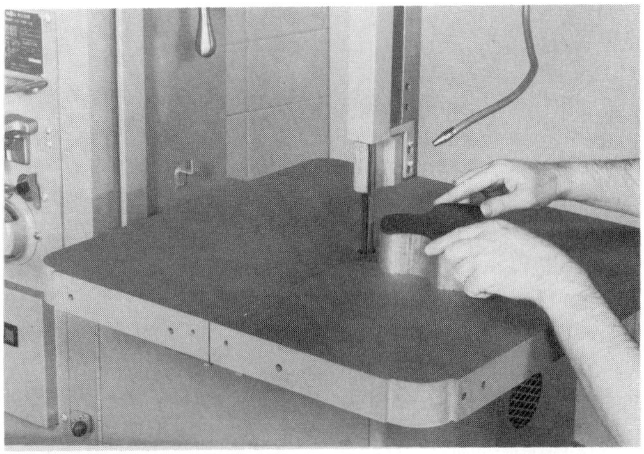

FIGURE 20–28
Band filing has proven to be seven to nine times faster than hand filing. (Credit: DoAll Company)

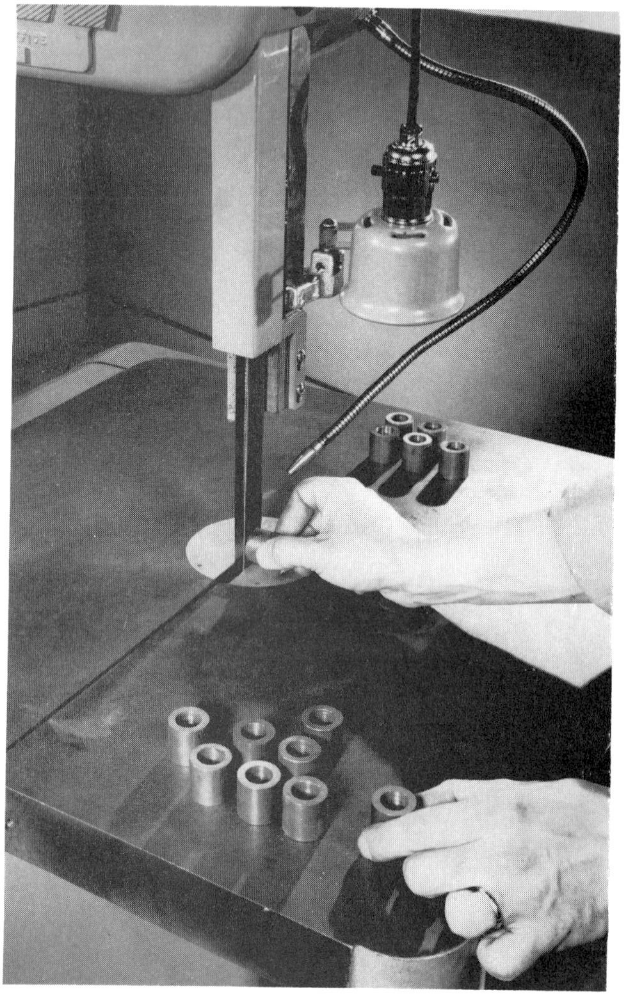

FIGURE 20-29
Aluminum oxide polishing bands are used for smoothing surfaces after filing. (Credit: DoAll Company)

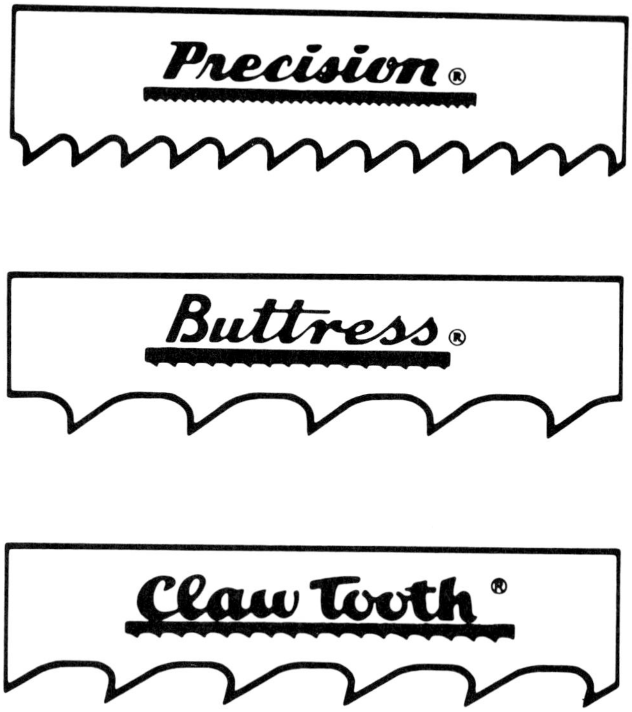

FIGURE 20-30
Bandsaw blade tooth forms. (Credit: DoAll Company)

Contour Bandsaw Blades

A continuous loop of steel, generally high-carbon steel that contains hardened teeth with a soft flexible back, is the cheapest and most widely used to be found in schools and industry. The shape, profile, and gullet of these teeth are illustrated in Figure 20-30.

Precision or regular tooth: This is considered the most versatile-type blade. There is a noticeably rounded gullet that combines with a zero rake angle and ample back clearance angle to leave a smooth edge finish. It is recommended for a variety of work from cutoff to contoured design work.

Buttress or skip tooth: This tooth design is similar to the regular tooth except that greater clearance has been provided by spacing the teeth further apart. It is recommended for cutting thick, heavy sections or for making deep cuts in nonferrous metals, woods, plastics, and other soft materials.

Claw or hook tooth: In application, the claw tooth blade is similar to the buttress tooth blade. While the gullet design is similar to the buttress tooth form, the claw tooth has a positive 10° rake angle. This design enables the teeth to penetrate the workpiece faster and with lower feed pressure; however, it will not leave so smooth a finish as the buttress tooth form.

Pitch and Set

Like the hacksaw blade, pitch and set play a role in blade selection. Pitch will be determined by the thickness of the material. The amount of set, clearance provided by offsetting the teeth, follows three common patterns (Figure 20-31):

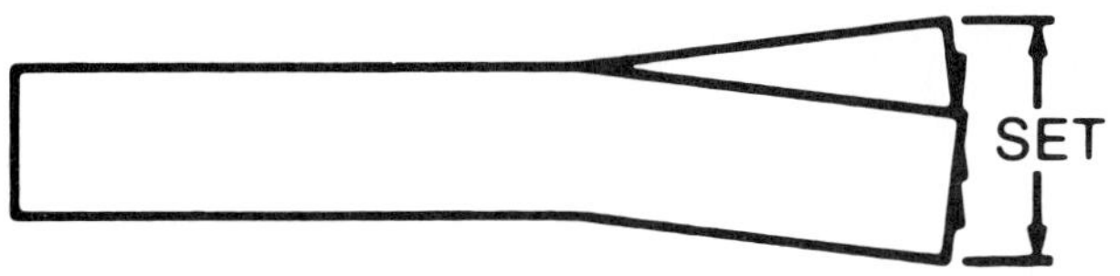

FIGURE 20-31
Common set patterns used for bandsaw blades.
(Credit: DoAll Company)

Raker set teeth follow a pattern where one tooth is bent to the right, the next to the left, and the third remains on center. It is designed for general-purpose work and is well suited for contour sawing.

Wave set teeth bear a gradual curving of teeth that are offset from one side of the blade to the other. These blades are recommended to be used on cross sections of workpieces that vary, such as tubing, angles, or pipes.

Straight set teeth have a steady right, left, right, left tooth-to-tooth bend. This type of blade is recommended for the free cutting of nonferrous metals, plastics, woods, as well as tubing and sheet metal.

Contour Bandsawing

In order to successfully use the contour bandsaw, the operator will have to become familiar with a number of operations such as welding, annealing, grinding, and mounting the proper saw blade for the material involved. One of the main concerns in any operation, however, is that of safety. Do not use too much force when cutting and keep fingers clear of the running blade.

Procedure for Mounting a Saw Blade

1. Open the upper door.
2. Open the lower door.
3. Loosen the tension on the underside of the filler plate and slide this section out from the table.
4. Loosen the blade tension handwheel and keep turning it to lower the upper wheel pulley, if necessary, to remove any saw blade.
5. Select and mount the proper saw guide in the upper and lower saw guide.
6. Lower the upper saw band carrier to bring both saw band carriers in closer alignment.
7. Mount the looped saw blade over the upper and lower pulley wheels with the teeth pointing in a down direction toward the table.
8. Carefully place the back of the blade between the saw guides and adjust the tension on the upper pulley wheel to keep the saw blade taut.
9. Start the machine to see whether the saw blade is properly tracking. If the blade is extended too far beyond the backup bearings, the upper wheel should be adjusted. (Loosen the locking nut on the inside of the adjustment knob in the center of the pulley wheel and turn the adjustment knob until the saw band rides close to but does not touch the backup bearing when it is not cutting. Lock this position by tightening the inner nut.)
10. Once the proper adjustment has been made, stop the machine.
11. Tighten the saw blade using the band tension indicator as a guide (Figure 20-32).
12. Close the upper and lower carrier housing doors.
13. Slide the filler plate back in the table and lock in place.

The saw blade has now been properly mounted to do its cutting.

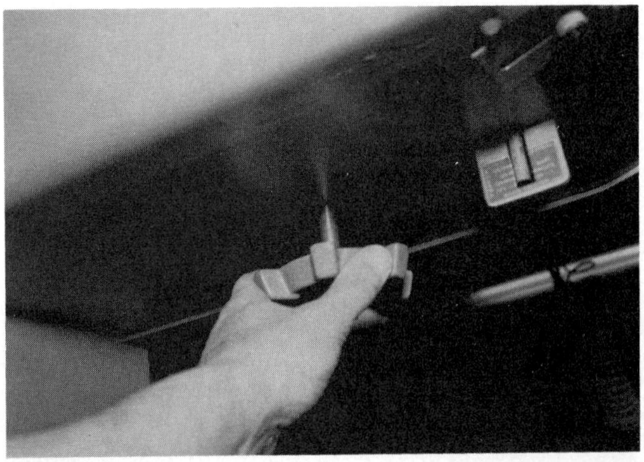

FIGURE 20-32
Tighten the saw blade using the band tension indicator as a guide.

FIGURE 20-33
Leave about $\frac{1}{4}$-in. (6-mm) clearance between the workpiece and the saw guide.

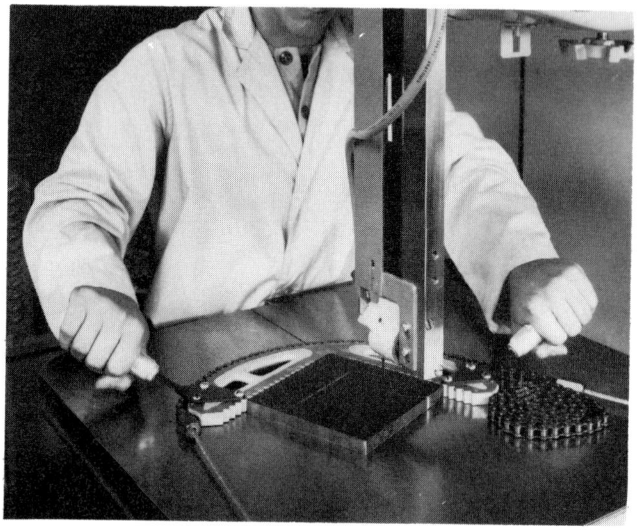

FIGURE 20-35
Using a guiding wooden block or a work-holding jaw to direct the workpiece during cutting. (Credit: DoAll Company)

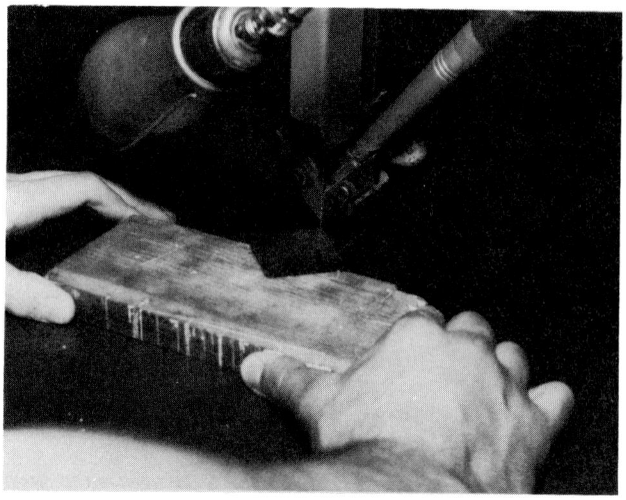

FIGURE 20-34
Never use bare hands to push against the workpiece.

Procedure for External Cutting

1. Appropriate safety glasses or face shield must be worn before attempting any cutting on the contour bandsaw.

2. Consult the job selector for correct blade and cutting speed.

3. Position the bandsaw blade with proper tension.

4. Lower the saw guide to within $\frac{1}{4}$ in. (6 mm) of the highest or thickest part of the workpiece to pass through and lock it in that position (Figure 20-33).

5. With the workpiece laid out for cutting, secure it firmly in a holding device or some type of guidance system such as the work-holding jaw (Figure 20-34).

6. Manually set the band speed adjustment to its high- or low-speed range.

7. Start the bandsaw.

8. Dial in the band speed selector to its proper speed as indicated on the job selector wheel.

9. Adjust the air applicator.

10. With a pusher block or using the work-holding jaw, lightly exert pressure to bring the workpiece in contact with the blade (Figure 20-35). *Note:* If the workpiece surface is to be machined to a finish in another operation, cut no closer than $\frac{1}{32}$ in (0.8 mm) from the layout line.

11. Keep the fingers free from the moving blade but do not push the workpiece. Carefully feed the workpiece into the saw and allow the teeth to cut at an even pace.

12. Stop the machine at the end of the cut and brush off the saw table of metal chips.

UNIT 21 FILES

One of the least understood cutting tools is the file. Its origin probably dates back to a time when volcanic rock was used for its grating surface to sharpen cutting tools.

Filing is considered a hand operation that seems easy to do and is probably one of the most difficult to master. There is no better proof of this than to file a block of steel square. Filing a block square is a time-honored tradition that is practiced all over the world. In this process the top surface of a vise-held work block is filed flat so that when the operator is through the surface should be at 90° to its sides. It sounds easy; try it some time.

Objectives
After completing this unit you will be able to

1. identify a variety of files and recognize their function.
2. properly use a number of files to perform various filing operations.
3. properly maintain and store files.

Safety

No filing should ever be attempted without a tight-fitting handle. This is a hard and fast rule that has to be maintained and for a good reason. A file without a handle is dangerous and can leave the filer injured or maimed for life.

A file (Figure 21-1) consists of two basic parts: a blade or body where teeth are cut into and a tang for insertion into a wooden or plastic handle. As a file is pushed forward in cutting, the tang can be driven into the palm of the hand if it is not covered with a handle, striking nerves that can cripple a hand for life. Nor is it any better to hold a file above an exposed tang. If the file were to be jammed or hit a moving force in its forward stroke, say in filing on a spinning chuck on a lathe, the tang can be driven into the body with just as disastrous or lethal results. Under no circumstances should a file ever be used without a handle.

Another common mistake is to drive a file tang into the handle by hammering. This can split the handle and must be avoided. The proper way is to insert the file tang as far as it will go into the file handle, and then by continuously striking the handle on something solid like a heavy steel block or concrete floor, the tang will seat itself tightly (Figure 21-2).

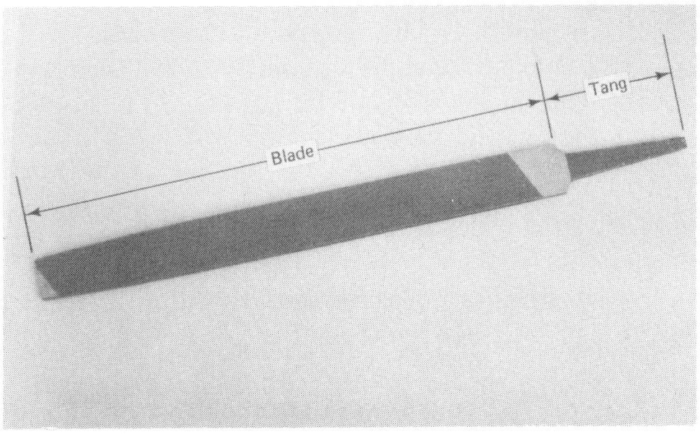

FIGURE 21-1
Basic parts of a file. (Credit: Philip Rapisarda)

FIGURE 21-2
Inserting a file in a handle. (Credit: Philip Rapisarda)

File handles also come with internal metal threads that are designed to screw onto the tang of a file. This type need not be impacted as the type previously mentioned.

Types and Purpose

Files are classified and named according to length, teeth cut, and cross-sectional area.

Length. The length of a file usually does not include the tang; however, on some Swiss pattern files the tang is included (Figure 21–3). Normally, hand files will come in lengths from 4 to 20 in. (100 to 500 mm), with general-use files being 6, 8, 10, and 12 in. (150, 200, 250, and 300 mm) long. For finer work, files 4 and 6 in. (100 and 150 mm) long are used.

Teeth Cut. There are four basic teeth structures or file cuts: single cut, double cut, rasp, and curved (Figure 21–4).

A single-cut file has a row of teeth cut on an angle of 65° to 85° across the width of the face and is used on hard metal such as in sharpening tools and finish filing. A single-cut file is considered one of the best tools to break the sharp edge on sheet metal cuts.

Double-cut files have two rows of teeth cut diagonally across the face. The first row, called the overcut, is set on an angle of 45°, while the second cut or upcut is cut deeper on an angle between 75° and 80° across the width. This crisscross pattern will cut faster than the single-cut file but will not leave the surface as smooth.

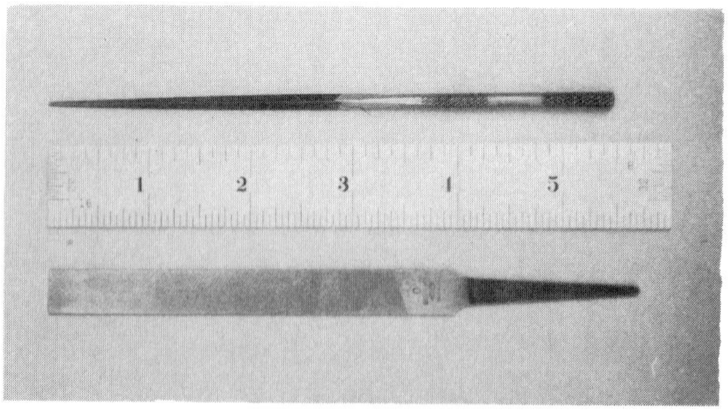

FIGURE 21–3
While length on American files excludes the tang, Swiss pattern files are measured overall. (Credit: Philip Rapisarda)

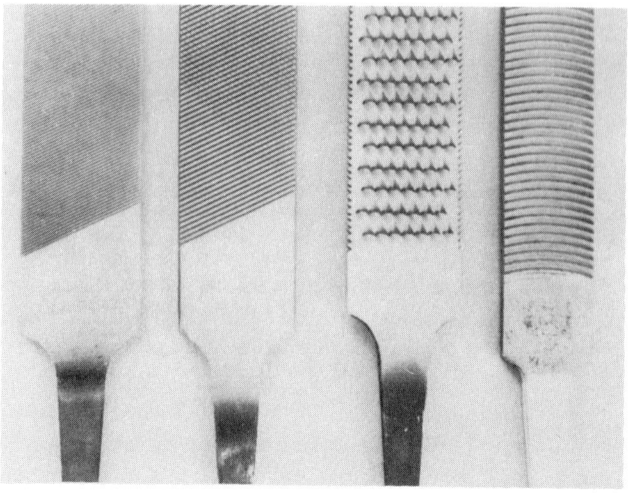

FIGURE 21–4
The four basic teeth cuts include single-cut, double-cut, rasp, and curved tooth files. (Credit: Philip Rapisarda)

The rasp file has a surface that is sharp to the touch due to a series of independent, single pointed teeth that are not connected to each other. This form allows quick cutting without clogging and is used on very soft material such as wood.

Curved-cut files have a series of contours resembling small arcs that are undercut to form cutting edges. A file of this sort is fast cutting for material such as lead, babbitt, or aluminum.

Grades of Cut. There are six common grades of cuts or teeth spacing, each affecting the finish of the single-cut and double-cut files.

Rough cut contains 20 teeth per in. (25 mm).

Course cut contains 25 teeth per in. (25 mm).

Bastard cut contains 30 teeth per in. (25 mm).

Second cut contains 50 to 60 teeth per in. (25 mm).

Dead smooth contains 100 or more teeth per in. (25 mm).

While the bastard, second-cut, and smooth-cut files are more frequently used, the others being reserved for special situations, it should be noted that the teeth spacing is not uniform. The spacing increases with the file's length. Thus, a 6-in. (152 mm) second-cut file has more teeth per in. (25 mm) than a 10-in. (254 mm) file of the same cut.

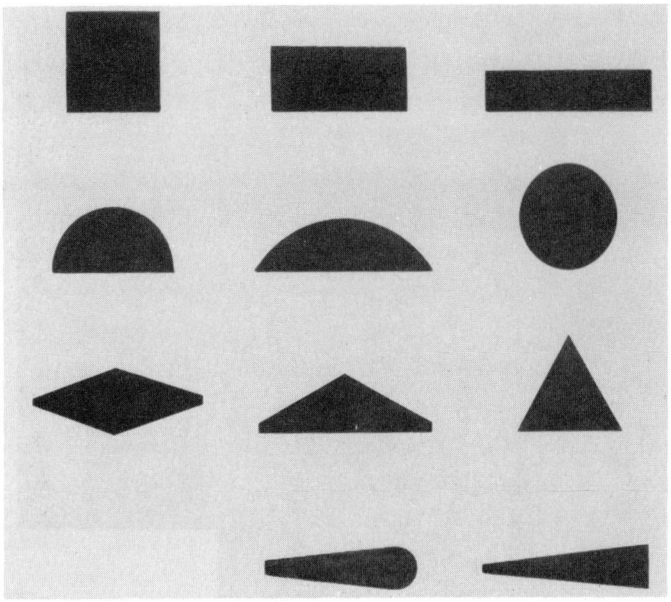

FIGURE 21–5
The cross-sectional area may change but they are all based on four geometric shapes: square, triangle, circle, and rectangle. (Credit: The Cooper Group)

Shapes of Files. Although files come in many shapes for special purposes, there are four basic cross-sectional areas of a file: the square, triangle, circular, and rectangular (Figure 21–5). These geometric patterns are modified, but they still remain the common shapes involved.

Basic Types. Files are divided into three broad groups: machinist's file, Swiss pattern/jeweler's files, and special-purpose files.

Machinist's files are general-purpose files and most often come in standard shapes, the most popular being the mill file (Figure 21–6), which comes in single-cut teeth and is for the most part rectangular. It comes in 6, 8, 10, and 12 in. (150, 200, 250 and 300 mm) long. The mill file is used on nonferrous metals such as brass and bronze as well as lathe work. The mill file derives its name from the fact that it was first used to sharpen mill saws.

Swiss pattern files, also known as needle files, are widely used by die makers and fine craftspeople for doing small, intricate filing and are sold in sets that are 5 or 6 in. (127 or 152 mm) long (Figure 21–7).

FIGURE 21–6
Mill file. (Credit: The Cooper Group)

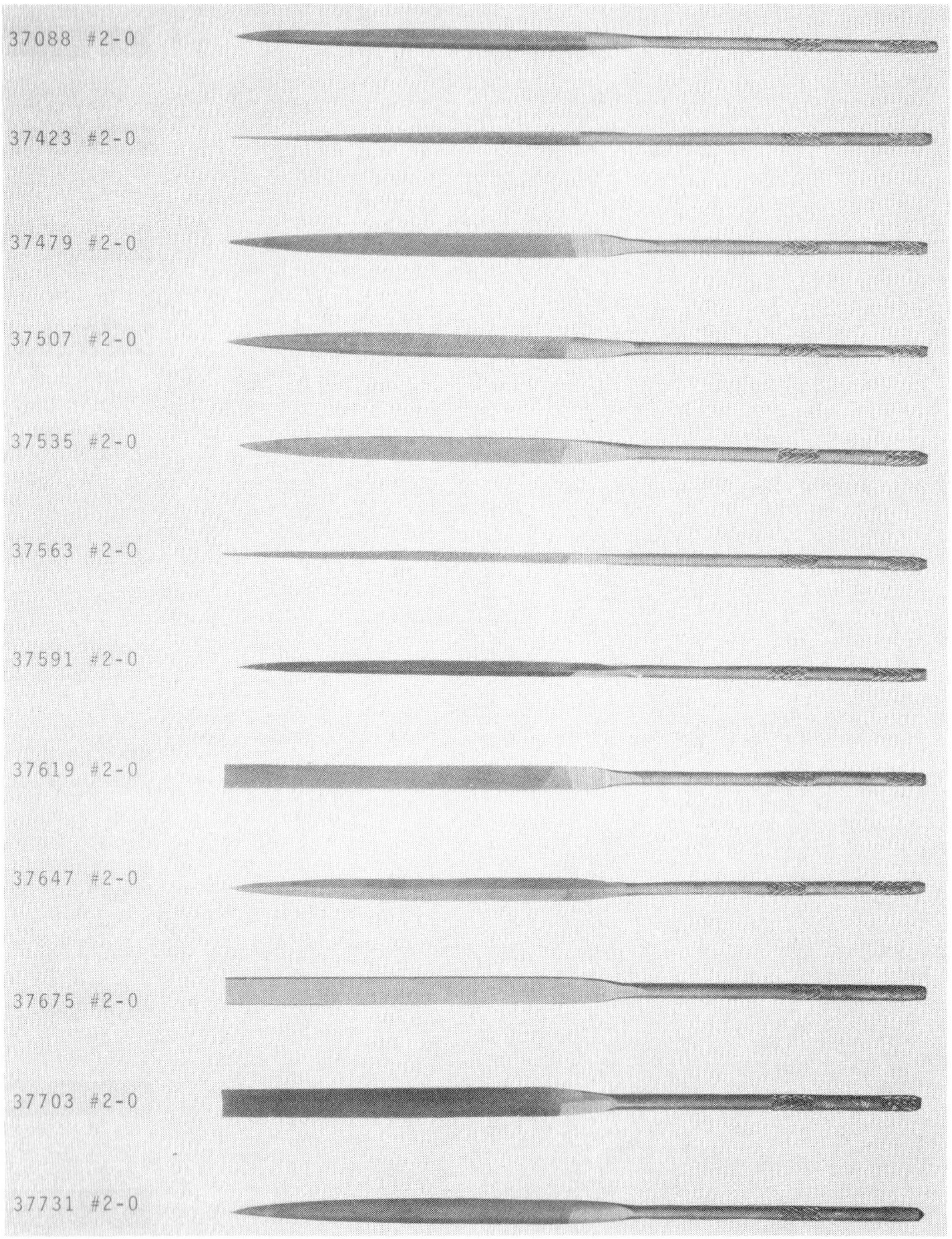

FIGURE 21-7
Swiss pattern files, also called needle files, are designed for fine, delicate work. (Credit: The Cooper Group)

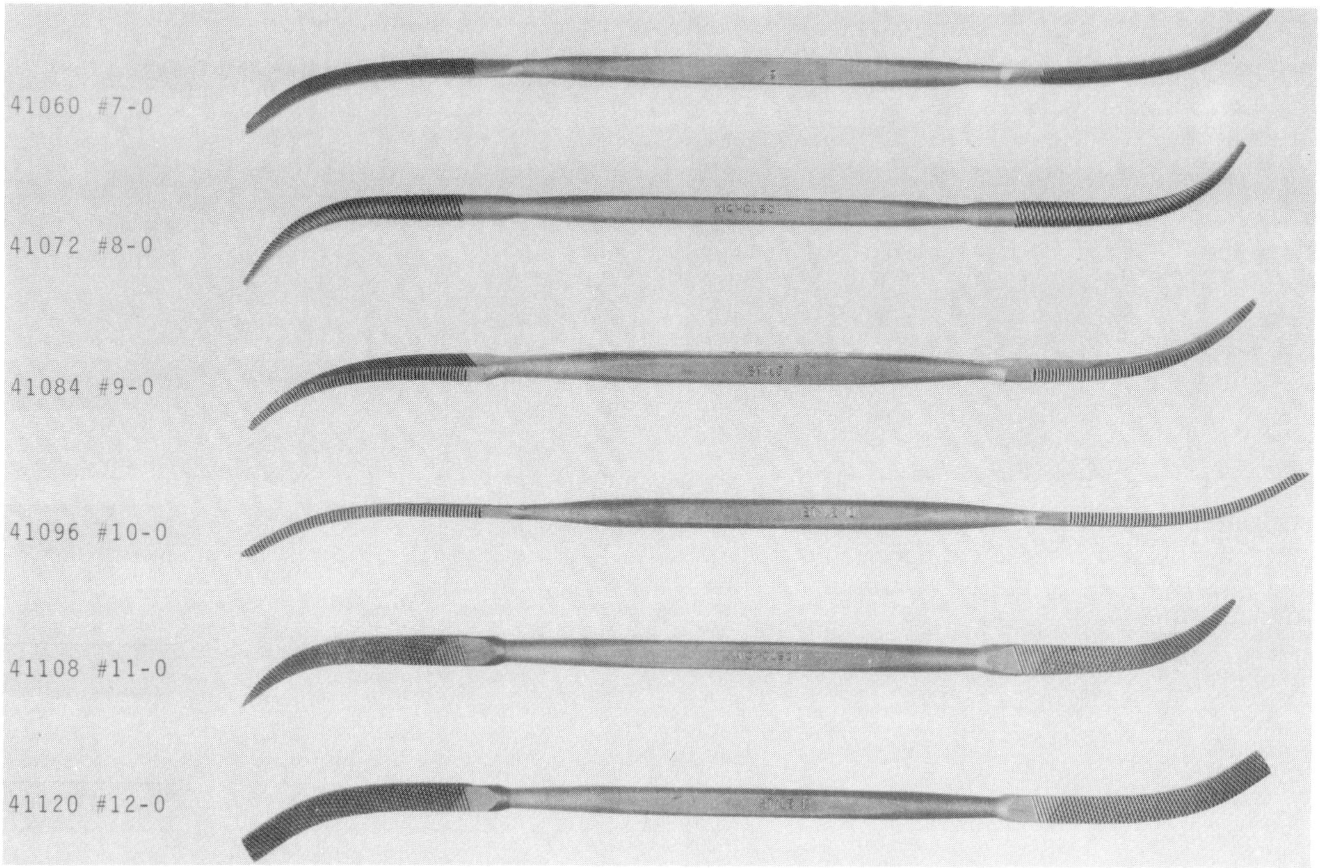

FIGURE 21-8
Riffler files are contoured to reach casting and die sinking recesses.
(Credit: The Cooper Group)

Special-purpose files, such as rifflers (Figure 21-8), are used to reach special contours as those found in die sinking. Their curved endings make them ideal to reach awkward die recesses.

Methods of Filing

As hand filing is usually done on a bench vise, certain considerations such as vise height, workpiece hold, and body position must be met.

The height of the vise should match the elbow height of a bent arm. If a filer is too short for an available vise, a platform should be provided to meet the elbow height requirement.

The workpiece should not extend too high above the vise jaws. Failure to do so will lead to vibration and chatter, leaving a tail of file marks on the work surface. About 1 in. (25 mm) above the jaws would be a good position. Furthermore, the workpiece should be protected against vise marks such as is caused by serrated teeth found on the jaws. A set of soft cover plates, usually made of copper, brass, or aluminum (Figure 21-9), are designed to prevent this. Wood may also be used if soft covers are not available. In filing very thin stock, the material should be sandwiched between and raised slightly above two blocks of wood to prevent bending or chatter.

The body position of a right-handed person (reverse directions if left handed) should be balanced with feet spread from 24 in. (609 mm) apart, the left foot in front of the right foot, enabling the filer to stroke the arms and shoulders freely.

In heavy pressure filing the proper grip has the handle grasped in the right palm, thumb on top, with the fingers of the left hand curled under the

FIGURE 21-9
Soft cover plates are made of nonferrous material and are designed to protect workpiece surfaces from marring.

FIGURE 21-11
Light pressure filing is designed for control and accuracy in cutting.

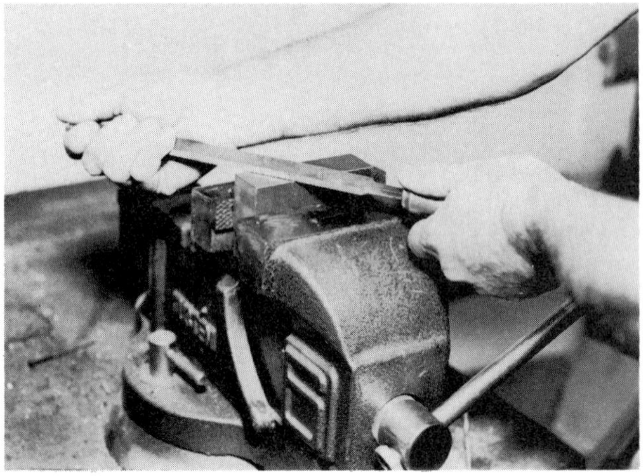

FIGURE 21-10
The grip on heavy pressure filing is designed for greater stock removal.

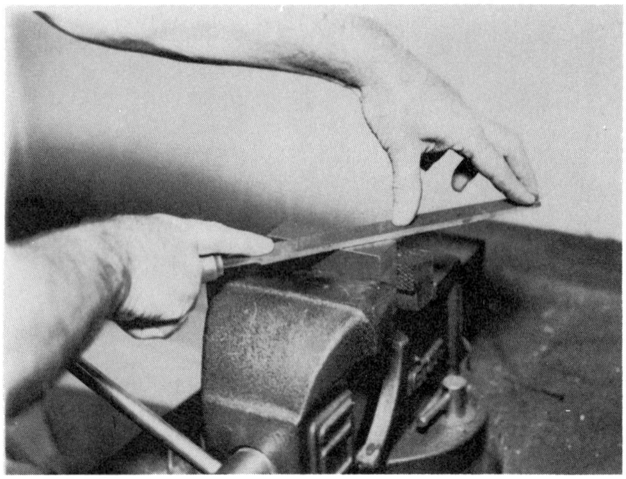

FIGURE 21-12
Flat filing. This position permits even pressure as well as a full stroke of the file.

front end of the file (Figure 21-10). Leaning forward as the file is stroked level, uniform pressure is applied to make sure the file cuts evenly, relaxing the pressure on the backward stroke to return the file to its original position. In filing hard materials (ferrous metals) the file should be raised slightly on the return stroke. In either case filing should be accomplished at not too fast a speed.

In light pressure filing (Figure 21-11) the tip of the file is held by the thumb and index finger, which renders greater control and accuracy in cutting.

Flat filing can be obtained by positioning and

spreading apart the thumb and fingers of the left hand on the end of the file (Figure 21-12). Pressure is distributed uniformly this way, tending to keep the file level, while permitting the full length of the file to be used.

It is important that a file be kept relatively clean or it will clog and cause scratches to be dug in the work surface. This condition is called "pinning" and can be prevented by cleaning the rows of teeth with a file card (Figure 21-13). If a file card fails to clean the teeth, the chips can be chased out with a piece of sharp pointed wire or a piece of copper. Rubbing a piece of chalk across a file's face prior to cutting is recommended to prevent a pinning condition. It should be noted that as a file wears pinning will lessen.

It is equally as important to keep the work surface clean by not fingering or touching the surface. As clean as the fingers may seem, they will leave a deposit of oil that can lead to dulling as the file slips on the oily surface. It is recommended that a clean, dry brush be used to wipe the surface clean.

When files are not in use, their teeth should be protected from damaging by banging or rubbing against other files or hardened tools. They should be stored separately (Figure 21-14), not thrown carelessly in a drawer.

FIGURE 21-13
A file card should be pushed parallel with the file's teeth to remove chips. (Credit: Philip Rapisarda)

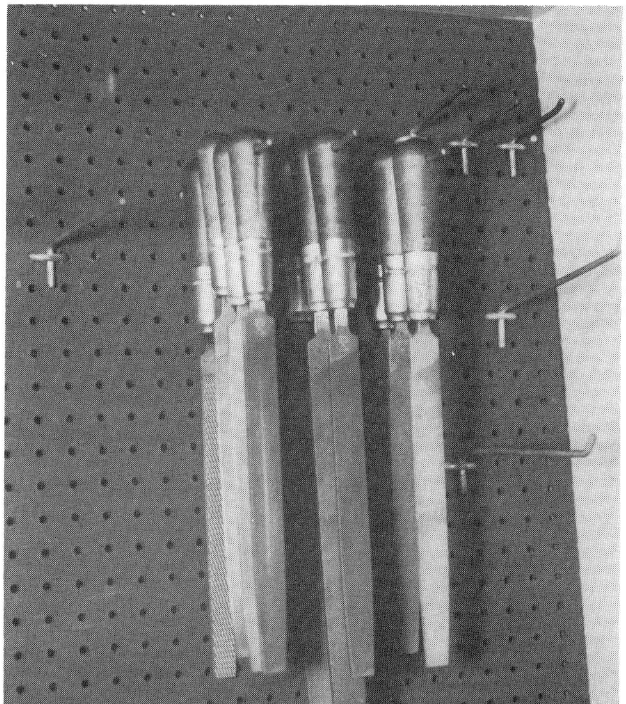

FIGURE 21-14
Files should be stored separately to avoid damage to the cutting edge. (Credit: Philip Rapisarda)

Draw Filing

When a job has been filed to size, it is completed by draw filing, an operation that produces a fine surface finish. A smooth file is used in draw filing, with a single-cut mill file giving good results. The file is grasped in both hands as they straddle the workpiece, extended thumbs being between $\frac{1}{2}$ and $\frac{3}{4}$ in. (13 and 19 mm) away from the workpiece (Figure 21-15). Holding the file at right angles to your work, cutting is achieved by a series of push/pull strokes that has a shearing or shaving effect on the surface. Uniform pressure has to be applied to avoid hollow spots and to keep the file level. To avoid pinning, it is recommended that different parts of the file be used. When one side of the file has been used, turn it over and use the other side. In situations where metal has to be removed rapidly, a flat bastard file is used, followed by a smooth file such as the mill file.

Draw filing will produce a wirelike edge on the sides of the surface being filed. To remove this, hold your file on an angle and use it lightly across each side.

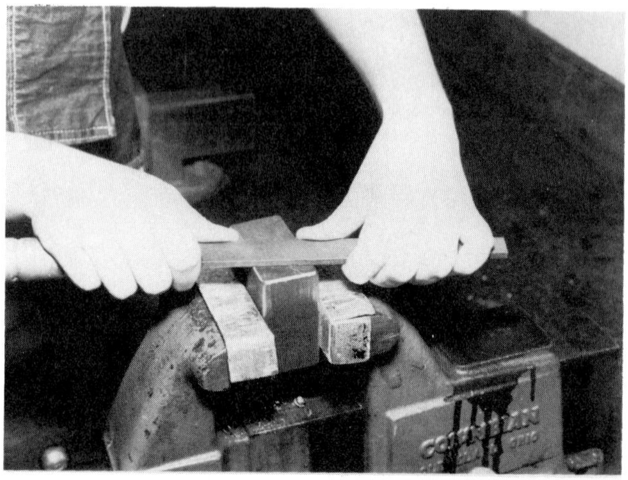

FIGURE 21-15
Draw filing requires a push/pull technique, with the file grabbed close on either side of the work. (Credit: Philip Rapisarda)

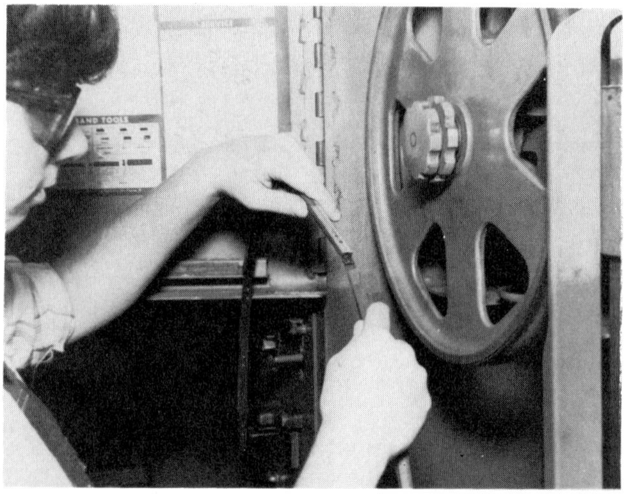

FIGURE 21-16
Preparing to mount a band file on a contour bandsaw.

Power Filing

There are three common filing techniques used in industry that come under the heading of power filing: band filing, machine filing, and rotary filing or burring.

Band filing (Figure 21-16) is considered the fastest and most accurate means of filing. Performed on a vertical bandsaw, a number of file cuts are possible, rendering coarse, fast removal of metal to fine cut files that will leave work smooth.

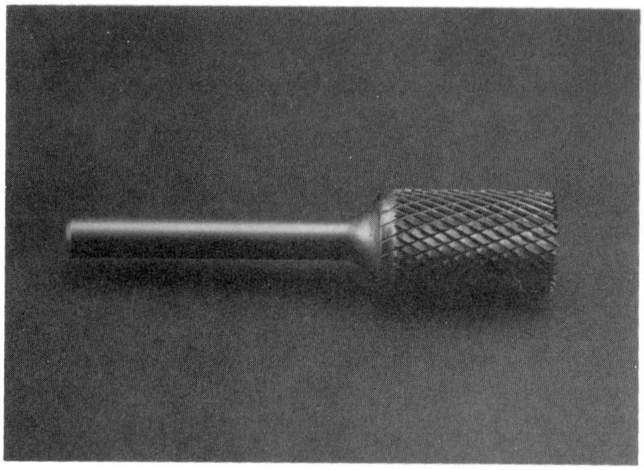

FIGURE 21-17
A carbide rotary burr. (Credit: DoAll Company)

A filing machine is often used by die makers to complete the shaping and contours and other tools. A much smaller version of band filing, machine filing, employs special files that are used with an up and down reciprocating motion. The table upon which the work rests is adjustable for angular setting, while a roller guide is adjusted behind the file to give it correct friction during use. Failure to properly adjust this roller guide will cause the file to break.

Rotary files and burrs are small cutters that are manufactured in many sizes and shapes (Figure 21-17). They are made of high-speed steel, solid carbide or steel spindles bearing carbide tips with ground or chiseled cutting edges. Generally used through flexible shafting or driven by air or electric portable tools, these cutting edges are found useful in reducing soft materials such as wood or plastics to the more commonly used metals that include cold-rolled steel, brass, or aluminum. The hand-chiseled cutting edges of the rotary file prevents them from heating up too quickly and are used extensively against abrasive or tough materials such as forgings, cast iron, and stainless steel. Burring tools are composed of ground, uniform teeth and flutes. This construction prevents them from clogging too easily and are, therefore, more efficient than rotary files.

UNIT 22 BENDING AND FORMING METALS

Bending of metal can be defined as changing a flat surface into another direction. Long before there were modern power presses to form and bend metal, there were any number of hand tools, devices, and operations from which they emerged. Thin flat materials required one type of tooling, while thin tubing required another. The ability to form a radius bend on thin tubing without causing it to crease would perhaps require special devices or, lacking tooling, improvising. Since many ferrous and nonferrous metals under $\frac{1}{4}$ in. (6 mm) are frequently cold formed, the various ways to accomplish this will be examined.

Objectives
After completing this unit you will be able to

1. identify and use various hand tools for bending metals.
2. use various techniques for bending metal.
3. calculate material allowances for making 90° bends.

Safety

No task in bending or forming metals should be undertaken without performing the following basic steps:

1. Always wear approved safety glasses.
2. Deburr and remove all sharp edges before starting.
3. Return all unused stock to storage.
4. Be careful in hammering not to hit your hand or fingers.
5. Learn to use any mechanical devices properly and avoid getting caught in the mechanism.

Basic Hand Forming Tools

The basic hand tools used in bending metal by hand are usually a bench vise, a ball-peen hammer, a wrench of some sort, a form block, and quite often something to hit against materials such as aluminum or hard phenolic (a plastic resin used by the electrical industry as a nonconducting insulating material) to eliminate the hammer marks. In situations where the metal thickness is light enough, a soft-faced mallet (Figure 22-1) can be used. Mallets come in rawhide, plastic, rubber tipped, and synthetic material. In order to be effective, however, a mallet should not bounce when it strikes. It should land with a dead blow.

Angular bends on flat metal are developed by folding or hammering the metal near the bend line. Another factor to be considered is the blank length. For making angular bends, the blank length should be longer than the bent form. How long? You should allow an amount equal to one-half the material thickness for each right-angle bend; therefore, if you are going to use $\frac{1}{8}$ in. (3.17 mm) material and make two right-angle bends, you should make the blank $\frac{1}{8}$ in. (3 mm) longer than the finished length required. The formula for developing a 90° bend allowance can be seen in Figure 22-2. This formula is carried out one step further in Tables 22-1 and 22-2 where 90° bends for common sheet metal thicknesses have been calculated.

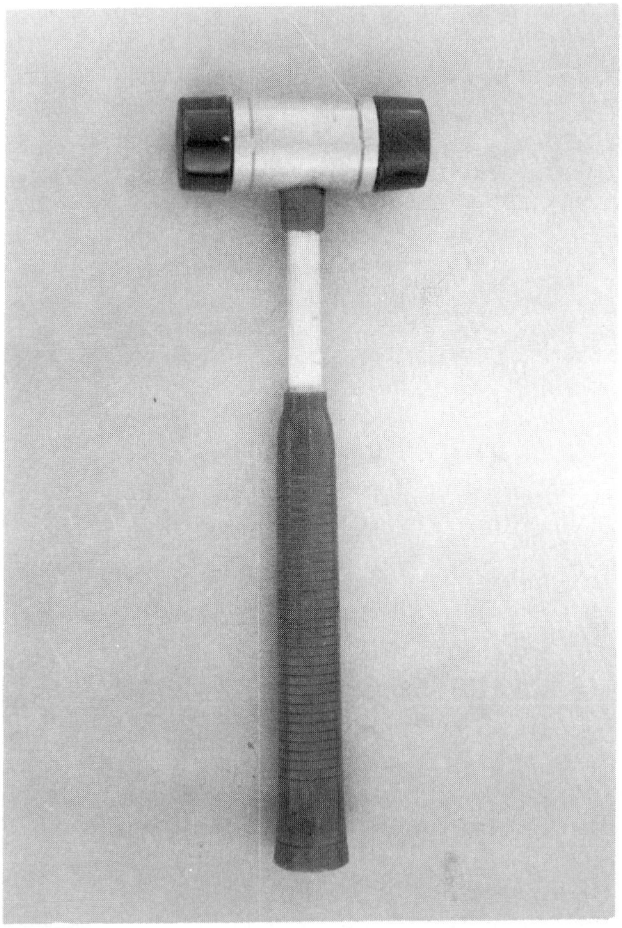

FIGURE 22-1
Plastic-tipped mallet. Mallets also come in wood, rubber, rawhide, and synthetic materials.

T = Stock thickness
R = Inside radius of bend
E = Bend allowance
Y = $T/3$, when R is less than $2T$
Y = $T/2$, when R is $2T$ or larger
E = $1.5708\,(R + Y)$

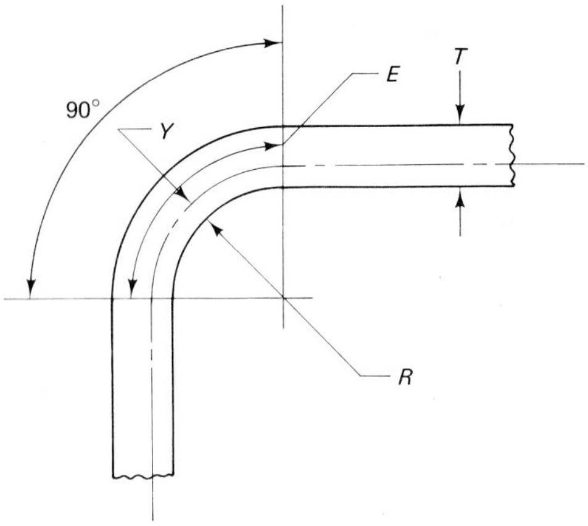

FIGURE 22-2
Formula for calculating 90° bend allowances.

TABLE 22-1
Bend allowances for 90° bends

Stock thickness	*Inside radius of bend in inches*									
T	$\frac{1}{64}R$	$\frac{1}{32}R$	$\frac{1}{16}R$	$\frac{3}{32}R$	$\frac{1}{18}R$	$\frac{3}{16}R$	$\frac{1}{4}R$	$\frac{5}{16}R$	$\frac{3}{8}R$	$\frac{1}{2}R$
0.015	0.033	0.061	0.110	0.159	0.209	0.307	0.405	0.503	0.601	0.798
0.032	0.041	0.066	0.123	0.172	0.222	0.320	0.418	0.516	0.614	0.810
0.062	0.057	0.082	0.131	0.186	0.245	0.344	0.442	0.540	0.638	0.834
0.093	0.074	0.098	0.147	0.196	0.255	0.368	0.466	0.564	0.663	0.859
0.125	0.090	0.114	0.164	0.215	0.267	0.373	0.491	0.589	0.687	0.884
0.156	0.106	0.131	0.180	0.229	0.278	0.384	0.500	0.614	0.712	0.906
0.187	0.123	0.147	0.196	0.245	0.295	0.394	0.510	0.624	0.736	0.933
0.250	0.155	0.180	0.229	0.278	0.327	0.425	0.529	0.643	0.756	0.982
0.312	0.188	0.212	0.262	0.311	0.360	0.458	0.556	0.662	0.776	1.002
0.375	0.221	0.245	0.294	0.344	0.393	0.491	0.589	0.687	0.797	1.043

TABLE 22-2
Bend allowances for 90° bends

Stock thickness T	*Inside radius of bend in millimeters*									
	0.40R	*0.79R*	*1.59R*	*2.38R*	*3.18R*	*4.76R*	*6.35R*	*7.94R*	*9.53R*	*12.7R*
0.38	0.84	1.55	2.79	4.04	5.31	7.80	10.29	12.77	15.27	20.27
0.81	1.04	1.68	3.12	4.37	5.64	8.13	10.61	13.11	15.60	20.57
1.57	1.45	2.08	3.33	4.72	6.22	8.74	11.27	13.72	16.21	21.18
2.36	1.88	2.49	3.73	4.98	6.48	9.35	11.84	14.33	16.84	21.81
3.18	2.29	2.90	4.17	5.46	6.78	9.47	12.47	14.96	17.45	22.45
3.96	2.69	3.33	4.57	5.82	7.06	9.75	12.70	15.60	18.08	23.01
4.75	3.12	3.73	4.98	6.23	7.49	10.01	12.96	15.85	18.69	23.70
6.35	3.94	4.57	5.82	7.06	8.31	10.80	13.44	16.33	19.20	24.94
7.92	4.78	5.38	6.65	7.90	9.14	11.63	14.12	16.81	19.71	25.45
9.53	5.61	6.22	7.47	8.74	9.98	12.47	14.96	17.45	20.24	26.49

Making Right-Angle Bends

1. Study the drawing or sketch to understand what material and bends are required.

2. Allow one-half the material thickness for each bend to develop blank length or consult Tables 22-1 and 22-2.

3. Lay out the bend line on the workpiece.

4. Put your safety glasses on.

5. Insert the material vertically in a bench view, clamping it against a form block about 1 in. (25 mm) above the jaws so that the bend line is even with the top of the form block.

6. Check the workpiece edge and form block with a square to see that they are at right angles to each other.

7. If the material is light enough and long enough, bend the top down by hand.

8. With a piece of hard phenolic or aluminum to buffer the hammer blows, strike the metal behind the bend line (Figure 22-3), distributing the blows to even out the folding process. *Caution:* Do not hammer the area excessively for you will thin it out, stretching the metal.

FIGURE 22-3
When it is necessary to prevent hammer marks, use a soft buffer material such as aluminum to hit against.

FIGURE 22-4
Completing an acute angle bend.

Angular Bends on Wide Sheet Stock

Wide sheet stock can be bent in the same manner as described in the right-angle outline except that the material will have to be sandwiched between steel—angle iron may very well do—or hardwood blocks.

1. Lay out the bend line.
2. Place the material between the steel blocks with the layout lines even with the top edge of the clamping blocks and lock in the bench vise.
3. Fold the top edge of the material down by hand.
4. Finish the bending operation by striking along the bend line with a hammer and phenolic or aluminum. If the material is light enough, a weighted mallet can do the job.

Making Bends Other than 90°

Acute angles, as angles less than 90° are called, can be formed by first holding the material as described in the preceding outline, then completing the bend between the jaws of a bench vise as shown in Figure 22-4 until the angle checks out with a protractor.

Making Angles Greater than 90°

Obtuse angles, as angles greater than 90° are called, can be formed against a matching form block with the angle already cut on it or be hammered against a straight edge form block as previously described until the angle has been achieved experimentally and checked out with a protractor.

Increasing a Bent Angle

Angles that have been overbent or require to have their opening increased without marring their appearance can be changed in the following manner:

1. Set the protractor to the angle desired.
2. With one hand seat and hold the angular opening against a length of rod.
3. With a mallet strike the top of the corner uniformly along its length.
4. Check the results with the protractor until the angle is obtained.

Twisting Metal

Metal can be twisted to present decorative patterns by means of a wrench. Short twists can be made by extending the metal vertically from the jaws of a vise and twisting it by means of a wrench (Figure 22-5). Longer twists can be obtained by clamping the material horizontally in a bench vise, slipping a pipe over the width of the material to prevent distortion and buckling during twisting, and applying the torque of a heavy wrench to achieve the twists (Figure 22-6). Blank lengths can be developed by previously twisting similar pieces experimentally.

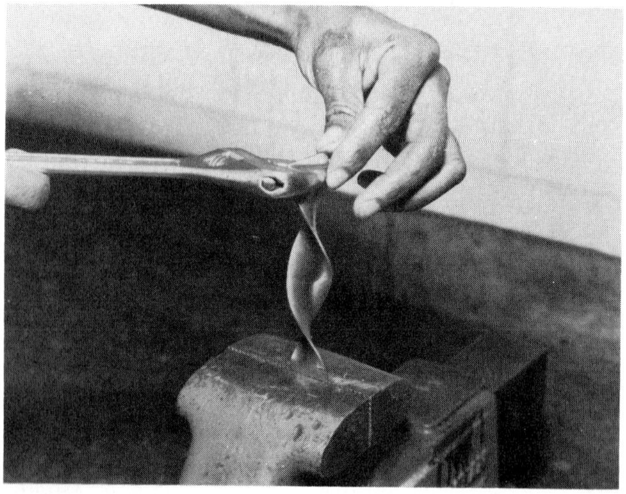

FIGURE 22-5
Different size wrenches can be used to twist metal.

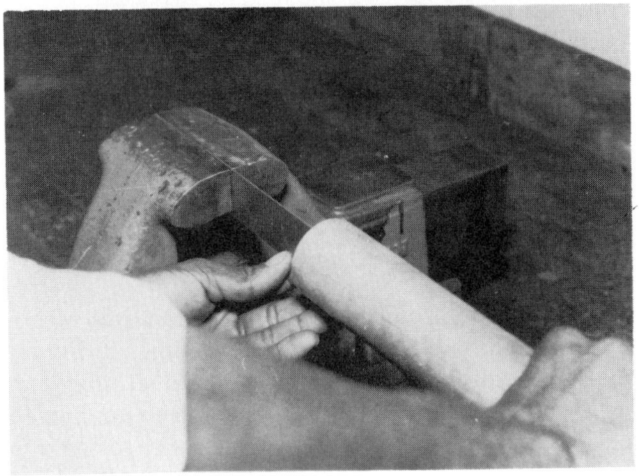

FIGURE 22-6
Long lengths of metal should be inserted within a heavy tube or pipe to prevent distortion and buckling during twisting.

FIGURE 22-8
Bending jig.

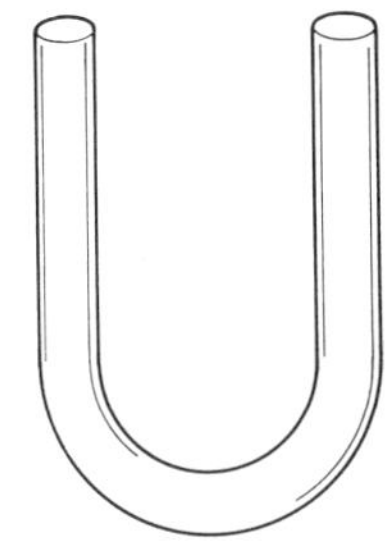

FIGURE 22-7
Bending fork. (Credit: Paul Rapisarda)

FIGURE 22-9
The curve of a scroll is layed out as a pattern on paper first.

Making Curved Bends

Resembling a flat circular spiral or unwound clock spring, scrolls are curved bends of metal that can be formed by the use of a bending fork (Figure 22-7) or a bending jig (Figure 22-8). The bending fork consists of a metal block with pins that can be set within a series of holes to accommodate the various metal thicknesses. Since there is no set way of curving the scroll, it will have to bend a little at a time and be checked frequently with the layout pattern as shown in Figure 22-9. The blank length can be calculated by forming soft wire on the layout scroll pattern and then measuring it when straightened.

Closed curves can also be accomplished by forming metal around a rod or pipe of the same desired diameter. Clamp an edge of metal as shown (Figure 22-10) and bend the metal toward the curving surface of the rod. As the curve is formed, the metal is moved in and around the rod until the curve is complete.

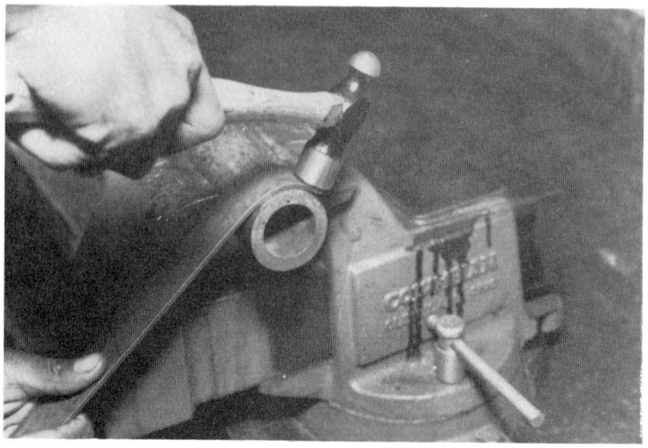

FIGURE 22-10
The bend of a curve begins by clamping the metal against the round form.

FIGURE 22-11
Insertion through a slotted pipe or tubing can aid in the bending process.

Still another method of closing a curve is to form the metal about a slotted pipe. A small opening is made in the pipe of the diameter required to allow the material to be inserted. Once so placed, the metal is pulled around as previously described (Figure 22-11).

Forming Small Rings

Very tight, small closed rings can be made by wrapping soft wire about a round rod as shown in Figure 22-12. When the rings have been formed, they can be cut by a hacksaw and then soldered or brazed closed. A tapered rod can also be used to make a long spiral coil.

Bending Tubing

Wooden form blocks can be used for bending thin tubing or pipe. Figure 22-13 shows how the block is grooved to accept the contour of the material. By clamping the form block and a bracing block to the work bench, tubing can be drawn to the radius of the bend involved. In situations where the bend may be too small and metal becomes kinked (uneven ridges), it is advised that the tubing be filled with wet sand or melted lead to firm it up. When the bend is completed, empty out the tubing or its contents.

Bending Machine

One hand-operated machine can do a multitude of bends and forms with wire, sheet metal, rods, tubing, channel metal, as well as special shaped material. The hand-operated bender shown in Figure 22-14 is ca-

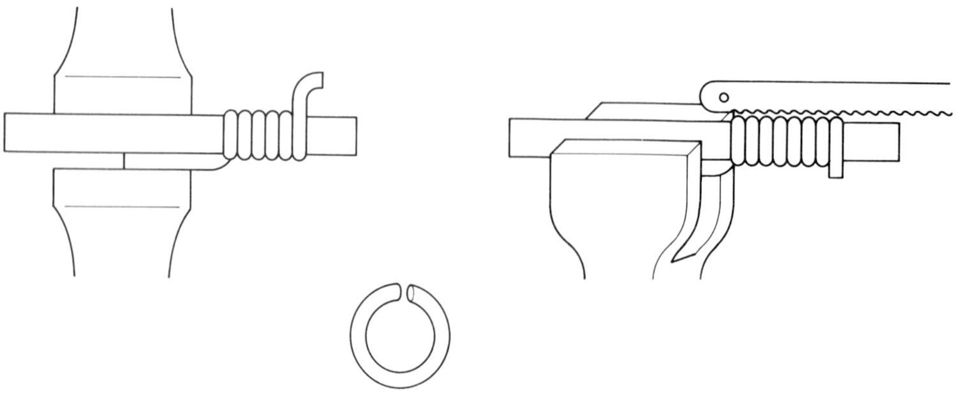

FIGURE 22-12
Making small rings.

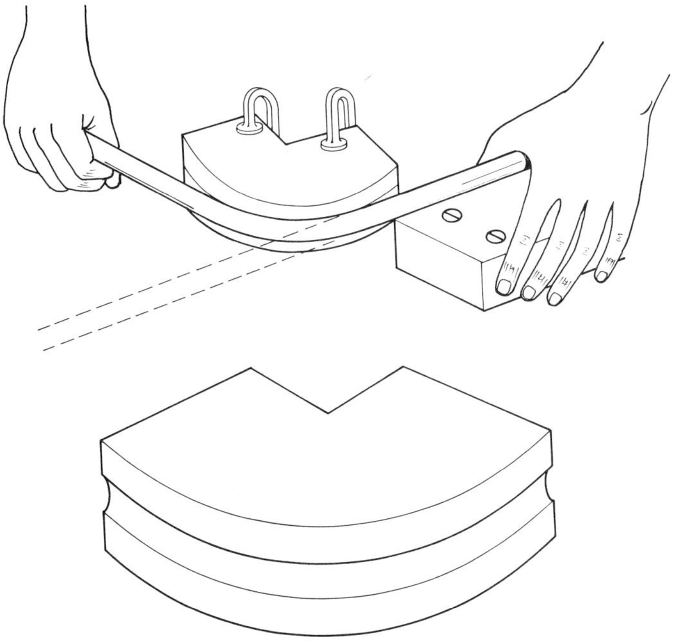

FIGURE 22-13
Wooden form blocks can aid in the bending of light tubing.

pable of developing any number of bending operations, such as circles, loops, and spirals, and zero radius bending. With special tooling added, a variety of different shapes of metals can be accommodated.

Consisting of a circular base and a holed mounting plate upon which a form, radius, and locking pins may be positioned, an operating arm is pivoted about the radius pin bearing a form rolling or shaping device.

One of its operations may best be explained with the forming of a circle. Because most metals "spring back" (open up) after they have been formed, a radius collar (form) is used slightly smaller than the required circle diameter. To determine the amount of spring-back involved in a particular metal will require experimentation, however. All metals have different reactions. To bend the circle, the exact length of the blank has to be precut before forming.

As with any machine, proper instructions should be given on its operation before attempting any sort of bending.

FIGURE 22-14
Di-acro bending machine. (Credit: Diacro Corporation)

Procedure for Bending a Circle

1. Cut the material to exact blank length.
2. Position the radius collar on the plate.
3. Clamp the material against the radius collar with the locking pin.
4. Set the forming nose of the operating arm against the material.
5. Rotate the operating arm, stretching and forming the radius, until the forming nose reaches the end of the material.
6. Relocate the material and reclamp it with the locking pin at a point near the unbent portion where a radius has been formed.
7. Complete rotating the operating arm so that the forming nose reaches the end of the material.

UNIT 23 METAL FASTENING

With the development of metal came the required means of holding or binding metals together. There are several ways of fastening metals, some very simple, some very permanent, and some unique. All of them are useful in modern technology where the demands of new and better ways of doing things or accomplishing travel have had to be explored and conquered. An overview of the subject matter will be presented at this time as some of the subjects are more fully covered in different parts of the book.

Objectives
After completing this unit you will be able to

1. identify and properly use various threaded fasteners.
2. identify and properly use various nonthreaded fasteners.
3. identify and properly use various metal bonding fasteners.
4. identify various permanent metal fasteners.

Threaded Fasteners

One of the most unique ways of holding metals together is by means of a threaded fastener. Probably originating with Archimedes, the noted Greek mathematician and sometimes called the father of experimental science whose contributions included the lever, the pulley, and hydrostatics, the formation of the screw or thread has evolved over the centuries. Originally constructed as a unit for drawing bilge water out of the holds of ships belonging to his king, Hiero of Syracuse, Archimedes' device consisted of a tube wrapped around a cylinder in a spiral rotation. Submerging the lower end and rotating the cylinder on its axis, water would be forced upward through the tube and emptied out at the other end.

The screw has developed from the hand-cut grooves found on suits of armor during the Middle Ages to very accurately ground, precise threads with specific angles. Today screws and threads comprise the most widespread method of fastening metal.

Threads can be made externally or internally. Nuts and screws are good examples of this. Think about it. You can use and reuse the threaded fasteners without damaging either member. Threads are principally made from a variety of metals—steel, aluminum, brass, copper alloys, stainless steel—to serve a variety of purposes such as tensile strength and temperature changes. They are also finished with coatings and platings for corrosion resistance. Threaded fasteners are also made from nylon and plasticlike materials that are light in weight and resist moisture and corrosion. These fasteners also offer a disadvantage in combating temperature, stress, and shear conditions. We will concentrate on some of the more widely used screws used in metalworking and their applications.

FIGURE 23-1
A bolt and screw example.

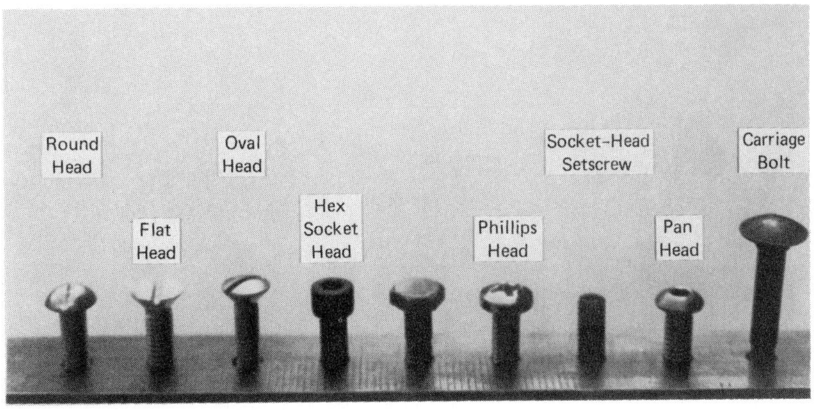

FIGURE 23-2
Screw head types.

Bolts and Screws

Figure 23-1 illustrates the difference between a bolt and a screw. Is there a difference? The answer lies in the application of the threaded fastener. If a threaded fastener is used to assemble parts and is normally inserted through a hole to be tightened or released by means of a nut, it is considered a bolt. If, on the other hand, a threaded fastener is able to mate with or make its own thread and can be tightened or released by turning the head of the fastener, it is considered a screw. Consequently, the same threaded fastener can be considered a screw or a bolt, depending on its application.

Screw Heads and Types (Figure 23-2)

Machine screws: The UNF and UNC machine screw ranges in diameter size from 0 (0.060) to $\frac{3}{8}$ in. (0.375) and in lengths from $\frac{1}{8}$ to 3 in. They are made in a number of head styles including round head, oval head, flat head, fillister head, and are also available with a Phillips head drive (Figure 23-3).

Machine bolts: Bolts in the UNF and UNC thread series range in diameter from $\frac{1}{4}$ in. (0.250) to 6 in. and in length from $\frac{1}{2}$ to 30 in. ISO metric bolts and screws vary in diameter from 1.6 to 68 mm and 5 to 500 mm in length, generally with square or hexagonal heads and nuts to match.

FIGURE 23-3
Machine screws used in industry.

Cap screws: In situations that demand an assembly where a screw is inserted through one piece so that it can thread into the bottom piece, allowing the head of the screw to press against the top piece as it locks and binds both pieces together, a cap screw is used. Cap screws are available with UNF, UNC, and metric threads and different types of heads and drives. This would include flat head, hex head, socket head, fillister head and round head (Figure 23-4).

FIGURE 23-4
Socket head cap screws.

FIGURE 23-5
The use of stud bolts can be found on an automotive engine block.

Stove bolts: Cut with a UNC thread, stove bolts are made with slotted flat or round heads to be driven by a screwdriver. Used in general assembly work, stove bolts were originally named for their value in metal stove assemblies.

Stud bolt: Threaded on both ends, a stud bolt is headless. It is used on heavy assemblies such as a cast-iron pump housing or an automobile engine. The shorter of the threaded ends is screwed tight into the cylinder block, for example, which allows the cylinder head to be placed over the upright studs where it is locked in place by means of tightened nuts (Figure 23-5).

Carriage bolt: Made with a square neck beneath its solid domed head, the carriage bolt is designed to hold against wood in wood-to-wood or wood-to-metal assemblies. Its square neck is made to sink into the wood when the bolt is tightened with a nut. Made in the UNC thread series ranging in diameter from $\frac{1}{4}$ to $\frac{1}{2}$ in. and as long as 6 in. in length, the carriage bolt can be purchased with a black coating.

Thread cutting screw: A boon in industry where they are used in heavy sheet metal and nonferrous assemblies, the thread cutting screw is made with hardened teeth. These screws are able to cut threads in previously drilled or punched out holes.

Setscrews: A variety of setscrews are made for locking collars, gears, or pulleys against shafts. They are classified by their point and head styles. Their points are case hardened and may vary from flat to cone point to a cup point. Their heads may be styled with a square head or a slotted headless type or may feature a socket head (Figure 23-6).

Nuts and Washers

The nuts and washers shown in Figure 23-7 play an important role in mechanical fastening. Often made in hexagonal or square shapes, the thread series of the nut must match the thread series of the bolt or screw involved. Nuts are manufactured by four gen-

FIGURE 23-6
Socket head setscrews.

FIGURE 23-7
A collection of nuts and washers.

eral methods of making blanks: cold punched, cold forged, hot forged, and milled from a bar. They are also made with a number of secondary operations that may include drilling, slotting, trimming, shaving, heat treating, and surface finishes to dimensional accuracy and special requirements.

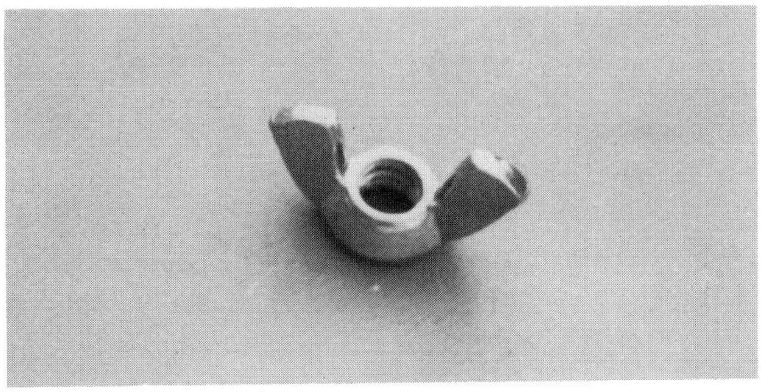

FIGURE 23-8
A wing nut.

Machine screw nut: Considered a standard hex nut, the machine screw nut serves as a multipurpose fastener. It is made for the UNF, UNC, and metric thread series.

Jam nut: Thinner than the machine screw nut, the jam nut is used to back up or lock the machine screw nut. It is also used where space requirement leave no room for the conventional nut height to be used.

Castle nut: Slotted across the top and into its machined diameter to render a castlelike profile, the castle nut is designed to be locked in place. Aligning one of the slots with a hole in the bolt after tightening, a cotter pin is slipped in place to prevent the nut from turning. The castle nut is widely used on automobile wheels and wheel bearings.

Self-locking nuts: Mechanical fasteners in the form of self-locking nuts that feature collars or plastic inserts are commercially available. Other self-locking nuts contain distorted threads to provide a friction grip that will not jar or vibrate loose.

Wing nut: Designed to be used by turning it with the thumb and index finger, the wing nut is used on light assembly work requiring fast and easy adjustments or removal (Figure 23-8).

Acorn nut: Manufactured to house or cover the end of a screw, the acorn nut lends appearance of good look as well as protection against sharp edges that may be exposed.

Washers

Washers serve several basic functions in mechanical fastening. They provide a bearing surface, distribute the load over a wider area, protect the adjoining surface, or may also lock assembled parts. These, too, are manufactured from various materials, heat treated, and made to withstand light- to heavy-duty applications.

Flat washer: Primarily used to provide a wider bearing surface, flat washers are measured by the hole size.

Lock washer: Resembling a split ring, the lock washer serves to prevent movement or loosening from vibration when tightened against by a nut or a screw head. Due to the fact that the split edges are made to bite into the surrounding surface, the lock washer should not be placed between a flat washer and the tightening member. To do so would only defeat the purpose of the lock washer (Figure 23-9).

FIGURE 23-9
A collection of lock washers.

FIGURE 23-10
Tooth-type lock washer.

Tooth-type lock washer: The tooth-type lock washers are in wide use. Fabricated with external, internal, or a combination of both types, hardened teeth are made to dig into the bearing surface to prevent slippage or loosening due to shock and vibration. This type of washer is in wide use in the automotive industry (Figure 23-10).

Preassembled nut and lock washer: Fitted loosely with a lock washer below the nut, preassembled nuts and lock washers are used to speed up the production on assembly lines (Figure 23-11).

Nonthreaded Fasteners

A number of nonthreaded fasteners are used in metalworking, some are of a reusable nature, some are very permanent.

Pins

Dowel Pins. Dowel pins (Figure 23-12) are manufactured in both a hardened and unhardened state. They are straight, accurately ground, and come in a number of sizes. The hardened style is widely used in the diemaking trade to prevent shifting or lateral movement between mating parts such as a punch and die, for example. Dowel pins are tapped or press fitted into reamed holes.

Tapered Pins. Also designed to prevent shifting of assembled parts, a tapered pin is often used to position gears and pulleys to shafts. The American standard taper pin tapers $\frac{1}{4}$ in. per foot and is used in a reamed hole bearing the same size. Unlike the dowel pin, the tapered pin can tighten its fit by being driven deeper in the hole (Figure 23-13).

FIGURE 23-11
Preassembled nut and lock washer.

FIGURE 23-12
Dowel pins.

FIGURE 23-13
A taper pin.

Keys

A key is a metal insert fitted between the shaft and the crank, pulley, or gear hub to provide alignment and the means for a positive drive. The key is made to fit mating slots that are found partially cut in the hub and in the shaft. The slot in the shaft is called the keyseat. The slot cut into the hub is referred to as the keyway (Figure 23-14).

A Woodruff key is one of the most commonly used keys in the machine tool industry. A flat sectioned disk in a shape that is made to fit in a semicircular keyseat, the Woodruff key has one-half of its height extending above the shaft to engage the hub above it.

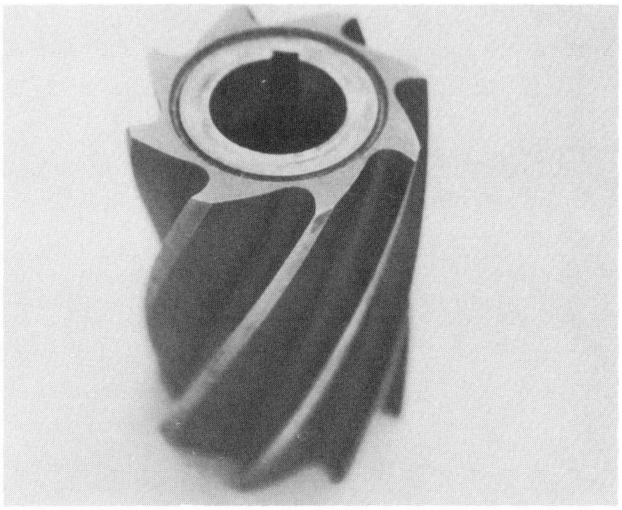

FIGURE 23-14
A keyway is cut into a cutter hub for alignment purposes.

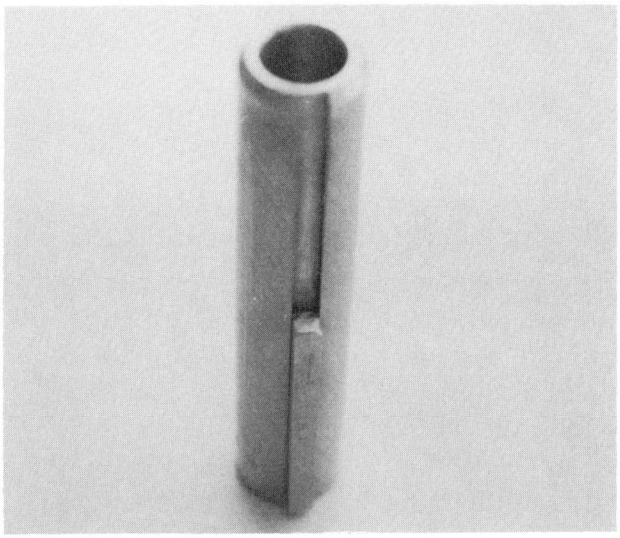

FIGURE 23-15
Using a key in a tailstock spindle assembly.

The square key is one of the simplest type of keys used. Made to seat half in the shaft and half in the hub, the square key may be slightly tapered on its upper surface for easier insertion (Figure 23-15).

The gib-head key is often used in place of the square key because its construction lends itself to easier insertion and removal.

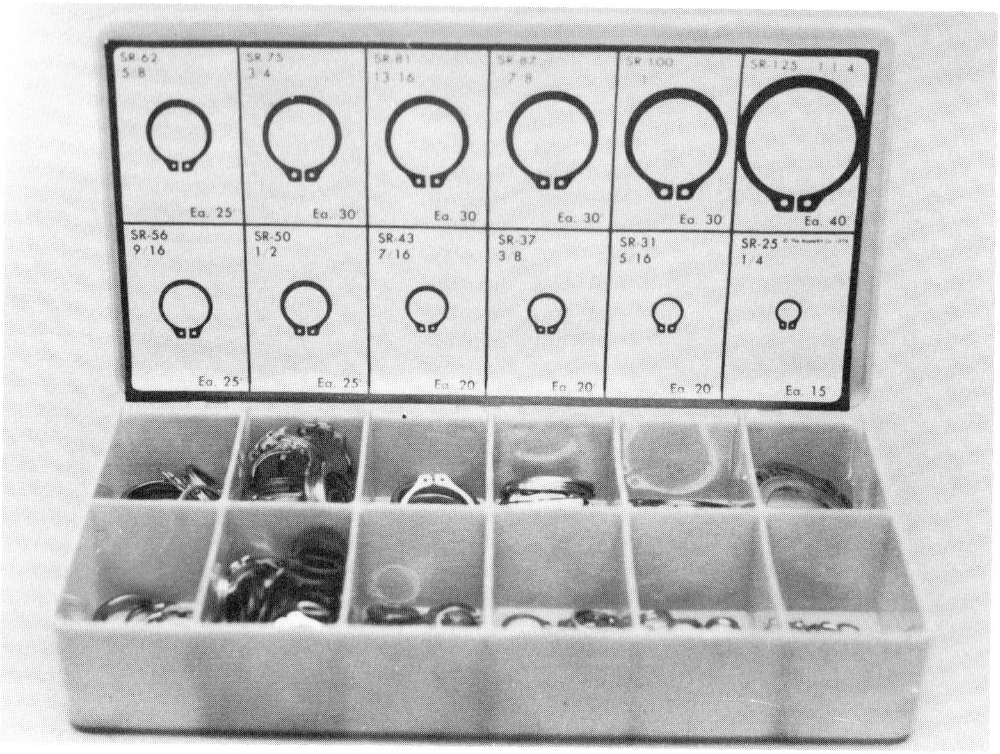

FIGURE 23-16
A set of retaining rings.

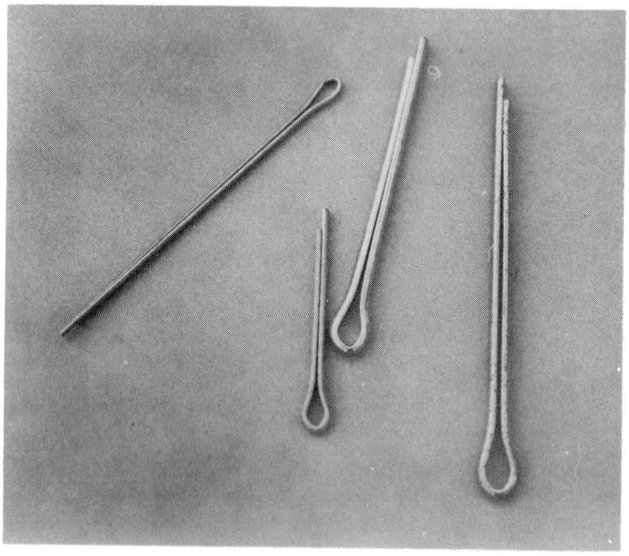

FIGURE 23-17
Cotter pins.

Retaining Rings

Another nonthreaded fastener used in industry is the retaining ring (Figure 23-16). Employing special tools, many are made to fit and seat in grooves and recesses. They can also be purchased as self-locking fasteners. Retaining rings are used for axial assemblies, end-play take-up, and radial assemblies. Hardened and tempered, retaining rings are designed to eliminate bulky fasteners, reducing product cost and weight in the process.

Cotter Pins

Made of heavy wire, a cotter pin (Figure 23-17) somewhat resembles a hair pin. Its use lies in preventing slippage or turning between a nut and a bolt. It is inserted into the hole of a bolt that has been tightened and aligned with the slot of the castle nut. Locking by spreading the legs should be done when the head of the cotter pin can be fully located within the slot and merely making contact with the outside edge of the nut.

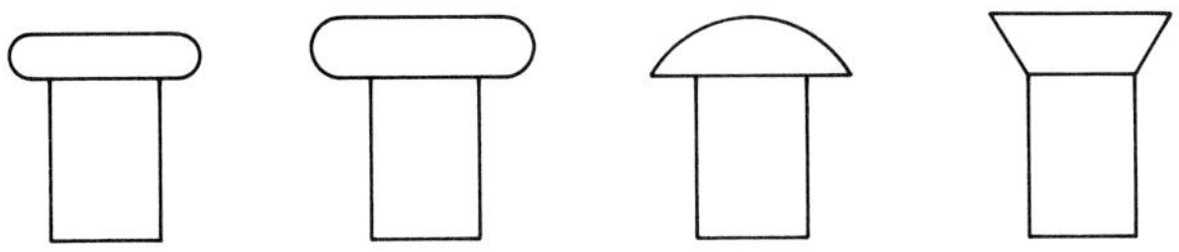

FIGURE 23-18
Rivets come in a variety of shapes, length, and materials. (Credit: Paul Rapisarda)

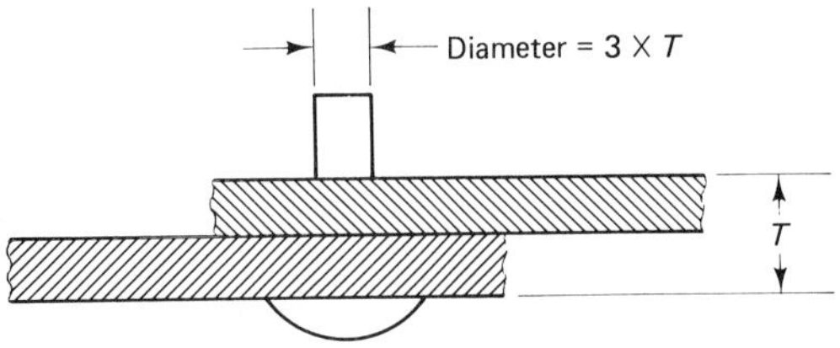

FIGURE 23-20
A rivet's diameter should be about three times the thickness of the surfaces being fastened.

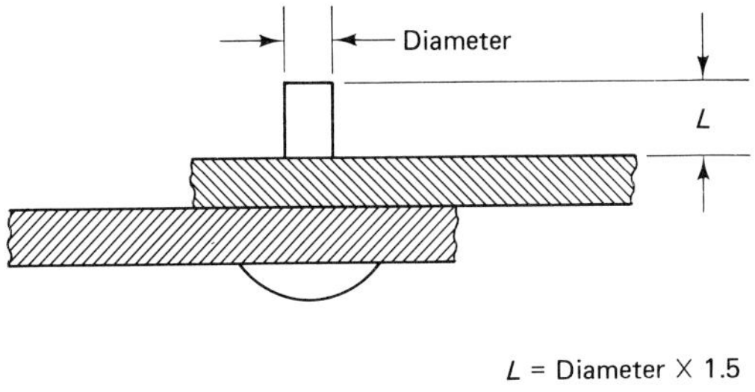

FIGURE 23-19
A rivet's length should be about $1\frac{1}{2}$ times longer than the rivet's diameter.

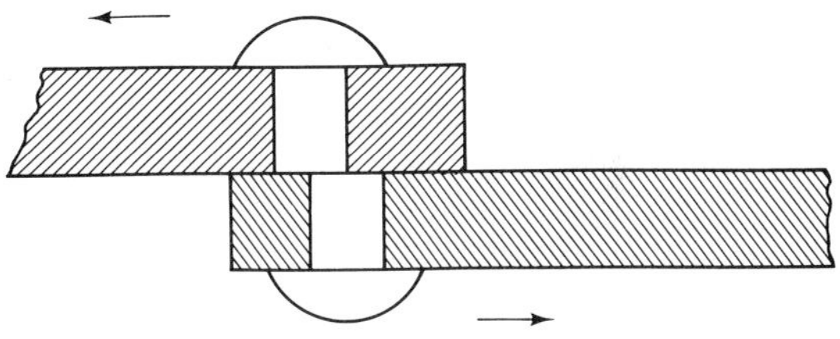

FIGURE 23-21
Plates pulling apart due to rivets being too small for the holes.

Riveting

One of the permanent means of binding metal together is by means of another type of nonthreaded fastener. It is called a rivet and is inserted as a straight pin in a hole between metal surfaces to be joined. Its end is flared, hammered, or otherwise spread to complete the shaping and closing of the rivet. Rivets come in a variety of metals that range from nonferrous metals such as copper, brass, and aluminum to soft steel. They come in a variety of head sizes and shapes (Figure 23-18), diameters, and lengths from $\frac{1}{4}$ to 3 in. (6.35 to 76.2 mm).

A rivet's holding power is based on three basic factors: rivet size, hole size, and hole spacing or pitch as it is referred to. The shape of a rivet's head plays a role in figuring the length of the rivet involved. There are many formulas to accommodate specific head shapes and materials, but one general rule is: rivet length should be about $1\frac{1}{2}$ times longer than the body diameter of the rivet (Figure 23-19).

A rivet's body should be about 3 times the thickness of the surfaces to be fastened and fit closely in the hole provided. This is illustrated in Figure 23-20. When the rivet's body size is too small, it may result in the plates being pulled apart (Figure 23-21).

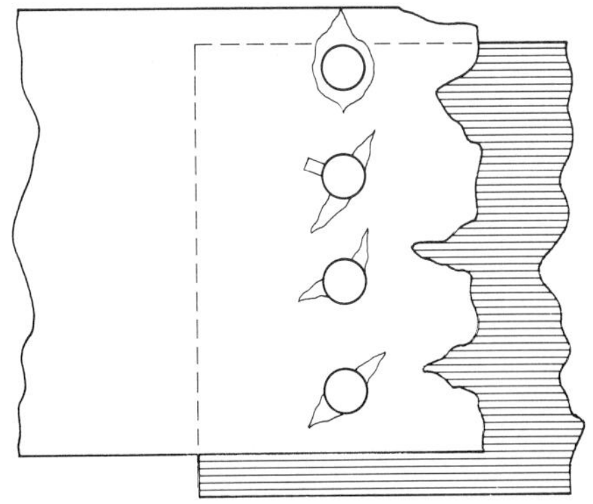

FIGURE 23-22
Bursting is caused by placing the rivets too close to the edge. (Credit: Paul Rapisarda)

Another consideration is the hole location. When a rivet is placed too near the edge, the plate is weakened and suddenly gives way, bursting (Figure 23-22). To prevent this, the center distance of a

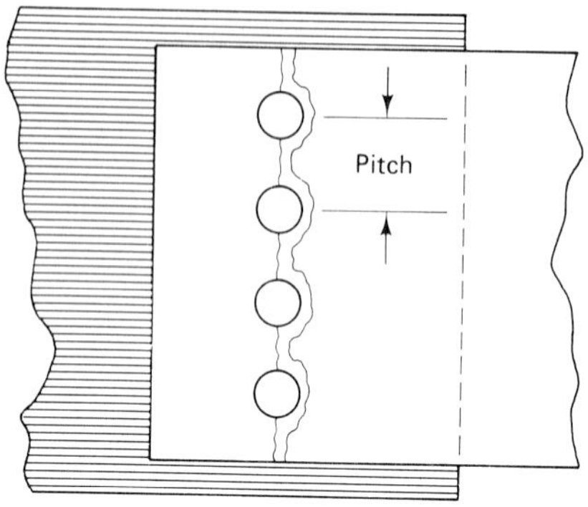

FIGURE 23–23
Tearing is caused by an incorrect pitch (center distance) of the rivets. (Credit: Paul Rapisarda)

rivet's hole to the edge of the plate should be $1\frac{1}{2}$ times the rivet's body diameter. Incorrect pitch or center spacing of the hole's location will lead to tearing (Figure 23–23). This can be prevented by the following formula: Pitch is equal to 1.6 times plate thickness plus $1\frac{1}{4}$ in. (31 mm).

Many situations in machine shops and industry will require the use of blind rivets or Pop rivets, as trade marked by the USM Fastener Company, so called because they are inserted and fastened on one side only. These rivets are available in either a tubular shank or a partial hole shank. Both contain a mandrel or pull stem that will fracture upon closing the surfaces.

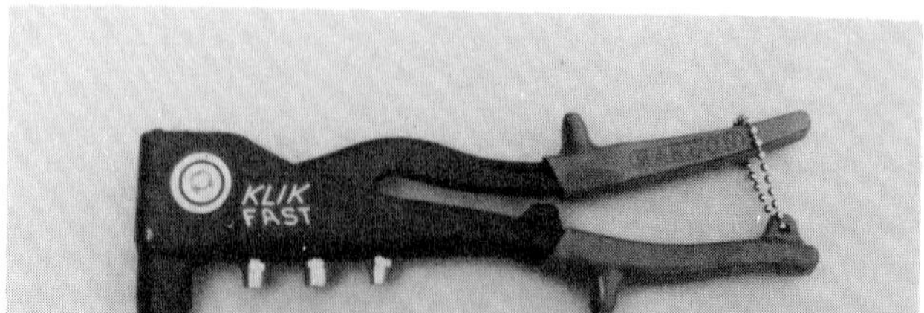

FIGURE 23–24
A clinching tool for blind rivets.

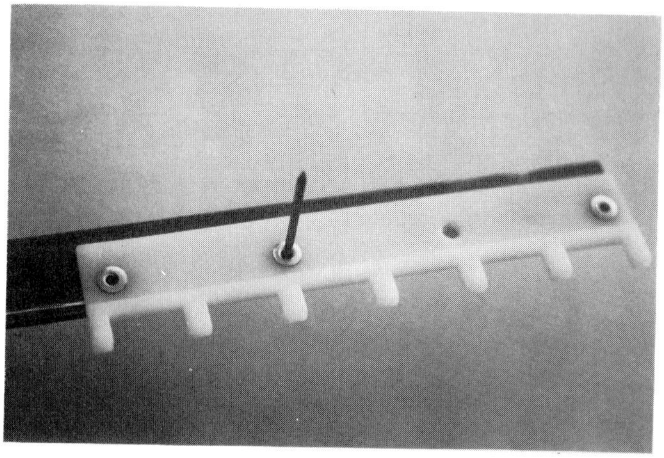

FIGURE 23–25
The use of tubular blind rivets.

Blind rivets can be secured by use of special riveting machines or closed by hand using a clinching tool (Figure 23–24). Clinching takes place by pulling up on the mandrel, which compacts the bottom of the rivet against the underside of the surface, breaking off (Figure 23–25).

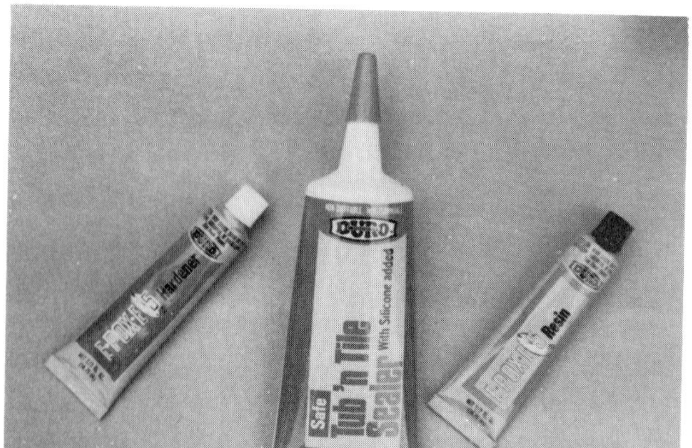

FIGURE 23–26
Epoxies are used for mending any number of household items.

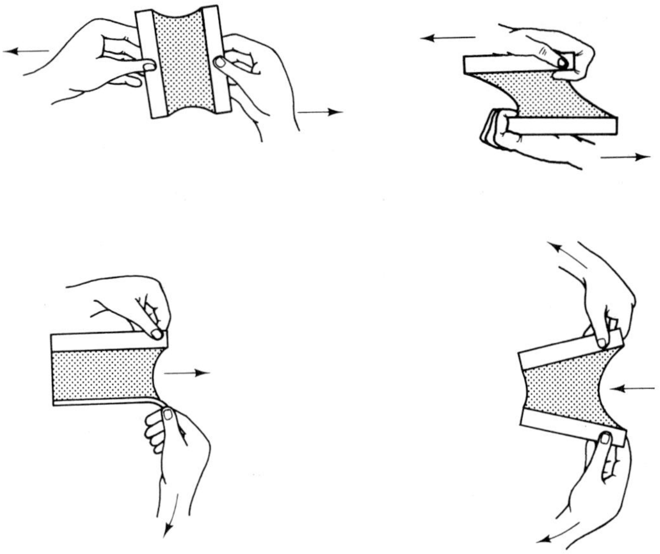

FIGURE 23-27
Stress forces to which adhesive bonds are subjected.

Cements

Bonding by glue and similar cements has been known for centuries. What is relatively new is the use of structural (load bearing) adhesives, which were pioneered by the aircraft industry. Increased properties have made these adhesives applicable for use in modern design. For instance, epoxy unites with the surface being bonded, which enables the "wonder glue" to mend broken pottery and other items around the house (Figure 23-26).

Stress Forces

In order for adhesives to be considered for structural or design application, they must stand up to stress conditions in use. These four conditions are (1) shear, (2) tension, (3) peel, and (4) cleavage (Figure 23-27).

Generally, adhesives fall in the categories of thermosetting, thermoplastic, or a combination of both. Thermosetting adhesives are a permanent type. They harden with heat and pressure and their state will not change. Epoxy, polyurethane, phenolic, and polyester fall in this category. Thermoplastic adhesives are softened by heating, harden upon cooling, and can be resoftened and set again. They lack the load stress of the thermosettings, however. Thermoplastic adhesives are known to creep; they do not stay set. In order to add some flexibility to thermosetting adhesives to avoid hardening as quickly, silicone or rubber will be added.

The use of adhesives as metal fasteners presents advantages and disadvantages. On the positive side, adhesives are neater, can act as insulating materials, can absorb vibration (damping) to minimize fatigue, can join diverse metals, and can minimize corrosion by not trapping moisture. On the negative side, structural adhesives have to be applied to very clean surfaces, rate very poorly in peel and cleavage stress, require special tooling so that curing (hardening by chemical reaction) can take place, and finally, there is no way of testing without destroying the bond.

In summary, there is no adhesive that will satisfy all conditions that includes strength to environmental change. They all have to be considered on individual merit.

Safety

Adhesives are classified by the temperature at which they harden. Below 68°F (20°C) they are classified cold setting. From 87 to 212°F (30 to 100°C) they are classified hot setting. Most structural adhesives fall in this category. Adhesives are available in solvents, paste, powder, tape form, syrups, sheet, and film, and they should not be used without safety considerations. Solvents come in liquid form and can be rolled on or sprayed on but should only be used in well-ventilated areas. Other adhesives can be harmful to the skin or eyes. In all cases manufacturer's directions should be followed in preparation and use.

Welding

The uniting of metal into a permanent bond by means of high temperature, compression, or sometimes a combination of both is called welding. While there are many processes, there are two broad categories that the welding processes fall into: fusion and pressure welding. Only a few processes will be examined here.

Fusion, the joining together by melting and then cooling to form a permanent bond, is best exemplified by arc welding (Figure 23–28). Other processes falling into the fusion type of welding will include gas welding (Figure 23–29), thermit welding, and laser welding.

Thermit welding is a process that involves a chemical reaction that takes place when iron and aluminum are superheated (thermit) to be poured into a mold surrounding the parts to be welded. The aluminum frees the iron of oxygen, and when the thermit combines with the joints already heated to a plactic state, a very solid weld takes place. Rails and repairing heavy machinery are done in this manner.

A number of welding applications are possible using laser technology. Heavy metals can be deeply penetrated by focusing a laser beam upon the workpiece to be fused. Vaporization that begins at the focus point burrows a column deep within the base

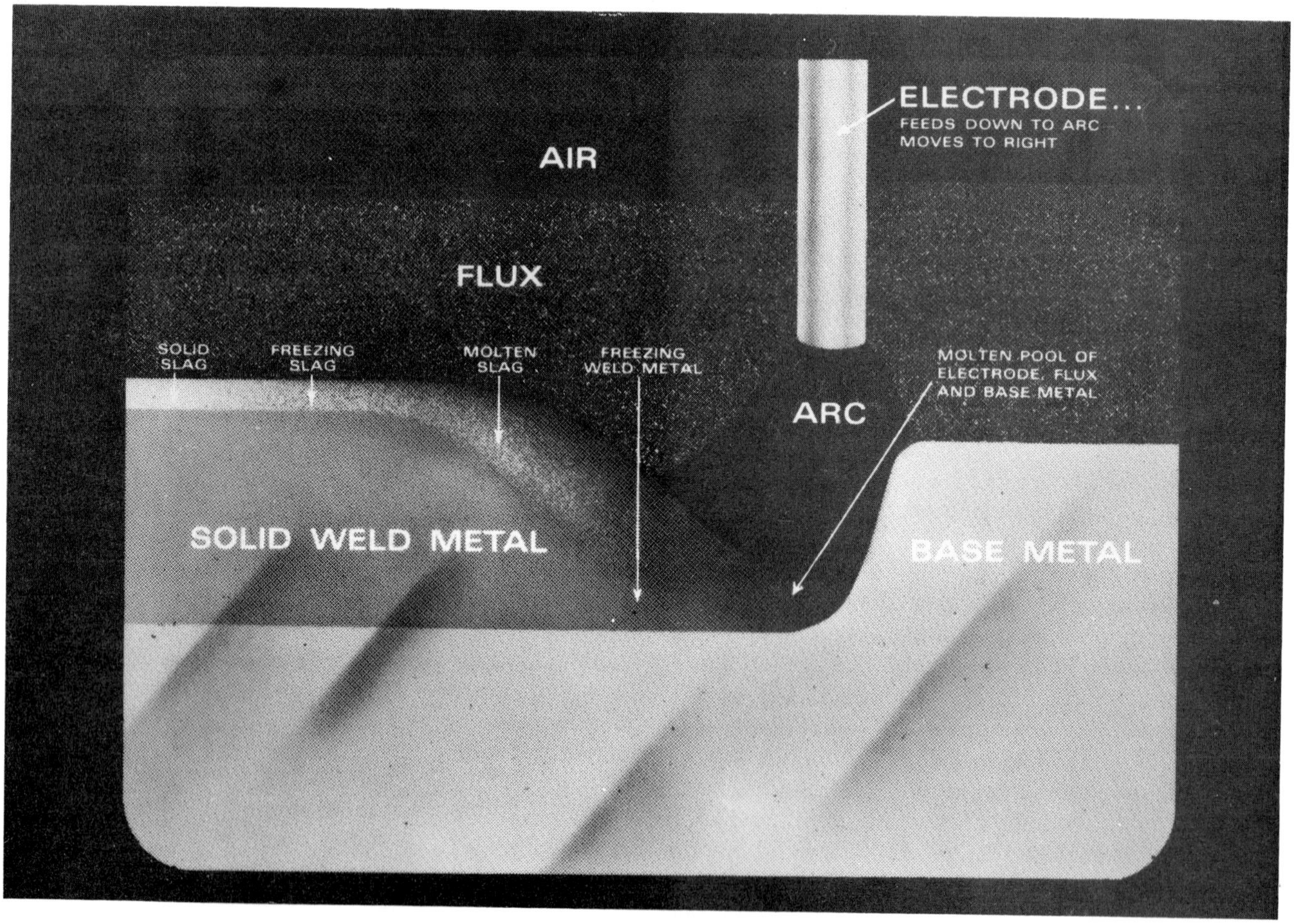

FIGURE 23–28
Arc welding is one of the most widely used methods of fusion welding. (Credit: The Lincoln Electric Corporation)

FIGURE 23–29
Oxy-acetylene welding is one of the most common forms of gas welding.

metal. Fusion takes place as the focused beam moves on, congealing a pool of molten metal that surrounds the vaporized column. Lasers make the ideal system for automated welding joining thin metals by low-power amplification. Each pulse of burning takes place within a few millionths of a second and in a small spot. Localized heating and rapid cooling results in minimizing distortion.

Although the blacksmith's art was not fusion, it was a form of pressure welding. A good hammer weld would not reveal its join and would have about 75% of the metal's strength. This type of pressure welding is still performed today as forge welding. Other forms of pressure welding will include resistance welding, which is accomplished by intensifying heat by not allowing the current flow to complete the passage. Butt welding, spot welding, and seam welding are but three examples.

UNIT 24 CUTTING EXTERNAL THREADS WITH A DIE

An outside screw thread is formed by a helical groove cut around a cylinder or a cone. The first standardized thread form was the Whitworth thread, which had an included angle of 55° and rounded top and bottom surfaces. The thread would later be modified in America to have a V-shaped included angle of 60°, with adjoining top and bottom flat surfaces (Figure 24–1). By an act of Congress, this thread was adopted and is known as the American (National) screw thread. While still in use today, the American National thread system has been slightly modified to become an international standard that is recognized by America, Canada, and the United Kingdom and is known as the Unified Thread System. In this system the bottom of the threads are rounded while the tops can either be flat or rounded. It is this development that allows an interchangeability between threads produced by the countries just mentioned.

Objectives
After completing this unit you will be able to

1. identify the four basic thread forms used in industry.
2. identify the various means of hand cutting external threads.
3. prepare and cut an external thread using a hand cutting tool.

Types of Threads

While there are as many as nine common thread forms, there are basically four thread forms used in American industry: the 60° V-type, the square, buttress, and acme threads (Figure 24–2). Each thread is designed to withstand the usage involved.

60° V-Type. Found in American national standards as well as in the metric standard thread, it is commonly used for screws, nuts, bolts, and other threaded fasteners for holding parts together. Due to its included flat surfaces between threads, this type thread is more durable and can take more handling abuse without injury than can the sharp V-thread. The sharp V-thread form has greater sealing capacity, wedging closed when tightened, but may tear or gall when releasing its hold. The sharp-V thread is rarely used except on steam fittings that require a seal.

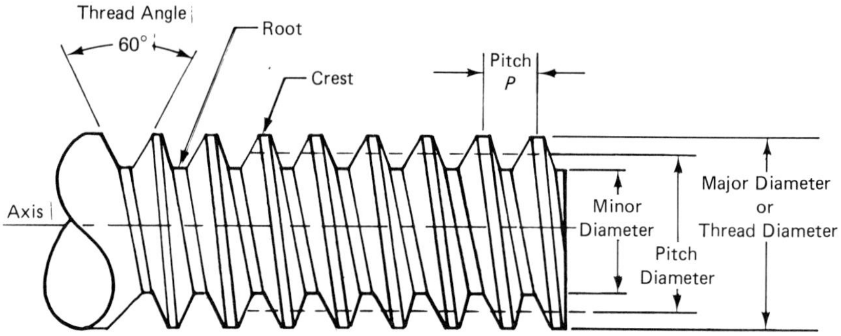

FIGURE 24–1
The American National screw thread.

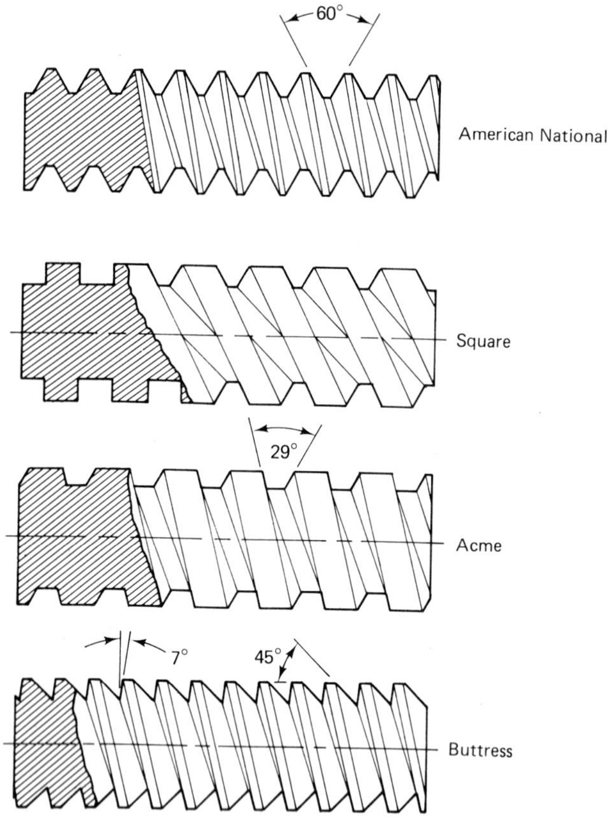

FIGURE 24–2
Four basic thread forms.

Square Thread. The sides of a square thread are perpendicular to the center of the axis and all surfaces are square with each other. This leaves less friction to overcome and will transmit all motion nearly parallel to its axis. It is difficult to cut and is found mostly in bench vises, clamp screws, house jacks, and similar devices that require force and power.

Buttress Thread. Resembling a saw tooth, the buttress thread combines the strength of a V-thread with the efficiency of a square thread. It is useful when applying pressure from one direction such as on jack screws, breech blocks on large guns, and weight-lifting devices.

Acme Thread. The profile of this thread contains a slight taper that enables the mating thread to easily engage it. This makes it ideal to be used as feed screws on precision machine tools. The lead screw on a lathe is a good example. The acme thread can also be easily reconditioned and, in general, has replaced the square thread.

Types of Dies

A screw is an example of an external thread and can be cut by one of two methods: using hand tools or by machines. Hand tools will require the use of a die. Basically, a die consists of a block of steel that has a hole with threads or teeth cut within, causing some dies to resemble a nut. The block is made of special tool steel and hardened so that it may cut external threads on a rod.

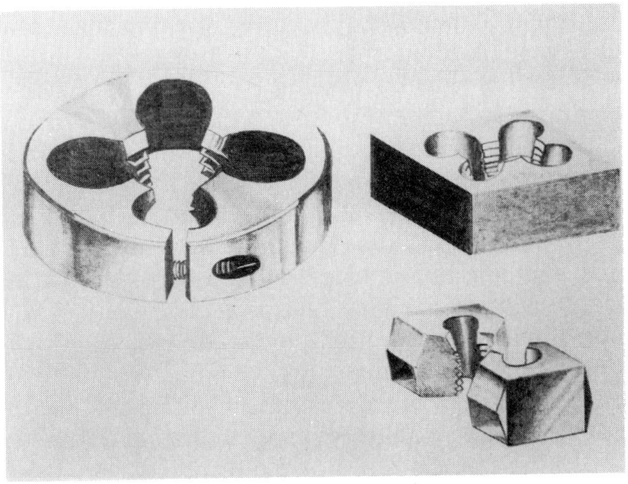

FIGURE 24-3
Three basic types of dies: (A) round split adjustable die, (B) adjustable two piece die, and (C) solid rethreading die.

There are three basic types of dies: round split adjustable die, two-piece die, and a solid rethreading die (Figure 24-3). The round split adjustable die and the two-piece die can be set to control the fit a resultant thread will have with its mating part. The solid rethreading die is used to clean up or restore threads that have been previously cut or damaged.

Work Preparation

The preparation for cutting a thread begins by securing a die. A solid rethreading die meets little resistance in cutting and will accept any wrench to turn it. While a two-piece die is held by a collet and guide (Figure 24-4), the chief method of holding a round split adjustable die is by means of a stock, a holder designed to clamp and hold the setting of the die; hence, the term stock and die.

American machine screws are identified by combining two basic dimensions—major diameter and threads per inch (Figure 24-5). Metric threads are identified by the letter *M* followed by the major diameter in millimeters and combined with the pitch (center-to-center distance of the threads) in millimeters. (See Unit 25.) This identification is stamped or etched on the die's face. The thread's major diameter will be the size of the rod to be used in cutting the thread. When the size falls below a fractional size for American machine screws, it is identified by a gage number from 0 to 12. Each number will represent the rod's diameter (see Tables 25-2 and 25-3). Each rod to be threaded will have its end chamfered (relieved on an angle) to ease the cutting action.

Other identification on the die's face will include whether a die is a left-hand thread or not. Ordinarily, threads are right handed, that is, they are cut and tightened in a clockwise direction. Most threads are right handed and there is no identification to this effect. When the die cuts left-handed threads (they cut and are tightened in a counterclockwise motion), it is designated LH on the die's face.

A stock and die is available in a screw plate set (Figure 24-6) that contains a series of taps and dies with holders to cut a variety of external and internal threads.

Cutting the Thread

Using a round split adjustable die requires some understanding of how it works. The adjustment screw

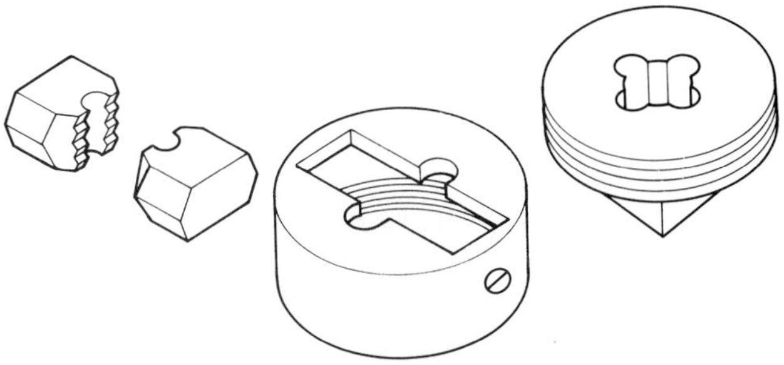

FIGURE 24-4
Two-piece holding system. Dies may be adjusted for depth of cut.

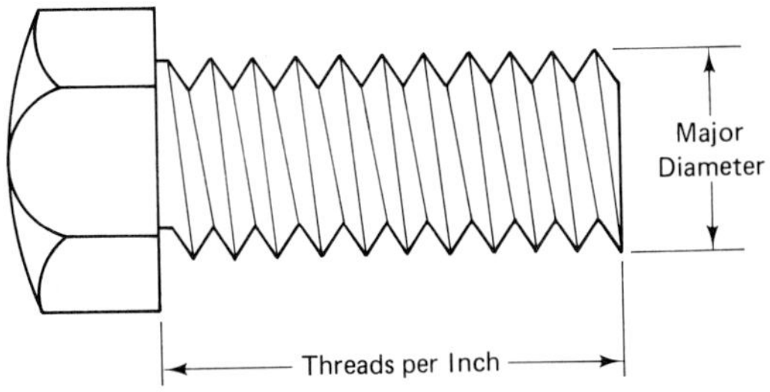

FIGURE 24-5
American Standard threads are identified by combining two basic dimensions—major diameter and number of threads per inch.

FIGURE 24-6
A screw plate set. (Credit: DoAll Company)

controls the depth of cutting. By turning the screw clockwise (Figure 24-7), the die opens up, resulting in a shallow cut. By turning the adjustment screw counter clockwise (Figure 24-8), the teeth come closer together, resulting in a deeper thread cut. Tension screws on the side of the stock keep the die in place and maintain the adjustment. A series of grooves intersects the threads to form cutting teeth and also allows the chips and waste to escape. A cross section of the die reveals that one end of the opening is funneled, deeply tapered, with very little taper on the other end (Figure 24-9).

Cutting a thread with this tool can be accomplished by two means: one is by adjusting the depth of the die's teeth to a thread already cut, the other by trial and error. The last method will require inspection and checking against mating members.

Adjusting Die Depth to a Thread Already Cut

1. Secure the threaded rod in a bench vise without damaging the threads.
2. Screw the die onto the threaded rod.
3. Turn the adjustment screw until the die teeth fit snugly against the threaded rod.

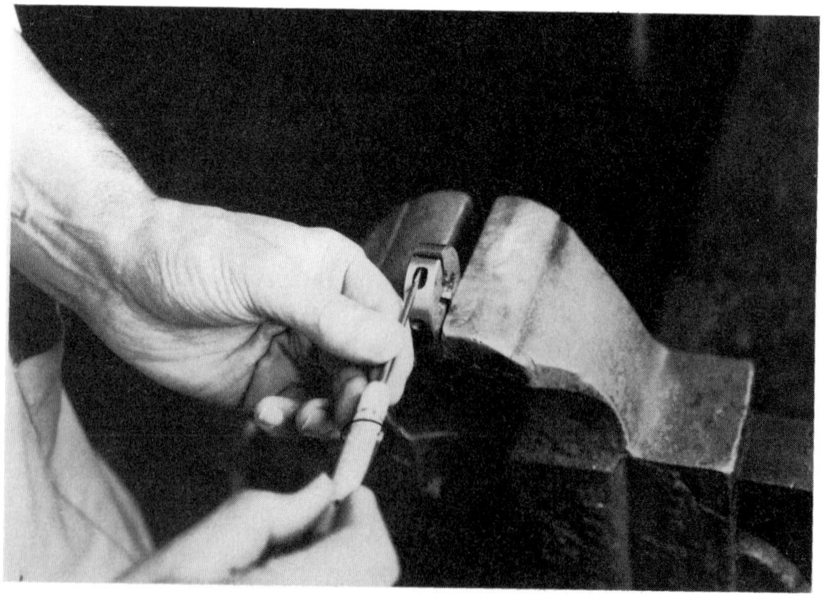

FIGURE 24-7
Adjusting a round split die to force the teeth apart, resulting in a shallow cut, a tighter mating fit. (Credit: Philip Rapisarda)

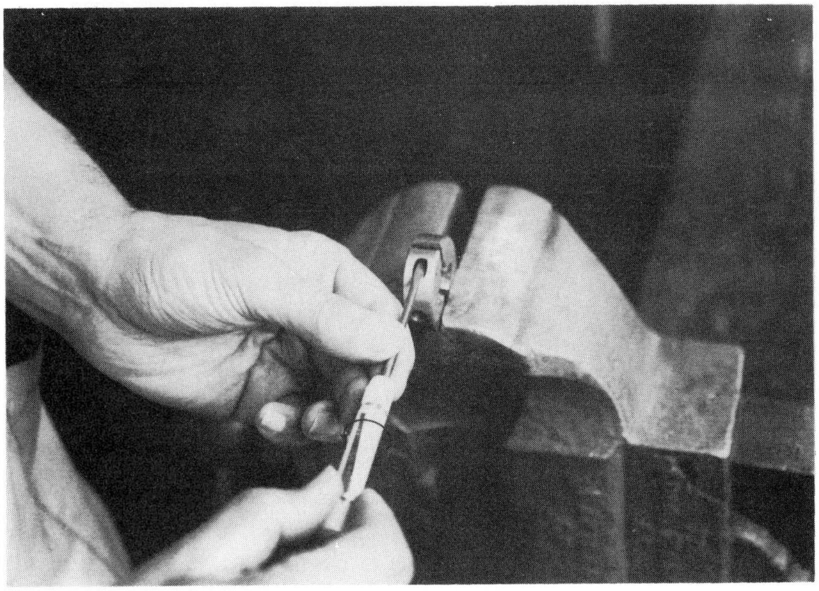

Figure 24-8
Adjusting a round split die to allow the teeth to come closer together, resulting in a deeper cut, looser mating fit. (Credit: Philip Rapisarda)

Threading a Rod with a Round Split Adjustable Die

1. Insert the proper size die in the stock, funneled opening facing you, and tighten the tension screws against it.
2. Place the proper size rod with one end chamfered in an upright position in a bench vise, extending it no more than 2 in. (50 mm) above the jaws.
3. Lubricate the rod (Figure 24-10).
4. Place the stock held with the funneled opening squarely on top of the workpiece.
5. Press down evenly on the ends of the stock and turn it clockwise about two or three turns (Figure 24-11).
6. Examine the position of the stock to see if the threads have been started squarely. If not, apply pressure on the desired side for several turns to straighten the thread cut and check again.
7. Progressively advance and back off the die to break the chips and ease the cutting for several threads.
8. Unscrew the die to check the thread against a mating part.

FIGURE 24-9
A cross section of the die reveals that one side is funneled, deeply tapered, to help start and ease the cutting action. (Credit: Philip Rapisarda)

FIGURE 24-10
In addition to having the end chamfered, the rod should be well lubricated when threading. (Credit: Philip Rapisarda)

FIGURE 24-12
Threading to a shoulder is accomplished by first cutting with the large tapered opening coming as close as $\frac{1}{16}$ in. (1.58 mm) away from the shoulder. (Credit: Philip Rapisarda)

FIGURE 24-11
In starting to thread, hold the die stock level and at right angles to the workpiece, applying even pressure on the ends of the stock. (Credit: Philip Rapisarda)

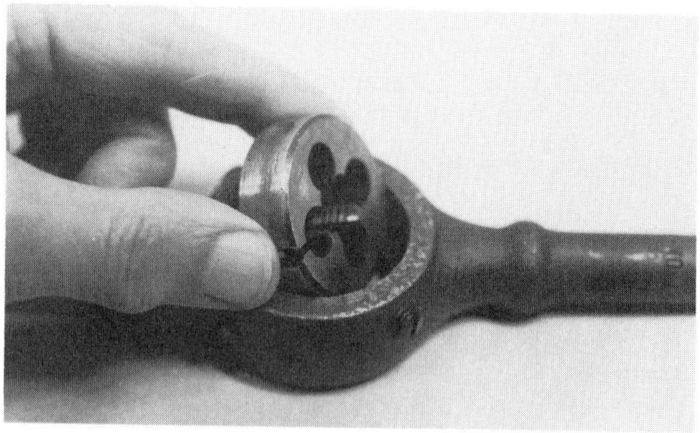

FIGURE 24-13
Upon threading near a shoulder, flip over the die to have the short taper opening face you, reinsert, tighten, and complete the job. (Credit: Philip Rapisarda)

9. Make any necessary die opening adjustment if necessary.

10. Cut a few more threads and recheck with the mating part.

11. Once the adjustment is correct, relubricate the rod occasionally during the operation and complete the thread.

Threading to a Shoulder (Figure 24-12)

1. Cut to about $\frac{1}{16}$ in. (1.6 mm) from the shoulder as previously described.

2. Unwind the die from the thread.

3. Remove the flip over the die (Figure 24-13).

4. Reinsert it so that the tapered opening has been reversed.

5. Recut the thread until the die almost touches the shoulder.

UNIT 25 CUTTING INTERNAL THREADS WITH A TAP

A tap is a hardened steel tool designed to cut threads within a hole. Technically, it is considered cutting internal helical grooves. A good example of this would be a nut.

A tap (Figure 25-1) consists of three major parts. The cutting portion, called threads or teeth, has a series of flutes running along its entire length. Flutes are channels that serve two purposes: (1) They act as a waste and lubrication recess and (2) they help sharpen the teeth. The unthreaded portion is called the shank and adds strength to the construction. It is on the shank that a tap's information will be found (Figure 25-2). The square on the end of the tap is designed for holding and turning purposes.

Objectives
After completing this unit you will be able to

1. calculate the tap drill size for tapping a hole.
2. know how to tap a hole by hand.
3. know how to remove a broken tap from a hole.

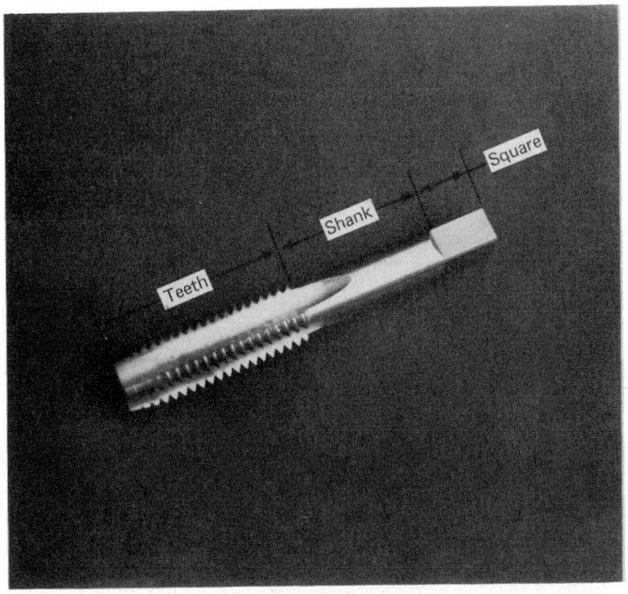

FIGURE 25-1
A basic tap consists of three major parts: teeth, shank, and square for holding.

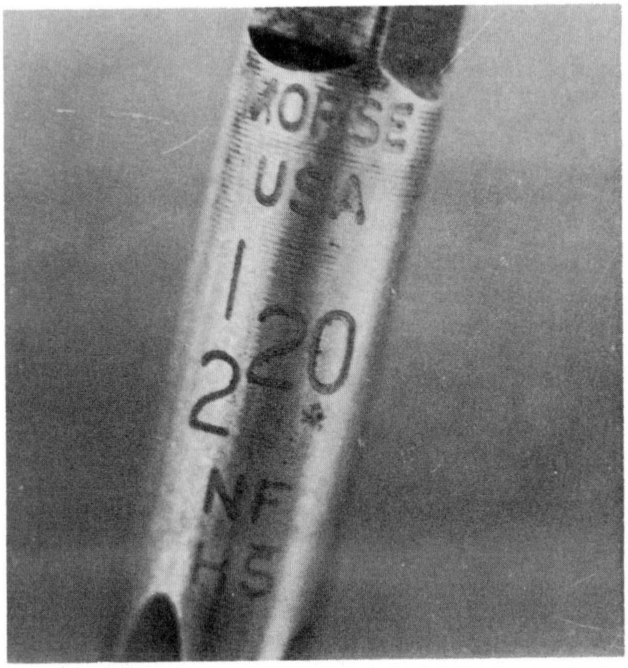

FIGURE 25-2
The shank contains a tap's information.

Types of Taps

Hand Taps. Initially, these taps were made for use as hand tools and still are. While originally made by machine cut, today most taps have ground threads with hand taps coming in two, three, and four flutes. To cover a variety of hole situations, a set of hand taps (Figure 25-3) are used. Consisting of three basic taps, each hand tap differs as to the amount of taper involved before the full thread diameter is reached. It is the taper that eases the cutting action in tapping. Each tap is identified as follows:

Taper tap: Also referred to as a starting tap, this tool has an 8- to 10-thread taper and is used in open hole threading (Figure 25-4A).

Plug tap: Widely accepted as a general-purpose tap, this tool has a 3- to 5-thread taper and is used when threading a portion of a hole not necessarily close to its bottom (Figure 25-4B). It can also be used instead of a taper tap to start a hole.

Bottoming tap: This cutting tool has a 1- to $1\frac{1}{2}$-thread taper. Used primarily to thread close to the shoulder of a blind hole (Figure 25-4C), a bottoming tap should only be used after first threading as far as you can go with a plug tap.

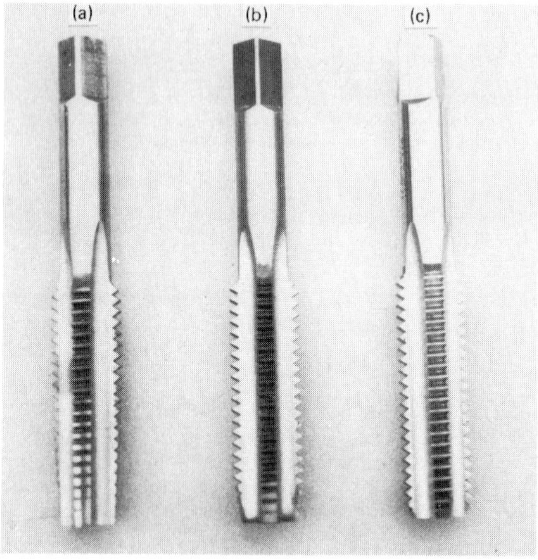

FIGURE 25-3
A set of hand taps consists of (A) taper tap, (B) plug tap, and (C) bottoming tap.

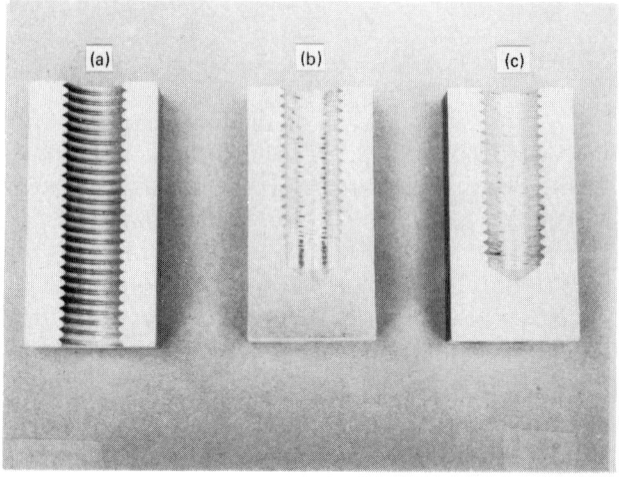

FIGURE 25-4
The type of hole dictates a tap's use. (A) A taper tap is used on an open hole, (B) a plug tap is used for general purpose tapping, and (C) a bottoming tap is used when threading close to the shoulder of a blind hole.

Machine Taps

Another type of tap, the machine tap, is widely used as a production tool (Figure 25-5). In addition to the straight flutes, one production tap is made with an angular flute at the point (Figure 25-6). This spiral pointed tool is called a gun tap and "shoots" the chips out ahead of its shearing action and is used for through holes. Other machine taps might include spiral fluted taps (Figure 25-7), which are designed to lift the chips clear of the blind holes they are used on.

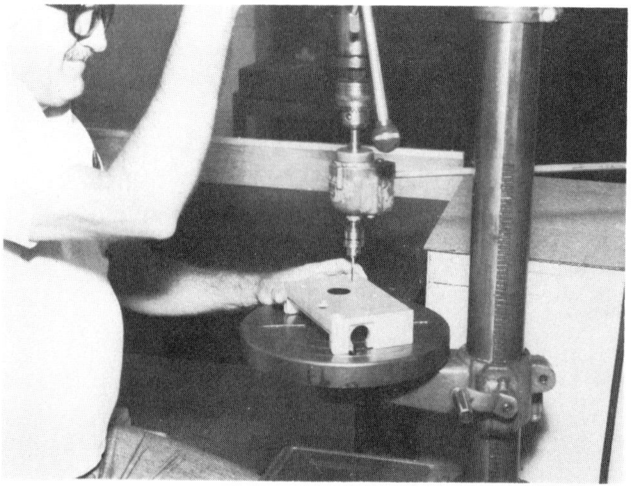

FIGURE 25-5
Tapping heads are designed to turn machine taps under power.

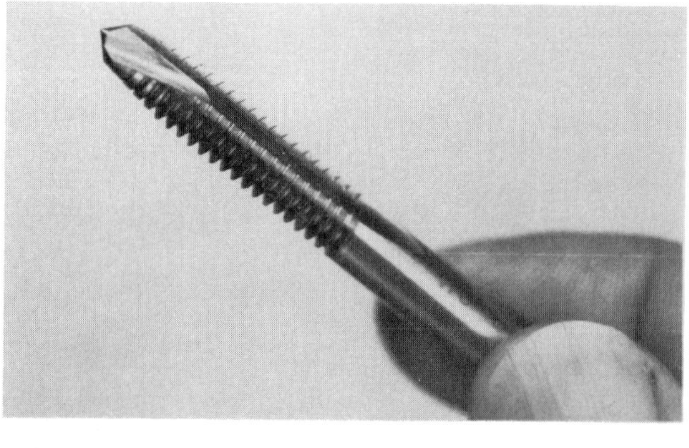

FIGURE 25-6
"Gun" taps have angular pointed flutes up front that shear the metal as it "shoots" out the chip ahead of its cutting action.

Pipe Taps and Threads

The American National Standard pipe thread is designed for use in assembly of threaded joints where a compound or sealer will not be objectionable to pressure tighten the joint. Four types of threads are used:

American National Standard pipe threads (NPT)

American National Standard straight pipe threads (NPSC)

American National Standard dryseal taper pipe threads (NPTF)

American National Standard dryseal straight pipe threads (NPSF)

Employing a taper thread form equal to $\frac{3}{4}$ in. per foot (Figure 25-8), the NPT thread is the most widely used fastener in commercial valve and pipe fittings. The external thread is always tapered. On jobs requiring high-pressure joints, the internal thread is also tapered. The basic pipe dimensions are shown in Table 25-1.

The NPSC thread is for internal application only. It is designed to mate with NPT threads and be used as pipe couplings (straight internal threads).

Dryseal threads (NPTF) have the same general form as the NPT thread, that is, an angle of 60° between the sides of the thread, but differ in the width of the profile section between the inside and outside thread. The dryseal thread is designed for interference between mating parts when turned hand tight. When tightened with force, it eliminates the need for a sealer, making it ideal to be used for gas or liquid containers.

The NPSF thread is designed as a coupling to assemble with an interference fit when mated with a dryseal external taper pipe.

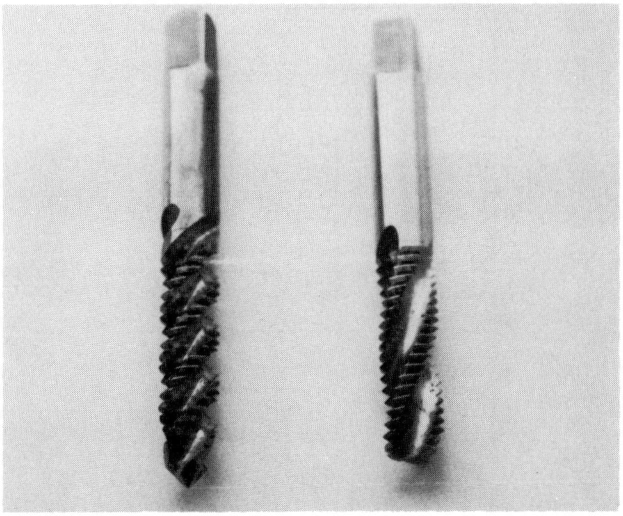

FIGURE 25-7
Spiral fluted taps are designed to lift the chips clear of the hole.

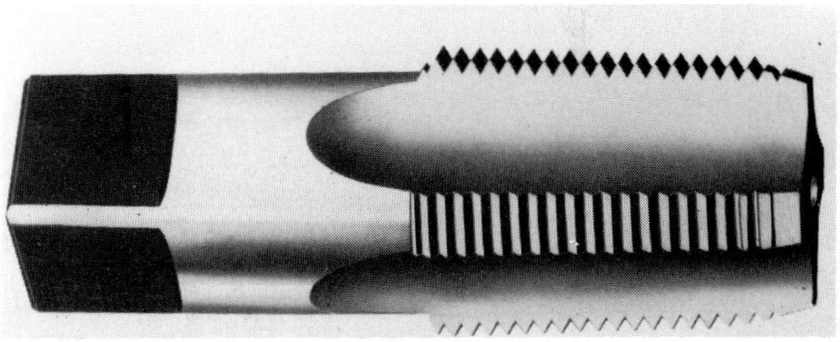

FIGURE 25-8
Taper pipe tap. (Credit: Union/Butterfield)

Tap Drill Sizes

A tap removes the material that forms the thread. In order to undertake tapping, a hole size has to be determined to be drilled. This is called the tap drill size. But will it be too big or too small? It has been found that a full thread engagement between mating parts, that is, 100% thread depth (Figure 25-9) is not the answer. Engineering tests have shown that a 100% thread depth is only 5% stronger than a 75% thread depth, but it requires three times the power to turn the tap, greatly increasing the chances of

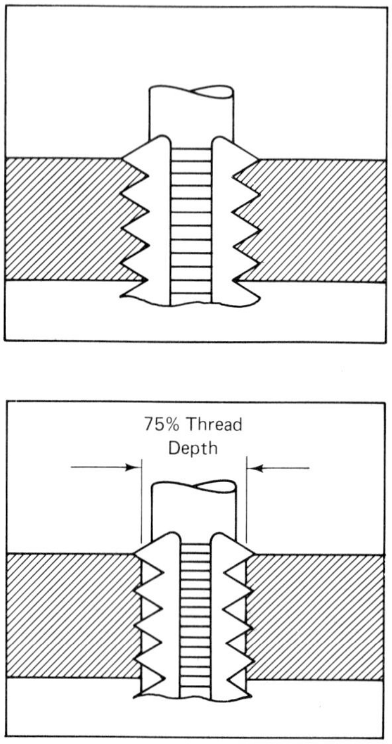

FIGURE 25-9
A 75% thread depth is used in standard tap drill sizes. A 100% thread depth is only 5% stronger but requires three times the power to turn the tap.

breaking a tap; consequently, a tap drill that gives you a standard 75% thread depth is generally used.

One quick way of obtaining this information is to refer to tap drill tables such as Tables 25-2 and 25-3, which list the hole sizes for common American National and Unified threads.

Another means of calculating any tap drill is by use of a simple formula:

$$\text{Tap drill} = \text{MD} - \frac{1}{N}$$

$$\text{MD} = \text{major diameter}$$

$$N = \text{number of threads per inch}$$

Example: Find the tap drill size for a 3/8-16NC tap.

$$\text{Tap drill} = \text{MD} - \frac{1}{N}$$

$$= \frac{3}{8} - \frac{1}{16}$$

$$= \frac{6}{16} - \frac{1}{16}$$

$$= \frac{5}{16}$$

TABLE 25-1
NPT national pipe thread and tap drill sizes

Tap size	*Threads per inch*	*Actual outside diameter*	*Actual inside diameter*	*Tap drill size*
$\frac{1}{8}$	27	0.405	0.270	$\frac{11}{32}$
$\frac{1}{4}$	18	0.540	0.364	$\frac{7}{16}$
$\frac{3}{8}$	18	0.675	0.494	$\frac{19}{32}$
$\frac{1}{2}$	14	0.840	0.623	$\frac{23}{32}$
$\frac{3}{4}$	14	1.050	0.824	$\frac{15}{16}$
1	$11\frac{1}{2}$	1.315	1.048	$1\frac{5}{32}$
$1\frac{1}{4}$	$11\frac{1}{2}$	1.660	1.380	$1\frac{1}{2}$
$1\frac{1}{2}$	$11\frac{1}{2}$	1.900	1.610	$1\frac{23}{32}$
2	$11\frac{1}{2}$	2.375	2.067	$2\frac{3}{16}$
$2\frac{1}{2}$	8	2.875	2.468	$2\frac{5}{8}$

TABLE 25-2
Tap drill sizes (American National and Unified fine thread form, NF and UNF)

			Tap Drill		*Clearance Hole Drill*			
					Close fit		*Free fit*	
Screw size	*Threads per inch*	*Major diameter (decimal equivalent)*	*Drill size*	*Decimal equivalent*	*Drill size*	*Decimal equivalent*	*Drill size*	*Decimal equivalent*
0	80	0.060	$\frac{3}{64}$	0.0469	52	0.0635	50	0.0700
1	72	0.073	53	0.0595	48	0.0760	46	0.0810
2	64	0.086	50	0.0700	43	0.0890	41	0.0960
3	56	0.099	45	0.0820	37	0.1040	35	0.1100
4	48	0.112	42	0.0935	32	0.1160	30	0.1285
5	44	0.125	37	0.1040	30	0.1285	29	0.1360
6	40	0.138	33	0.1130	27	0.1440	25	0.1495
8	36	0.164	29	0.1360	18	0.1695	16	0.1770
10	32	0.190	21	0.1590	9	0.1960	7	0.2010
12	28	0.216	14	0.1820	2	0.2210	1	0.2280
$\frac{1}{4}$	28	0.250	3	0.2130	F	0.2570	H	0.2660
$\frac{5}{16}$	24	0.3125	I	0.2720	P	0.3230	Q	0.3320
$\frac{3}{8}$	24	0.375	Q	0.3320	W	0.3860	X	0.3970
$\frac{7}{16}$	20	0.4375	$\frac{25}{64}$	0.3906	$\frac{29}{64}$	0.4531	$\frac{15}{32}$	0.4687
$\frac{1}{2}$	20	0.500	$\frac{29}{64}$	0.4531	$\frac{33}{64}$	0.5156	$\frac{17}{32}$	0.5312
$\frac{9}{16}$	18	0.5625	$\frac{33}{64}$	0.5156	$\frac{37}{44}$	0.5781	$\frac{19}{32}$	0.5938
$\frac{5}{8}$	18	0.625	$\frac{37}{64}$	0.5781	$\frac{41}{64}$	0.6406	$\frac{21}{32}$	0.6562
$\frac{3}{4}$	16	0.750	$\frac{11}{16}$	0.6875	$\frac{49}{64}$	0.7656	$\frac{25}{32}$	0.7812
$\frac{7}{8}$	14	0.875	$\frac{13}{16}$	0.8125	$\frac{57}{64}$	0.8906	$\frac{29}{32}$	0.9062
1	12	1.000	$\frac{15}{16}$	0.9375	$1\frac{1}{64}$	1.0156	$1\frac{1}{32}$	1.0312

TABLE 25-3
Tap drill sizes (American National and Unified coarse thread form, NC and UNC)

			Tap Drill		*Clearance Hole Drill*			
					Close fit		*Free fit*	
Screw size	*Threads per inch*	*Major diameter (decimal equivalent)*	*Drill size*	*Decimal equivalent*	*Drill size*	*Decimal equivalent*	*Drill size*	*Decimal equivalent*
1	64	0.073	53	0.0595	48	0.0760	46	0.0810
2	56	0.086	50	0.0700	43	0.0890	41	0.0960
3	48	0.099	47	0.0786	37	0.1040	35	0.1100
4	40	0.112	43	0.0890	32	0.1160	30	0.1285
5	40	0.125	38	0.1015	30	0.1285	29	0.1360
6	32	0.138	36	0.1065	27	0.1440	25	0.1495
8	32	0.164	29	0.1360	18	0.1695	16	0.1770
10	24	0.190	25	0.1495	9	0.1960	7	0.2010
12	24	0.216	16	0.1770	2	0.2210	1	0.2280
$\frac{1}{4}$	20	0.250	7	0.2010	F	0.2570	H	0.2660
$\frac{5}{16}$	18	0.3125	F	0.2570	P	0.3230	Q	0.3320
$\frac{3}{8}$	16	0.375	$\frac{5}{16}$	0.3125	W	0.3860	X	0.3970
$\frac{7}{16}$	14	0.4375	U	0.3680	$\frac{29}{64}$	0.4531	$\frac{15}{32}$	0.4687
$\frac{1}{2}$	13	0.500	$\frac{27}{64}$	0.4219	$\frac{33}{64}$	0.5156	$\frac{17}{32}$	0.5312
$\frac{9}{16}$	12	0.5625	$\frac{31}{64}$	0.4844	$\frac{37}{64}$	0.5781	$\frac{19}{32}$	0.5938
$\frac{5}{8}$	11	0.625	$\frac{17}{32}$	0.5312	$\frac{41}{64}$	0.6406	$\frac{21}{32}$	0.6562
$\frac{3}{4}$	10	0.750	$\frac{21}{32}$	0.6562	$\frac{49}{64}$	0.7656	$\frac{25}{32}$	0.7812
$\frac{7}{8}$	9	0.875	$\frac{49}{64}$	0.7656	$\frac{57}{64}$	0.8906	$\frac{29}{32}$	0.9062
1	8	1.000	$\frac{7}{8}$	0.8750	$1\frac{1}{64}$	1.0156	$1\frac{1}{32}$	1.0312

Tap drill sizes for ISO metric screw threads are called out in Table 25-4. Tap drill sizes for metric taps can be calculated in much the same manner as the U.S. Standard threads but with a slight difference. Metric threads are designated by the letter *M* and are identified by a nominal diameter (major diameter) in millimeters followed by a pitch value in millimeters. A typical thread callout such as an M 3 × 0.50 would indicate M is a metric thread, 3 is the nominal or major diameter in millimeters, and 0.50 is the thread pitch value in millimeters.

The tap drill formula for metric threads would be used in the same manner as the American National and Unified threads.

Example: Find the tap drill size for an M 3 × 0.50 tap.

$$\begin{aligned} \text{Tap drill} &= \text{MD (mm)} - \text{pitch (mm)} \\ &= 3.00 - 0.50 \\ &= 2.50 \text{ mm} \end{aligned}$$

Work Preparation

In preparing to tap a hole, one must have the answers to certain questions as well as the tools to do the job. The first question to be answered would be the thread size. American Standard screws are mea-

TABLE 25-4
ISO metric tap drill sizes

Nominal diameter (mm)	*Thread pitch (mm)*	*Tap drill size (mm)*	*Clearance hole drill*	
			Close fit (mm)	*Free fit (mm)*
1.60	0.35	1.25	1.70	2.00
2.00	0.40	1.60	2.20	2.60
2.50	0.45	2.05	2.70	3.10
3.00	0.50	2.50	3.20	3.60
3.50	0.60	2.90	3.70	4.10
4.00	0.70	3.30	4.30	4.80
5.00	0.80	4.20	5.30	5.80
6.00	1.00	5.00	6.40	7.00
8.00	1.25	6.80	8.40	10.00
10.00	1.50	8.50	10.50	12.00
12.00	1.75	10.20	13.00	15.00
14.00	2.00	12.00	15.00	17.00
16.00	2.00	14.00	17.00	19.00
20.00	2.50	17.50	21.00	24.00
24.00	3.00	21.00	25.00	28.00
30.00	3.50	26.50	31.00	35.00
36.00	4.00	32.00	37.00	42.00
42.00	4.50	37.50	43.00	48.00
48.00	5.00	43.00	50.00	56.00
56.00	5.50	50.50	58.00	66.00
64.00	6.00	58.00	66.00	74.00
72.00	6.00	66.00	74.00	82.00
80.00	6.00	74.00	82.00	91.00
90.00	6.00	84.00	93.00	101.00
100.00	6.00	94.00	104.00	112.00

sured by combining the major diameter and the number of threads per inch (see Figure 24-5). The size of the tap must match the size of the thread. This information will be listed on the tap's shank (Figure 25-2).

Thread measurements can be obtained by inserting the screw's or tap's diameter through a drill plate gage (Figure 25-10) or measuring with a micrometer.

The number of threads can be solved by a few simple methods. One would be to count the number of threads along a given length (Figure 25-11); another would be to match profiles of screws and taps whose identification is known. The third method involves the use of a screw pitch gage (Figure 25-12). By dropping a blade that matches the thread's profile, the number of threads involved will be the number stamped on the blade.

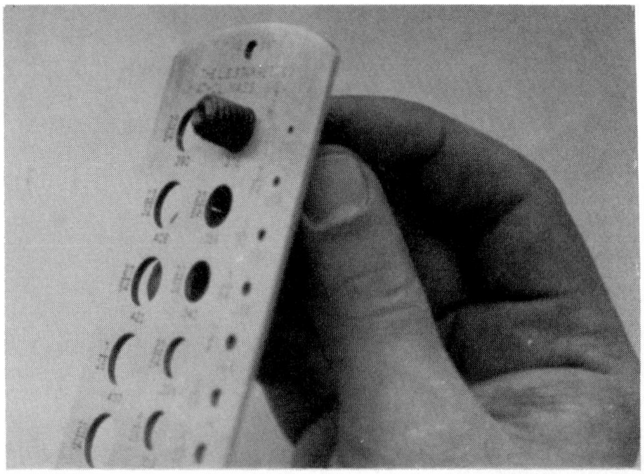

FIGURE 25-10
A screw's major diameter can be quickly obtained by insertion through a drill plate gage.

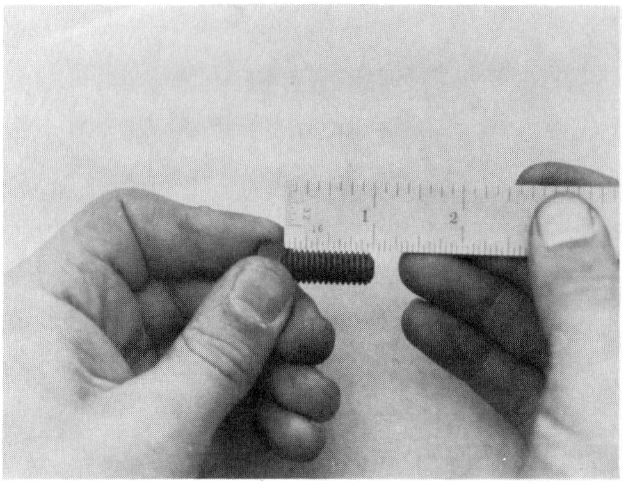

FIGURE 25-11
A screw's number of threads can be counted over 1 in. length by use of a 6-in. rule.

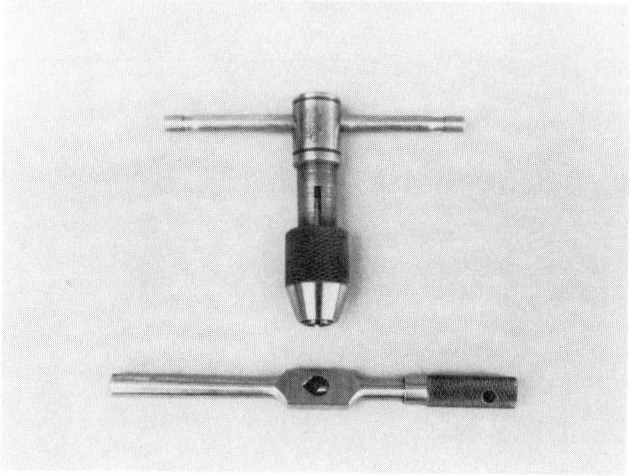

FIGURE 25-13
The principal means of holding a tap in hand tapping—T-handle and long handle tap wrenches.

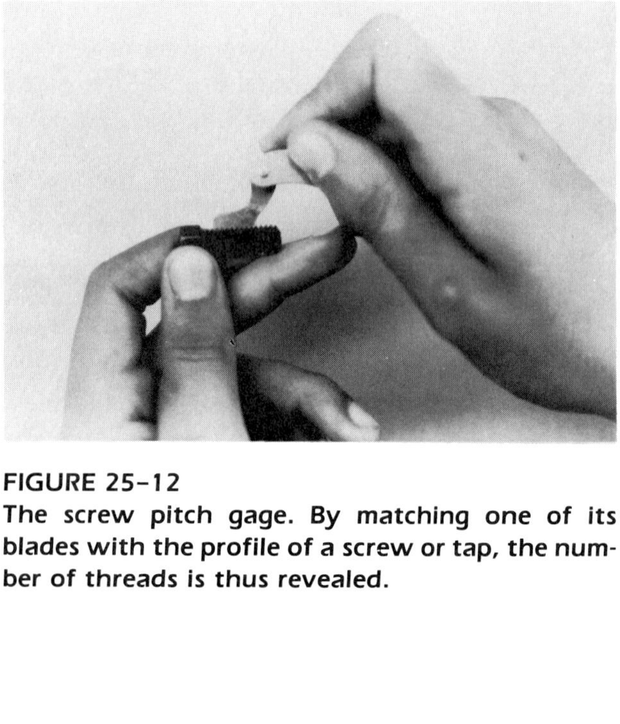

FIGURE 25-12
The screw pitch gage. By matching one of its blades with the profile of a screw or tap, the number of threads is thus revealed.

Pitch. It should be understood that the number of threads per inch plays a role in computing the mathematical thread spacing or center-to-center location of threads measured parallel to the axis. By dividing one inch by the number of threads ($\frac{1}{N}$), the resultant answer will be the pitch. In the metric system this value need not be computed, it is given in millimeters.

Tap Wrenches. Taps are held and turned by two principal types of hand wrenches: the T-handle tap wrench and the straight-handle tap wrench (Figure 25-13). The T-handle tap wrench is adjustable and comes in different sizes as does the straight-handle tap wrench. Its length affords greater leverage, however, and should not be used on small taps. In general, it should be turned with caution.

Lubrication. Cutting fluids such as chlorinated sulfur-based oil for steels and a mixture of kerosene and lard oil for nonferrous metals should be used frequently to avoid over heating, dulling the tap, ease the cutting action, and produce a quality thread. Some materials such as gray cast iron, plastics, hard rubber, and fiber may be tapped dry.

Tapping

Taps are made of a high-grade hardened steel. This plus its construction makes it fragile and can break very easily. For this reason safety glasses must be worn in undertaking tapping. Both tap and workpiece must be fastened securely before starting to avoid the problems of a broken tap, which is difficult to get out and may ruin your workpiece.

The design of a four-fluted hand tap is such that as the tap is turned, a groove is cut, forming a chip that curls bigger and bigger as the tapping continues, creating more and more resistance that is felt by the person tapping. If the tapping is not

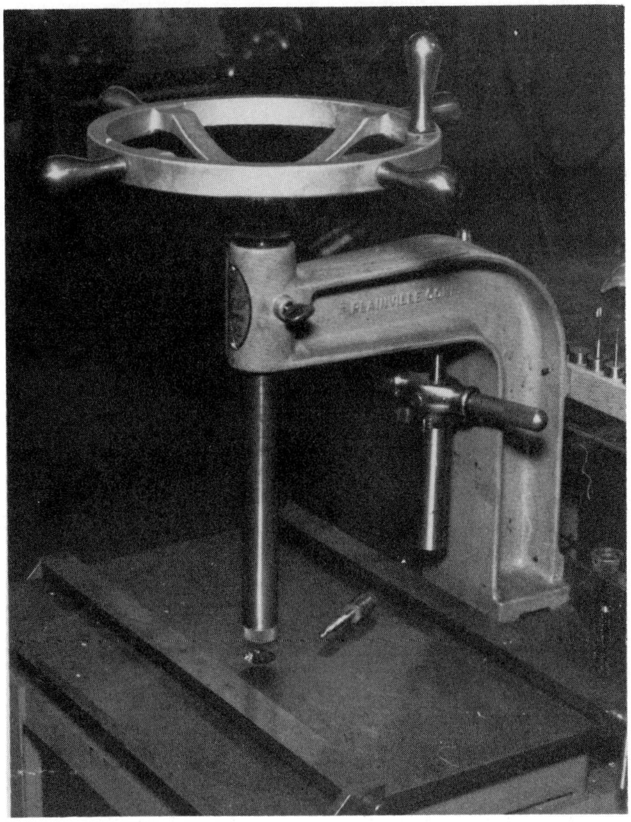

FIGURE 25-14
A tapping fixture allows a tap to be held at right angles to the workpiece.

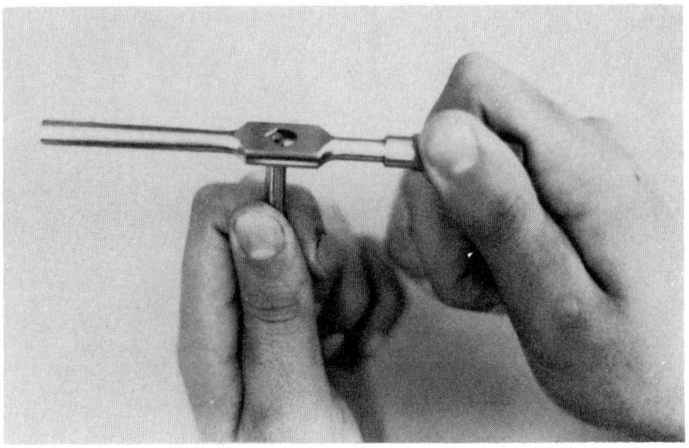

FIGURE 25-15
Tighten the wrench against a tap's square.

FIGURE 25-16
A tap should be checked to see that it has been started square.

discontinued and the tap backed off, it will snap the cutting tool. While hand tapping with a four-fluted tap will require a process of advancing and backing off the cutting tool, there is no set rule as to how far to advance the tap before backing it off to clear the chip. This will depend on the material, the sharpness of the tap, and the lubricant involved.

Being able to tap square (at right angles to the workhole) is best achieved when the tap drill hole remains clamped directly under the drill chuck, whereby the tap can be supported and turned at 90° to the work. A tapping fixture (Figure 25-14) will also do the job.

To tap square in a bench vise requires more skill and patience. General tapping can be achieved in the following manner:

1. Wear your safety glasses before starting.
2. Clamp your workpiece in a bench vise so that the flat surface is level and extends slightly above the jaws.
3. Tighten a taper or plug tap in a tap wrench (Figure 25-15).
4. Lubricate the tap with an adequate cutting fluid.
5. Cup the tap holder in the hand palm and insert the tap vertically.
6. Start tapping clockwise by applying a steady downward pressure in making two or three turns.
7. With a small square, check to see that the tap has been started squarely (Figure 25-16) by checking two positions 90° apart.

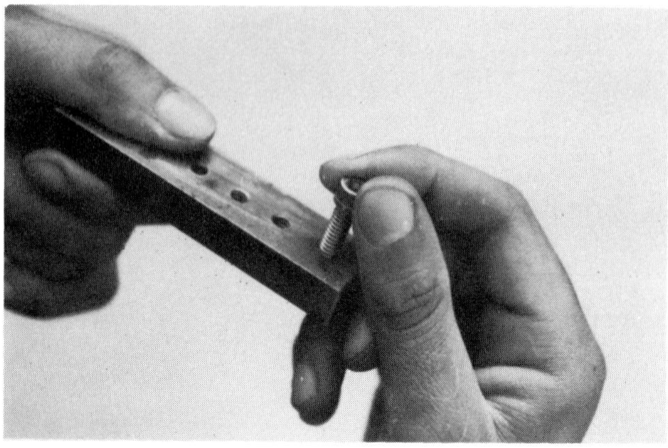

FIGURE 25-17
Mating the screw with the tapped hole.

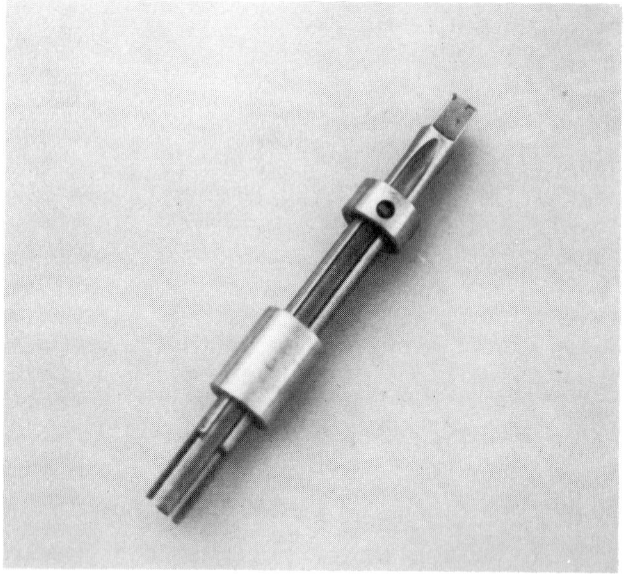

FIGURE 25-18
A tap extractor.

8. If the tap is out of square, remove it from the hole. Restart the tapping, moving the cutting tool in the direction required but be careful not to exert too much pressure in doing this.

9. With the tap started squarely, turn the wrench with both hands for about one-half to a full turn.

10. Now back off the tap by reversing it one full turn. You should be able to hear and feel the chip breaking.

11. Apply more lubricant.

12. Advance the tap to the point where it last cut and advance it another one-half turn beyond it.

13. Now back off the tap a full turn. If the tap can not be backed off smoothly, breaking the chip in the process, it is recommended that the tap be rocked counterclockwise and clockwise in small motions until the tap can be moved freely.

14. Back out the tap occasionally to clear the chips as it is lubricated and continue to advance and back off the tap as before until the thread depth has been reached.

15. Lastly, try the mating screw to see that the threads are free of any waste material (Figure 25-17).

Removing Broken Taps

Due to its fragile nature, a broken tap is one of the hazards confronting any person who does any tapping. The consequences of a broken tap are many, such as ruining a workpiece, loss of a particular hole in the workpiece, or having to make the workpiece again. All may turn out to be time consuming, frustrating, or expensive. What follows are some of the methods used in removing or eliminating a broken tap.

The Tap Extractor

The tap extractor (Figure 25-18) is a tool that can be used with a tap holder. It has four prongs, the lengths of which are adjustable, to fit into the flutes of a tap and will come in a number of sizes to fit a variety of taps. Once inserted, a backup collar is slid down its body to rest above the tap to support and strengthen the grip of the inserted prongs. This will enable the user to unscrew the broken tap.

For all its simplicity, two things must be remembered. In order to be used, the flutes must be clear of chips and waste, and once the prongs are inserted, care must be exercised during use.

Using a Tap Extractor

1. Put safety glasses on before attempting to use a tap extractor.

2. Tighten the proper tap extractor in a tap wrench (Figure 25-19).

3. Slide the prongs into the flutes of the broken tap as far as they will go (Figure 25-20).

4. Slide the backup collar against the top of the tap.

5. Turn the tap wrench carefully in a counterclockwise direction for right-handed threads, clockwise for left-handed threads.

Note: If movement is difficult, it may be necessary to move the tap wrench in small advancing and backing off motions until the tap can be freed.

Removal by Annealing
When broken taps are made from carbon steel, not high-speed steel, they may be removed by annealing. Once annealed, drilling or breaking up the tap in small pieces may be possible. One of the drawbacks to this method is that hard spots may remain to hinder any drilling.

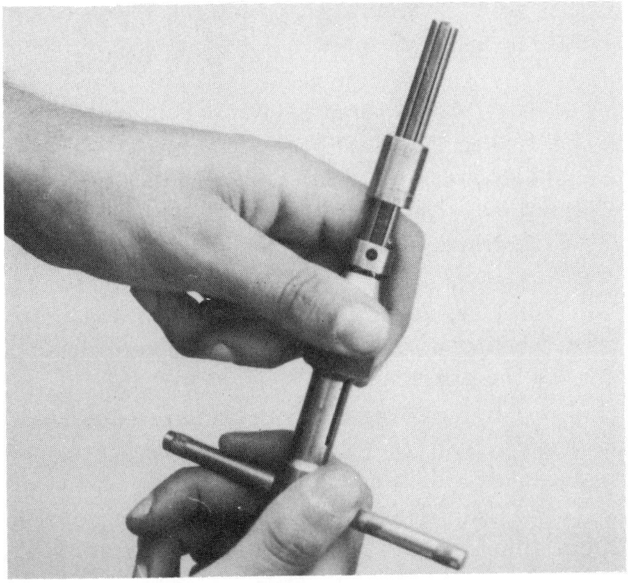

FIGURE 25-19
Tightening a tap extractor in a T-handle tap wrench.

Procedure

1. Put safety glasses on before attempting to anneal the tap.

2. Using a torch, heat up the broken tap until it turns a bright red color and then allow it to cool to room temperature *slowly.*

3. Select a drill smaller than the space between the flutes and attempt to drill through.

4. Enlarge this hole to come closer to the diameter created by the flutes.

5. Using a pin punch, break the sections that remain and remove the pieces.

Note: If drilling becomes difficult, breaking off small pieces with a pin punch may be the only means of removal.

Removal by Acid
Acid is another means of tap removal. This method uses a solution of diluted nitric acid injected into the hole to loosen the tap.

Procedure

1. Prepare to mix a batch of acid by first putting on a face shield.

2. In a beaker carefully dilute about one part of nitric acid with five parts of water.

3. Inject the mixture into the hole.

4. Remove the loosened tap with a pair of needle-nose pliers or with a tap extractor.

5. Wash out any remaining acid with water to discontinue the acid's eating action on the workpiece.

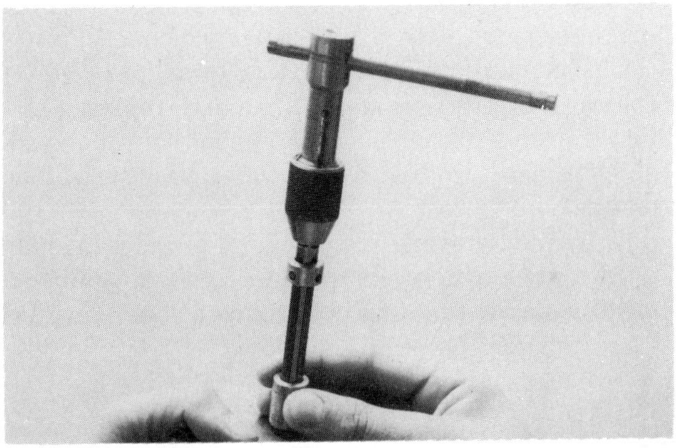

FIGURE 25-20
Insert the prongs into the flutes of the broken tap as far as they will go.

FIGURE 25–21
An EDM machine can pulverize a broken tap.

Thermal Disintegration
Broken taps may also be removed by thermal disintegration. Broken taps that are placed in an EDM machine (Figure 25–21) are destroyed by an electric arc that is sent through a cutting tool placed above the tap. A small gap left between the surfaces creates an energy discharge that pulverizes the tap in a controlled manner without destroying its surrounding surface. (See Unit 61.)

UNIT 26 OFF-HAND GRINDERS

For centuries, grindstones were used to sharpen knives, weapons, and similar tools. At one time, it was common practice to have vendors come around neighborhoods with a foot-pedal-driven grindstone to sharpen knives and scissors (Figure 26–1). Today, the old-fashioned grindstone has been replaced by a power-driven abrasive wheel and enjoys wide use in shaping and grinding round, flat, and other shaped contours.

While there are many general-purpose grinders used for removal of sharp edges, welding joints, or casting flaws, this unit will concern itself with off-hand grinders found in toolrooms.

Objectives
After completing this unit you will be able to

1. change a grinding wheel on a pedestal grinder.

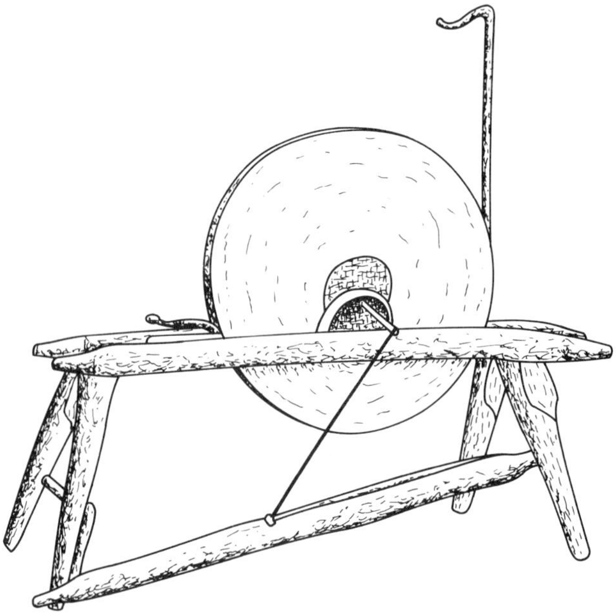

FIGURE 26–1
A foot-pedal-driven grindstone. (Credit: Paul Rapisarda)

2. use the different methods of dressing a wheel.
3. set up a pedestal grinder safely.

General Safety

1. Always wear appropriate safety glasses in any grinding area (Figure 26–2).
2. Always remove rings, jewelry, or wrist watches before attempting any grinding.
3. Never use gloves or rags to hold your work.
4. Disconnect the machine from any power source before attempting to adjust the machine or change the grinding wheels.
5. In removing any spindle nut, wedge a piece of wood to hold the wheel while turning the wrench in the same direction as the wheel cuts.
6. Only mount grinding wheels whose maximum rpms match *but never exceed* those of the grinder.
7. Always stand to one side when turning on a grinder. A shattering wheel can kill anyone in its path (Figure 26–3).

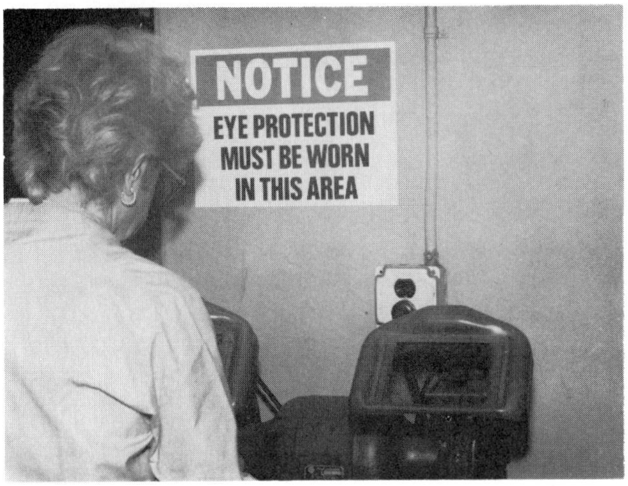

FIGURE 26-2
Safety glasses should always be worn in a grinding area.

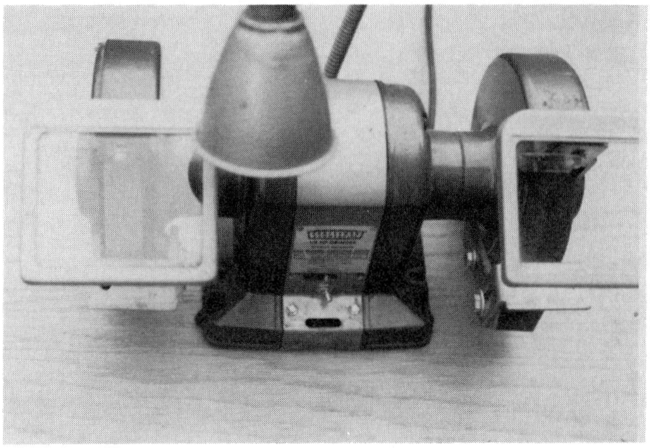

FIGURE 26-4
A light-duty bench grinder.

FIGURE 26-3
Always stand clear of the grinder when turning it on.

Off-hand Grinders

Off-hand grinding is not a precise method of removing metal. Toolrooms are generally equipped with a pedestal grinder and in some cases a bench grinder as well. In this method cutting tools such as drills or tool bits are held by hand and fed against a rotating abrasive wheel (off-hand grinding) to shape or sharpen them.

The bench grinder (Figure 26-4) is a smaller, light-weight version of the pedestal grinder. It is designed to be mounted on top of a workbench. Its electric motor drives an extended spindle upon whose ends are mounted abrasive wheels for light metal removal. Wire wheels or buffing wheels can also be mounted for light polishing purposes.

The pedestal grinder (Figure 26-5) is a heavy base, floor model off-hand grinding machine. A threaded spindle extends from either side of a heavy-duty motor. The left side has left-hand threads, while the right side has right-hand threads. It is important to understand this as it may present a problem in mounting or removing the grinding wheel. The nuts holding the wheels in place can be removed by turning them in the same direction as the wheels do when cutting. A tool rest is located in front of the wheel and slightly below its centerline. A spark guard is provided to protect the operator against a stream of fiery hot sparks, and in front of the spark guard is an eye shield. End covers can also be found to protect against abrasive dust or pieces of broken wheel that may occur.

Due to the fact that heat will be generated when grinding metal, a water trough or container should be filled to cool the work by dipping it often in the water (Figure 26-6).

Types of Wheels

As discussed in Unit 57, every grinding wheel has two major elements—the abrasive grain and the

FIGURE 26-5
A pedestal grinder.

bond. The grain size will determine the cutting, while the bond acts as the cement that binds and supports the grain in its cutting action.

The composition of the wheels will determine the type of metal to be ground. Steels and alloy steels will require one type of abrasive wheel; cast iron and nonferrous metals will require another. Generally, most pedestal grinders use heavy-duty, vitrified bonded (special clays baked in electric furnaces) aluminum oxide abrasive wheels. A coarse and a fine wheel are used for general-purpose offhand grinding. The coarse wheel with a 46-grit surface (see Unit 57, grain size) is designed to remove metal rapidly, while a fine wheel of 60 to 80 grit is used for touching up or smoothing the surface left by the coarse wheel.

Aluminum oxide wheels are intended to grind soft steel and steels used to make tools such as carbon steel, high-speed steel, and metals of this nature. Nonmetallic materials and nonferrous metals such as aluminum, brass, bronze, etc. must never be

FIGURE 26-6
The water container should be filled before doing any grinding.

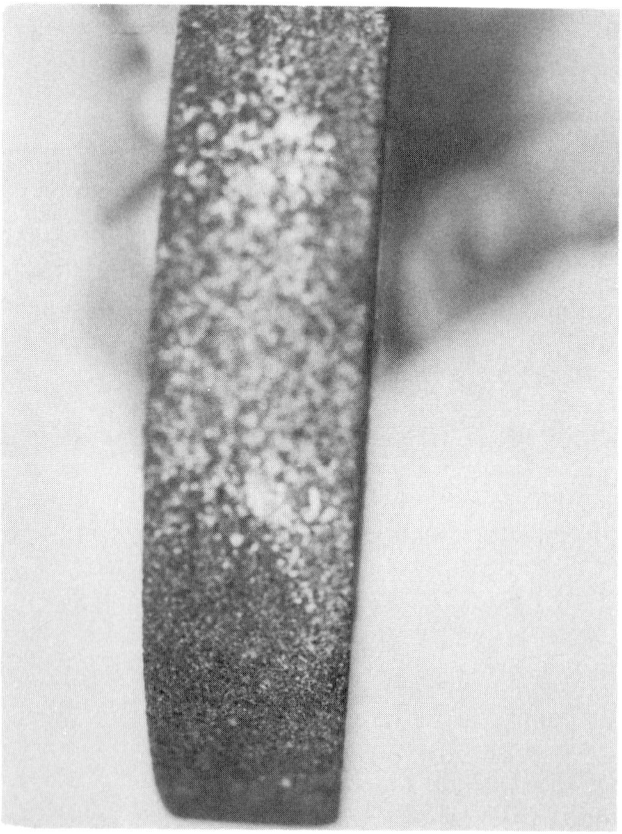

FIGURE 26-7
A loaded grinding wheel is one whose surface has been clogged with aluminum or other nonferrous metals.

used against wheels designed to grind steel and steel alloys. They will clog the wheels (Figure 26-7), leaving them unusable. Wheels composed of silicon carbide should be used on nonmetallic materials, nonferrous, and many types of cast iron.

Wheel Use

Safety flanges and safety washers should be used when mounting abrasive wheels. Safety flanges are large metal washers and should measure from one-third to one-half the wheel diameter. They should also be recessed so that its outer edge presses against the safety washer. The safety washer acts as a blotter and is placed between the safety flange and the wheel. When tightened against, it acts as a cushion against the abrasive surface and creates a better friction drive (Figure 26-8).

Wheels should be checked for cracks before mounting by testing. The wheel should be held on a wooden handle or dowel and tapped lightly with wood (Figure 26-9). A ringing sound indicates the wheel is good; a dull sound indicates the wheel is cracked and unusable.

The center hole of the grinding wheel is also important. It should match the spindle diameter. If it is too large, a reducing bushing should be inserted in the wheel's center hole.

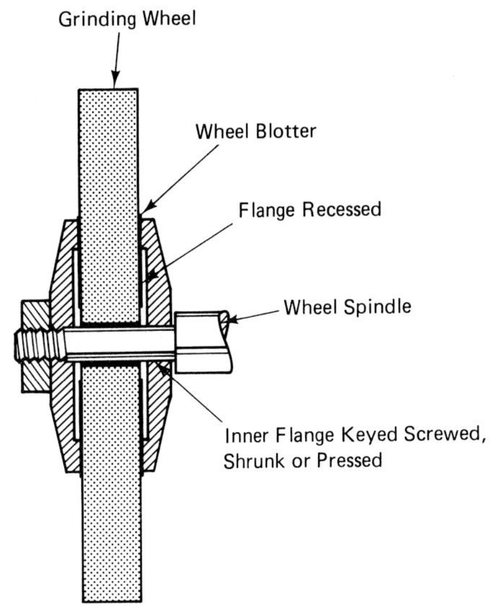

FIGURE 26-8
Cross section of a wheel mounted on a grinding spindle. The safety washer acts as a blotter between the safety flange and the abrasive wheel, forming a gripping surface. (Credit: Bay State Abrasives)

FIGURE 26-9
Testing a grinding wheel by tapping it with wood. A ringing sound indicates the wheel is good; a dull sound indicates the wheel is cracked and unusable.

Dressing the Wheel

Any wheel that is newly mounted or one that has been used for a while will have to be "trued" and resurfaced. The act of resurfacing is to remove all hollows and grooves, to unclog the abrasive pores, and expose newer and sharper cutting abrasives. The term "truing" implies to make the wheel run more uniform, more even in roundness, the removal of high spots or wheel bounce; the business of resurfacing a wheel is referred to as dressing the wheel.

There are three types of wheel dressers: mechanical, abrasive, and diamond.

A Huntington wheel dresser, also called a star dresser (Figure 26-10), is widely used as a mechanical dresser. Composed of a series of steel discs and star-shaped wheels, the device is placed firmly against the rotating front surface of the wheel and slowly moved from side to side as the wheel is trued. Care should be taken not to go beyond the width of the wheel. This will create unevenness of wear.

There are two types of abrasive dressers, the abrasive wheel dresser and abrasive stick wheel dressing tool. The abrasive wheel dresser is usually made of silicon-carbide and trues the grinding wheel with a shearing action when pressed against it. The abrasive stick wheel dressing tool is rolled across the toolrest as a steel tube with hard abrasives is pressed firmly against the front of the rotating wheel (Figure 26-11).

A diamond point wheel dressing tool has a short diamond point nib mounted in a holding device that can also be used to dress a pedestal grinder. This dressing device should be held slightly lower than the wheel's centerline, while running the diamond slowly across the face of the wheel, holding it at a 30° angle in doing so to avoid chatter.

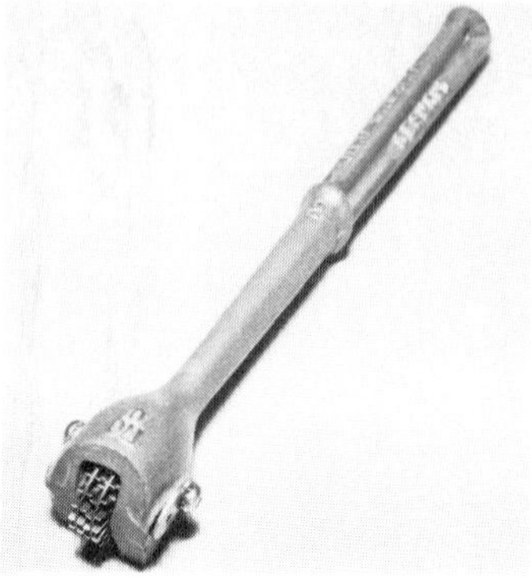

FIGURE 26-10
A Huntington wheel dresser is also called a star dresser.

FIGURE 26-11
Wheel dressing abrasive sticks. (Credit: DoAll Company)

Pedestal Grinder Setup

1. Wear appropriate safety glasses before attempting adjustments.
2. Disconnect the grinding machine from any power source.
3. Clean off any spindle prior to mounting a wheel.
4. Test the wheels for cracks before mounting (see Figure 26-9).
5. Check the wheel's center hole to see that it matches the grinder's spindle diameter. If not, insert a correct size bushing in the wheel's center hole.
6. Check to see that each abrasive wheel has a safety washer.
7. Check the safety flanges to see they are free of burrs, allowing firm seating against the safety washer.
8. Slip on the abrasive wheel and safety flanges. Firmly tighten but *do not overtighten* the spindle nut.
9. Adjust the spark guard not more than $\frac{1}{8}$ in. (3 mm) away from the wheel.
10. Adjust the toolrest not more than $\frac{1}{8}$ in. (3 mm) away from the wheel (Figure 26-12).
11. Turn the wheel by hand to see that it clears the toolrest freely.
12. Tighten the end covers in place.
13. Reconnect the pedestal grinder to its power source.
14. Stand out of line with the wheel's rotation and turn on the grinder.
15. Let the grinder run for at least one full minute before attempting any dressing or grinding.

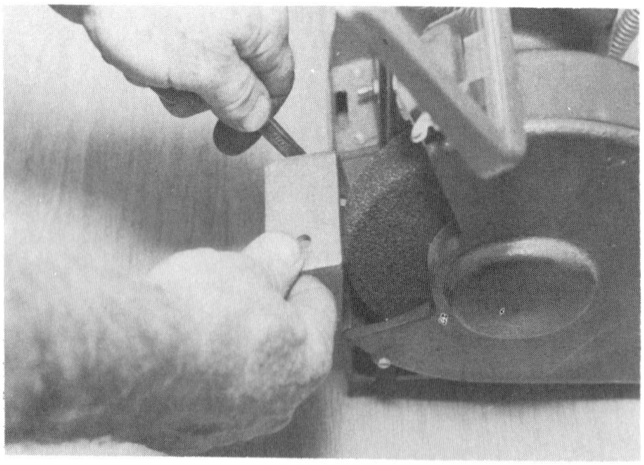

FIGURE 26-12
The toolrest should be adjusted not more than $\frac{1}{8}$ in. (3 mm) away from the grinding wheel.

SECTION REVIEW

Unit 20

1. Name the five common methods of cutting material in the metalworking industry.
2. In which direction should the teeth of a hacksaw blade be pointing in use?
3. Besides hardness, name two factors in determining the selection of a hacksaw blade.
4. Describe the type of work on which the following hacksaw blades should be used: (a) 14-pitch blade, (b) 18-pitch blade, (c) 24-pitch blade, and (d) 32-pitch blade.
5. Name the two basic types of power hacksaws.
6. What type of cutoff wheels does an abrasive saw use?
7. Describe the cutting blade used on a cold saw and the material from which it is made.
8. What is the purpose of the lead-in rollers used on the horizontal bandsaw?
9. List the three main parts of a contour bandsaw.
10. What is the purpose of the job selector found on the door that encloses the upper pulley wheel?
11. Briefly explain the friction cutting method.
12. How are heavier metals cut using the friction cutting technique?
13. What are the three common patterns to be found in the set when making a contour bandsaw blade selection?

Unit 21

1. What are the two main parts of a file?
2. Under what conditions can a file be used without a file handle?
3. What are the four basic teeth structures or file cuts?
4. What are the four basic cross-sectional areas from which all files are shaped?
5. What is another name for Swiss pattern files?
6. At what height should the bench vise jaws be for filing?
7. How should very thin material be held in a vise for filing?
8. What is the name of the condition that is said to cause scratches in work surfaces during filing?
9. What is the name of the tool used in cleaning files of chips and waste?
10. What is the name of the operation where the file is held flat while at 90° to the work and stroked in a series of push/pull movements?

Unit 22

1. How may the bending of metal be defined?
2. What basic hand tools are generally used to bend metal by hand?
3. In developing blank lengths involving right-angle bends, how much material thickness for every bend should be added to the overall length?
4. What two conditions develop if the metal is hammered excessively when bending it?
5. What are angles less than 90° called?
6. What are angles greater than 90° called?
7. What two conditions are eliminated by twisting metal supported through a long pipe that slips over the metal's width?

Unit 23

1. Who is credited with having originated the threaded fastener?
2. Name a simple example of an external and internal thread.
3. What is the purpose of a setscrew and how is it classified?
4. Name three basic functions served by washers.
5. What is the purpose of a dowel pin?
6. What is a key and what is its purpose?
7. What is the result of placing rivets too close to the edge of the lapping plates?
8. Name two basic types of structural adhesives.

9. What is the manner in which structural adhesive bonds are tested?

10. Name three stress conditions that an adhesive has to withstand in order to be considered for structural or design application.

11. Name two metallic means of permanently bonding metal together.

Unit 24

1. What is considered to be the first standardized thread form?

2. Name four basic thread forms used in American industry.

3. For what purpose are threading dies used?

4. Name two means of cutting an external thread.

5. Name three basic types of dies.

6. What two basic dimensions are used to identify the American National screw thread?

7. In which direction is a die turned to cut a left-hand thread?

8. In using an adjustable round split die, in which direction is an adjustment screw turned in order to achieve a shallow cut?

9. (a) What is the name of the die used for restoring threads and (b) what is the name of the holder used in turning it?

10. What is the name of the set that contains a series of hand tools to cut a variety of internal and external threads?

Unit 25

1. What are the three major parts of a tap?

2. What is the name of the recess channel that collects and holds lubrication and chip waste?

3. What are the names of the principal wrenches used in hand tapping?

4. What is the name of the hole required for tapping and what percentage thread depth does it generally render?

5. What are the names of the three taps used in a set of hand taps?

6. What is another name for a general-purpose tap?

7. What is another name for a spiral pointed tap?

8. What is the name given to the tap designed to lift the chips clear of the hole?

9. What is the name of the tool that, by measuring the spacing from thread to thread, determines the number of threads involved?

10. What expression is used to denote that the cutting tool is threading at 90° to the workhole?

Unit 26

1. How is off-hand grinding defined?

2. Name two types of grinding wheels used on a bench or pedestal grinder.

3. Why are coarse- and fine-grained wheels used on a bench or pedestal grinder?

4. Why must brass, bronze, or aluminum not be used on ferrous grinding wheels?

5. If the nuts on the threaded spindles contain right-hand and left-hand threads, how are the wheels removed?

6. How close should the tool rest be to the wheel? Explain why.

7. What is provided on the pedestal or bench grinder to protect the operator's hands against a stream of fiery hot sparks?

8. How are wheels tested for cracks before mounting?

9. What precautions should be observed when starting a grinder and explain why.

10. Name three types of wheel dressers.

11. What is another name for the Huntington wheel dresser and how is it classified?

Section 7

CUTTING PRINCIPLES

UNIT 27 PHYSICS OF METAL CUTTING

Research has shown that the theory stated so often in technical publications that metal cutting tools sever by pushing the metal apart, citing the wood splitting axe as an illustration, is simply not true. Like other technologies that have been used by civilizations without fully understanding the basic principles involved, such as casting and forging, the use and understanding of how metal cutting tools work has evolved slowly. The findings of many metallurgical researchers in industry and in different countries will be used in the following text.

Objectives
After completing this unit you will be able to

1. understand the basic principles of metal cutting.
2. identify the three chips formed during machining operations.
3. understand the movement of the chip over the tool face.

The Dynamics of Cutting

Extensive research has shown that the cutting action of any cutting tool is the result of applying concentrated forces to the cutting edge(s) generated by the machine tool (Figure 27-1). As the cutting tool continues to travel against the workpiece, the many tons of pressure (psi) at the cutting edge causes the work material to compress and deform in front of the tool face. Tremendous heat is generated in the cutting zone due to the resistance of the material to the tool face. The heat from this friction continues to build until a plasticlike state has developed in the path of the cutting tool advancement, a plastic state pliable enough to flow the material under pressure. One of two conditions exist as the tool continues to advance: There will be either a rupture or a plastic flow of material in the direction of the tool face.

As the chip is formed due to the deformation of material in the cutting zone, its thickness will be determined by the shear angle (Figure 27-2). This shear angle is a determining factor in generating temperature and will affect the cutting tool life. The greater the chip thickness, the greater the force required to form the chip. This will result in a higher temperature in the cutting zone. The opposite is also true. The thinner the chip, the smaller the force required to form it. A light chip will generate less heat in the cutting zone.

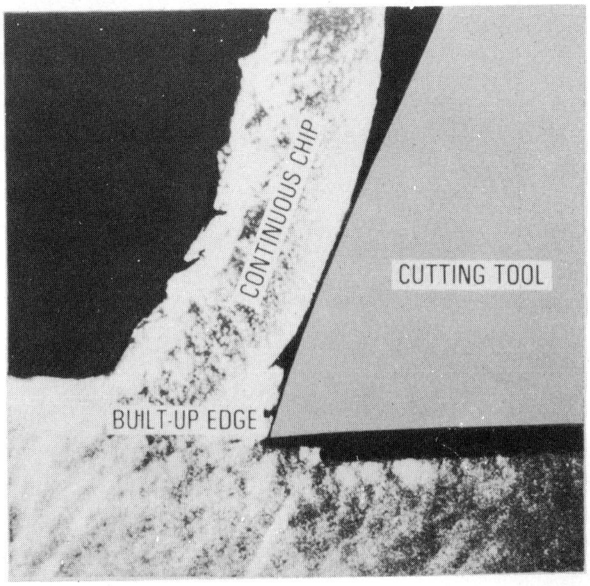

FIGURE 27-1
Basic chip formation. (Credit: Cincinnati Milacron Inc.)

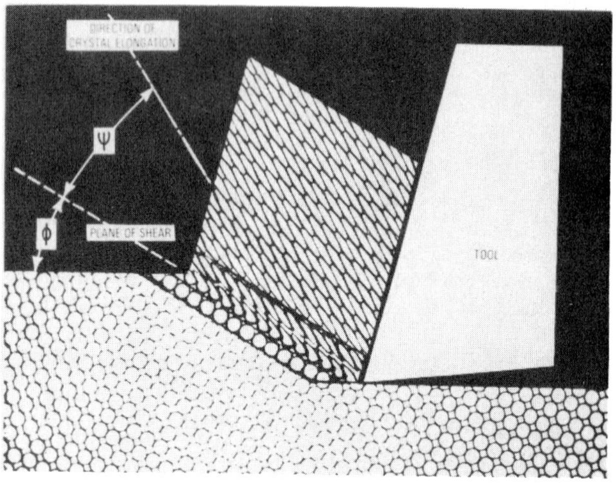

FIGURE 27-2
The deformation of the material in the cutting zone makes the formed chip thicker than the layer of metal being removed. Chip thickness is determined by the shear angle. (Credit: Cincinnati Milacron Inc.)

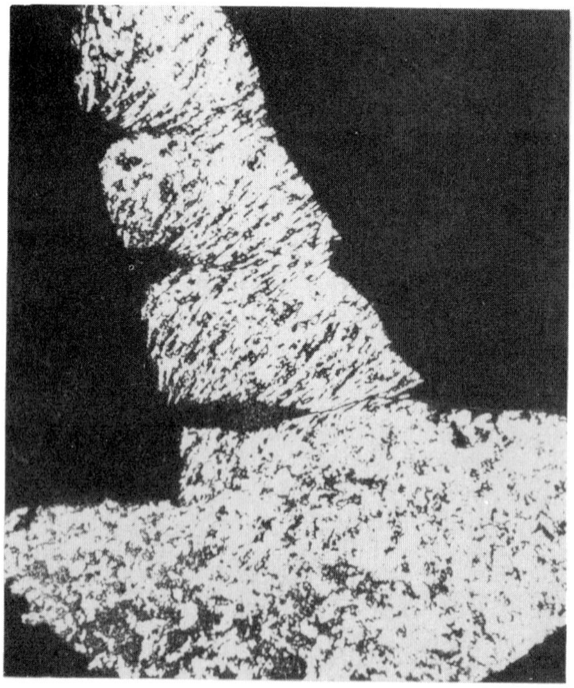

FIGURE 27-3
The discontinuous chip is segmented. (Credit: Cincinnati Milacron Inc.)

The type of material is also a contributing factor to the amount of heat generated in the cutting zone. Aluminum alloys, mild steel, and other ductile metals tend to be plastic and deform easily at a relatively low temperature. Alloy steels, titanium, stainless steel, and other difficult to machine metals have an increased resistance to compression and deforming ahead of the cutting tool. These metals require greater forces to overcome them. This greater force results in higher temperatures in the cutting zone.

Types of Chips

Chips formed by the cutting tools on lathes, milling machines, and similar machine tools are classified into three basic types: the discontinuous chip, the continuous chip, and the continuous chip with a built-up edge. All will exist under certain conditions.

The Discontinuous Chip

The discontinuous or segmental chip (Figure 27-3) is often the result of machining brittle materials such as cast iron and hard bronze. Other operating conditions such as a low cutting speed, a small rake angle, a large chip thickness, or vibration of the cutting tool may also be contributing factors.

In this situation the chip is produced by the advancing tool, which compresses the material briefly and causes it to escape freely along the free space above the cutting edge. As the material tends to flow further and further ahead of the tool, it finally ruptures as more stress is applied and is broken up in small segments that easily clear the cutting zone. The cycle is repeated indefinitely (Figure 27-4).

Generally, tool failure is experienced with discontinuous chips due to the wear and rounding over of the cutting edge. And while a poor finish is usually the case, a good finish can be produced if the chip segments are small.

The Continuous Chip

The continuous chip (Figure 27-5) is much in evidence in the machine shop where spiral lengths of tightly coiled ribbons of metal can be seen in the areas where aluminum alloys and free machining steel are being machined on a lathe.

Metal is made up of many crystalline grains. Ductile materials offer a good plastic flow due to the many crystallographic slip planes, which enables a chip to withstand greater distortion without fracturing or rupturing. A continuous chip is formed as the metal ahead of the cutting tool is compressed and

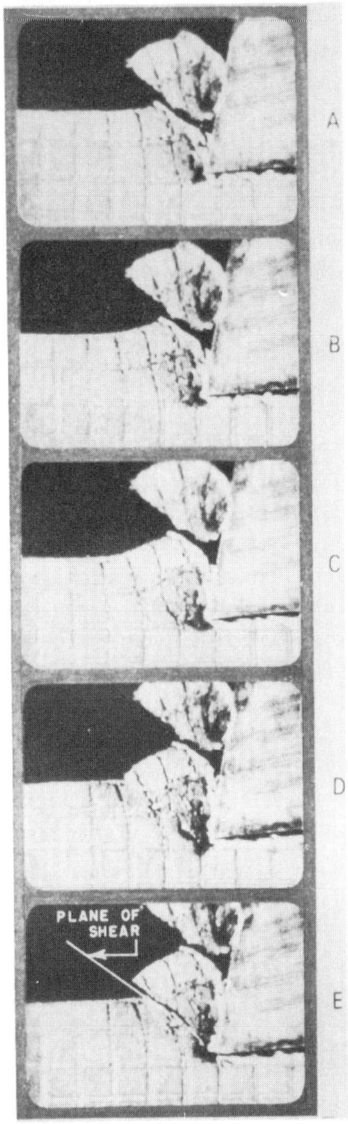

FIGURE 27-4
The formation of a typical discontinuous chip. (Credit: Cincinnati Milacron Inc.)

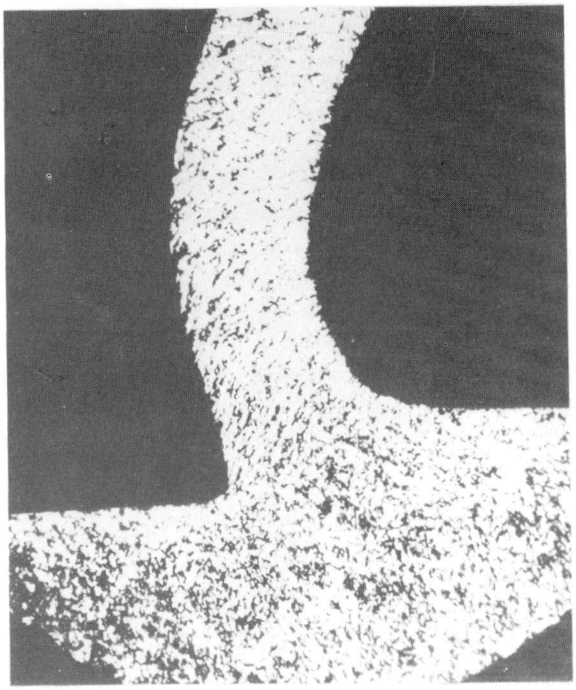

FIGURE 27-5
The continuous chip is able to be formed in long ribbonlike curls or short curls. (Credit: Cincinnati Milacron Inc.)

continues compressed as it separates from the metal, escaping in a steady flow along the face of the tool. The chip ceases to elongate and compress, becoming work hardened as it starts to curl. This is evidenced by the shiny layer of material found on the underside of the chip. The cycle is repeated indefinitely, leaving a good work surface finish.

The continuous chip can be produced using the following recommendations:

1. Ductile work material.
2. High cutting speed.
3. Sharp cutting edge.
4. Small chip thickness with fine feed.
5. Large rake angle.
6. Use of cutting fluid to keep the tool point and work surface cool.
7. Minimize resistance to chip flow by
   (a) a highly polished tool face
   (b) use of cutting fluid to reduce friction
   (c) use of a cutting tool that will have a low coefficient of friction such as cemented carbide.

Note: While the continuous chip sounds like an effective method of machining, its use in steel turning tools is questionable. Long continuous chips are a safety hazard to the operator, are difficult to handle, and may twist around to clog automatic turning machines. Chipbreakers, which are ground into the turning tool, are designed to produce short, broken chips. A continuous figure-nine-shaped chip is considered ideal when machining steel. Broken chips take up less space, permit a better flow of cutting fluid to the cutting edge, and are disposed of more economically.

The Continuous Chip with Built-up Edge

The continuous chip with built-up edge (Figure 27-6) is frequently formed on low-carbon steel that is cut with high-speed steel cutting tools without the aid of cutting fluid.

As the metal is compressed ahead of the cutting tool, it forms a chip that begins to slide off the cutting tool face. Due to the high pressure and high temperature in the cutting zone, as well as the high friction between the compressed layer and the tool face, the chip does not slide off cleanly. Rather, it "welds" part of itself to the face of the tool along the cutting edge while the remaining chip barely manages to shear off and escape. As the cutting action continues, additional particles of compressed material build up, moving the line of shear further and further away from the cutting edge. Eventually, the enlarged buildup is fractured or torn away, particles of which find their way and bond themselves to both the chip and the work material.

The recurring condition of building up and tearing away occurs rapidly during cutting. The results are a rough finished workpiece that is covered with many particles of built-up edge. This type of chip making also shortens the tool life due to cratering and abrasion.

Cratering is a depression or wear zone on the tool face that exists just behind the cutting edge. As the cutting continues, cratering enlarges in size, extending outwardly to the cutting edge until the point weakens and breaks off, causing the tool to fail completely.

Abrasion is the result of the tool flank coming in contact with the fragments that are freed from the workpiece and chip.

Summary of Findings

There are two basic processes involved in all metal cutting operations, the formation of a chip and the movement of the chip over the tool face. Both operations result in heat due to the deformation and friction of the metal. The discontinuous chip may be formed with reasonable tool life and good finish when machining brittle metals; excessive tool wear and poor finish will result when machining ductile metals at low speeds. The continuous chip will deform metal continuously without rupturing due to low friction between the chip and the tool face. The continuous chip with built-up edge operates at high friction between the chip and tool face resulting in the worst surface finish of the three (Figure 27-7).

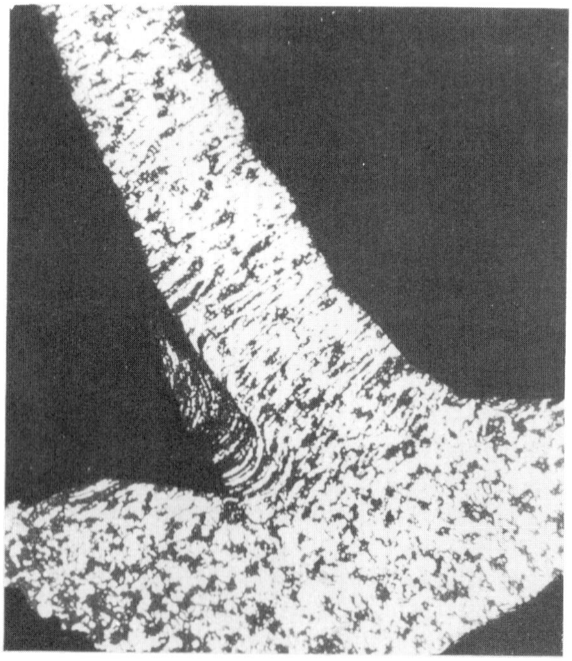

FIGURE 27-6
The continuous chip with built-up edge accumulates or "welds" metal to the surface of the cutting tool. (Credit: Cincinnati Milacron Inc.)

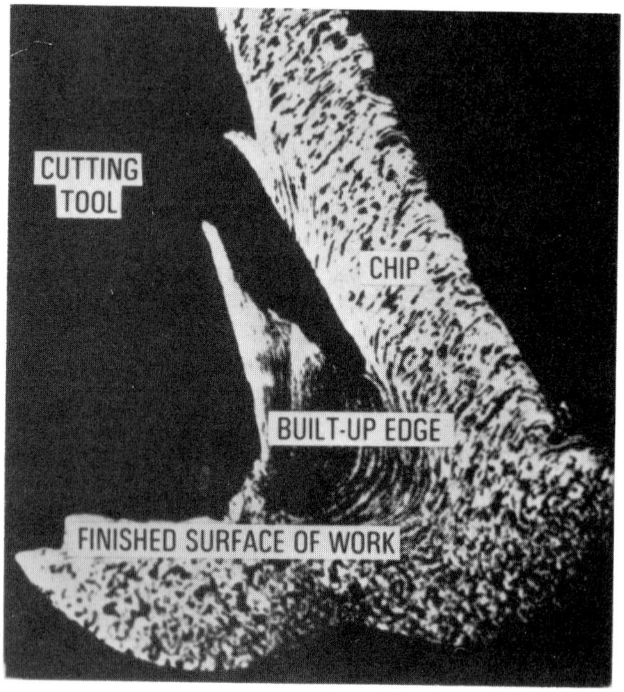

FIGURE 27-7
A continuous chip with built-up edge being formed adjacent to the tool face. (Credit: Cincinnati Milacron Inc.)

While the conditions that cause the three different types of chips have been examined, it should be noted that by changing the chip thickness, temperature, cutting fluid, and similar operating conditions, it is possible to change one type of chip for another without changing the material.

UNIT 28 MACHINABILITY OF METALS

Machinability of metals is a technical term that refers to the ease or difficulty with which metals may be machined. There are many factors that are taken into account such as cutting speed, tool wear, finish, accuracy, cutting fluid, and power requirement.

Objectives
After completing this unit you will be able to

1. know the factors that affect the machinability of metals.
2. explain what a metallurgical chipbreaker is.
3. explain the difference between a positive and negative rake and how they are used to improve tool design.
4. understand the purpose of the different cutting fluids.

Machinability Rating

The machinability of work materials varies with their physical makeup and chemical composition. Table 28–1 is a partial listing and is composed to show the percent of the relative cutting speeds in surface feet per minute (sf/min), using B1112 cold-drawn steel, also referred to as mild or free machining steel, as the standard rated at 100%. It will also be noted that the chemical composition is listed under the identical SAE and AISI numbers. The values listed involve the use of a high-speed steel toolbit obtained on various experimental data and average production work.

As a means of comparison, if tool steel is rated at 50% and brass is rated at 200%, it can be readily seen that the tool steel (low machinability) will take twice as long to machine while the brass (high machinability) would machine twice as fast as the standard cold-drawn steel (B1112) under the same operating conditions.

TABLE 28–1
Machinability ratings of SAE and AISI cold-drawn carbon steel

SAE and AISI No.	*Machinability rating (%)*	*BHN*	*Cutting speed (sf/min)*	
			HSS	*Carbide*
B 1111	95	163	155	500
B 1112	100	167	169	500
B 1113	135	167	180	600
C 1117	90	137	140	450
C 1118	85	143	150	500
C 1119	100	137	165	500
C 1132	75	183	130	500
C 1137	70	197	130	500
C 1140	70	170	130	500
C 1141	70	212	130	500
C 1144	80	217	130	500
C 1145	65	187	130	500
C 1146	70	187	130	500
C 1151	65	207	90	250

Microstructure

The inner composition of steel, or its microstructure, has a direct bearing on the cutting tool's performance, effecting penetration, surface finish, and the ability to maintain a sharp cutting edge. Machinability is a direct function of the metal's tensile strength. High-carbon steels alloyed with tungsten and molybdenum, for instance, are difficult to machine and therefore have a low machinability rating. These carbide-forming elements provide a resistance to wear and cutting. The machining quality in steel can be improved by annealing. It can also be improved with the addition of other elements in the alloying of some metals to make them free machining metals.

Free machining metal is the result of adding one or more contaminants at the mill to give small broken chips, use lower power consumption, while obtaining a better surface finish and prolonging a longer tool life (Figure 28–1). To avoid long chips from fouling the cutting tool, the sulfur (manganese sulfides) content is raised during the alloying of steel, for instance, to make it more machinable. Sulfur is added to ductile materials such as stainless steel, copper, and some steels. The sulfur is a modest embrittler but does not bother the mechanical properties of the ductile materials much. It tends to act as a "metallurgical chip breaker."

FIGURE 28-1
Samples are taken from each blast furnace cast for spectrographic and chemical analysis. (Credit: Bethlehem Steel Corporation)

The following is a list of alloying agents used as metallurgical chip breakers.

Alloying agent	*Metal*
Sulfur or lead	Steel
Lead	Brass
Lead and bismuth	Aluminum
Sulfur or selenium	Stainless steel

Low-Carbon (Mild) Steel

One of two contaminants may be added to low-carbon steel to make it free machining: sulfur or lead. The composition of steel is iron plus approximately up to 2% carbon, generally averaging 0.18 to 0.95% carbon, plus additional elements. As the carbon content increases, so does the presence of pearlite—layers of ferrite and iron carbides (cementite). The addition of sulfur does much to change the state of the cementite from brittleness into spheroid balls, a rounded or globular form of carbides that are considerably easier to machine.

Lead does not mix in a true solution (insoluble) in ductile material such as aluminum, copper, iron, or their alloys. It can be found as leaded globules at the grain boundaries and tends to lubricate and helps break up the chip.

Cast Iron

There are four different types of cast iron, two of which will be discussed here. All are defined by their microstructure.

Gray cast iron is granular in form, contains large graphite flakes that are formed during the cooling stage of solidification, and ranges in carbon content between 3 and 4%. Its graphite content acts as a lubricant and produces short chips without the aid of further alloying or processing. The heavy concentration of carbon will bring with it a greater pearlite condition. The finer pearlite spacing will result in a decrease in machining efficiency and shorten tool life. Machinability may be increased by converting the carbides that exist in the pearlite layers to a flake graphite condition by annealing.

When white cast iron is cooled rapidly during solidification, the carbon in the molten iron remains united with the iron and forms layers of iron carbides, which is a hard and brittle compound. Since the iron carbides saturate the microstructure, white cast iron is hard and brittle and features a white crystalline surface. The hardness of white cast iron is controlled by the amount of iron carbides produced during solidification.

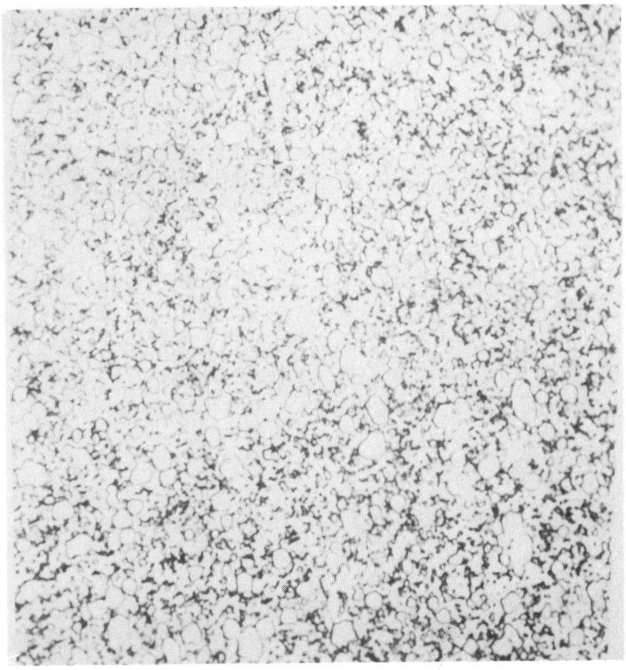

FIGURE 28–2
Microstructure of high-speed steel that has been annealed by spherodizing for ease of machinability. (Credit: Latrobe Steel Company)

Alloy Steel

Alloy steels are steels that have other elements besides carbon added to the iron to change or impart different properties. They may have nickel, chromium, or vanadium added to improve toughness, strength, and wear resistance, for example. As such, alloy steels are generally difficult to machine and may undergo different normalizing and annealing processes to promote machinability (Figure 28–2). Other elements that may be specified to improve machinability include the use of nitrogen, tellurium, or bismuth. Sulfur or selenium may be added to 303 stainless steel to make it as machinable as 302 stainless.

Nonferrous Alloys

Pure aluminum is soft and ductile. It can be alloyed with a number of elements such as copper, manganese, magnesium, silicon, and zinc to increase more desirable properties such as strength, ductility, corrosion resistance, and weldability to name but a few. Machinability of aluminum alloys is governed by heat treatment and tempering that they undergo as well as with the addition of bismuth, lead, and tin.

Copper and copper-base alloys are available in various degrees of hardness, referred to as temper by the copper industry, such as hard, half-hard, quarter-hard, and spring. Copper, which machines as a long, tough, string material, is often modified with the addition of small amounts of other elements to produce different cutting characteristics. Consider 360 yellow brass as an example. When made as an alloy of 65% copper and 35% zinc, the result when machined is a cut that is long and stringy. By substituting 3% lead with a like amount of copper, a free cutting brass is produced. 360 yellow brass is machined to produce a brittle, easily broken chip. It has a machinability rating of 100% and is used as the standard to which all other copper and copper-base alloys are compared. Tellurium and selenium in proper amounts are also used to increase machinability.

Cutting Tool Geometry

Cutting tools for turning and similar types of operations are made from a variety of materials as well as a number of different shapes. The contour or tool shape will vary with the type of work material and may have a strong influence on the efficiency of the cutting tool. From a standpoint of tool wear and surface finish, the tool should have a low coefficient of friction.

There are two principal causes of cutting tool failure:

1. Excessive temperature at the cutting edge
2. Gradual tool wear that develops as flank, nose, or crater wear (Figure 28–3).

The greatest limitation to the speed of machining is the heat generated at the point of contact between the tool and the work. *Nose wear,* as the result of the cutting action is called, affects the quality of the work surface finish. Using a larger nose radius will produce a better finish and will permit a faster feed rate, if the factors of work material and cutting fluid are not considered. The heat generated must be carried off by either the cutting fluid or the chip itself. The efficiency of the cutting fluid will directly affect the temperature of the work and the tool. The chip will be affected by the geometry of the tool, the sharpness of the cutting edge, and the depth of cut.

When there are free machining alloys or moderately machinable alloys available, chip contact is minimized due to the rapid chip breaking action. Here the requirement is a cutting fluid whose function is to act as a coolant. Rake and clearance angles can be held to a minimum. The small rake angle will break the chip almost immediately, while the minimum clearance angle provides a greater support for the tool's cutting edge.

Metals that have not had elements added to act as metallurgical chipbreakers will require a lubricant to reduce friction and prevent the metal from "welding" itself to the face of the tool. Unless the cutting fluid can act as a coolant as well, the cutting speed will have to be reduced. Rake angles will have to be increased to start the cutting and removal of the chip as rapidly as possible. A highly polished tool face will help reduce friction, prevent *crater wear* (surface depression), and will help the chip slide off as rapidly as possible.

Flank wear, the tool wear just below the surface of the cutting edge, is another result of gradual tool breakdown. As some of the tool's strength is reduced due to increased temperatures at the cutting edge during machining, some of the tool material manages to slowly plasticize, sticking to the tool surface below the cutting edge. As this plastic deformation of tool material becomes greater, it will greatly reduce or eliminate the clearance angle along the flank area. As the rubbing between the tool and work continues, temperatures at the cutting edge are increased until eventually the tool fails completely.

Lack of tool strength may also lead to fracturing or chipping of the cutting edge. Tiny cracks or stress marks may be brought about by excessive pressure acting on the cutting tool. These conditions may be brought about by poor tool geometry, material selection, the resharpening of the cutting tool, and by poor machining conditions that can lead to vibration and chatter.

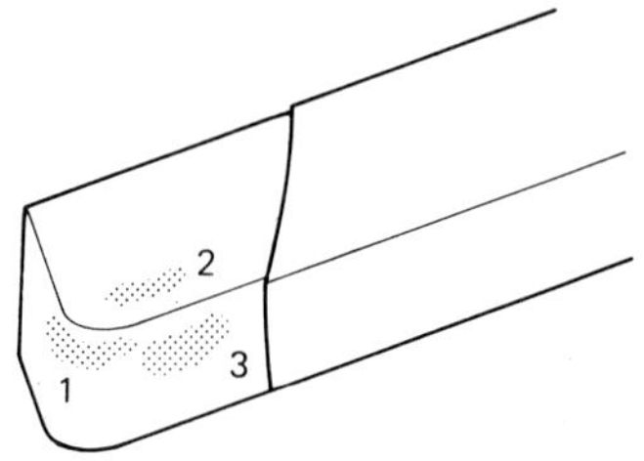

FIGURE 28–3
The tool wear of a cutting tool as indicated by (1) nose wear, (2) crater wear, and (3) flank wear.

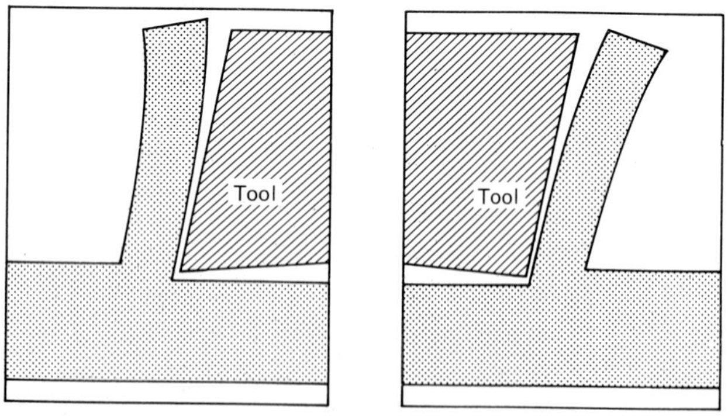

FIGURE 28–4
Tool life is affected by the direction of the rake angle. (A) The chip slides along the tool face when a positive rake is ground; (B) the chip slides up the tool face when using a negative rake.

Overcoming Tool Failure by Design

There are several methods used for overcoming tool failure such as the use of positive and negative tool rake angles (Figure 28–4).

A positive rake angle can be identified by the slide of the chip along the tool face. If the chip slides down the face of the toolbit when the point of the cutting tool and cutting edge contact the work, the cutting tool is considered having a positive rake angle. Larger rake angles tend to decrease the cutting force and cutting temperatures, while at the same time it will benefit tool life or permit a higher cutting speed. It will all depend on the selection of the cutting tool and work material. Cemented carbide cutting tools, for example, which are harder and more brittle than high-speed steel, require larger side rake angles for easier chip escape.

Negative rake angles are sloped in the opposite direction as the positive rake angles. In simplified terms, if the chip slides up the face of the toolbit when the point of the cutting tool and cutting edge contact the work, the cutting tool is considered having a negative rake angle. It is recommended that negative rake angles be used on heavy interrupted work surface cuts and for turning very hard materials.

Other advantages offered by negative rake angles are:

1. Strengthens the nose of the tool.
2. Strengthens the cutting edge to enable it to withstand heavier cutting and shock loads.
3. Economic advantage when using negative rake indexable insert and tool holder. Top and bottom of the inserts can be used, offering twice the cutting edges.

Sources of Heat

There are three principle sources of heat generated during metal cutting:

1. Compression and deformation of the metal while forming a chip.
2. The thickness of chip that is formed by the shear angle of the cutting tool.
3. The high friction created as a chip begins to slide off the tool face.

The heat generated during the three stages of metal cutting vary in temperature according to the metal. During the compression and deformation of metal while forming a chip, the heat generated in the cutting zone is high enough to almost melt the metal. This plasticlike state is pliable to flow the material under pressure. This temperature, while registering one level for ferrous alloys and another level for nonferrous alloys, contributes greatly to shortening tool life. The heat of deformation, allied with the heat developed by the formation of the chip thickness (chip distortion) and the friction developed by the chip sliding over the tool face and the workpiece on the tool, will affect the surface finish, rate of metal removal, and accuracy of the workpiece.

The capacity of the cutting tool material is another contributing factor toward maintaining cutting efficiency. High-speed steel cutting tools maintain their cutting edges at temperatures up to 1000°F (537°C) even though the cutting tool has turned a dull red. This property is known as "red hardness" and is used as a measuring level at which cutting edges perform. As the temperature begins to range between 1000 and 1100°F (537 and 593°C), the cutting tool loses its cutting edge.

In contrast to this, cemented carbide cutting tools are productive even at temperatures ranging up to 1700°F (927°C). Cemented carbide tools are many times harder than high-speed steel tools, have high compression strength, and some grades are quite brittle. They also have a low heat conductivity and can be used at cutting speeds several times as fast as high-speed steel.

Surface Finishes

Surface finish is affected by many factors encountered in machining—dull tools, speeds and feeds, heat generated during the machining operation, and the condition of the machine. While not singling out any particular condition as being the most important, experiments by metallurgist Hans Ernst have been conducted to focus on the role that heat plays on surface finish.

A cylindrical surface of aluminum alloy was turned at three different temperatures, 200 (93), 75 (24), and −60°F (−51°C) without the use of any cutting fluid. It was found that the surface of the aluminum alloy ranged from a ragged, torn appearance at the higher temperature to a very shiny, smooth appearance that was obtained with the supercooled [−60°F (−51°C)] workpiece.

A second test was conducted with the temperature of the aluminum alloy work surface being machined at a temperature of 75°F (24°C), using a very low cutting speed to minimize any heat increase. It was found that a large built-up edge was formed merely by machining the aluminum material using a low-temperature and low-speed technique.

During the turning operation, a cutting fluid void of any lubricating qualities, was used. A few drops of carbon tetrachloride were placed between

FIGURE 28–5
A high-pressure jet of coolant is used in this machining operation. (Credit: DoAll Company)

FIGURE 28–6
Coolants can also be effectively used through mist spraying.

the tool face and the chip. The effect was immediate and dramatic, producing an all together different chip formation. Gone was the fragmented chip; what was left behind was a bright and smooth finish.

The analysis of the experiment led to the conclusion that the coolant effect of a cutting fluid plays a major role in producing better workpiece finishes.

The Role of Cutting Fluids

Some of the advantages provided by the cutting fluids have been cited throughout this unit. There are other properties that require closer examination such as coolant and lubricant.

Proper use of cutting fluids requires the use of a coolant or a lubricant. Cutting fluids made to reduce or remove heat from the cutting zone are known as coolants. Cutting fluids formulated to reduce the frictional heat generated in the cutting zone are called lubricants.

Coolant

A coolant is effective if it can penetrate to the heat in the cutting zone and has the capacity to carry off or eliminate the heat. This will depend on the work material, the difficulty of the operation, the contour of the cutting tool, and the properties of the fluid. Water is an excellent heat conductor and coolant. It increases in temperature only half as much as a like volume of oil and is capable of dissipating (scattering) heat better than two times faster than oil. Water will also absorb a great deal more heat than oil before vaporizing into a gas (vaporization).

Despite its excellent properties, water by itself will cause machines, tools, and ferrous metals to rust. Water-soluble oil, an emulsion of soluble oil and water plus the addition of a rust inhibitor, is widely used as a coolant. Whatever lubricating properties are to be found in the soluble oil are small. This type of coolant is used for light to heavy machining operations of ferrous materials and may be dispensed as a circulating fluid (Figure 28–5) with high-pressure jets and also through mist spraying (Figure 28–6). Additional advantages of water-soluble oil coolants are:

1. Higher machining speeds can be used.
2. Cutting tool remains cooler.
3. Economical to buy and use.
4. Results in less odor and cleaner work conditions.
5. Poses no fire hazard.

Properly applied high-quality coolants can dramatically reduce labor, overhead, and tooling costs. It does not pay to use inexpensive low-quality coolants on expensive machine tools.

Chemical Cutting Fluids

All coolants are synthetic. In general usage, the term "synthetic" implies any coolant or lubricant concentrate that does not contain petroleum oil. In fact, the major source of raw materials for all types of coolants, including pure synthetics, are petrochemicals made from crude oil and natural gas.

There are three principal types of chemical cutting fluids manufactured: (1) true solution, (2) plain surface-active, and (3) heavy-duty surface-active types.

1. True solution fluids are rust inhibitors that provide good cooling but very little lubrication. These fluids are generally clear solutions and are used primarily to combat rusting and for rapid cooling during grinding operations. These fluids may have dyes added for identification purposes and are mixed in correct proportions with deionized water. Depending on the application, this mixture may include 1 part solution to as much as 250 parts deionized water. *Note:* It is recommended that deionized water be used in mixing fluid concentrates. Severe water hardness and the presence of other minerals can result in rusting, staining, or corrosion of both machines and workpieces.

2. Plain surface-active fluids contain additives that improve the wetting action by lowering the surface tension of the water. These fluids have good cooling and moderate lubrication characteristics. Other additives are mixed to improve lubricity and cooling. Plain surface-active fluids can be found in use when high-speed turning or milling with carbide tools.

3. Heavy-duty surface-active fluids cool effectively but have other compounds such as sulfur, chlorine, or phosphorous added that offer superior high-pressure lubrication.

The following benefits are found in synthetic cutting fluids:

1. Nontoxic.
2. Good rust and corrosion control.
3. Nonflammable.
4. Nonsmoking.
5. Excellent cooling properties.
6. Excellent penetration.
7. Effective lubricity.
8. Long-range stability and product durability.

Synthetic coolants may offer longer tool life for operations that require more cooling than lubricity. These coolants are not a cure all for all operations, however. There is no one perfect coolant. To make a choice, one should consider the cutting fluid as a tool. Many chemical cutting fluids are not recommended by manufacturers for use on zinc, cadmium, lead, or magnesium alloys, for example. Synthetic cutting fluids are more susceptible to mold growth than soluble oils. To maintain integrity and stability of the cutting fluid, good housekeeping is in order. Not only will it require proper selection and control of fluid concentration, but also prevention and elimination of contaminants as well as adequate filtration. To prevent bacterial growth will require more than the use of good water. There are no short cuts, no buttons to push if the quality of synthetic coolants is to be maintained.

Lubricant

There are a number of cutting fluids that are used as lubricants, the principle being the use of straight mineral oil (cutting oil). Mineral oils are used in combination with such additives as lard oil, castor oil, or chemically active elements (sulfur and/or chlorine) to form compounded cutting oils.

Lubricants are used to reduce friction between the flowing chip and the cutting tool face, lowering the heat generated in the cutting zone. Approximately one-third of the tool heat generated is caused by the chip sliding up the face of the tool and the tool flank sliding over the cut surface. The other two-thirds is caused by internal friction, the resistance of metal shearing. The tougher film produced by compound cutting oils prevents the chip from "welding" itself to the tool face. Straight mineral oils are recommended for light machining on metals of high machinability such as aluminum, brass, magnesium, and free machining steels. Heavy-duty compounds are recommended for difficult-to-machine metals when extra lubrication is needed and for form grinding from the solid. They are also recommended for work materials when extreme accuracy and fine finish are a requirement.

Additional benefits include:

1. Less frictional heat is generated (Figure 28–7).
2. Lowers the horsepower requirement for machining.
3. Lowers the amount of pressure required to form a chip (Figures 28–8 and 28–9).
4. Deeper depths of cut are possible.
5. Increases cutting capabilities.
6. Prolongs tool life.
7. Maximizes production rates (Figure 29–10).

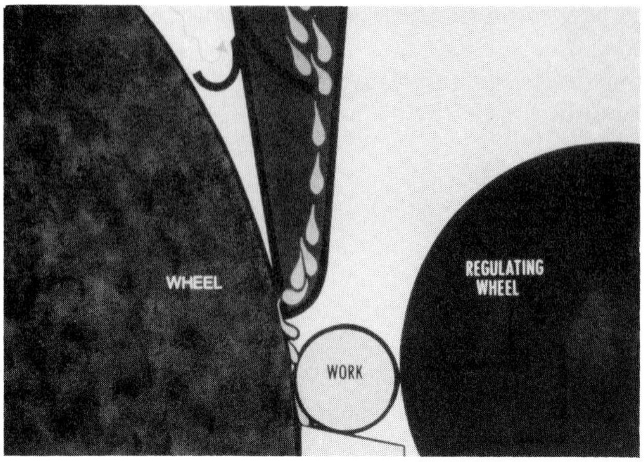

FIGURE 28-7
Not only does the cutting fluid generate less frictional heat in this grinding operation, but it keeps the wheel clean. (Credit: Master Chemical Corporation)

FIGURE 28-9
It is usually desirable to use a ring nozzle system with a face mill. (Credit: Master Chemical Corporation)

FIGURE 28-8
Cutting fluid in proper volume is supplied to both sides of the cutter in this slab milling operation. (Credit: Master Chemical Corporation)

FIGURE 28-10
Cutting fluid for grinding operations should be applied in large quantities and under very little pressure. (Credit: Landis Tool)

Note: The handling and storage of cutting oil fluids may be hazardous to the machine operator. Though some cutting fluids may contain bactericide to prevent contamination and spread of bacteria, care should be taken not to come in contact with the fluid. Not only are unpleasant odors caused by the bacteria growth, but a serious skin disorder (dermatitis) may also result from it. The bacterial life cycle may also attack the fluid concentrate, weakening it to leave a deposit of acid and salts. This condition may result in extensive rusting and corrosion of both the material being machined and the moving machine parts. Not only should high-quality cutting fluids be used, but they should be kept clean with good housekeeping maintenance.

SECTION REVIEW

Unit 27

1. What was the original theory concerning the cutting action of metal cutting tools?
2. Explain briefly how the cutting action of any cutting tool occurs according to the presently accepted theory of cutting.
3. What determines the thickness of chip during cutting?
4. Name three types of chips.
5. What is cratering and what is its result?
6. Name three reasons why long continuous chips should not be used during production machining.
7. Name three reasons why chip breakers should be used instead.
8. What causes tool abrasion?

Unit 28

1. How is machinability of metals defined?
2. What two factors affect the machinability of metals?
3. How are free machining metals produced?
4. What alloying agent or agents are added to the following metals to act as metallurgical chip breakers? (a) Steel, (b) brass, (c) aluminum, and (d) stainless steel.
5. How does lead react in ductile materials such as aluminum, copper, iron, and their alloys and what are the benefits if any?
6. What are the two principle causes of cutting tool failure?
7. Name three principle sources of heat generated during machining.
8. How do the terms "coolant" and "lubricant" differ?
9. On what type of materials should tools with negative rakes be used?
10. Name four factors that affect surface finish in machining.
11. Name two problems that may arise from bacteria growth in the handling and storage of cutting fluids.

Section 8

MACHINE TOOL PROCESSES

Machine tool is a term given to a machine that can produce parts used for itself or other machines such as lathes, milling machines, shapers, planers, grinders, and drill presses. All are classified as machine tools. Even in an age when machine tools are becoming more sophisticated through the use of computers and other advanced technologies, machine tool processes remain the methods of efficiently and intelligently operating machine tools. These processes become part of the necessary skills needed by people in the machine tool trade that support the engineering staff.

UNIT 29 THE DRILL PRESS

Drilling is perhaps one of civilization's oldest mechanical means of using tools, estimated to have taken place some time after the use of the hammer. When one considers that the idea of rotating or twisting a stick against a stationary piece of wood to start a fire, for example, one can see that the idea of rotating or drilling a tool-like device is very ancient.

In the area of machining, the drill press is one of the most basic and widely used machine tools. Not only is the drill press used for producing straight holes in metal and other materials, but it can be used for other operations such as reaming, boring, counterboring, countersinking, spot-facing, and tapping (Figure 29–1).

> Reaming is an operation of accurately finishing a hole to size by using a multiedge cutting tool.
>
> Boring is an operation of enlarging a hole using an adjustable single-point cutting tool.
>
> Counterboring is an operation of enlarging the top of a previously drilled hole.
>
> Countersinking is an operation of cutting a cone-shaped enlargement in the end of a previously drilled hole.
>
> Spot-facing is an operation of smoothing and squaring an area around an uneven surfaced hole.
>
> Tapping is an operation of cutting internal threads in a previously drilled hole by means of a tap.

A more detailed explanation of these operations will be given in the next several units.

Objectives
After completing this unit you will be able to

1. identify at least six basic operations that can be performed on a drill press.
2. identify five types of drilling machines and their purpose.
3. name the different parts of a sensitive drill press and explain their function.

Safety

One of the dangers involved in the use of the drill press is the fact that it is a common shop machine tool. The average on-looker believes there is no skill involved in drilling a hole, that it can be done with complete disregard for safety or equipment. Unfortunately, if one is not careful in drilling, not only can the workpiece be destroyed but the operator can also be injured. The ability to work safely on this machine tool is a basic necessity.

One of the keys to working safely is to develop positive work habits such as not handling moving machine parts or sharp edged tools. The same applies to not wearing loose clothing that can get easily caught around machinery and keeping the work area free from chips, tools, and oil spills.

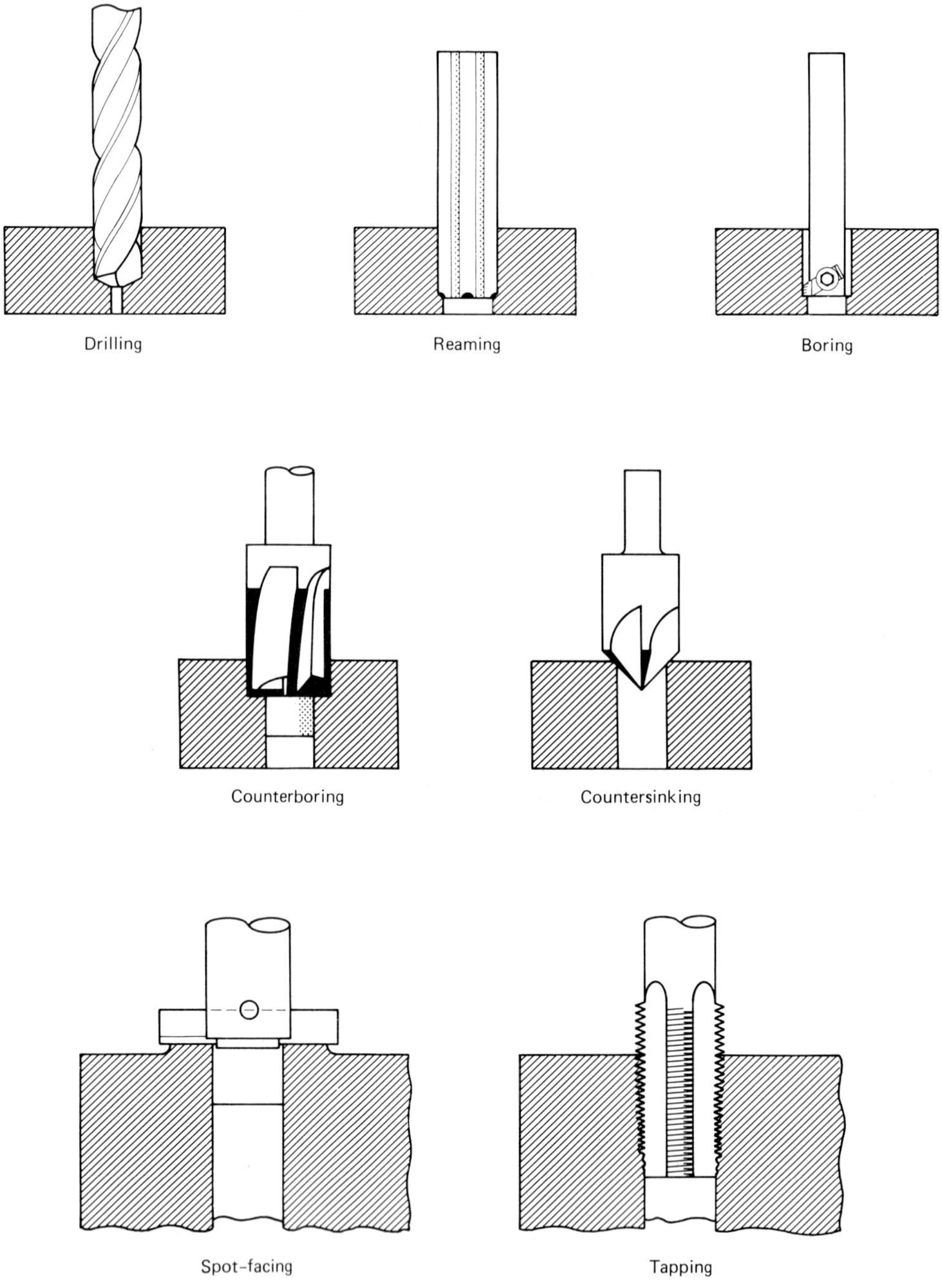

FIGURE 29-1
Common drill press operations.

In working on and around a drill press, the following rules of safety should be observed:

1. Always wear approved safety glasses to protect your eyes.

2. Remove all rings and jewelry before attempting to work; these can get caught in the machine to cause a serious injury.

3. Never wear loose fitting clothing and roll up any long sleeves above the elbow to prevent them from getting caught in the machine.

4. Keep any long hair under a cap or held in place with a hair net to prevent it from becoming caught in a revolving drill.

5. Do not wear gloves while operating a drill press or any machine tool. Gloves can get easily caught, drawing in fingers or the hand itself to cause a serious injury to the hand.

6. Never leave a chuck key in a drill chuck at any time (Figure 29-2). Any chuck key so left can be thrown out with great force once the drill press spindle is switched on.

7. Never attempt to remove a drill chuck from the drill press spindle with a cutting tool held in its jaws. The operator's hand can be punctured or maimed this way (Figure 29-3).

8. Always set the safety clamp on the column below the table. This will eliminate any deep or sudden drop of the table if the column clamp is not tightened.

9. Remove all objects and tools from a drill press table before attempting to move it up or down to a new position (Figure 29-4).

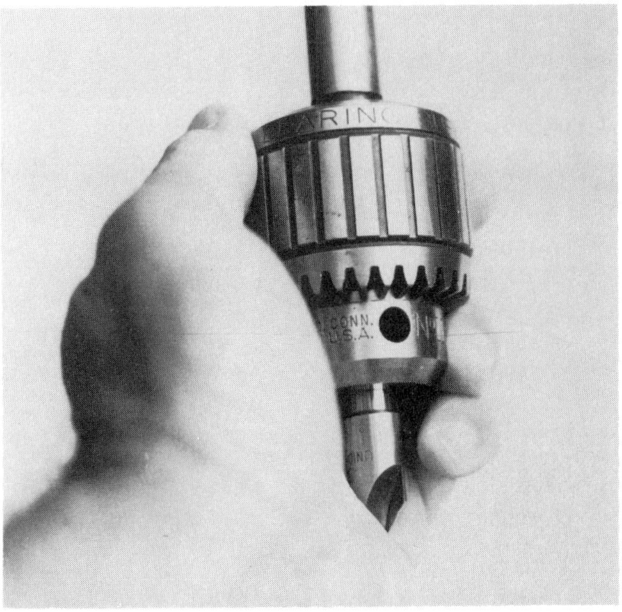

FIGURE 29-3
Never remove the drill press chuck with a sharp tool held in its jaws. It can puncture or maim an operator's hand. (Credit: Philip Rapisarda)

FIGURE 29-2
Never leave a key in the drill press chuck. It can fly out with great force when the drill press is started.

FIGURE 29-4
Remove all objects from the table before attempting to raise or lower the table.

FIGURE 29-5
A bench-type sensitive drill press is used for light general work.

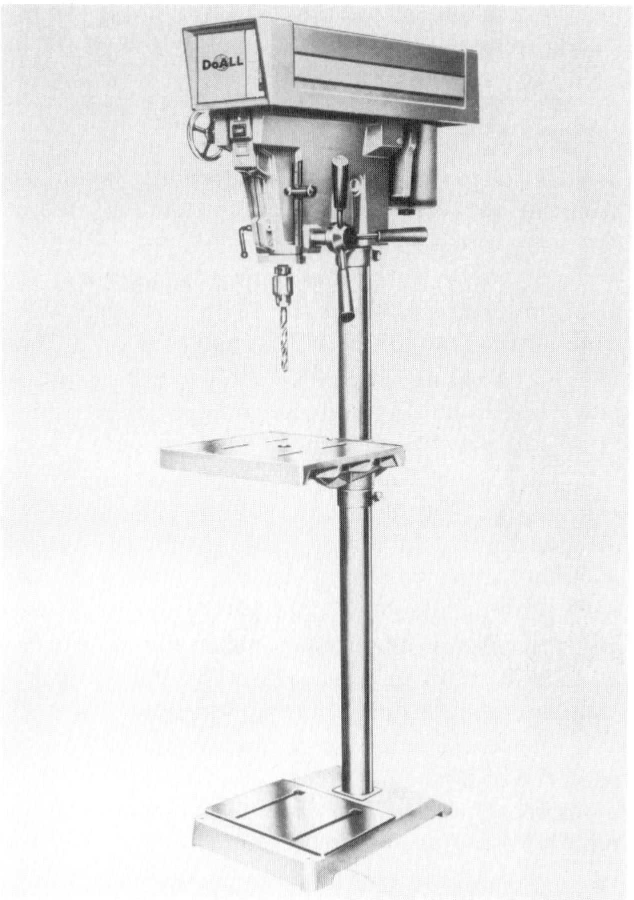

FIGURE 29-6
A floor model sensitive drill press can be used for drilling and tapping a variety of materials. (Credit: DoAll Company)

10. Never stop the drill chuck by hand even when it is coasting to a stop. Chips collected around the chuck jaws can severely cut the hand.

11. Never attempt to change any pulley belt unless the machine has been stopped and disconnected from its electrical source.

12. Always clamp a table stop or barrier on the left side of the table behind the drill vise to prevent the vise from spinning with the drill's rotation.

13. Always clamp the drill vise to the table before attempting any large hole drilling.

14. If attempting to drill a large hole in a large workpiece, be sure that parallels are used between the workpiece and table before clamping to the table.

15. Never attempt to clean a drill press table free of chips with bare hands nor use an air hose. Make sure the drill press spindle has stopped rotating and use a brush instead.

Types of Drilling Machines

There are many types of drilling machines used in the metalworking industry. The sensitive drill press can be found in school shops as well as in an industrial machine shop. Available in either the bench type (Figure 29-5) or the floor model (Figure 29-6), both can be used for a variety of operations as well as for drilling small holes and working on small workpieces.

The common parts to be found on a sensitive drill press are as follows:

Base: The foundation of the drill press that provides support as well as firmness when bolted to the floor or bench. It is also slotted to allow clamping of workpieces or devices to it.

Column: An upright cylindrical post that fits

into the base and supports the rest of the machine.

Table: Fastened to the column and locked, the table can be raised, lowered, or swiveled into a new position. Made either round or rectangular with a waste hole or slots for clamping purposes, the table provides the surface on which to rest the work or the drill vise.

Drilling head: The cast section that is mounted near the top of the column and contains all the working mechanisms for drilling.

Motor and step pulley: Located in the rear of the drilling head, this assembly is the means by which power is supplied.

Spindle: The tool holding shaft connected to step pulleys and linked to the motor pulley by means of a V-belt.

Quill: The housing that surrounds the spindle and is able to slide up or down to control the depth of drill penetration.

Feed handle: The means by which the quill is raised or lowered.

Drill chuck: The drill holding device fitted to the end of the spindle opening and has adjustable jaws to accommodate different diameter drills.

A variation to the pulley-driven sensitive drill press is the geared-head floor model (Figure 29–7). Equipped with a geared transmission that provides the drive for the spindle, this upright model is larger and heavier. These machines feature a power quill, a micrometer stop, as well as a positive speed selection to match the required drilling speeds. In addition to being able to easily raise and lower the head and table, some uprights are available with a coolant system. Because of its power and flexibility, it is also a reliable production machine tool, capable of accommodating larger and heavier workpieces.

Gang Drilling Machine

The gang drilling machine (Figure 29–8) consists of a group of independent drill presses placed side by side on a common table to perform a variety of operations such as drilling, tapping, and reaming on the same workpiece. The tools for these operations are guided into place by the drill bushings located in the jigs and fixtures—tooling that holds the parts to be machined. Once the particular drilling operation has been performed, the tooling is moved to another station under another drill to complete that stage of the operation.

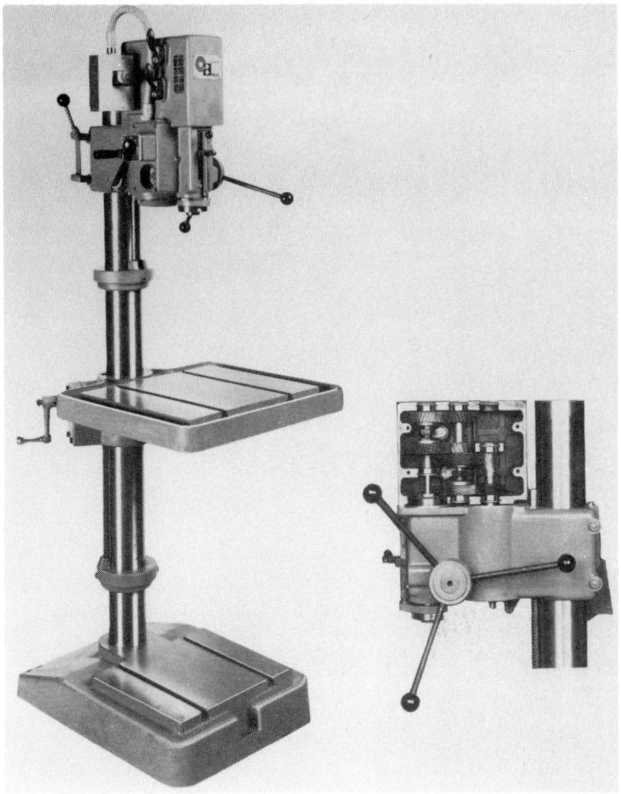

FIGURE 29–7
A geared-head floor model drill press can accommodate larger and heavier workpieces. (Credit: Clausing Industrial Inc.)

FIGURE 29–8
Gang drills consist of a series of spindles mounted on one base and are used to perform a variety of operations on a workpiece.

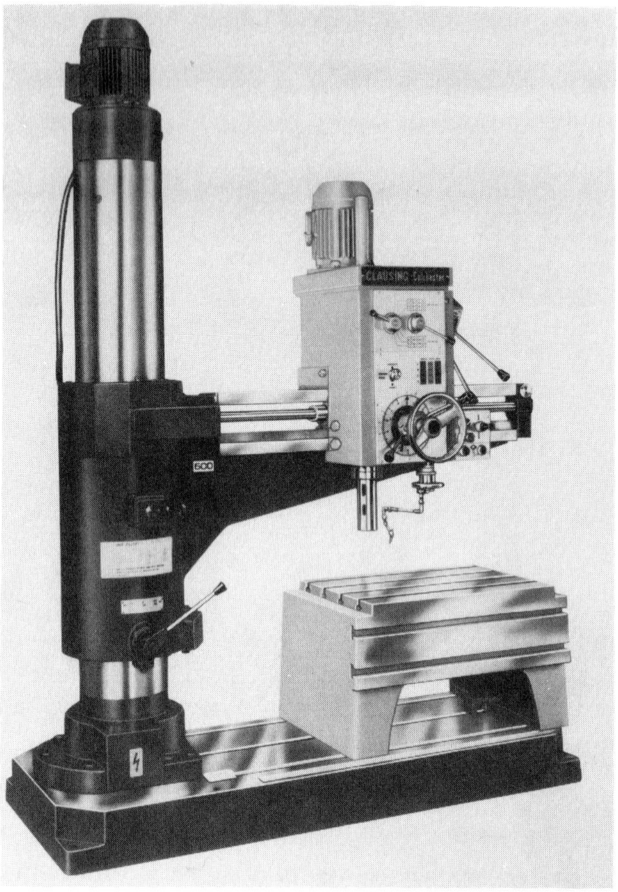

FIGURE 29-9
On a heavy-duty radial drilling machine, the spindle head is movable along an arm in a horizontal plane within its extension. (Credit: Clausing Industrial)

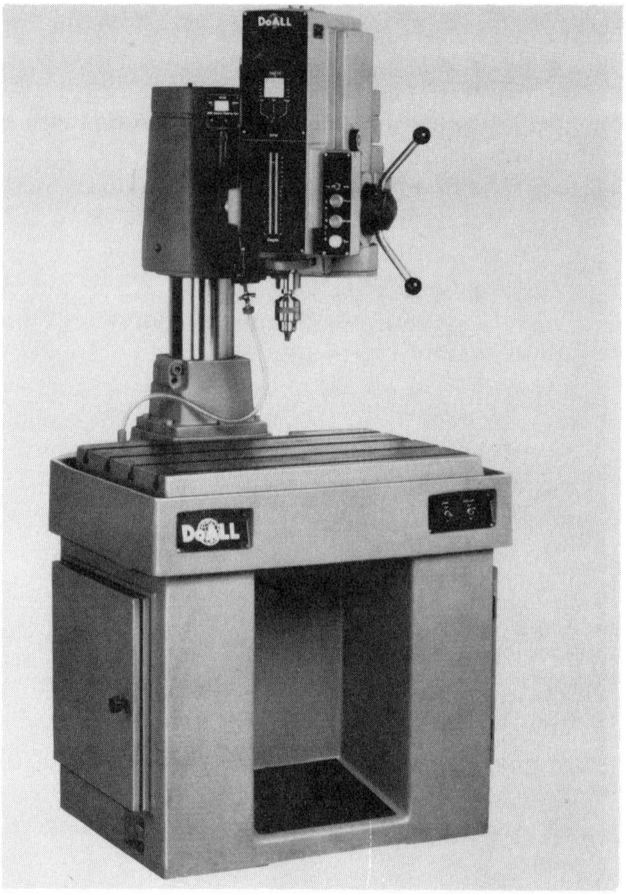

FIGURE 29-10
A toolroom size radial drilling machine is capable of handling a wide range of jobs including large workpieces. (Credit: DoAll Company)

Radial Drilling Machine

The radial drilling machine shown in Figure 29-9 is a rigidly constructed, cast-rion ribbed supported model and is capable of being used on heavy work and large castings. The spindle head is movable along a horizontal plane within its extension. This flexibility of reach allows its heavy-duty spindle head to carry out the different operations of drilling, tapping, reaming, and boring with ease.

The radial drilling machine is available in different sizes such as the model shown in Figure 29-10. Small enough for toolroom use, it features a geared headstock capable of multispeed ranges. The spindle head is mounted on a short radial arm that can pivot from the column. Both the arm and column can easily be raised or lowered and locked in position. This allows drilling and tapping of different locations on a workpiece being held in the same position.

Automatic Drilling Machine

The turret drilling machine has many tool holding devices that can be revolved and indexed (repositioned) so that the proper tool can be brought into use quickly. The turret drilling machine can be controlled manually or automatically through a numerical control system.

Other automated drilling machines, may have milling capabilities, multiple spindles, multiple-tool storage, as well as some type of tool changing mechanism. These machines are numerically controlled, with all movements from workpiece locations to drill changes to drill depths programmed. Once the workpiece has been clamped in position at the proper reference point, the operations are carried out automatically and with accuracy. Repeated hole locations can be held to a 0.001-in. (0.02-mm) tolerance (Figure 29-11).

FIGURE 29–11
An N/C machine tool that is capable of automated milling and drilling operations. (Credit: DoAll Company)

UNIT 30 CUTTING TOOLS AND ACCESSORIES

The cutting tools and the means by which they are held are a few of the important steps that must be undertaken for successful drilling. While the twist drill is usually associated with drill press work, there are a number of other cutting tools that can be held in the conventional drill press chuck.

Objectives
After completing this unit you should be able to

1. identify the parts of a twist drill.
2. identify and use four different size systems of twist drills.
3. identify and state the purpose of at least five different types of special drills.
4. identify and use three different types of tool holding devices.

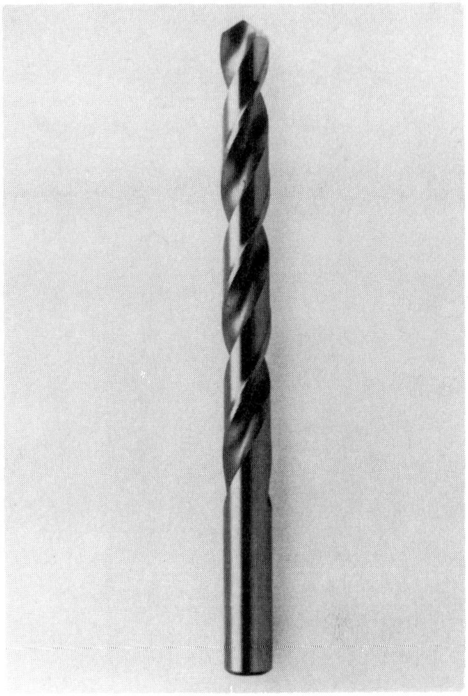

FIGURE 30–1
A general-purpose two-fluted twist drill. (Credit: DoAll Company)

The Twist Drill

The most common type of cutting tool used in the drill press is the two-fluted twist drill (Figure 30–1). The twist drill is generally made of high-speed steel. High-production drills are made with carbide tips and outperform the high-speed steel drill by approximately three to one. Not only can carbide drills cut hard materials, but they can outwear, outlast, and are more heat resistant than high-speed steel drills. For our purposes the high-speed steel drill will be explained since it is the one most commonly found in school shops and toolrooms.

Parts of a Twist Drill
The twist drill has three main parts: the shank, the body, and the point (Figure 30–2).

Shank. The holding portion of the drill can be either straight or tapered. Straight shank drills, generally drills up to $\frac{1}{2}$ in. (13 mm) in diameter, are held in a drill chuck. Tapered shank drills are held in an internally tapered spindle (Figure 30–3).

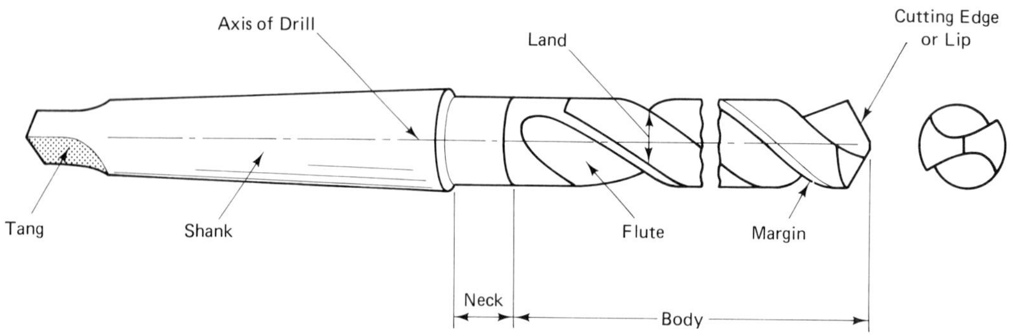

FIGURE 30-2
Parts of a twist drill.

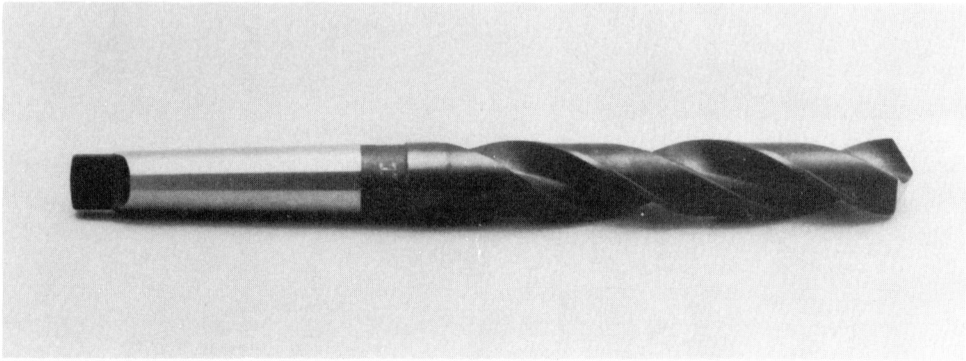

FIGURE 30-3
A tapered shank twist drill is held in an internally tapered spindle. (Credit: DoAll Company)

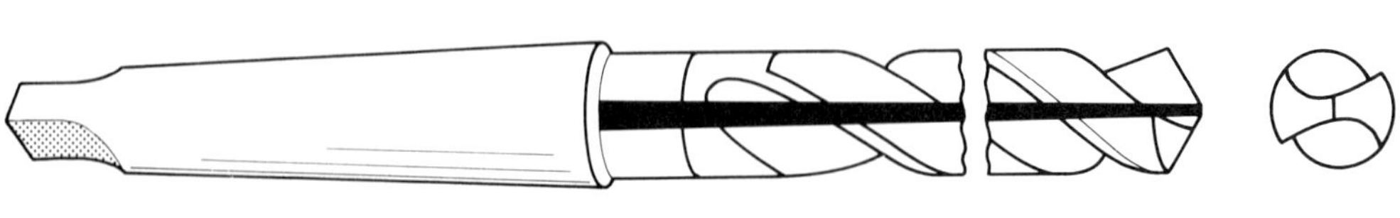

FIGURE 30-4
The web extends the length of the body and acts as the spinal column of a drill.

Body. That section between the shank and the point which is composed of the flutes, land, margin, and web. (a) *Flutes* are the spiral grooves that twist along the length of the body. They aid in eliminating chips, allow lubrication to enter, and help form the cutting edges. (b) The *land* is the surface between the flutes. The greatest portion of this surface is undercut and acts as a body clearance to reduce friction during cutting. (c) The *margin* is that narrow raised strip that runs along the edge of the flutes on the land. The margin helps form the diameter of the drill. (d) The *web* is the surface thickness between the flutes and becomes thicker as it extends the length of the body toward the shank. The web forms the drill's spinal column to give it strength (Figure 30-4).

Point. The conical section of a drill, also known as the drill point, is identified by several parts. (a) The cutting edges or lips extend angularly across the width of the drill to the margin (Figure 30-5). The cutting edges are uniform in length and form an included angle of 118° for general-purpose drilling. (b) The point of the dead center, known as the chisel edge, is formed by an angle of 120° to 135° with the cutting edge (Figure 30-6). (c) The outer edge of the drill directly behind the cutting edge formed at the margin is called the heel. (d) The lip clearance is that conical surface ground away between the lip and the heel to aid cutting (Figure 30-7). Depending on the hardness of material involved, this angle is generally 8° to 12°.

Types of Twist Drills

Twist drills are available in a variety of sizes and fall in four distinct measurement systems: fractional, number, letter, and metric. The sizes of these drill systems can be found in Table 30-1.

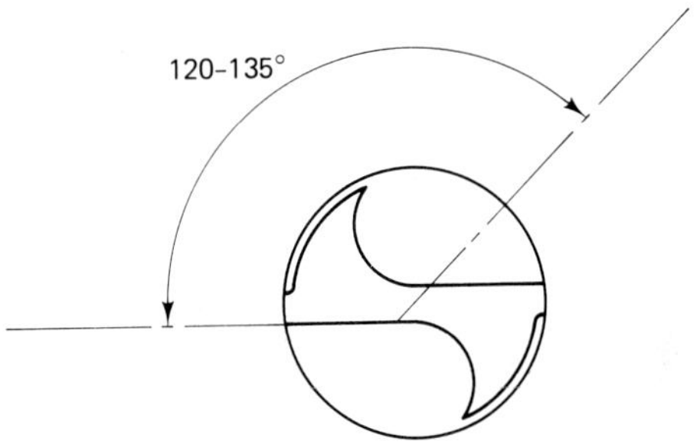

FIGURE 30-6
The point of the dead center is also known as the chisel edge.

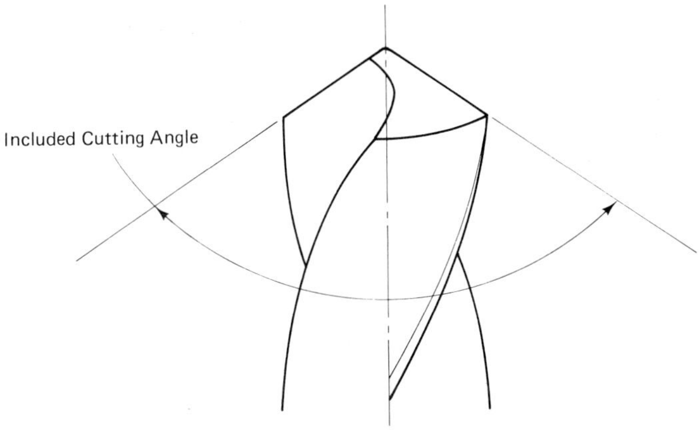

FIGURE 30-5
The included cutting angle is generally 118° for general-purpose drills.

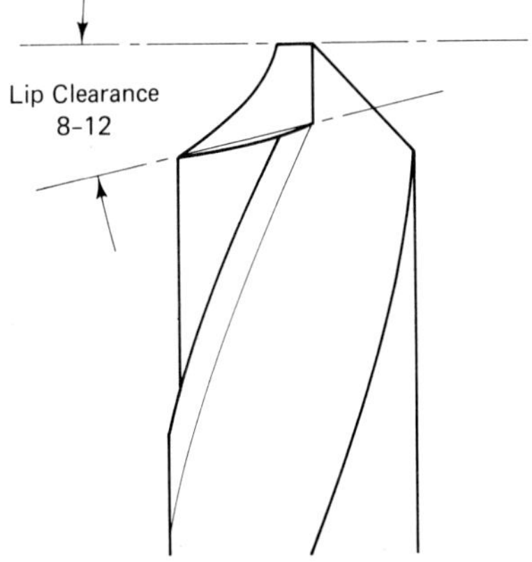

FIGURE 30-7
Lip clearance is generally 8° to 12°.

TABLE 30-1
Metric conversion chart (Credit: Union/ Butterfield)

Fractional sizes	*No. sizes*	*Decimal Equivalent (inches)*	*M.M. sizes*	*Metric Equivalent (mm)*
	97	.0059	0.15	
	96	.0063	0.16	
	95	.0067	0.17	
	94	.0071	0.18	
	93	.0075	0.19	
	92	.0079	0.20	
	91	.0083		
	90	.0087	0.22	
	89	.0091		
	88	.0095		0.241
		.0098	0.25	
	87	.0100		0.254
	86	.0105		0.267
	85	.0110	0.28	
	84	.0115		0.292
		.0118	0.30	
	83	.0120		0.305
	82	.0125		0.318
		.0126	0.32	
	81	.0130		
	80	.0135		0.343
		.0138	0.35	
	79	0.145		0.368
		.0150	0.38	
1/64		.0156		0.397
		.0157	0.40	
	78	.0160		0.406
		.0165	0.42	
		.0177	0.45	
	77	.0180		0.457
		.0189	0.48	
		.0197	0.50	
	76	.0200		0.508
	75	.0210		0.533
		.0217	0.55	
	74	.0225		0.572
		.0236	0.60	
	73	.0240		0.610
	72	0.250		0.635
		.0256	0.65	
	71	.0260		0.660
		.0276	0.70	
	70	.0280		0.711
	69	.0292		0.742
		.0295	0.75	
	68	.0310		0.787
1/32		.0312		0.794
		.0315	0.80	
	67	.0320		0.813
	66	.0330		0.838
		.0335	0.85	
	65	.0350		0.889
		.0354	0.90	
	64	.0360		0.914
	63	.0370		0.940
		.0374	0.95	
	62	.0380		0.965
	61	.0390		0.991
		.0394	1.00	
	60	.0400		1.016
	59	.0410		1.041
		.0413	1.05	
	58	.0420		1.067
	57	.0430		1.092
		.0433	1.10	
		.0453	1.15	
	56	.0465		1.181
3/64		.0469		1.191
		.0472	1.20	
		.0492	1.25	
		.0512	1.30	
	55	.0520		1.321
		.0531	1.35	
	54	.0550		1.397
		.0551	1.40	
		.0571	1.45	
		.0591	1.50	
	53	.0595		1.511
		.0610	1.55	
1/16		.0625		1.588
		.0630	1.60	
	52	.0635		1.613
		.0650	1.65	
		.0669	1.70	
	51	.0670		1.702
		.0689	1.75	
	50	.0700		1.778
		.0709	1.80	
		.0728	1.85	
	49	.0730		1.854

Fractional sizes	*No. sizes*	*Decimal Equivalent (inches)*	*M.M. sizes*	*Metric Equivalent (mm)*
		.0748	1.90	
	48	.0760		1.930
		.0768	1.95	
5/64		.0781		1.984
	47	.0785		1.994
		.0787	2.00	
		.0807	2.05	
	46	.0810		2.057
	45	.0820		2.083
		.0827	2.10	
		.0846	2.15	
	44	.0860		2.184
		.0866	2.20	
		.0886	2.25	
	43	.0890		2.261
		.0906	2.30	
		.0925	2.35	
	42	.0935		2.375
3/32		.0938		2.381
		.0945	2.40	
	41	.0960		2.438
		.0965	2.45	
	40	.0980		2.489
		.0984	2.50	
	39	.0995		2.527
	38	.1015		2.578
		.1024	2.60	
	37	.1040		2.642
		.1063	2.70	
	36	.1065		2.705
7/64		.1094		2.778
	35	.1100		2.794
		.1102	2.80	
	34	.1110		2.819
	33	.1130		2.870
		.1142	2.90	
	32	.1160		2.946
		.1181	3.00	
	31	.1200		3.048
		.1220	3.10	
1/8		.1250		3.175
		.1260	3.20	
	30	.1285		3.264
		.1299	3.30	
		.1339	3.40	

Fractional sizes	*No. sizes*	*Decimal Equivalent (inches)*	*M.M. sizes*	*Metric Equivalent (mm)*
	29	.1360		3.454
		.1378	3.50	
	28	.1405		3.569
9/64		.1406		3.572
		.1417	3.60	
	27	.1440		3.658
		.1457	3.70	
	26	.1470		3.734
	25	.1495		3.797
		.1496	3.80	
	24	.1520		3.861
		.1535	3.90	
	23	.1540		3.912
5/32		.1562		3.969
	22	.1570		3.988
		.1575	4.00	
	21	.1590		4.039
	20	.1610		4.089
		.1614	4.10	
		.1654	4.20	
	19	.1660		4.216
		.1693	4.30	
	18	.1695		4.305
11/64		.1719		4.366
	17	.1730		4.394
		.1732	4.40	
	16	.1770		4.496
		.1772	4.50	
	15	.1800		4.572
		.1811	4.60	
	14	.1820		4.623
	13	.1850	4.70	
3/16		.1875		4.762
	12	.1890	4.80	
	11	.1910		4.851
		.1929	4.90	
	10	.1935		4.915
	9	.1960		4.978
		.1969	5.00	
	8	.1990		5.055
		.2008	5.10	
	7	.2010		5.105
13/64		.2031		5.159

TABLE 30-1
Continued

Fractional sizes	*No. & letter sizes*	*Decimal Equivalent (inches)*	*M.M. sizes*	*Metric Equivalent (mm)*
	6	.2040		5.182
		.2047	5.20	
	5	.2055		5.220
		.2087	5.30	
	4	.2090		5.309
		.2126	5.40	
	3	.2130		5.410
		.2165	5.50	
7/32		.2188		5.556
		.2205	5.60	
	2	.2210		5.613
		.2244	5.70	
	1	.2280		5.791
		.2283	5.80	
		.2323	5.90	
	A	.2340		5.944
15/64		.2344		5.953
		.2362	6.00	
	B	.2380		6.045
		.2402	6.10	
	C	.2420		6.147
		.2441	6.20	
	D	.2460		6.248
		.2480	6.30	
1/4	E	.2500		6.350
		.2520	6.40	
		.2559	6.50	
	F	.2570		6.528
		.2598	6.60	
	G	.2610		6.629
		.2638	6.70	
17/64		.2656		6.747
	H	.2660		6.756
		.2677	6.80	
		.2717	6.90	
	I	.2720		6.909
		.2756	7.00	
	J	.2770		7.036
		.2795	7.10	
	K	.2810		7.137
9/32		.2812		7.144
		.2835	7.20	
		.2874	7.30	
	L	.2900		7.366
		.2913	7.40	

Fractional sizes	*Letter sizes*	*Decimal Equivalent (inches)*	*M.M. sizes*	*Metric Equivalent (mm)*
	M	.2950		7.493
		.2953	7.50	
19/64		.2969		7.541
		.2992	7.60	
	N	.3020		7.671
		.3031	7.70	
		.3071	7.80	
		.3110	7.90	
5/16		.3125		7.938
		.3150	8.00	
	O	.3160		8.026
		.3189	8.10	
		.3228	8.20	
	P	.3230		8.204
		.3268	8.30	
21/64		.3281		8.334
		.3307	8.40	
	Q	.3320		8.433
		.3346	8.50	
		.3386	8.60	
	R	.3390		8.611
		.3425	8.70	
11/32		.3438		8.731
		.3465	8.80	
	S	.3480		8.839
		.3504	8.90	
		.3543	9.00	
	T	.3580		9.093
		.3583	9.10	
23/64		.3594		9.128
		.3622	9.20	
		.3661	9.30	
	U	.3680		9.347
		.3701	9.40	
		.3740	9.50	
3/8		.3750		9.525
	V	.3770		9.576
		.3780	9.60	
		.3819	9.70	
		.3858	9.80	
	W	.3860		9.804
		.3898	9.90	
25/64		.3906		9.922
		.3937	10.00	
	X	.3970		10.084

Fractional sizes: 1/32nd & larger	Fractional sizes: 1/64ths	Decimal Equivalent (inches)	M.M. sizes	Metric quivalent (mm)
	Y	.4040		10.262
13/32		.4062		10.319
	Z	.4130		10.490
		.4134	10.50	
	27/64	.4219		10.716
		.4331	11.00	
7/16		.4375		11.112
		.4528	11.50	
	29/64	.4531		11.509
15/32		.4688		11.906
		.4724	12.00	
	31/64	.4844		12.303
		.4921	12.50	
1/2		.5000		12.700
		.5118	13.00	
	33/64	.5156		13.097
17/32		.5312		13.494
	35/64	.5469		13.891
		.5512	14.00	
9/16		.5625		14.288
	37/64	.5781		14.684
		.5906	15.00	
19/32		.5938		15.081
	39/64	.6094		15.478
5/8		.6250		15.875
		.6299	16.00	
	41/64	.6406		16.272
21/32		.6562		16.669

Fractional sizes: 1/32nd & larger	Fractional sizes: 1/64ths	Decimal Equivalent (inches)	M.M. sizes	Metric quivalent (mm)
		.6693	17.00	
	43/64	.6719		17.066
11/16		.6875		17.462
		.6890	17.50	
	45/64	.7031		17.859
		.7087	18.00	
23/32		.7188		18.256
	47/64	.7344		18.653
		.7480	19.00	
3/4		.7500		19.050
	49/64	.7656		19.447
25/32		.7812		19.844
		.7874	20.00	
	51/64	.7969		20.241
13/16		.8125		20.638
		.8268	21.00	
	53/64	.8281		21.034
27/32		.8438		21.431
	55/64	.8594		21.828
7/8		.8750		22.225
	57/64	.8906		22.622
29/32		.9062		23.019
	59/64	.9219		23.416
15/16		.9375		23.812
	61/64	.9531		24.209
31/32		.9688		24.606
		.9843	25.00	
	63/64	.9844		25.003
1		1.0000		25.400

Source: Credit of Union/Butterfield.

Fractional drills range in size from $\frac{1}{64}$ in. (0.40 mm) to as large as 4 in. (100 mm) in diameter. Drills up to $\frac{1}{2}$ in. (13 mm) diameter generally have straight shanks and are available in increments of $\frac{1}{64}$ in. (0.40 mm). Tapered shank fractional drills are produced starting at $\frac{1}{2}$ in. (13 mm) diameter.

Number drills are straight shanked and range in size from a very fragile diameter of 0.0059 in. (0.15 mm), identified as a No. 97 drill, to 0.218 in. (5.5 mm), identified as a No. 1 drill.

Letter drills range in size from A to Z, with letter A measuring 0.234 in. (5.90 mm), letter Z measuring 0.413 in. (10.50 mm).

Metric drills are measured in millimeters. The fragile diameter drills are manufactured in sizes that range from 0.04 to 0.99 mm. Each drill in this series increases in size by 0.01 mm. Straight shank metric drills are manufactured from 0.5 to 20 mm in. diameter. Tapered shank drills are available as large as 80 mm, as small as 8 mm.

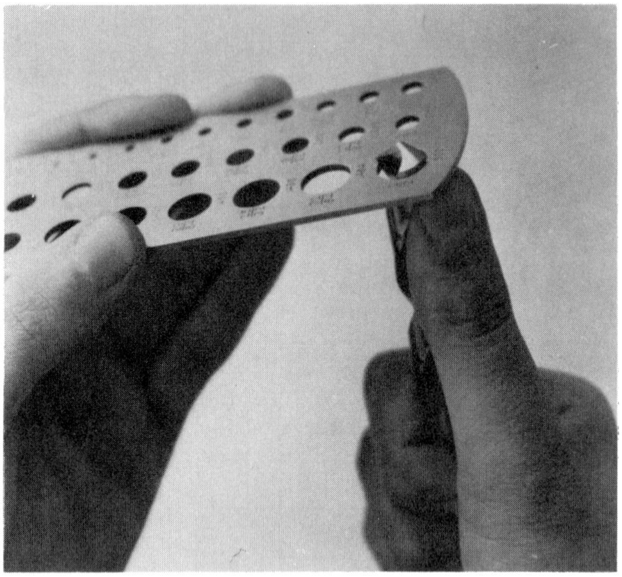

FIGURE 30-8
Measuring a drill's diameter using a drill plate.

FIGURE 30-9
Holes are included in an oil drill to allow fluids to cool the tool and flush out the chips. (Credit: Union/Butterfield)

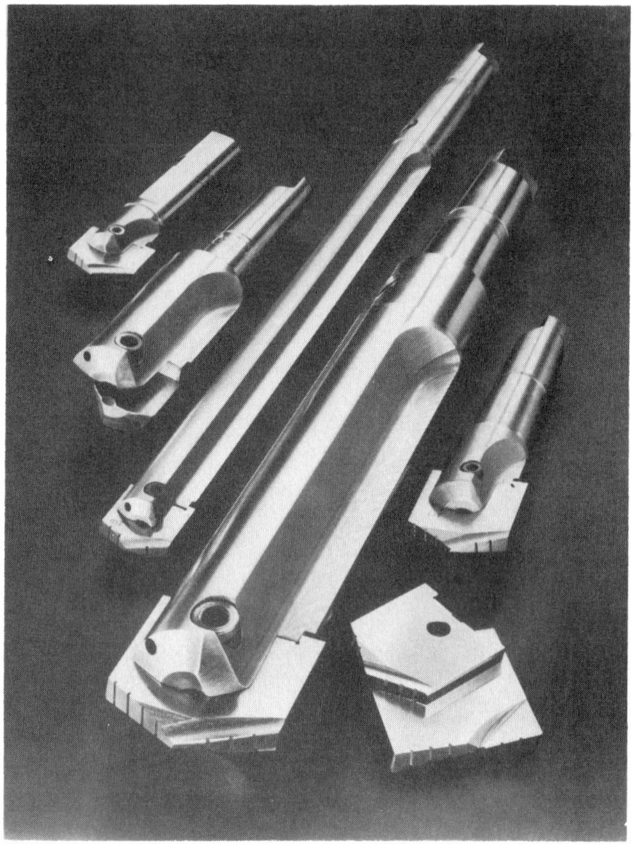

FIGURE 30-10
The cutting blades on spade drills are replaceable. (Credit: Union/Butterfield)

Drills in all four systems are measured across the margin using a micrometer or a drill plate gage (Figure 30-8).

Special Drills

There are many kinds of special drills, used for specific drilling operations, several of which will be examined at this time.

The *oil hole drill* (Figure 30-9) is primarily used on the turret lathe and screw machine. It is designed with two holes in the drill land that extend from the shank to the cutting edge to allow cutting fluids, compressed air, or oil to flow through. Oil hole drills are designed to cool the cutting edges, aid the cutting action, and flush away chips from the hole.

The *spade drill* (Figure 30-10) is available as a toolholder and an interchangeable blade for drilling large holes. The tool design makes a greater distribution of cutting fluid possible, reducing blade temperature in the process. This results in a more efficient chip flow and a larger lasting cutting edge. Due to the fact that a dull blade can be replaced without disturbing the setup, resetting the stops or increasing or decreasing the length of the drilling depth, the spade drill has wide use on automatic and numerically controlled machine tool operations. Spade drills are made in standard sizes ranging from 1 through 6 in. (25 to 150 mm).

The *straight fluted carbide drill* (Figure 30-11) is manufactured as a solid carbide drill in diameters up to $\frac{5}{32}$ in. (4 mm). On drills $\frac{3}{16}$ in. (4.80 mm) and over, the body is made of high-speed steel, straight fluted, and the cutting edge is carbide tipped. These drills are recommended for use in drilling hardened steel, such as die blocks, using a generous supply of coolant and a careful, steady handfeed.

Core drills (Figure 30-12) are available with three or four flutes. They are used to enlarge drilled, punched, or cored holes. The size of these holes can be as small as 60% of the diameter of the core drill. The principal advantages of a core drill is a design that combines maximum flute space with maximum rigidity. This results in an increased metal removal rate, improved finish, the hole accuracy.

The *high helix drill* (Figure 30-13) is designed for drilling deep holes in low tensile strength materials such as aluminum, magnesium, copper alloys, die cast materials, as well as free machining brass, bronze, and screw stock. Rapid chip removal is made possible by polished wide flutes and a faster helix angle (35° to 40°).

Tool Holding Devices

Cutting tools may be held directly in the spindle of a drill press or indirectly by means of a drill chuck, a tapered sleeve or other devices, such as a tap held in a tapping head. The most common methods of holding drills in the metalworking industry are the drill chuck, the tapered spindle, and tapered sleeve and socket.

Drill Chuck

The standard drill chuck (Figure 30-14) is the most widely used device for holding drills in the toolroom. It uses a chuck key for tightening its three jaws uniformly against the cutting tool's straight shank. The jaws can be made to open or close and will move together, free from tension, by rolling the outer sleeve by hand. Once the jaws have gripped the tool's straight shank, the chuck key should be in-

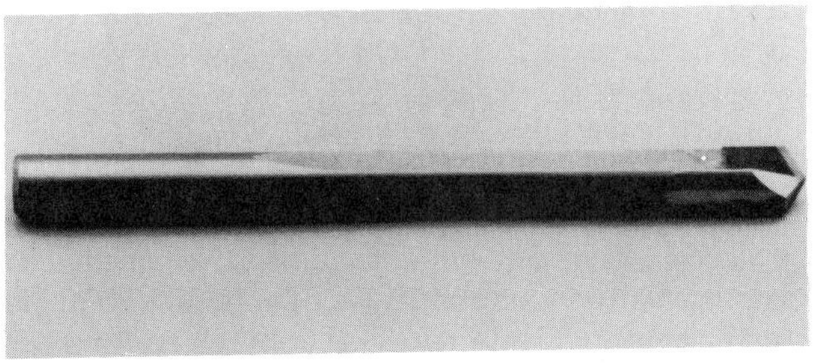

FIGURE 30-11
The straight fluted carbide drill can cut into hardened steel. (Credit: DoAll Company)

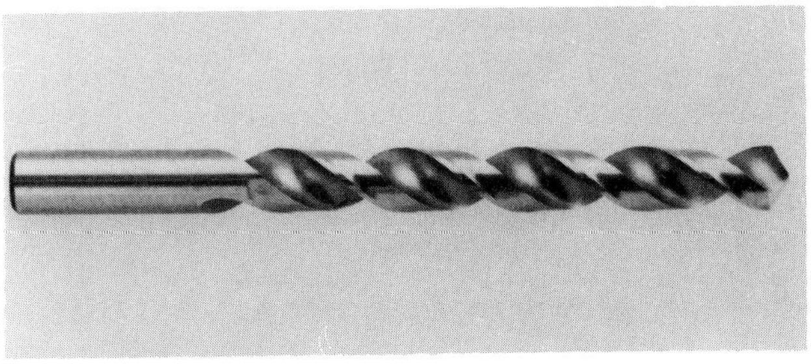

FIGURE 30-13
A high helix drill permits rapid chip removal. (Credit: DoAll Company)

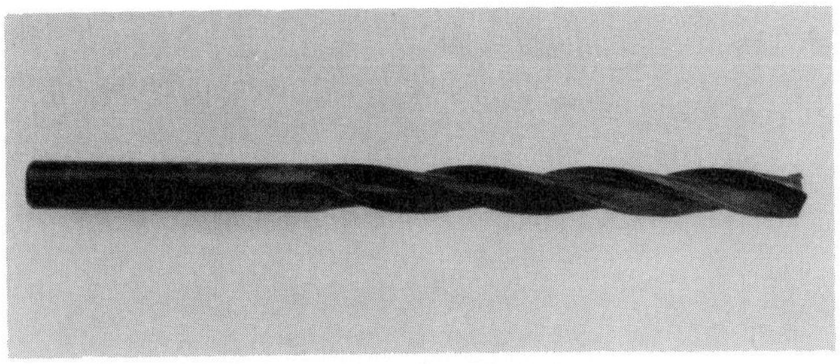

FIGURE 30-12
A core drill combines maximum flute space with maximum rigidity. (Credit: DoAll Company)

FIGURE 30-14
A drill chuck with key for tightening.

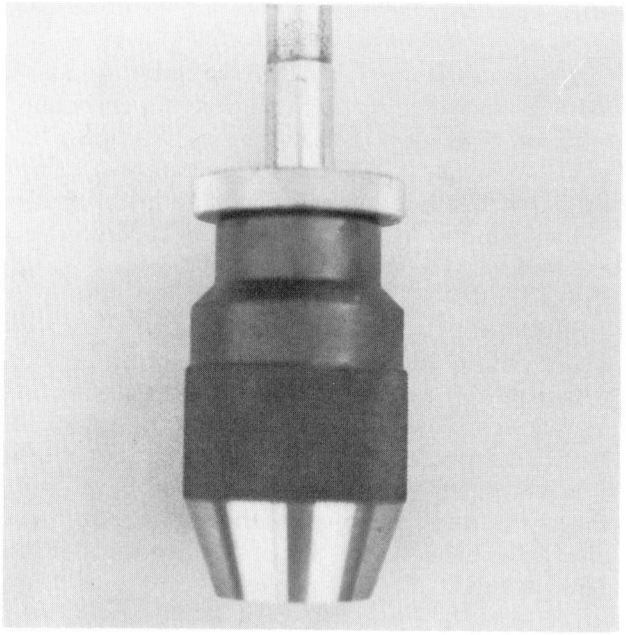

FIGURE 30-15
A keyless drill chuck requires hand tightening only.

serted in a chuck hole and turned against the geared teeth. Tightening should never take place in one hole only. This will strain the jaws. The drill chuck should be tightened using all three holes.

The keyless drill chuck (Figure 30-15) requires hand tightening only. Tightening takes place by inserting a straight shank drill into the opening of the jaws and turning the knurled outer sleeve until the drill is firmly held. This is a precision-type chuck, designed for smaller drills and light work. The hand tightening cannot match the torque obtained by a chuck and chuck key arrangement.

Sleeve and Sockets

Larger drill presses will use the holding power of a tapered drill shank that is inserted into a Morse tapered hole in its spindle. The hole size is related to the size of the drill press and may have a number 0 up to a number 7 internal Morse taper, depending on the drill press size. In order to hold a tapered shank drill directly, the shank's taper must match the spindle's hole taper.

If the drill shank taper is too small, a drill sleeve (Figure 30-16) is used to adapt the cutting tool shank to the drill press spindle taper. Conversely, if the taper on the cutting tool shank is too big, a drill socket is used. On these devices the drill is inserted in the socket hole and then the socket shank is fitted into the drill press spindle taper.

The tapered shank of the cutting tool as well as the internal taper of the drill press spindle must be clean before any mounting. Once cleaned, the tang of the tool is aligned with the slot within the spindle hole and while held with a soft cloth, inserted into the hole with a short snap.

To remove any tapered tool or device from the drill press spindle, a drift (Figure 30-17) is driven

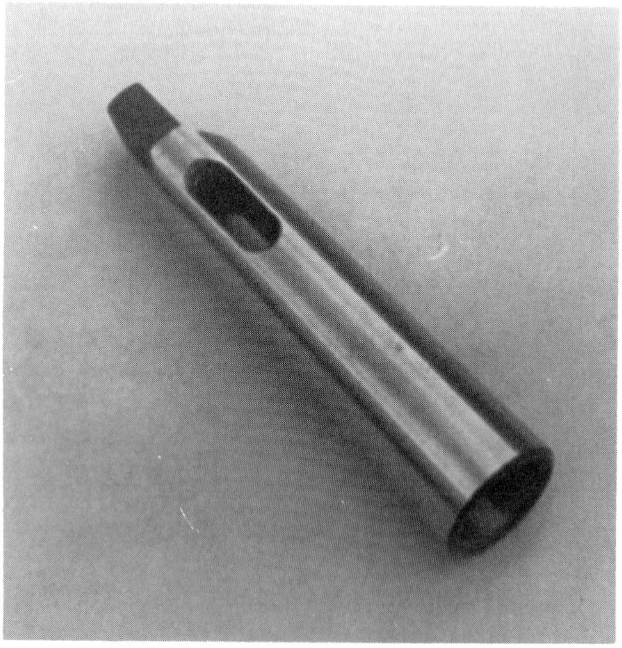

FIGURE 30-16
A drill sleeve is used to adapt the cutting tool shank to a tapered spindle.

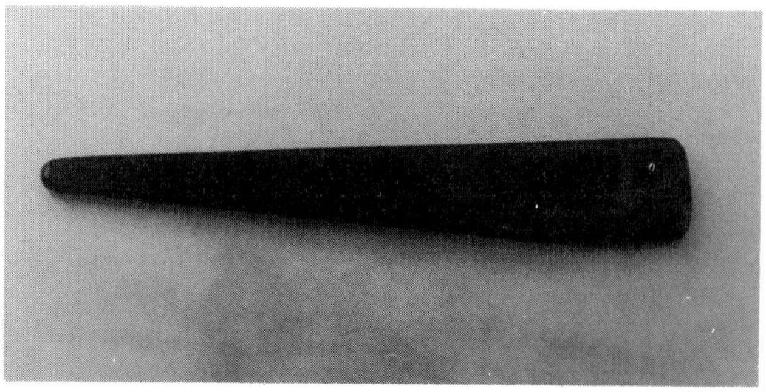

FIGURE 30-17
A drill drift is used for removing tapered shank drills.

against the shank end of the drill or holding device. An elongated slot is provided in the drill press spindle that allows the drift to be hammered when it is wedged between the slot and the end of the object. To prevent the cutting tool from striking the table upon release, a wooden block should first be placed below it.

UNIT 31 DRILLING SPEEDS AND FEEDS

Several factors affect the performance of any drill press operation. They are drill material, size of drill, material being cut, depth of hole, cutting fluid, and rigidity of machine and work setup. To not understand this relationship or to ignore it will result in poor efficiency, short tool life, and wasted time.

Objectives
After completing this unit you should be able to

1. calculate the proper revolutions per minute (rpm) to use on inch and metric drills.
2. select a proper feed to use for different materials on inch or metric drills.

Recommended Cutting Speed

One of the basic questions asked in using a cutting tool is: "How fast should the machine tool be set?" The cutting speed of a drill, referred to as surface speed or peripheral speed, is the distance one edge on the drill's circumference will travel on a straight line in one minute (circumference distance). Since the distance would be many inches (millimeters) long if extended, cutting speed represents the amount of surface covered in feet or meters per minute. This value is represented in technical tables as sf/min (surface feet per minute) or m/min (meters per minute). Tool life is affected by the recommended cutting speed; if the drill rotates too fast, as indicated by the blue chips forming, the tool is burned out; if it turns too slow, time is wasted. As suggested in Table 31-1, a range of starting cutting speeds is indicated for a number of commonly used materials. Since other variables such as tool construction and fixturing may enter the picture, it is better to start a little slower than the recommended speed and gradually increase. A more accurate cutting speed may be established as experience is gained, after trials indicate the job can be run faster.

TABLE 31-1
Suggested speeds for high-speed steel (HSS) and carbide drills

		CS (cutting speed)			
Material	*Drill*	*sf/min*	*m/min*	*Feed*	*Cutting Fluid*
Aluminum	HSS	200–300	60–90	Medium	Two-thirds lard oil
	Carbide	150–500	45–150		One-third kerosene
Brass (free cutting)	HSS	150–300	45–90	Heavy	Two-thirds lard oil
	Carbide	150–300	45–90		One-third kerosene
Bronze (high tensile)	HSS	70–100	21–30	Heavy	Two-thirds lard oil
	Carbide	200–300	60–90		One-third kerosene
Cast iron (medium)	HSS	50–100	15–30	Medium	Dry
	Carbide	115–140	35–42	Light	
Copper	HSS	70–100	21–30	Light	Soluble oil
	Carbide	200–300	60–90		
Tool steel	HSS	50–60	15–18	Medium	Lard oil or soluble oil
	Carbide	75–100	23–30		
Stainless steel (free machining)	HSS	30–80	9–24	Heavy	Sulphur-based oil
	Carbide	100–140	30–42		

TABLE 31–2
Inch cutting speeds

	Surface feet per minute (sf/min)												
	10	20	30	40	50	60	70	80	90	100	150	200	300
Diameter in inches	Revolutions per minute (rpm)												
$\frac{1}{16}$	611	1222	1833	2445	3056	3667	4278	4889	5500	6112	9167	12224	18336
$\frac{1}{8}$	306	611	917	1222	1528	1834	2139	2445	2750	3056	4584	6112	9167
$\frac{3}{16}$	204	407	611	815	1019	1222	1426	1630	1834	2038	3056	4074	6112
$\frac{1}{4}$	153	306	458	611	764	917	1070	1222	1376	1528	2292	3056	4584
$\frac{5}{16}$	122	244	367	489	611	733	856	978	1100	1222	1833	2445	3667
$\frac{3}{8}$	102	204	306	408	509	611	713	815	916	1018	1528	2037	3056
$\frac{7}{16}$	87	175	262	349	437	524	611	699	786	874	1310	1746	2619
$\frac{1}{2}$	76	153	229	306	382	459	535	611	688	764	1146	1528	2292
$\frac{5}{8}$	61	122	183	245	306	367	428	489	552	612	917	1222	1833
$\frac{3}{4}$	51	102	153	203	254	306	357	408	458	508	764	1019	1528
$\frac{7}{8}$	44	87	131	175	219	262	306	349	392	438	655	873	1310
1	38	76	115	153	191	229	267	306	344	382	573	764	1146

Revolutions per Minute

In using high-speed steel drills, a standard formula is used to establish the number of revolutions per minute the machine's spindle must turn to achieve effective drill cutting. By referring to Table 31–2, cutting speeds (CS) have been combined with the drill's diameter (D) to develop the number of rpm. The formula reads

$$\text{rpm} = \frac{\text{CS} \times 12}{3.1416 \times D}$$

$$= \frac{\text{CS} \times \overset{3.82}{\cancel{12}}}{\cancel{3.1416} \times D}$$

$$= \frac{\text{CS} \times 3.82}{D}$$

This is the formula on which technical tables are based; however, this formula has been rounded off for general calculations so that it reads

$$\text{rpm} = \frac{\text{CS} \times 4}{D}$$

Example: Determine the rpm required to drill a $\frac{1}{4}$-in. hole in machine steel (CS 100) using a high-speed steel drill.

Solution:

$$\text{rpm} = \frac{\text{CS} \times 4}{D}$$

$$= \frac{100 \times 4}{\frac{1}{4}}$$

$$= 100 \times 4 \times \frac{4}{1}$$

$$= 1600$$

Metric rpm Formula

The formula for metric drills is derived from the following information:

$$\text{rpm} = \frac{\text{CS (m)}}{\pi\, D \text{ (mm)}}$$

In order to make the formula work easier, all the values should be either in meters or millimeters. Since the drill's diameter is measured in millimeters, it is best to simplify the formula using millimeters as the standard measurement.

$$\text{rpm} = \frac{\text{CS} \times 1000}{3.1416 \times D}$$

$$= \frac{\text{CS} \times \overset{318.3}{\cancel{1000}}}{\cancel{3.1416} \times D}$$

$$= \frac{\text{CS} \times 318.3}{D}$$

Rounded off, the formula now reads

$$\text{rpm} = \frac{\text{CS} \times 320}{D}$$

Example: Determine the rpm required to drill a 5-mm hole in aluminum (CS 60) using a high-speed steel drill.

Solution:

$$\text{rpm} = \frac{\text{CS} \times 320}{D}$$

$$= \frac{60 \times 320}{5}$$

$$= 3840$$

On drill presses that do not have variable speed controls and exact speeds cannot be set, it is advisable to set the drill press to the closest speed *under* the calculated speed.

Recommended Feeds

As important as the speed is to the drilling process, so also is the feed—the distance a drill travels into the workpiece with each rotation. Feeds are governed by the hardness of the material being cut, the size of the drill, the rigidity of the machine and work setup, as well as the accuracy of the hole requirement and the coolant used.

TABLE 31-3
Suggested drill feeds

Drill Diameter			*Feed per revolution*					
			Light		*Medium*		*Heavy*	
in.	*mm*	*Material*	*in.*	*mm*	*in.*	*mm*	*in.*	*mm*
$\frac{1}{16}-\frac{1}{8}$	1.6–3	HSS	0.001	0.02	0.002	0.05	0.003	0.07
		Carbide	0.001	0.02	0.002	0.05		
$\frac{1}{8}-\frac{1}{4}$	3–6	HSS	0.002	0.05	0.004	0.10	0.006	0.15
		Carbide	0.002	0.05	0.004	0.10		
$\frac{1}{4}-\frac{1}{2}$	6–13	HSS	0.004	0.10	0.007	0.17	0.010	0.25
		Carbide	0.003	0.10	0.007	0.17		
$\frac{1}{2}-1$	13–25	HSS	0.007	0.17	0.010	0.25	0.015	0.38
		Carbide	0.005	0.13	0.009	0.23	0.012	0.30
Over 1	25+	HSS	0.015	0.38	0.020	0.51	0.025	0.64
		Carbide	0.008	0.20	0.011	0.30	0.015	0.38

As indicated in Table 31–3, the feed is increased as the drill's diameter is increased. Smaller feed values are recommended for hard materials such as alloy steels; the higher feed range values should be used for soft materials such as brass and aluminum. Too coarse a feed can result in drills chipping or breaking. Too light a feed can result in scraping or chattering noises, which can cause the cutting edges of a drill to become dull.

Carbide Twist Drills

The use of carbide twist drills require a rigid work setup and the use of a machine tool free of conditions that can lead to tool chatter. This means that the machine tool guides or moving parts should be free of wear and play. Additionally, there should be no vibration or end play in the spindle rotation.

Small drills measuring $\frac{1}{8}$ in. (3 mm) in diameter or less are made of solid carbide. Twist drills larger than this are generally carbide tipped. Carbide twist drills are not recommended for deep hole drilling nor are the use of extra long carbide drills. The lack of rigidity in these cases cause the cutting edges to chip, the end tips to break. Drill depths not more than six times the diameter of the twist drill is the recommended application guideline. Carbides are sensitive to temperature changes and if a coolant is to be used, it should be used uniformly throughout the operation. This is especially true when deep drilling. Carbides will crack if suddenly cooled after heating up (thermal shock).

Speed is critical to the performance of a carbide drill and should never be used at speeds less than the highest recommended for high-speed steel drills for the same operation. For a listing of general guides for high-speed steel and carbide twist drills, see Table 31–4.

UNIT 32 DRILLING HOLE METHODS

There are a number of hole operations that can be performed in drilling. The most obvious are the holes produced in the drill press. Holes can also be produced using a portable hand drill. In this unit the different methods of drilling holes will be examined.

Objectives
After completing this unit you should be able to

1. identify and explain the proper use of several work holding devices.
2. drill a hole to an accurate layout.
3. properly drill a hole smaller than $\frac{1}{2}$ in. (13 mm) in diameter.
4. properly drill a hole larger than $\frac{1}{2}$ in. (13 mm) in diameter.

Drilling Safety

There is a tendency for an operator to force or ram the feed of a drill as the cutting point starts to break through the bottom of a workpiece (Figure 32–1). The result is drill breakage or the drill does not get a chance to cleanly cut the material; consequently, the thin angular slope wedges itself into the flutes

TABLE 31–4
Metric cutting speeds

	Meters per minute (m/min)												
	6	*12*	*18*	*24*	*30*	*36*	*42*	*48*	*54*	*60*	*66*	*72*	*78*
Diameter in millimeters	*Revolutions per minute (rpm)*												
2	955	1910	2865	3820	4775	5729	6684	7639	8594	9549	10505	11460	12415
3	637	1273	1910	2545	3185	3820	4456	5093	5730	6366	7002	7639	8276
4	477	955	1430	1910	2385	2865	3428	3820	4297	4775	5252	5729	6207
5	382	764	1145	1530	1910	2292	2674	3056	3438	3820	4201	4584	4966
6	318	635	955	1275	1590	1910	2228	2546	2865	3183	3501	3820	4138
7	273	545	820	1090	1365	1637	1910	2183	2456	2728	3001	3274	3547
8	239	475	715	955	1195	1432	1671	1910	2149	2387	2626	2865	3104
9	212	425	635	850	1060	1273	1485	1698	1910	2122	2334	2546	2759
10	191	350	520	695	870	1146	1335	1528	1719	1910	2100	2292	2483
15	127	255	380	510	635	764	891	1019	1146	1273	1400	1528	1655
20	95	190	285	380	475	573	668	764	859	963	1050	1146	1241
25	76	150	230	305	380	458	535	611	688	764	840	917	993

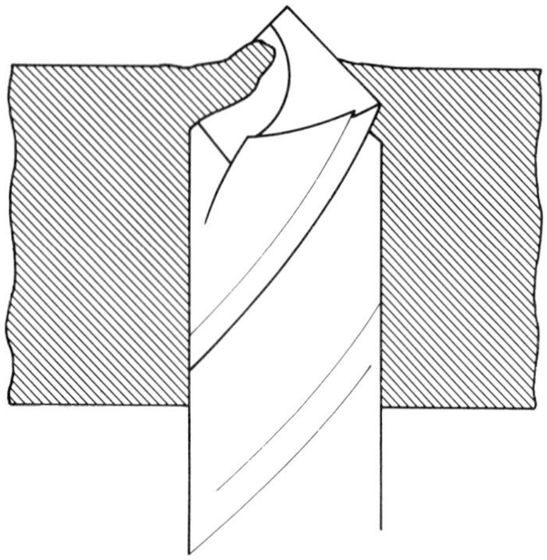

FIGURE 32-1
Excessive cutting pressure can cause the workpiece to spiral upward or spin with the rotation of the drill.

FIGURE 32-3
Flat-type drill vise with clamping bolts.

FIGURE 32-2
Clamping can prevent spinning or upward spiraling of the workpiece.

of the drill. This can cause the workpiece to be lifted as it spirals upward within the "twin alleys" provided by the twisting flutes. Forcing the feed into the material can also lead to the workpiece rotating in the same direction as the drill's rotation. These conditions can be avoided by easing up on the feeding pressure as the drill begins to break through, allowing the hole to be cut cleanly, and clamping the drill vise or workpiece to the table (Figure 32-2).

Work Holding Devices

The drilling of holes requires various holding devices, the most common of which is the flat-type drill vise (Figure 32-3). The drill vise is used for a variety of shapes ranging from flat, round, to irregular. A variation of this type is the angle drill vise (Figure 32-4), which permits the vise to be set on an angle. While there are a number of vises available, such as the contour vise that contains a series of free moving segments that adjust automatically to the shape of the workpiece, it is advisable to use one that is provided with the means for clamping or bolting it to the table.

For holding round work, V-blocks are recommended. Hardened blocks that have been deeply grooved and ground to a 90° included angle are

FIGURE 32–4
An angle drill vise will permit a workpiece to be drilled on an angle. (Credit: DoAll Company)

FIGURE 32–5
V-block supported workpiece clamped to the drill press table.

used, sometimes clamped to the drill press table in pairs (Figure 32–5) and sometimes not when clamping to the table is not required (Figure 32–6).

A variation of the V-block is the bench block (Figure 32–7). This, too, can be used for clamping round work in its V-groove for drilling, as well as general all-around machine shop use.

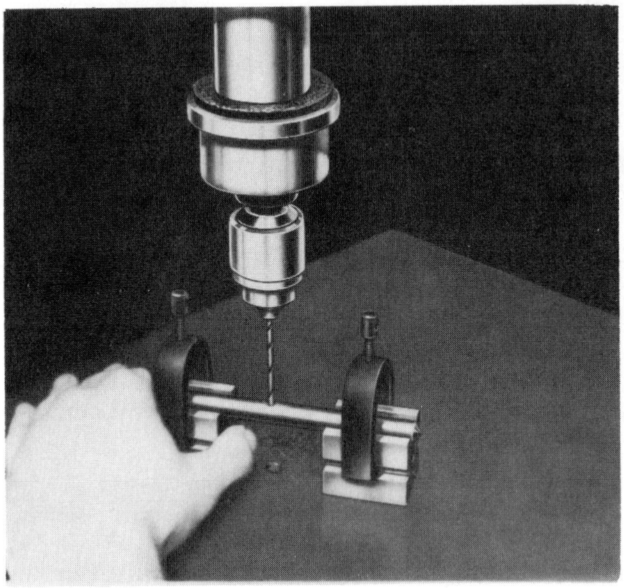

FIGURE 32–6
Hand positioned work held in a pair of V-blocks. (Credit: L.S. Starrett Company)

FIGURE 32–7
A bench block is a utility-type tool that will accept round and flat work for drilling.

FIGURE 32-8
A clamp and strap set that can be used for holding down work.

FIGURE 32-9
Step blocks can be the means that support the end of a clamp in holding work down.

FIGURE 32-10
T-bolts are used in connection with hold-down straps.

FIGURE 32-11
To be properly used with the clamp, the bolt should be used as close as possible to the workpiece.

Another means of holding work down is by use of the clamp and strap system (Figure 32-8). Step blocks (Figure 32-9) are one means used to raise the strap or clamp (either term is acceptable) to a height level or slightly higher than the workpiece, while a T-bolt is used to secure the set up (Figure 32-10). As illustrated in Figure 32-11, the bolt should be placed as close as possible to the workpiece. The holding pressure should be applied to the workpiece, not the step block. Drilling can also be carried out by having the work clamped to an angle plate (Figure 32-12). Larger fixtures, tooling devices, and workpieces can be mounted against slotted angle plates shown in Figure 32-13).

Cutting Fluids

Heat is the biggest factor that affects any drilling operation. The heat generated by friction between the drill and the workpiece during cutting can result in premature wear, breakdown, and shortened tool life. To reduce this heat, cutting fluids or coolants are used.

Cutting fluids will produce a better finish, and prevent the drill from tearing the surface of the hole being machined. At the same time, the cutting fluid can lower the tool force on the workpiece, which can result in greater hole accuracy. Because of its capacity to act as a lubricant between the drill and the chip, generally higher cutting speeds can be used (Figure 32-14) when coolants are used.

FIGURE 32-12
Work can be clamped to an angle plate for drilling purposes.

FIGURE 32-13
Heavy tool fixturing can be bolted to slotted angle plates. (Credit: DoAll Company)

FIGURE 32-14
Cutting fluids can produce a better finish as well as lower the cutting tool force.

There are many fluids and coolants on the market. Their uses will depend on the machinability of the material involved as well as the cutting tool. The requirements for using a carbide drill would be different than the conditions that exist when using a high-speed steel drill.

The ideal cutting fluid should have the following characteristics:

1. Act as a coolant for the drill and the workpiece.
2. Lubricate to reduce friction.
3. Increase work surface finish.
4. Help remove chips.
5. Prevent rust.
6. Not be harmful to the operator.
7. Not be damaging to the work or machine surface.

Center Hole Drilling

The method of center drilling is better covered in Unit 43 but needs to be mentioned here because of the important role it plays in spotting a hole location.

There are two types of center drills (Figure 32-15). In addition to the plain style where the countersink has a 60° angle, the bell style center drill has a secondary angle above the 60° slope. The diameter where the two angles intersect is called the bell diameter. The bell style center drill is designed to produce a hole that protects the edges of the center hole in production where the part is frequently handled for different machining operations.

Due to the fact that the chisel edge of a drill is generally larger than the center punch mark made on the work surface, a center drill is used. To be effective, center drills should pick up the center punch mark where it exists and not be forced into a false

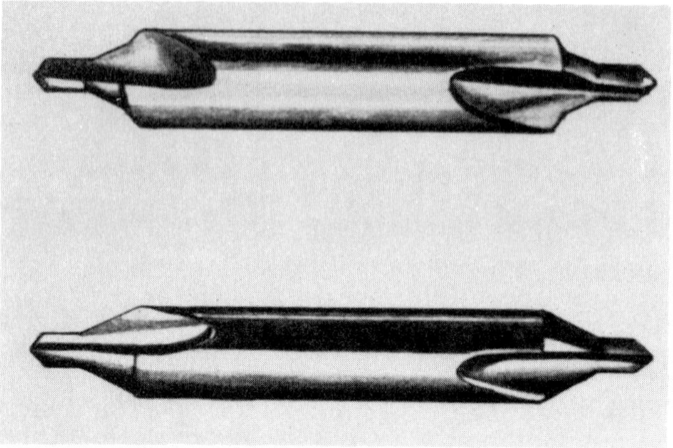

FIGURE 32-15
A plain center drill (top) is used for general machine shop work, while the bell type (bottom) is designed to produce a protected center-drilled hole for production.

FIGURE 32-17
Begin the drilling operation by using a small center drill.

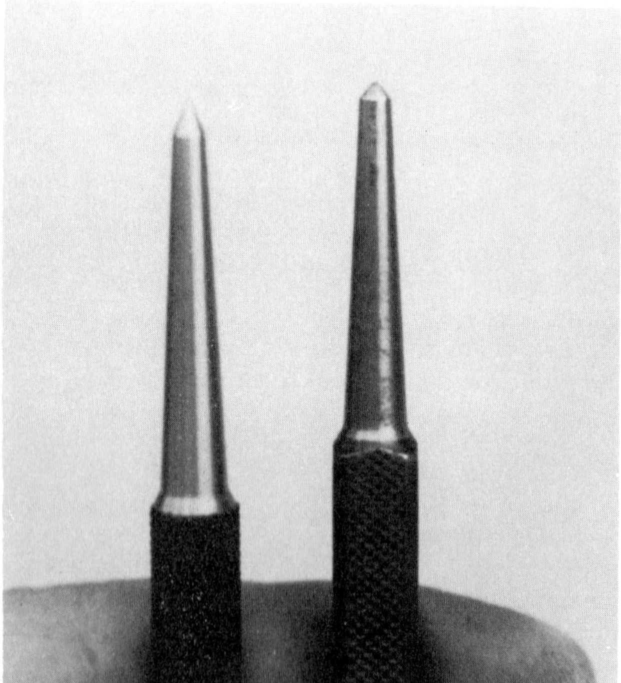

FIGURE 32-16
Note the angle difference between a prick punch and a center punch.

setup location. For this reason, the workpiece or vise should not be clamped to the table when center drilling. The workpiece should be allowed to move so that the center punch mark can be located by the rotating center drill.

Procedure

1. Carefully indent the hole's center location by using a prick punch.
2. Check the punch mark location and correct if necessary.
3. Enlarge the indent by using a center punch (Figure 32-16) being careful to hold the punch as vertical as possible.
4. Mount the workpiece in a vise or set it on the drill press table but do not clamp it in place.
5. Mount a small size center drill in a drill chuck, extending it not more than $\frac{1}{2}$ in. (13 mm) Figure 32-17).
6. Set the drill press speed to approximately 1000 to 1200 rpm.
7. Start the drill press spindle.
8. Lower the center drill into the center punch mark and allow the workpiece to "float" as it centers itself with the center drill point.
9. Continue drilling until a cone-shaped hole has been produced by the center drill's tapered section.

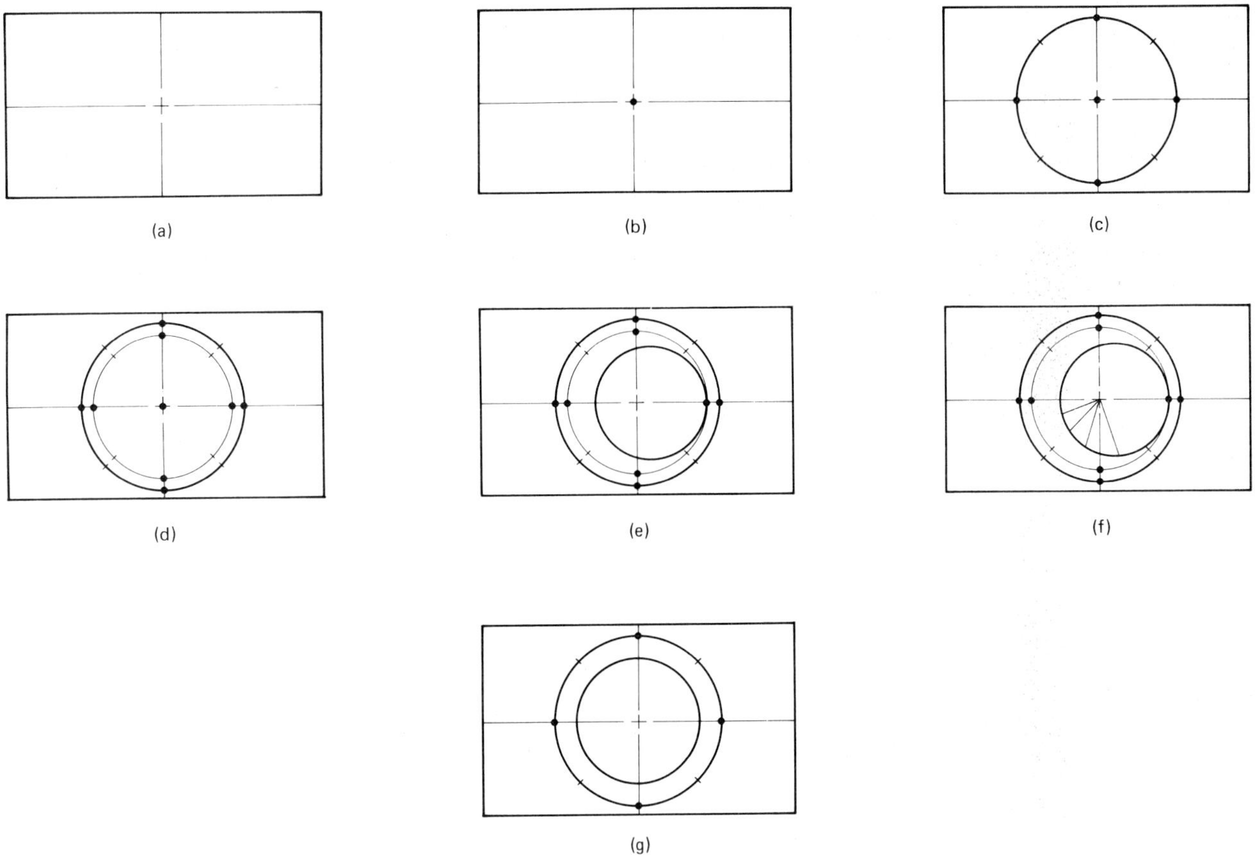

FIGURE 32–18
The procedure for drawing a drill to a layout.

Layout Drilling

It is much easier to layout a hole location accurately than to drill it accurately in the drill press. Due to the fact that a drill may be started off center, location accuracy would be impossible. It would then be necessary to move the drill point so that the location would be concentric with the layout. This is illustrated in Figure 32–18.

Procedure

1. Coat the clean surface of the workpiece with layout dye.
2. Layout the intersecting lines of the hole location (Figure 32–18a).
3. While viewing the workpiece through magnification, lightly prick punch the intersection of the lines (Figure 32–18b).
4. Set a pair of dividers for the radius of the intended hole diameter.
5. Scribe the hole location (Figure 32–18c).
6. Set the dividers to draw a concentric test circle $\frac{1}{16}$ in. (1.6 mm) smaller than the hole size.
7. Scribe the test circle.
8. Prick punch four points on the circumference of the test circle and eight points on the hole diameter (Figure 32–18d).
9. Provide a deeper indentation for center drilling by center punching the center location of the hole.
10. Center drill to a depth where a narrow cone-shaped hole is produced.
11. Replace the center drill with a proper size drill.
12. Determine the correct speed for the twist drill size and set the drill press accordingly.

13. Start the machine and drill to a depth equal to 60% of the drill diameter.

14. Examine the shallow impression for concentricity with the inner test circle (Figure 32-18e).

15. If it is not concentric, draw the circle over by cutting a few V-grooves with a round nose chisel on the side toward which the drill must be moved (Figure 32-18f).

16. Start the drill press and lower the drill into the spotted and grooved hole. The drill will be moved toward the direction of the grooves. *Note:* The hole location cannot be moved once the drill cuts to its full diameter.

17. Examine after drilling a shallow depth. If it is not concentric with the inner test circle, it may be necessary to add a few more grooves until the drill point is moved to the center (Figure 32-18g).

18. Upon successful location, complete drilling to depth.

Hardened Drill Plate Method

Layout drilling can be accurately completed by using a drill plate as a guide. Made of oil hardening stock (Figure 32-19), a blank is drilled and reamed to size, hardened and tempered, then placed and clamped exactly on the circumference of the hole laid out. The hardened drill plate hole size can be the actual desired diameter or can act as a pilot hole for any drill to follow. For any surface that lends itself to clamping (as opposed to an irregular shaped casting, for example), this method is far superior to the method using a chisel to move over the hole location (Figure 32-20).

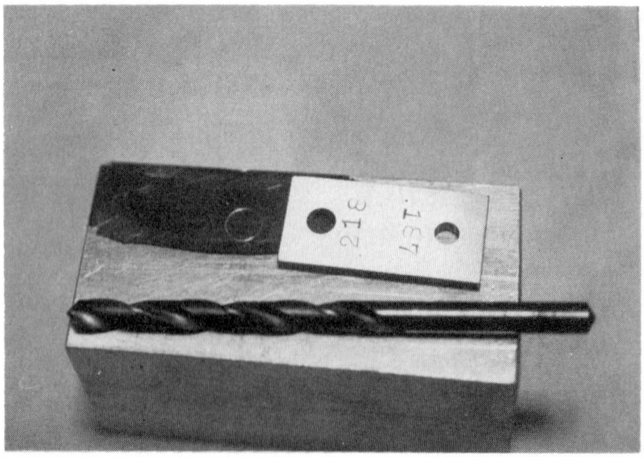

FIGURE 32-19
A hardened drill plate guide is used for drilling a layout accurately.

Drilling Holes under $\frac{1}{2}$ in. (13 mm)

In drilling holes up to $\frac{1}{2}$ in. (13 mm), general safety precautions should be observed although it may not be necessary to clamp the drill vise directly to the table. A table stop on the left side of the table or a barrier (Figure 32-21) should be provided that permits a small amount of room for vise movement. This space will permit ease in picking up the hole location without forcing the drill into a false position.

Procedure

1. Clear the drill press table.

2. Tighten the workpiece in a drill vise to rest firmly on a set of parallels supporting it.

3. Tighten a proper size center drill in the drill chuck.

4. Lower the center drill just above the hole location and lock the spindle.

5. Clamp a table stop on the left side of the table or a barrier about $\frac{1}{8}$ in. (3 mm) away from the drill vise.

6. Set the speed for center drilling.

7. Start the drill press spindle.

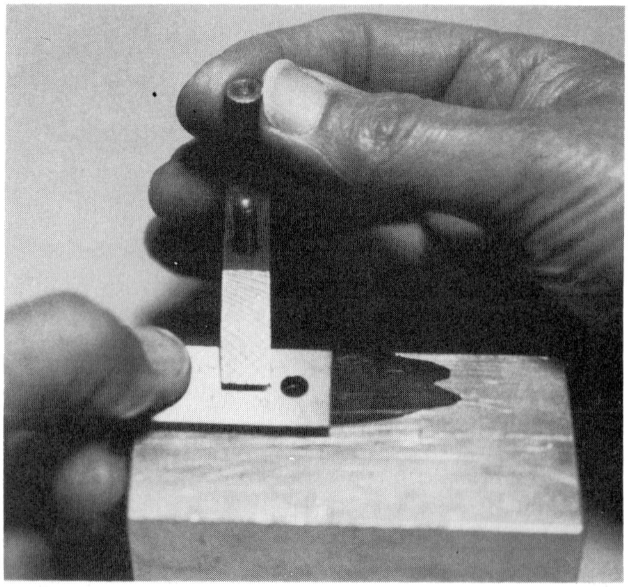

FIGURE 32-20
The drill plate guide is carefully placed and clamped directly over the layout hole.

FIGURE 32-21
Always clamp a table stop or barrier to the table before attempting any drilling.

FIGURE 32-22
In drilling large holes, the vise or workpiece must be clamped to the table.

8. Unlock the spindle.
9. Drill the center location to produce a cone-shaped hole one-half the diameter of the center drill.
10. Stop the drill press spindle.
11. Replace the center drill with the proper size drill and tighten it in the drill chuck.
12. Set the spindle speed for the drill's diameter.
13. Apply cutting fluid and drill the workpiece.
14. Complete the operation, being sure to ease up on pressure as the drill breaks through the workpiece.

Drilling Large Holes

In drilling holes larger than $\frac{1}{2}$ in. (13 mm) diameter, there are a few rules to keep in mind. First, the workpiece or drill vise must be clamped to the drill press table (Figure 32-22), and second, a pilot hole or smaller hole should be drilled. Cutting fluids and proper spindle speeds will of course enter the picture, too.

While a drill as large as $\frac{1}{2}$ in. (13 mm) diameter can be used directly after having the hole started with a center drill, holes over $\frac{1}{2}$ in. (13 mm) diameter will require a pilot hole. The cutting action of a large drill must be examined to understand why a pilot hole should be used first.

In addition to not being sensitive enough to pick up a center punch mark as the chisel point of a drill (web) gets thicker as it increases along the drill's body length, a greater drilling pressure will be required to cut the material. While thinning the web section can reduce the cutting pressure, a pilot hole is still required.

The size of the pilot drill should never be much more than the web of the final size drill to be used. There is a tendency for the drill to work harden the surface of the hole, which causes the next drill to wear at the center of the pilot hole. This is especially true in tool and alloy steels. If a pilot hole is much larger than the web of the following drill, a distorted hole opening, out of roundness, or drill chatter may occur.

Once the pilot hole has been accurately located and drilled, drilling the large size hole can be carried out as previously described while working within the rules of safety:

Workpiece or vise clamped to the table.
Proper spindle speed.
Using an adequate cutting fluid.

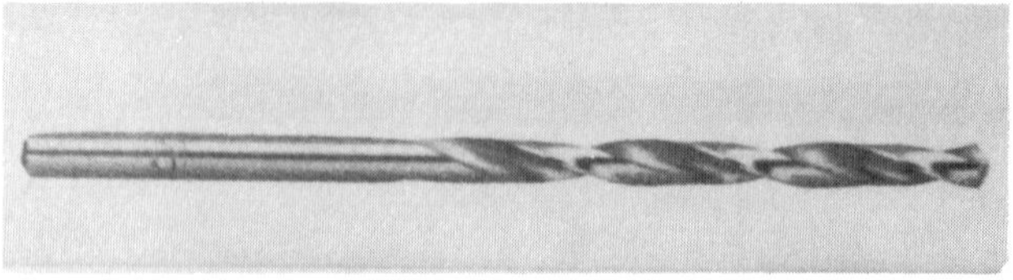

FIGURE 32-23
A low helix drill can be used to drill sheet metal.
(Credit: DoAll Company)

Drilling Sheet Metal

Ordinarily, sheet metal holes are punched or pierced by hand or machine tools. Holes can be produced on a drill press, but due to the thinness of the metal, it can be dangerous if certain precautions are not taken. The holes produced look ragged and torn. Another problem to contend with is that the area below the drilling point is pushed down with pressure. Therefore, sheet metal will not only have to be clamped down, but it should also be supported with a piece of wood or some other material to get a decent looking hole. Standard ground twist drills will not produce a clean hole in thin material, however. Drills should have a short flat ground along the length of both cutting edges to make them suitable for this type of operation. Other recommendations would include the use of a low helix drill or a straight fluted drill (Figure 32-23).

UNIT 33 DRILL PRESS OPERATIONS

Due to its versatility as a machine tool, there are many operations such as reaming, counterboring, countersinking, tapping, and spot-facing that are still performed safely and efficiently on the drill press.

Objectives
After completing this unit you should be able to

1. identify and explain the difference between a hand reamer and a machine reamer.
2. machine ream a hole.
3. countersink and counterbore holes.
4. hand tap a hole in the drill press.

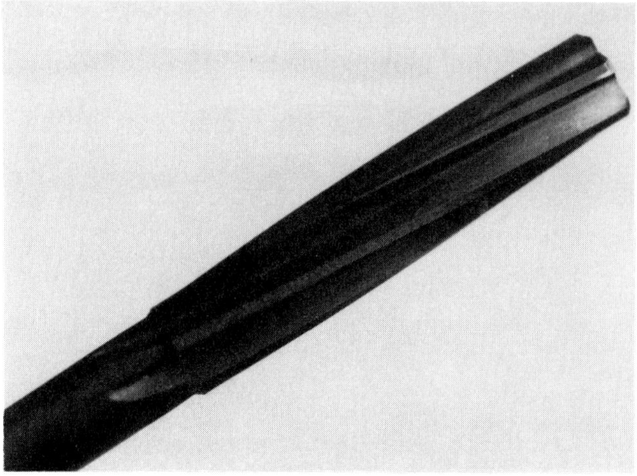

FIGURE 33-1
A reamer is used to enlarge a hole clean, smooth, and to exact size.

Reamers

A reamer is a rotating cutting tool that is used to enlarge and finish a previously drilled hole clean, smooth, and to exact dimension. Containing straight or helical cutting edges formed by grooves or flutes along its body (Figure 33-1), reamers are available made of high speed steel, cast alloys, or may be carbide tipped.

Reamer Terms
Most reamers consist of three principal parts: angle of chamfer, body, and shank (Figure 33-2).

The *angle of chamfer* refers to the entering part of the reamer that does the actual cutting. It is ground uniformly on every flute to provide clearance behind the chamfered cutting edge, specifically on rose-type reamers.

The *body* refers to the full diameter portion of the reamer that is fluted along its entire length to form the cutting teeth. Each tooth on the body consists of a land (the portion between each flute) and a margin. The margin is adjacent to the cutting edge and is an unrelieved portion of the land that extends the length of the flute. Friction during cutting is reduced by a relief (body clearance angle) that is ground behind the margin.

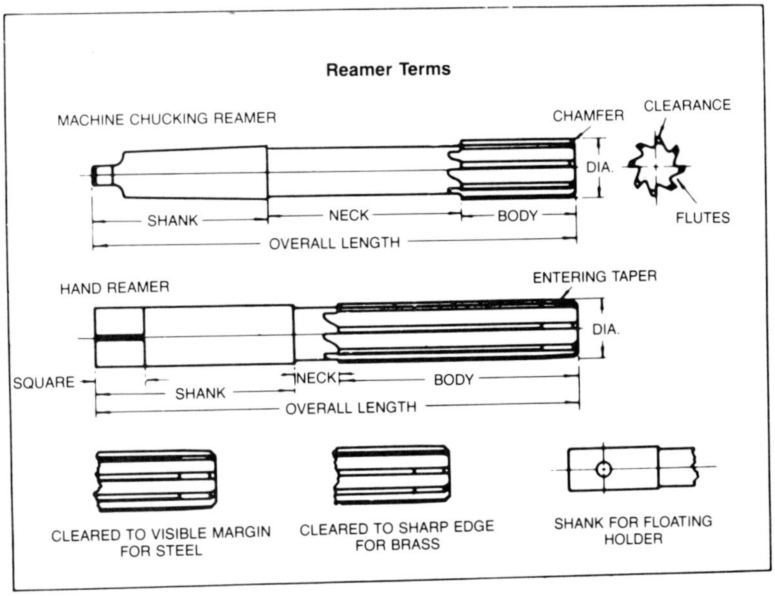

FIGURE 33-2
Reamer terms. (Credit: DoAll Company)

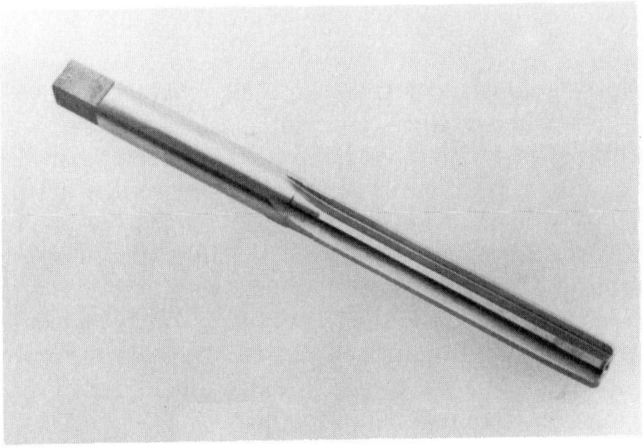

FIGURE 33-3
Straight fluted solid hand reamer. (Credit: DoAll Company)

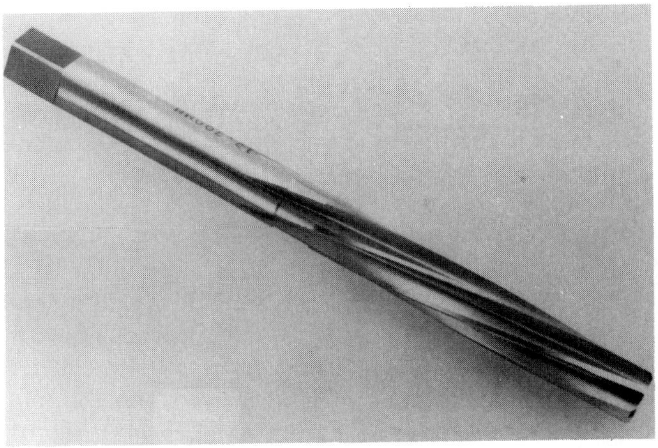

FIGURE 33-4
Helical fluted solid hand reamer. (Credit: DoAll Company)

The *shank* is that portion of the reamer by which it is held and driven. It may be straight (round) or tapered, and shanks on hand reamers contain a square on the end for holding a tap wrench.

Types of Reamers

There are many types of reamers in use and they broadly fall into two general categories: hand reamers and machine reamers.

Hand Reamers. Hand reamers are designed to be turned by a suitable wrench and contain a square on the end of the shank for this purpose (Figure 33-3). They should never be used under power. Hand reamers are used for reaming light amounts of material, no more than 0.005 in. (0.13 mm), to bring a hole to exact size. They are available in either the solid or expansion type and may come with either straight or helical flutes. Solid hand reamers are generally made from high-speed steel (Figure 33-4). One of the

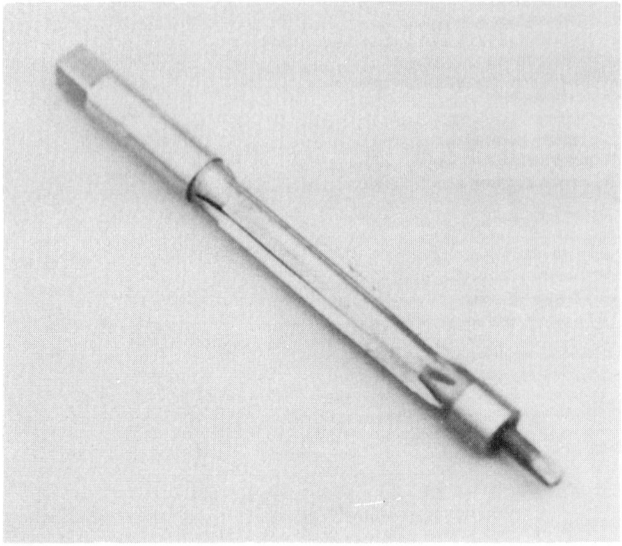

FIGURE 33-5
Expansion-type straight fluted hand reamer will break if expanded too far.

FIGURE 33-7
If the flutes twist away in a counterclockwise direction, it is a left-hand helix.

biggest causes of breakage of expansion-type hand reamers is expansion of these tools beyond the manufacturer's recommendation (Figure 33-5).

Machine Reamers. Machine reamers are made to be used in a drill press, lathe, vertical milling machine, and other machine tools capable of drilling operations. Machine reamers are available with straight shanks for tightening in a drill press chuck or a standard Morse tapered shank to fit the machine's spindle directly (Figure 33-6). While straight fluted reamers cut freely in many materials, other types of helical flutes generally perform better. Helical reamers with a left-hand helix are recommended for interrupted cuts. A helical reamer with a left-hand helix can be identified by looking at the end of the reamer (Figure 33-7). If the flutes twist away from the viewer in a counterclockwise direction, it is a left-hand helix. If the flutes twist away in a clockwise direction, it is a right-hand reamer (Figure 33-8). Only a few of the more commonly used machine reamers will be discussed.

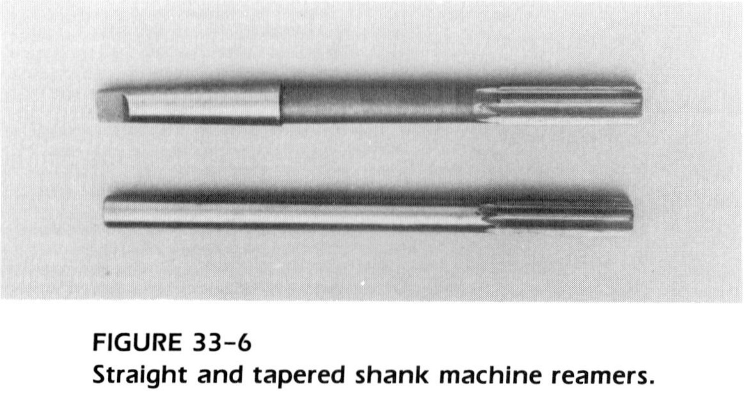

FIGURE 33-6
Straight and tapered shank machine reamers.

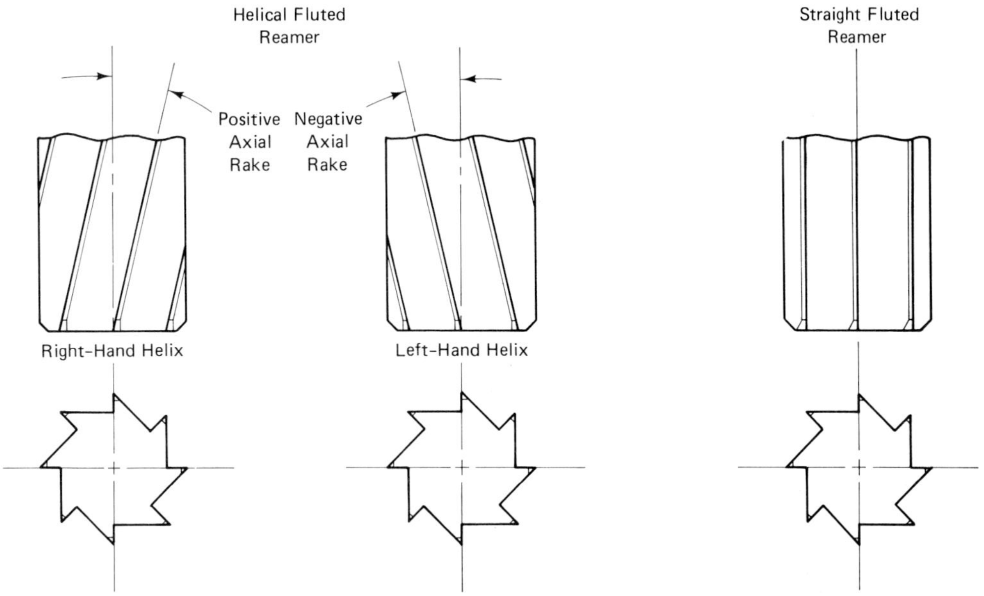

FIGURE 33-8
The cutting structure of a reamer.

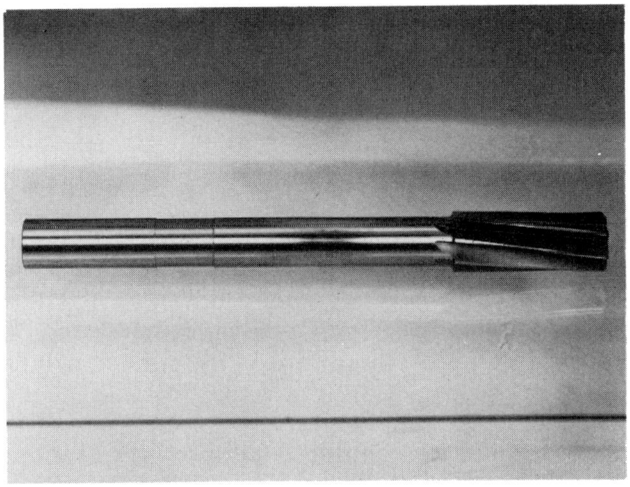

FIGURE 33-9
Rose reamers do all their cutting on the chamfered cutting edges on the end of the reamer. (Credit: DoAll Company)

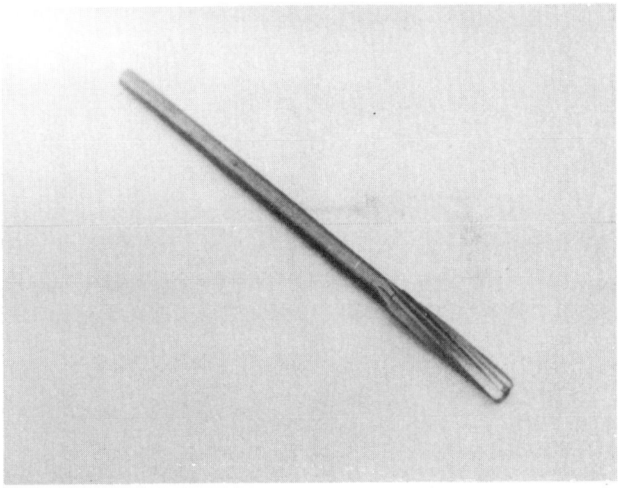

FIGURE 33-10
Helical fluted machine reamer.

Rose reamers are available with either straight (Figure 33-9) or helical flutes along their lengths, straight or tapered shanks, and are used primarily to remove metal quickly to bring a hole close to finish size. The cutting edges on the reamer are located on the 45° chamfer on the end of the cutter. This is the only section of the flutes that is ground to do any cutting.

Fluted machine reamers (Figure 33-10) are available with straight or helical flutes and are designed to cut on both sides and on the end of the tool. These reamers are available in both straight or tapered shanks and in production are used after the rose reamer to cut a hole smooth, clean, and to exact size.

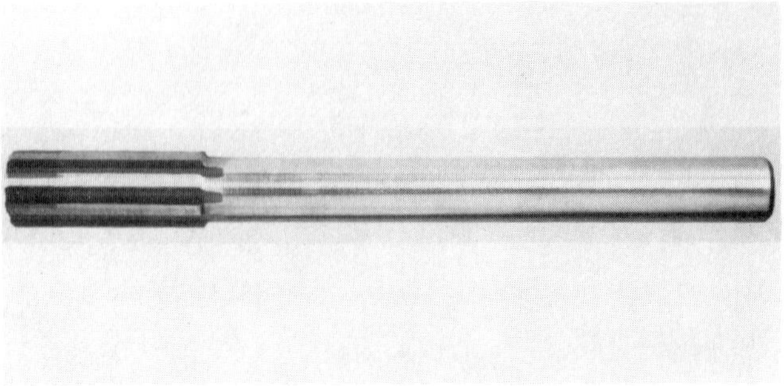

FIGURE 33-11
Chucking reamers are designed to be used in turret lathes, screw machines, and drill presses. (Credit: DoAll Company)

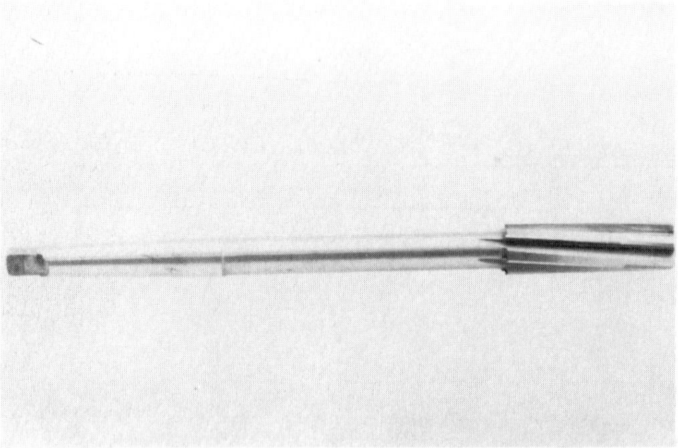

FIGURE 33-12
Due to their long life and free cutting action, carbide chucking reamers are economical to use. (Credit: DoAll Company)

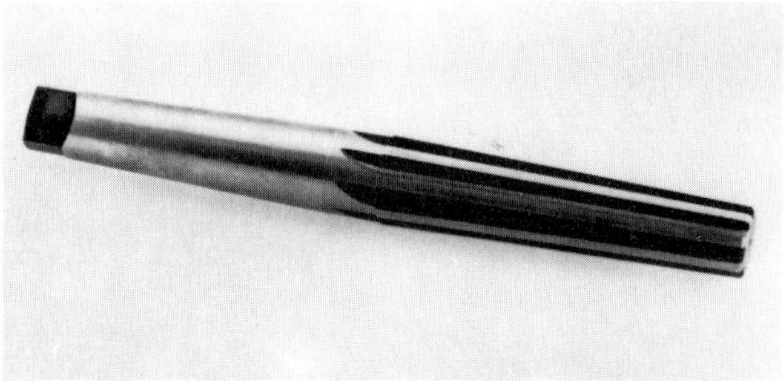

FIGURE 33-13
A tapered shank, Morse taper finishing reamer. (Credit: DoAll Company)

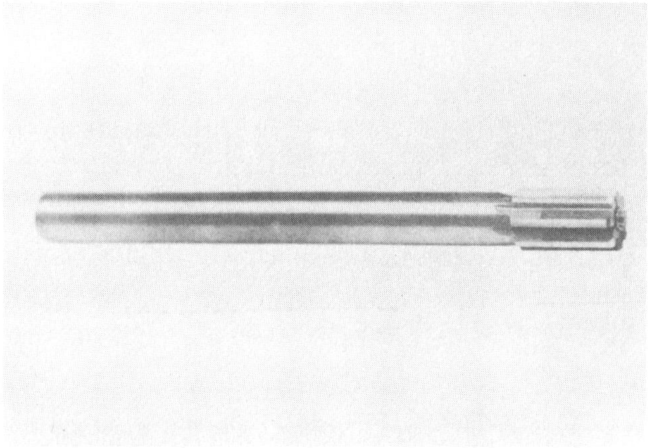

FIGURE 33-14
An expansion chucking reamer can be enlarged slightly and ground to size. (Credit: Union/Butterfield)

Chucking reamers are designed to be used in turret lathes, screw machines, and drill presses. While the straight fluted are carbide tipped (Figure 33-11), the flutes are staggered or unevenly spaced around the circumference of the body to prevent chatter, which is caused by the machine tool's cutting forces and slippage. Also available with tapered shanks and helical flutes (Figure 33-12), carbide-tipped chucking reamers remain abrasion resistant, maintain a free cutting action even in difficult to ream material, and leave a smooth hole.

Morse taper reamers (Figure 33-13) are available with helical or straight flutes and are used for finish reaming tapered holes to fit Morse standard taper shanks such as sockets, sleeves, and machine tool spindles. The finishing reamer should only be used to clean up a few thousandths of an inch (millimeters) that have been left by the roughing reamer.

The *expansion chucking reamer* (Figure 33-14) is available in both taper and straight shanks. It is designed to give longer life to a finishing reamer. A recessed adjusting socket head screw at the bottom of the cutting tool is the means by which the dull blades are expanded oversize and then reground to size. Turning the screw will only expand a 1-in. (25 mm) reamer about 0.005 in. (0.13 mm). The recessed screw also makes possible the reaming to the bottom of a blind hole.

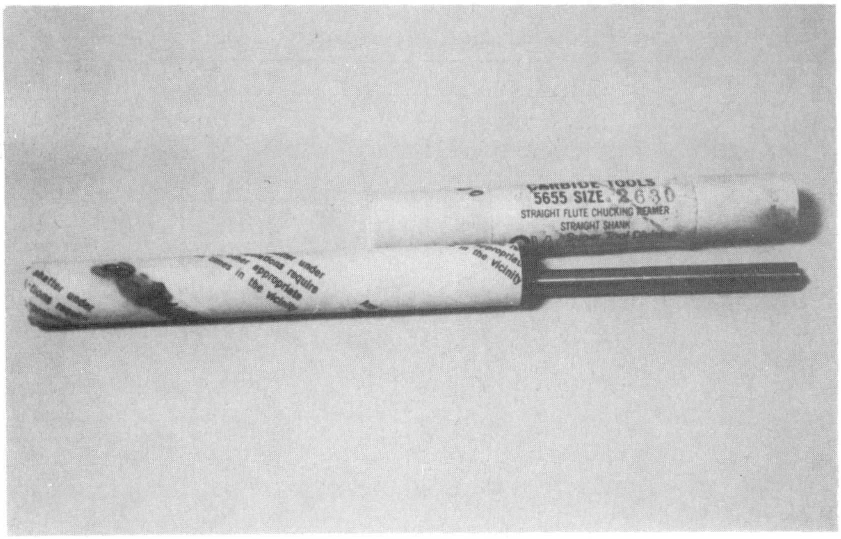

FIGURE 33-15
Reamers should be stored separately when not in use.

Reamer Handling
Reamers are sharp cutting tools. To stay sharp, they must be handled carefully. During use, reamers should not be turned backwards for any reason. This will only ruin the cutting edge. The dropping of reamers on hard surfaces such as concrete floors, machine tool tables, metal plates, etc. should also be avoided. Cutting edges can only get broken or burred this way.

Nor should reamers be neglected when storing. In order to produce smooth, accurate holes, the cutting edges of these tools should be protected in separate containers such as wooden or plastic tubes or heavy-duty envelopes when not in use (Figure 33-15). Furthermore, a film of oil should be applied before storage to prevent rust.

Reaming Allowances

Hole finishes will be affected by the type of reamer being used as well as material left in the hole to be cut. While the minimum stock allowance will improve the finish, too little can lead to cutting trouble. Rather than cut it, the edges of the reamer can be deflected by the material.

Any number of technical publications list the amount of material to be left in a hole for reaming. No two seem to be alike. General guides do exist and they are good starting points, such as in Table 33-1.

For machine reaming in steel or most ferrous metals, for example, it is not uncommon to leave $\frac{1}{64}$ in. (0.4 mm) for stock removal. As a general rule, $\frac{1}{32}$ in. (0.8 mm) stock should be left for reaming holes larger than $\frac{1}{2}$ in. (13 mm) in diameter.

For hand reaming a range of 0.002 to 0.005 in. (0.50 to 0.13 mm) material is sufficient, although it is recommended never to leave more than 0.005 in. (0.13 mm).

TABLE 33-1
Suggested stock removal for machine reaming

Hole size		*Stock removal*	
in.	*mm*	*in.*	*mm*
$\frac{1}{4}$	6	0.010	0.25
$\frac{1}{2}$	13	0.015	0.38
$\frac{3}{4}$	19	0.017	0.43
1	25	0.020	0.50
$1\frac{1}{4}$	31	0.022	0.55
$1\frac{1}{2}$	38	0.025	0.63
$1\frac{3}{4}$	44	0.027	0.69
2	50	0.030	0.76

TABLE 33-2
Suggested reaming speeds and feeds for high speed steel reamers

	CS (cutting speed)			
Material	*sf/min*	*m/min*	*Feed*	*Coolant*
Aluminum	130–200	39–60	0.0015 in. (0.04 mm)	Soluble oil
Brass (Free cutting)	130–200	39–60	to	Soluble oil
Bronze (Hard)	50–70	15–21	0.004 in.	Soluble oil
Cast iron (Soft)	50–100	15–30	(0.10 mm)	Dry or compressed air
Machine steel	50–70	15–21	Per flute	Soluble oil
Magnesium	170–270	52–82		Soluble oil
Tool steel	30–40	9–12	Per revolution	Soluble oil
Stainless steel	20–30	6–9		Soluble oil

Reaming Speeds and Feeds

Reaming speeds and feeds are based on the material being cut as well as the material and size of the reamer involved. Rule of thumb suggests a reamer speed of about 50% of that used for drilling. As a starting point for using high-speed steel reamers, a conservative listing of speeds and feeds can be found in Table 33-2, where it will be found that the coolants chosen for reaming may differ from that used for drilling. A reamer will cut at a much slower rate and higher feed than a drill of equal size. The fluid used is to produce a better finish rather than cooling. Carbide reamers are far superior to high-speed steel reamers not only for their long life and resistance to wear and abrasion but can operate at speeds $1\frac{1}{2}$ times that of high-speed steel reamers.

Reamer chatter is one of the causes of producing a flawed hole finish. The rapid vibration of the cutting tool produces a wavelike surface in the hole. Chatter can be eliminated by reducing the reamer speed, increasing the feed or reducing the lead angle. In some cases chatter may be caused by using a straight fluted reamer. A helical fluted reamer should be used instead. A straight fluted reamer will eliminate chatter if the cutting edges used are staggered or unevenly spaced around the circumference of the reamer body.

Reaming a Straight Hole

Lack of rigidity is one of the causes of poor finishes and chatter marks. Before attempting any drilling and reaming, the vise-held workpiece should be firmly clamped to the drill press table and not moved during the entire operation. As with many other hole producing operations, tool alignment is important to a successful conclusion.

Procedure

1. Clamp the work-mounted vise firmly to the drill press table.
2. Tighten a proper size drill for reaming allowance in the drill chuck.
3. Drill the hole. *Note: Do not move the hole location at this time.*
4. Tighten a proper size reamer in the drill chuck.
5. Set the spindle speed for the work material and type and size of the reamer being used.
6. Start the drill press spindle.
7. Lubricate the reamer freely if the work material requires it.
8. Slowly lower the reamer until the end of the reamer begins cutting.
9. Carefully feed the reamer to complete the cutting.

Counterboring

A counterbore is a multicutting edge tool used for the purpose of enlarging the top of a previously drilled hole to a limited depth (Figure 33-16). It is a flat bottom cutting tool guided into the hole by means of a pilot. Counterboring may be used to recess items such as bolt head screws, springs, fillister head screws, or hex head screws. Counterboring will produce holes that are wider and deeper than the size of socket head cap screws it must often accommodate.

Counterbores are available in a variety of styles. Some counterbores are supplied with solid pilots as shown in Figure 33-17, while other counterbores are available with replaceable pilots.

FIGURE 33-16
A counterbored hole is drilled to a limited depth.

Procedure for Counterboring a Cap Screw Hole

1. Fasten the workpiece in a drill vise.
2. Drill the body size hole to accommodate the screw to be used.
3. Mount the proper size counterbore in the drill chuck.
4. Set the spindle speed to approximately one-quarter of that used for drilling.
5. Lower the counterbore pilot into the drilled hole and lock the spindle.
6. Set the drill press depth stop for the thickness of the screw head.
7. Release the spindle and raise the counterbore pilot out of the hole.
8. Start the drill press spindle.
9. Apply cutting fluid.
10. Counterbore to the required depth as set by the drill press depth stop.

FIGURE 33-17
A set of counterbores with solid pilots. (Credit: DoAll Company)

Spot-Facing

Spot-facing is an operation performed for the purpose of smoothing and truing the surface around the top of a hole to provide an area to seat washers, nuts, or cap screws (Figure 33-18). Made of high-speed steel or with carbide tip inserts, spot-facing tools are made with interchangeable pilots and can be used as counterbores as well (Figure 33-19). Like counterbores, the work must be securely held and the spindle speed set to about one-quarter that used for drilling holes of similar size.

Boring

Boring is an operation performed on an undersized hole in order to produce a hole to size. The hole will be smooth, straight, and concentric. Generally, a single or double-edged cutting tool is used in drill press work.

Since the accuracy of a machined hole depends on the type of hole making process, boring is better performed on a vertical milling machine where a rigid setup and a power feed quill are possible. In a modern machine shop, boring on a drill press is seldom done.

FIGURE 33–18
A spot-faced surface can be used to seat washers, nuts, or cap screws.

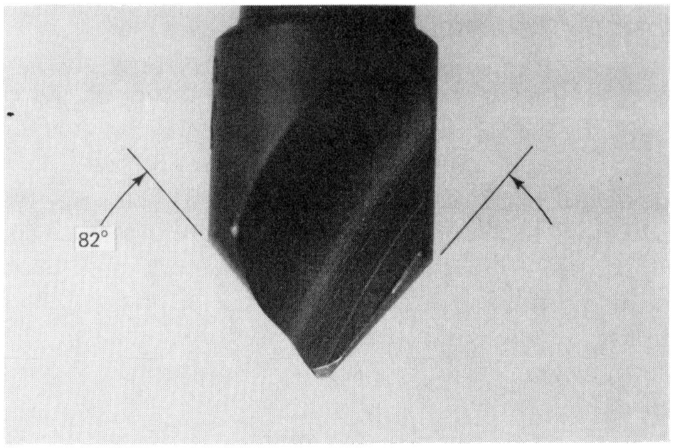

FIGURE 33–20
An 82° countersink is used for enlarging or producing cone-shaped holes.

FIGURE 33–19
Spot-facing tools can also be used as counterbores.

FIGURE 33–21
Countersinks can also be used as deburring tools.

Countersinking

A countersink is used for enlarging or producing cone-shaped holes (Figure 33–20). One of its uses is to remove burrs left on holes previously drilled (Figure 33–21). Its chief purpose is to recess holes to accommodate flat head screws flush with or slightly below the work surface (Figure 33–22). Countersinks are available with various included angles such as 60°, 82°, 90°, 100°, 110°, and 120°.

Countersinking is also used to remove material from a hole that will be tapped. In order to eliminate an unwanted burr caused by the initial thread (first-thread burr), the hole is countersunk deeper, widening the opening slightly wider than the major diameter of the thread (Figure 33–23).

The cutting speed will vary with the type and size of the countersink involved. As a general rule, three fluted high-speed steel countersinks should be run about one-quarter to one-half the drilling speed. Carbide countersinks may run at 60 to 70% of the drilling speed.

To Countersink a Hole for a Flat Head Screw

1. Measure the diameter of the flat head screw with a rule.

2. Tighten an 82° countersink in the drill chuck.

3. Set the spindle speed between one-quarter and one-half that used for drilling.

4. Place the workpiece on the drill press table.

5. Lower the countersink into the hole to center it if clamping is required. If using a pilot-type countersink, slip fit the pilot into the hole first, then clamp the workpiece.

6. Raise the countersink slightly and start the drill press spindle.

7. Slowly hand feed the countersink into the hole, raising it periodically to check the diameter being cut with the rule or compare it directly by placing the head of an inverted screw into the hole.

8. Complete countersinking until the top of the hole is slightly larger than the diameter of the screw head.

Drill Press Tapping

The subject of tapping, both by hand and under power, has been covered in previous units. What is required at this time is to explore the operation of hand tapping in a drill press. Maintaining alignment of tap and hole is the key to a successful operation. Once the tap drill has been used, the hole location should not be moved.

Procedure for Hand Tapping

1. Mount the center-punched workpiece on parallels in a drill vise.

2. Tighten a small center drill in the drill chuck.

3. Extend the drill press spindle about 3 in. (76 mm) and lock it.

4. Place the workpiece under and close to the center drill.

5. Set the spindle speed for center drilling.

FIGURE 33-22
Holes should be countersunk deep enough to keep flat-head screws flush or slightly below the surface.

FIGURE 33-23
A countersink was used to remove material to eliminate the first thread burr.

FIGURE 33-24
A stub center is used to guide a hand tap or reamer perpendicular to the hole.

6. Release the spindle and start the drill press.

7. Lower the center drill and allow the workpiece to move as it centers itself with the center drill point.

8. Once a cone-shaped hole has been produced, stop the drill press and with the center drill still inside the hole, lock the spindle.

9. Clamp the drill vise to the table.

10. Remove the center drill and mount the correct tap drill.

11. Drill the hole to the correct tap drill size for the tap to be used. *Note: Do not move the hole location at this time.*

12. Replace the tap drill with a stub center (Figure 33-24).

13. Fasten a long-handle tap wrench on the end of the tap to be used.

14. Apply a suitable lubricant to the tap and place it in the drilled hole.

15. Lower the drill press spindle so that the stub center can fit into the tap's center hole.

16. Keeping light pressure against the tap's center hole, turn the tap wrench clockwise to start the tapping.

17. Complete the tapping in the usual manner: applying lubricants, maintaining light pressure against the tap center hole, while advancing and reversing the tapping rotation.

UNIT 34 SHARPENING A DRILL

Proper drill sharpening and drill economy plays an important role in the metalworking industry. Twist drills supplied by the manufacturer are properly sharpened on special drill sharpening grinders. These drills are designed to wear evenly and cut equally on both cutting edges, thereby assuring a long drill life. This unit deals with the proper reconditioning of a twist drill.

Objectives
After completing this unit you should be able to

1. name five characteristics found in a properly ground drill.
2. identify the three causes for drill failure.
3. identify four types of web thinning.
4. grind a general-purpose twist drill.

Drill Cutting Characteristics

In order to remove metal quickly and efficiently a drill has to be properly sharpened for cutting. This design will combine accurate drill pointing, web thinning, and ample clearance without reducing strength or increasing wear in order to achieve the following:

- Control of the chip formation.
- Control of the size and shape of the chip.
- Control of the chip as it slides up the flutes.
- Control the size, quality and straightness of the drilled hole.
- Control the amount of burr produced on either side of the drilled hole.
- Increase the cutting lip strength.
- Reduce the rate of wear at the cutting edges.
- Reduce the amount of drilling pressure required.
- Reduce the amount of heat generated.
- Permit a wider range of speeds and feeds for more efficient drilling.

Causes of Drill Failure

Wear
Wear is one of the main causes of drill failure and generally starts at the corners of the drill. As the outer tips begin to round, wear takes place at the

chisel point and cutting lips, leaving a conical surface with no clearance. The drill tends to rub rather than cut in the hole. In this condition greater power and force is required to operate the drill. The result is a heat buildup around the cutting edges and a rapid increase in wear. The more the drill is used, the greater the wear. The drill will eventually become completely dull and will not cut unless it is resharpened. In reconditioning the drill, the worn section must be removed by grinding.

Web Thinning

The lack of web thinning is another cause of drill failure. Due to the fact that the web becomes thicker as it progresses along the length of the drill toward the shank, the chisel edge length increases as the drill point is resharpened and presents drilling problems (Figure 34-1). The wider than normal web requires greater pressure to penetrate the work surface and will also generate more heat, shortening drill life. Along with difficulty in spotting holes, increased chisel edges can cause oversize diameters.

Web thinning can be done in a conventional tool and cutter grinder or it can be done as an offhand grinding operation. There is also a special web thinning grinder available. There are a few general rules to keep in mind when web thinning.

- In all cases thinning should be done by removing material evenly and uniformly along both lips of the drill.
- Thinning should be kept central with axis of the drill, with thinning kept to a minimum so as not to weaken the drill (Figure 34-2).
- A narrow soft wheel should be used so as not to burn the drill edges when grinding.

Examples of web thinning are shown in Figure 34-3.

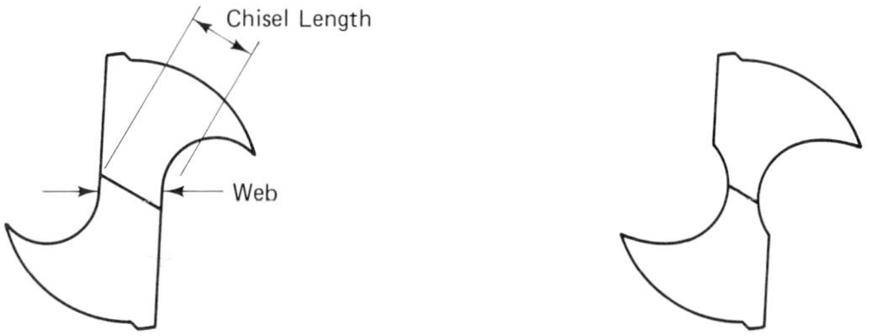

FIGURE 34-1
The chisel point length increases as the drill point is resharpened.
(Credit: Union/Butterfield)

Drill Diameter	*Percentage of drill diameter* H.S.S.	*Percentage of drill diameter* Carbide
$\frac{1}{8}$ in. (3 mm)	20	16
$\frac{1}{4}$ in. (6 mm)	17	15
$\frac{1}{2}$ in. (13 mm)	14	13
1 in. (25 mm)	12	13
1+ in. (25+ mm)	11	12

FIGURE 34-2
General recommended web thicknesses.

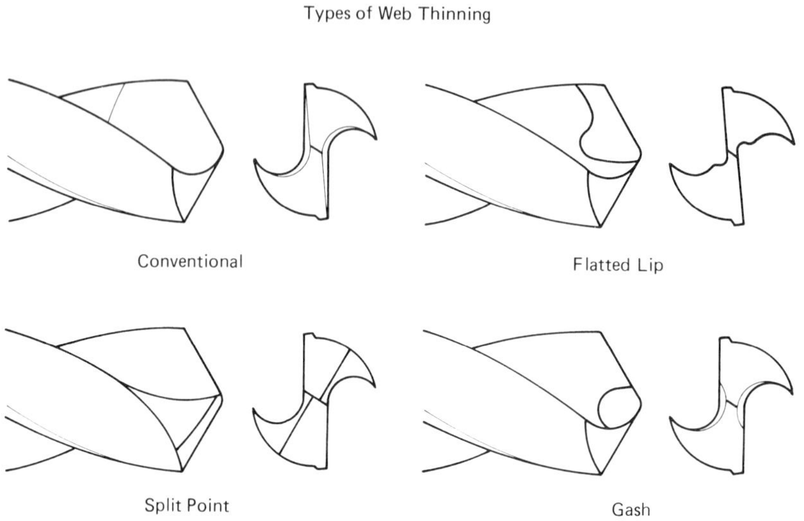

FIGURE 34–3
Types of web thinning: (a) Conventional—stock removal follows flutes, (b) flatted lip—flatting the back of the flute to eliminate the drill from digging in, (c) split-point-type thinning—designed for deep oil hole drilling and will also minimize skidding and walking, (d) gash-type thinning—produced by two half-round grindings near the center of the drill to reduce webbing. (Credit: DoAll Company)

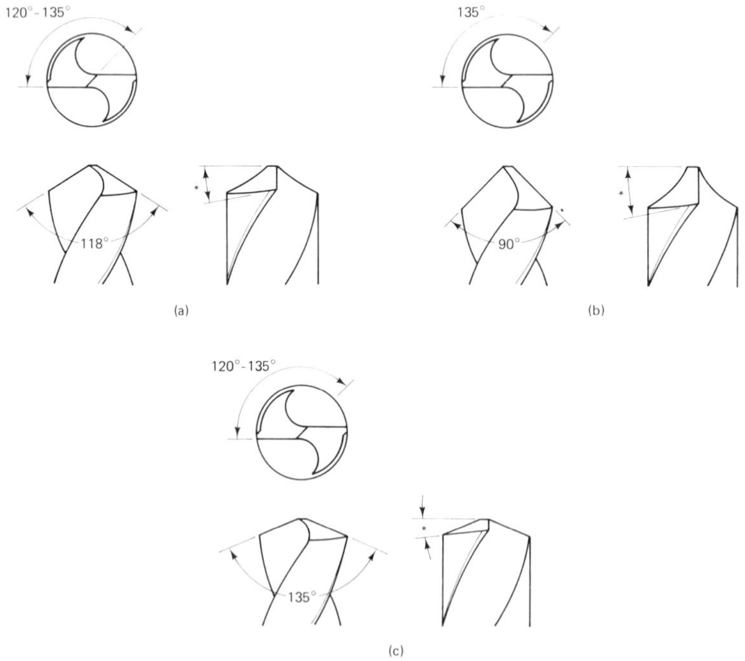

FIGURE 34–4
(A) 118° drill point is used for general-purpose applications, (B) 90° drill points are used on softer or ductile materials, (C) 135° point angle is used on hard materials, highly alloyed steels.

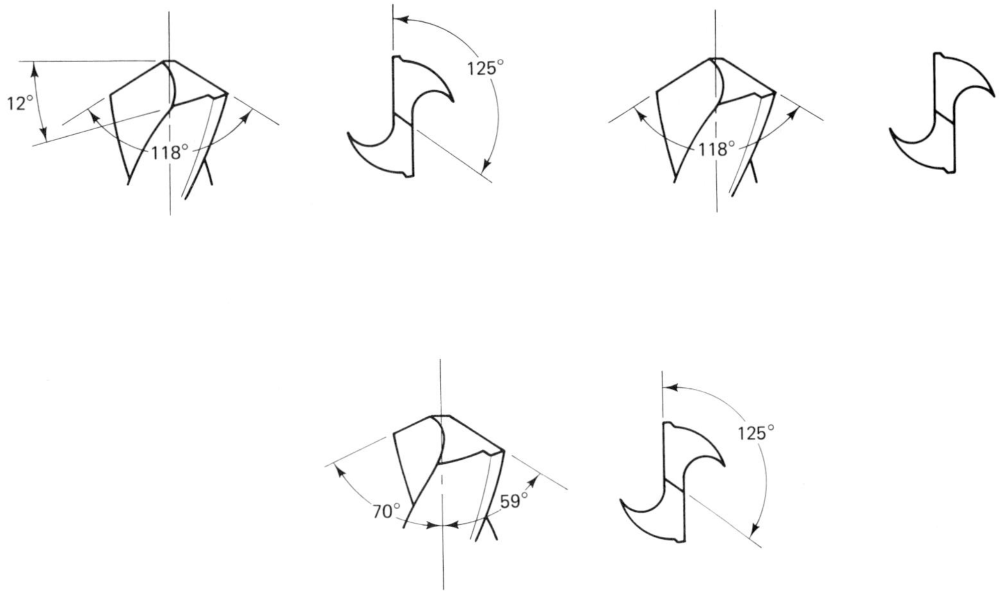

FIGURE 34–5
Uneven lip length and uneven angles will result in oversize, inaccurate holes. (Credit: Union/Butterfield)

Drill Pointing

The third cause of drill failure is incorrect point regrinding. As indicated in Unit 30, the point is the conical section at the end of the twist drill and consists of several parts: the lips or cutting edges, lip clearance, the heel, and chisel point.

There are three general drill point angles in use, each designed for various hardness or softness of materials (Figure 34-4).

An 118° point drill is used for general-purpose drilling and will include a lip clearance of 8° to 12°. Too much lip clearance will weaken the cutting edge; too little will require greater drilling pressure, generate heat, and will wear out the cutting edges quickly.

An 135° point drill is used for tough or hard materials such as alloyed steels. The flatter angle requires less power to cut and reduces drill strain. Since the cutting edges will require as much support as possible, lip clearance of between 6° and 8° is recommended.

A 90° point drill is used for drilling soft cast iron, hard rubber, bakelite, plastics, and similar nonferrous materials. The recommended lip clearance for this type of drill is between 12° and 15°.

In any reconditioning, it is important that the cutting edges be uniform in length. Failure to grind the lips or angles evenly will result in an inaccurate hole as indicated in Figure 34-5.

FIGURE 34–6
Drill pointing grinding machine.

To Sharpen a Twist Drill

A twist drill can be looked upon as a round tool having sharp cutting edges with relief or clearances provided to prevent it from binding while drilling and still stand up against pressure and wear. There are point grinding machines available to recondition twist drills that are worn or have its cutting edges damaged in any way (Figure 34-6). It can also be done using a drill grinding attachment (Figure 34-7).

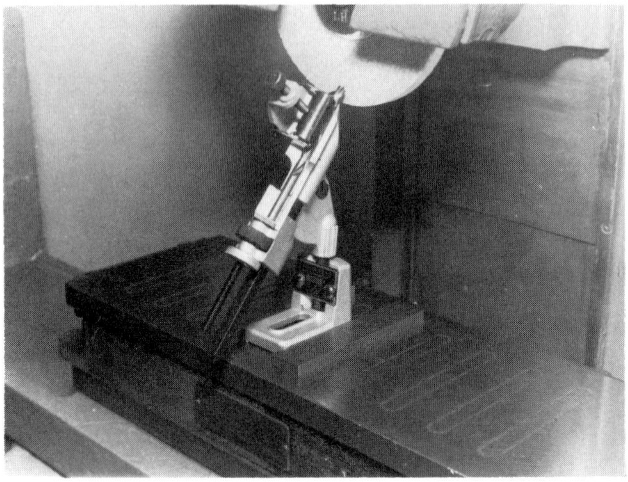

FIGURE 34-7
Drill grinding attachment.

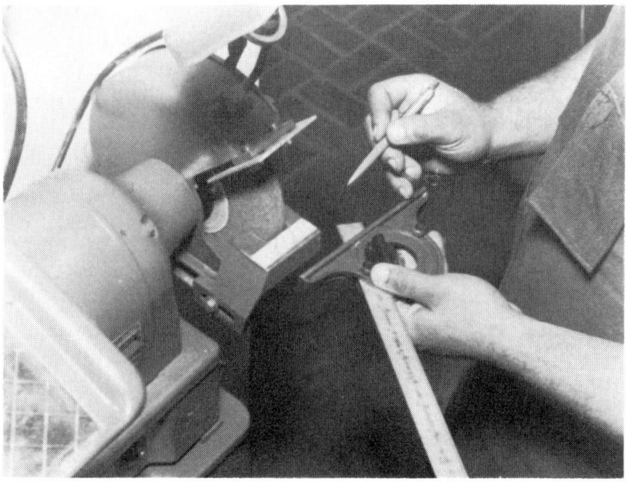

FIGURE 34-9
Draw a few spaced lines at 59° to the wheel face.

FIGURE 34-8
Adjust the tool rest not more than $\frac{1}{8}$ in. (3 mm) from the wheel face.

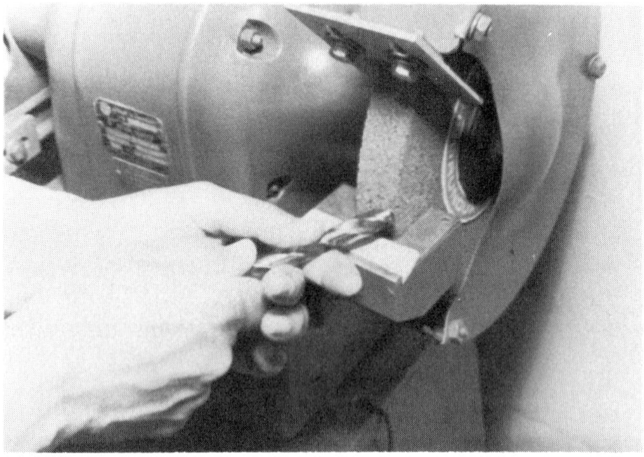

FIGURE 34-10
Hold the drill in both hands with the back hand lower than the front hand.

In the machine shop it is very common for a machinist to touch up or repoint a twist drill by hand against a pedestal grinding wheel. The machinist should be prepared by using methods of safety, a clean dressed abrasive wheel, and gages for measuring the accuracy of the ground drill.

Procedure for Grinding a General-Purpose Twist Drill

1. Appropriate safety glasses should be worn before attempting any drill grinding.

2. Examine the twist drill for wear or damaged cutting lips.

3. Examine the pedestal grinding wheel and dress it, if necessary, to clean and smooth the front face of the wheel.

4. Adjust the toolrest to leave a gap not more than $\frac{1}{8}$ in. (3 mm) from the wheel face (Figure 34-8).

5. Check the water trough and fill it if necessary.

6. Using masking tape, lay down a strip on the toolrest to draw a few spaced lines at a 59° angle to the wheel face (Figure 34-9).

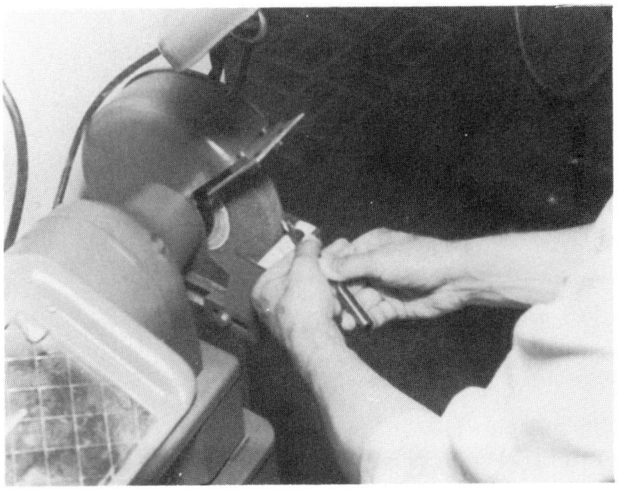

FIGURE 34-11
Align the body of the drill by eye with one line on the tape.

FIGURE 34-12
Measuring lip length and angle using a drill point gage.

7. The drill should be held in both hands, with the back hand on the shank slightly lower than the front hand (Figure 34-10).

8. Hold the cutting edge level and parallel to the wheel face, aligning by eye the main body of the drill with one line on the tape (Figure 34-11).

9. Lightly make contact with the rotary wheel and slowly lower the back hand while keeping the drill in contact with the grinding wheel. *Do not twist the drill.*

10. Quench in water to prevent over heating and softening of the cutting edge.

11. Rotate the drill half a turn and repeat the process on the other cutting edge.

12. Check the drill point angle and length of cutting lips against a drill point gage for accuracy (Figure 34-12).

13. Repeat steps 7 through 12 until any worn sections have been removed and both cutting edges are uniform in length and angle.

Special Drill Grinding

A number of special drill grinding operations can be performed using off-hand grinding techniques. While the step-by-step procedure will not be discussed, it is worthwhile examining a few of the drills and see for what purpose they can be used.

Flat-bottomed drills (Figure 34-13) are used for producing holes to accommodate springs and other flat-bottomed hardware. To do so, the point of the drill must be removed by grinding both lips flat and uniform. Lip clearance must be provided as with a standard twist drill. A flat bottom drill of this type cannot be used unless a standard twist drill of the same size precedes it, with the bottom of the hole later leveled by the use of the flat bottom drill (Figure 34-14).

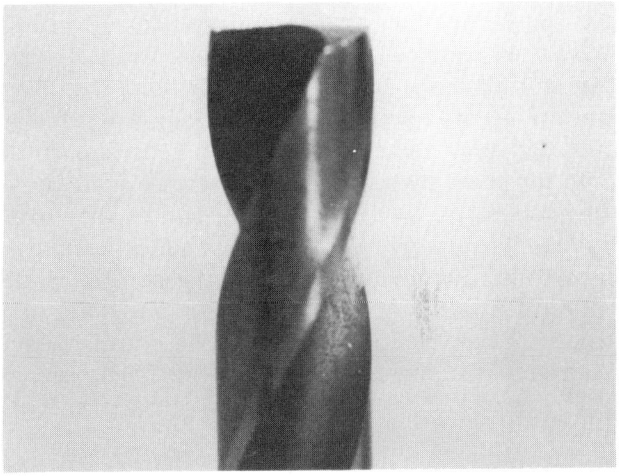

FIGURE 34-13
A flat-bottomed drill is used for producing holes to accommodate springs and similar flat-bottomed hardware.

What about the drilling of materials such as brass or bronze? Should drills be treated differently for these copper-base alloys? The answer is defi-

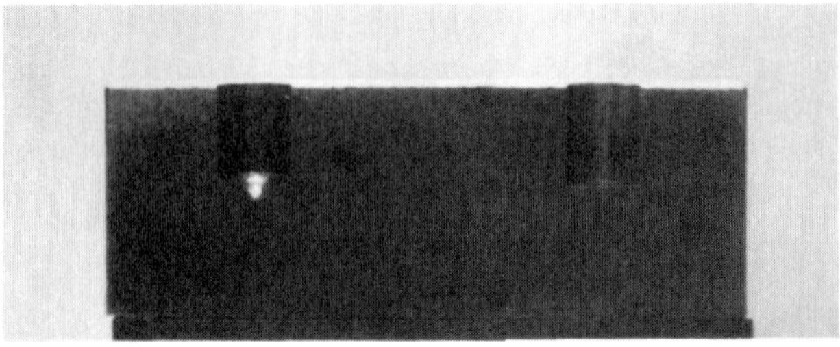

FIGURE 34–14
A standard sized drill must first be used before using the flat bottom drill.

nitely "yes!" Drills should be ground to prevent digging in and breaking by producing a neutral rake or helix angle at the cutting edge (Figure 34–15). Unless a drill has been prepared in this manner, it should not be used for drilling brass or bronze.

This twist drill has been expertly sharpened on a tool grinder for use as a sheet metal drill (Figure 34-16). Notice how two secondary back-off grinds have been produced, reducing some of the land, leaving just a narrow cutting edge with ample lip clearance. The chisel point has been reshaped to leave an included spur point. When using a general-purpose drill for sheet metal drilling, the chisel point tends to push metal away, deforming the surface even though it is supported underneath, before it will cut. The construction of a general-purpose drill is designed for deep hole drilling, making long flutes necessary. Sheet metal drills should be sturdier, have shorter flutes, and a heavier web construction.

FIGURE 34–15
A neutral rake or helix angle should be ground on drills used for brass or bronze to prevent the drill from drawing in.

FIGURE 34–16
A twist drill reground for sheet metal drilling.

SECTION REVIEW

Unit 29

1. Hole drilling is the chief operation performed on a drill press. Name five operations that also can be carried out.
2. Name two common types of sensitive drill presses to be found in a metalworking shop.
3. Why is it dangerous to leave a chuck key in a drill chuck at any time?
4. Why is it dangerous to remove a drill chuck from the drill chuck spindle with a cutting tool held in its jaws?
5. Describe the difference between a gang drill machine and a turret drill press.
6. For what is a radial drill press used?
7. Briefly describe the function of the following parts found on a standard drill press: (a) base, (b) table, (c) spindle, and (d) drill chuck.

Unit 30

1. Of what material is the twist drill found in the machine shop generally made?

2. Name the three principal parts of a twist drill.
3. What are the two types of drill shanks used?
4. State the three purposes of drill flutes.
5. What is the surface between the flutes called and what is its function?
6. Where is the twist drill's web located and what is its function?
7. Name the four measurement systems used to identify the twist drill sizes.
8. Name two methods of measuring drills.
9. Name three common methods of holding drills.
10. Briefly describe how to tighten a standard drill chuck.
11. How is a tapered shank cutting tool removed from the drill press spindle and what precautions should be taken to prevent damaging the drill press table?

Unit 31

1. How may the cutting speed of a drill be defined?
2. What are the dangers involved in not achieving the recommended cutting speed in drilling?
3. How may feed be defined and what are the factors affecting feed?
4. What two general conditions will eliminate the use of carbide drills?
5. To what hole depth can a carbide drill be used?
6. What is the danger of suddenly cooling a carbide drill that has been heated up?

Unit 32

1. What two conditions may result when forcing the feed of a drill as the cutting point starts to break through the bottom of a workpiece?
2. List four common methods of holding work for drilling purposes.
3. Where should the holding pressure be applied when using bolts and step blocks?
4. Name five characteristics a cutting fluid should have.
5. What is a bell-type center drill and what is its purpose?
6. Briefly describe how a drill plate can be used to locate the center of a hole location.
7. When is it necessary to clamp a drill vise directly to the table?
8. List three important rules that should be observed in order to drill a large hole accurately.
9. How large should the size of a pilot hole be?

Unit 33

1. What is the purpose of a reamer?
2. How are hand reamers identified?
3. Name two different types of hand reamers.
4. As a rule of thumb, at what speed should a machine reamer be used?
5. Why should reamers not be turned backwards when being used?
6. How should reamers be stored when not being used?
7. What is the purpose of a counterbore?
8. What is the purpose of spot-facing?
9. What is the purpose of countersinking?
10. What is the key to a successful hand tapping operation in a drill press?

Unit 34

1. List three causes for drill failure.
2. List three conditions that can result due to lack of web thinning.
3. List the drill point angle and lip clearance used for general-purpose drilling.
4. List the drill point angle and lip clearance used for very hard materials such as alloyed steels.
5. List the drill point angle and lip clearance used for drilling soft cast iron and nonferrous materials.
6. What is the purpose of using a flat bottom drill?
7. Why should drills be ground with a negative rake or helix angle at the cutting edges when drilling brass or bronze?
8. List seven favorable conditions that will result from a properly sharpened drill.

Section 9

THE LATHE

UNIT 35 THE ENGINE LATHE

One of the earliest lathes—the "tree lathe"—was in use before the screw cutting lathe. Work suspended between two trees was revolved by means of a rope that was wrapped around it. One end of the rope was tied to a movable branch overhead; the other end was pulled on by an assistant in a back and forth motion, permitting the operator to work the wood by a hand tool.

By 1740, a screw cutting lathe was introduced in France. It was rotated by turning a hand crank connected to the headstock. While the lead screw was geared to the spindle, it could only cut a thread with one particular pitch. It was not until the Industrial Revolution, about 1797, that Henry Maudslay introduced the basic principles of the screw cutting engine lathe. His design allowed gears that connected the headstock spindle with the lead screw to be changed. This permitted different screw pitches to be cut.

The invention of the engine lathe made possible the removal of material or chips from a rotating workpiece with a stationary cutting tool. The cutting tool could move across the length or across the diameter of the workpiece by mechanical or manual means. None of the machine tools developed could match the lathe for versatility. It is hard to imagine to what degree industrial progress could have been made without the lathe. Many special-purpose machinery such as the jeweler's lathe (Figure 35-1), heavy-duty production lathe (Figure 35-2), or a hydraulic tracer lathe (Figure 35-3) are evidence to the lathe's flexibility. Many operations and special equipment used in automated production processes could not have been developed without the lathe.

The toolroom lathe (Figure 35-4) remains one of the most widely used machine tools in industry, schools, and machine shops. This unit will deal mainly with the toolroom lathe.

Objectives
After completing this unit you will be able to

1. identify and explain the function of the different parts of the lathe.
2. determine and set the lathe to run at its required speed.
3. determine and set the proper feed for the cut required.

Safety

While it is important to understand how to operate a lathe before machining with it, the key to using it is to make it work for the individual safely. There is no reward for working hard on a lathe any more than it would be to work dangerously. While there can be a great deal of satisfaction in completing an opera-

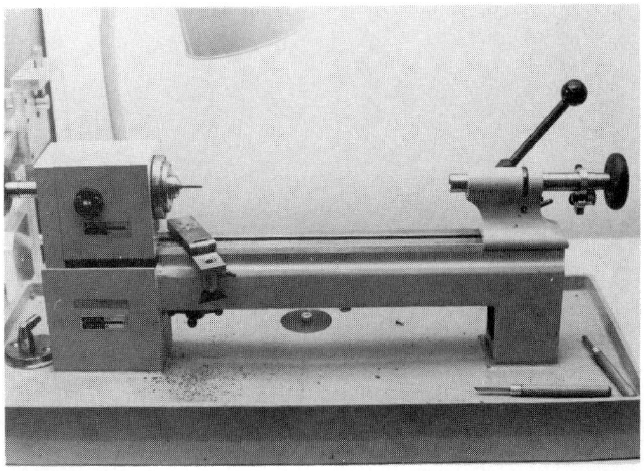

FIGURE 35-1
A jeweler's lathe: designed for very light, sensitive work.

FIGURE 35–2
A long, heavy-duty 20-in. (750-mm) production lathe. (Credit: Clausing Industrial Inc.)

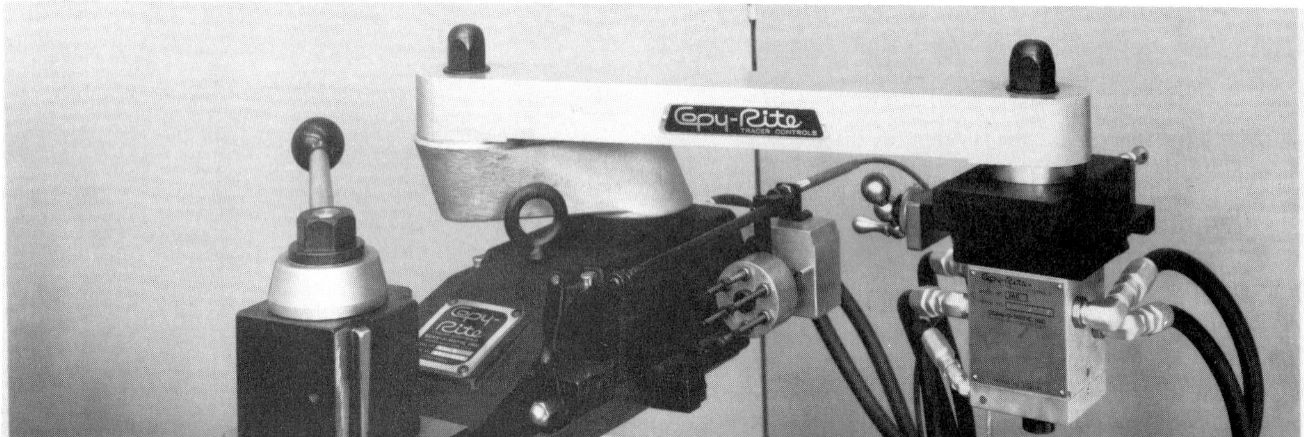

FIGURE 35–3
A hydraulic tracer attachment can follow a pattern or a model to reproduce an exact copy of the workpiece. (Credit: LeBlond Makino Machine Tool Company)

FIGURE 35-4
An inch/metric toolroom lathe with digital readout unit. (Credit: HLV® DR is a trademark of Hardinge Brothers, Inc.)

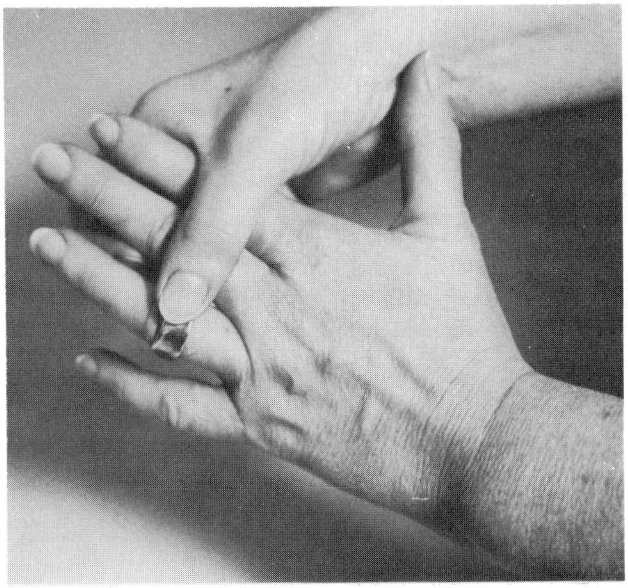

FIGURE 35-5
Never wear rings or other jewelry when operating a lathe.

tion successfully, the greatest reward anyone can achieve is to be able to approach a lathe, work on it, and be able to walk away from it without damaging the equipment or maiming one's self. The following few rules can help the individual reach that goal.

Lathe Safety

1. Always wear appropriate safety glasses before attempting to work on a lathe.

2. Be thoroughly familiar with the operating mechanisms.

3. Never attempt to operate a lathe without proper supervision and permission to do so.

4. Never wear loose clothing, rings, watches, or bracelets (Figure 35-5).

5. Never leave the chuck key in the chuck at any time.

6. Do not attempt to stop a moving chuck, drive plate, or workpiece with your hand.

7. Never attempt to measure moving surfaces. Always stop the lathe first.

8. Always establish a lathe stop to prevent running the cutting tool into the lathe chuck (Figure 35-6).

9. Butt the carriage against a set lathe stop and by turning the chuck by hand, examine to see if there is enough clearance between the chuck jaws and cutting tool, and whether the jaws will chew up the compound rest or dovetail.

10. Never tighten a toolbit in the holder by swinging the wrench in an arc in front of the cutting tool point.

11. Never attempt to oil, clean, or adjust the machine while it is in operation.

12. Never attempt to clean the workpiece or lathe with a rag while the machine is in motion. The rag along with the hand may be drawn into the machine this way.

13. Always clean the floor area of oil, grease, or dirt before starting the lathe operation.

14. Always fasten a drive plate or lathe chuck securely before attempting an operation.

15. Never attempt to remove chips by hand. Use a brush or a pair of standard pliers when the lathe is not in motion.

FIGURE 35-6
Establish a stop to prevent running the cutting tool into the lathe chuck.

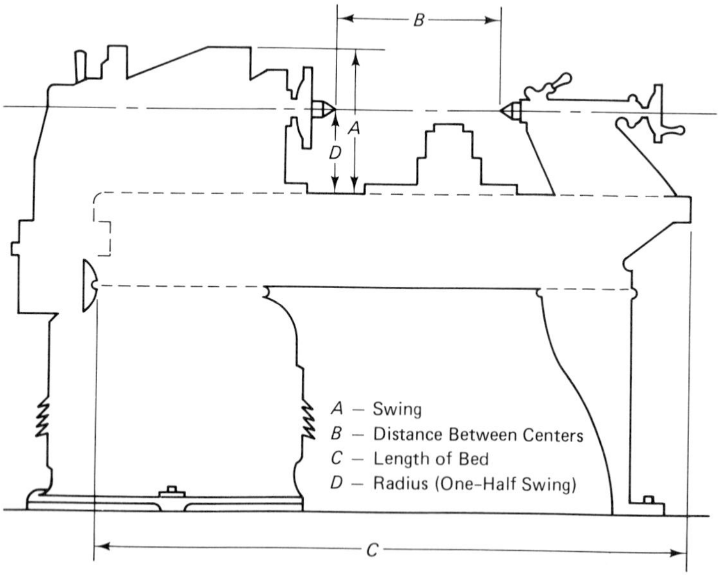

FIGURE 35-7
The size of a lathe is determined by the swing over and the length of the bed. (Courtesy of South Bend Lathe, Inc.)

Size of a Lathe

The size and capacity of an engine lathe or lathe, as it is commonly called in the machine shop, is generally determined by some manufacturers by the maximum diameter of work that can be turned or swung over the bed and the length of work turned between centers (Figure 35-7). The length of the lathe is determined by the length of the bed. This is often used by some manufacturers who determine the size and capacity of a lathe as being the largest work diameter that can be swung over the bed and the length of the lathe bed. European manufacturers classify the size of the lathe by the radius (one-half the swing), measured from the center axis of the headstock spindle down to the bed. In practice, the swing over the cross slide is found to be less than the swing over the bed and distance between centers is found to be less than the total length of the bed.

Manufacturers produce lathes in varying sizes, ranging from 9 to 30 in. (225 to 750 mm) in swing and from 16 to 160 in. (400 to 4000 mm) between centers. As an example, a medium-duty toolroom lathe may have a 13-in. (325 mm) swing and a 40-in. (1000 mm) distance between centers.

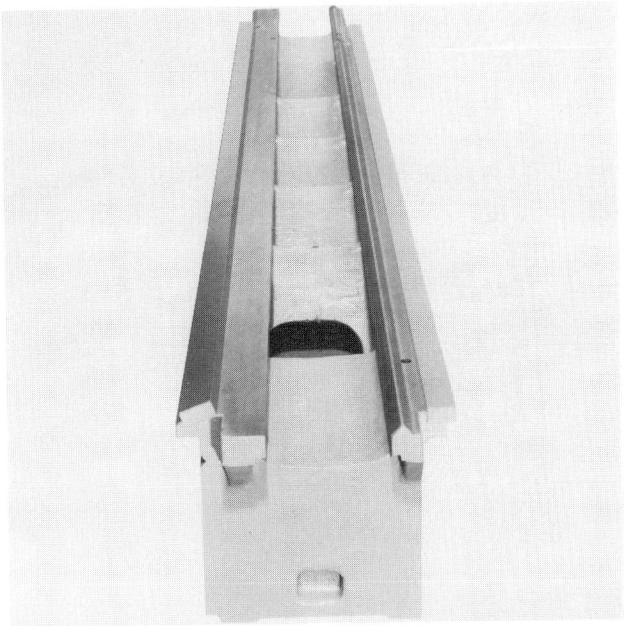

FIGURE 35-8
The bed and ways of a well-constructed bed.
(Credit: LeBlond Makino Machine Tool Company)

Parts of a Lathe

The main parts of a lathe are the bed, the headstock, the carriage, the quick-change gearbox, the tailstock, and their chief component parts.

Bed

The *bed* of the lathe is the machine's foundation and is the traffic roadway for transporting mechanisms and cutting tools. The bed shown in Figure 35-8 is ruggedly cast with its ways as one unit. Its outer construction consists of two outer ways (flats and vees) that, while they are hardened and ground to assure accurate alignment of the moving parts, are replaceable. On other lathes the bed may consist of two outer inverted V-ways whose accuracy is matched by two inner bearing surfaces, one flat and one inverted V-way (Figure 35-9). While one particular lathe model may be explained, the instructions given may apply to almost any general lathe.

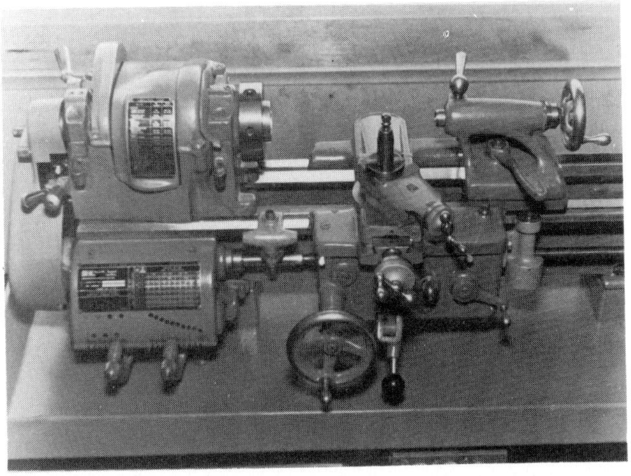

FIGURE 35-9
The bed of this lathe contains three inverted V-ways and one flat bearing surface.

Headstock

The *headstock* is the housing or casting on the upper left side of the lathe and is accurately aligned with respect to the ways and bed, as well as the other mechanisms that travel on this "roadway." Located within the headstock is the *headstock spindle,* a long

FIGURE 35-10
A cam lock spindle nose. (Credit: Leblond Makino Machine Tool Company)

FIGURE 35-11
The gearing mechanism transmits power from the motor on a geared-head lathe. (Credit: LeBlond Makino Machine Tool Company)

tubular shaft that is suspended by bearings. The spindle's inner hole is tapered in front to grip and drive tooling whose surface bear an equal taper. The outer construction, which extends from the headstock, called the spindle nose (Figure 35-10), is designed to have work holding devices mounted on it.

Headstock spindles can be rotated by the gearing mechanism, which transmits power from the motor in the form of controlled speeds on gear-driven lathes (Figure 35-11).

On most lathes the *feed reverse lever* is located on the headstock. By moving the feed reverse lever, the direction of cutting travel is reversed.

Quick-Change Gearbox

In front and below the headstock housing is the *quick-change gearbox,* which is responsible for selecting the lathe's feed as well as for changing the pitch for thread cutting. Gears within the box are selected by positioning tumble plungers, which will affect the rate the feed rod advances the lathe when it is to operate in the automatic turning mode. It will also affect the rate that the lead screw revolves for thread cutting operations. These plungers should not be forced into position or moved while the lathe is in operation (Figure 35-12).

FIGURE 35-12
The quick-change gearbox permits fast settings of feed and thread cutting. (Credit: LeBlond Makino Machine Tool Company)

FIGURE 35-13
The main parts of an apron. (Credit: LeBlond Makino Machine Tool Company)

Carriage

The *carriage* is an assembly that moves along the ways between the headstock and the tailstock. The carriage consists of four main parts: the saddle, the cross slide, the apron, and the compound rest.

The *saddle* is an H-shaped casting that is machined and accurately fitted to move along the ways of the bed.

The *apron* is bolted to and is found on the front bottom section of the saddle. The apron houses the gearing for carriage movement as well as control devices for cross feeding and thread cutting operations. A safety interference bar or mechanism on the back of the apron prevents a dual engagement of levers. The lead screw and half-nut lever (for thread cutting) cannot be engaged at the same time that the lathe is set to run when the feed control lever is in place.

The *apron handwheel* (Figure 35-13) is responsible for moving the carriage as it turns a gear in mesh with a long geared tooth rack. The carriage can be moved manually with the apron handwheel or automatically by engaging the carriage feed mechanism.

The *cross slide* sits atop and spans across the width of the saddle. It is controlled by turning the cross-feed handle to provide cross movement for the cutting tool in machining operations made in a direction 90° to the axis of the spindle.

The *compound rest* sits on top of the cross slide and consists of two major parts: the base and the slide. The base of the rest is marked off in degrees, allowing the compound rest to be set to an angular position. The slide is T-slotted and is used for supporting the cutting tool. Movement of the compound rest is achieved by hand turning the compound rest handle that, like the cross-feed handle, is measured using a graduated micrometer collar reading in thousandths of an inch or hundredths of a millimeter. The compound rest makes it possible to cut short tapers as well as adjusting the cutting tool to different angles.

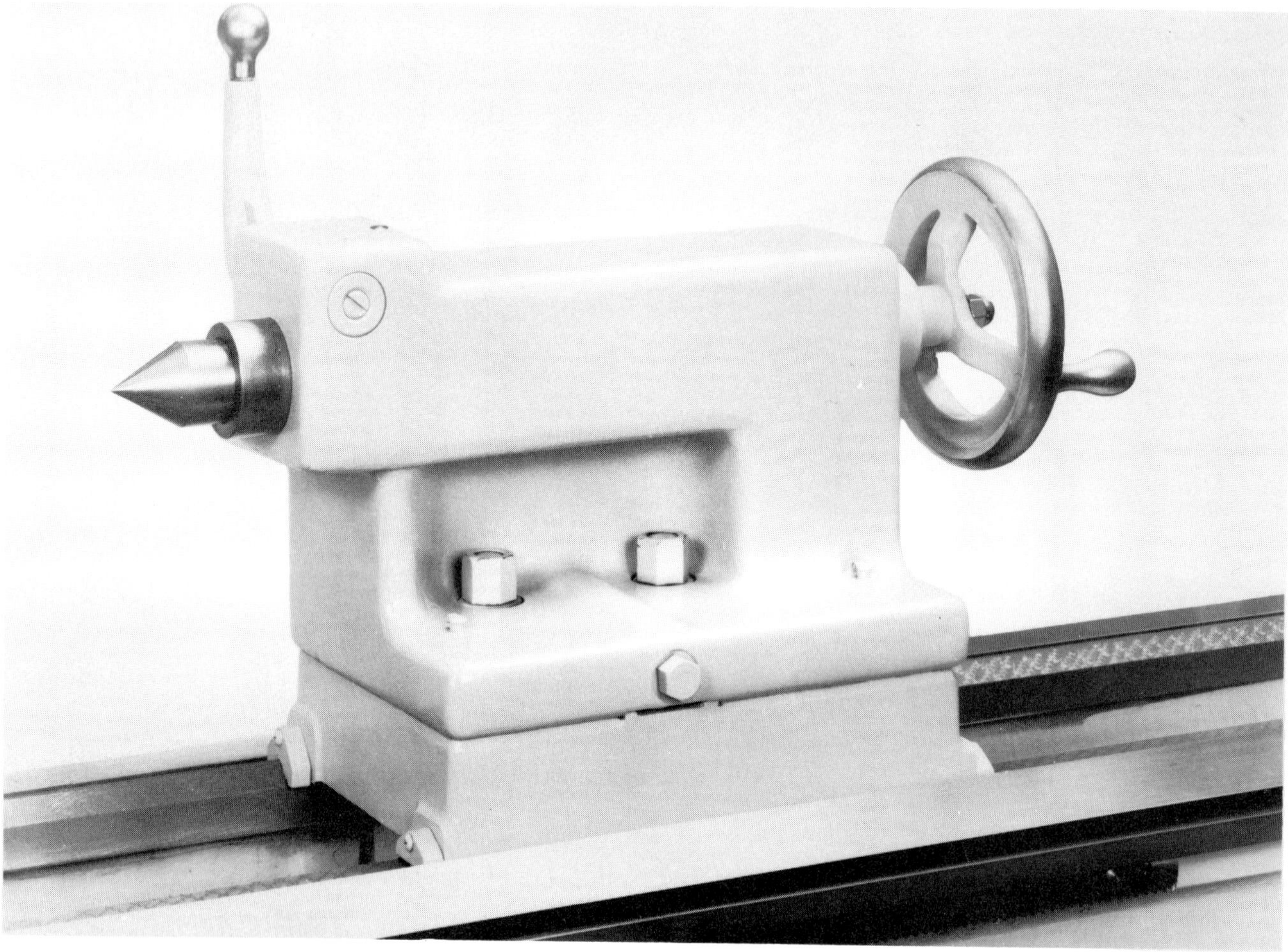

FIGURE 35-14
The tailstock of an engine lathe. (Credit: LeBlond Makino Machine Tool Company)

Tailstock

Located on the right-hand side of the lathe is the *tailstock* (Figure 35-14). The tailstock consists of two sections that can be slid and adjusted toward or away from the lathe's center axis. It comes equipped with a self-holding internally tapered spindle. The spindle can hold drill chucks, drills, reamers, and other tools that contain a Morse taper shank. The tailstock handwheel is turned to advance or retreat the spindle, making it possible to hand feed cutting tools during drilling and reaming operations. The tailstock spindle can also be locked in a fixed position by means of a binding lever or clamp nut.

Types and Uses of Accessories

To add to the flexibility of the lathe, a variety of accessories are available to hold and support work during machining. Some of these include three- and four-jaw chucks, drive plates, lathe centers, and mandrels. Many of these will be explained in greater detail in units to follow.

Cutting Tools

The most commonly used cutting tools in the machine shop are made of high-speed steel. Other materials include carbide, cast alloys, ceramics, cubic bo-

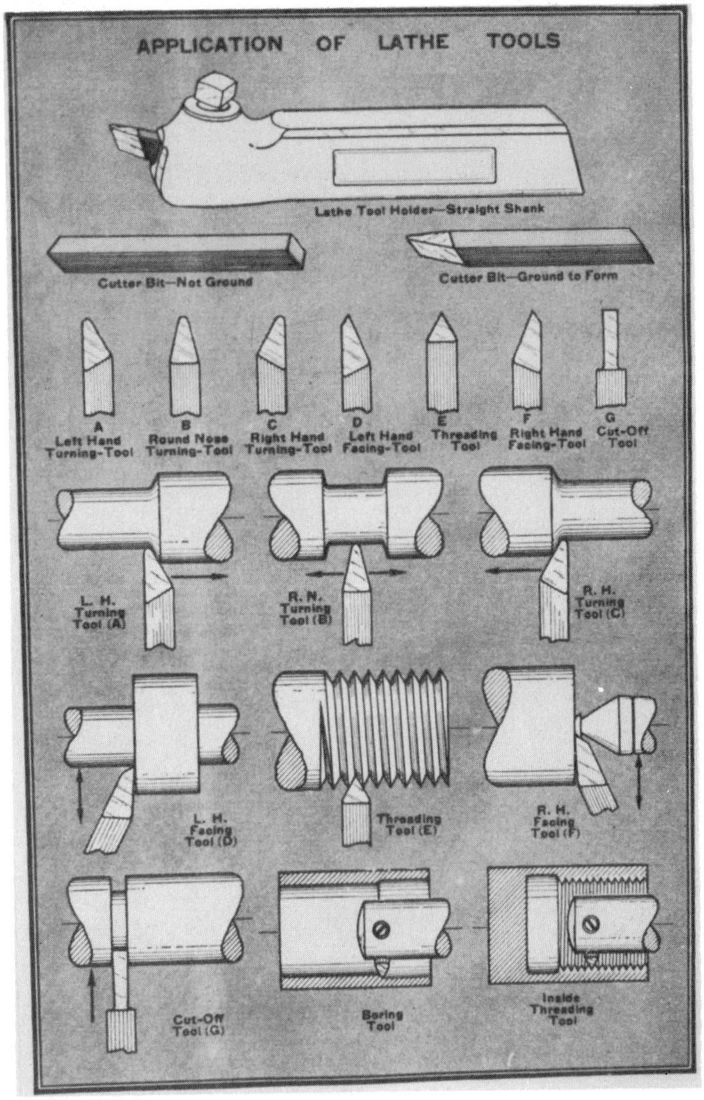

FIGURE 35–15
Application of lathe tools. (Courtesy of South Bend Lathe, Inc.)

ron nitride, and diamond. High-speed steel toolbits will be mostly discussed here.

Toolbits are ground in a specific way to include various angles, not only to perform as long as possible but are designed to cut in a particular manner. When a toolbit is ground with the cutting edge on the left side and cuts from the tailstock toward the headstock, it is considered a right-hand cutting tool. If it is ground with the cutting edge on the right side, it cuts from the headstock toward the tailstock and is considered a left-hand cutting tool. Some of these cutting tools are shown in Figure 35–15.

Types of Toolholders

High-speed toolbits used in school shops and in many toolroom lathes are held in *standard toolholders* that hold the toolbit at an angle of about 16.5°. In turn, these toolholders are held in a round tool post. Three general types of toolholders are shown in Figure 35–16. While the right- and left-hand holders will require cutting tools that allow as much machining room as possible, the straight toolholder can be used for general machining and will accept either a right- or a left-hand cutting tool.

The *straight shank toolholder* shown in Figure

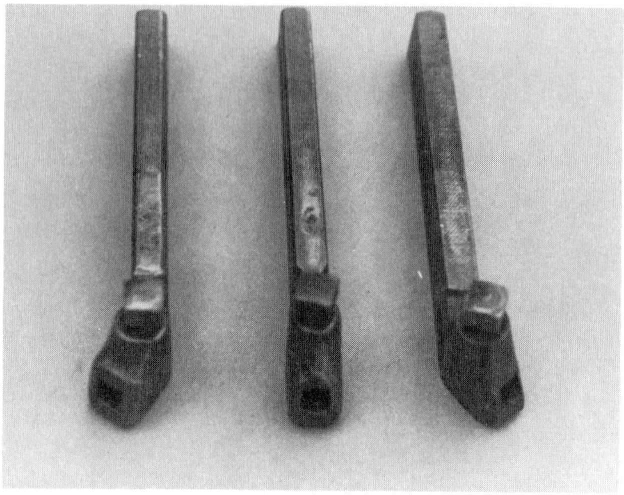

FIGURE 35–16
Left, straight, and right-hand toolholders.

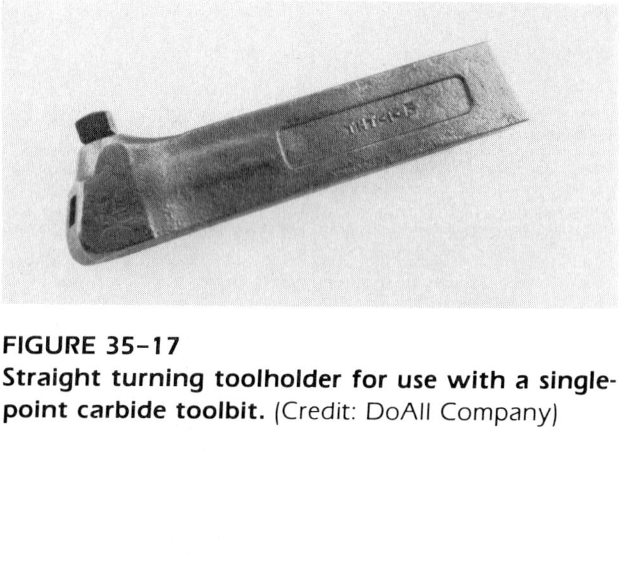

FIGURE 35–17
Straight turning toolholder for use with a single-point carbide toolbit. (Credit: DoAll Company)

35–17 does not have an inclined tool hole. The toolbit is held in a horizontal position and is used with carbide toolbits as well as cast alloy toolbits.

The *cutoff toolholder* uses a long narrow cutoff blade whose cutting edge is normally produced with side clearance to prevent it from binding when used as a parting tool, cutting to a shoulder or grooving. The blade is held in the toolholder by means of a cam lock or locking nut that holds it firmly against a slotted surface (Figure 35–18).

FIGURE 35–18
The blade in the cutoff toolholder can be used for parting or grooving operations.

The *four-way turret tool post* (Figure 35–19) can hold as many as four separate tools or toolholders at the same time. Unlike the round tool post that has a wedge under the toolholder to center the tool, the turret-type tool post requires shims to adjust the tool height. By loosening the handle, the tool post can be indexed to any of 12 different positions and locked in place.

The *quick-change tool post* shown in Figure 35–20 uses preset tool blocks that slide onto the dovetail of the tool post and are locked in place by means of a clamping lever. This rigid mounting will permit heavy depths of cut to be taken that are free of vibration and chatter. Each type of tool will require a special holder, as shown in Figures 35–21 and 35–22. The tools are centered by means of a knurled knob or nut on the toolholder.

Lathe Speeds and Feeds

Speeds

In order to machine metal in a lathe efficiently without wasting time or avoiding premature breakdown and tool wear, the proper speeds and feeds must be used. The speed at which a lathe should be turning is referred to as the number of revolutions the spindle will turn in one minute (rpm). To calculate the rpm, the cutting speed (CS) of the material must be known. Cutting speed refers to the amount of mate-

FIGURE 35-19
The four-way turret tool post can hold as many as four separate tools or toolholders at the same time. (Credit: LeBlond Makino Machine Tool Company)

FIGURE 35-20
The quick-change tool post uses preset tool blocks that can be accurately set and rigidly set in place.

FIGURE 35-21
A quick-change multipurpose toolholder. (Credit: DoAll Company)

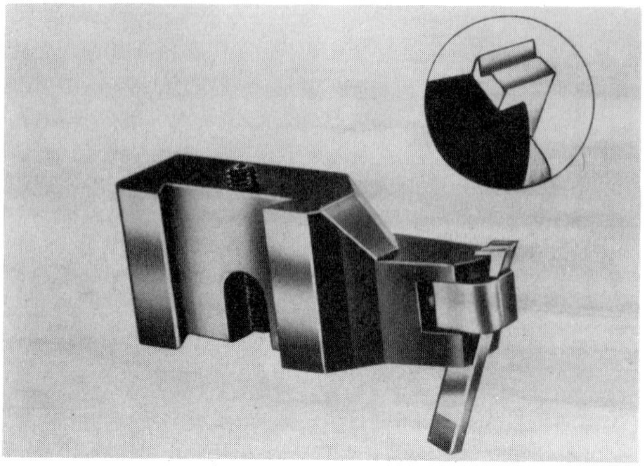

FIGURE 35–22
The quick-change box toolholder provides a rigid tool setup. (Credit: DoAll Company)

rial removal recommended for effective machining. In some books this is expressed by the letters ft/min. (feet per minute) or sf/min (surface feet per minute). It refers to the amount of material in feet or meters that goes past a cutting point in one minute's time. This can be seen if the material were stretched out in one continuous length. Working with all this information is the diameter (D) of the workpiece involved. The formula is expressed

$$\text{rpm} = \frac{\text{CS} \times 4}{D}$$

where cs = cutting speed

d = diameter of workpiece

Example
Calculate the rpm required to turn a 2-in.-diameter piece of tool steel (CS 50)

$$\text{rpm} = \frac{\text{CS} \times 4}{D}$$

$$= \frac{50 \times 4}{2}$$

$$= 100$$

For metric lathes the formula reads

$$\text{rpm} = \frac{\text{CS} \times 320}{D}$$

Example
Calculate the rpm required to turn a 30-mm-diameter piece of aluminum (CS 90)

$$\text{rpm} = \frac{\text{CS} \times 320}{D}$$

$$= \frac{90 \times 320}{30}$$

$$= 960$$

TABLE 35–1
Recommended average cutting speeds and feeds using high-speed steel (HSS) lathe tools

	Turning, facing and boring					
	Rough-cut feed		*Finish-cut feed*		*Cutting screw threads*	
	.007–.015 in.	*0.17–0.38 mm*	*.003–.006 in.*	*0.07–0.15 mm*		
Material	*SFPM*	*M/min*	*SFPM*	*M/min*	*SFPM*	*M/min*
Aluminum	200	61	300	90	50	15
Brass	150	45	200	60	50	15
Bronze	90	27	100	30	25	8
Cast iron	60	18	80	24	25	8
Machine steel	90	27	100	30	35	11
Tool steel	50	15	80	24	25	8

FIGURE 35-23
In order to machine under power on a lathe, the feed change lever must be properly positioned.

FIGURE 35-24
The quick-change gearbox is set by positioning the tumbler levers using the column and row location.

While conditions will vary in machining, the preceding formulas represent average lathe speeds. Cutting speeds for high-speed steel toolbits are listed in Table 35-1.

Feeds

The feed of a lathe is the distance a cutting tool will travel along the workpiece's length with each revolution. Generally, machining will be done in two operations: rough and finish turning. The rate of feed will be different for each operation. The purpose of rough turning is to remove excessive stock quickly. Finish turning is a smoothing operation that brings the workpiece down to size and produces a good finish. The rate of feed for rough and finish cuts will vary with the size of the lathe, but as a general guide 0.007 to 0.015 in. (0.17 to 0.38 mm) can be used for roughing cuts, while 0.003 to 0.006 in. (0.07 to 0.15 mm) can be used for finishing cuts.

Setting Feeds

The quick-change gearbox is the key to setting feeds for a lathe. An index plate attached to the front of the gearbox is lined to form a grid pattern, indicating by rows and columns the feed and pitch or number of threads that can be cut, as well as the gear shift lever position. Setting the lathe will require that the tumbler levers and/or setting lever, depending on the style lathe used, be placed or positioned in the proper column and row box location to match the box location of the required feed.

Procedure for Setting the Feed

1. Set the feed change lever into the proper location if need be (Figure 35-23).
2. Locate the feed desired as listed on the index plate attached to the quick-change gearbox (Figure 35-24).
3. Position the tumbler lever into the hole directly under the column location of the desired feed.
4. Move the tumbler lever into the proper gear shifting hole as indicated on the index plate. On geared lathes a lever is positioned in place.

As a measure of safety, turn the lathe chuck by hand to see that the gears are fully engaged. Do not turn on the power unless the gears have properly meshed. The gears inside a quick-change gearbox can be damaged very easily, and can put a lathe out of commission.

UNIT 36 WORK HOLDING DEVICES

The most popular means of holding work for machining in a lathe are the various chucks that are used. They are handy, quick to use, and easy to change. Their capacities range from holding simple round stock to odd-shaped workpieces. While there are several types of lathe chucks, this unit will concern itself with the more commonly used work holding devices.

Objectives
After completing this unit you should be able to

1. mount and remove a three-jaw chuck.
2. mount and remove a four-jaw chuck.
3. mount and remove a spring collet chuck.
4. mount and remove a plain mandrel.

Types of Chucks

The *three-jaw universal chuck* (Figure 36–1) is extensively used for machining round and hexagonal work. A set of jaws are provided that are fitted to an inner scroll thread, which enables all these jaws to move at the same time and come to a common center. While the jaws on the chuck shown are designed to hold small diameters and to grip internal openings such as rings, another set of jaws are provided that are stepped in just the opposite direction. These jaws can hold large diameters of varying sizes. For accuracy, new chucks can run a workpiece true within 0.003 in. (0.07 mm) or better; however, the accuracy will not be maintained as the scroll thread wears.

The *four-jaw independent chuck* (Figure 36–2) was designed to be flexible. Unlike the three-jaw universal chuck where the jaws move together, each jaw on the four-jaw chuck is independent. Not only can each jaw be loosened or tightened individually, but it is also reversible. This feature allows round, square, octagonal, and odd-shaped workpieces to be set up or trued with a high degree of accuracy.

The *six-jaw chuck* (Figure 36–3) features six opposed jaws that enable it to grip soft metals and tubes firmer with less pressure and less distortion. In addition, these workpieces can be extended further from the jaws than are normally held in the conventional three-jaw chuck and still be accurately machined. The six-jaw chuck is sensitive enough to hold many workpieces that are ordinarily machined while held in spring collets.

The *draw-in collet chuck* (Figure 36–4) is one of the handiest of all the chuck types and the most accurate in holding small parts. Collets, the gripping devices, are made in round, square, hexagonal, and other shapes.

To use the spring collet chucks, a long tubular handwheel is slipped through the back end of the headstock spindle (Figure 36–5). This draw bar, as it

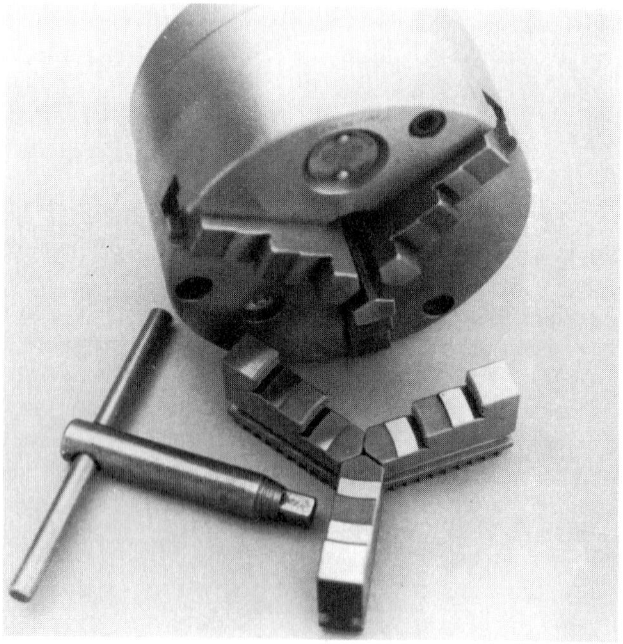

FIGURE 36–1
The three-jaw universal chuck is used to hold round and hexagonal work.

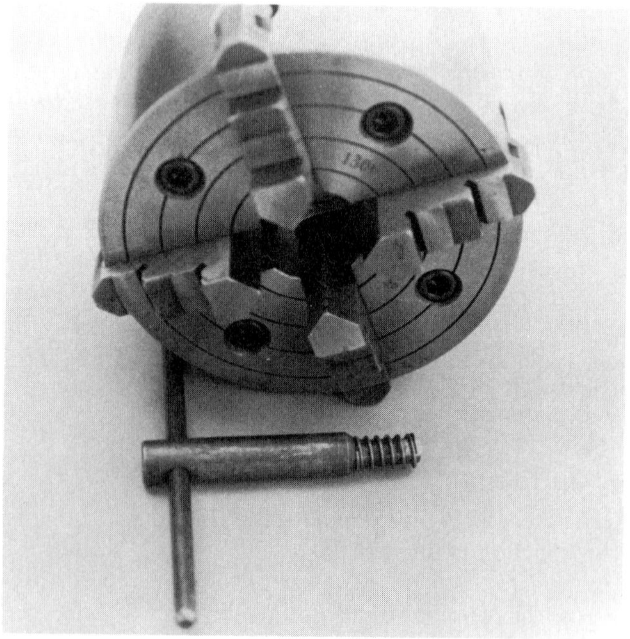

FIGURE 36–2
The four-jaw independent chuck can be used to hold conventional as well as odd-shaped workpieces.

is called, has its opening threaded. A tapered collet sleeve or adapter is fitted into the taper of the headstock spindle and is keyed to accept the collets to be used. Tightening the workpiece is accomplished by mating the threads of the collet and draw bar and by locking the headstock spindle. The handwheel is then turned to draw the collet tightly around the workpiece.

Releasing is accomplished by locking the headstock spindle, unscrewing the draw-bar handwheel a few turns, and hitting the back of the handwheel with an open hand to release the collet from the spindle taper. To maintain the accuracy built into the system, workpieces should not vary more than 0.002 in. (0.05 mm) than the collet size opening.

The Jacobs collet chuck (Figure 36-6) uses an impact tightening handwheel rather than a draw-bar handwheel. A set of tapered steel bars, rubber flex collets, are forced against or released from the workpiece by turning the handwheel. Due to the nature of the collets, they have about a $\frac{1}{8}$-in. (3-mm) wider gripping range than the spring collets.

FIGURE 36-3
The six-jaw chuck permits precision machining usually performed in spring collets. (Credit: DoAll Company)

FIGURE 36-5
The draw-in collet is tightened around the workpiece by turning a long tubular handwheel that is positioned through the back end of the spindle.

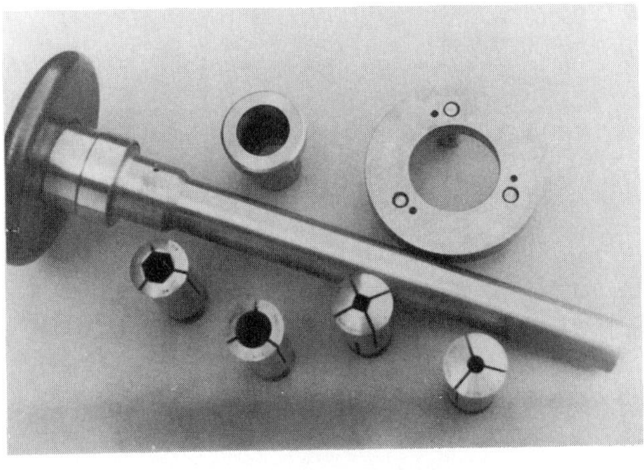

FIGURE 36-4
The draw-in collet chuck with varied spring collets.

FIGURE 36-6
The Jacobs collet chuck uses rubber flex collets and closes with an impact tightening handwheel.

Mounting and Removing Chucks

The use of chucks in a lathe requires careful setups and removal, either condition requiring safety to prevent injuring the operator as well as damaging the lathe. While there are three styles of locking devices—the cam lock, the standard tapered spindle, and the threaded spindle—they all have one thing in common. Each system requires clean mounting surfaces to assure safety and accuracy. Additionally, it is recommended that the lathe be disengaged from its electrical power to prevent any accidental turn on while chucks are mounted or removed.

Cam Lock Lathes

The cam locking lathe uses a short taper (type D-1) and a number of cam locks against extended studs to tighten its chuck on the spindle. Both the spindle bearing surfaces and cam locking studs (Figure 36–7) should be thoroughly cleaned before attempting any mounting.

Alignment of registration marks on the spindle and cam locks indicate the open position for releasing and mounting the chuck. Before attempting any mounting, be sure these registration marks are lined up and visually inspect the holes to see if the locks are clear for entry.

Cam Lock Lathe Mounting Procedure

1. Place the chuck on a wooden block or chuck cradle to prevent personal injury or damage to the lathe (Figure 36–8).
2. Set the lathe spindle in neutral.
3. Turn the spindle by hand to align the cam lock holes with the studs.
4. Slide the chuck into the cam lock holes.
5. Turn the cam lock wrench clockwise to tighten each stud.

Cam Lock Removal Procedure

1. Place a wooden block or chuck cradle under the chuck.
2. Set the lathe spindle in neutral.
3. Turn the cam lock wrench counterclockwise to align the registration marks on the spindle surface and cam locks.
4. Loosen the chuck by giving it a sharp blow with a lead hammer or plastic mallet (Figure 36–9).
5. Slide the freed chuck out on the wooden block or chuck cradle.

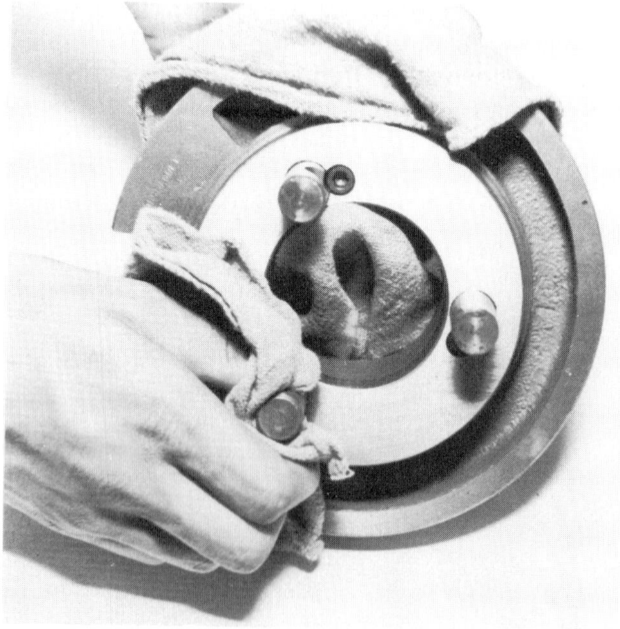

FIGURE 36–7
Spindle bearing surfaces and cam studs should be thoroughly cleaned before a chuck is mounted.

FIGURE 36–8
A lathe chuck should be supported by a wooden block or cradle when attempting to mount or dismount a chuck.

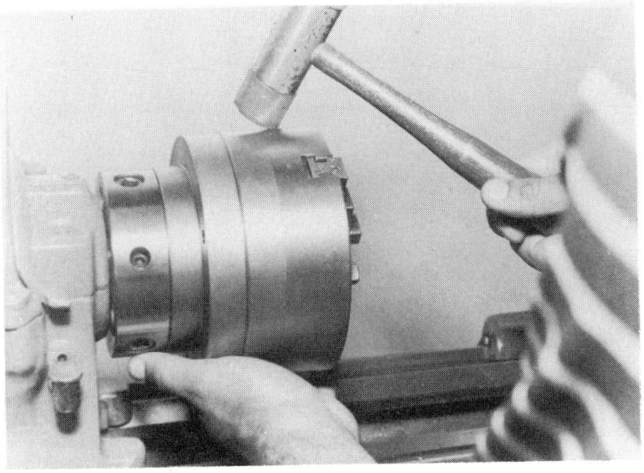

FIGURE 36-9
Strike the chuck to free it from the cam lock holes.

FIGURE 36-10
The type L tapered spindle nose.

Tapered Spindle Nose Lathes

The tapered spindle nose is designed with a steep taper (type L) and a key to align any mounting chuck or spindle accessory (Figure 36-10). A threaded lock ring holds the mounted device on the taper and is tightened with a spanner wrench.

Tapered Spindle Nose Lathe Mounting Procedure

1. Wipe clean the tapered nose surface and tapered back surface of the chuck.
2. Rest the chuck on a wooden block or cradle.
3. Align the chuck's keyway with the tapered nose key.
4. Slide the chuck onto the nose, forcing it against the spindle with one hand, threading the lock ring with the other.
5. Set the lathe to its lowest spindle speed.
6. With the spanner wrench hooked onto the lock ring, strike the end of the wrench in a counterclockwise direction with the hand to lock the chuck in place (Figure 36-11).

FIGURE 36-11
To lock the ring on a taper spindle nose, strike the end of the wrench in a counterclockwise direction.

Tapered Spindle Nose Lathe Removal Procedure

1. Place a wooden block or cradle under the chuck.
2. Set the lathe to its lowest spindle speed.
3. Hook the spanner wrench onto one of the locking ring grooves, positioning the handle vertically.
4. Holding the wrench and locking ring with one hand, strike the end of the wrench clockwise by hand to jar the ring loose (Figure 36-12).

FIGURE 36-12
To loosen the hold, strike the end of the wrench in a clockwise direction.

FIGURE 36-14
Jaws should be mounted in sequence in a three-jaw chuck if it is to run true.

FIGURE 36-13
Once loosened, the ring can be turned by hand.

5. Once loosened, turn the spanner wrench by hand until the ring can be turned by hand (Figure 36-13).

After the chuck has been removed from the spindle, the back opening should be stuffed with a cloth before storing to keep it free from chips and dirt.

Threaded Spindle Lathes
The threaded spindle lathe is one of the oldest types of lathes in use. Before any chuck mounting is attempted all mating threads, both on the spindle and in the back of the chuck, should be cleaned. In addition, the shoulder on the spindle should be clean and free from burrs. Any nicks on the shoulder or the edge of the chuck hole that fits against it will prevent the chuck from running true. A few drops of oil on the spindle and the chuck is ready to be threaded in place.

Using the Three-Jaw Chuck

Before attempting to work in a three-jaw chuck, the jaws should be checked for coordination of movement. When the jaws close, do they come to a common center? Do the numbers on the jaws match the numbers stamped alongside the slots? As a means of assuring accuracy, chuck manufacturers identify and fit the jaws and chuck so they will form one complete unit. The jaws are ground for a particular chuck and should never be interchanged with another chuck. It is also necessary that the jaws be inserted and threaded in sequential order or they cannot hold work properly (Figure 36-14).

Procedure for Assembling Jaws

1. Clean the jaws and jaw slots thoroughly.
2. Insert a chuck wrench and turn it to move the chuck scroll so that its leading edge can be seen in the middle of slot 1 (Figure 36-15).
3. Turn the chuck wrench counterclockwise to move the front edge of the scroll so that it just clears the slot opening.

FIGURE 36-15
Turn the chuck wrench so that the leading edge of the scroll can be seen in the middle of the slot.

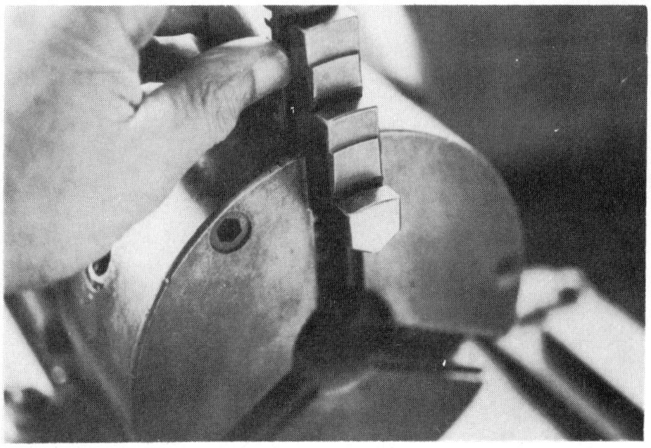

FIGURE 36-16
Push down on the jaw and turn the chuck wrench until the scroll engages the first jaw thread.

4. Insert jaw 1 into slot 1, pushing it down with one hand, turning the chuck wrench clockwise a few turns with the other to engage the first jaw thread (Figure 36-16).

5. Stop turning the chuck wrench and test the jaw engagement by pulling it up and down. If the jaw cannot be slid out, it has been properly engaged; otherwise, start the steps over again.

6. Insert jaw 2 into slot 2 and repeat steps 2 to 5.

7. Insert jaw 3 into slot 3 and repeat steps 2 to 5.

8. Close the jaws to test whether they come to a common center.

Some chucks are equipped with reversible top jaws. The jaws are actually a two-part construction, with stepped top halves fastened to the grooved lower half by means of cap screws (Figure 36-17). At times these jaws are left in a soft state in order to machine them for gripping special contours. In order to take full advantage of their stepped design, the screws have to be removed, the parts thoroughly cleaned and then reversed before being refastened. Tightening has to be done uniformly in order to prevent uneven gripping.

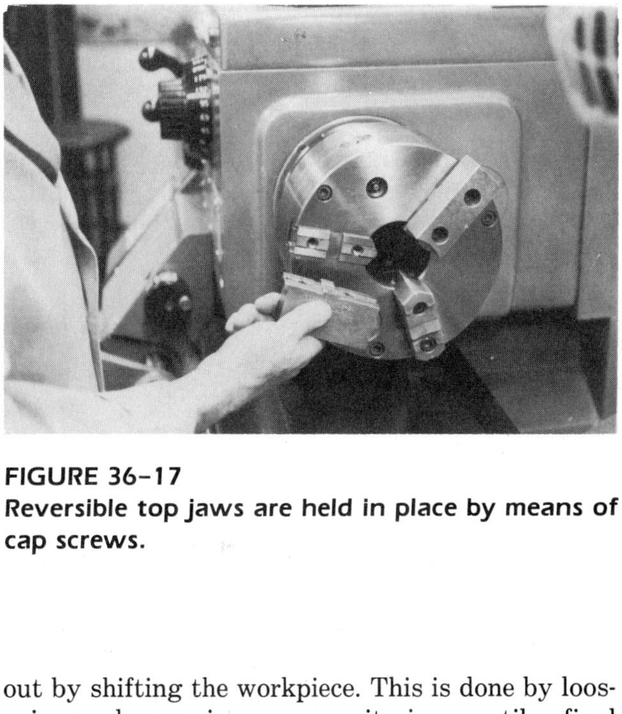

FIGURE 36-17
Reversible top jaws are held in place by means of cap screws.

Mounting Work in a Four-Jaw Chuck

Upon first setting up in a four-jaw chuck, the operator will find the workpiece running erratically. To make it run true, the wobble will have to be taken out by shifting the workpiece. This is done by loosening and snugging up opposite jaws until a final setting can be tightened. A number of methods can be used to make the workpiece run concentrically, depending on the accuracy involved. Regardless of the method, the concentric grooves that circle the face of the chuck should be used as an aid. These grooves are evenly spaced and will do much to approximate the position of the chuck jaws or position the diameter of round work. Working on a four-jaw chuck will require patience, understanding, and some common sense.

Procedure for General Setup

1. Measure the diameter of the workpiece.
2. Open two jaws wide enough to accept the diameter, using one concentric ring in the chuck face as a guide.
3. Open the opposite two jaws in the same manner.
4. Place the workpiece in the four-jaw chuck.
5. Lightly tighten the jaws on the workpiece.
6. Adjust all four jaws so that they bear the same position relative to the concentric ring in the chuck face.
7. Apply any of the following methods for general concentricity.
   (a) The chalk method
      (1) Clear the compound rest of the tool post.
      (2) Move the lathe carriage to bring the compound rest close to the workpiece for hand support.
      (3) Set the lathe to run at a moderate speed.
      (4) Start the lathe.
      (5) Slowly feed a piece of chalk held in both hands against the rotating workpiece.
      (6) Stop the lathe.
      (7) Check the workpiece to see if a white stripe has circled the workpiece. If it does, the workpiece is running true.
      (8) If the chalk mark appears on the surface, it will mark the high side (Figure 36–18).
      (9) Loosen the jaw opposite the white stripe (low side) about $\frac{1}{8}$ turn.
      (10) Tighten the jaw near the white stripe (high side).
      (11) Erase the white stripe and repeat the process until a white line can be seen circling the workpiece.
   (b) The surface gage method
      (1) Place a surface gage with extended scriber point on the compound rest.
      (2) Set the lathe spindle in neutral.
      (3) Rotate the lathe chuck by hand.
      (4) Advance the crossfeed handle to slowly feed the surface gage's scriber close to the rotation to find the high side.
      (5) Rotate the chuck by hand until the low side is nearest the scriber point.
      (6) Loosen the jaw nearest the low side about $\frac{1}{8}$ turn.
      (7) Tighten the jaw opposite the low side to move the workpiece closer to center (Figure 36–19).
      (8) Repeat steps 3 through 7 until the workpiece is running true.

FIGURE 36–18
Using the chalk method to true up work.

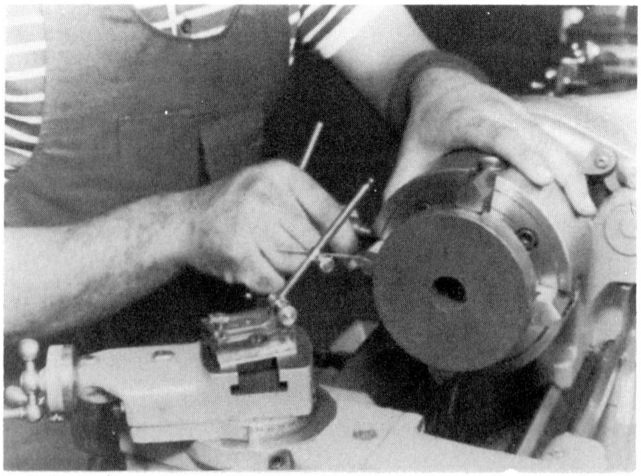

FIGURE 36–19
Tighten the high side to move it away from the scriber point.

Dial Test Indicator Method

The most accurate method of truing work in a four-jaw chuck is done using a dial test indicator. With it, one can achieve a very high degree of accuracy. Once the workpiece has run true by one of the preceding methods, mount a dial test indicator in the tool post. When the indicator is first used, there will be a wide rise and fall of the indicator's needle. The difference between the two readings will indicate the run out. The maximum reading will be considered the high side; the minimum reading will be considered the low side. The objective is to get a steady reading on the dial, with little or no movement of the needle.

Procedure for Truing Large Round Work

1. Mount the workpiece in a four-jaw chuck and true it using the chalk or surface gage method.
2. Mount a dial test indicator in the tool post or the compound rest (Figure 36–20).
3. Set the dial test indicator's contact point near the center of the workpiece (Figure 36–21).
4. Set the lathe spindle in neutral.
5. Advance the crossfeed handle until the indicator's contact point registers about 0.020 in. (0.50 mm).
6. Slowly rotate the lathe chuck by hand (Figure 36–22).
7. Identify the high- and low-side readings indicated by the dial needle (Figure 36–23) and note the difference.

FIGURE 36–20
Mount a dial test indicator in the tool post or on the compound rest.

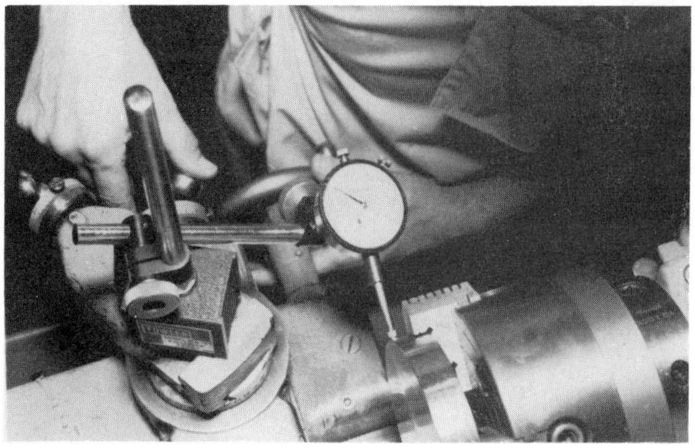

FIGURE 36–21
Set the indicator's contact point near the center of the workpiece.

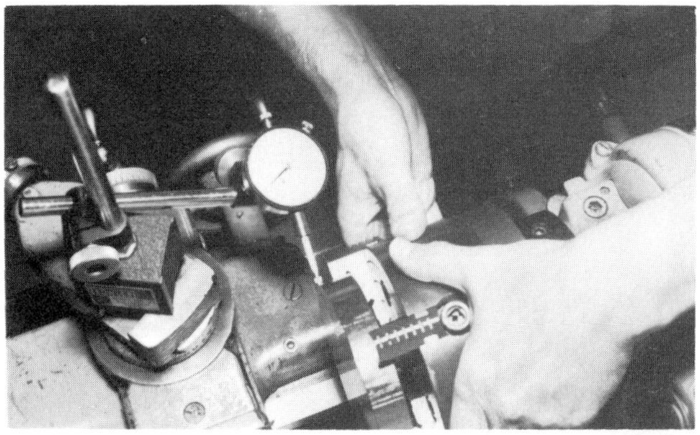

FIGURE 36–22
Slowly turn the lathe chuck by hand to register the workpiece against the indicator's contact point.

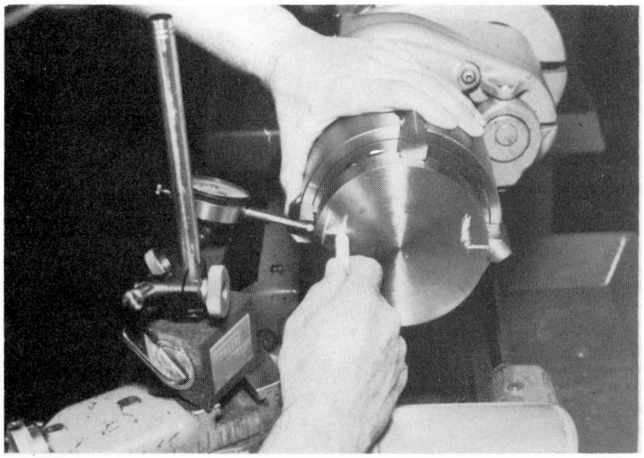

FIGURE 36–23
Identify the high and low sides.

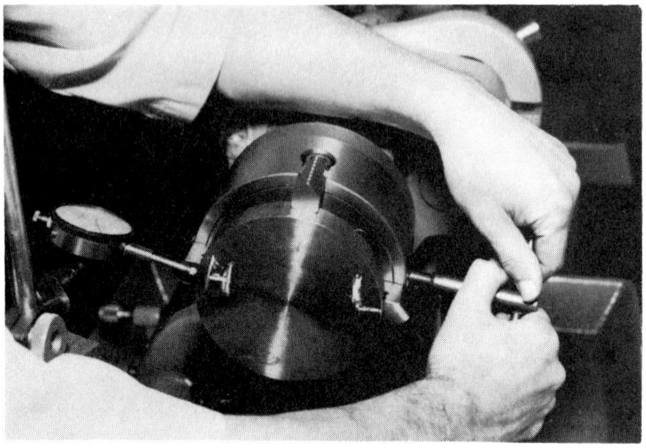

FIGURE 36–24
Barely loosen the low side so that the high side can be moved when tightened.

8. Hold the high- and low-side jaws level, with the contact point testing against the high side.

9. Barely loosen the opposite jaw or low side (Figure 36–24).

10. Snug up the high-side jaw to move the workpiece half the difference between the two indicator readings.

11. Continue to adjust these two opposite jaws until the runout has disappeared. Always align the two opposite jaws first before proceeding to the next set of jaws.

12. Take a new reading on the other two jaws and repeat the process until the runout has disappeared.

13. Tighten all four jaws evenly.

14. Recheck the indicator readings by slowly rotating the lathe spindle by hand to see that the locked position has not been disturbed.

FIGURE 36–25
A steady rest is fastened to the lathe bed to support long and slender work held in a chuck or between centers. (Credit: LeBlond Makino Machine Tool Company)

Steady Rest

A steady rest is a mechanical device that is fastened to the inner ways of the lathe bed and is used to support long and slender work held in a chuck or between lathe centers (Figure 36-25). The three jaws of the steady rest are adjustable and may be tipped with bronze, plastic, or rollers to prevent scoring and overheating the workpiece. The jaws are turned to lightly contact against the workpiece and locked in place to prevent the workpiece from yielding to cutting pressure (springing) during machining. Care must be taken when using the steady rest. If used with a chuck, the steady rest should be placed as close as possible to the end of the workpiece. Alignment with the chuck is also important. Misalignment should not be greater than 0.001 in. (0.02 mm) in 12 in. (300 mm).

Follower Rest

The follower rest (Figure 36-26) is mounted directly to the carriage and travels along with it, supporting the workpiece along its entire cut. Constructed with two adjustable plain bronze or antifrictional roller jaws, one jaw is positioned directly opposite the cutting tool. The other jaw is positioned above the workpiece, preventing it from yielding to cutting pressure during machining. Both jaws may be interchanged without affecting capacity. The follower rest offers maximum stability when machining long slender shafts between centers.

FIGURE 36-26
A follower rest travels with the lathe carriage to offer maximum stability for the entire length of the cut. (Credit: LeBlond Makino Machine Tool Company)

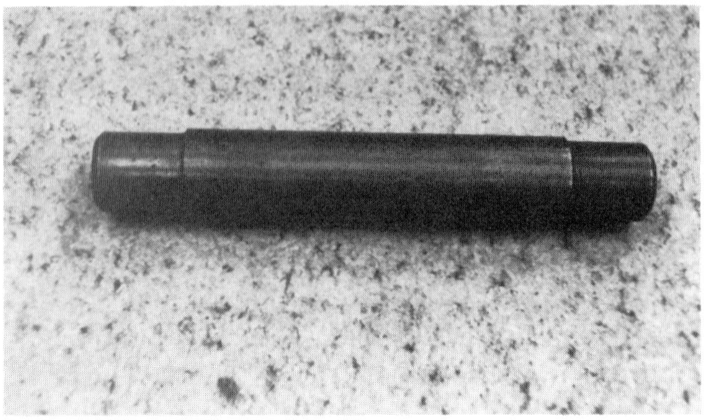

FIGURE 36-27
A plain mandrel is a solid shaft that is slightly tapered to fit in an exact dimensioned hole to prevent the workpiece from slipping during machining.

FIGURE 36-28
An arbor press is a device that is used for pressing or removing an arbor or mandrel from a workpiece hole.

The Plain Mandrel

The machining of pulleys, gear blanks, and similar type work where outside dimensions have to run concentrically with the inner hole are usually performed on a mandrel. A mandrel is a spindle or shaft that has been made for mounting and machining bored work between lathe centers. There are several types of mandrels to accommodate a variety of work, including a threaded mandrel, a gang mandrel, an expansion mandrel, and the most common type, the plain mandrel (Figure 36-27).

The plain mandrel is a solid shaft that is slightly tapered along its diameter and contains a center-drilled hole on each end. The holes are concentric with the outside taper and provide a good bearing surface for the lathe centers. A mandrel can be made in the shop or is commercially available hardened and tempered. It has a taper of about 0.006 to 0.008 in./ft (0.15 to 0.20 mm/m) with the large end stamped with the mandrel's size.

The mandrel is designed to be pressed fit tightly in an exact dimension hole that has been bored or reamed to prevent the workpiece from slipping during machining. It is pushed into the workpiece hole, small end first, by using an arbor press (Figure 36-28). Before mounting, it is important that the mandrel and hole surface be lubricated to prevent "freezing," ruining the work when removing it. Once assembled, a proper lathe dog is fastened to the larger end and prepared for machining between lathe centers.

UNIT 37 FACING

Facing is one of the basic lathe operations. It is defined as machining to resurface the end of a workpiece. The machined surface, in turn, is called the face of the workpiece. Facing is performed for a number of reasons such as establishing a reference surface (face) for measurement, provide a surface 90° to the machined (turned) diameter, and also to machine the workpiece to length. Facing may be performed with work held in a chuck, collet, or between centers.

Objectives
After completing this unit you will be able to

1. set up a lathe cutting tool for facing.
2. set up a workpiece to face the ends.
3. face a workpiece to an accurate length.

Setting Up the Facing Tool

The test of a true faced surface is to be able to place a straight edge across it without seeing any light escaping beneath it. In order to face the end of a workpiece flat, the facing tool has to be mounted exactly on center. If the facing tool is placed slightly below center, a stub of metal will be produced; if the facing tool is placed slightly above center, the stub of metal just mentioned is crushed in machining, and the result is a bump on the middle of the surface.

There are two methods used in facing the ends of a workpiece: the facing tool can be fed from the center outward or may be fed from the circumference toward the center. Roughing cuts may be taken feeding from either the center outward or from the circumference toward the center. If excessive stock has to be removed from the length, it is advisable to feed the tool sideways, making a series of step cuts. Much time can be saved using short step cuts instead of long radial cuts. The feed rate for roughing cuts will vary from 0.005 to 0.015 in. (0.12 to 0.38 mm).

Finishing cuts are made from the center outward, using a feed rate of 0.002 to 0.005 in. (0.05 to 0.12 mm). To assure a smooth finish, the end of the cutting tool should be slightly rounded and honed with an oilstone. The point should be placed against the workpiece so that a small angle is created between the edge of the facing tool and the workpiece end. In this manner the edge of the cutting tool will begin to cut when moving from the center outward. The surface will then be smoothed by the rounded point that completes the depth of cut. The finish face cut can be further enhanced by the use of a little cutting oil.

Depth of Cut

There are two ways of controlling the depth of cut when facing. One method is to use the compound rest so that its slide is set parallel to the lathe ways (Figure 37-1). With this method, the facing tool is at 90° to the lathe axis and is fed straight into the work by turning the compound rest feed handle. The depth of cut is measured directly on the micrometer collar by the number of graduations that are moved past the index line. This value would usually be in thousandths of an inch or millimeters, in half-thousandths of an inch or half-millimeters for 1 : 1 lathes.

The second method has the compound rest set at 30° (Figure 37-2). In this position the depth of cut is exactly half the amount of compound feed. For example, if the compound rest feed handle is turned 0.030 in. (0.75 mm), the depth of cut (side movement) will be 0.015 in. (0.38 mm). On 1 : 1 lathes, the depth of cut would measure half as much. An illustration of this 30-60 compound rest setup is shown in Figure 37-3. The compound rest set at this angle will also provide more working room when working between centers.

The depth of facing cut will vary with the lathe speed and material to be cut. As a general rule, finishing cuts should not exceed 0.015 in. (0.38 mm) in depth. Roughing cuts should not exceed 0.060 in. (1.50 mm) in depth. In either case the carriage should be locked (Figure 37-4) to prevent the facing tool from moving during machining.

Measurement

Semiprecise facing lengths should be layed out with an adjustable square or similar instrument and then indented along this line with center punch marks. Facing should then be carried out until these marks are cut in half.

Semiprecise measurements are made by taking

FIGURE 37-1
Setting the compound rest with its slide set parallel to the lathe ways.

FIGURE 37-2
Setting the compound rest 30° to the right.

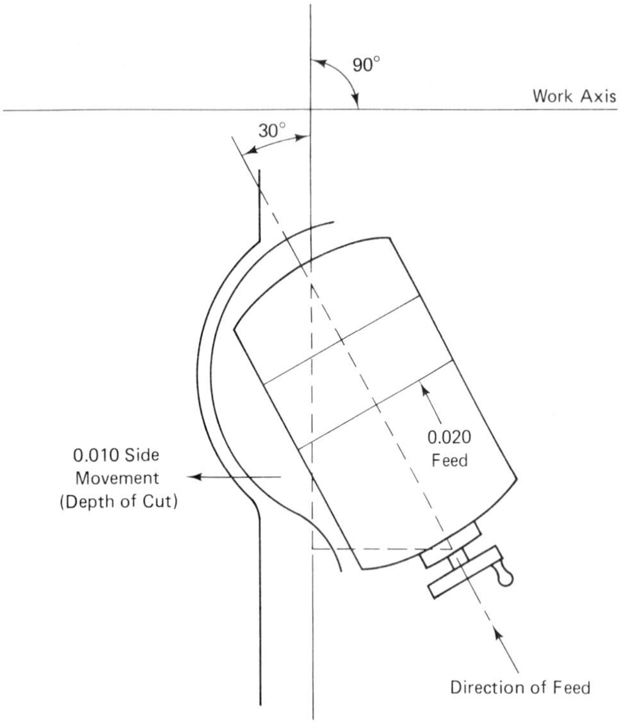

FIGURE 37-3
Feeding the facing tool at a 30° angle advances the depth of cut (side movement) to one-half the amount of feed.

FIGURE 37-4
To prevent the facing tool from moving during machining, the lathe carriage should be locked.

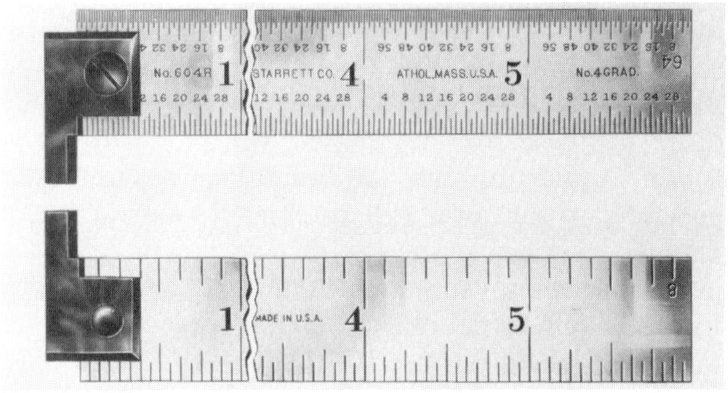

FIGURE 37-5
A hook rule is a good semiprecise measuring tool to use when facing to length. (Credit: L. S. Starrett Company)

trial cuts and then measuring with a hook rule (Figure 37-5). More accurate measurements can be made using a dial caliper or micrometer, spacing permitting.

Facing in a Chuck

Since facing is generally followed by turning, judgment will have to be used as to how far to extend the workpiece beyond the lathe chuck jaws. As a general rule, the workpiece should not extend more than three times its diameter and the workpiece should be securely tightened (Figure 37-6).

Procedure

1. Set the compound rests 30° to the right or parallel to the lathe ways (Figure 37-7).
2. Extend the workpiece from the chuck jaws not more than three times its diameter and tighten it securely.
3. Select a honed right-hand facing tool.
4. Extend the facing tool about $\frac{1}{2}$ in. (13 mm) from a toolholder designed to work from the center outward and tighten the toolholder setscrew (Figure 37-8).
5. Set the toolholder on the left side of the compound rest (Figure 37-9).
6. Snap in a tailstock center.

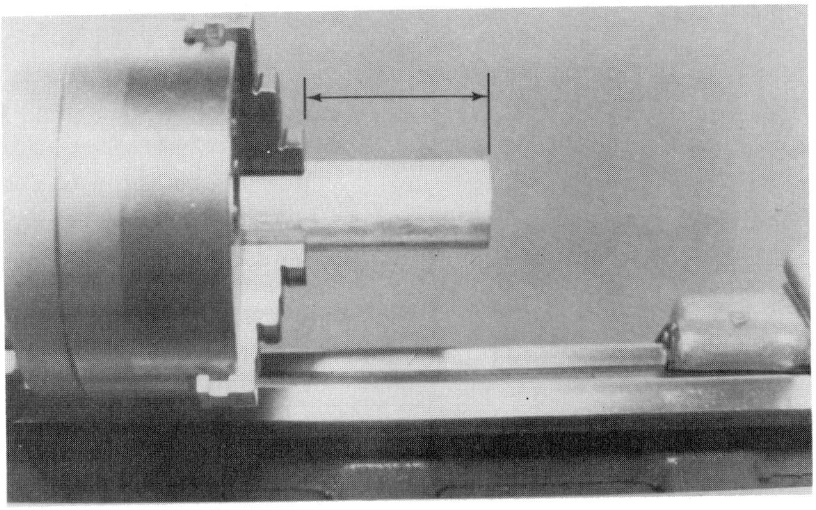

FIGURE 37-6
A workpiece should not extend beyond the chuck jaws more than three times its diameter.

FIGURE 37-8
Extend the facing tool about $\frac{1}{2}$ in. (13 mm) from the toolholder and tighten.

FIGURE 37-7
The facing tool should be set up so that depth of cuts can be controlled by the compound rest.

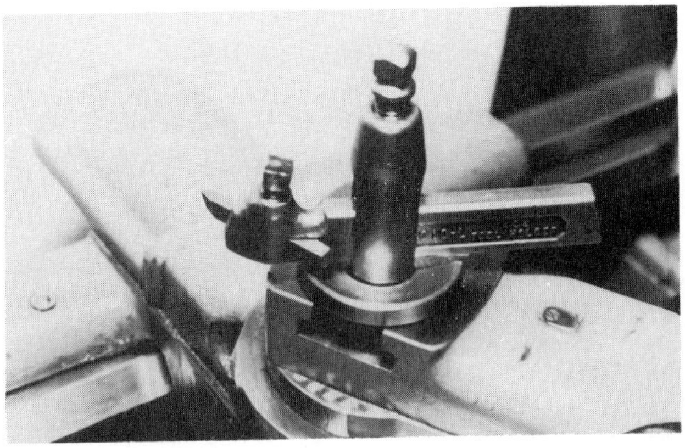

FIGURE 37-9
For less damage to the compound rest, set the toolholder on the left side of the rest.

7. Adjust the toolholder until the point of the facing tool is at the same height as the tailstock center point (Figure 37-10) and lightly tighten the tool post screw.

8. Move the lathe carriage to position the facing tool near the workpiece.

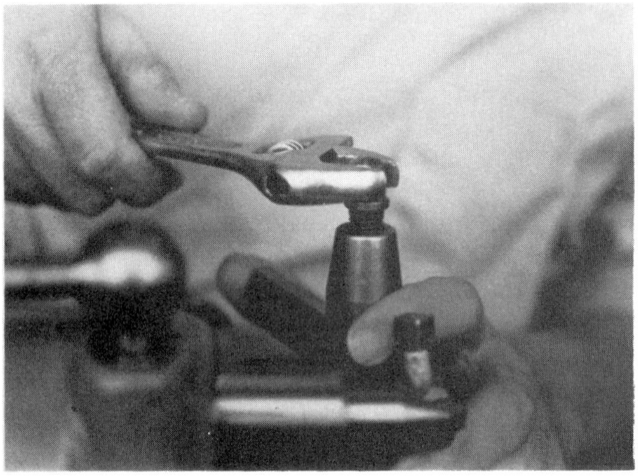

FIGURE 37-10
Adjust the cutting tool's height to match the lathe center's height.

9. Tap the toolholder lightly until the side of the facing tool forms a small angle with the workpiece (Figure 37-11).

10. Tighten the tool post screw securely to keep this position.

11. Lightly tighten the carriage lock about two fingers tight.

12. Set the correct lathe speed for the diameter and type of material to be cut.

13. Start the lathe and feed into the workpiece near the center using the compound rest feed handle.

14. Using both hands, slowly turn the cross-feed handle toward the center until the metal chip has dropped off at the center.

15. Reverse the rotation by slowly feeding away from the center using a steady even turn to complete the facing.

16. Repeat steps 13 to 15 if facing is not completed.

Facing Between Centers

To successfully face between centers, the following conditions must be met: alignment of centers, an undercut half center, proper work setup, lathe speed, and facing tool setup. While roughing cuts may be

FIGURE 37-11
Place the edge of the facing tool at a small angle to the workpiece.

started from the center or circumference and cut in the opposite direction, finishing cuts should be started at the center and fed outward. This will prevent the center hole from being burred in. It will also be found that with the exception of a few steps, the operations will be similar to facing in a chuck.

Procedure

1. Place the compound rest 30° to the right.
2. Snap in a half center in the tailstock spindle with its flat vertical and facing the operator (Figure 37-12).
3. Select a honed right-hand facing tool.
4. Extend the facing tool about $\frac{1}{2}$ in. (13 mm) from a toolholder designed to work from the center outward and tighten the toolholder setscrew.
5. Set the toolholder on the left side of the compound rest.
6. Adjust the toolholder until the point of the facing tool is at the same height as the tailstock center point and lightly tighten the tool post screw.
7. Extend the tailstock spindle $2\frac{1}{2}$ to 3 in. (62 to 75 mm) beyond the tailstock.
8. Fill the right-hand side workpiece hole with antifrictional compound.
9. Mount the workpiece between centers.
10. Lock the tailstock clamping nut.
11. Adjust the tailstock handwheel for proper mounting tension.
12. Turn the apron handwheel to bring the facing tool near the end of the workpiece.
13. Tap the toolholder lightly until the side of the tool forms a small angle with the workpiece.
14. Tighten the tool post screw securely to keep this position.
15. Lightly tighten the lathe carriage lock.
16. Set the correct lathe speed for the diameter and type of material to be cut.
17. Turn the crossfeed handle clockwise to move the facing tool as close as possible to the center of the work.
18. Set the depth of cut using the compound rest feed handle.
19. Using both hands, slowly feed away from the center using a steady, even rate to complete the facing (Figure 37-13).

UNIT 38 TURNING AN ACCURATE DIAMETER

One of the requirements of a machinist is to be able to machine work to size as well as produce a diameter that is uniform and parallel throughout its length. Parallel turnings will depend on machining conditions as well as skill on the part of the operator. If work is to be turned between lathe centers, for example, the centers must be perfectly aligned. Time is also a factor. Time should not be wasted in ma-

FIGURE 37-12
The half center should be inserted with its flat vertical and facing the operator.

FIGURE 37-13
Slowly feed away from the center using two hands.

chining an accurate diameter if the accuracy is not required. While the general practice is to complete a diameter to size in two cuts—a roughing cut and a finishing cut—the operation should be completed with as few cuts as possible.

Objectives
After completing this unit you should be able to

1. understand the depth of cut measurement using the cross-feed micrometer collar.
2. set up a rough turning toolbit for maximum rigidity.
3. complete an accurate diameter using the trial cut method.

Graduated Micrometer Collars

There is a graduated micrometer collar that circles the cross-feed handle and is designed to assist the operator in setting the depth of cut for material removal. Lathes come equipped with micrometer collars that have a 1 : 1 relationship (Figure 38–1) or a 2 : 1 relationship (Figure 38–2). It is the pitch or spacing between the threads of the feed screw that controls the length of movement. All technical explanations aside, when a depth of cut is taken using the crossfeed handle with a 1 : 1 micrometer collar, the workpiece diameter is reduced by the number of graduations advanced past the index line. Essentially, if the micrometer collar is turned five graduations past the reference line, the workpiece diameter is reduced five-thousandths of an inch (0.005 in.). On a metric lathe the 1 : 1 micrometer collar is graduated in measurements of two-hundredths of a millimeter (0.02 mm). If the micrometer collar has a 2 : 1 relationship, as is the case with many of the older lathes, a depth of cut feed represented by the movement of five graduations will reduce the workpiece diameter ten-thousandths of an inch (0.010 in.). On a metric lathe the material removed from the diameter by the advancement of five graduations would be 5 × 0.02 mm × 2 or 0.20 mm.

Setup of Cutting Tools

Rigidity and safety must be considered in setting up cutting tools in a lathe. Generally, if the toolbit is used to cut toward the headstock, a straight or left-hand toolholder should be used. The toolbit should not extend more than about $\frac{1}{2}$ in. (13 mm) beyond the toolholder and be clamped tightly (Figure 38–3).

The length that the cutting edge extends beyond the center of the tool post is important, too. Allowing the toolholder to extend too far is a poor setup. This may permit spring and vibration that will cause chattering and lead to inaccurate machining. What, then, is a good setup? The expression in the machine shop is to "catch the tool short." If the toolholder can be tightened by placing a finger's width between the holder's screw and the post, not only will it provide a rigid setup but it allows room for tightening the toolholder's clamping screw (Figure 38–4).

FIGURE 38–1
Lathe crossfeed micrometer collar indicating a 1 : 1 feed.

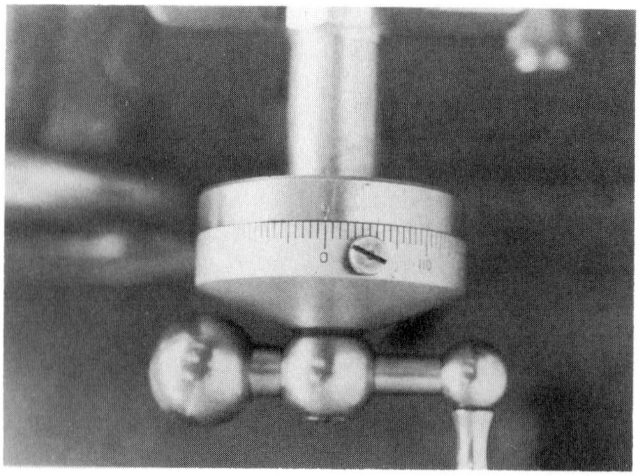

FIGURE 38–2
Lathe crossfeed micrometer collar indicating a 2 : 1 feed.

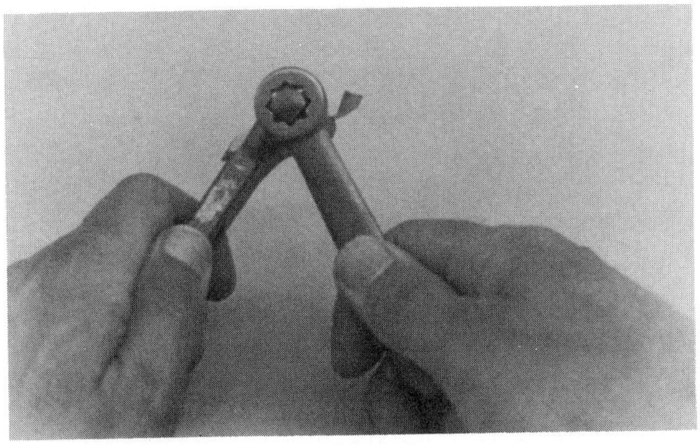

FIGURE 38–3
The toolbit should not extend more than $\frac{1}{2}$ in. (13 mm) beyond the toolholder and be tightened securely.

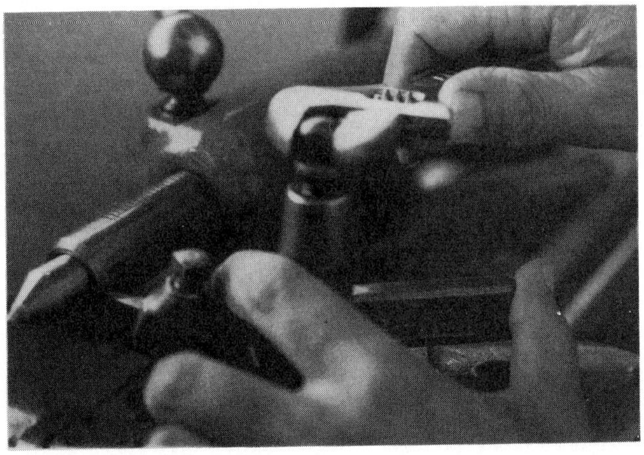

FIGURE 38–4
Allow a finger's width between the toolholder's screw and the tool post.

The other safety factor to consider is the position of the tool post. As a rule, the tool post should be placed at the left-hand end of the T-slot in the compound rest. Additionally, the dovetail slide of the compound rest should not be exposed more than it has to be (Figure 38–5). It should be covered by the upper portion of the compound rest as much as possible. Doing so will prevent the dovetail slide from being chewed up by the hardened jaws of a chuck or damaged by the rotation of a lathe dog.

FIGURE 38–5
The dovetail of the compound rest should not be exposed more than it has to. It is bad machining practice and can lead to being chewed up by hardened chuck jaws.

Taking a Trial Cut

The key to accurate machining is control. One of the methods used in controlling any cut is to take a trial cut. The machining is done on a very short distance on the end of the workpiece and is measured before the cut is allowed to run its course. There is no need for anyone operating a lathe to discover an undercut has been made in completing the cut. With a trial cut the operator can either accept the depth of cut undertaken or the cutting tool can be reset to cut deeper or shallower. The trial cut method is followed in sound machining, whether to establish the initial cut or complete the final cut.

Procedure

1. Appropriate safety glasses must be worn before doing any machining on a lathe.
2. Set up the workpiece and cutting tool for turning.
3. Set the lathe for proper speed and feed for the material being machined.
4. Start the lathe spindle.
5. Move the carriage to position the cutting tool about $\frac{1}{8}$ in. (3 mm) from the end of the workpiece.
6. Turn the crossfeed handle clockwise to feed the toolbit into the work to lightly groove the surface (Figure 38–6).

FIGURE 38-6
Lightly groove the work surface to establish the crossfeed setting.

FIGURE 38-8
Take a trial cut for $\frac{1}{4}$ in. (6 mm) along the length of the work.

FIGURE 38-7
Do not touch the crossfeed handle until the trial cut has to be taken.

FIGURE 38-9
Measure the trial cut diameter to calculate the amount of material that remains to be removed.

7. *Do not move the crossfeed handle setting* (Figure 38-7) and move the carriage to the right so that the toolbit clears the end of the workpiece.

8. Turn the crossfeed handle clockwise to remove 0.010 (0.25 mm) from the workpiece diameter.

9. Advance the carriage in automatic feed to take a trial cut for a $\frac{1}{4}$ in. (6 mm) along the length of the work (Figure 38-8).

10. Disengage the automatic feed and stop the lathe.

11. Turn the carriage handwheel to clear the toolbit past the end of the work.

12. Measure the trial cut diameter using a micrometer or similar instrument and calculate the amount of material that remains to be removed (Figure 38-9).

13. Reset the cutting tool as required.

14. Take another trial cut $\frac{1}{4}$ in. (6 mm) long and stop the lathe.

15. Measure the diameter and reset depth of cut if necessary.

16. If accurate, machine the workpiece to length.

Rough Turning

Turning on a lathe is done in two basic operations: rough and finish cutting. Rough turning is intended to reduce the workpiece diameter quickly. The lathe operation is geared to efficiently reduce the diameter so that the workpiece is left slightly larger than the finish size, about $\frac{1}{32}$ in. (0.80 mm) larger. Depth of cuts should be made as deep as the machine, material, and cutting tool will take. While normally one roughing cut is taken from the workpiece, excessive material removal, such as $\frac{1}{2}$ in. (13 mm) or greater, may require two or three cuts. Lathe speeds are set according to the diameter and material being used. Roughing cuts are set to move the cutting tool from 0.007 to 0.015 in. (0.17 to 0.38 mm) per revolution.

For general rough turning, the toolbit should be set on center or slightly above center. The toolbit should be positioned with a lead angle so that the edge of the cutting tool contacts the workpiece first before the point does (Figure 38-10). Once the toolbit has been properly set against the workpiece, rough turning a diameter can be carried out.

Procedure

1. Mount a general-purpose turning tool in a straight toolholder or left-hand toolholder.
2. Position the toolholder to the left side of the tool post.
3. Lightly tighten the toolholder a finger's width beyond the tool post and approximately even with or slightly above lathe center (Figure 38-11).
4. Adjust the toolholder so that the edge of the toolbit forms a lead angle with the workpiece and secure this position as tight as possible with two hands.
5. Set the lathe speed for the material and diameter being cut (see Table 35-1).
6. Set the quick-change gear box for a rough-cut feed, approximately 0.007 to 0.015 in. (0.17 to 0.38 mm) per revolution.
7. Mark off the length to be turned and set the point of the toolbit on this mark.
8. Tighten a lathe stop against the carriage at this length (Figure 38-12).

FIGURE 38-11
Position the point of the toolbit even with or slightly above lathe center.

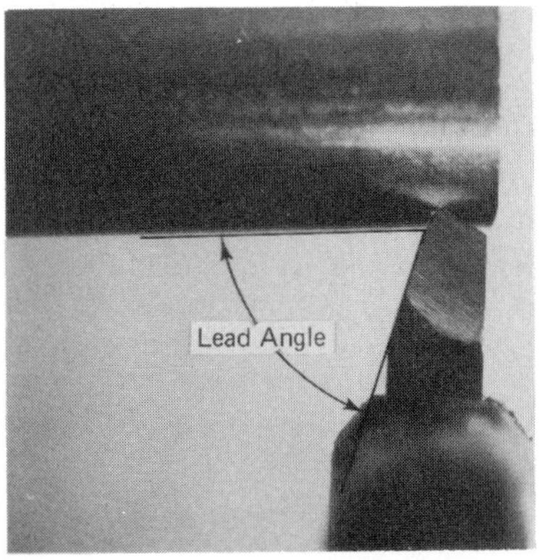

FIGURE 38-10
Position the rough turning tool so that a lead angle is formed between the cutting edge and the workpiece.

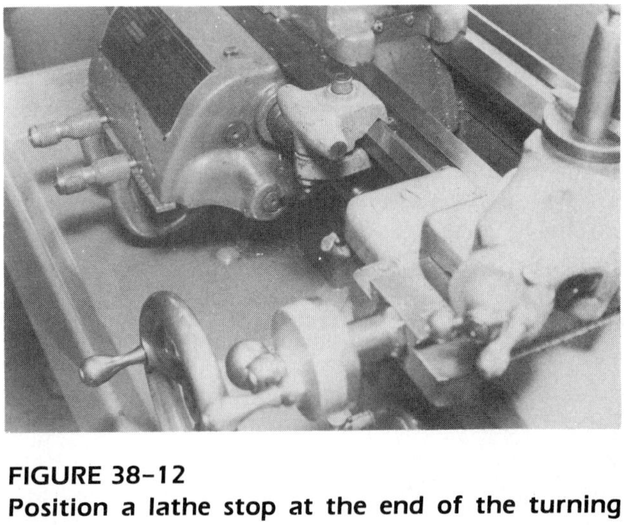

FIGURE 38-12
Position a lathe stop at the end of the turning length.

9. Back off the cutting tool by turning the crossfeed handle counterclockwise.

10. Turn the carriage handwheel to position the toolbit to the right and approximately $\frac{1}{8}$ in. (3 mm) near the end of the workpiece.

11. Start the lathe spindle.

12. Turn the crossfeed handle clockwise to lightly groove the rotating workpiece.

13. Move the carriage to the right so that the toolbit point clears the workpiece by about $\frac{1}{4}$ in. (6 mm).

14. Turn the crossfeed handle clockwise approximately 0.010 in. (0.25 mm).

15. Engage the lathe in automatic feed to take a trial cut about $\frac{1}{4}$ in. (6 mm) long.

16. Disengage the automatic feed and stop the lathe. (*Do not move the crossfeed handle setting.* It has now been established as a starting point for any depth of cut to be taken.)

17. Turn the carriage handwheel until the cutting tool clears the right-hand end of the workpiece.

18. Measure the diameter of the trial cut and calculate how much material must be removed to leave approximately $\frac{1}{32}$ in. (0.8 mm) for a finish cut.

19. Turn the crossfeed handle clockwise to remove the desired amount on the micrometer collar.

20. Take another trial cut.

21. Check the diameter, readjust the depth of cut if necessary, and take the roughing cut for the required length.

Finish Turning

Finish turning should not be attempted unless the workpiece is near room temperature. In taking roughing cuts, there is a tendency to overheat and cause the diameter of the workpiece to grow larger. If a finish cut is taken in this state, the diameter can change in size, depending on the material, by as much as 0.002 in. (0.05 mm) on a 1-in. (25 mm) diameter piece of work when it cools off. There is also a tendency that the workpiece surface may work harden, which will require a deeper trial cut to penetrate it.

Finish cuts are turned at higher speeds and finer feeds than are used for rough turnings. The toolbit should also be sharp and have a radius on the point to smooth out the work surface. There will be times when a roughing tool will remain sharp enough to be repositioned and used as a finish tool.

Procedure

1. Insert and tighten a finish turning tool in the toolholder.

2. Mount the toolholder in the tool post and tighten it with the toolbit at lathe center height.

3. Set the lathe speed and feed for finish turning. The speed can be as much as $\frac{1}{3}$ higher than the speed used for rough turning, depending on the material. A feed of 0.003 to 0.005 in. (0.07 to 0.12 mm) per revolution is recommended (Figure 38–13).

4. Mount the workpiece.

5. Establish a lathe stop.

6. Move the carriage to the right-hand end of the workpiece and prepare to take a trial cut.

7. Start the lathe spindle.

8. Turn the crossfeed handle clockwise to lightly groove the rotating workpiece.

9. Move the carriage to the right until the toolbit clears the end of the workpiece.

10. Turn the crossfeed handle clockwise approximately 0.015 in. (0.36 mm) (Figure 38–14).

11. Engage the lathe in automatic feed to take a trial cut about $\frac{1}{4}$ in. (6 mm) long.

12. Disengage the automatic feed and stop the lathe. (*Do not move the crossfeed handle.* It has now been established as a depth of cut starting point.)

13. Turn the carriage handwheel until the cutting tool clears the right-hand end of the workpiece.

FIGURE 38–13
A finish feed of 0.003 to 0.005 in. (0.07 to 0.12 mm) per revolution is recommended.

14. Measure the trial cut diameter and calculate how much material remains to be removed for a finish cut diameter.

15. Turn the crossfeed handle clockwise to remove the desired amount on the micrometer collar.

16. Start the lathe and engage it in automatic feed to take another trial cut about $\frac{1}{4}$ in. (6 mm) long.

17. Stop the lathe and recheck the diameter for size.

18. If the diameter is to the finish size, complete the cut to its required length.

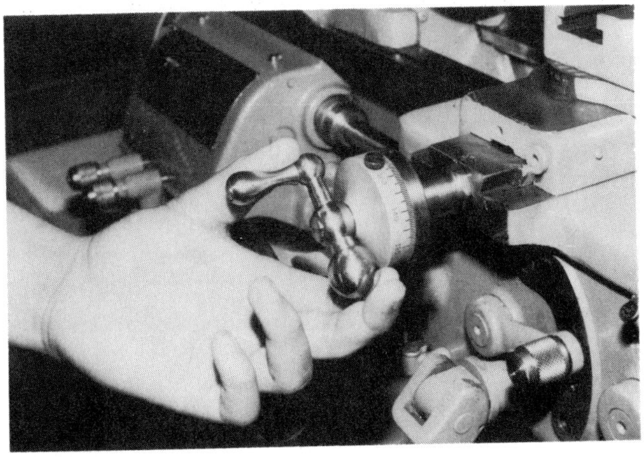

FIGURE 38-14
To penetrate a surface that has been work hardened, a trial cut of 0.015 in. (0.12 mm) deep is recommended.

Shoulder Turning

Shoulder turning is similar to facing. The difference lies in definition. While facing involves machining the end of the workpiece, shouldering involves the machining that begins as one diameter and suddenly increases in overall size. Examples of common types of shoulders can be seen in Figure 38-15. A regular turning tool may be used for both rough and finish cuts on longer shoulder cuts. On short shoulder lengths, the turning tool is used to finish the entire length.

Once a frame of reference is established, such as the face, shouldering of lengths or steps can be accomplished in either of two ways. One way is to layout the required shoulder length with a scriber and adjustable square, center punch the mark, and finish machining the shoulder by splitting this line. A more precise method involves the use of a micrometer carriage stop (Figure 38-16), which has a movable spindle that can be adjusted to 0.001 in. (0.02 mm) movement.

Procedure to Machine a Shoulder Square

1. Upon completing the facing operation, layout the shoulder length required by using a scriber and adjustable square or similar measuring tool.

2. Lightly groove this mark with the point of a facing tool.

3. Slide the micrometer carriage stop so that its extended spindle butts against the lathe carriage and lock it in place.

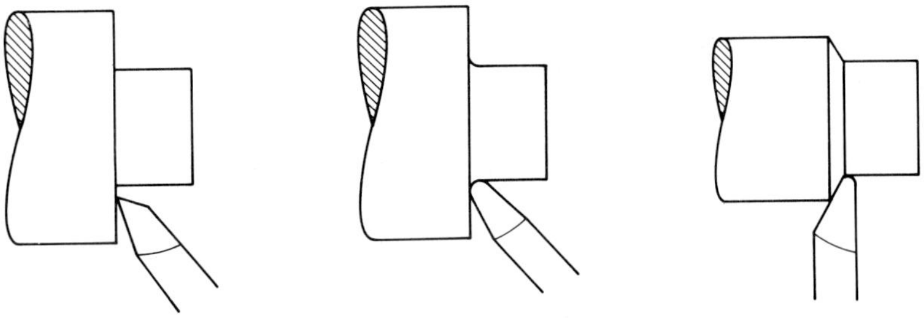

FIGURE 38-15
Three common types of shoulders. (Credit: Paul Rapisarda)

FIGURE 38-16
A micrometer lathe stop will permit carriage movements of very small measurements.

4. Move the carriage handwheel so that the toolbit clears the workpiece and prepare to rough and finish the small diameter. (*Note:* The facing tool need not be changed if the point is sharp and contains a small radius.)

5. Turn the crossfeed handle clockwise to set the rough depth of cut.

6. Start the lathe and engage the power feed lever.

7. Disengage the power feed within $\frac{1}{8}$ in. (3 mm) of the stop and, by hand, carefully cut within $\frac{1}{32}$ in. (0.80 mm) of the finish length.

8. Move the carriage handwheel to position the toolbit clear of the workpiece.

9. Advance the crossfeed handle to take the finish cut and note the micrometer collar reading.

10. Engage the power feed.

11. Disengage the power feed within $\frac{1}{8}$ in. (3 mm) of the stop and, by hand, carefully cut until the carriage lightly butts against the stop.

12. Face the shoulder by turning the crossfeed handle counterclockwise, using two hands, so that the facing tool cuts from the center to the outside.

13. If the shoulder has not been cut to length, return the crossfeed handle to the same setting as in step 9 or adjust the micrometer carriage stop by turning the knurled nut. Carefully butt the carriage against the stop and machine the shoulder as in step 12.

Machining a Filleted Shoulder

There are times when shoulders will have to be machined so as to add increased strength to the part. Filleted or radius shoulders are often machined on parts that will undergo heat treating. The rounded surface does much to avoid cracks and fracturing at this point.

Procedure

1. Layout the shoulder length required by using a scriber and adjustable square or similar measuring tool.

2. Lightly groove this mark with the point of a facing tool.

3. Slide the micrometer carriage stop so that its spindle butts against the lathe carriage and lock it in place.

4. Move the carriage handwheel so that the toolbit clears the workpiece.

5. Turn the micrometer stop's knurled nut to extend the spindle the radius to be cut. For example, if the radius to be cut is 0.060 in. (1.50 mm), extend the spindle 0.060 in. (1.50 mm).

6. Prepare to rough and finish cut the small diameter.

7. Mount and tighten the correct radius toolbit; set it to center.

8. Loosen the micrometer carriage stop and slide it to the left.

9. Set the lathe to run at one-half the turning speed.

10. Apply layout dye to the small diameter as close as possible to the unfinished shoulder.

11. Start the lathe and feed the radius toolbit until it just removes the layout dye.

12. Slowly turn the carriage handwheel with both hands until the tool starts to cut the unfinished shoulder.

13. Slide the micrometer stop so that its spindle butts against the carriage and lock it in place.

14. Turn the micrometer stop's knurled nut to back out the spindle to one-half the amount of the radius being cut. [For an 0.060-in. (1.50-mm) radius, back it out 0.030 in. (0.75 mm).]

15. Use cutting fluid.

16. Turn the carriage handwheel slowly with both hands to feed the cutting tool sideways to cut the shoulder.

17. Repeat steps 15 and 16 until the shoulder has been cut to length.

Beveled Shouldering
Angular shoulders are produced for a number of reasons, such as eliminating weight, sharp edges, or similar handling problems. They are also produced to improve appearance, increase strength and reduce size.

In contrast to the long angular shoulder, which is cut by using the point of a turning tool fed in a compound rest set to the required angle, the short beveled or angular cut can be produced by using the edge of a cutting tool. The cutting edge of the toolbit is set to a required angle by a protractor (generally 30° to 60°, usually 45°) and held against the rotating workpiece until the desired angle has been cut. If the material proves too difficult to machine or chatter develops, it is recommended that the compound rest method be used to produce the beveled cut.

Procedure for Beveling a 45° Shoulder

1. Layout the length of the small diameter.
2. Rough and finish turn the small diameter to size.
3. Mount and tighten a side cutting tool in the toolholder.
4. Set the cutting tool to center and lightly tighten it in the tool post.
5. Using the head of an adjustable square against the workpiece, tap the side of the toolholder with a tool post wrench until the side of the cutting edge of the toolbit matches the 45° angle.
6. Tighten the tool post screw securely.
7. Apply layout dye to the small diameter as close as possible to the shoulder.
8. Start the lathe spindle.
9. Turn the crossfeed handle until the point of the cutting tool barely removes the layout dye.
10. Using two hands, turn the carriage handwheel to slowly feed the cutting tool against the shoulder.
11. Apply cutting fluid and continue feeding the edge of the cutting tool using two hands until the required size has been machined.

UNIT 39 ALIGNING LATHE CENTERS

It is important that the headstock and tailstock be in alignment when working between centers in order to be able to machine parallel diameters. If the alignment is not exact, the resultant work will be tapered, with different diameters on both ends of the workpiece. The misalignment of the tailstock will also lead to other machining problems, such as not being able to drill a hole in a lathe with any degree of accuracy. It is with the elimination of this misalignment that this unit will deal.

Objectives
After completing this unit you should be able to

1. align the lathe centers using a visual method.
2. align the lathe centers using the trial cut method.
3. align the lathe centers using a test bar and a dial indicator.

Construction of the Tailstock

Four methods are generally used for aligning lathe centers. The key to aligning the center axis of the lathe centers lies with the construction of the tailstock. It is constructed of two halves, the baseplate and the tailstock body (Figure 39-1). The whole construction is tightened in place to the lathe ways by a clamping nut. The tailstock body is designed to

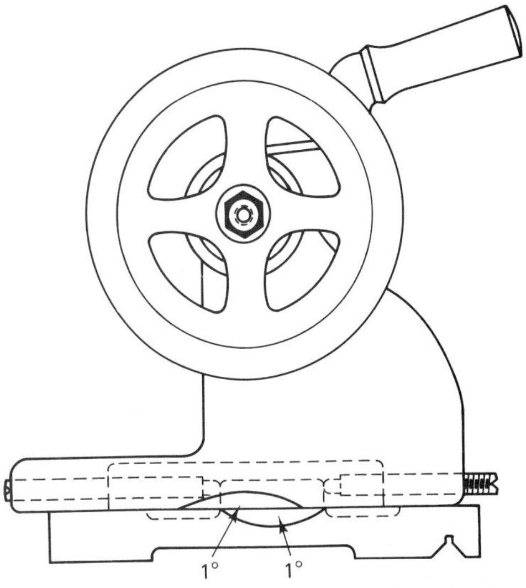

FIGURE 39-1
The tailstock is constructed of two sections: the tailstock body and base.

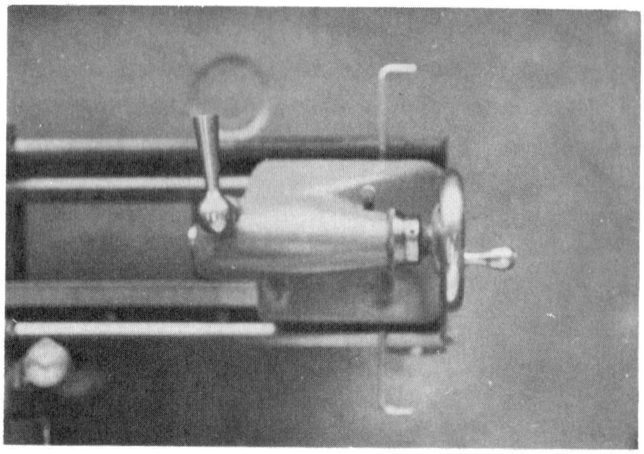

FIGURE 39-2
Adjustment screws permit the tailstock body to move laterally.

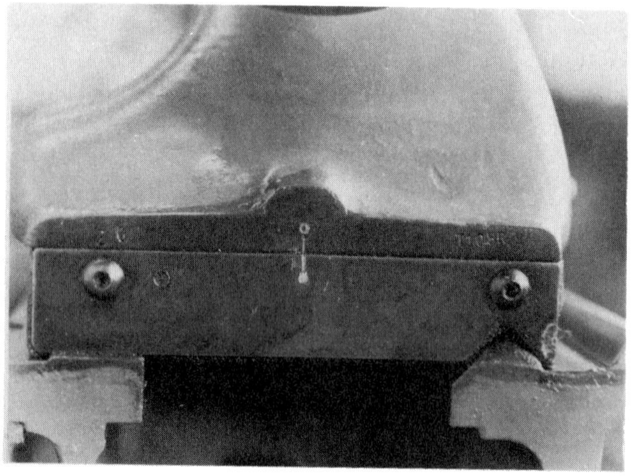

FIGURE 39-3
Matching lines help to visually align the lathe centers.

move laterally, toward or away from the operator. It is like being able to slide the top slice of bread on a sandwich. Adjustment screws on either side of the tailstock body make lateral movements possible (Figure 39-2). Once the tailstock clamp nut is loosened, one adjustment screw is loosened while the opposite side screw is tightened. It will be found that the tailstock body moves in the opposite direction by tightening the adjustment screw.

Due to the adjustments required in some of the operations that follow, one must be careful not to overloosen or overtighten adjustment screws or the tailstock spindle binding lever. The term "snug-up"

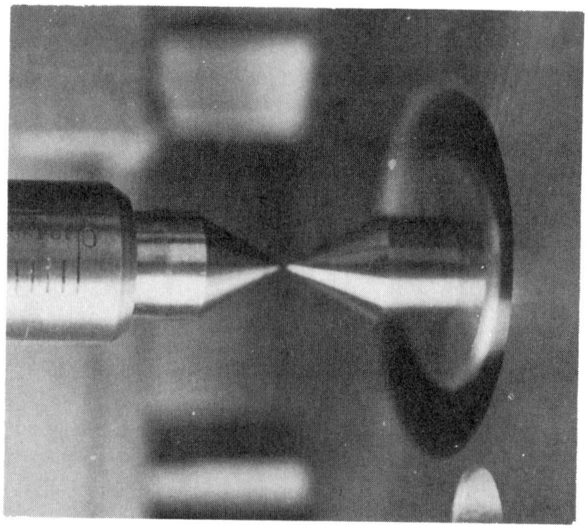

FIGURE 39-4
Positioning the centers about $\frac{1}{16}$ in. (1.6 mm) apart will help to visually align them.

means exactly that. Overtightening a tailstock spindle binding lever will raise the dead center in the tailstock spindle. This will cause a continuous error in measurements. Overloosening and then taking up the slack with the adjustment screw will cause a wide shift in lateral movement. Experience will show that snugging-up might mean to use light hand pressure, such as pulling on a wrench or lever with two fingers.

Visual Tailstock Alignment

The back of some tailstocks are graduated with two or more lines for visual alignment. Generally, one line is located on the tailstock body and one line is located on the base. When the two lines match so as to form one continuous line, the tailstock is considered to be visually aligned (Figure 39-3). Another visual alignment method is to physically match the headstock and tailstock centers, leaving a small gap between the points when matching them (Figure 39-4). Since either method is a general indication, it is not very accurate and should not be used for precision machining.

Procedure

1. Loosen the tailstock clamping nut (Figure 39-5).
2. Snug-up the clamping nut by hand (Figure 39-6).

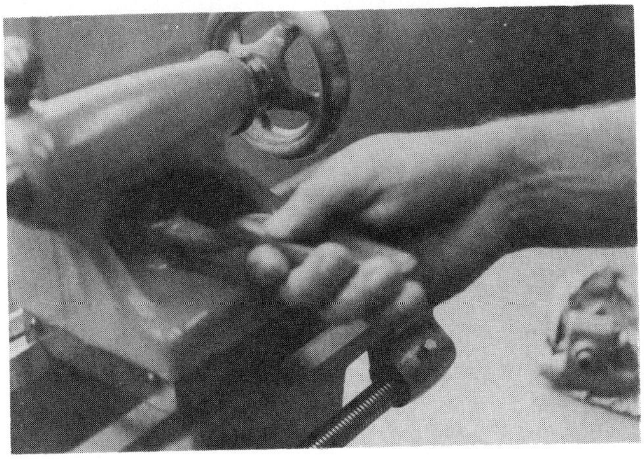

FIGURE 39–5
To begin tailstock adjustment, loosen the tailstock clamping nut.

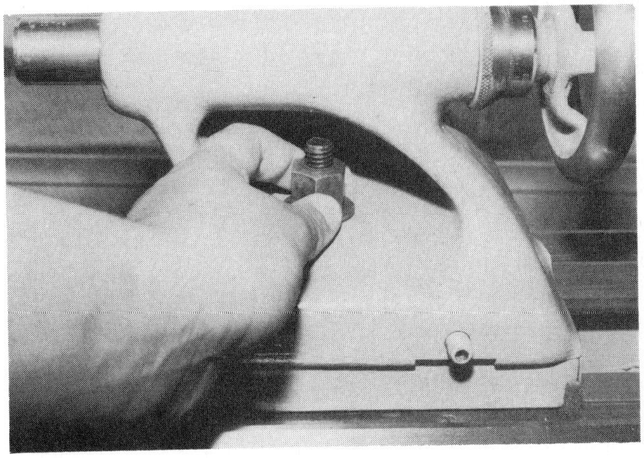

FIGURE 39–6
Hand tighten the tailstock clamping nut before making any more adjustments.

3. Loosen the adjustment screw opposite the direction of movement required for alignment (Figure 39–7).
4. Tighten the opposite side adjustment screw (Figure 39–8).
5. Repeat the process until the mating lines match up.
6. Lock the loosened adjustment screw to hold the position.
7. Check to see that the opposite adjustment screw has no play.

Trial Cut Method of Alignment

The trial cut method involves taking two cuts between centers. The size and length of the stock is not important as the method can be normally done on any workpiece being machined. Having set a turning tool as previously explained in Unit 38, proceed as follows:

1. Wear appropriate safety glasses before attempting machining in the lathe.
2. Start the lathe spindle.
3. Using the automatic feed, take a trail cut from the workpiece at the tailstock end, machining a true diameter about $\frac{1}{4}$ in. (6 mm) long.
4. Disengage the automatic feed and stop the lathe.
5. Note the reading on the crossfeed micrometer collar.

FIGURE 39–7
Adjustment screws on either side of the tailstock body can be loosened at the same time.

FIGURE 39–8
Tightening the adjustment screw affects the tailstock lateral movement.

6. Back off the cutting tool from the workpiece by moving the crossfeed handle counterclockwise a few turns.

7. Move the carriage to position the cutting tool about 1 in. (25 mm) away from the lathe dog.

8. Start the lathe spindle.

9. Move the crossfeed handle to the same micrometer collar reading as it was when completing the initial trial cut (Figure 39–9).

10. Using the automatic feed, cut a length about $\frac{1}{2}$ in. (13 mm) long.

11. Stop the lathe and measure both diameters with a micrometer or similar instrument.

12. If the diameters are the same, the lathe centers are in line. If the diameters are different, move the tailstock one-half the measured difference between the diameters as indicated in (a) or (b):

(a) Move the tailstock toward the operator if the diameter at the tailstock end is larger.

(b) Move the tailstock away from the operator if the diameter at the tailstock is smaller.

13. Recut the two sections again and if the diameters are still different when measured, readjust the tailstock (Figure 39–10).

Test Bar and Dial Indicator Method

The test bar method is the most accurate method as it uses a dial test indicator to show any error there is in misalignment.

Procedure

1. Mount a parallel test bar between centers, allowing no end play.

2. Snug-up the tailstock spindle binding lever using light hand pressure (Figure 39–11).

3. Tighten a dial test indicator in the tool post (Figure 39–12).

4. Position the test indicator's contact point on the center of the test bar (Figure 39–13).

5. Turn the crossfeed handle until the contact

FIGURE 39–10
Recut the two sections again if the taper correction is not made.

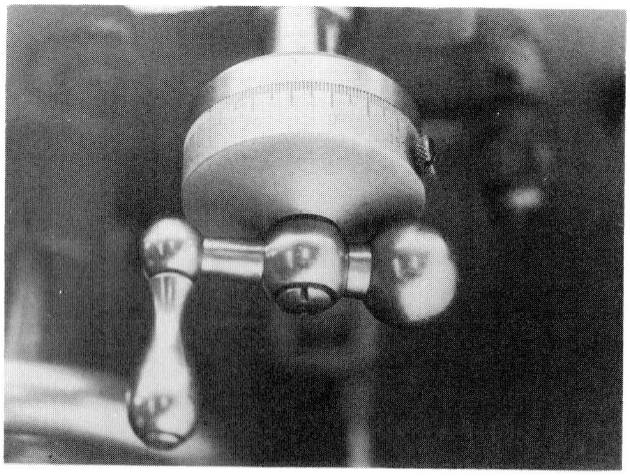

FIGURE 39–9
The depth of cut reading for the headstock should be the same as was taken on the tailstock end.

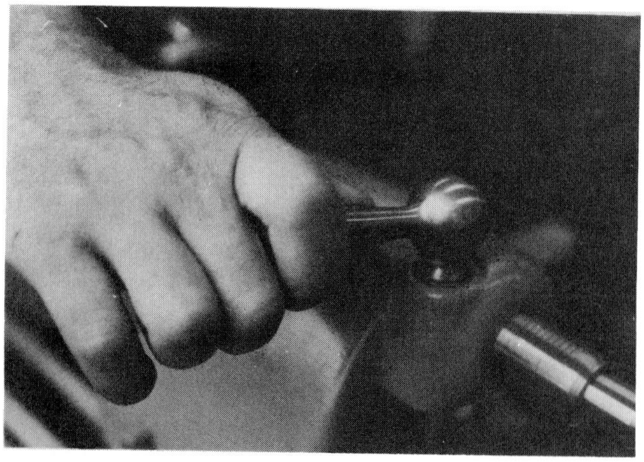

FIGURE 39–11
Snug-up the binding lever with light hand pressure.

point registers against the tailstock end of the bar, no more than $\frac{1}{4}$ of a revolution, and set the dial face on zero.

6. Turn the carriage handwheel to move the indicator across the length of the workpiece (Figure 39-14).

7. If the reading remains the same, the lathe centers are in alignment (Figure 39-15). If there is a difference in reading, the tailstock will have to be moved according to directions indicated in (a) or (b):

(a) If the reading starts to climb, the total reading at the end of the workpiece is the amount the tailstock will have to be moved toward the operator. Place the indicator contact point against the tailstock end of the test bar with a low reading registering and proceed to adjust the tailstock as previously explained.

(b) If the dial test reading starts to fall off when moving toward the headstock, place the indicator contact point to register a low reading against the headstock end. Move the carriage handwheel to slowly advance the indicator toward the tailstock. The total difference between the ends is the amount the tailstock will have to be moved away from the operator.

FIGURE 39-12
Mount a dial test indicator in the tool post.

FIGURE 39-13
Position the indicator on the tailstock end of the workpiece.

FIGURE 39-14
Slowly move the indicator along the length of the work.

FIGURE 39-15
If there is no difference in reading at both ends, the lathe centers are in alignment.

UNIT 40 MOUNTING WORK BETWEEN CENTERS

In order to machine work between lathe centers, the workpiece will have to be mounted so that a drive plate can drive the lathe dog clamped assembly. This would mean that not only are holes in the ends of the workpiece to be center drilled to proper depth, but a proper bearing and heat resisting center must be provided for the tailstock end hole. A hardened dead center using an antifrictional compound or a revolving ball bearing center is generally used in this end. It is the hardened point, ball bearing center that offers the greatest heat-resistant bearing support.

Objectives
After completing this unit you should be able to

1. prepare a lathe for working between centers.
2. set up a workpiece for machining between centers.

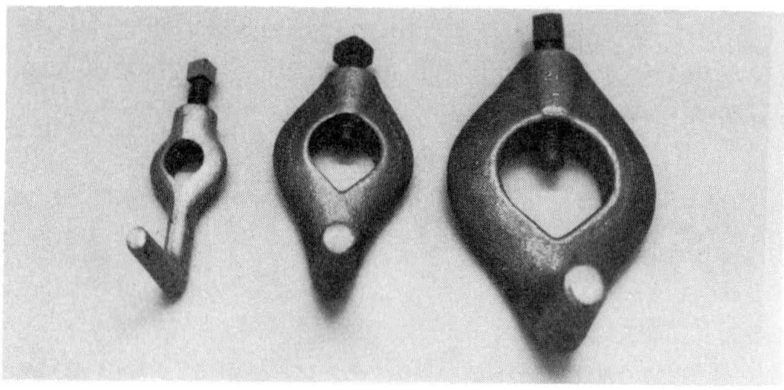

FIGURE 40–1
Different size lathe dogs.

FIGURE 40–2
A lathe dog featuring a safety headless setscrew.

The Pros and Cons to Machining Between Centers

There are advantages and disadvantages to machining between lathe centers. On the positive side the work is supported on both ends, permitting the operator to take heavier cuts. Another advantage is being able to remachine a workpiece accurately once it has been removed from the lathe. This can only happen as long as the centers are not disturbed. Cutting an unfinished thread that has been removed for inspection is a good example. Still another advantage is being able to turn a shaft end for end to complete a concentric machining operation.

One of the principal disadvantages is not being able to use a cutoff tool to part the workpiece between centers. If attempted, the cutoff tool would bind and break, ruining the workpiece as well.

Lathe Accessories

A lathe dog is a clamping device fastened to a workpiece that is to be machined between centers. It will turn the work when its bent tail is inserted into a hole or slot in the drive plate. Lathe dogs come in different sizes (Figure 40–1) and with different types of screws. Many feature the square head setscrew that is turned by a wrench, while others contain the headless safety setscrew (Figure 40–2). It is advisable that when fastening a lathe dog to a workpiece that a small piece of copper or soft brass be placed under the hardened setscrew to prevent marring or damaging the surface of the workpiece.

The clamp lathe dog adjusts with a set of screws and may be used to hold round, square, hexagonal, or rectangular shapes. In selecting the size of a lathe dog, it is best to use the smallest size that will fit the workpiece.

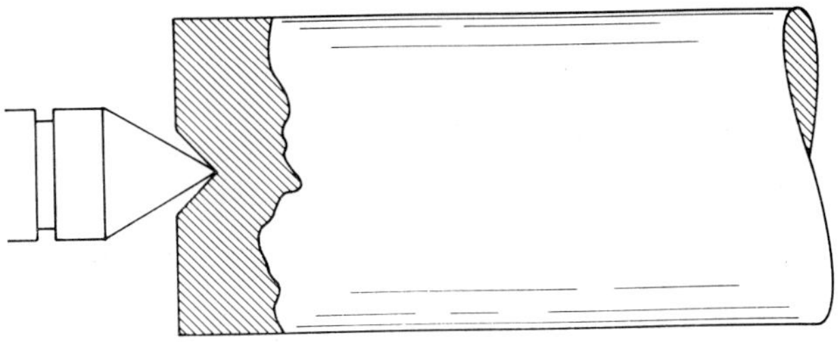

FIGURE 40-3
Shallow center drilling will give no bearing support.

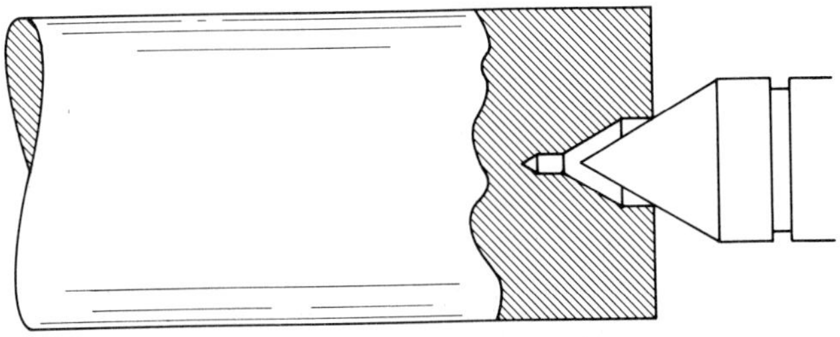

FIGURE 40-4
Center drilling too deep will give a false bearing support.

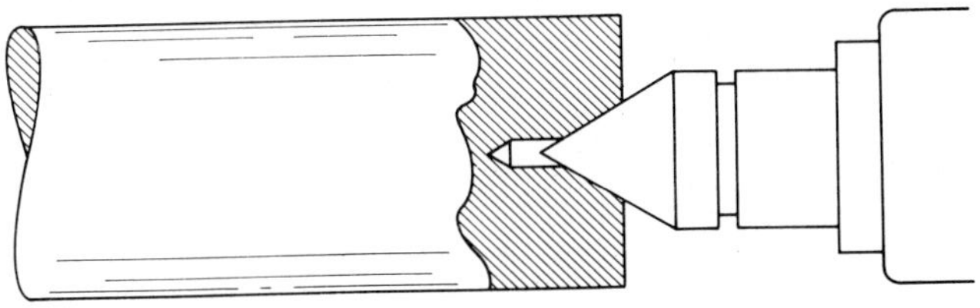

FIGURE 40-5
Correct bearing support.

The drive plate provides the torque against the lathe dog when working between centers. The drive plate is smaller and is limited in number of slots or holes within which the lathe's dog tail is inserted and differs from a face plate. Face plates are provided with a series of holes and/or slots so that work of varying shapes and sizes may be bolted or strapped to it for drilling or machining purposes. Because machining may cause the workpiece to shift weight, counterweights generally have to be clamped to the faceplate to provide a balanced turning.

FIGURE 40-6
Power should be disconnected before attempting to clean a lathe.

Work Preparation

The depth of a center-drilled hole is an important consideration in working between centers. Unless proper bearing contact between the hole and the center is made and maintained, the results can be ruinous to both the center and the workpiece. Too shallow in center drilling will result in little or no support (Figure 40-3). Too deep a hole will also result in a false bearing and will not allow the 60° taper on the hole and center to make contact (Figure 40-4). Figure 40-5 shows the correct depth that will allow full support and contact between the mating surfaces. Unless there is a specified diameter that the center-drilled opening should be, rule of thumb suggests drilling to a depth equal to $\frac{3}{4}$ the length of the slope of the center drill being used. This entry should assure proper support and contact of mating surfaces.

Lathe Preparations

Since work will be mounted between 60° center points, all bearing surfaces of the spindle openings and centers should be examined, cleaned, and tested for accuracy. For safety reasons all cleaning of spindle openings should be done while the lathe has been disconnected from the electrical source to assure no accidental running while cleaning (Figure 40-6).

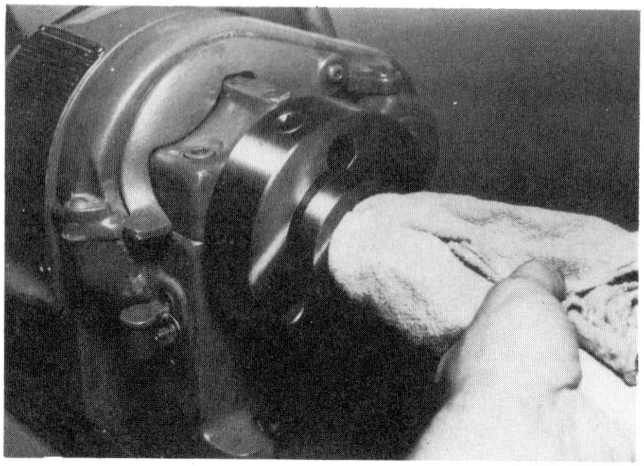

FIGURE 40-7
Because sharp chips can accumulate, never clean the headstock spindle by hand.

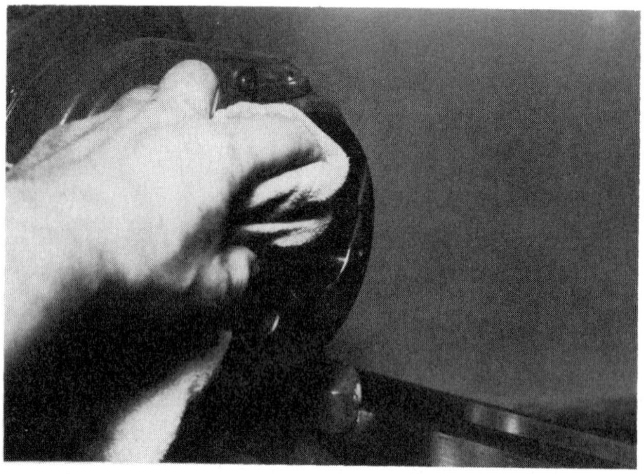

FIGURE 40-8
Clean the headstock spindle nose before mounting the drive plate.

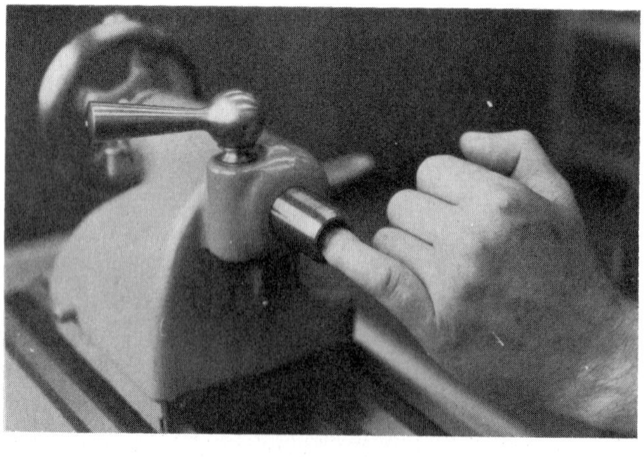

FIGURE 40-9
Clean the tailstock spindle taper.

Procedure

1. Clean the headstock spindle taper by pushing through a cloth wrapped around a stick (Figure 40-7).
2. Clean the headstock spindle nose (Figure 40-8).
3. Clean the tailstock spindle taper (Figure 40-9).
4. Examine the lathe centers for burrs and sharp edges (Figure 40-10).
5. Use an oilstone to remove raised surfaces on the hardened centers, a file on soft centers.
6. Insert the drive plate in the headstock and lock in place (Figure 40-11).
7. Insert the headstock center with a hard quick push.
8. Insert the hardened tailstock center in the same manner (Figure 40-12).

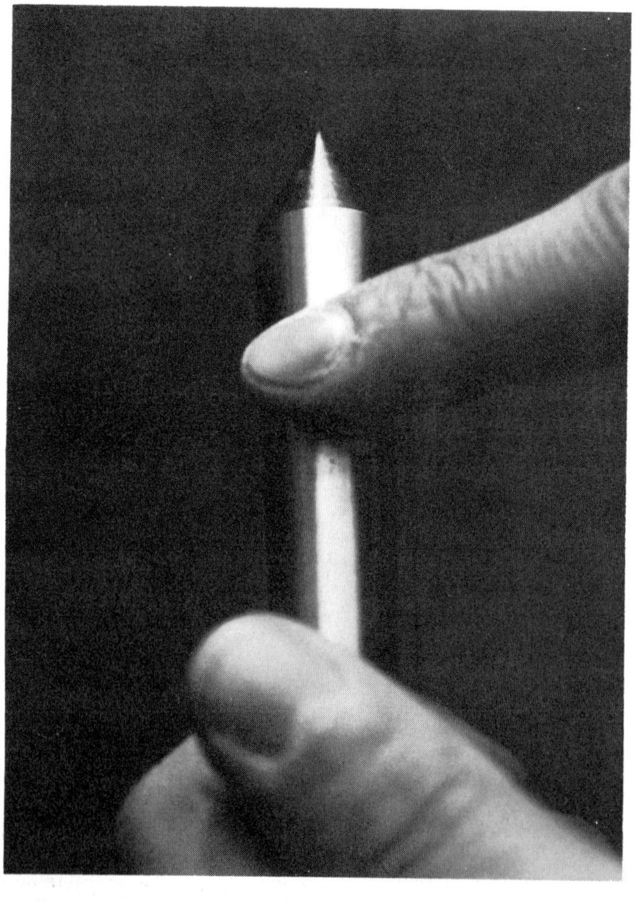

FIGURE 40-10
Examine the lathe centers for burrs or raised edges.

It is important that the headstock center run true. This center rotates with the workpiece and is called the live center. The live center requires no lubrication and need not be of a hardened type; however, hardened centers are more accurate and give longer life. Any inaccuracy in setup will result in a taper when machining. To check a live center for trueness, proceed as follows:

1. Disengage the headstock from the drive.

2. Mount a dial test indicator in the tool post and adjust it until the indicator's contact point is set on center.

3. Lightly touch above the point area with the indicator's contact point and turn the drive plate by hand (Figure 40-13).

FIGURE 40-11
Lock the drive plate in place.

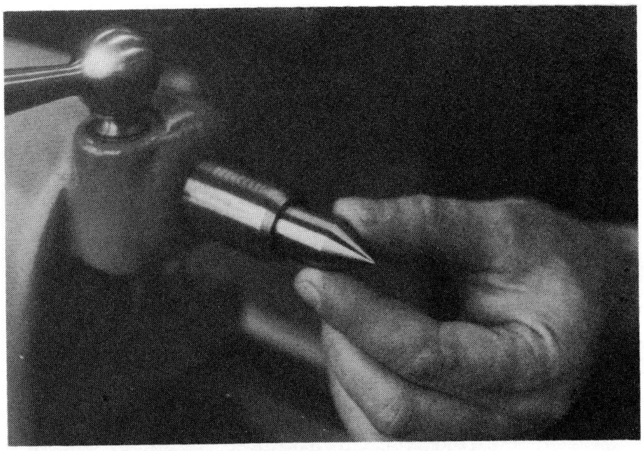

FIGURE 40-12
Insert the tailstock center with a hard quick snap.

If the runout is more than 0.001 in. (0.02 mm), remove the center, examine it for burrs, clean and then reinsert, and retest until it runs true.

Types of Tailstock Centers

There are several styles of tailstock centers such as the hardened center, the carbide-tipped center (Figure 40-14), the half center, and the revolving center. When the tailstock center is of this hardened-type

FIGURE 40-13
Turn the drive plate by hand to indicate the center for trueness.

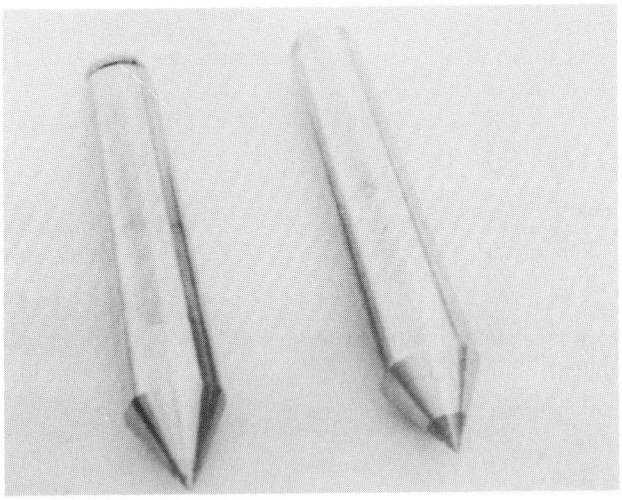

FIGURE 40-14
All hardened and carbide-tipped lathe centers.

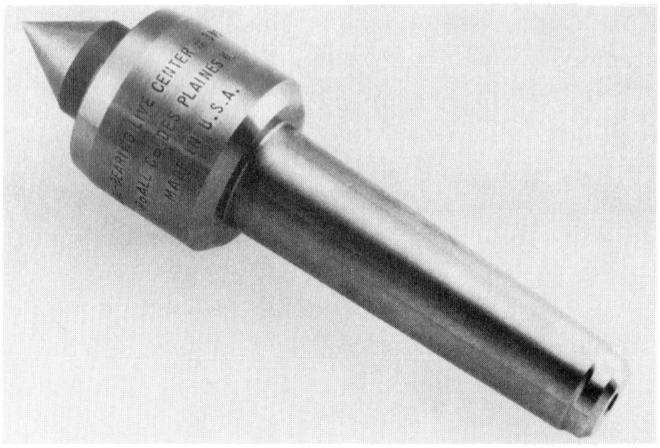

FIGURE 40–15
A ball bearing lathe center.

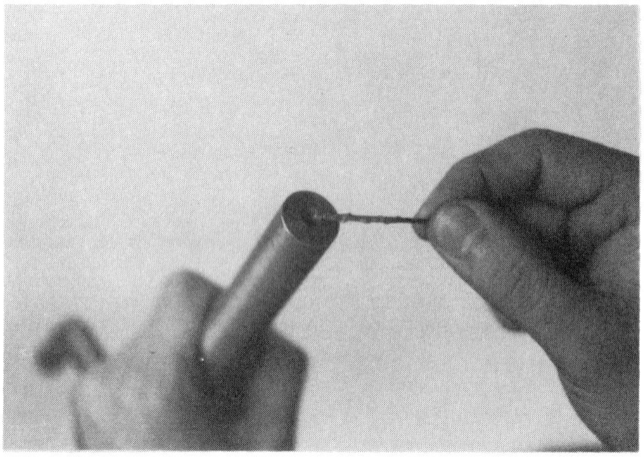

FIGURE 40–17
Apply an antifrictional compound to the tailstock end hole.

surface, it will be stationary in the spindle and is called the dead center. The dead center will require a high heat-resistant lubricant when used. An exception to using a lubricant is the ball bearing center (Figure 40–15). Centers equipped with sealed ball bearings or tapered roller bearings rotate with the workpiece and are lubricated for life by the manufacturer.

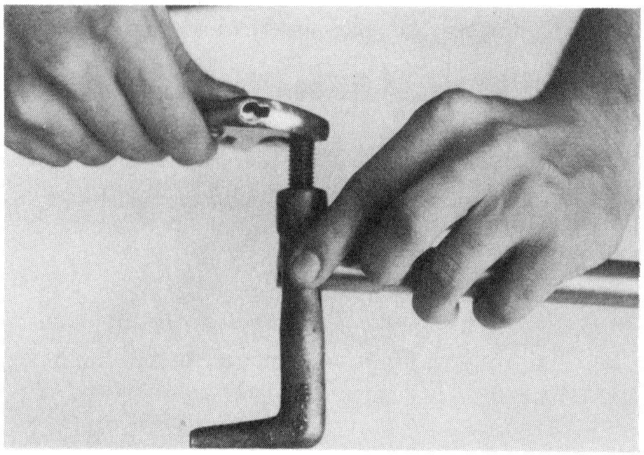

FIGURE 40–16
Clamp a lathe dog on the left-hand end of the workpiece.

Work Mounting Procedure

Heat, caused by the friction of the cutting action, is the greatest danger when working between centers. It is the heat that will expand the workpiece and cause it to bind on the centers, preventing machining. For this reason the workpiece should be carefully mounted and properly lubricated. With the lathe and workpiece prepared as previously explained, proceed as follows when using a dead center:

1. Loosen the tailstock clamping nut.
2. Extend the tailstock spindle about $2\frac{1}{2}$ to 3 in. (62 to 75 mm) beyond the tailstock.
3. Clamp a lathe dog on the left-hand end of the workpiece (Figure 40–16).
4. Fill the opposite center hole with an antifricitional compound (Figure 40–17).
5. Position the assembly on the live center, making sure the dog tail does not bind in the drive plate slot or hole.
6. Holding the workpiece by the left hand, slide the tailstock so that the center enters the lubricated hole.
7. Tighten the tailstock clamping nut.
8. Position the drive plate by hand so that the lathe dog tail rests horizontally.
9. Lift the dog tail up while adjusting the tailstock handwheel with the other hand to engage the center firmly into the hole.

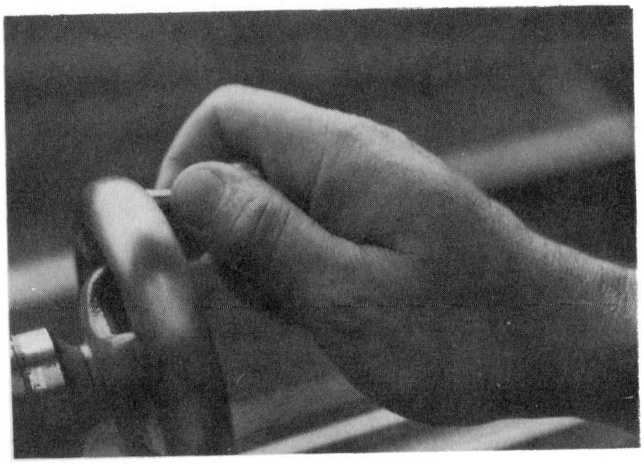

FIGURE 40–18
Adjust the tailstock handwheel to allow the dog tail to drop from its own weight.

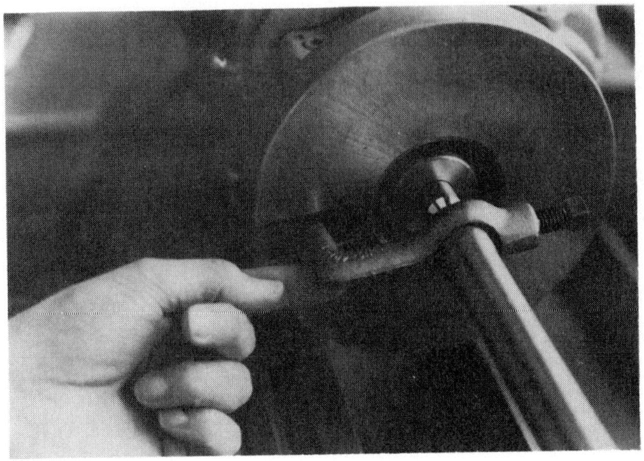

FIGURE 40–20
When the tension is correct between centers, the dog tail should drop from its own weight.

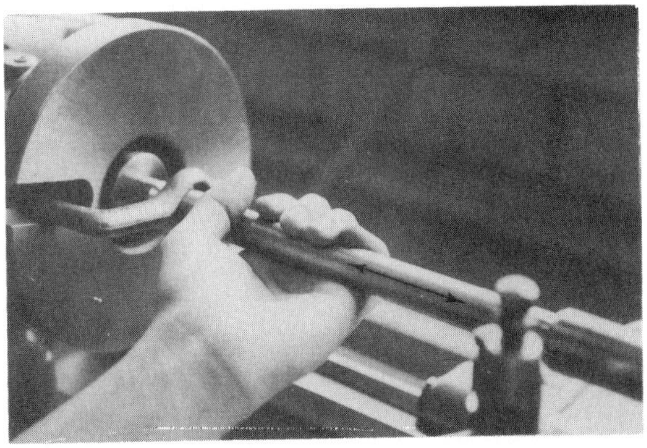

FIGURE 40–19
Move the workpiece between centers to check for end play.

10. Carefully back off the tailstock handwheel until the dog tail drops from its own weight (Figure 40–18).

11. Tighten the tailstock spindle binding lever using light hand pressure.

12. Check the workpiece for motion by moving it from side to side between centers. There should be no end play (Figure 40–19).

13. Repeatedly hold the tail of the dog up in the slot to see that it drops from its own weight. Click! Click! Click! (Figure 40–20).

14. Move the carriage to the left-hand end of the cut to position a lathe stop (Figure 40–21). If a full-length cut is taken, turn the drive plate by hand to see that the lathe dog clears the compound rest.

FIGURE 40–21
When positioning a lathe stop, make sure the lathe dog clears the compound rest.

Using a Revolving Center
There are not as many steps to setting up to work between centers when a ball bearing center is used. The steps would be as follows.

Procedure

1. Loosen the tailstock clamping nut.

2. Extend the tailstock spindle about $2\frac{1}{2}$ to 3 in. (62 to 75 mm) beyond the tailstock (Figure 40–22).

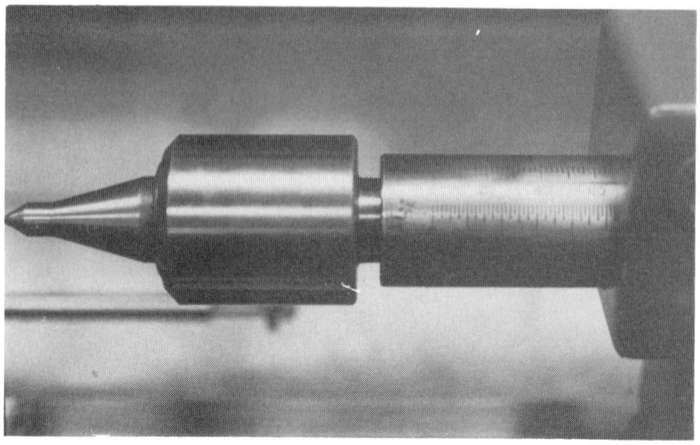

FIGURE 40-22
Extend the tailstock spindle about $2\frac{1}{2}$ to 3 in. (62 to 75 mm) beyond the tailstock.

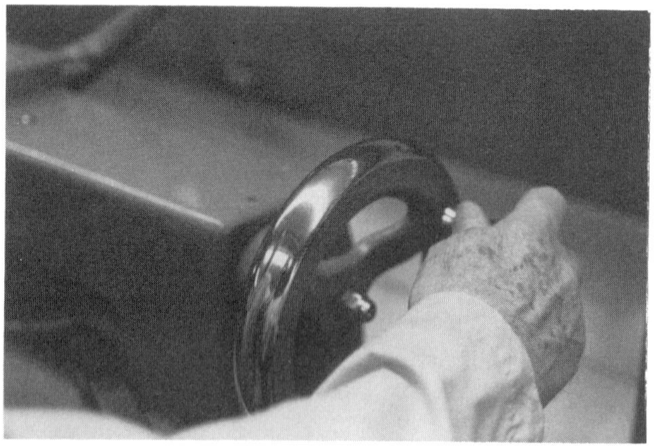

FIGURE 40-23
Adjust the tailstock handwheel until the center revolves with the workpiece without slippage.

3. Clamp a lathe dog on the left-hand end of the workpiece.

4. Slide the tailstock until the revolving center enters the workpiece.

5. Tighten the tailstock clamping nut.

6. Start the lathe.

7. Adjust the tailstock handwheel until the center begins to turn with the workpiece without slippage (Figure 40-23).

8. Lock the tailstock spindle binding lever with light hand pressure.

9. Position a lathe stop as explained in step 14 previously.

UNIT 41 TAPERS AND MACHINING TAPERS

A taper is defined as a gradual increase or decrease in diameter of a workpiece measured along its length. The end of a sharpened pencil (Figure 41-1) is a common example of a taper. Tapers exist in many forms in industry. They may be either external or internal in construction. Either way, tapers are efficient and offer a precise alignment and interchangeability of tools and machine parts. Self-holding tapers provided on shanks of certain types of cutting tools, such as twist drills and reamers, can be found in most any machine shop. Holding devices, such as drill chucks, will also come with tapered shanks (Figure 41-2). Releasing tapers, which can be found on quick-change spindles (Figure 41-3), are also widely used in production machining.

FIGURE 41-1
A common example of a taper.

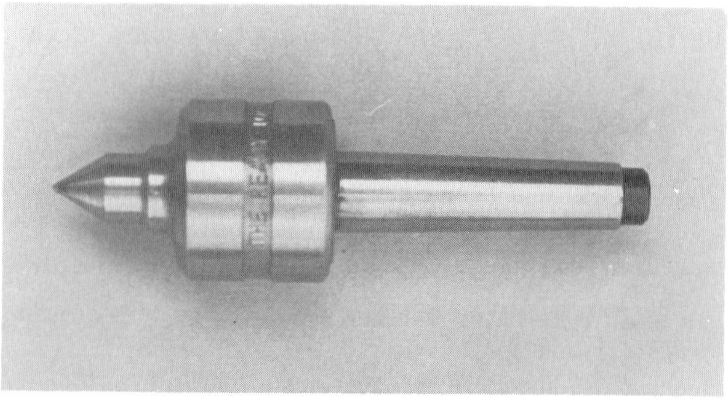

FIGURE 41-2
Self-holding tapers are commonly found on tools such as lathe centers.

Objectives
After completing this unit you should be able to

1. identify the different types of tapers and how they are measured.
2. machine a taper on a lathe using the compound rest.
3. machine a taper between lathe centers.
4. machine a taper using the taper attachment.

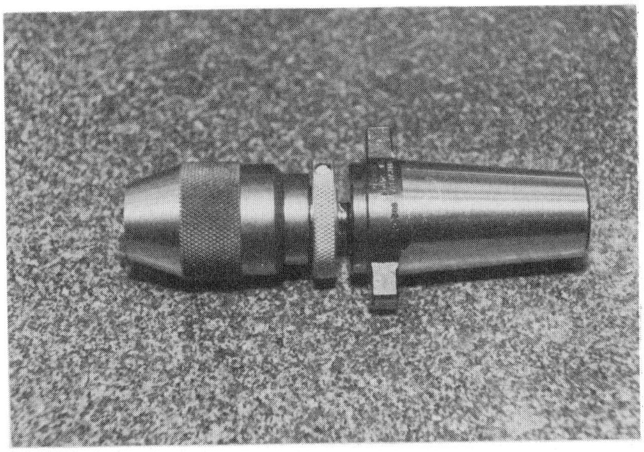

FIGURE 41-3
An example of a self-releasing taper found on a quick-change spindle.

Types of Tapers

There are four basic parts to a taper: length of taper, large diameter, small diameter, and degree of slope (Figure 41-4). On the inch taper the degree of slope may be expressed as either taper per inch or taper per foot. On metric tapers this degree of slope is expressed as a ratio of one millimeter per unit of length. For example, a part with a taper of 1 mm in a length of 10 mm would be expressed as a ratio of 1 : 10 and would be indicated on a print as TAPER = 1 : 10.

On machines and tools, these tapers are generally classified by the American Standard Association as being self-holding, steep taper, or self-releasing.

Self-Holding Tapers

Self-holding tapers are common in the machine tool industry. The friction-type resistance of the taper permits the seating of mated members to be positive and firmly gripped. The Morse Standard Taper (Table 41-1) tapers approximately $\frac{5}{8}$ in. per foot and varies in dimensions ranging in size from 0 through 7. Morse tapers may be found on a number of tools and machine tool parts such as lathe tailstock and headstock spindles. Tapered shanks of twist drills are only made to Morse taper specifications.

The Brown and Sharpe taper series also falls in the category of self-holding tapers as does the $\frac{3}{4}$ in. per foot machine taper series. Tapering approximately $\frac{1}{2}$ in. per foot, except for its No. 10 taper, which measures 0.5161 in. per foot, Brown and

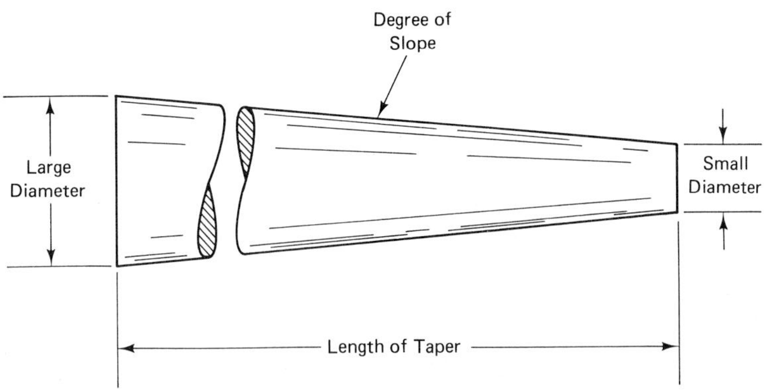

FIGURE 41-4
The basic parts of a taper. (Credit: Paul Rapisarda)

TABLE 41-1
Standard dimensions of self-holding tapers

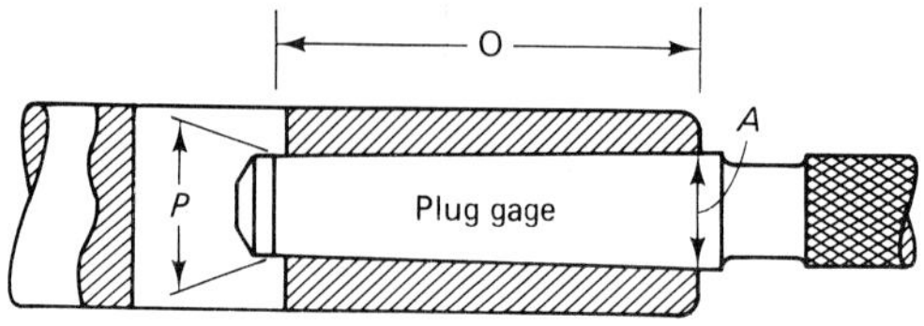

Number of taper	Taper per foot	Diameter at gage line (*A*)	Diameter at small end (*P*)	Length (*O*)	Origin
0	0.6246	0.3561	0.252	2	Morse taper series
1	0.5986	0.475	0.369	$2\frac{1}{8}$	
2	0.5994	0.700	0.572	$2\frac{9}{16}$	
3	0.6023	0.938	0.778	$3\frac{3}{16}$	
4	0.6233	1.231	1.020	$4\frac{1}{16}$	
$4\frac{1}{2}$	0.624	1.500	1.266	$4\frac{1}{2}$	
5	0.6315	1.748	1.475	$5\frac{3}{16}$	
6	0.6256	2.494	2.116	$7\frac{1}{4}$	
7	0.624	3.270	2.750	10	
0.239	0.502	0.2392	0.200	$\frac{15}{16}$	Brown and Sharpe taper series
0.299	0.502	0.2997	0.250	$1\frac{3}{16}$	
0.375	0.502	0.3752	0.3125	$1\frac{1}{2}$	
200	0.750	2.000	1.703	$4\frac{3}{4}$	$\frac{3}{4}$ in. tpf series
250	0.750	2.500	2.156	$5\frac{1}{2}$	
300	0.750	3.000	2.609	$6\frac{1}{4}$	
350	0.750	3.500	3.063	7	
400	0.750	4.000	3.516	$7\frac{3}{4}$	
450	0.750	4.500	3.969	$8\frac{1}{2}$	
500	0.750	5.000	4.422	$9\frac{1}{4}$	
600	0.750	6.000	5.328	$10\frac{3}{4}$	
800	0.750	8.000	7.141	$13\frac{3}{4}$	
1000	0.750	10.000	8.953	$16\frac{3}{4}$	
1200	0.750	12.000	10.766	$19\frac{3}{4}$	

Sharpe tapers can be found on arbors and collets as well as tapered shanks of cutting tools such as reamers and end mills. These tapers are used in the spindles of milling machines, grinding machines, and all Brown and Sharpe machines.

While shank-held friction caused by the wedging of mated members provides some of the drive, larger size tapers depend on a tang to prevent slippage inside a fitted socket or spindle of a machine. Tapered shank twist drills, for example, cannot be removed from a tang drive hole unless freed from this grip by a drill drift. Headstock lathe centers are freed by ramming the small end of the headstock center with the length of a nonferrous metal bar. Larger diameter tapers are generally key driven with the shank held in place by means of a draw-in bolt.

Steep or Self-Releasing Tapers

While the angle of the self-holding taper is but 2° to 3°, a steep or self-releasing taper such as a milling machine arbor (Figure 41-5) will have an included angle of more than 16°. This design will provide an easier release. The positive drive is provided by two opposite sides set of keys on the spindle that locate in slots in the tool or arbor to prevent slippage. The arbor or tool is held in place securely by means of a drawbolt.

FIGURE 41-5
A milling machine arbor is an example of a steep taper.

FIGURE 41-6
A type D-1 spindle nose is used with a cam-lock chuck.

Other examples of the self-releasing taper are the type L and type D-1 lathe spindle nose. The type L spindle nose has a $3\frac{1}{2}$ in. per foot taper and uses a threaded ring and key for its positive drive. The type D-1 spindle nose (Figure 41-6) uses a very short 3 in. per foot taper. The holding device—chuck or driveplate—is positioned on the spindle nose by means of studs and secured in place by means of three cam locks.

Nonstandard Tapers

The Jarno taper remains one of the easier tapers to calculate and measure. The taper per foot is always 0.600 in. The tapers range in size from 2 to 20, with each number indicating the size of the large diameter in eighths of an inch, the small diameter in tenths of an inch, and the length of the taper indicated by the size number divided by 2. For example, a No. 6 Jarno taper is $\frac{6}{8}$ in. diameter at the large end, $\frac{6}{10}$ or 0.600 in. at the small end, and $\frac{6}{2}$ or 3 in. in length.

The Jarno taper can be found on various machine tools but in a limited capacity. The resistance to making the system more widespread lies in the economics of replacing all the many millions of existing tapered tools, machine tools, and parts already made to existing taper systems.

Standard Taper Pin

Tapered pins are used in holes that are usually small in diameter and comparatively deep. They are used to accurately locate two different work parts with one another. These pins are cut with $\frac{1}{4}$ in. per foot taper and range in size at the large diameter from 0.062 to 0.706 in.

Taper Calculations

In order to cut a taper between centers, it is first necessary to calculate the amount of offset. Tapers in the inch system are based on the 12-in. length as being the standard and are specified as either taper per inch (tpi) or taper per foot (tpf). This taper represents the difference between the large diameter and the small diameter over the 12-in. length.

Calculating the tpf

$$\text{tpf} = \frac{(D - d) \times 12}{\text{TL}}$$

where D = large diameter

d = small diameter

TL = taper length

Example: calculate the taper per foot for the workpiece shown in Figure 41-7.

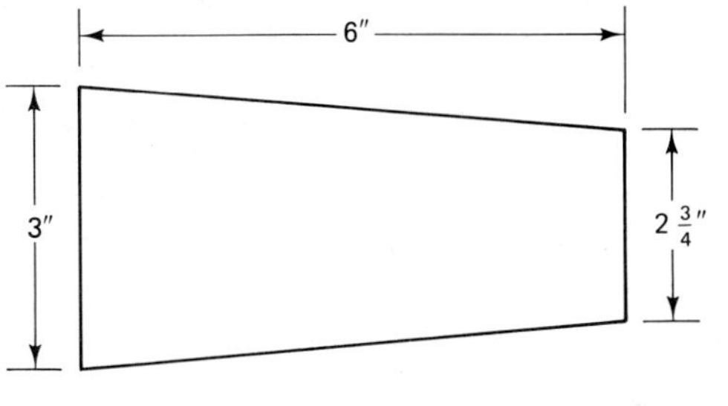

FIGURE 41-7
A taper dimensioned for the inch system.

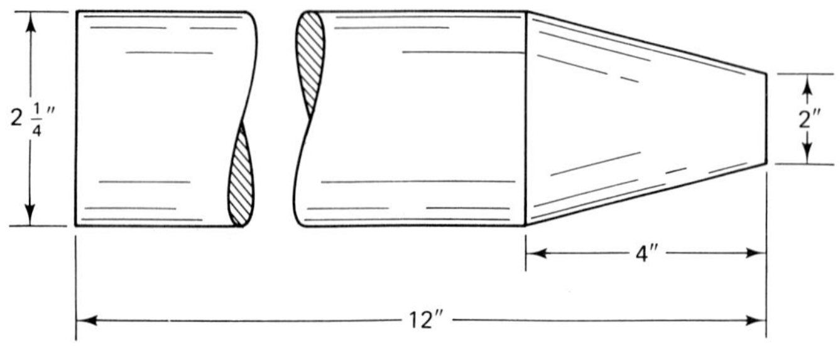

FIGURE 41-8
A workpiece that is tapered part of its overall length.

Solution:

$$\text{tpf} = \frac{(D - d) \times 12}{\text{TL}}$$

$$= \frac{(3 - 2\tfrac{3}{4}) \times 12}{6}$$

$$= \frac{\tfrac{1}{4} \times 12}{6}$$

$$\text{tpf} = \tfrac{1}{2} \text{ in.}$$

When the taper involved represents a short distance of the total length of the workpiece as in Figure 41-8, the tailstock offset will have to take into consideration the overall length of the work. The formula in that case would be

$$\text{Tailstock offset} = \frac{\text{OL}}{\text{TL}} \times \frac{(D - d)}{2}$$

where OL = overall length
TL = taper length
D = large diameter
d = small diameter

Example: Calculate the tailstock offset for the workpiece shown in Figure 41-8.

Solution:

$$\text{Tailstock offset} = \frac{\text{OL}}{\text{TL}} \times \frac{(D - d)}{2}$$

$$= \frac{12}{4} \times \frac{(2\tfrac{1}{4} - 2)}{2}$$

$$= 3 \times \frac{\tfrac{1}{4}}{2}$$

$$= 3 \times \tfrac{1}{8}$$

$$= \tfrac{3}{8} \text{ in.}$$

The tailstock offset can be calculated if the taper per foot is known. The simplified formula reads

$$\text{Tailstock offset} = \frac{\text{TPF} \times \text{OL}}{24}$$

If the taper per foot is not known, it will have to be calculated as previously explained. This information is then used with the formula just stated.

Example: Calculate the tpf and tailstock offset for the workpiece shown in Figure 41-8.

Solution:

$$\text{tpf} = \frac{(D - d) \times 12}{\text{TL}}$$

$$= \frac{(2\tfrac{1}{4} - 2) \times 12}{4}$$

$$= \frac{\tfrac{1}{4} \times 12}{4}$$

$$= \tfrac{3}{4} \text{ in.}$$

$$\text{Tailstock offset} = \frac{\text{tpf} \times \text{OL}}{24}$$

$$= \frac{\tfrac{3}{4} \times 12}{24}$$

$$= \frac{\tfrac{3}{4}}{2}$$

$$= \tfrac{3}{8} \text{ in.}$$

Tailstock offset can also be calculated if the taper per inch is known. If unknown, a simplified formula can be used to calculate this value.

$$\text{tpi} = \frac{\text{tpf}}{12}$$

$$\text{Tailstock offset} = \frac{\text{tpi} \times \text{OL}}{2}$$

Example: calculate the tpi and tailstock offset for the workpiece shown in Figure 41–8.

$$\text{tpi} = \frac{\text{tpf}}{12}$$

$$= \frac{\frac{3}{4}}{12}$$

$$\text{tpi} = \tfrac{1}{16}\ \text{in.}$$

$$\text{Tailstock offset} = \frac{\text{tpi} \times \text{OL}}{2}$$

$$= \frac{\frac{1}{16} \times 12}{2}$$

$$= \tfrac{1}{16} \times 6$$

$$= \tfrac{3}{8}\ \text{in.}$$

Metric Taper Calculations

An explanation of the metric taper is illustrated in Figure 41–9, which shows a taper = 1 : 10. The large diameter (D) is equal to the small diameter (d) plus the amount of taper. Since the workpiece tapers 1 mm for every 10 mm of length, the dimension at a point 10 mm away from the small diameter (d) will be $d + 1$ (1 mm longer).

Redimensioning a second drawing, it can now be seen that the unit length of taper represented by 10 mm in Figure 41–10 is now represented by the letter k. Since the amount of taper for the unit length is 1 mm, the amount of taper per millimeter of unit length = $1/k$.

The total amount of taper will be the taper per millimeter ($1/k$) × total length of the taper (1).

$$\text{Total taper} = \frac{1}{k} \times 1 \quad \text{or} \quad \frac{1}{k}$$

$$D = d + \text{total amount of taper}$$

Therefore $$D = d + \frac{1}{k}$$

Example: Calculate the large diameter (D) for a 1 : 60 taper having a small diameter of 30 mm and a length of 180 mm on a workpiece 360 mm long.

Solution:

Since the taper is 1 : 60, $k = 60$

$$D = d + \frac{1}{k}$$

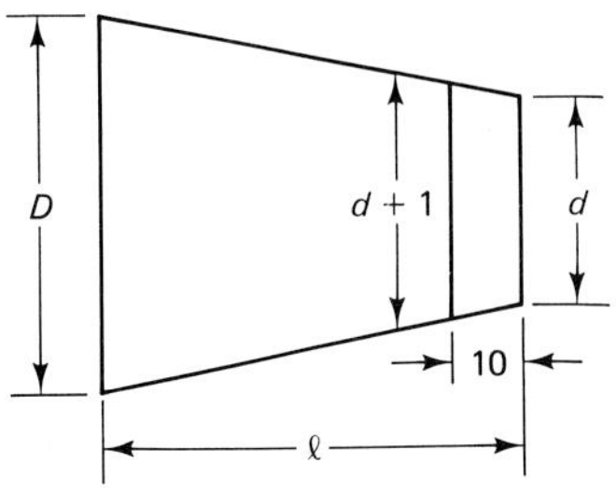

FIGURE 41–9
Dimensions of a metric taper.

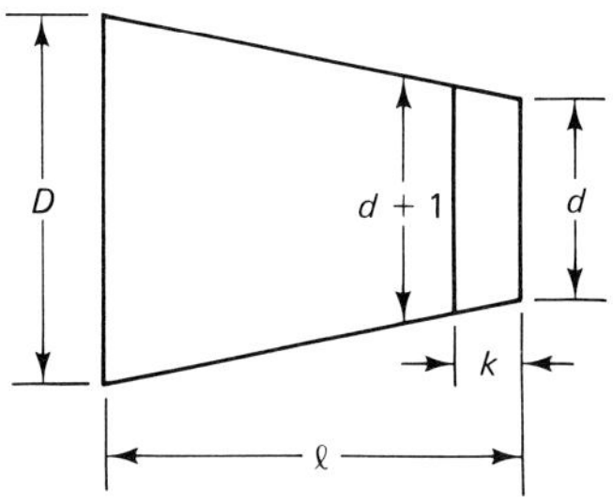

FIGURE 41–10
Identifying the basic dimensions of a metric taper.

$$= 30 + \frac{180}{60}$$

$$= 30 + 3$$

$$= 33\ \text{mm}$$

Metric Tailstock Offset Calculations

The formula for calculating tailstock offset for machining metric tapers is as follows:

$$\text{Tailstock offset} = \frac{D - d}{2 \times l} \times L$$

where D = large diameter

d = small diameter

l = taper length

L = overall length of workpiece

Example: calculate the tailstock offset for the workpiece problem just solved.

Solution:

$$\text{Tailstock offset} = \frac{D - d}{2 \times l} \times L$$

$$= \frac{33 - 3}{2 \times 180} \times 360$$

$$= \frac{3}{360} \times 360$$

$$= 3 \text{ mm}$$

Machining the Taper

Besides using the edge of a cutting tool as a form tool, there are three basic methods of machining a taper in a lathe: by using the compound rest, by offsetting the tailstock, and by using the taper turning attachment. The first two methods can be used on any standard lathe; the last method involves special equipment. The first rule to remember in machining a taper is that the cutting edge of the toolbit must be mounted exactly on center to turn a true taper. Failure to do so will result in a false taper.

Setting the Compound Rest

The compound rest has a limited travel and is used for cutting relatively short steep tapers. The compound rest is mounted on a base that can swivel and is graduated in degrees so it can be set to any angle to cut a taper. The lathe's toolbit is fed by hand by turning the compound rest handle. By using the compound rest, it acts like a path or a guide that directs the line of cut.

To set the compound rest to any required angle the angle of the taper must be examined. Is the taper indicated with an included angle or is it indicated with an angle given from one side? The secret of the setup is to place the compound rest along the same centerline as the print's centerline was drawn and then move the compound rest to favor the cutting side. This movement will either be half of the degrees if the print's taper is shown as an included angle or to the given angle if the print's taper is shown with a given angle from one side.

Cutting a short steep taper can be done in the following manner using a turning tool with a small radius on the end of the toolbit.

Procedure

1. Loosen the compound rest and prepare to position it to use
   (a) The included angle method: Place the compound rest on the same centerline of the proposed taper, then move the compound rest to half the included angle so that it favors the cutting side (Figure 41-11).
   (b) The given angle method: Place the compound rest on the same line of the proposed taper, then move the compound rest to the given angle so that it favors the cutting side. The edge of the compound rest will be parallel to the proposed taper (Figure 41-12).
2. Lock the compound rest.
3. Set the turning tool on center so only the point does the cutting.
4. Set the lathe to its proper turning speed.
5. Start the lathe spindle.
6. Face off the workpiece if necessary.

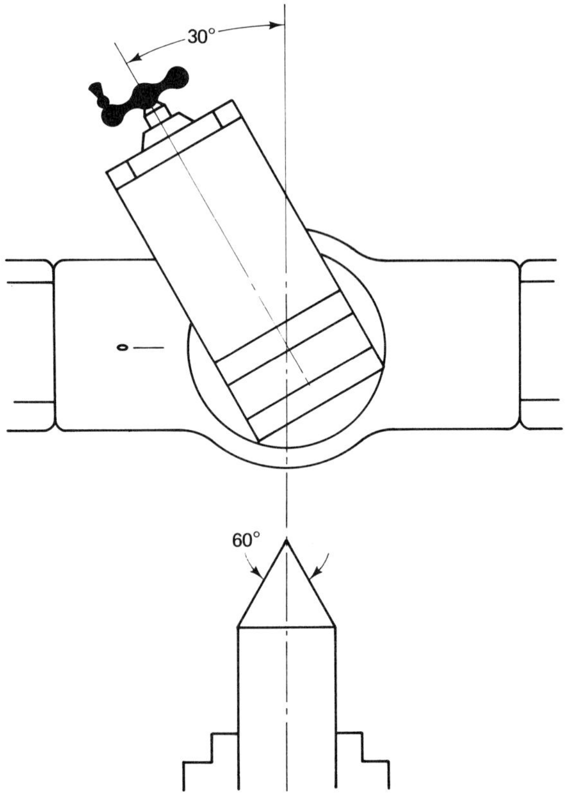

FIGURE 41-11
Setting the compound rest using the included angle method. (Credit: Paul Rapisarda)

7. By turning the crossfeed handle and the carriage handwheel, advance the toolbit close to the workpiece without touching it.

8. Withdraw the compound rest so that its dovetail slide extends about the same length of the tapered section to be cut.

9. Lock the carriage.

10. Turn the compound rest handle to position the toolbit so it overlaps the edge of the workpiece about $\frac{1}{8}$ in. (3 mm).

11. Turn the crossfeed handle so that the toolbit makes light contact with the revolving workpiece.

12. Back off the cutting tool by reverse turning the compound rest handle counterclockwise.

13. Advance the crossfeed handle to set the depth of cut.

14. Take a short trial cut to check the small diameter of the taper and the angle.

15. Machine the workpiece by turning the compound rest handle (Figure 41–13).

16. Upon completing the cut, return the cutting tool beyond the front edge of the workpiece and repeat steps 13 and 15 to complete the machining.

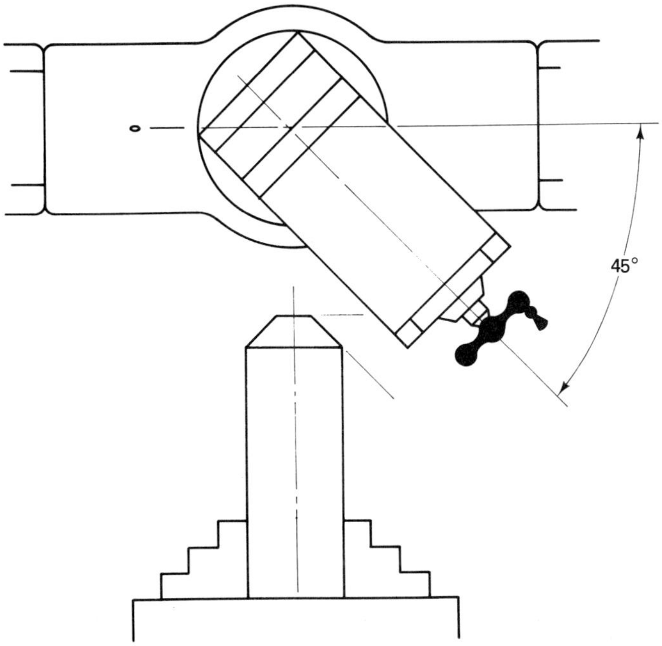

FIGURE 41–12
Setting the compound rest using the given angle method. (Credit: Paul Rapisarda)

Offsetting the Tailstock

The tailstock is offset or set over, as it is also called, when long gentle tapers are to be cut between centers. This method can only be used to cut external tapers. As mentioned in Unit 39, the construction of the tailstock makes it possible to offset the tailstock centers out of alignment with the headstock by adjusting a set of screws on either side of the tailstock base. When the tailstock center is offset beyond the center axis of the headstock so that it is closer to the operator (Figure 41–14A) to create a setup such as in Figure 41–14B, the small diameter of the taper occurs at the tailstock end (Figure 41–14C). When the tailstock center is moved away from the operator beyond the center axis of the headstock, the small diameter of the cut begins at the headstock end.

Adjusting the Tailstock

The amount of tailstock offset can be measured using three different methods, each varying in degree of accuracy: using a visual method, using a dial indicator, using the crossfeed micrometer collar.

Using the Visual Method

The visual method is a semiprecise method of offsetting the tailstock. Movement of the tailstock can be measured with a rule held between the centers (Figure 41–15) or by measuring the distance between the alignment marks on the back of the tailstock base with a rule.

FIGURE 41–13
Using the compound rest to turn a taper.

Procedure

1. Loosen the tailstock clamping nut.
2. Snug-up the clamping nut by hand.
3. Loosen the adjustment screw in the base of the tailstock on the opposite side of the direction of movement required.
4. Tighten the other side until the upper section of the tailstock has been moved the required length.
5. Tighten the opposite adjustment screw to lock the upper section of the tailstock in place.

FIGURE 41-14A
Moving the tailstock center closer to the operator . . .

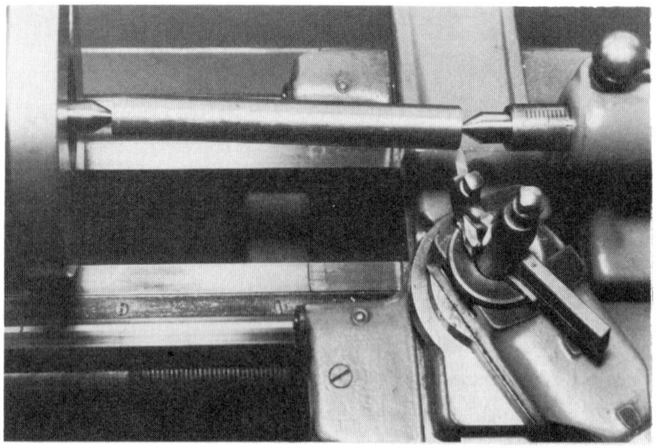

FIGURE 41-14B
. . . to create this setup . . .

FIGURE 41-14C
. . . will result in a taper with the small diameter in the tailstock end.

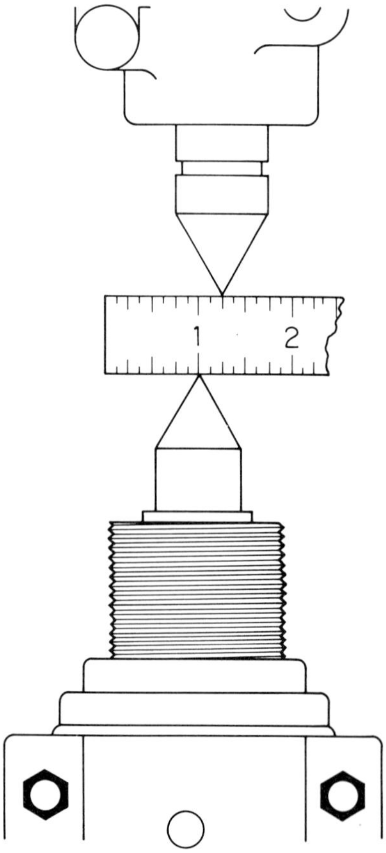

FIGURE 41-15
Measuring the tailstock offset by rule.

Using a Dial Indicator

The use of a dial indicator is the most accurate means of measuring tailstock offset. The dial indicator may be either held in a tool post or used in connection with a magnetic base fastened to the top of a compound rest. Once mounted, the indicator is positioned against the tailstock spindle to measure its movement. When the offset is less than the indicator range, the following steps should be followed.

Procedure

1. Extend the tailstock spindle about 2 in. (50 mm) and lock the tailstock spindle binding lever.
2. Mount a dial indicator in the toolpost or on a magnetic base holder with the indicator's probe in a horizontal position and on center.
3. Turn the crossfeed handle counterclockwise several turns.
4. Turn the crossfeed handle clockwise to remove any backlash and keep turning to bring the indicator's plunger into contact against the tailstock spindle until the needle moves about one-half revolution.
5. Set the indicator's bezel to zero.
6. Set the crossfeed micrometer collar to zero.
7. Loosen the tailstock clamping nut.
8. Snug-up the clamping nut by hand.
9. Proceed to offset the tailstock by loosening and tightening adjustment screws in the base of the tailstock as previously explained.

When the offset is more than the indicator's range, the following steps should be followed.

Procedure

1–8. Follow the steps outlined for offsetting less than the indicator's range.

9. Turn the crossfeed handle to move the indicator needle the amount of offset required.
10. Adjust the tailstock spindle by loosening and tightening the adjustment screws in the base of the tailstock until the needle registers zero.
11. Tighten both adjustment screw, making sure the zero reading does not change.

There is still another method. Gage blocks or an adjustable parallel can be set to the required offset. This spacer is placed between the tailstock spindle and dial indicator. The indicator is advanced against the gage, set to a reading, the spacer is removed, and the tailstock spindle is adjusted to the indicator reading.

Using the Crossfeed Micrometer Collar

The third method of offsetting involves the use of the crossfeed micrometer collar to register the amount of tailstock spindle movement.

Procedure

1. Tighten a toolholder in the tool post with its back end facing the tailstock spindle.
2. Rotate the crossfeed handle counterclockwise several turns.
3. Turn the crossfeed handle clockwise to remove the backlash and keep turning it until a paper feeler is pinched between the tailstock spindle and the toolholder.
4. Set the micrometer collar to zero.
5. Back off the crossfeed handle a few turns and advance it by turning it clockwise to remove the backlash.
6. Keep advancing the crossfeed handle and position it on the graduation line represented by the zero minus the offset amount.
7. Adjust the tailstock spindle by loosening and tightening the tailstock adjustment screws until a paper feeler can be pinched between the spindle and the toolholder.

Taper Turning Attachment

There are two types of taper attachments available: the plain taper attachment (Figure 41–16) and the telescopic taper attachment. They both operate using the same principle. A slide, which is independent to the ways, can be angled and acts as a guide to control the cross-slide movement. Machining a taper on a workpiece takes place as the toolbit cuts along the guided path.

There are many advantages to using the taper turning attachment.

- The lathe centers can remain aligned and the center holes are not distorted during machining.
- Any workpiece length can be machined.
- Workpieces can be machined being held in a chuck or between centers.
- Tapers can be machined on either external (turning) or internal (boring) surfaces.

FIGURE 41–16
This plain taper turning attachment is capable of producing a maximum taper of 3 $\frac{1}{2}$ in./ft (300 mm/m). (Credit: LeBlond Makino Machine Tool Company)

- A wider range of taper sizes (as much as $3\frac{1}{2}$ in./ft) can be machined.
- The taper turning attachment can be used with an offset tailstock to machine tapers too steep for either method.
- A power feed may be used.
- Tapers can be produced set to either degrees and millimeters for metric lathes or degrees and inches per foot for the inch system lathes.

The operation of the taper attachments are similar except for the depth of cut mechanisms. In order to use the plain taper attachment, the cross-feed screw, which governs the cross-slide movement, has to be disconnected. This is done by removing the screw that holds the crossfeed nut to the cross slide. All depth of cuts must be taken by using the compound rest handle. A binding screw is then tightened to connect the guide block (sliding shoe) and the slide extension of the cross slide.

On the telescopic taper attachment, the cross slide and crossfeed nut are left alone. The depth of cut is made by using the crossfeed handle.

Procedure for Using a Taper Turning Attachment

1. Tighten any looseness between the guide block and guide bar.
2. Adjust the compound rest gibs if need be.
3. Position and tighten the compound rest about 30° to the cross slide.
4. Clean and oil the guide bar and guide block.
5. Loosen the lock screws and adjust the

guide bar to the required angle of taper.

6. Tighten the lock screws.

7. Mount a turning tool and set it up on center.

8. Mount the workpiece.

9. Mark the tapered section to be cut.

10. Move the carriage handwheel to position the turning tool near the center of the tapered section to be cut.

11. Lock the guide bar anchor bracket to the lathe V-way. *Note:* If using a plain taper attachment, the following steps should be followed at this point.

(a) Remove the screw that connects the cross slide and crossfeed nut.

(b) Tighten the binding screw to connect the cross-slide extension with the guide block.

(c) Insert a plug to help keep the binding screw hole free of dirt and chips. Depth of cut can only be taken now using the compound rest feed.

12. Turn the carriage handwheel to move the turning tool clear of the right-hand end of the workpiece by about $\frac{1}{2}$ in. (13 mm).

13. Take a light trial cut for about $\frac{1}{16}$ in. (1.50 mm) long, stop the lathe, and check the taper for size.

14. Set the depth of roughing cut about 0.06 in. (1.50 mm) oversize and machine the taper.

15. Check the taper for fit. (See earlier section on checking fits.)

16. Readjust the taper attachment if necessary, take another light trial cut, and recheck the fit.

17. Finish turning to size.

Taper turning attachments may be set accurately to a given taper by mounting a dial indicator in the tool post as well as an angled workpiece or taper plug gage between centers. The guide bar is then positioned as the indicator is moved along the length of the test piece. When there is no movement on the needle, the guide bar is locked in place.

For boring operations the side opposite the surface used for turning should be indicated for taper attachment setup.

Checking Tapers

There are several methods used for checking external tapers. These include a standard micrometer, a taper micrometer, a sine bar, a shadowgraph, and a ring gage. All but the standard micrometer will require special equipment or special setups to measure a taper in a lathe. Measurement with a standard micrometer is as follows.

Procedure for Inch Tapers

1. Mark a length of the tapered section to be measured with layout dye.

2. With spindle extended, butt a micrometer lathe stop against the carriage and lock in place.

3. Place the lathe in neutral.

4. Advance a mounted, sharp-pointed toolbit toward the workpiece as it is being turned by hand.

5. Cut a shallow line into the surface.

6. Back off the micrometer stop exactly 1 in.

7. Move the carriage handwheel to butt the carriage against this stop and lightly cut the second line.

8. Measure the large diameter on the line with the edge of the spindle and anvil of a standard micrometer (Figure 41-17).

9. Measure the small diameter of the taper in the same manner (Figure 41-18).

10. Subtract the difference in reading. Measurement will indicate the taper per inch of the workpiece. Workpieces that are not held in the lathe can be checked in an upright position on a surface plate with a standard micrometer, a set of dowel pins, and two sets of gage blocks.

FIGURE 41-17
Measure the large diameter of the taper with the tips of the micrometer anvil and spindle.

Procedure

1. Place the tapered workpiece in a vertical position.

2. Place a set of gage blocks of exact height (say from 0.500 to 1.000 in.) on either side of the upright taper.

3. Place a set of uniform size dowel pins on the level of the gage blocks.

4. Using a standard micrometer, measure across the full width of the dowel pins for the first diameter measurement of the taper.

5. Build up a set of gage blocks exactly 1.000 in. higher than the first set.

6. Place the set alongside the upright taper.

7. Measure the tapered diameter at this level as before. The difference in reading will indicate the taper per inch of the part.

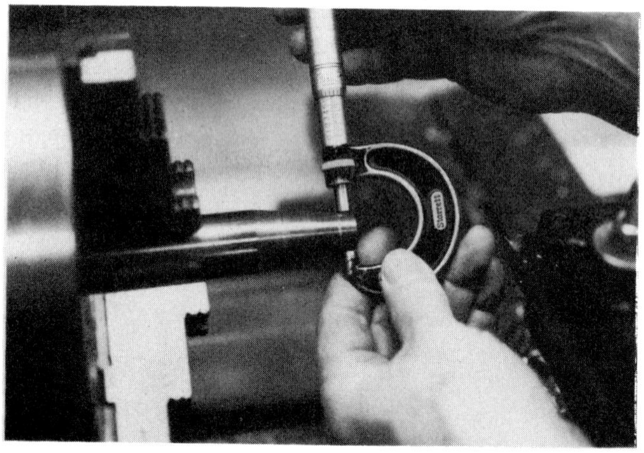

FIGURE 41-18
Measure the small diameter of the taper, 1 in. away, in the same way. The difference in reading will be the taper per inch.

Checking a Metric Taper

1. Mark a tapered section of work with layout dye.

2. Scribe two lines on the workpiece whose distance apart will match the ratio number of the taper. For example, if the taper is specified 1 : 10, the lines would be 10 mm apart.
(*Note:* Where workpiece length permits, measurement can be checked easier if the lines scribed apart are drawn two or three times the length of the tapered section. For instance, the 10-mm line in the preceding problem would be marked off 30 mm. While the difference in diameters would also be proportional—three times greater—the measurement would be more accurate.)

3. Measure the large diameter of the line with the edge of the spindle and anvil of a standard metric micrometer.

4. Measure the small diameter of the taper in the same manner.

5. Subtract the difference in reading. The difference should be one millimeter for each unit of length.

Fitting an External Taper

Procedure

1. Cover the entire surface of the taper with chalk or a light coating of Prussian blue.

2. Slip a ring gage onto the surface as far as it will go.

3. Carefully rotate the ring gage counterclockwise by hand about one-half turn.

4. Remove the ring gage.

5. Rub marks will reveal where the taper high spots are. (These marks indicate where any correction in taper cutting have to be made.) Uniform rub marks along the entire length will reveal a good fit.

Fitting an Internal Taper

Internal tapers are checked in a manner similar to checking external tapers; only a plug gage is coated to be tested inside the taper.

UNIT 42 SURFACE FINISHING IN A LATHE

Different surface finishes are possible in a lathe, either to alter or improve the surface quality. These operations might include filing, polishing, knurling, or grinding. Each operation is produced by applying a particular method. Each will require that a safe approach be used.

Objectives
After completing this unit you should be able to

1. use a file to remove small amounts of metal to improve surface quality.
2. produce a good surface finish by polishing.
3. produce a good pattern finish by knurling.
4. set up a lathe grinder for cylindrical grinding.

Filing

In a lathe filing is generally restricted to light metal removal such as touching up a tapered surface to ensure proper fit or perhaps removing a burr or sharp edge. It can also improve surface finish by removing tool marks. Care must be taken, however, not to remove too much material from a diameter or the part will be left out of round. Filing is the poorest method of bringing a diameter to size.

There are three factors to be considered when filing in a lathe: (1) type of file, (2) lathe speed, and (3) most important of all safety. There are a few files recommended for lathe work such as a single-cut, mill bastard file or a long-angle lathe file that can effectively cut clean without producing file marks. If working between centers, the workpiece should be mounted properly as explained in Unit 40. Also a left-handed technique should be used when working as it is safer. Right-hand filing has resulted in a greater number of accidents, such as exposing the knuckles or left hand to a rotating lathe chuck or spinning drive plate or lathe dog.

Before attempting any filing, the lathe ways should be covered with paper to prevent metal filings from imbedding themselves on the wear surfaces. The compound rest should also be cleared of the tool post and moved as far to the right as possible to provide filing room.

Procedure

1. Appropriate safety glasses must be worn before attempting any work in the lathe.
2. Select a file with a snug-fitting handle.
3. Clean the file and apply chalk to prevent it from clogging (Figure 42-1).
4. Cover the lathe ways with paper.
5. Set the lathe to run at about two times the normal turning for the material's diameter.
6. Start the lathe spindle.
7. Hold the top of the file with the fingers of the right hand to use a left-hand technique (Figure 42-2).

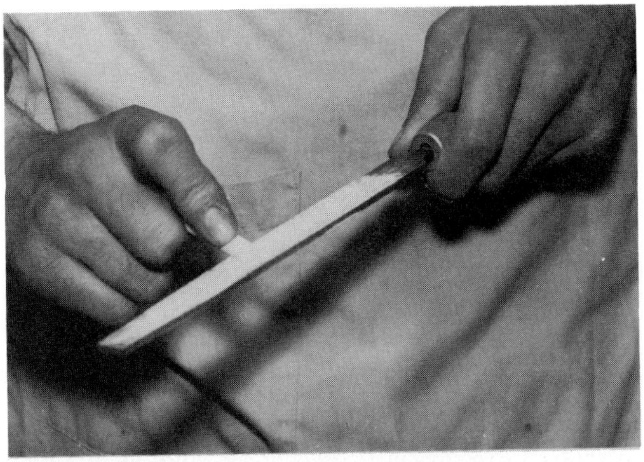

FIGURE 42-1
Applying chalk to the file to prevent clogging.

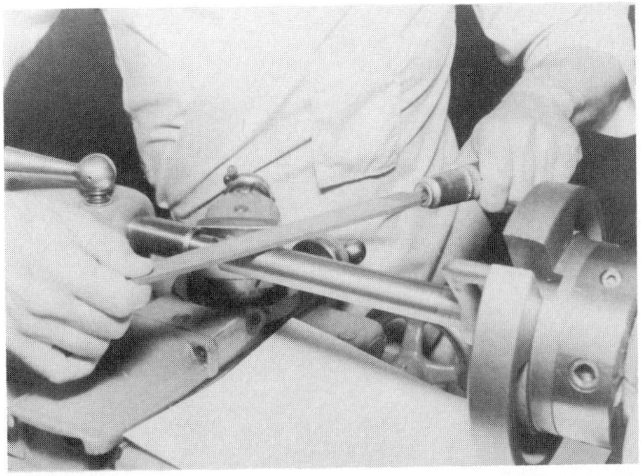

FIGURE 42-2
A left-hand filing technique provides much more safety than right-handed filing.

8. File across the moving part on an angle, using a light steady stroke.

9. Release the pressure on the return stroke, not necessarily lifting it.

10. Each successive pass should overlap the previous stroke by one-half the file's width.

11. Clean the file and chalk it frequently.

Polishing

Once a diameter has been filed to size, it can be polished. Polishing can be accomplished with abrasive cloth that will range in grit size (Figure 42-3). For rough polishing use abrasive cloth between 80 and 100 grit, while for finish polishing a grit of 120 or higher is recommended. Aluminum oxide abrasive cloth is recommended for ferrous metals. Silicon carbide abrasive cloth is recommended for nonferrous metal. Worn abrasive cloth mixed with a little oil will also produce a smooth gleaming finish. Polishing is intended to heighten the finish of a surface, not remove excessive metal. Polishing should never be used to take the place of a toolbit or a file. To do so is time consuming and will produce a workpiece that is out of round. And like filing, polishing will heat up the workpiece, enlarging its diameter. It is recommended that the workpiece cool off before attempting to measure.

Polishing can be carried out a few ways. As with filing, the lathe ways should be protected with paper against the falling abrasive that will collect. The lathe should also run at about two times the speed used for filing. A narrow 1-in. (25-mm) strip torn from the length of a sheet of abrasive cloth, approximately 10 in. (250 mm) long can be used by holding the cloth against the rotating workpiece, pulling on the ends with your hands. The abrasive cloth will resemble an elongated letter *U*, which is then moved slowly from side to side (Figure 42-4). Precautions should be taken, however, that the abrasive cloth does not wrap around the workpiece.

FIGURE 42-4
Care must be taken when polishing that the abrasive cloth does not wrap around the workpiece.

FIGURE 42-3
Abrasive cloth of varying grit sizes are used for polishing in a lathe.

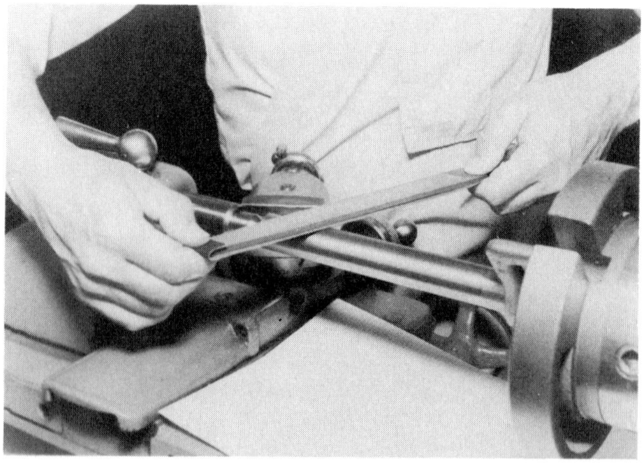

FIGURE 42-5
A left-hand polishing technique eliminates much of the danger of hitting a rotating drive plate or lathe chuck.

The abrasive cloth can also be held on the bottom of a flat file (Figure 42-5). Curling the end of the cloth over the tip of the file and pinching it with the fingers of the right hand, the other end is stretched and held near the handle with the left hand. The diameter can be polished using long strokes in a manner similar to the left-hand filing technique.

FIGURE 42-6
Knuckle joint knurling tools featuring straight and diamond pattern rollers.

FIGURE 42-7
A revolving head knurling tool that can be used for producing different textured work. (Credit: DoAll Company)

Knurling

Knurling is a machine operation that is used to change the finish and appearance of a workpiece and provide a better gripping surface. A knurling tool with one set of hardened rollers (Figure 42-6) or one with three pairs of rollers (Figure 42-7) imprints its pattern onto the surface of a workpiece. The result of these imprints or depressions is a diamond-shaped or straight-line pattern that can be controlled to produce a fine, medium, or coarse texture (Figure 42-8). Knurling will enlarge the diameter of a workpiece (Figure 42-9).

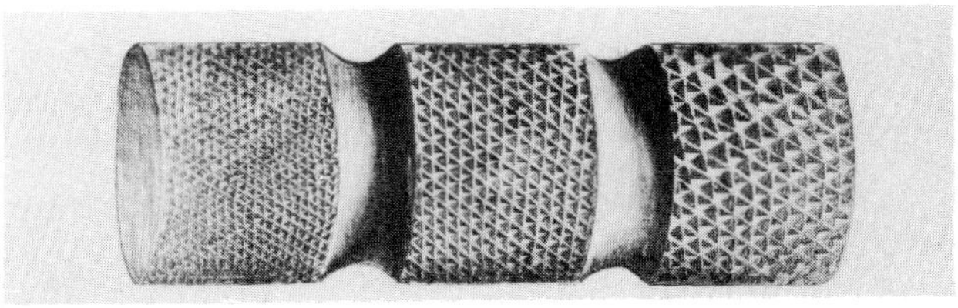

FIGURE 42-8
Samples of fine, medium, and coarse knurling.

FIGURE 42-9
In addition to providing a gripping surface, knurling has enlarged the body diameter.

There are several conditions or requirements that must be met to successfully knurl. While the knurling tool is self-centering, the rollers must be set against the workpiece so that the full width of the rollers are tracking. Not centering the tool will produce an uneven knurl because more pressure is placed on one roller than the other. Not only must plenty of oil be used during the operation, the lathe is also made to run at one-quarter of the recommended turning speed of the material. The last requirement concerns itself with a damaged knurl that makes itself known with the formation of rings. To avoid these rings, never disengage the automatic feed until the depth of the knurl has been completed. If more than one pass has to be made to complete the depth of the impression, stop the lathe and change the rotational direction of the lead screw.

Procedure

1. Set the lathe to run at one-quarter the recommended turning speed of the material.
2. Extend the knurling tool only far enough from the tool post to make a rigid setup.
3. Place the center of the knurling tool's floating section to match the height of the tailstock center point and snug up.
4. Mount the workpiece to run between lathe centers. *Note:* Work held in a chuck should be supported by a rotating tailstock center whenever possible.
5. Set the knurling tool at 90° to the workpiece and tighten it in that position. *Note:* If the knurling rolls are worn, it is advisable to set the knurling tool on a slight angle so that the edge of the rolls start the pattern.
6. Traverse the carriage to both extremes to assure that one-half of the knurling roll widths overlap the edges of the workpiece.
7. Set the quick-change gearbox for a 0.020- to 0.030-in. (0.50- to 0.75-mm) feed.
8. Start the lathe and test the knurling by lightly imprinting against the front edge by gradually turning the crossfeed handle.
9. Stop the lathe to check that the knurling pattern is uniform. If a true diamond pattern has formed, it is correct. (If the full width of the rollers are not tracking, recheck the center height and try again on another section.)
10. Move the knurling tool to the front end of the work, overlapping this edge by half the roller width.
11. Oil the section to be knurled freely.
12. Feed the knurling tool in 0.025 in. (0.64 mm). *Note:* The lathe may be started before feeding the knurling tool into the work 0.025 in. (0.64 mm) to produce the knurled pattern.
13. Start the lathe and engage the automatic feed to knurl the required length. *Note:* Do not disengage the automatic feed during knurling as rings will be formed in the pattern.
14. Stop the lathe and reverse the feed to deepen the knurl as it travels back to the front end of the work.
15. Repeat the passes until the diamond pattern is to a point.

The Tool Post Grinder

The tool post grinder (Figure 42–10), so named because it is mounted on the compound rest in place of the tool post, is capable of cylindrical and taper grinding. Equipped with internal grinding attachments, it can also be used for grinding holes and bushings. As in other forms of cylindrical and internal grinding, the tool post grinding wheel will best cut to accuracy and finish by having the workpiece rotate in the opposite direction to the grinding wheel's turning. Tool post grinding will also require that the lathe ways be protected against the abrasive particles that collect.

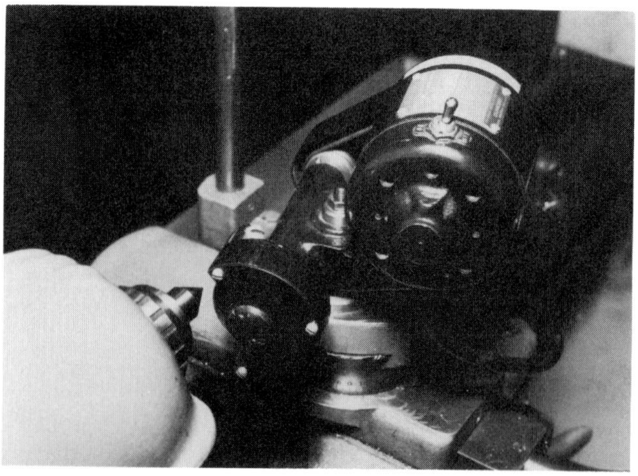

FIGURE 42–10
A tool post grinder is capable of cylindrical and taper grinding.

Setup for Grinding a Lathe Center

1. Insert the lathe center to be ground in the headstock spindle.

2. Rotate the compound rest to 30° with the centerline of the lathe.

3. Mount a dial test indicator in the tool post to test the accuracy of the 30° setup.

4. Adjust the indicator till the indicator contact point is set on center (Figure 42–11).

5. Lightly touch the lathe center's surface and by turning the compound rest feed handle, move the indicator along the length of the taper.

6. If the 30° setup is not accurate, reset the compound rest and repeat step 5.

7. Protect the lathe ways and the upper part of the lathe carriage with cloth or similar material.

8. Clear the compound rest of any tool post.

9. Mount the tool post grinder and adjust the center of grinding spindle to the lathe center's height.

10. Mount a proper grinding wheel so it can be dressed and trued.

11. Place a shallow pan of water on the lathe carriage, directly under the grinding wheel.

12. Set the lathe to turn at a slow speed.

13. Start the lathe so that the center turns in reverse.

FIGURE 42–11
Set the indicator's probe on the same height as the lathe center.

14. Start the grinder and by turning the crossfeed handle, touch the center until it sparks lightly.

15. Lock the carriage.

16. Carefully turning the crossfeed handle, take a 0.001-in. (0.02-mm) grinding depth of cut.

17. Turn the compound rest feed handle slowly to move the grinding wheel along the face of the center.

18. Check the angle ground by placing a center gage level with the lathe's centerline. (Placing the center gage above or below the centerline will give a false reading.)

19. Finish grinding the center.

20. Polish the center with a piece of worn abrasive cloth and a little oil at a high spindle speed if a high finish is desired.

UNIT 43 DRILLING AND TRUING CENTER HOLES

A common method of machining metal in a lathe is to support it between centers. This will require both work and lathe preparation. Since work will be supported between 60° centers, one mounted in the tailstock and the other in the headstock spindle, holes will have to be drilled in the ends of the workpiece to match the taper point of each center. (See Unit 40 for a fuller explanation.)

Other situations will arise when work containing damaged center holes will have to be repaired or remachined. The damaged center hole in the end of a shaft is a good example. The center hole will have to be trued before it will permit the shaft to be mounted and run true.

Objectives
After completing this unit you should be able to

1. drill a center hole accurately.
2. true an existing center hole.

Types of Center Drills and Sizes

The cutting tool used to machine the proper tapered hole to fit the lathe center is a combination center drill and countersink, commonly called a center drill (Figure 43–1). By construction, one can see that the pilot of the center drill leads to a 60° tapered diameter that reaches its full size on the body. It is the body size of the center drill that dictates the size. Center drills used in the machine shop generally are numbered from 1 to 8, although smaller sizes such as 00 and 0 are available, and range in size from $\frac{1}{8}$ to $\frac{3}{4}$ in. (3 to 9 mm) in diameter. For example, a number 6 center drill has a $\frac{1}{2}$-in. (13-mm) body diameter and a $\frac{7}{32}$-in. (5.50-mm) drill diameter. A full listing of center drills as well as recommended sizes for different work diameters can be found in Table 43-1.

Work Holding Methods

In addition to being able to center drill in the drill press (see Unit 32), there are several work holding methods used in the lathe. In all methods the workpiece is revolved while the center drill is fed into the end surface by a drill chuck mounted in the tailstock. These methods would include the three-jaw chuck (Figure 43-2), the four-jaw chuck, and the

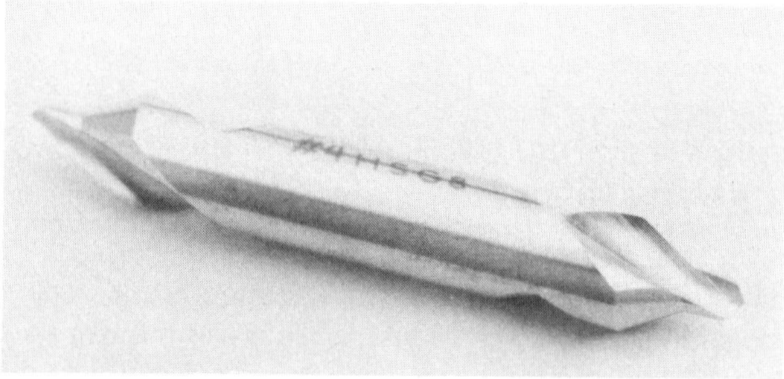

FIGURE 43-1
The main parts of a center drill.

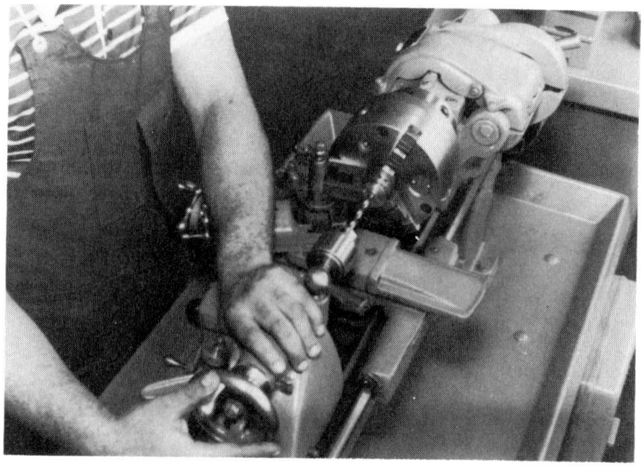

FIGURE 43-2
Drilling work held in a three-jaw chuck.

TABLE 43-1
Center drill sizes

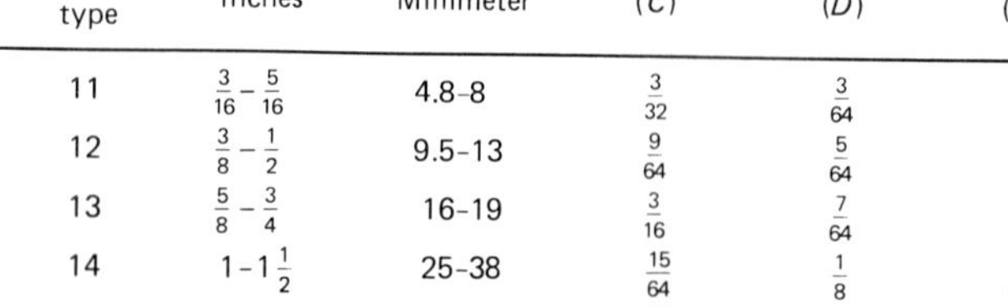

Center drill sizes

Size		Work diameter (W)		Diameter of countersink	Drill point diameter	Body size
Regular type	Bell type	Inches	Millimeter	(C)	(D)	(F)
1	11	$\frac{3}{16}$-$\frac{5}{16}$	4.8-8	$\frac{3}{32}$	$\frac{3}{64}$	$\frac{1}{8}$
2	12	$\frac{3}{8}$-$\frac{1}{2}$	9.5-13	$\frac{9}{64}$	$\frac{5}{64}$	$\frac{3}{16}$
3	13	$\frac{5}{8}$-$\frac{3}{4}$	16-19	$\frac{3}{16}$	$\frac{7}{64}$	$\frac{1}{4}$
4	14	1-1$\frac{1}{2}$	25-38	$\frac{15}{64}$	$\frac{1}{8}$	$\frac{5}{16}$
5	15	2-3	50-75	$\frac{21}{64}$	$\frac{3}{16}$	$\frac{7}{16}$
6	16	3-4	75-100	$\frac{3}{8}$	$\frac{7}{32}$	$\frac{1}{2}$
7	17	4-5	100-125	$\frac{15}{32}$	$\frac{1}{4}$	$\frac{5}{8}$
8	18	6 and over	150 and over	$\frac{9}{16}$	$\frac{5}{16}$	$\frac{3}{4}$

spring collet chuck (Figure 43–3). The Jacobs collet chuck (Figure 43–4), which uses a handwheel to tighten or release a set of rubber flexed collets, is another method used for holding small round stock. Larger workpieces, such as shafting, are held by a chuck on one end and supported by means of a steady rest on the other.

Speeds and Feeds

Center drills are often broken due to the sensitivity of the center drill's pilot (the drill diameter). There are several reasons why this breakage occurs:

- The lathe centers are not aligned.
- Feeding the center drill too fast for the lathe speed involved.
- Uneven work surface prevents uniform tool cutting.

In center drilling the lathe should be set to run at the proper speed for the size of the center drill being used. Generally, a speed of 1200 to 1500 rpm is recommended for most center drills. The feed of the center drill is very important. The pressure that builds up from too fast a feed for the lathe's speed will cause the drill point to break. The center drill should be carefully fed in and backed out frequently, applying a few drops of cutting fluid during the process.

FIGURE 43–3
Center drilling work held in a spring collet chuck.

An uneven surface that is produced by a saw-cut should be faced or spotted by a center hole before center drilling. This is done by using the point of a mounted facing toolbit. Using two hands, the toolbit point is forced into the center of the work by carefully moving the carriage handwheel until a small chip curls off the end of the toolbit (Figure 43–5).

FIGURE 43–4
Uneven surfaces should be faced or the hole spotted before center drilling.

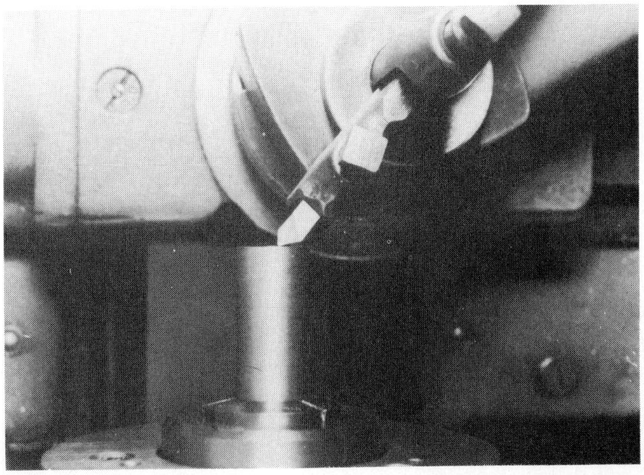

FIGURE 43–5
The toolbit should be set to center whether hole spotting or facing.

Center Drilling Holes

The result of using a center drill in the end of a workpiece is a tapered shape opening. Besides preparing the workpiece for mounting between lathe centers, center drilling is used as the first step in drilling a hole in a lathe. It enables the operator to drill the center locations accurately without having to lay out the locations first.

Procedure

1. Appropriate safety glasses must be worn before doing any machining in a lathe.
2. Mount the workpiece in a three-jaw chuck, extending it about $\frac{1}{2}$ in. (13 mm) beyond the jaws, never more than three times its diameter, and tighten it.
3. Tighten a facing toolbit in the toolholder, extending it no more than $\frac{1}{2}$ in. (13 mm) from the holder.
4. Position the cutting tool for a facing operation and tighten (Figure 43–5).
5. Set the lathe speed for the diameter and material to be cut.
6. Start the lathe and (a) face the end surface clean, or (b) using two hands to move the carriage handwheel, force the point of the toolbit into the center of the workpiece to machine a small spotting center.
7. Stop the lathe and turn the crossfeed handle to back out the facing tool to clear the area for drilling.
8. Clean the tapered shank of a lathe drill chuck.
9. Clean the tailstock spindle taper.
10. Insert the drill chuck with a short sharp snap (Figure 43–6).
11. Mount a proper size center drill in the drill chuck, extending it not more than $\frac{1}{2}$ in. (13 mm) from the chuck jaws.
12. Set the lathe speed for approximately 1200 to 1500 rpm.
13. Slide the tailstock toward the workpiece, locking it when the center drill is close to the workpiece.
14. Lightly tighten the tailstock spindle clamp to create a slight drag when turning the tailstock handwheel.
15. Start the lathe spindle.
16. Slowly turn the tailstock handwheel, using two finger pressure, to feed the center drill in the work (Figure 43–7).
17. Back out the center drill frequently by reversing the handwheel rotation to remove the chips and apply cutting fluid (Figure 43–8).
18. Continue center drilling until the countersunk opening is the correct size for the workpiece diameter (Figure 43–9).

FIGURE 43–6
Insert the drill chuck in a clean spindle with a short snap.

FIGURE 43–7
Carefully feed the center drill into the end of the workpiece.

FIGURE 43-8
Back out the center drill frequently to remove the chips.

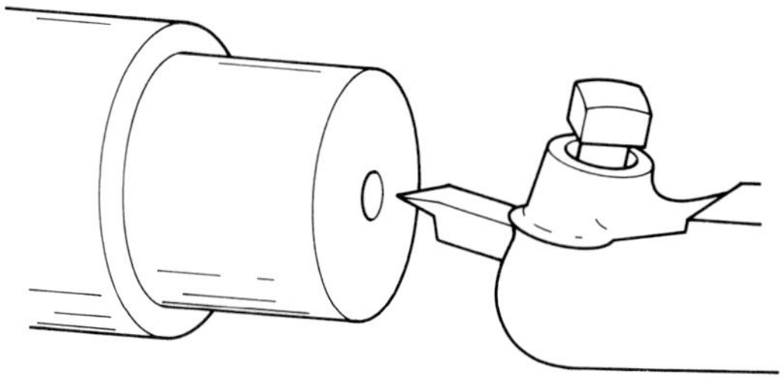

FIGURE 43-10
A 60° spotting tool is used for truing damaged center holes.

FIGURE 43-9
Stop center drilling when the countersink opening is the correct size for the workpiece diameter.

FIGURE 43-11
A special ground countersink bearing a flat, single cutting edge may also be used for truing a damaged center hole.

Truing a Center Hole

There are a few ways of truing a damaged center hole. One method will require the use of a 60° shaped spotting tool (Figure 43-10). The toolbit must have sufficient side clearance ground on it to prevent rubbing in the center hole. The other cutting tool can be a special 60° countersink that contains one flat, single cutting edge (Figure 43-11). Both methods require that the lathe centers be in line.

With the workpiece running true in a three- or four-jaw chuck, proceed as follows:

Using a Spotting Tool

1. Tighten the spotting tool in a level holding toolholder (carbide toolholder).
2. Mount the spotting tool at 90° to the workpiece hole.
3. Set the lathe to run at about 200 rpm.
4. Start the lathe spindle.
5. Slowly turn the carriage handwheel to feed the toolbit into the center of the hole.

FIGURE 43-12
Back out the cutting tool to clear the area for drilling.

6. Carefully turn the crossfeed handle to gradually feed the cutting edge against the damaged surface until it has been smoothed.

7. Stop the lathe and clear the toolbit from the hole by moving the carriage handwheel.

8. Turn the crossfeed handle counterclockwise to back out the cutting tool, clearing the area for drilling (Figure 43-12).

9. Clean the tailstock spindle.

10. Clean the shank of a proper drill chuck.

11. Insert the drill chuck into the tailstock spindle with a snap.

12. Mount a sharp 60° countersink in the drill chuck.

13. Turn the tailstock handwheel to extend the spindle about 1 in. (25 mm) beyond the tailstock.

14. Slide the tailstock toward the headstock and lock it close to the center hole.

15. Using cutting fluid, lightly drill the center hole to restore it to its true 60° tapered opening.

Using a Special Countersink

The method of using a special countersink is similar to the spotting tool method, only no other cutting precedes the actual countersinking. To prevent deflection of the cutting tool, the tailstock spindle clamp should be lightly tightened before feeding the cutting tool into the work hole.

Procedure

1. Set the lathe to run at about 600 rpm.
2. Clean the tailstock spindle.
3. Clean the drill chuck shank.
4. Insert the drill chuck in to the tailstock spindle with a sharp snap.
5. Insert the special countersink into the drill chuck, extending it not more than $\frac{1}{2}$ in. (13 mm) from the jaws, and tighten it.
6. Turn the tailstock handwheel to extend the spindle about 1 in. (25 mm) beyond the tailstock.
7. Slide the tailstock toward the headstock, locking it close to the workpiece.
8. Lightly tighten the tailstock spindle clamp to create a slight drag when turning the tailstock handwheel.
9. Start the lathe spindle.
10. Apply cutting fluid.
11. Slowly turn the tailstock handwheel to feed the cutting tool into the damaged center hole.
12. Continue feeding until the center hole has been cleaned and runs true.

UNIT 44 DRILLING, BORING, REAMING, AND TAPPING

There is a relationship between drilling, boring, reaming, and tapping. They all require the cutting of a hole. In performing these operations in a lathe, the workpiece will be generally held in the headstock in some fashion. With the exception of the boring tool, which is held in the tool post, the cutting tool for drilling, reaming, and tapping will be held in a tailstock drill chuck or tailstock spindle and fed from a clamped-down tailstock. Generally, the workpiece hole will be machined concentric with the outside diameter.

Objectives
After completing this unit you should be able to

1. drill different size holes to depth in a lathe.
2. bore holes to accurate dimensions.
3. identify different types of reamers.
4. ream a hole to an accurate size.
5. tap a hole square in the lathe.

Work Preparation

Before attempting any drilling, the workpiece should be faced to provide a flat, square surface so that the center drill can get started. The lathe speed can be calculated using the basic formula,

$$\text{rpm} = \frac{\text{CS} \times 4}{D}$$

The tailstock must be aligned properly to do any drilling. With the tailstock spindle clean, snap in a drill chuck in a slightly extended spindle and tighten the center drill. Slide the tailstock toward the headstock, clamping it down on the lathe ways with the center drill about 1 in. (25 mm) away from the workpiece (Figure 44-1).

Methods of Holding Drills

There are several methods of holding twist drills for cutting. The most popular method is using a straight shank drill and tightening in a drill chuck for usage. Generally, twist drills of $\frac{1}{2}$ in. (13 mm) diameter or larger have a lathe dog fastened to the shank to prevent its twisting inside the drill chuck (Figure 44-2). The tail of the dog rides along the top of the compound rest as the drill is fed into the workpiece.

The other general method of holding a twist drill in the tailstock is by means of a tapered shank. Drills up to $\frac{1}{2}$ in. (13 mm) diameter are usually made with a straight shank. Above this diameter they are generally, though not always, made to the Morse taper system, a system that uses an internal holding system by mating tapered surfaces. The exception to this rule are drills larger than $\frac{1}{2}$ in. (13 mm) diameter that have their shanks turned down to fit a straight shank drill chuck.

Tapered shank drills are mounted directly into the tailstock spindle (Figure 44-3) and are released

FIGURE 44-2
Use a lathe dog to prevent the drill from twisting inside the drill chuck.

FIGURE 44-1
Clamp the center drill a short distance away from the workpiece before drilling.

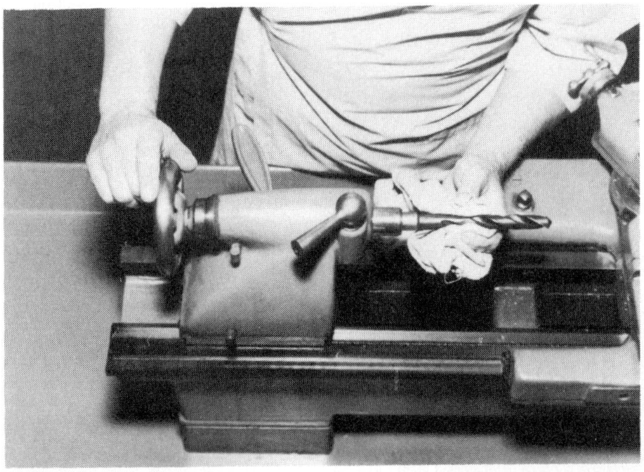

FIGURE 44-3
Tapered shank drills should never be inserted bare handed.

by turning the tailstock handwheel counterclockwise until an internal ram forces the drill free. Like the larger straight shank drills, a lathe dog should be fastened to tapered shank drills to prevent twisting in the tailstock spindle and damaging the tapers on both surfaces.

Drilling Speeds and Feeds

Drilling speeds for lathe work can be calculated by using the basic formula,

$$\text{rpm} = \frac{\text{CS} \times 4}{D}$$

It should be understood that the diameter of the drill and not the workpiece is the information required. This calculation can be speeded up by referring to Table 31–2. For center drilling it is recommended that the lathe be slowed to three-fourths of the calculated rpm.

Unlike production drilling in the drill press where the rate of feed is calculated to determine the production expectancy and the life of the tool, drilling in a lathe involves a steady rate of feed where the tailstock handwheel is turned just fast enough to allow the chips to curl out of the flutes. Cutting fluid or lubrication is freely used and withdrawal of the drill to clear it from chips is recommended. Smaller drills should not be fed into the workpiece using heavy handed pressure when turning the tailstock handwheel.

Drilling to Depth

The depth of a hole on a print is measured from the reference surface to the bottom of the hole at the full diameter, not the point of a drill. This fact should be kept in mind when drilling to a depth. In this type of drilling, a graduation on the tailstock spindle is used as a reference line to gage the depth that the hole is being drilled.

Procedure

1. Appropriate safety glasses must be worn before any machining.
2. Drill into the workpiece just beyond the widest diameter of the drill behind the point.
3. Stop the lathe.
4. Loosen the tailstock.
5. Turn the tailstock handwheel to set the spindle to the 1-in. (25-mm) line.
6. Lock the tailstock spindle clamp.

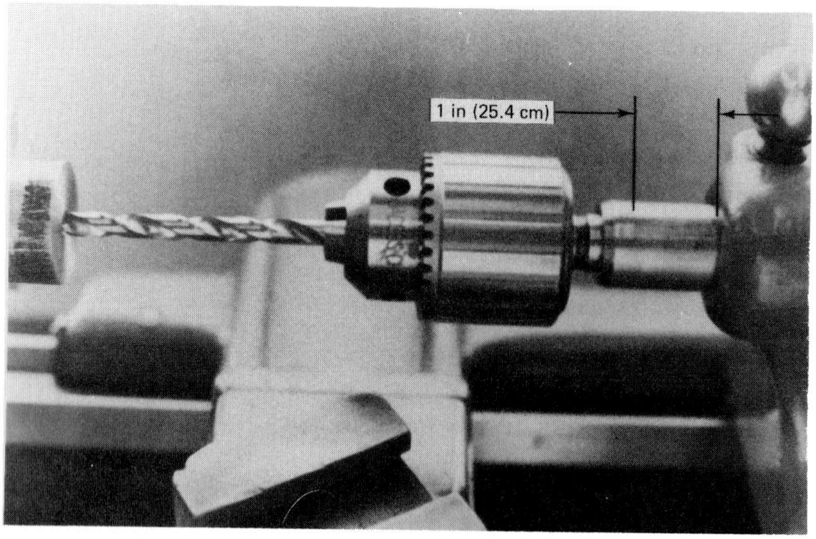

FIGURE 44–4
Use the 1-in. (25-mm) mark on the spindle as a reference line to measure the drilling depth.

7. Slide the tailstock toward the workpiece so that the drill's full diameter behind the point is even with the face of the workpiece (Figure 44-4).

8. Lock the tailstock with this reference line established.

In the event that neither the tailstock spindle nor the handwheel is graduated, depth of drilling can be carried out by establishing your own reference line.

Procedure

1. Drill into the workpiece just beyond the widest diameter of the drill behind the point.

2. Stop the lathe.

3. Crank the tailstock handwheel to adjust the drill's full diameter behind the point to be even with the workpiece's face.

4. Mark the spindle next to its opening with a pencil line.

5. Measure the distance this mark moves in drilling by using a rule held against the tailstock spindle opening.

Boring

Boring is an operation that may be performed in a lathe in order to enlarge holes or to clean and true the surface of the hole itself. In a lathe boring is performed with a single-point cutting tool, generally a small-sized toolbit held in a special boring bar fastened in the tool post. These tools are made to be used in boring tool holders (Figure 44-5) as well as in dovetail locking tool holders (Figure 44-6). In using any boring tool, it is important that the boring bar be as solid and heavy as possible, that it does not extend too far beyond the tool holder and that the boring bar enters the hole freely without obstruction. An overextended boring bar will lead to vibration and chatter, producing a poor machined surface and inaccurate work.

The best way of explaining how a boring tool should be used is to go through the steps of boring a hole in steel hex bar stock that is $1\frac{1}{4}$ in. (32 mm) long so it can be hand tapped $\frac{3}{4}$-16 NF.

Procedure

1. Appropriate safety glasses must be worn before any machining.

2. Mount the workpiece in a lathe chuck.

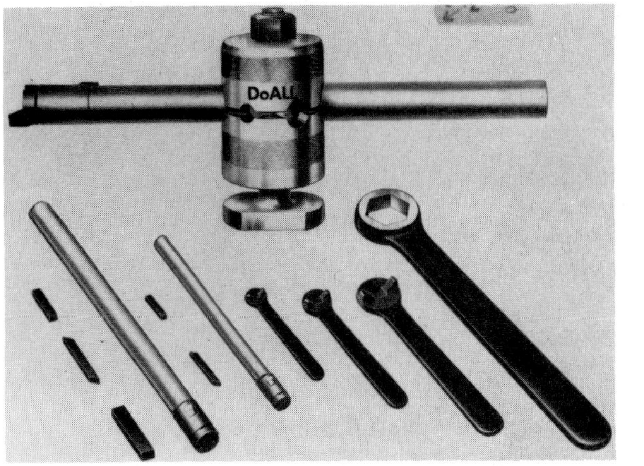

FIGURE 44-5
A three-bar boring tool set. (Credit: DoAll Company)

FIGURE 44-6
A dovetail locking toolholder and boring bar.

3. Drill the tap hole size $\frac{1}{16}$ in. (1.60 mm) undersize to an $\frac{11}{16}$ in. (17.50 mm) diameter hole.

4. Select a proper size boring bar.

5. Extend and tighten the boring bar tool $1\frac{1}{2}$ in. (38 mm) long beyond the holder to clear the length of the hole by about $\frac{1}{4}$ in. (6 mm) (Figure 44-7).

6. Set the cutting edge on center, leaving the holder level with the axis of the hole.

7. Wind the lathe carriage handwheel so that the boring tool enters the full length of the cut to check for clearance.

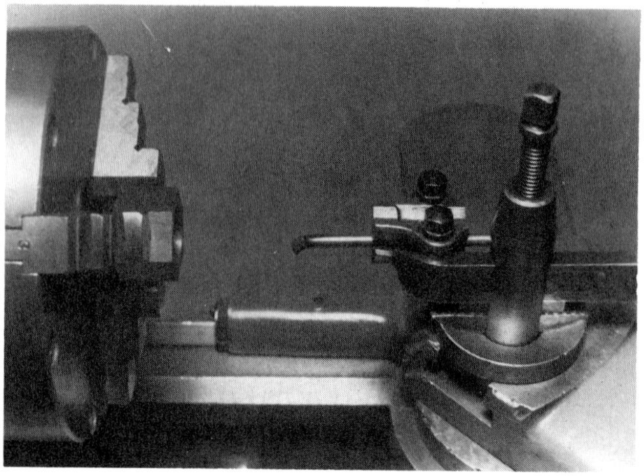

FIGURE 44-7
For rigidity purposes a boring bar cutter should extend beyond the holder about $\frac{1}{4}$ in. (6 mm) longer than the length of the cut involved.

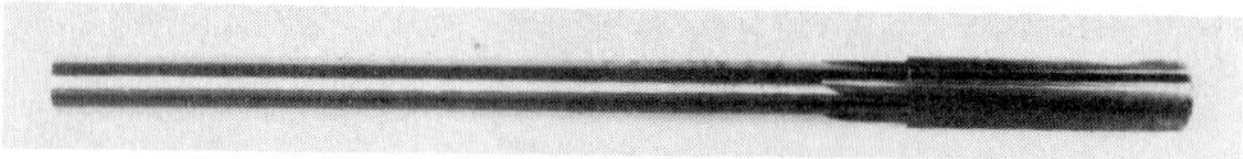

FIGURE 44-8
The straight fluted rose reamer cuts only on the front chamfered edges of its teeth. (Credit: DoAll Company)

8. Calculate the lathe speed to cut a $\frac{3}{4}$ in. (19 mm) diameter hole.

9. Start the lathe and bring the boring tool in light contact with the diameter of the hole.

10. Move the boring bar beyond the front edge of the work surface.

11. Take a trial cut, using a 0.005 to 0.010 in. (0.12 to 0.25 mm) depth of cut and advance the carriage slowly by hand.

12. Stop the lathe upon completing a $\frac{1}{4}$ in. (6 mm) wide cut.

13. Measure the diameter using a vernier caliper or inside micrometer.

14. Set the cross-feed handle to remove the difference between the diameter just cut and hole size, leaving 0.010 to 0.020 in. (0.25 to 0.50 mm) for a finish cut.

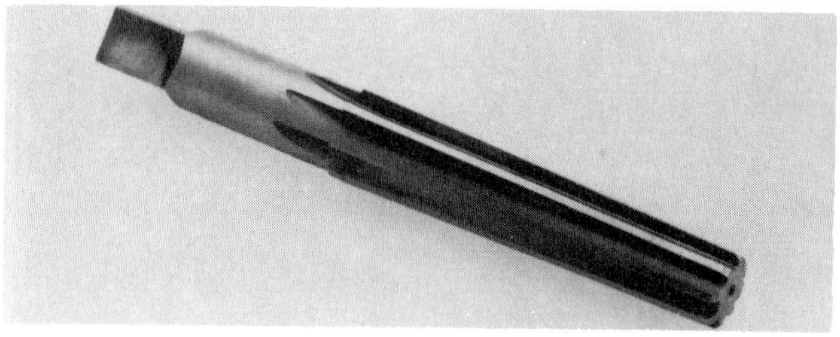

FIGURE 44-9
A tapered shank Morse taper reamer. (Credit: DoAll Company)

15. Start the lathe and take the roughing cut.

16. Upon completion of the cut, stop the lathe and move the boring bar clear of the hole without disturbing the cross-feed handle.

17. Set the lathe for a fine feed.

18. Take a trial cut and measure the diameter.

19. Set the lathe to take the depth of finish cut and complete the boring operation.

Types of Reamers

As previously discussed in Unit 33, a reamer is a cutting tool whose edges are made to machine a hole accurately within very close limits. Machine reamers can be had in a number of styles that can be used for either roughing or finishing a hole. Only a few types will be discussed here.

The rose reamer (Figure 44-8) can come with straight or helical flutes along its length, straight or tapered shank, and is used primarily to remove metal quickly to bring a hole close to finish size. The cutting edges on the reamer are located on the 45° chamfer on the end of the cutter. This is the only section of the flutes that is ground to do any cutting.

Fluted machine reamers (see Figure 33-10) are available with straight or helical flutes and are designed to cut on both the sides and on the end of the tool. These reamers are available in both straight or tapered shanks (Figure 44-9) and in production are used after the rose reamer to cut a hole smooth, clean and to exact size.

Reaming a Hole

In general, reaming in a lathe will be similar to reaming in a drill press. The amount of material to be left in a hole will depend on the metal involved and how the hole was produced. Drilled and bored holes will require more material than holes previously rough reamed.

Lathe reaming speeds should be about two-thirds of the required drilling speed for the same material. Running the lathe too fast will result in an oversize hole. As a general guide, reamers should be brought to the edge of the hole before starting and fed slowly into the work, using plenty of lubrication in the process.

Hand reamers should be started straight and kept true in the hole. To help alignment in a lathe, a center is used. The lathe center is used to support the square shank and is advanced against the reamer as the square is turned by a suitable wrench. Again, lubrication should be used freely.

Tapping a Hole Square

Hand tapping in a lathe where the workpiece hole has not been disturbed after machining provides an excellent means for tapping square. This can be accomplished with plenty of lubrication and the normal advancing and backing-off of the tap with the center being used to keep the tap square in the hole. In order to produce the $\frac{3}{4}$-16NF tapped hole previously discussed in the boring operation, the following steps must be followed.

Procedure

1. Wear appropriate safety glasses.
2. Chamfer the end of the hole larger than the tap diameter.
3. Clear the compound rest of any tooling and cover it with a cloth (Figure 44-10).
4. As a means of "locking" the headstock, set the lathe to its lowest speed.
5. Select the proper size tap with a center hole in the end of the square shank.
6. Insert a hardened lathe center in the tailstock spindle.
7. Extend the tailstock spindle to an inch (25 mm) length.
8. Place the tap in the workpiece hole and slide the tailstock until the center enters the end of the tap (Figure 44-11).
9. Lubricate the tap (Figure 44-12).
10. Grip the tap's square shank with a suitable wrench and place the other hand on the tailstock handwheel.
11. Turn the tap wrench and handwheel clockwise with uniform motion for about one-quarter turn (Figure 44-13).
12. Stop advancing the tap and prepare to back it out.
13. Unlock the tailstock but do not remove the center (Figure 44-14).

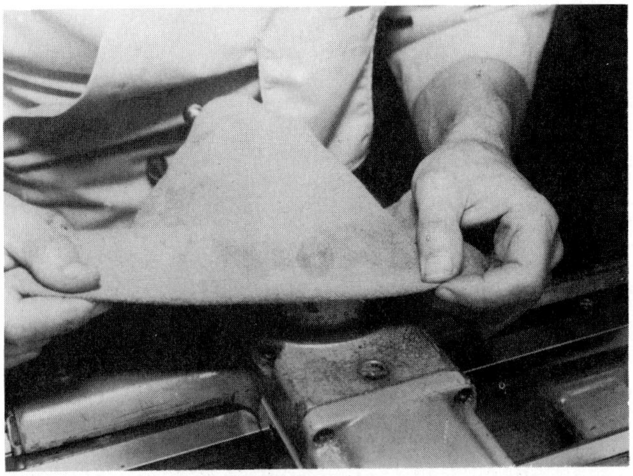

FIGURE 44-10
Clear the compound rest and cover it with a cloth.

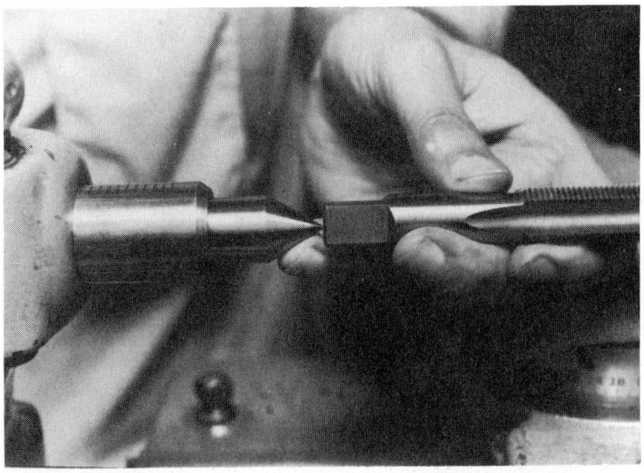

FIGURE 44-11
The lathe center should support the end of the tap.

FIGURE 44-12
Lubricate the tap before starting the tapping operation.

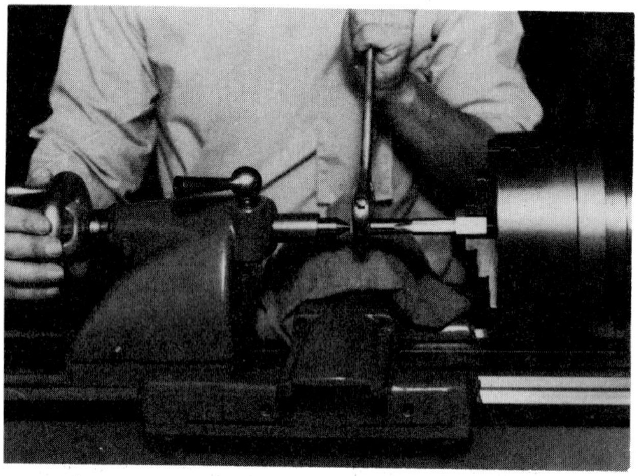

FIGURE 44-13
The handwheel and tap wrench should be turned together with uniform motion.

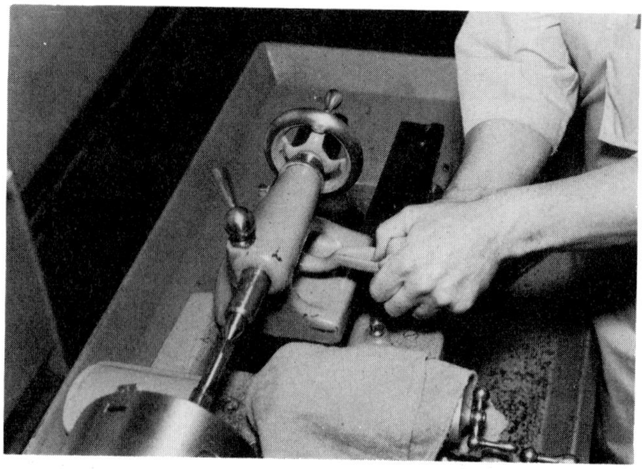

FIGURE 44-14
Unlock the tailstock but do not remove the center.

FIGURE 44-15
Back off the tap to clear it of chips.

14. Back off the tap by turning it counterclockwise to clear it of chips formed (Figure 44-15). The tailstock will move with it.

15. Relubricate the tap.

16. Slide the tailstock toward the workpiece (Figure 44-16).

17. Thread the tap into the hole by hand.

18. Lock the tailstock.

Repeat the original step of uniformly turning the tap wrench and handwheel (step 11) until the tap begins to bind again. *Caution:* Do not force the tap's advancement when it begins to bind for it will lead to breakage. Repeat the above steps until the hole has been threaded (Figure 44-17).

Die Chasing

External threads can be cut in a lathe by means of a single-point tool that is ground to match the profile of the thread being cut (Figure 44-18), or it can be cut by chasing. Single-point tools complete the

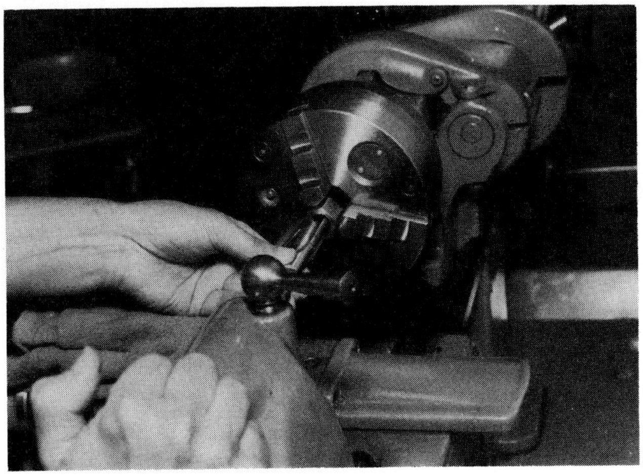

FIGURE 44–16
Slide the tailstock center toward the workpiece to again support the tap.

FIGURE 44–18
Cutting a thread using a single-point tool.

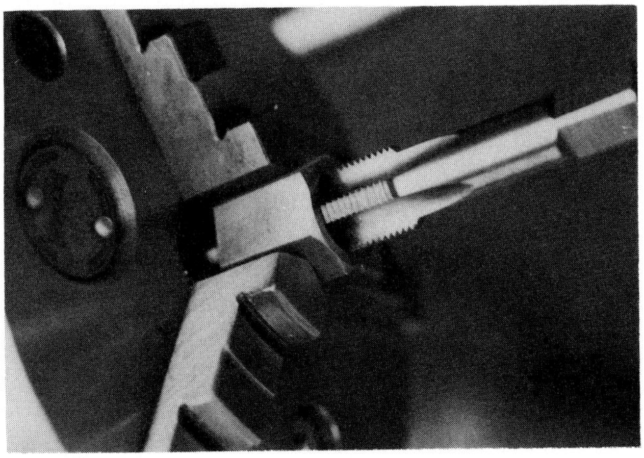

FIGURE 44–17
When the tap can be turned freely in the hole, the tapping operation is complete.

FIGURE 44–19
The releasing mechanism in this die chasing head permits close shoulder threading. (Credit: Union/Butterfield)

threading operations through a series of cuts whereby the cutting tool is synchronized with the rotating workpiece to enter the path of the spiral involved precisely time after time.

Chasing is somewhat similar to cutting threads with a stock and die. A series of cutting teeth (chasers) are engaged to thread the workpiece. One type of chasing is performed with a stationary die head (Figure 44–19). This style of thread cutting is used in industry on hand-operated turret lathes and single-spindle automatic lathes. Stationary die heads can also be mounted in the tailstock of an engine lathe. These tools can be equipped with four or more chasers to cut right- or left-hand threads, straight or taper, and in a variety of diameters, depending on the capacity of the tool. The self-opening die head is placed against the workpiece and set by means of a closing handle. Once the threads are cut, deeper and deeper by successive teeth, the die head opens at the completion of the forward travel. The cutting tool is then automatically withdrawn, thus eliminating the need to reverse the spindle rotation.

UNIT 45 CUTTING 60° SCREW THREADS

Cutting a thread in a lathe requires an understanding of the technical terms associated with the screw thread and the ability to measure it. Properly done, there should be no guesswork involved in obtaining a thread measurement with reasonable accuracy.

The actual process of cutting the thread involves making a series of cuts. The threading tool is fed a little at a time, not plunged in one complete cut. The manner in which the cutting tool is fed, the depth of cut, and correct shape of the cutting tool all play a role in the process of cutting a thread.

Objectives
After completing this unit you should be able to

1. calculate the thread form measurements for an external Unified thread.
2. calculate the thread form measurements for an external ISO metric thread.
3. set up a lathe to cut an external 60° Unified thread.
4. set up a lathe to cut an external 60° ISO metric thread.
5. reset a threading tool.

Thread Terms

A screw thread may be defined as a helical groove formed around the inside or outside of a cylinder or cone. Although there are a variety of shapes or profiles of thread forms, the American National screw and its later slightly modified counterpart, the Unified thread, are the most widely used thread forms in commercial use throughout the world. This thread form is illustrated in Figure 45-1 and will be used for the thread cutting operations that follow. Herewith is a listing of the more important terms:

External thread: The thread on the outside surface of a fastener such as a bolt or a screw.

Internal thread: The thread on the inside surface of a hole diameter. A nut would be a prime example.

Major diameter (MD): The largest diameter to be found on an external or internal thread, replacing former thread terms such as OD (outside diameter) and full diameter.

Minor diameter: The smallest diameter to be found on an external or internal thread, replacing former terms such as "core diameter," "root diameter," and "inside diameter."

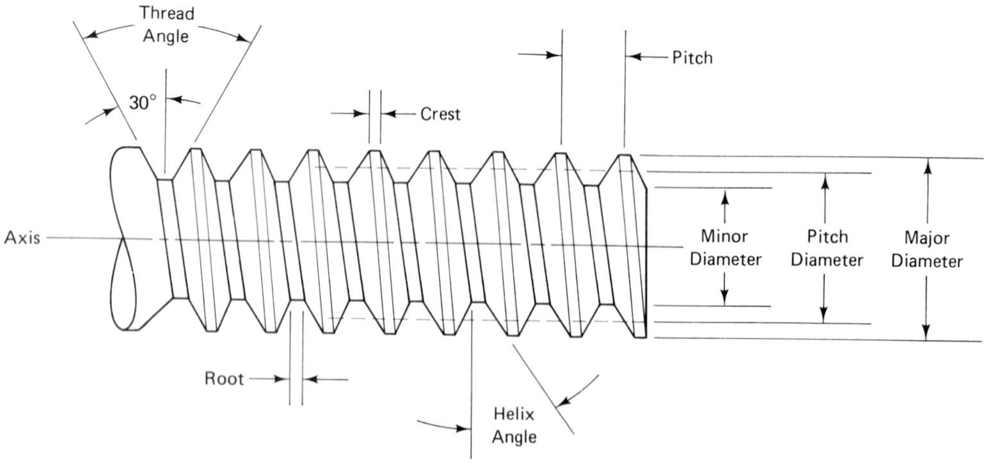

FIGURE 45-1
Main parts of a thread.

N: The number of threads per inch of a threaded section. This term does not apply to metric threads.

Pitch (P): The distance from a point on one thread to a corresponding point on the thread next to it, measured parallel to the axis, is called pitch. This term does not apply to metric threads.

Pitch diameter (PD): An invisible diameter that is formed in the thread's profile between the major and the minor diameters as it passes through a point where the width of the thread and groove are equal.

Root: The bottom surface of a thread that lies between the joint's adjacent threads.

Crest: The top surface of the two sides of a thread.

Flank: The side of a thread that connects the crest with the root.

Depth of thread (D): The distance between the crest and root measured perpendicular to the axis.

Lead: The distance a screw thread advances axially in one complete turn. On a single-thread screw, the lead and pitch are identical; on a double-thread screw, the lead is twice the pitch; a triple-thread screw, the lead is three times the pitch etc.

Lead angle: The angle made by the helix of the thread at the pitch diameter with a plane perpendicular to the axis.

Right-hand thread: A thread wound or advanced in a clockwise direction.

Left-hand thread: A thread wound or advanced in a counterclockwise direction.

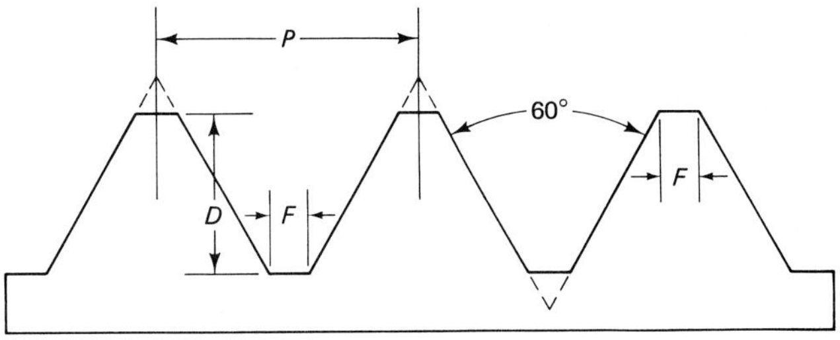

$$D = 0.6495 \times P \text{ or } \frac{0.6495}{N}$$

$$F = 0.125 \times P \text{ or } \frac{0.125}{N}$$

(a)

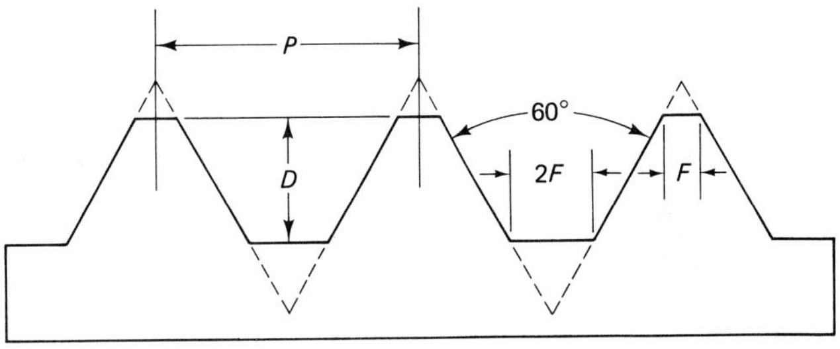

D (external) = $0.54127 \times P$
FC = $0.125 \times P$
FR = $0.250 \times P$

(b)

FIGURE 45–2
Comparison of (a) American National and (b) ISO metric thread.

The ISO Metric Thread Form

There are a number of similarities and differences between the ISO metric thread and the American Standard or Unified thread form. As listed in Table 25–4, there are 25 thread sizes ranging in diameter from 1.6 to 100 mm. While both thread forms have a 60° included angle, the ISO metric thread has a difference in depth of thread, which affects the size of the root and crest of the thread. As can be seen in Figure 45–2, the illustration shows how the root of the metric thread is wider than the Unified thread form. The depth of the metric thread is not as deep, which tends to produce a stronger thread for the diameter and pitch involved. The American Standard thread form has a uniform crest and root. The root of the ISO metric thread is one-fourth the pitch ($0.250 \times P$).

Thread Calculations

In order to cut a thread in a lathe, a series of calculations have to be performed. The pitch, depth, minor diameter, width of flat, and, if a thread micrometer is to be used, the pitch diameter will have to be determined.

Example

Find the pitch, depth, width of flat, minor diameter, and pitch diameter for a 1″–8 NC thread.

$$P = \frac{1}{N} = \frac{1}{8} = 0.125 \text{ in.}$$

$$\text{Depth } (D) = 0.64952 \times P = 0.64952 \times 0.125 = 0.081 \text{ in.}$$

$$\text{Width of flat} = \frac{P}{8} = \frac{0.125}{8} = 0.0156 \text{ in.}$$

$$\text{Minor diameter} = \text{major diameter} - (D + D) = 1.000 - (0.081 + 0.081) = 1.000 - 0.16238 = 0.838 \text{ in.}$$

$$\text{Pitch diameter} = \text{MD} - D = 1.000 - 0.081 = 0.919 \text{ in.}$$

Example

Calculate the pitch, depth, minor diameter, width of crest and root, and pitch diameter for a M 6.3×1.0 thread.

$$\text{Nominal diameter (MD)} = 6.3 \text{ mm}$$

$$P = 1.0 \text{ mm}$$

$$D = 0.54127 \times 1 = 0.54 \text{ mm}$$

$$\text{Minor diameter} = \text{MD} - (D + D) = 6.30 - (0.54 + 0.54) = 5.22 \text{ mm}$$

$$\text{Width of crest} = 0.125 \times P = 0.125 \times 1 = 0.125 \text{ mm}$$

$$\text{Width of root} = 0.250 \times P = 0.250 \times 1 = 0.25 \text{ mm}$$

$$\text{Pitch diameter} = \text{MD} - D = 6.3 - 0.54 = 5.76 \text{ mm}$$

Machining the Pitch Diameter

The problem facing anyone cutting a thread is fit. How deep does the operator cut with the threading tool, and when is it time to stop cutting? Since the class of fit of any thread is controlled by the dimension of the pitch diameter, it remains for the operator to achieve this measurement by machining.

A simple means of understanding the amount of metal to remove in cutting a thread would be to visualize a threaded blank as a solid blank with two distinct diameters. The minor diameter forms a step with the major diameter (Figure 45-3). The measured distance from the major diameter to the minor diameter on one side is called the single depth. The single depth represents the amount of metal removed, or depth of threading tool penetration, on one side of the thread. The total amount of metal removal would be considered the double depth. What an operator will be doing is grooving a machined blank with a threading tool, using the single-depth or double-depth calculations. The threads cut will be measured with a tool such as a thread micrometer. The measurement obtained will be the pitch diameter.

The double-depth value can be found on the back of a center gage as well as in Table 45–1. An operator will have three methods of cutting the thread. The cutting tool can be advanced straight in using the crossfeed handle, or it can be advanced angularly using the compound rest handle.

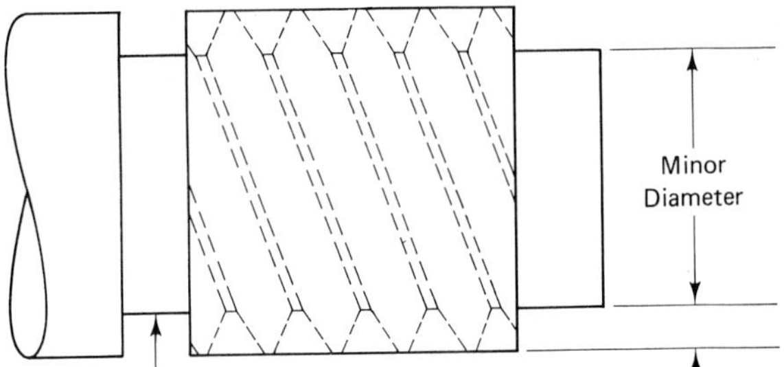

FIGURE 45–3
Illustration showing the amount of metal that is removed to cut a thread.

Lathe Operation Setup

There are several steps to be taken in cutting a thread besides putting the lathe in a thread cutting mode. These would include safety, assuring workpiece accuracy, gearing to cut a required screw pitch, tool setup, and lathe speed.

1. Whether a workpiece is machined between centers or held in a chuck, room should be provided at the end of the thread for an operator to back off the threading tool without running into the rotating

TABLE 45–1
60° National form thread cutting information

	Crossfeed setting		*29° Compound rest setting*	
Threads per inch	*Single depth*	*Double depth*	*Single depth*	*Double depth*
4	0.1625	0.325	0.1858	0.3716
6	0.108	0.216	0.1235	0.247
7	0.093	0.186	0.106	0.212
8	0.081	0.162	0.092	0.184
9	0.072	0.144	0.082	0.164
10	0.065	0.130	0.074	0.148
11	0.059	0.118	0.0674	0.1348
13	0.050	0.100	0.057	0.114
14	0.0464	0.093	0.0525	0.105
16	0.0405	0.081	0.046	0.092
18	0.036	0.072	0.041	0.082
20	0.0325	0.065	0.037	0.074
24	0.027	0.054	0.0308	0.0616
28	0.0231	0.0463	0.0264	0.0528

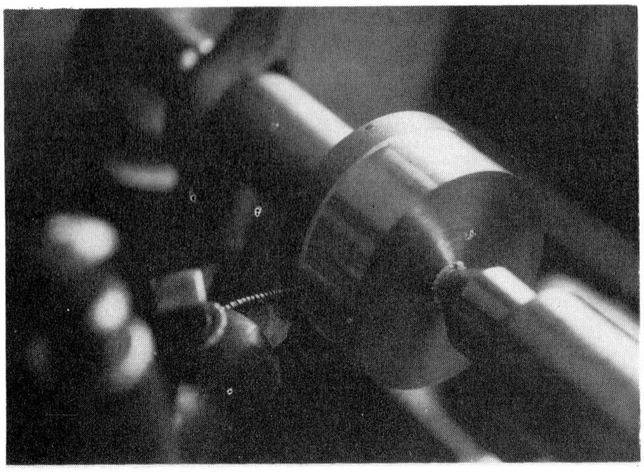

FIGURE 45–4
Do not disturb the concentricity (trueness) of the work setup before cutting the thread.

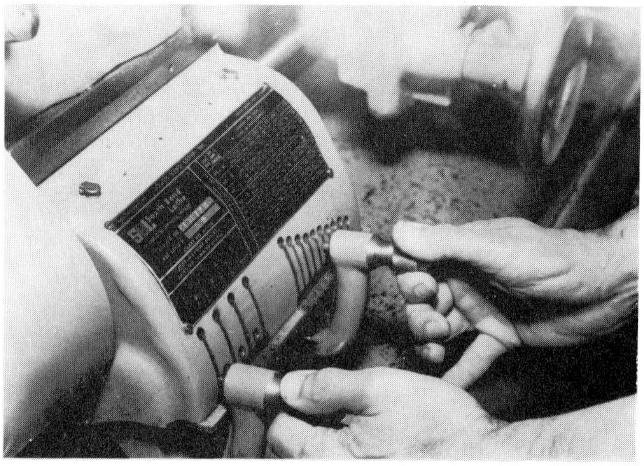

FIGURE 45–5
Setting the quick-change gearbox to cut a thread.

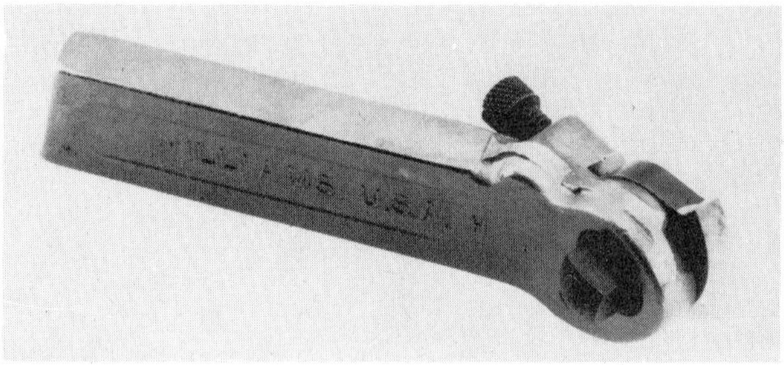

FIGURE 45–6
A formed threading tool. (Credit: DoAll Company)

chuck or spinning lathe dog. To assure accuracy, the major diameter should be machined prior to the thread cutting operation (Figure 45–4). If working between centers, always mark off the drive plate slot with chalk where the lathe dog's tail enters. This will prevent any false setup if the workpiece is removed and has to be set back for additional machining. Setting up in a drive plate slot other than the original will cause a new series of threads to be cut, ruining or cross threading what has already been cut.

2. Modern lathes use a quick-change gearbox to set the lathe for the required pitch of the thread to be cut (Figure 45–5). By moving tumbler levers or similar arms, positive carriage movement is achieved by positioning the gearing between the spindle and the lead screw.

3. Beginners should bear in mind that since cutting threads is really a problem of coordination that involves disengaging the carriage feed and cutting tool from the workpiece at the same time, lathe speed is important. More advanced operators can cut a thread at about one-half or higher the rpm required for turning the material involved.

The Cutting Tool

If a formed threading tool (Figure 45–6) is not used, a toolbit will have to be ground to the shape or form of the screw thread being cut. A center gage with included angles of 60° notched along its edges is used in checking the accuracy of the threading tool's included angle.

The toolbit should not be ground to a point, but rather have a flat equal to the thread's root (0.125P for Unified threads, 0.250P for metric threads); fine-pitched threads will require a rounding of the point with an oilstone. Front clearance on a threading tool should be between 8° and 10° to cut freely without interfering with the helix angle. Generally, the top of the toolbit is flat and ground without slopes and slants, although a side rake may be used when cutting into steel.

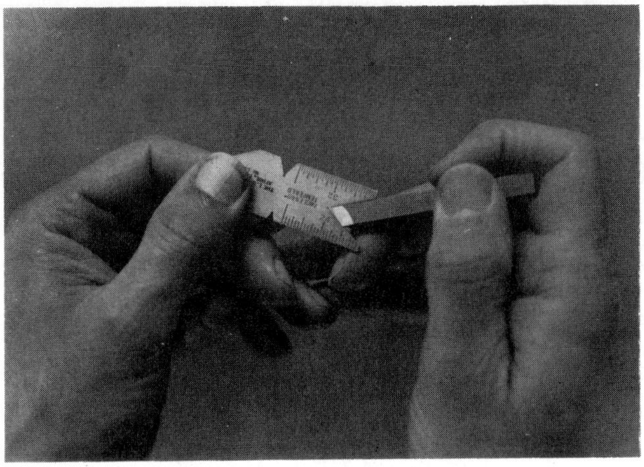

FIGURE 45-7
Checking the angle of a thread cutting tool.

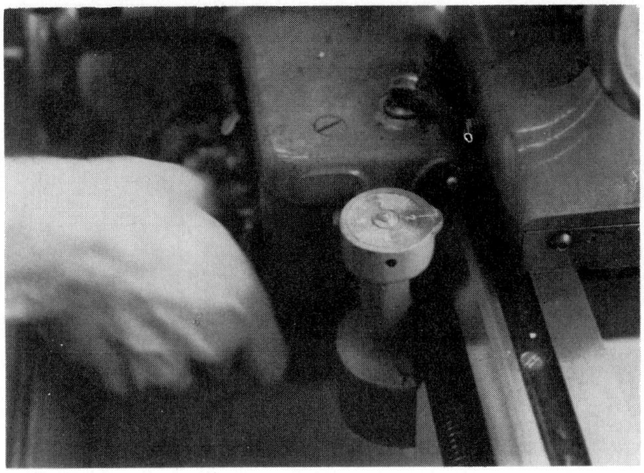

FIGURE 45-8
Using a thread dial indicator as a guide to engage the thread cutting tool.

While the conventional threading tool is shaped with its angle near midpoint, the toolbit should be ground by using the right side of the toolbit as part of the 60° included angle (Figure 45-7). This will provide more room for thread cutting as the toolholder is swung away from the left side of the workpiece.

Thread Chasing Dial

A thread chasing dial, also called a thread dial indicator by some manufacturers (Figure 45-8), is a simple clocklike device attached to the carriage or in some cases built into the apron. The thread chasing dial enables the operator to enter the threading tool in the same groove everytime if engaged properly.

A small gear located on the lower half of the thread dial indicator meshes with the lead screw, causing the graduated dial on the upper portion to rotate (Figure 45-9). The dial is divided into eight graduated lines, four of which are numbered. Every movement of the carriage or lead screw causes the graduated lines to move past a reference mark. When the lathe is set for thread cutting, the carriage is put into motion by engaging a split-nut lever that closes a split nut onto the lead screw.

The ability to engage, disengage, and reengage the split-nut lever so that the tool enters the originally cut groove every time speeds up the thread cutting process. With the use of the thread chasing dial, one does not have to reverse the spindle rota-

FIGURE 45-9
Cutaway of a thread dial indicator meshing with the lead screw.

Threads per inch	When to engage half nut	Reading on dial
Even-numbered threads	Engage at any line on dial	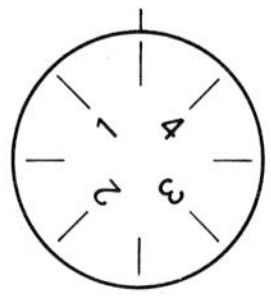
Odd-numbered threads	Engage at numbered line on dial	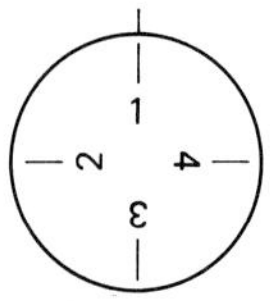
($\frac{1}{2}$) Fractional numbered threads	Engage at odd-numbered line on dial	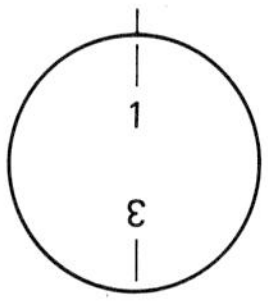
($\frac{1}{4}$) Fractional numbered threads	Engage at starting line on dial each time	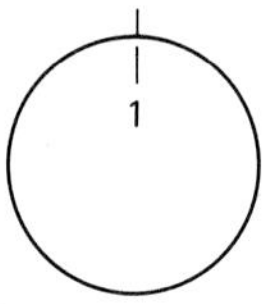
Multiple of lead screw threads	Engage at any time	Use of dial not needed

FIGURE 45–10
Thread chasing dial guide.

tion to come back to the original cut except for special situations. A thread chasing dial guide can be seen in Figure 45–10. With the exception of fractional and special threads, the split nut is engaged on any line when cutting even numbered threads; odd threads on numbered lines only.

Positioning the Compound Rest

For more efficient cutting, the compound rest should be swung counterclockwise, positioned and locked at 29°. (For left-hand threads, the compound rest should be turned clockwise and locked at 29°.) It has been found that by advancing the cutting tool angularly, the left side of the toolbit will remove most of the chip while shaving the right side, leaving a cleaner cut. Another factor in its favor is that the micrometer collar on the compound rest handle acts as a counter in taking depth of cut readings.

Cutting Tool Setup

Once the compound rest has been positioned, the threading tool can be set up.

Procedure

1. Lock the toolholder near the left side of the compound rest.
2. Set the 60° threading tool within the toolholder, locking it with about $\frac{1}{2}$ in. (13 mm) extended (Figure 45–11).
3. Set the cutting tool even with the height of the lathe center and snug-up the tool post screw (Figure 45–12).
4. Place a center gage against the centerline of the workpiece or tailstock spindle.
5. Lay a sheet of white paper beneath the cen-

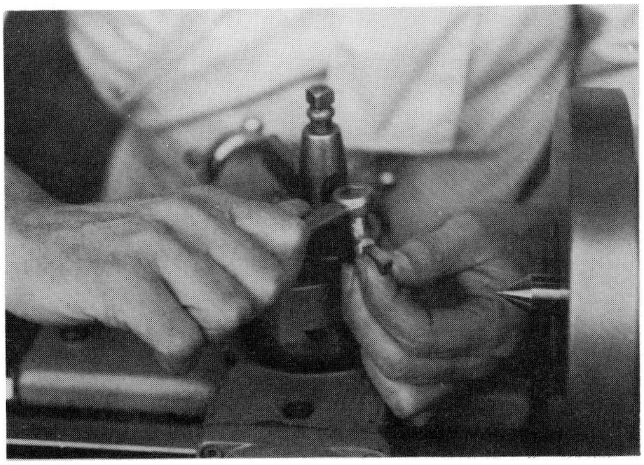

FIGURE 45-11
Extend the threading tool not more than $\frac{1}{2}$ in. (13 mm) from the toolholder.

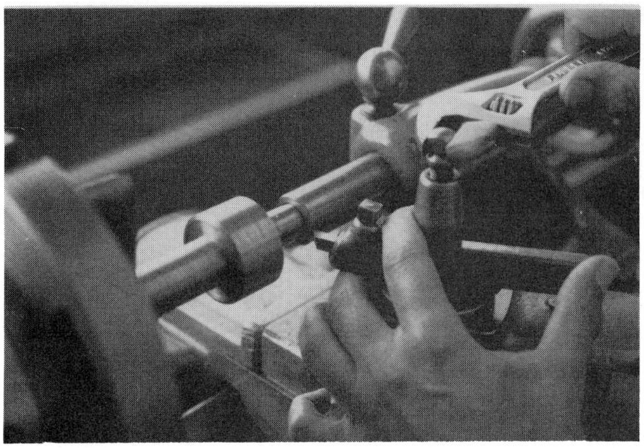

FIGURE 45-12
Snug-up the tool post screw so that the cutting tool height can be ajdusted.

ter gage to better see how well the threading tool lines up with the center gage.

6. Lightly tap the side of the toolholder to direct the toolbit's cutting edge against the surface of the V-notch in the gage (Figure 45-13).

7. Once the alignment is achieved, tighten the tool post screw securely. (*Note:* Do not jam the threading tool into the V opening to assure alignment.)

8. Recheck the height of the toolbit against the lathe center to make sure it has not changed.

Depth Settings

The double-depth value found on the back of a center gage is calculated feeding the cutter straight into the workpiece. To accomplish the same depth using the compound rest set at 29° will require a slightly longer travel of the cutting tool. This can be seen in Table 45-1 where the single-depth and double-depth settings can be found for crossfeed setting (straight in cut) and for 29° (angular cut). For example, the double-depth setting for an 8-thread per inch screw with the cutter fed straight in (crossfeed setting) would be 0.162 in.; fed at 29°, the setting would read 0.184 in. The depth setting for 60° ISO metric threads can be found in Table 45-2.

Thread Cutting Operation

One of the prime requirements of thread cutting is safety. Beginners should work on surfaces that are undercut or machined behind the threaded area to a level below the minor diameter. The undercut should be wide enough to permit the cutting tool to complete the thread cut and still allow for its withdrawl without damaging the adjoining surface or shoulder. It is also recommended that several practice runs of engaging the split-nut lever (while using the thread chasing dial as a guide) and disengaging the split-nut lever (while backing out the threading tool) be attempted before actual cutting takes place.

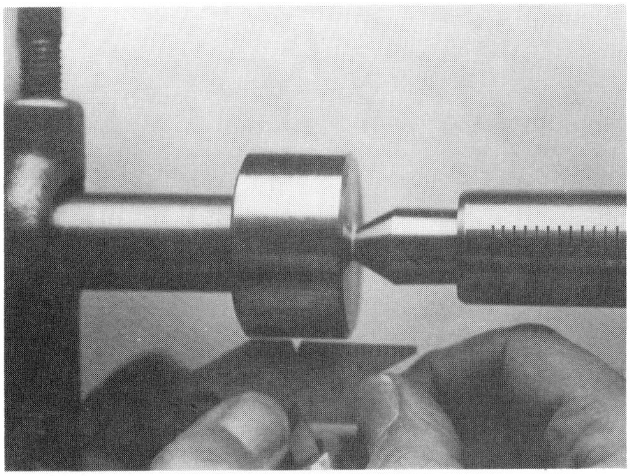

FIGURE 45-13
Place the toolbit's cutting edge square against the workpiece using a center gage.

TABLE 45-2
ISO metric thread cutting information

	0° Crossfeed setting		*29° Compound rest setting*	
Pitch (mm)	*Single depth*	*Double depth*	*Single depth*	*Double depth*
0.35	0.19	0.38	0.21	0.42
0.4	0.21	0.42	0.24	0.48
0.45	0.24	0.48	0.27	0.54
0.5	0.27	0.54	0.31	0.62
0.6	0.32	0.64	0.37	0.74
0.7	0.37	0.74	0.43	0.86
0.8	0.43	0.86	0.49	0.98
1.0	0.54	1.08	0.62	1.24
1.25	0.67	1.34	0.77	1.54
1.5	0.81	1.62	0.93	1.86
1.75	0.94	1.88	1.08	2.16
2.0	1.08	2.16	1.24	2.48
2.5	1.35	2.70	1.55	3.10
3.0	1.62	3.24	1.85	2.70
3.5	1.89	3.78	2.16	4.32
4.0	2.16	4.32	2.47	4.94
4.5	2.44	4.88	2.78	5.56
5.0	2.71	5.42	3.09	6.18
5.5	2.98	5.96	3.40	6.80
6.0	3.25	6.50	3.71	7.42
6.5	3.52	7.04	4.02	8.04
7.0	3.79	7.58	4.33	8.66
7.5	4.06	8.12	4.64	9.28
8.0	4.33	8.66	4.95	9.90

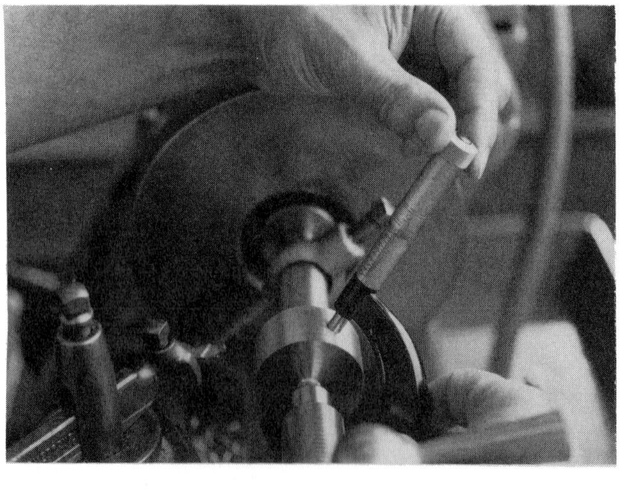

FIGURE 45-14
The major diameter should measure 0.002 in. (0.05 mm) below the basic size.

Procedure

1. Wear appropriate safety glasses before attempting to operate a lathe.

2. Check the workpiece to see that it is running true.

3. Measure the major diameter. It is good practice to have it measure 0.002 in. (0.05 mm) below the basic size (Figure 45-14).

4. Put the lathe in a thread cutting mode, which will allow the split-nut lever to be engaged.

5. Set the quick-change gearbox for the required number of threads being cut (pitch in millimeters if using metric).

6. Set the lathe to run at one-half the regular turning speed.

7. Start the lathe, and using the V edge of the cutting tool, chamfer the front and back of the

FIGURE 45-15
To avoid a sharp edge burr, chamfer the front and back of the thread area.

FIGURE 45-17
Lock the crossfeed micrometer collar on zero.

FIGURE 45-16
Coat the surface with layout dye.

FIGURE 45-18
Lock the compound rest micrometer collar on zero.

thread area just below the minor diameter (Figure 45-15).

8. Coat the threading area with layout dye (Figure 45-16).

9. To remove all backlash, reverse the crossfeed handle a few turns, and just as quickly advance the cutting tool toward the workpiece so that the tool is close but not touching, leaving the crossfeed handle in about the 9 o'clock position.

10. Lock the zero line on the micrometer collar to match the index line (Figure 45-17).

11. Reverse the compound rest handle a few turns and carefully advance the threading tool to just touch the inner edge of the workpiece.

12. Lock the micrometer collar zero line to match the index line (Figure 45-18).

13. Using the thread chasing dial as a guide, engage the split-nut lever to take a trial cut.

14. At the end of the cut, disengage the split-nut lever with one hand while at the same time backing off the threading tool by turning the crossfeed handle with the other hand.

15. Stop the lathe and check the pitch with a screw pitch gage (Figure 45-19).

FIGURE 45-19
Check the trial cut with a screw pitch gage.

FIGURE 45-21
Engage the split-nut lever using the thread chasing dial reading as a guide.

FIGURE 45-20
Advance the compound rest handle for a 0.005-in. (0.12-mm) depth of cut.

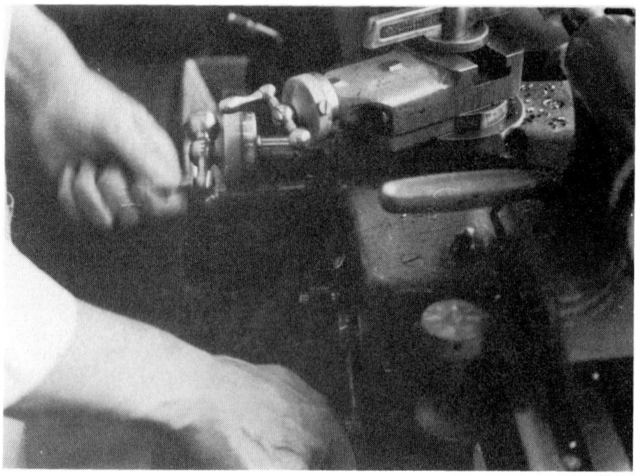

FIGURE 45-22
Disengage the split-nut lever and back out the threading tool at the same time.

16. Move the threading tool in front of the diameter to be threaded and turn the crossfeed handle clockwise to bring it back to its original zero setting.

17. Advance the compound rest handle to take a 0.005-in. (0.12-mm) depth of cut (Figure 45-20).

18. Start the lathe and apply cutting fluid to the workpiece.

19. Using the thread chasing dial reading as a guide, engage the split-nut lever to start the cutting (Figure 45-21).

20. Disengage the split-nut lever at the completion of the cut with one hand while at the same time turning the crossfeed handle counterclockwise a few turns with the other hand (Figure 45-22).

21. Move the threading tool to the right, lining it up near the front edge of the workpiece (Figure 45-23).

22. Carefully turn your crossfeed handle clockwise to advance the threading tool close to the workpiece without touching it for visual alignment.

FIGURE 45-23
Move the lathe carriage to the right and position the threading tool near the front edge of the workpiece.

FIGURE 45-25
Turn the compound rest handle to take another 0.005-in. (0.12-mm) depth of cut.

FIGURE 45-24
Turn the crossfeed handle to reset the micrometer collar to its original zero reading.

FIGURE 45-26
Apply cutting fluid.

23. Position the threading tool by moving the carriage to the right beyond the start of the thread.

24. Carefully turn the crossfeed handle to reset it to a zero reading with the index line (Figure 45-24).

25. Take another 0.005-in. (0.12-mm) depth of cut using the compound rest handle (Figure 45-25).

26. Apply cutting fluid (Figure 45-26).

27. Engage the split-nut lever as before, using the thread chasing dial as a guide (Figure 45-27).

Steps 20 to 27 are repeated, taking 0.005-in. (0.12-mm) depth of cuts until the operator is within 0.007 to 0.010 in. (0.17 to 0.25 mm) of completing the double-depth requirement. The pitch diameter should be checked, having first removed any burrs from the top of the threads with a file, as well as cleaning any chips and cutting fluid (Figure 45-28). Completion of the thread will require careful measurement and light cuts of 0.001 to 0.002 in. (0.02 to 0.05 mm) deep.

FIGURE 45-27
Engage the split-nut lever using the thread chasing dial reading as a guide.

FIGURE 45-28
Threads should be clean of chips and cutting fluid when measuring their accuracy.

Resetting the Threading Tool

Completing a shallow cut thread or one that has been interrupted due to premature removal from the lathe or tool sharpening can present a problem in resetting the threading tool. Not only does the tool have to be placed square against the body of the threads, but the tool must enter exactly into the same groove without leaving a ridge.

Procedure

1. Appropriate safety glasses must be worn before attempting any work on a lathe.
2. Set up the lathe and workpiece for thread cutting as previously explained.
3. Start the lathe spindle.
4. To remove all backlash, reverse the crossfeed handle a half of a turn, and then advance the cutting tool near the workpiece but do not touch it.
5. Using the thread chasing dial as a guide, engage the split-nut lever on the correct line and stop the lathe with the threading tool within the length of the unfinished threads. (*Note:* Do not disengage the split-nut lever.)
6. Reverse the compound rest handle half a turn and then carefully advance the threading tool to visually align it slightly to the right of center of the partially cut thread.
7. Advance the crossfeed handle until the threading tool just touches the right side of the thread.
8. Hold the crossfeed handle and carefully set the micrometer collar zero line to match the index line.
9. Repeat the operation for the compound rest feed handle.
10. Reverse the crossfeed handle, moving it counterclockwise a few turns to withdraw the threading tool from the thread groove.
11. Disengage the split-nut lever.
12. Move the carriage to position the threading tool to the end of the workpiece.
13. Turn the crossfeed handle clockwise to its zero line setting.
14. Start the lathe and take a trial cut without advancing the compound rest handle.
15. Upon disengagement, move the carriage to the right until the toolbit clears the end of the workpiece.
16. Reset the crossfeed handle to its zero reading.
17. Apply cutting fluid to the workpiece.
18. Continue the screw cutting operation to its finished size, taking depth of cuts using the compound rest handle as previously explained.

Converting the Lathe for Cutting Metric Threads

Metric threads can be cut on a standard quick-change gear lathe by using conversion gears in the gear train of the lathe. Since the lead screw of a standard lathe is designed to cut the pitch of a screw in inches, the lead screw has to be converted to cut the pitch of a metric thread, which is measured in millimeters. A relationship has to be established to determine the spacing (in threads per inch) in a metric thread. To do so, it will be necessary to understand the relationship between the inch and the centimeter.

$$1 \text{ in.} = 2.54 \text{ cm}$$

It can be readily seen that a ratio of 1 : 2.54 exists. In order to cut a metric thread, a set of conversion gears with a ratio of 1 : 2.54 will have to be used. Multiplying by 50 to get a whole number, these gears are

$$\frac{1}{2.54} \times \frac{50}{50} = \frac{50 \text{ teeth}}{127 \text{ teeth}}$$

In practical terms this means that a driving gear with 50 teeth is used as the spindle gear, while a follower, 127-tooth gear, is placed on the lead screw.

Using a Standard Quick-Change Gear Lathe to Cut a 2.5-mm Thread

Procedure

1. Mount the 50-tooth gear on the spindle.
2. Mount the 127-tooth gear on the lead screw.
3. Convert the 2.5-mm pitch to threads per centimeter.

$$10 \text{ mm} = 1 \text{ cm}$$

$$P = \frac{10}{2.5} = 4 \text{ threads/cm}$$

4. Set the quick-change gear box to cut 4 threads/in.
5. Set up the lathe for thread cutting as previously explained.
6. Start the lathe and take a trial cut.
7. Stop the lathe at the end of the cut, withdraw the threading tool but do not disengage the split-nut lever. *The split-nut lever should not be disengaged until the thread has been cut to depth.*
8. Reverse the lathe spindle rotation to move the threading tool clear of the workpiece.
9. Stop the lathe.
10. Check the lightly grooved thread cut with a metric screw pitch gage.
11. Complete the thread to the required depth as listed in Table 45-2.

Cutting a Left-Hand 60° Thread

Left-hand threads are often used for special holding requirements such as on the spindle of a toolroom surface grinder. Due to the fact that the rotation of the spindle or grinding wheel is in the opposite direction to the tightening of a left-hand nut, a locking action takes place in use.

As will be seen, the cutting of a left-hand thread is very similar to cutting the conventional right-hand thread.

Procedure

1. Set the quick-change gearbox to cut the number of threads required.
2. Set the lathe in a thread cutting mode, which will allow the split-nut lever to be engaged.
3. Position the feed reverse lever so that the lead screw will rotate in the opposite direction as for cutting a right-hand thread.
4. Position the compound rest 29° to the *left* (Figure 45-29).
5. Position the threading tool to the left of the compound rest, at lathe center height and at 90° to the workpiece.
6. Start the thread cut from the left side of the workpiece so that it advances toward the tailstock.
7. Complete the thread by taking depth of cuts similar to cutting any 60° thread, always cutting from left to right.

Cutting the Tapered Section Thread

Workpiece sections may be machine tapered using the offset method or a taper turning attachment.

FIGURE 45–29
The compound rest should be swung 29° to the left when cutting left-hand threads.

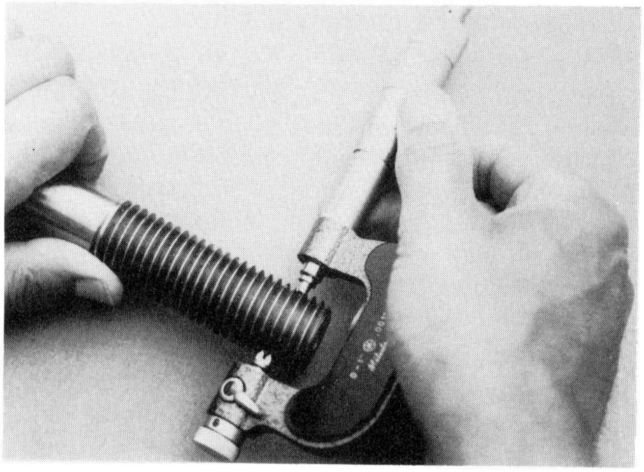

FIGURE 45–31
Checking the pitch diameter using a thread micrometer.

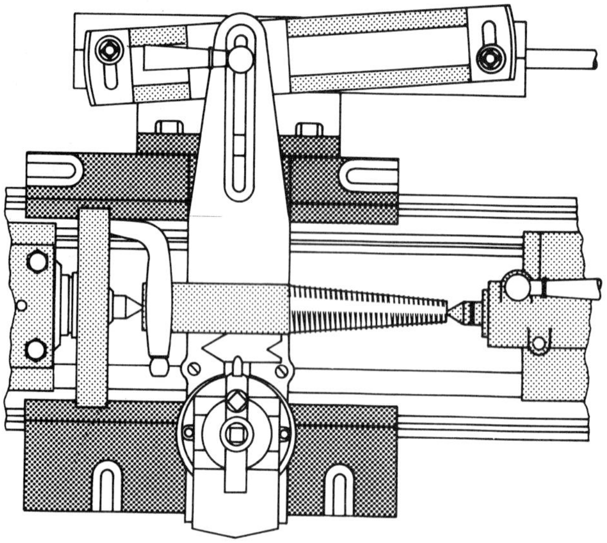

FIGURE 45–30
The toolbit should be placed square against the workpiece body and not against the tapered section.

Cutting a thread on these sections is no different than cutting a conventional right-hand thread. The only exception will be in the threading tool setup. It is important to understand that the toolbit should be placed square against the axis of the workpiece and *not* against the tapered section (Figure 45–30).

Measuring Thread Methods

There are a number of devices and tools used for measuring internal and external threads. Among them are the thread plug gage, the thread ring gage, and the roll thread snap gage, all of which have been covered in the fixed and comparison gage units.

Depending on how well equipped a toolroom is, measuring thread accuracy is generally done in one of several ways. The optical comparator (Figure 13–22) enlarges the profile of a thread and precisely compares its shadow to an equally scaled and enlarged detailed drawing.

The thread comparator visually indicates tolerance limits between set guides while also checking pitch diameter, lead, and thread angle error.

The screw thread micrometer (Figure 45–31) comes with measuring ends that include 60° points and grooves designed for a specific thread range. By placing the grooved anvil on the bottom thread and making a measuring contact directly opposite with the pointed spindle as in using a regular micrometer, the reading shown will be the pitch diameter.

Endorsed by the Bureau of Standards and the National Screw Thread Commission, the three-wire method is considered one of the most accurate methods of checking the pitch diameter. It is used by measuring with a micrometer over the diameters of hardened and highly accurate, uniform rods. Two rods are placed in adjoining thread grooves to enable the micrometer's anvil to seat against them while above, straddled directly opposite them, a third rod is set in a groove (Figure 45–32).

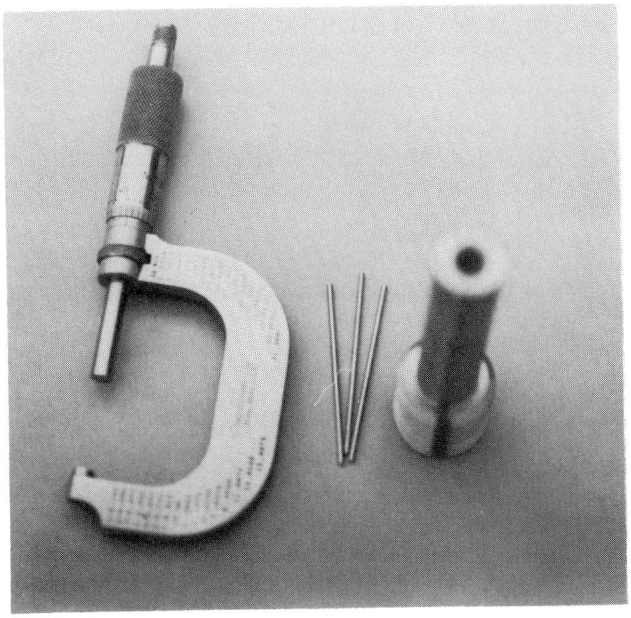

FIGURE 45-32
Preparing to check a thread using the three-wire system.

Using the Three-Wire Method

The accuracy for checking the pitch diameter of a standard single-thread screw is based on the best wire being used and elimination of errors in cutting the form of the thread, such as incorrect lead and thread angles. Since measurement is made over the seated wires, excessive contact pressure should also be avoided. For most practical requirements in cutting a 60° thread, the following formulas are recommended.

Wire size:

Smallest size wire = 0.50518 × pitch

Best size wire = 0.57735 × pitch

Largest size wire = 1.01036 × pitch

Measurement over wire:

$$M = D - (1.5155 \times P) + (3 \times W)$$

where M = over wire measurement

D = major diameter

P = pitch

W = wire size

Example:

Find the pitch diameter of an 1″-8 NC thread.

Procedure

1. Using the best size wire, calculate it as follows:

$$\begin{aligned}\text{Best size wire} &= 0.57735 \times \text{pitch}\\ &= 0.57735 \times \tfrac{1}{8}\\ &= 0.57735 \times 0.125\\ &= 0.07216\end{aligned}$$

2. With the best size wire selected, determine the measurement over wires.

$$\begin{aligned}M &= D - (1.5155 \times P) + (3 \times W)\\ &= 1.000 - (1.5155 \times P) + (3 \times 0.07216)\\ &= 1.000 - 0.18943 + 0.12648\\ &= 1.0270\end{aligned}$$

3. With the overall measurement known, determine the pitch diameter.

$$\begin{aligned}\text{PD} &= M - 3W + (0.86603 \times P)\\ &= 1.02705 - (3 \times 0.07216) + (0.86603 \times 0.125)\\ &= 1.02705 - (0.21648) + (0.10825)\\ &= 0.9188\end{aligned}$$

UNIT 46 CUTTING ADDITIONAL SCREW THREADS

Thread cutting operations on a lathe may also include the development of thread forms such as a square, Acme, or multiple lead threads. The square and Acme threads are designed to impart or transmit power, while the multiple-start threads are used in situations that may require the rapid advancement of travel along the thread and are often used in the construction of machines.

Objectives
After completing this unit you will be able to

1. set up and cut an external multiple thread.
2. set up and cut an external square thread.
3. set up and cut an external Acme thread.

Multiple Threads

On a single-start thread, the lead or the distance a nut advances on a straight line along the axis in one complete turn is identical to the pitch. The pitch is always a straight-line dimension of one point on a thread to a corresponding point on the thread next to it, measured parallel to the axis. Multiple-start threads may have a relationship of two, three, or four pitches to one lead. As illustrated in Figure 46–1, a single-start thread has one groove; a double-start thread has two grooves; a triple-start thread has three grooves, and a quadruple-start thread has four grooves.

The depth of thread cut on a multiple-start thread is not as deep as that of a single-start thread and produces a more pleasant appearance. The multiple-start thread also has less holding power and less tightening force than the single-start thread.

Several methods are used to cut multiple-start threads:

1. By using an accurate multislotted drive plate or face plate.
2. By using the spindle gear in the lathe's drive train for indexing.
3. By using the thread chasing dial to cut double-start threads with an odd-numbered lead.
4. By using the compound rest to index the threading tool.

Cutting a 60°, 16 tpi Double-Thread

Procedure

1. Select an accurately slotted drive plate or face plate. If neither is available, a special indexing plate may be used.
2. Fasten a lathe dog to the workpiece and mount it between centers, marking the dog tail slot or hole entry with chalk.
3. Set the lathe and cutting tool for cutting a single-start thread.
4. Set the quick-change gearbox to cut 8 tpi. (The lead of this thread is $\frac{1}{8}$ in.)
5. Calculate the depth of thread required for 8 tpi.
6. Start the lathe and feed the compound rest handle to take a trial cut.
7. Cut the first thread to within 0.001 in. of half the depth required for 16 tpi.
8. Note the reading on the compound rest micrometer collar.
9. Back out the threading tool using the compound rest handle but *do not move the crossfeed handle setting.*
10. Remove the workpiece and dog assembly from the lathe centers.
11. Rotate the workpiece so that the dog tail enters a slot or hole 180° opposite the entry first used.

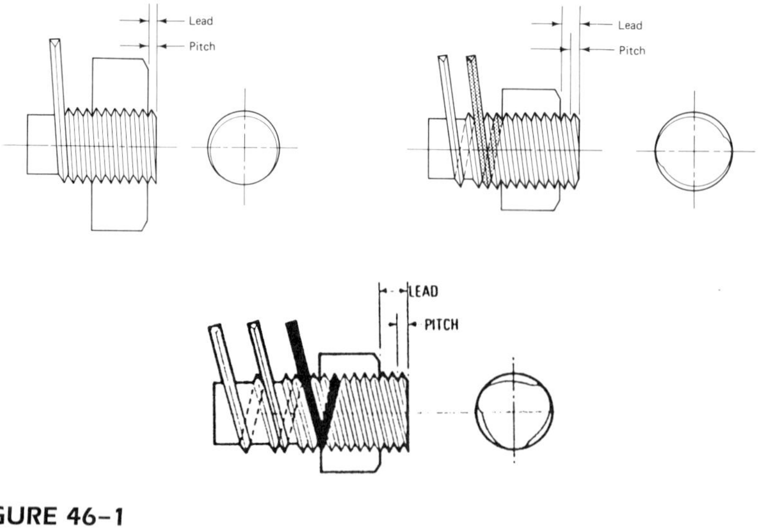

FIGURE 46–1
An illustration of a single-, double-, and triple-start threads. (Credit: Kostel Enterprises Limited)

12. Complete cutting the thread as explained in steps 6 and 7.

OR

1. Examine the gear train to see which tooth on the spindle gear is exactly between two teeth of the intermediate gear.

2. With chalk, mark both the spindle gear tooth and the space on the intermediate gear as well as any corresponding tooth and space on the lead screw gear (Figure 46-2).

3. Count the number of teeth required for a half revolution of the spindle gear, beginning with the first tooth next to the marked tooth, and mark this tooth with chalk.

4. Disengage the gearing that connects to the lathe spindle.

5. Turn the lathe spindle by hand so that the spindle gear is rotated 180° to line the upper tooth mark with the chalk mark on the intermediate gear.

6. Reengage the intermediate gear.

7. Start the lathe.

8. Since the crossfeed handle was not moved, reset the compound rest handle until the threading tool just touches the rotating workpiece.

9. Reset the micrometer collar to zero if necessary.

10. Cut the second thread to the calculated depth (0.001 in. deeper than the first thread).

FIGURE 46-2
Gear teeth should be properly marked before attempting any indexing.

11. Leave the crossfeed and compound rest handles at the final depth of cut setting and reset the gear positioning that was used to cut the first thread.

12. Take a cut through the first thread to assure that both threads are cut to the same depth.

Using the Thread Chasing Dial

The thread chasing dial may be used to cut double-start threads, including fractional or odd-numbered leads, without changing gears or rotating the workpiece in the following manner.

Procedure

1. Set up the lathe for thread cutting.

2. Take a trial cut along the workpiece by engaging the split-nut lever on a numbered line (Figure 46-3).

3. Reverse the crossfeed handle to withdraw the threading tool from the workpiece at the end of the cut.

4. Disengage the split-nut lever to move the carriage and position the threading tool to clear the end of the thread.

5. Return the crossfeed handle to zero (Figure 46-4).

FIGURE 46-3
The split-nut lever should be engaged on the number line for one cut of the double-start thread.

FIGURE 46-4
Reset the crossfeed handle to zero.

FIGURE 46-5
To index the threading tool, set the compound rest parallel to the lathe ways.

6. Engage the split-nut lever on an unnumbered line. This cut will be located in the center of the two adjoining grooves of the first thread.

7. Continue cutting the thread to the calculated depth of cut, alternating each pass by engaging the split-nut lever first on a numbered line and then on an unnumbered line. *Note:* Fractional or odd-numbered leads can be cut using the thread chasing dial, too, only the split-nut lever must be engaged on the even-numbered lines for one cut (2 or 4) and engaged on the odd-numbered lines for the second cut (1 or 3).

Using the Compound Rest Method

The compound rest may be used to cut multiple threads when it is set parallel to the lathe ways, and the threading tool is indexed by turning the rest handle. This leaves the depth of cut to be taken by advancing the crossfeed handle.

Procedure

1. The $\frac{1}{8}$-in. lead in all these problems has been based on a value of $\frac{1}{16}$-in. pitch. Convert this pitch to thousandths: $\frac{1}{16} = 0.0625$.

2. Set the compound rest parallel to the lathe ways (Figure 46-5).

3. Set up the lathe for thread cutting.

4. Set the quick-change gearbox to cut 8 tpi.

5. Back off the compound rest handle a few turns and then advance the threading tool one-quarter turn clockwise to remove any backlash.

6. Set the micrometer collar to zero.

7. Turning the crossfeed handle, carefully advance the threading tool to just touch the inner edge of the workpiece.

8. Set the crossfeed micrometer collar to zero.

9. Calculate the depth of thread required (0.6495/16 = 0.041).

10. Using the crossfeed handle, advance the threading tool by taking 0.005 in. depth of cuts with each pass until the thread has been cut to depth.

11. Move the carriage at the completion of the first thread and clear the threading tool from the workpiece to prepare cutting the second thread.

12. Advance the compound rest handle clockwise 0.0625 in.

13. Repeat steps 7 through 10 to complete the thread cutting operation.

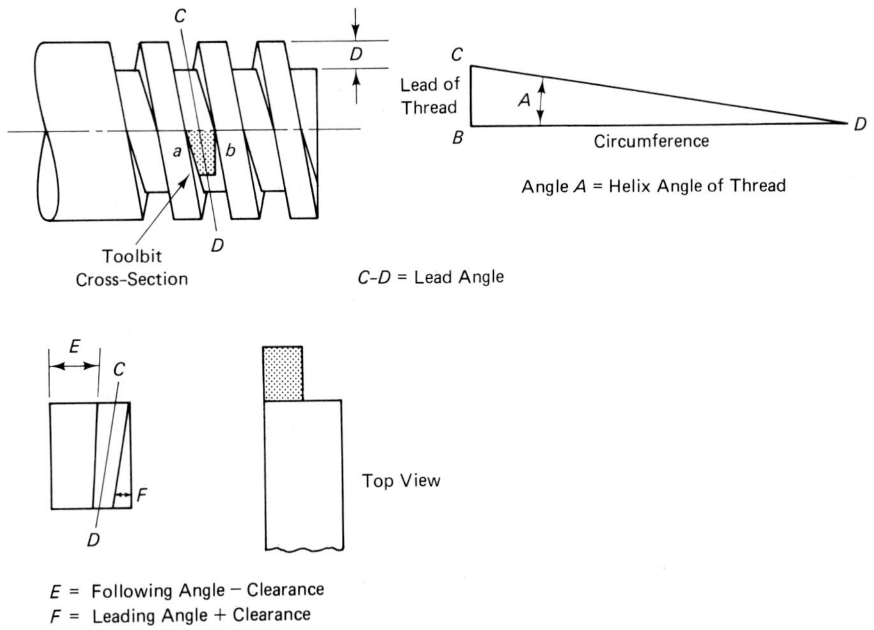

FIGURE 46-6
The design of a square thread tool.

Square Threads

The square thread was once widely used in machine design and construction, primarily to transmit maximum power and motion. Bench vise screws and house jackscrews are common applications. Acme threads have largely replaced the use of the square thread. Due to the difficulty of cutting square threads with hand cutting tools, the replacement of existing square threads is done by cutting them on a lathe. Once the tool of the right shape has been produced and it contains a sharp cutting edge, the operation is no more difficult than cutting a conventional 60° thread.

The Square Threading Tool

At first appearance the square thread tool looks very much like a stubby cutting-off tool. The similarity ends there because the sides of the threading tool are slanted to the shape or helix angle of the thread (Figure 46-6) and must be provided clearance to prevent rubbing during cutting. It is the grinding of these sides that determine the success of the threading tool.

The slant of the threading tool will depend on two things:

1. The slant (helix angle) varies for each different lead of thread for a given diameter. The greater the lead of the thread, the greater will be the slant of the tool.

2. The helix angle changes for each different diameter of thread for any given lead. The larger the diameter, the less will be the slant of the tool.

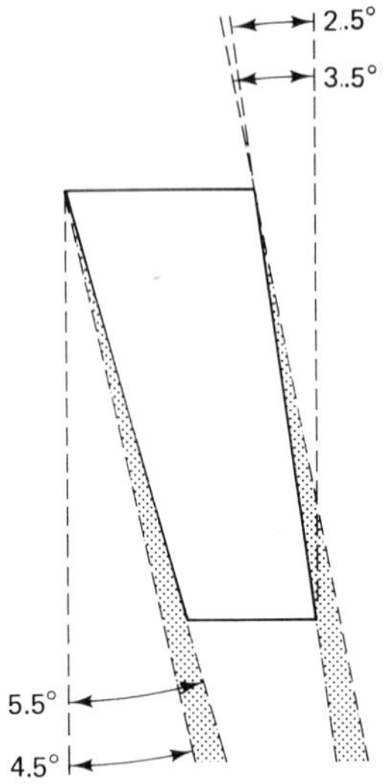

FIGURE 46–7
Illustration showing the clearance angles on the leading and following sides of a square threading tool.

The curvature of the thread demands that the threading tool be made thinner toward the bottom, in addition to the slant it must have. This surface is referred to as the leading side of the tool. It will also be seen that the following side, while slanting in the same general direction, will also be made thinner near the bottom. The amount of helix angle or slant on either side for a square thread can be calculated by using the right triangle illustrated in Figure 46–7. One side represents the lead of the thread, while the side adjacent represents the circumference of either the major or minor diameter of the thread. The angle between the side adjacent (circumference) and the hypotenuse represents the helix angle of the thread.

Square Threading Tool Calculations

$$\text{Tan of leading angle} = \frac{\text{lead of thread}}{\text{circumference of minor diameter}}$$

$$\text{Tan of following angle} = \frac{\text{lead of thread}}{\text{circumference of major diameter}}$$

Clearance

In practice, square threading tools are given a 1° clearance on each side in addition to the slant angles, providing a smaller cross section at the bottom. For grinding the leading side of the tool, add 1° clearance to the calculated slant angle. For grinding the following side, subtract 1° clearance from the calculated thread angle.

Example

Find the leading and following angles of a threading tool required for cutting a 1 1/8-4 square thread.

Solution

$$\text{Lead} = 0.250 \text{ in.}$$

$$\text{Major diameter} = 1.125 \text{ in.}$$

$$\text{Minor diameter} = \text{major diameter} - \text{lead}$$
$$= 1.125 - 0.250$$
$$= 0.875 \text{ in.}$$

$$\text{Tan of leading angle} = \frac{\text{lead of thread}}{\text{minor diameter circumference}}$$
$$= \frac{0.250}{0.875 \times \pi}$$
$$= \frac{0.250}{0.875 \times 3.1416}$$
$$= \frac{0.250}{2.7489}$$
$$= 0.09094$$

$$\text{Tan of leading angle} = 5°11'$$

$$\text{Toolbit leading side clearance} = 5°11' + 1°$$
$$= 6°11'$$

$$\text{Tan of following angle} = \frac{\text{lead of thread}}{\text{major diameter circumference}}$$
$$= \frac{0.250}{1.125 \times \pi}$$
$$= \frac{0.250}{1.125 \times 3.1416}$$
$$= \frac{0.250}{3.5343}$$
$$= 0.07073$$

$$\text{Tan of following angle} = 4°3'$$

Toolbit following side

$$\text{clearance} = 4°3' - 1°$$
$$= 3°3'$$

Cutting a Square Thread

There are two schools of thought concerning the cutting of a square thread. On this particular screw the width, depth, and space between the threads are equal. While the depth of cut is important, the objective of the operation is to develop a smooth finished thread. In order to fit the mating nut easily, there are those who suggest grinding the width of the tool 0.002 in. (0.05 mm) wider. Still others would suggest cutting the thread using two toolbits: one for roughing, using a cutting tool 0.015 in. (0.38 mm) undersize, and another cutting tool 0.002 in. (0.05 mm) oversize for a finishing cut. The choice of methods remains with the machinist.

Procedure

1. Align the lathe centers if necessary and mount the work.
2. Cut a small shoulder on the right-hand side of the work, approximately $\frac{1}{16}$ in. (1.6 mm) wide, and equal to the minor diameter in size. This will help determine when the threading tool has reached the full depth of cut.
3. If space allows, undercut a recess at the end of the thread section to the minor diameter to provide room for the threading tool run out.
4. Calculate the leading and following angles for the threading tool.
5. Grind the toolbit(s) to size of width, as suggested in the opening paragraph.
6. Set the compound rest at 30° to the right. This will permit any side movement to reset the threading tool if necessary.
7. Mount the threading tool near the left side of the compound rest, on center and square with the work.
8. Set the quick-change gearbox for the required number of tpi.
9. Calculate the single depth of the thread as follows:

$$\frac{P}{2} \text{ or } \frac{0.500}{N}$$

10. Start the lathe spindle.
11. Turn the crossfeed handle clockwise and just touch the work diameter with the threading tool.
12. Set the crossfeed micrometer collar to zero but *do not move the crossfeed handle.*
13. Turn the crossfeed handle clockwise to take a 0.003-in. (0.08-mm) depth of cut.
14. Using the thread chasing dial as a guide, engage the split-nut lever to take a trial cut.
15. Stop the lathe at the end of the cut and disengage the split-nut lever.
16. Check the thread with a screw pitch gage.
17. Using the crossfeed handle, advance the threading tool toward the work 0.002 to 0.007 in. (0.05 to 0.18 mm) for each cut, depending on the thread size and work material.
18. Start the lathe spindle.
19. Apply cutting fluid with each pass and cut the thread to depth.

Cutting an Acme Thread

The operation of cutting an Acme thread is similar in some ways to the operation of cutting a square thread. (1) A small section on the right side of the work may be machined to the minor diameter to indicate when the full depth of cut has been made. (2) A recess at the end of the thread length may also be machined to the minor diameter to allow for threading tool runout. (3) All depth of cuts are taken using the compound rest.

Toolbits ground for cutting Acme threads are made to fit an Acme thread gage (Figure 46–8). The gage contains a 29° included angle as well as an angular side for setting up the threading tool square with the work. A series of notches can also be found on the gage to measure the different widths at the end of the Acme threading tool and are numbered to correspond to the number of threads per inch being cut.

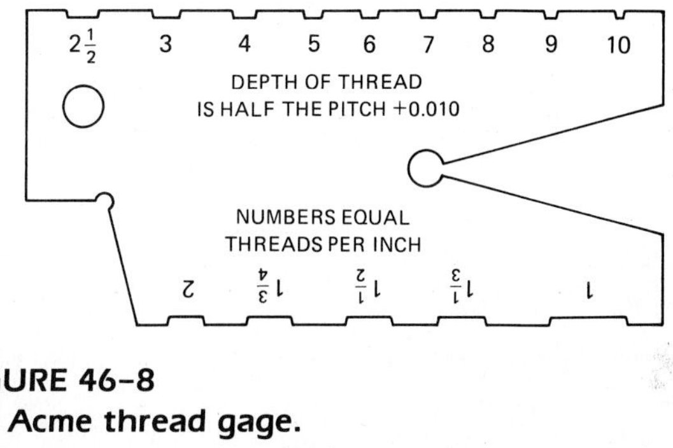

FIGURE 46–8
An Acme thread gage.

Procedure

1. Calculate the minor diameter of the thread as follows, where D = depth and N = tpi:

$$D = \frac{0.500}{N} + 0.010$$

$$\text{Minor diameter} = \text{major diameter} - 2D$$

2. Mount the work securely in the lathe.

3. Cut a small shoulder on the right side of the work, approximately $\frac{1}{16}$ in. (1.6 mm) wide and equal to the minor diameter in size.

4. Undercut a recess at the end of the thread length to the minor diameter to provide room for the threading tool to runout.

5. Grind a toolbit to fit the Acme thread gage, allowing enough side clearance to prevent the cutting edge from rubbing during use.

6. Grind a flat on the point of the tool until it fits the notch on the gage that corresponds to the threads per inch required. *Note:* If a gage is not at hand, the width of the flat may be calculated as follows and measured with a micrometer.

$$\text{Width of point} = \frac{0.3707}{N} - 0.0052$$

7. Set the compound rest 14.5° to the right.

8. Set the quick-change gearbox to the required number of threads per inch.

9. Set the proper spindle speed.

10. Mount the threading tool.

11. Set the Acme toolbit to center height and square with the work using the Acme thread gage (Figure 46–9).

12. Start the lathe spindle.

13. Turn the compound rest handle clockwise and just touch the work diameter with the threading tool.

14. Set the compound rest micrometer collar to zero.

15. Turn the compound rest handle clockwise to take a 0.003-in. (0.08-mm) depth of cut (Figure 46–10).

16. Using the thread chasing dial as a guide, engage the split-nut lever to take a trail cut.

17. Stop the lathe.

18. Check the threads cut with a screw pitch gage.

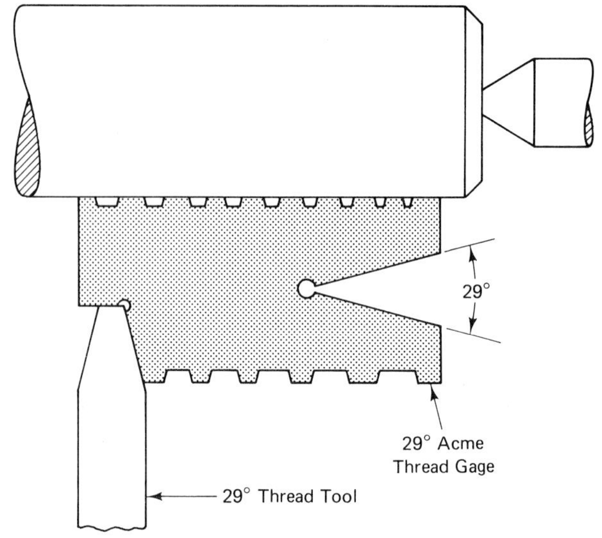

FIGURE 46–9
The threading tool is set square against the work using an Acme thread gage.

FIGURE 46–10
All depth of cuts for Acme threads are taken using the compound rest.

19. Start the lathe spindle.

20. Apply cutting fluid with each pass, advancing the threading tool 0.002 to 0.007 in. (0.05 to 0.18 mm) for each cut using the compound rest handle and thread to depth. *Note:* Some machinists prefer to use a square threading tool ground 0.005 in. (0.13 mm) smaller than the root of the Acme thread to cut the thread to depth. The Acme toolbit is then used to finish the sides. The results are a more accurate and smoother thread.

UNIT 47 SHARPENING A TOOLBIT

Due to the characteristics of the machinability of the metal and the properties of the cutting tool material, lathe cutting tools will have to be designed (tool geometry) and ground to maintain efficiency and produce a quality finish. Improperly ground and dull tools are wasteful and dangerous to use. Lathe cutting tools will have to combine proper clearance angles, cutting angles, and point radius to produce a sharp cutting edge that is sufficiently supported and wear resistant, yet tough enough to resist chipping. There is no one design that will fit all machining requirements. The design requirement for machining ferrous materials may be all together different than machining nonferrous materials.

Objectives
After completing this unit you should be able to

1. identify five main types of cutting tool materials.
2. identify the various surfaces ground on a toolbit.
3. grind a general-purpose cutting high-speed steel toolbit.

Cutting Tool Materials

Advances in the cutting tool technology over the years has seen the wide use of different types of lathe toolbit materials such as high-speed steel, cemented carbide, cast alloys, ceramics, ceramic–metal combinations (cermets), etc. By far, the most advance cutting tool materials to be introduced were the superabrasives: cubic boron nitride and diamonds. These remarkable materials and their performances will be covered in detail in Unit 72.

There are several factors that influence the tool design and cutting conditions when machining with a single-point cutting tool. These would include

- Rigidity of the workpiece
- Rigidity of the cutting tool
- Rigidity of the machine tool
- Cutting fluid and coolants

Working hand in hand with these factors are the properties of the cutting tool material itself. In order for any cutting tool to be successful, it must contain four important properties:

- High hardness (strength)
- Wear resistance (resistance to abrasion)
- Shock resistance (resistance to impact)
- Heat resistance (softening at high temperatures)

High-Speed Steel Toolbits

High-speed steel is a heavily alloyed steel whose principal elements of tungsten and molybdenum are combined with elements of cobalt, vanadium, and chromium in varying levels to produce carbide forming and wear-resistant properties. High-speed steel has the ability to machine at temperatures as high as 1050°F (577°C) and still retain its hardness and cutting edge when it has cooled to room temperature.

There are two general types of high-speed steel toolbits, designated by the AISI as being a tungsten-base (T) and molybdenum-base (M) alloy. The T1 type is a commonly used toolbit in the machine tool trade and has an 18–4–1 composition. This means that it contains 18% tungsten, 4% chromium, and 1% vanadium.

A molybdenum-base toolbit classification would be an M10 type whose composition would be 8–4–2. This toolbit would contain 8% molybdenum, 4% chromium, and 2% vanadium.

The toolbits just mentioned are of a general-purpose variety. Within both groups, each grade is identified by a suffix number, i.e., M2, M6, T4, T15, etc. It is best to go to a manufacturer's recommendation as to what grade is best suited for the job involved.

Hard-Cast Alloys

Operated at cutting speeds more than two times that of high-speed steel, hard-cast alloy toolbits (stellite) are nonmagnetic and do not contain enough iron to be considered a steel. Stellite is an alloy containing varying levels of cobalt (as much as 35%), tungsten (4 to 24%), chromium (25 to 35%), carbon (1 to 4%), and other elements added for special properties (Figure 47–1).

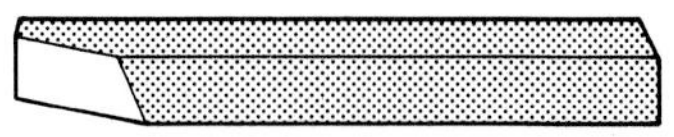

FIGURE 47–1
Hard-cast alloy toolbits can operate at cutting speeds twice that of high-speed steel.

Stellite toolbits reach full hardness when cast and are not heat treated. In terms of cutting properties, the tool has excellent wear resistance and will not lose its temper when machining at red hot temperatures. It is also more brittle and weaker in tension than high-speed steel. To prevent chipping and breaking during cutting, stellite toolbit cutting edges should be well supported and ground with just enough clearance to permit cutting. Unlike high-speed steel toolbits, which are cooled by quenching in water during grinding, no water should be used to quench stellite toolbits. They will crack due to thermal shock.

Cemented Carbides

Cemented carbide toolbits are available in two types: the throwaway insert type (Figure 47–2) and the brazed to the steel shank type (Figure 47–3). These cutting tools are made from pulverized tungsten carbide or a mixture of two or more carbides and fused (sintered) with a binder of cobalt to be molded into shape under high pressure. The cutting tool is brittle, porous, and extremely hard. It also has low heat conductivity, as evidenced by the fact that it can be used at cutting speeds 10 to 30 times faster than those used with high-speed steels in mod-

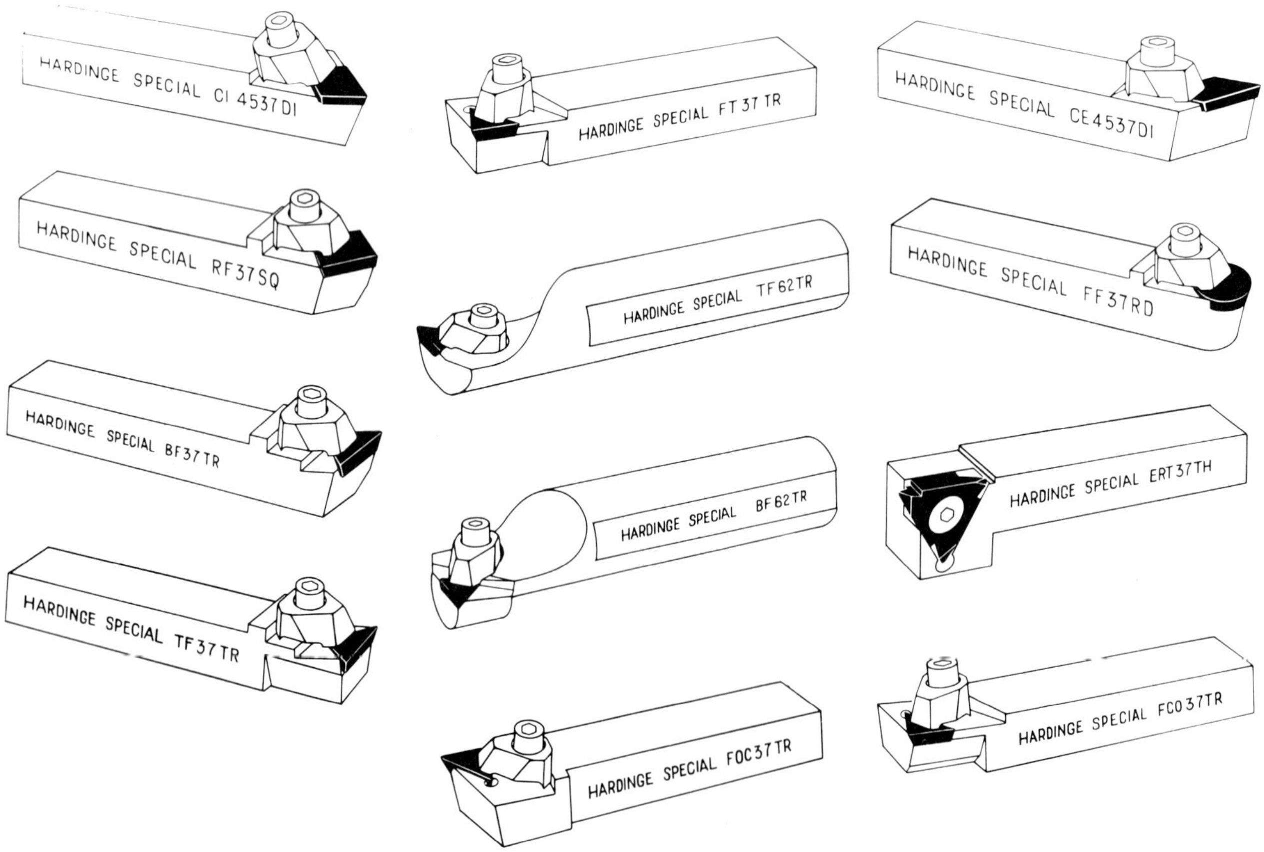

FIGURE 47–2
Throwaway carbide inserts prove economical to use as they eliminate resharpening and machine downtime. (Credit: Hardinge Brothers, Inc.)

ern, well-supported machine tools.

Tungsten carbide lathe toolbits are sharpened with a special grade of silicon carbide or diamond grinding wheel (Figure 47–4), and due to its high compression strength, it need not be ground with excessive clearance angles. Like the stellite toolbit, cemented carbide toolbits should have just enough clearance to allow the tool to cut freely.

FIGURE 47–3
A cemented carbide toolbit is a widely used production cutting tool.

Ceramic Cutting Tools

Ceramic cutting tools have been developed that are composed of cemented oxide materials. The materials may consist primarily of aluminum oxide, silicon oxide, or magnesium oxide. These powdered oxides are formed into shape by cold or hot pressing and sintered at temperatures near 3000°F (1649°C). While these oxides may be used separately or together, metallic binders such as titanium or chromium oxides may also be used in small quantities to improve impact strength. Ceramic cutting tools are produced as throwaway inserts and are available in a variety of shapes and styles, such as square, round, triangular, or rectangular. These cutting tools can be turned in special holders until all the cutting edges have been used and thrown away. Ceramic cutting tools are also available to be clamped in mechanical toolholders and as tool shanks that have ceramic tips cemented to them by means of an epoxy glue. The bonding process frees the ceramic from any clamping stresses set up in mechanical toolholders.

Ceramic cutting tools are crystalline in structure, brittle, but will retain their high hardness up to 1400°F (769°C). They do not weaken in strength until about 2200°F (1099°C). They are stronger than cemented carbides at elevated temperatures and enjoy a longer tool life at these temperatures on some materials. Due to the brittleness, ceramic cutting tools have a low resistance to both mechanical and thermal shock at room temperature. They should only be used in equipment that are solid in setup and free from vibration. Light finishing cuts are possible at high cutting speeds to improve surface quality of relatively hard materials, often eliminating the need of a grinding operation.

FIGURE 47–4
Cemented carbide toolbits are sharpened on an adjustable tool grinder using silicon carbide or diamond cutting wheels.

General Cutting Tool Terms

The *base* is the bottom support surface of the cutting tool shank.

The *face* is the surface upon which the cutting chip slides as it separates from the workpiece.

The *flank* may be one or two surfaces that lie below or adjacent to the cutting edge.

The *nose* is the connection between the side-cutting and end-cutting edges.

The *nose radius* is the contoured tip of the cutting tool and will affect the finish of the cut as well as the life of the tool.

The *point* is the end part of the tool, which has been ground to produce the face and cutting edges.

The *shank* is the main body portion of the toolbit, one end of which may be ground, shaped, or used for insert cutter support.

Cutting Tool Angles

The design of a single-point tool is shaped by the angles of seven sides or surfaces that are ground into the toolbit. These surfaces are angled to obtain a good performing cutting tool as well as a long wearing tool. The angles will vary with the material being cut.

Side-Cutting Edge Angle. The angle formed by the cutting edge and the side of the tool shank and will generally measure between 10° and 20°. Excessive angles ground over 30° will lead to tool chatter.

End-Cutting Edge Angle. The clearance angle between the cutting tool and the finished workpiece surface that is formed by the end-cutting edge of the tool and a line at right angles to the tool shank.

Side Relief Angle. A clearance angle between the flank of the tool and the surface of the workpiece. Generally measuring 6° to 10°, the angle is ground on the flank of the tool to permit the toolbit to travel lengthwise into the workpiece without rubbing.

End Relief Angle. A clearance angle ground on the end of the toolbit below the end-cutting edge. This angle will vary with the hardness of the material, usually ground about 10° when measured in the toolholder. Harder materials may require greater tool support and will be ground to a smaller angle.

Side Rake Angle. The angle that slopes away from the cutting edge of the toolbit to the opposite side. This angle controls the chip direction and free cutting behind the cutting edge. It will also vary in degree of slope. Little or no side rake is used for hard materials such as cast iron or hard bronze; as much as 20° to 30° is used for certain types of aluminum.

Back (Top) Rake. This angle slopes away from the top of the cutting edge to the back of the toolbit and is designed to give a slicing action. It will direct the chip away from the nose of the cutting tool. In toolholders where the toolbits are held in a square hole with an inclined angle of 20°, a back rake is normally provided (Figure 47-5) and need not be ground. A back rake of 5° to 15° is recommended when machining aluminum.

Angle of Keeness. This included angle is produced by the formation of the side rake and the side relief angles. This wedge will vary, depending on the hardness of the material.

FIGURE 47-5
The incline of the toolholding position provides a normal back rake.

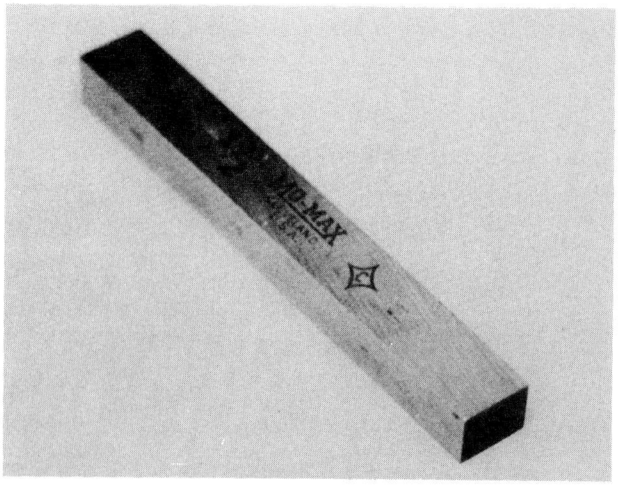

FIGURE 47-6
Toolbits come in long square blanks when they are new.

TABLE 47-1
Recommended average tool angles for single-point lathe toolbits

	End relief		*Side relief*		*Side rake*		*Back rake*	
Material	*HSS*	*Carbide*	*HSS*	*Carbide*	*HSS*	*Carbide*	*HSS*	*Carbide*
Aluminum	8°-15°	6°-10°	8°-15°	6°-10°	15°-20°	10°-20°	5°-15°	0°-10°
Brass	8°-10°	6°-8°	10°-12°	6°-8°	0°-5°	8°-5°	0	0°-5°
Bronze	8°-10°	6°-8°	8°-10°	6°-8°	0°-5°	8°-5°	0	0°-5°
Cast iron	6°-8°	5°-8°	8°-10°	5°-8°	8°-12°	6°-7°	8°	0°-7°
Copper	7°-10°	7°-10°	7°-10°	7°-10°	20°-25°	15°-25°	10°	4°-8°
Machine steel	8°-10°	5°-10°	8°-10°	5°-10°	12°-18°	6°-7°	8°-15°	0°-7°
Tool steel	6°-8°	5°-8°	7°-9°	5°-8°	8°-12°	6°-7°	8°	0°-7°
Stainless steel	8°-10°	5°-8°	8°-10°	5°-8°	15°-20°	6°-7°	8°	0°-7°

To Grind a General-Purpose HSS Toolbit

The high-speed steel toolbit remains one of the most widely used lathe cutting tools in school shops. When they are new, toolbits come in long square blanks (Figure 47-6) and their clearance angles are ground on some form of pedestal grinder, using off-hand grinding techniques. The average angles for high-speed toolbit are listed in Table 47-1.

As will be seen, the grinding of many sides on a toolbit will require holding the toolbit against the grinding wheel at three different angles at the same time. And because high-speed steel toolbits are expensive, it is helpful if the beginner practices on cold-roll steel blanks. A second suggestion would be to draw the cutting edge line on the toolbit in pencil. If the surface is too smooth, use a grease pencil. While the line will not last long as grinding is carried out, it will help in establishing the surface against the wheel.

Procedure

1. Wear appropriate safety glasses and remove all jewelry before attempting any grinding.
2. Dress and true the grinding wheel of a pedestal grinder (see Unit 26).
3. Fill the water container.
4. Layout the side-cutting edge line in pencil so that it extends about $\frac{1}{2}$ in. (13 mm) long and is one-fourth the width of the toolbit.

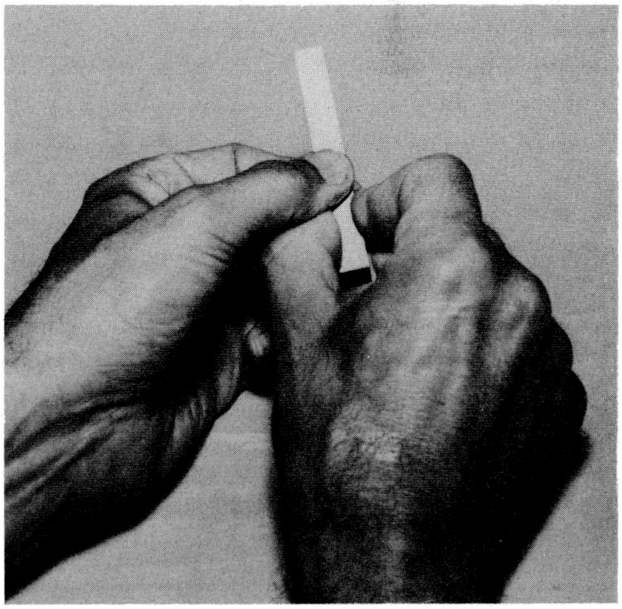

FIGURE 47-7
Grip the toolbit in both hands when off-hand grinding.

5. Grip the toolbit in both hands (Figure 47-7).
6. *Approach* the rough grinding wheel holding the blank level so that the side-cutting edge line is parallel to the face of the wheel (Figure 47-8).
7. *Elevate* the front end of the toolbit about $\frac{1}{4}$ in. (6 mm) to provide front clearance.

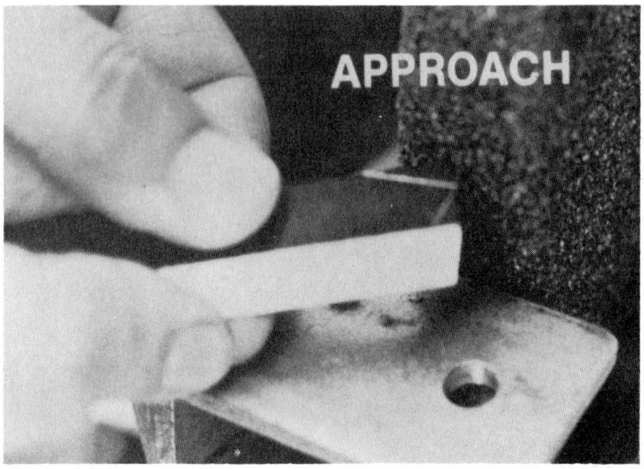

FIGURE 47–8
Approach with the side to be ground parallel to the wheel face. (Credit: Photocom Productions)

FIGURE 47–10
Completion of the side-cutting edge.

FIGURE 47–9
Quench the toolbit to prevent overheating.

FIGURE 47–11
Checking the side relief angle.

8. Now *rotate* the toolbit clockwise about 10° for the side relief angle.

9. Holding the *approach, elevate* and *rotate* positions against the rotating face of the wheel, move the toolbit from side to side to prevent grooving the wheel.

10. Quench it to prevent overheating! (Figure 47–9).

11. Repeat the *approach, elevate,* and *rotate* positioning, grinding and quenching often, until the cutting edge angle has been ground (Figure 47–10).

12. Using a toolbit grinding gage, check the side relief angle with the toolbit held in a toolholder (Figure 47–11).

13. Correct any angle error by either increasing or decreasing the angle of *rotation* and continue grinding until the side-cutting angle is complete.

14. To grind the end-cutting edge angle, *approach* the wheel holding the toolbit so it forms an angle slightly less than 90° (Figure 47–12).

15. *Elevate* the front end to help generate the end relief angle (front clearance).

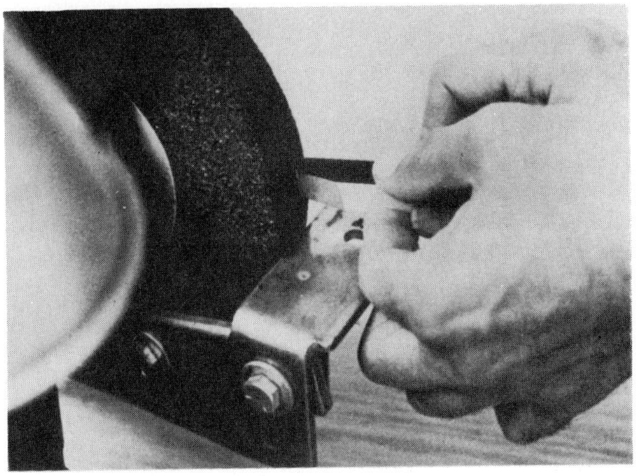

FIGURE 47-12
Starting to grind the end-cutting edge angle. (Credit: Photocom Productions)

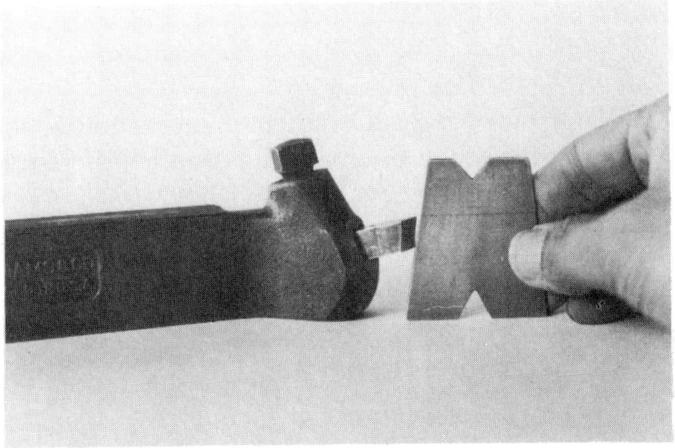

FIGURE 47-13
Checking the end relief angle.

16. At the same time, *rotate* it counterclockwise about 10°.

17. Holding the *approach, elevate,* and *rotate* positions against the wheel, move the toolbit from side to side as it is ground against the wheel face.

18. Quench it often to prevent overheating.

19. As before, check the end relief angle with the toolbit held in the toolholder (Figure 47-13).

20. If the front clearance is not right, increase or decrease the *elevation* positioning in touching it up.

21. The side rake is achieved by holding the top of the toolbit at a 45° angle to the wheel and *rotate* the toolbit so as to form a 14° slope.

FIGURE 47-14
Checking the angle of keeness. (Credit: Photocom Productions)

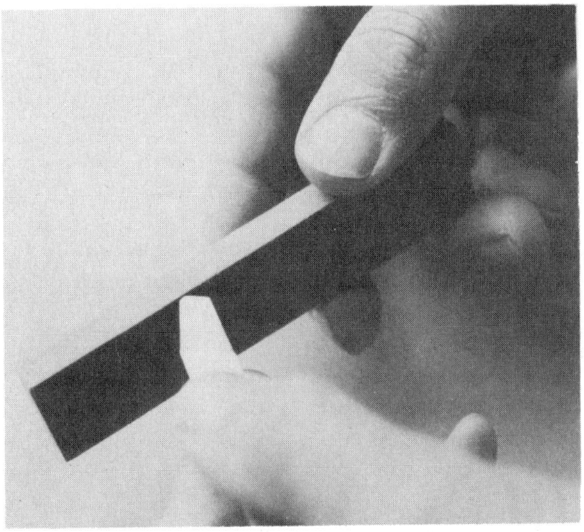

FIGURE 47-15
Hone the cutting edge to increase its tool life.

22. Hold the position against the wheel until the side rake is ground.

23. Quench the toolbit.

24. Check the angle of keeness (Figure 47-14).

25. Grind a slight radius on the point of the toolbit while still maintaining the same clearance angles already ground.

26. With an oilstone, hone the radius of the toolbit to increase the tool life and produce a better finish (Figure 47-15).

SECTION REVIEW

Unit 35

1. How is the size of an engine lathe generally determined by manufacturers?
2. List the five main parts of a lathe.
3. State the function of the following parts: (a) headstock spindle, (b) saddle, (c) apron handwheel, (d) cross slide, and (e) compound rest.
4. List five materials used for lathe cutting tools in the machine shop.
5. Explain the difference between rpm and cutting speed (CS).
6. How may a lathe's feed be defined?
7. What is the purpose of rough turning?
8. What is the purpose of finish turning?

Unit 36

1. Name five different types of lathe chucks.
2. What safety requirements must be met to mount and remove chucks in a lathe?
3. Why is a wooden block or chuck cradle used when mounting or removing a cam lock lathe?
4. How is the chuck mounted and fastened to the headstock spindle of a tapered spindle nose lathe?
5. Briefly describe how to remove a chuck from a threaded spindle lathe.
6. How should the jaws of a three-jaw chuck be assembled and tested for alignment?
7. For what purpose is a four-jaw chuck used?
8. Name three different semiprecise methods of truing up work in a four-jaw chuck.
9. What is the most accurate method of truing work in a four-jaw chuck and to what degree of accuracy can it achieve?

Unit 37

1. How may facing be defined as a lathe operation?
2. Name three reasons facing may be performed during machining.
3. What is the result of not placing a facing tool exactly on center?
4. What are the two ways of controlling the depth of cut when facing?
5. How far should work extend beyond the chuck jaws when facing?
6. What are the two methods used for facing the ends of a workpiece held in a chuck?
7. What conditions must be met to be able to face between centers?
8. Briefly describe how a facing cut can produce a smooth finish.
9. Why should finishing cuts be started at the center and fed outward when facing between centers?

Unit 38

1. Name two types of graduated collars found on the crossfeed of a lathe and state the difference between them.
2. What type of toolholder is recommended when using a toolbit advancing toward the headstock?
3. How far should the toolbit be extended from the toolholder for general machining?
4. How far should the toolholder extend beyond the tool post to eliminate vibration and chatter?
5. What is the purpose of a trial cut?
6. Briefly describe how the toolbit should be positioned for rough turning.
7. In rough turning, how much material should be left before attempting a finish cut?
8. Why is it important that the workpiece be near room temperature before attempting a finish cut?
9. How may shouldering be defined?
10. What is the purpose of machining a filleted shoulder?
11. What is another term for beveled shouldering and list several reasons why it is machined?

Unit 39

1. What is the result of machining between lathe centers that are not aligned?
2. Name at least three methods of aligning lathe centers.
3. Briefly describe how to test alignment of centers using the most accurate method.

Unit 40

1. List three advantages to working between lathe centers.
2. What safety procedure must be followed when cleaning the spindle opening of a lathe?
3. Why is the depth of a center-drilled hole an important consideration when working between centers?
4. How is marring prevented when fastening a lathe dog to a workpiece?
5. Explain the difference between a live center and a dead center.
6. Name four types of centers used in the tailstock.
7. Briefly explain how to test a live center for accuracy.
8. How are raised surfaces or burrs removed from lathe centers?
9. Why should excessive binding of the tailstock center be avoided?
10. Describe the type of tension that should be found in a workpiece properly mounted between centers.

Unit 41

1. How is a taper defined?
2. What kind of taper is used on twist drills, reamers, and similar cutting tools?
3. What kind of taper is used on a milling machine arbor and what is its included angle?
4. How are type L and type D-1 tapers described and where are they used?
5. Give two examples of nonstandard tapers.
6. List four basic parts of a taper.
7. List the four different methods of producing a taper in a lathe.
8. Calculate the tpf and tailstock offset for the following problem: D = 1.5 in., d = 1.375 in., TL = 3 in.
9. How are metric tapers expressed?
10. Calculate the large diameter for a 1 : 30 taper having a small diameter of 15 mm and a length of 90 mm.
11. List six advantages to using the taper turning attachment.
12. List four different methods of checking external tapers.
13. Describe the procedure for fitting an external taper.

Unit 42

1. What three factors must be considered when filing in a lathe?
2. Why is a left-handed technique recommended when filing in a lathe?
3. What type of abrasive cloth is recommended for ferrous metal? Nonferrous metal?
4. What precaution should be observed when attempting to polish in a lathe?
5. At what speed should a lathe be turning for polishing with abrasive cloth?
6. How is knurling defined?
7. Briefly explain how the knurling tool should be set up for knurling.
8. How can a damaged knurl in the form of rings be avoided?
9. How can die bushings and holes be ground in a lathe?

Unit 43

1. List four work holding methods when center drilling in a lathe.
2. Generally, at what speed should a lathe be turning for center drilling?
3. How should a sawcut or uneven surface be treated before center drilling?
4. Name two different methods of truing a damaged center hole.

Unit 44

1. What is the relationship shared by drilling, reaming, boring, and tapping operations?
2. How is the depth of a drilled hole measured on a print?
3. How can the depth of a hole be checked while it is being drilled in a lathe?
4. Define the boring operation.
5. List three conditions that should be met before using a boring bar.
6. What is the purpose of a rose reamer?
7. What is the recommended speed for machine reaming in a lathe?
8. What is the result of reaming too fast in a lathe?
9. How is hand tapping square in a lathe accomplished?

Unit 45

1. Define a screw thread.
2. Define the following terms: (a) major diameter, (b) minor diameter, (c) crest, (d) pitch diameter.
3. In which direction should a left-hand nut be turned to tighten it?
4. How is pitch measured on an Unified thread? Metric thread?
5. Compare the similarities and differences between the ISO and Unified thread form.
6. How is class of fit controlled when cutting a thread?
7. Calculate the pitch, depth, and minor diameter of a $\frac{1}{2}$-20 NC thread.
8. Calculate the pitch, depth, and minor diameter of a M14 $\times$ 2 thread.
9. How is the included angle of a ground toolbit used for cutting a Unified thread measured?
10. How is the width of the flat end of a threading toolbit calculated for Unified and metric threads?
11. What is the purpose of a thread chasing dial?
12. At what angle should the compound rest be turned when cutting a 60° thread?
13. What is the purpose of the quick-change gearbox?
14. For what reason should work surfaces be undercut or machined behind the threaded area to a level below the minor diameter?
15. For what purpose is the front and back of the thread area chamfered?
16. At what point on the thread chasing dial should the split-nut lever be engaged for the following threads: 8, 10, 11, 13, 18, 20.
17. Briefly describe how to set up a lathe for cutting a 60° thread.
18. How should the compound rest be positioned to cut a left-hand 60° thread?
19. How should the threading tool be set to cut a thread on a tapered section?
20. What measurement is obtained when using the three-wire method on a 60° single-thread screw?

Unit 46

1. What is the purpose of a square or Acme thread?
2. What is the purpose of a multiple-start thread and give an example of where it is used.
3. Name four methods of cutting multiple-start threads.
4. What two factors affect the helix angle of a square thread?
5. How is the threading tool ground so that its sides do not rub against the sides of a square thread?
6. Name two methods of producing a smooth finished square thread.
7. How are toolbits ground for cutting Acme threads measured?
8. Calculate the width of the point of an Acme threading tool used to cut 6 threads per inch.
9. How is the compound rest positioned and set when cutting an Acme thread?

Unit 47

1. List five common materials used in the manufacture of cutting tools.

2. Name two materials used in the manufacture of advanced cutting tools.

3. List the factors that influence the tool design and cutting conditions affecting the single-point cutting tool.

4. What four properties must any cutting tool contain to be successful?

5. Briefly describe the manufacture of a cemented carbide cutting tool.

6. What two methods are used to sharpen tungsten carbide cutting tools?

7. What is the chief reason ceramic cutting tools have a low resistance to both mechanical and thermal shock?

8. What is the purpose of the following parts on a single-point cutting tool? (a) base, (b) face, (c) nose radius, and (d) shank.

9. Define the purpose of the following cutting tool angles: (a) back rake and (b) end-cutting angle.

10. What is the recommended side relief angle for a general-purpose high-speed steel toolbit?

11. What is the function of the side rake angle and where is it located on a single-point toolbit?

12. Generally, at how many angles is a high-speed toolbit held against a grinding wheel in order to grind one side?

13. How should the nose radius of a high-speed toolbit be finished in order to increase tool life and produce a better finish?

Section 10

MILLING MACHINES

UNIT 48 HORIZONTAL MILLING MACHINES

The horizontal milling machine was invented by Eli Whitney during the early part of the nineteenth century. This machine tool uses a process whereby the workpiece is moved past a rotating cutter. Material is removed when the workpiece is moved at a steady rate of feed, while the stationary cutter revolves at a relatively high speed. Capable of using one or more multitooth cutters of different shapes, sizes, and materials, the horizontal milling machine is an important machine tool. The use of production milling machines will be covered elsewhere in the book. This unit will concern itself with the horizontal milling machine used in the toolroom.

Objectives
After completing this unit you should be able to

1. recognize and know the difference between the two horizontal milling machines used in the toolroom.
2. recognize and know the purpose of the main operational parts of a horizontal milling machine.
3. recognize and know the purpose of six horizontal milling machine accessories.

Safety

As with any machine tool, safety is very important to any successful operation.

1. Always wear appropriate safety glasses or goggles when in a machine shop.
2. Make sure that the work table, vise, and surrounding area are clean before starting work.
3. Clean all machined surfaces before mounting vises, cutters, arbors, and other working parts (Figure 48–1).
4. Remove all burrs from a workpiece and mount it securely in the vise. Rigidity is one of the requirements to any milling operation.
5. Use a soft-faced mallet or lead hammer to seat the workpiece in the vise (Figure 48–2).
6. Set the workpiece on parallels that raise it high enough and safely to do any machining.
7. Check to see that the workpiece and vise can safely pass any supporting device before attempting any machining (Figure 48–3).
8. Always make sure that the cutting tool revolves in the correct direction before attempting any machining.
9. Keep fingers clear of a cutter or moving part at all times.

FIGURE 48–1
Always clean the bottom of any work holding base before mounting.

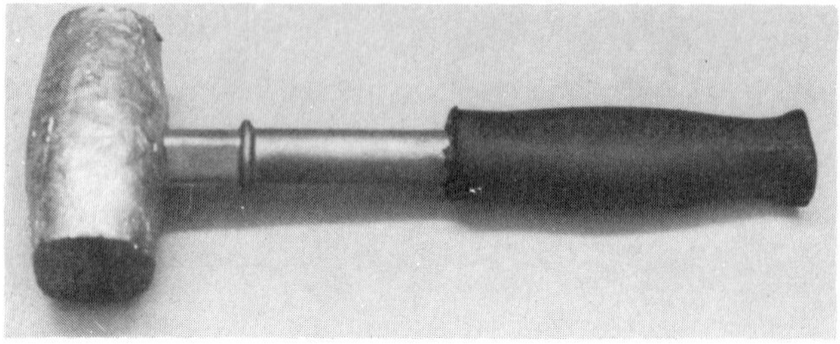

FIGURE 48-2
Use a lead hammer or a weighted mallet to seat work without marring it.

10. Do not attempt to change any belt or gears on any machine unless the machine has been stopped.

11. Loosen any quill, knee, column, table, or other necessary clamps when making setup adjustments.

12. Use only proper fitting wrenches to tighten or loosen working parts. Poor fitting wrenches can slip and damage the head of a driven nut. It can also injure the hand that slips with it.

13. Never remove any chips by hand. Always use a brush and then only when machine has stopped (Figure 48-4).

FIGURE 48-3
Check to see that there is ample clearance to allow work and the vise to pass safely.

FIGURE 48-4
Always use a brush to remove chips from the work areas.

FIGURE 48-5
Arbors should remain in an upright position when not in use to prevent warping.

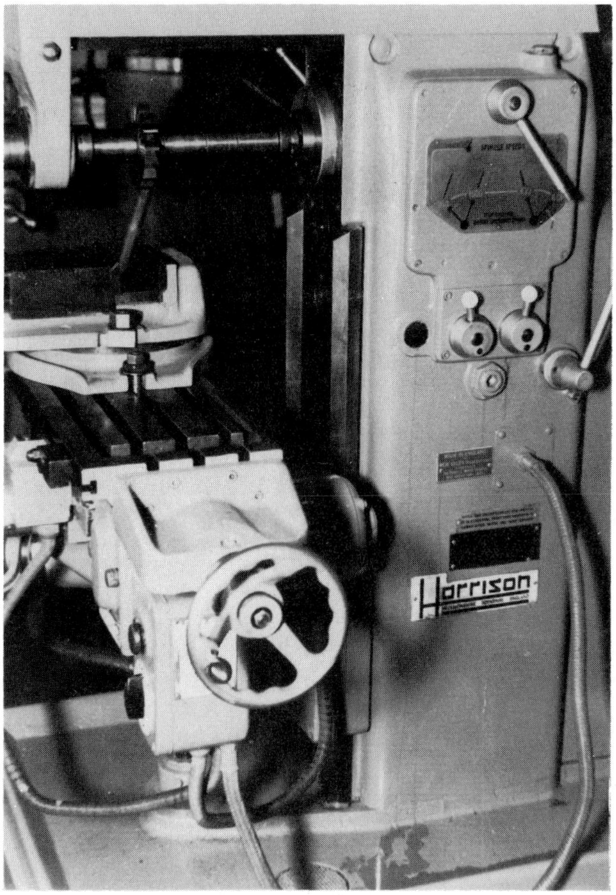

FIGURE 48-6
The plain knee and column-type milling machine.

FIGURE 48-7
The universal horizontal milling machine.

14. Apply a light film of oil to arbors when not in use and hang them in an upright position to prevent warping or bending (Figure 48-5).

15. Never attempt to run a milling machine unless authorized to do so by your instructor.

Types of Milling Machines

There are two types of horizontal-style milling machines used in the toolroom. These include the plain knee and column (Figure 48-6) and the universal knee and column (Figure 48-7).

The universal knee and column type can perform a variety of machining operations as can the plain knee and column type. They differ from each other only to the extent that the universal milling machine has a table swivel housing located between the table and the saddle. This housing permits the table to be moved and set at 45° in either direction to its normal axis, making it possible to mill helical grooves in gears, twist drills, and milling cutters.

Milling Machine Construction

The *base* serves as the very rigid and sturdy foundation to the machine and will also act as the reservoir basin for cutting fluids and coolants.

The *column face* is the machined upright section that acts to support and guide the knee when it is raised or lowered vertically.

The *knee* usually houses the feed mechanisms and is attached to the column face. The knee can be raised or lowered manually or automatically.

The *saddle* is mounted on the upper portion of the knee. The saddle may be moved toward or away from the operator (traversed) by means of the cross-feed handwheel, either manually or automatically.

FIGURE 48-8
The table handwheel is turned to move the table in a horizontal plane.

FIGURE 48-9
The crossfeed handwheel is turned to move the table crossfeed in a back-and-forth motion.

FIGURE 48-10
The arbor support is adjustable for use with varying size arbors.

The *table* is mounted in the saddle's guideways and supports the vise and work. By design, the table can travel longitudinally in a horizontal plane. When coupled with the knee and saddle movements, the table can be moved in three directions: traversely, longitudinally, and vertically.

The *table handwheel* is used to move the table horizontally in a side-to-side motion in front of the operator (Figure 48-8).

The *crossfeed handwheel* is used to move the table toward or away from the operator (Figure 48-9).

The *feed dial* is used to set the table feed movements.

The *spindle* provides the opening and drive for arbors and other attachments used on a milling machine.

The *overarm* is adjustable and provides alignment and support for the arbor and other attachments.

The *arbor support* is also adjustable and can be positioned and locked to support arbors and attachments of varying lengths (Figure 48-10).

The *elevating screw* enables the knee and table to be raised or lowered vertically, either manually or by automatic feed.

The *speed change dial* indicates the spindle speed and may be set by turning a crank. On some machines, this may require the use of two levers. One should always check to see if the machine speed can be changed while the machine is running or whether it has to be completely stopped.

Milling Machine Accessories

There are a number of accessories used on horizontal milling machines to greatly increase their work capacity.

The *vertical milling attachment* is mounted on

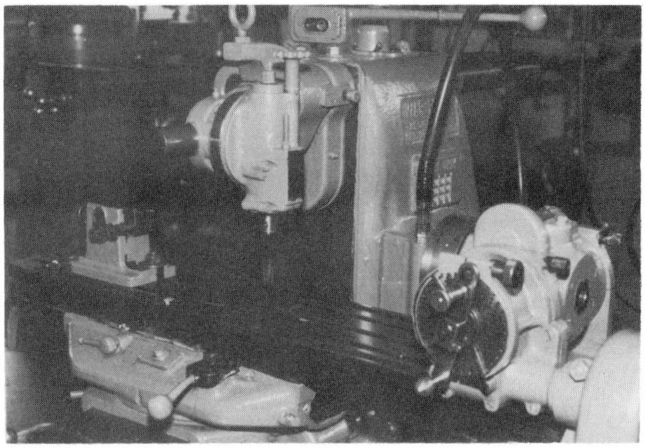

FIGURE 48-11
A mounted vertical milling attachment.

FIGURE 48-13
A shell end mill arbor permits vertical or horizontal face milling.

FIGURE 48-12
An indexing head being used for machining the head of a bolt.

FIGURE 48-14
A collet adaptor can be used for mounting drills or end mills.

the face or the column or on the overarm (Figure 48-11) and can convert a horizontal milling machine to accept vertical cutting tools. With it, operations such as end milling, face milling, T-slot milling, boring, and drilling are possible.

Also referred to as a dividing head, the *indexing head* (Figure 48-12) is a device that will permit work to be rotated so its periphery can be divided or sectioned into any number of equal parts and will hold the work in that position while it is being machined. Grooves, gear teeth, and bolt heads are common machining operations performed using an indexing head.

The *arbor* is the keywayed shaft used with spacers for mounting and positioning milling cutters.

The *shell end mill arbor* (Figure 48-13) is designed with a shank tapered for a draw-in bolt so that it may fit the spindle of a horizontal milling machine. This device allows shell mills to be mounted on it, greatly increasing metal removal efficiency.

The *collet adapter* (Figure 48-14) is used for holding single- and double-end end mills and is mounted in the vertical milling attachment.

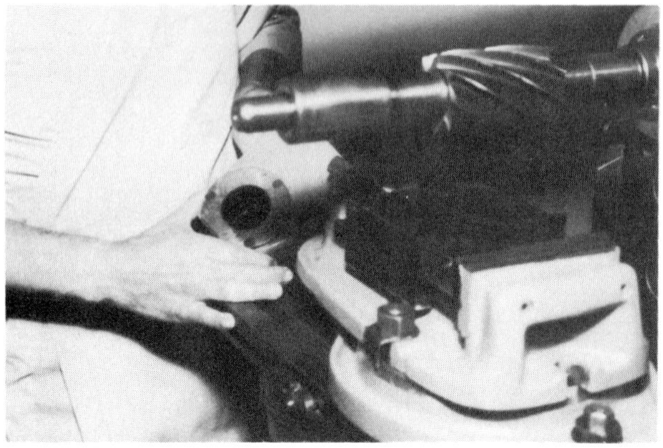

FIGURE 48–15
A swivel base vise can be set to any angle.

FIGURE 48–16
A plain milling vise cannot be repositioned.

FIGURE 48–17
The universal vise can be set to cut compound angles.

The main method of holding work in the milling machine is by means of the vise. The *swivel base vise* (Figure 48–15) is keyed at the bottom for quick and accurate mounting in the T-slots of the milling machine table and also contains hardened jaws. The 360° swivel base enables the vise to be positioned on any angle.

The *plain vise* differs very little from the swivel base vise except that it cannot be positioned to any angle (Figure 48–16).

The *universal vise* may be positioned to any degree due to its swivel base and in addition may be tilted at an angle of up to 90°. This dual setting permits compound angles to be machined (Figure 48–17).

UNIT 49 MOUNTING AND REMOVING ARBORS AND CUTTERS

The arbor is the main method by which cutters are held on a horizontal milling machine. Other types of holding devices are used, including the shell end mill arbor and various other collets and adaptors. This unit will deal mainly with the light-duty milling machine arbor and the common milling cutters used in the toolroom.

Objectives
After completing this unit you should be able to

1. identify and explain the purpose of at least seven standard milling cutters.
2. identify and explain the purpose of at least five special-purpose milling cutters.
3. mount and remove a milling machine arbor.
4. mount and remove a milling cutter.

Milling Methods

There are two common milling methods: peripheral milling and face milling.

In peripheral milling, the workpiece surface is machined by the outer rotation of the cutting teeth. The finished surface lies parallel to the axis of the cutter. Peripheral milling is generally carried out on the horizontal milling machine but can also be done on the vertical milling machine.

In face milling, the cutting teeth are located on the face of the cutter, generally at right angles to the axis of the cutter. Face milling can be done on both horizontal and vertical milling machines and is preferred over peripheral milling whenever possible.

Common Milling Cutter Terms

To better understand a milling cutter, an operator should be familiar with a few basic terms, illustrated in Figure 49–1, and the uses of each cutter.

Straight fluted cutter: Milling cutters may have straight or helical flutes. Straight fluted cutters have their cutting edges parallel to the axis of the cutter.

Hand of flute helix: Helical flutes twist away in one direction only. When viewed from either end of the cutter, a right-hand helix twists away from the viewer in a clockwise direction. A left-hand helix twists away from the viewer in a counterclockwise rotation.

Hand of rotation: Milling cutters are manufactured as either being right-handed or left-handed. If a milling cutter rotates counterclockwise when cutting, it is considered a right-handed cutter. If it rotates in a clockwise direction when cutting, it is considered a left-hand cutter.

Gullet: The gullet may be defined as the space between the teeth of a multitooth cutter. It is very important to the cutting action since it helps to form and free the chip from clogging. If a chip is unable to be set free, it packs into the gullet and will prevent the tool from cutting.

Heel: That portion of the tooth behind the land is called the heel.

Land: The surface directly behind the cutting edge is known as the land. The width of the land varies with different cutter diameters.

Rake angle: The effective rake angle is formed by the slope of the tooth face, which may be positive or negative. If the face leans forward at the tip, it is a positive rake angle; if it leans backward, it is considered a negative rake angle.

Relief: A radial relief is provided on milling cutters to prevent the heel of the tool from rubbing. Relief on side teeth is called peripheral relief.

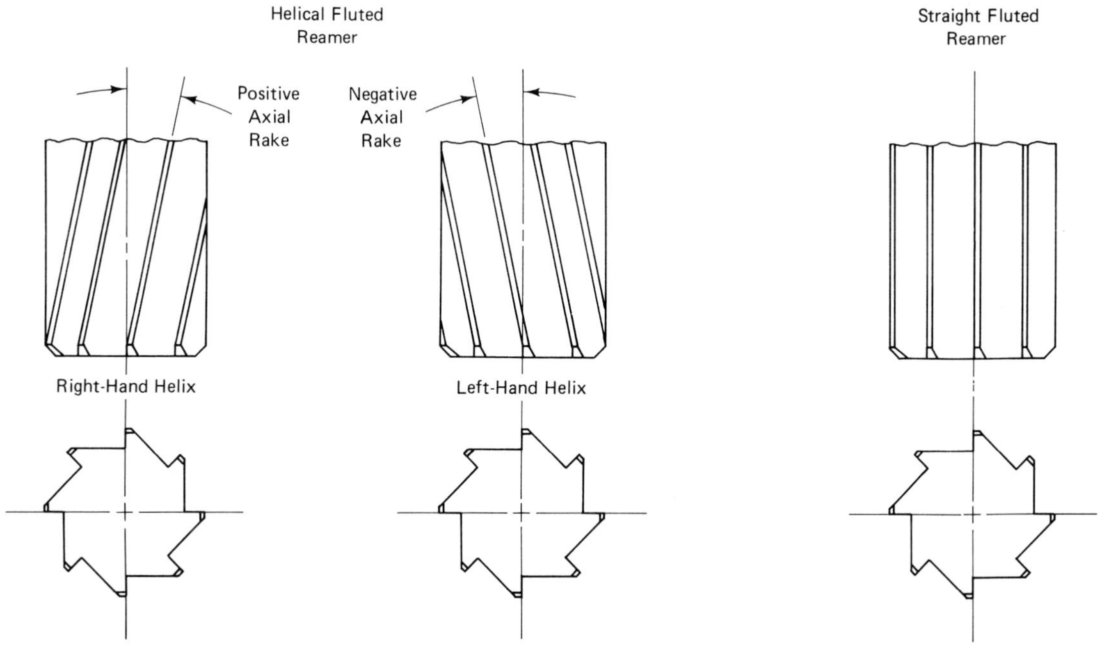

FIGURE 49–1
Common milling cutter terms.

FIGURE 49-2
Horizontal milling machine cutters. (Credit: Union/Butterfield)

FIGURE 49-3
A light-duty plain milling cutter. (Credit: Regal-Beloit Corporation)

Common Milling Machine Cutters

Cutters used on a plain horizontal milling machine come in many sizes, shapes, and contours. The shape of the cutter is generally a wheel around which teeth are staggered or spaced (Figure 49-2). Horizontal milling machine cutters, generally made of high-speed steel, have an inner hole that is ground to an accurate diameter and has a keyway slot. This allows the cutter to be mounted and keyed on an arbor. Many cutters used in production are made of tungsten carbide or have cemented carbide cutting teeth to meet higher production quotas, cut harder materials, and be revolved at faster speeds.

Plain Milling Cutters

Made with cutting teeth on the periphery only, plain milling cutters vary in width and tooth construction and are designed to mill a flat surface where the area involved is less than the width of the cutter. As a general classification, plain milling cutters are either light- or heavy-duty cutters.

Light-Duty Cutters. Light-duty cutters (Figure 49-3) generally are manufactured in widths less than $\frac{3}{4}$ in. (19 mm) and have straight teeth and undercut sides. Larger sizes have a left-hand helix that ranges between 15° and 25° and a radial rake of 10° to 20°. Light-duty cutters are generally used for moderate cuts in most materials.

Heavy-Duty Cutters. Heavy-duty cutters (Figure 49-4) are generally produced over 2 in. (50 mm) wide and have deep gullets for faster stock removal. These cutters are produced with a 12° rake and a 25° to 45° left-hand helix to provide a smooth finish. They are recommended for taking heavy cuts in mild steel and similar ferrous materials.

FIGURE 49–4
A heavy-duty plain milling cutter is provided ample chip clearance, permitting heavier cuts to be taken.

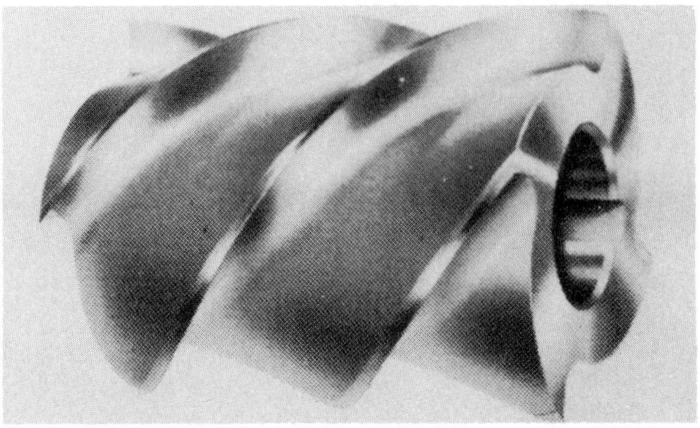

FIGURE 49–5
A high helix cutter is ideal for machining mild steel. (Credit: Union/Butterfield)

High Helix Cutters. High helix cutters (Figure 49–5) are manufactured to leave no tool marks as they run at high speeds. They are available in sizes as wide as 8 in. (200 mm) and numbering in teeth from 4 to 6. Their left-hand spiral/right-hand cut make them ideal for milling mild steel as well as for taking light cuts on thin materials.

Face Milling Cutters

The *face mill* (Figure 49–6) is generally made 6 in. (150 mm) in diameter or larger and is fitted onto the end of an adaptor. It can then be mounted in a vertical or horizontal milling machine and is used for milling large, flat surfaces. The cutting teeth located on the end and periphery of the tool are replaceable. They may be made of high-speed steel or cast alloy or may be tipped with carbide cutting edges. The face mill is designed to make roughing and finishing cuts in one pass. While the peripheral teeth do most of the cutting, the end teeth finish the surface created by the peripheral teeth.

The *shell end mill* is a smaller version of the face mill and is often used instead of a large end mill.

FIGURE 49–6
A face milling cutter is designed to make a rough and finishing cut in one pass.

FIGURE 49–7
A shell mill is a general-purpose milling cutter.
(Credit: Regal–Beloit Corporation)

FIGURE 49–8
A plain side milling cutter is well adapted for shallow cuts and slotting.

Mounted and fastened by a screw or a nut on a short stub arbor, its multiteeth are helical and are cut from solid stock (Figure 49–7).

FIGURE 49–9
Interlocking side milling cutter is used in straddle milling.

Side Milling Cutters

The *side milling cutter* (Figure 49–8) has cutting edges on the side as well as on the periphery. The side milling cutter can be used for general-purpose cutting, milling shallow cuts, and cleaning an upright surface, and when paired and spaced in tandem, side milling cutters can be used for straddle milling operations. Chip clearance is provided by the undercut area between the hub and the cutting teeth. Side milling cutters with inserts are also available in diameters of 8 in. (200 mm) or larger and in general come as wide as 1 in. (25 mm).

The *half side milling cutter* is used in straddle milling and is available as either right- or left-handed. This cutter is used for milling a side surface or when one side of the cutter is used. Both the half side milling cutter and the *interlocking side mill* (Figure 49–9) are manufactured with a 12° to 15° rake tooth and are capable of heavy-duty milling.

The *staggered tooth side milling cutter* (Figure 49–10) is provided with alternate right and left spiral cutting teeth and is capable of rapid stock removal. It is recommended for milling deep slots and

FIGURE 49–10
Staggered tooth side milling cutter.

grooves in nonferrous metals such as aluminum using high speeds and feeds. Staggered tooth cutters reduce vibrations and also produce a smoother cut.

Metal Saws

There are two styles of metal slitting saws: one made with plain concave sides and one made with side teeth.

The *plain concave slitting saw* (Figure 49–11) is ground concave on both sides for clearance, with its hub flat and the same thickness as the teeth. A less than 1° negative radial rake is used on teeth up to $\frac{3}{32}$ in. (2.4 mm) wide; wider teeth have a positive 10 radial rake angle. Since the saw is used for one size slitting and cutoff operations, the width of the teeth are held to very close tolerances. The metal slitting saw should only be used with the arbor key extending into adjoining collars on both sides of the cutter to prevent slippage and damage that would otherwise result from friction driving.

Metal slitting saws are also available with side teeth and staggered teeth. *Straight-tooth slitting*

FIGURE 49–11
A metal slitting saw.

FIGURE 49–12
A straight-tooth, side cutting metal saw.

saw (Figure 49–12) is produced with a positive undercut as well as side relief ground concave on both sides. This design gives a freer cutting action, eliminating binding and dragging when involved in deep slotting or cutoff operations.

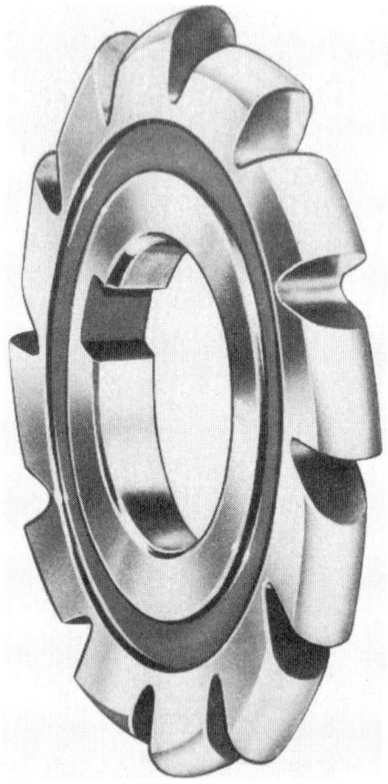

FIGURE 49–13
A convex cutter will produce a female half circle. (Credit: Regal–Beloit Corporation)

FIGURE 49–15
A corner rounding cutter. (Credit: Regal–Beloit Corporation)

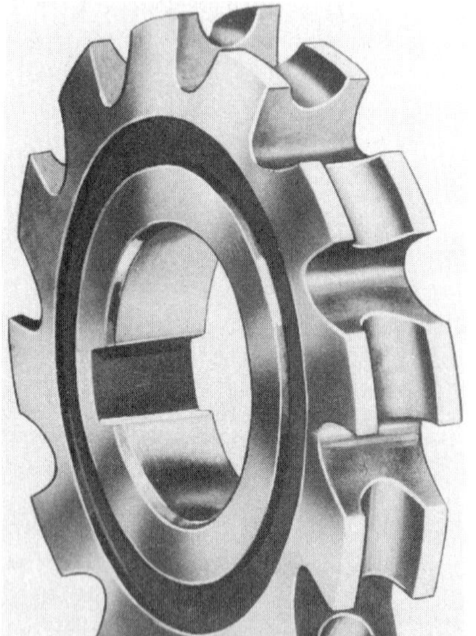

FIGURE 49–14
A concave cutter will produce a male half circle. (Credit: Regal–Beloit Corporation)

Special-Purpose Cutters

Milling machines use a number of common standard cutters that can be used for a wide variety of cutting applications. Still to be examined are a few of the specially designed tools for specific operations.

Convex and Concave Cutters. These form-relieved cutters are mounted on a horizontal milling machine and are designed for producing half circles or forms in workpieces, each opposite from the other. The convex cutter (Figure 49–13) produces a female half circle, while the concave cutter (Figure 49–14) produces a male half circle. They are both manufactured with a straight chip carrying space on the end of each tooth (gash), which permits resharpening without disturbing the form.

Corner Rounding Cutter. Designed with a right- or left-hand cut, the corner rounding cutter (Figure 49–15) is used for radius milling. These, too, are supplied with straight gash teeth and can be resharpened by grinding the surface only without destroying the contour.

FIGURE 49-16
A single-angle cutter. (Credit: Union/Butterfield)

FIGURE 49-17
A double-angle cutter. (Credit: Regal–Beloit Corporation)

Angle Cutters. Angle cutters are available with either a single or a double angle. Single-angle cutters (Figure 49-16) are available with either a right- or left-hand cut and with an included angle of 45° or 60°. The teeth are straight gashed and are capable of angular as well as side cutting. These cutters are used for cutting dovetails, ratchet wheels, and similar work.

Double-angle cutters (Figure 49-17) are often used for cutting V-grooves and similar milling operations. They are constructed with twin angular edges that converge. The included angle thus formed may be 45°, 60°, or 90° and is measured at a right angle to the axis of the cutter. When these angles are equal, the cutter is symmetrical and is identified by the included angle. Cutters are also available with each side having a different angle and they are considered unsymmetrical. These cutters are identified by the angles on each side of the cutter.

The *hob* (Figure 49-18) is an arbor-type milling cutter used for cutting gear teeth. Both the workpiece and the hob are rotated in a timed movement with each other. The hob, which has its cutting edge arranged in a helical thread, is fed into the workpiece to produce the gear tooth form. In hobbing, the gear tooth form produced on the workpiece is seldom like the cutting edge form on the hob.

FIGURE 49-18
A hob is used to cut gear teeth. (Credit: Union/Butterfield)

The *multiple-thread milling cutters* (Figure 49-19) are available in either arbor or shank type and can be used for both internal and external

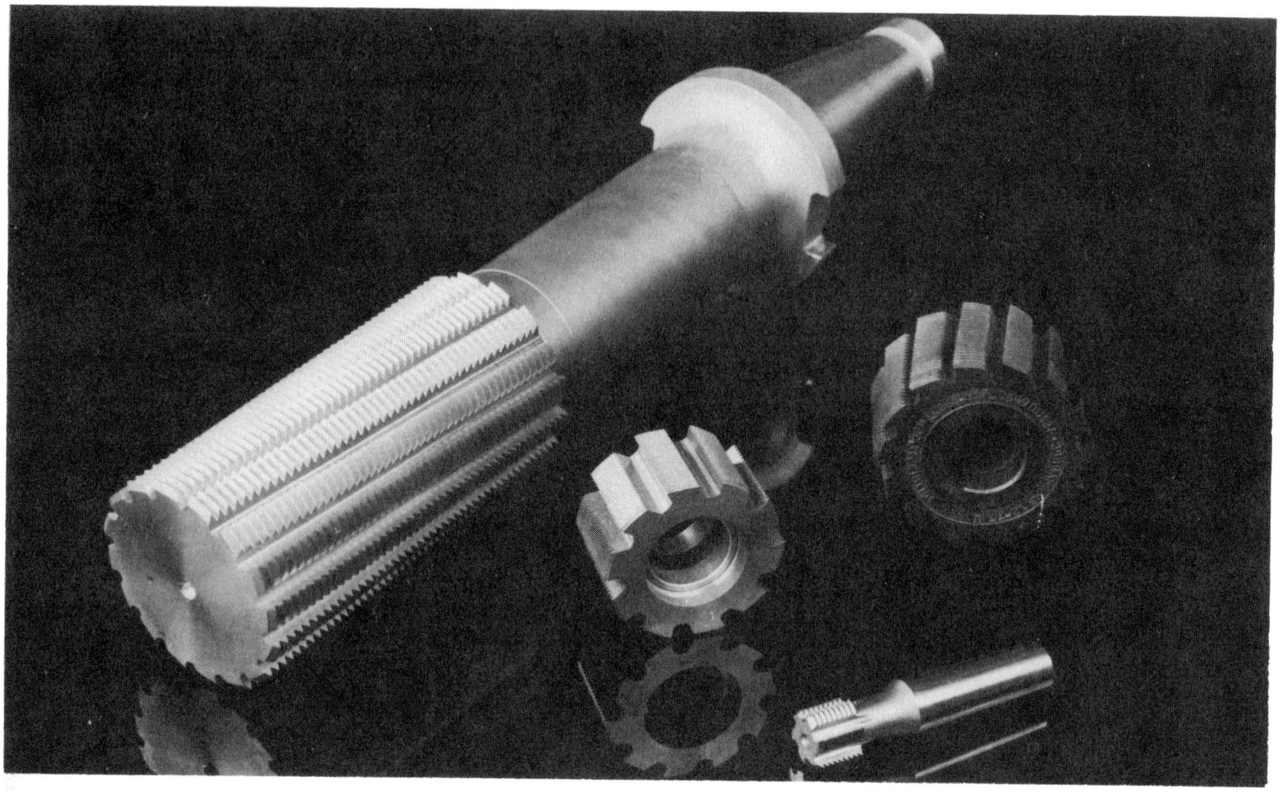

FIGURE 49-19
Multiple thread milling cutters. (Credit: Union/Butterfield)

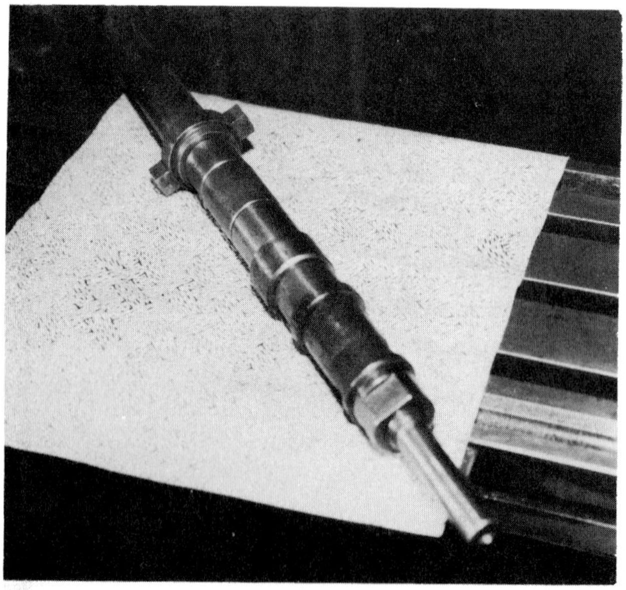

FIGURE 49-20
A horizontal milling machine arbor placed on a protected table.

thread milling. While they are available for all the standard thread types, they can be designed for straight or tapered threads and will thread an entire workpiece length in one complete revolution.

The Arbor Assembly

A milling machine arbor resembles a long shaft extending from the main spindle (Figure 49-20). One end is tapered and is held within the spindle by means of a draw bolt. The other end is threaded to fit an arbor nut while a sliding bearing bushing is made to fit within an overarm arbor support. Spacing collars, while not as large as the bearing bushing, are provided to allow the cutter and bearing bushing to be placed at any position along the arbor. Since the milling cutter hole is ground and can be keyed to be held accurately on the arbor, the spacing collars are held when tightened by the arbor nut. For this reason, the surface of the spacers and milling cutters must be free of chips and dirt, especially when assembled on the arbor.

Mounting an Arbor

Before attempting to mount an arbor, a few conditions should be met. The machine must be set at the lowest spindle speed, a good fitting wrench should be used and the operation should be done safely.

1. Shut off the machine.
2. Clean the tapered end of the arbor and remove any burrs.
3. Clean the tapered spindle opening.
4. Align the slots on the arbor's tapered end flange with the spindle's driving lugs and insert the arbor.
5. Thread the draw-in bar into the arbor's end.
6. Hand tighten the jam nut against the spindle's shoulder (Figure 49–21).
7. Complete the tightening of the jam nut using a wrench.
8. Turn on and engage the milling machine spindle to see if there is any arbor runout.

If there is any runout, the machine should be shut off and the arbor removed and recleaned before remounting. If runout still exists after it has been cleaned, the arbor should be replaced.

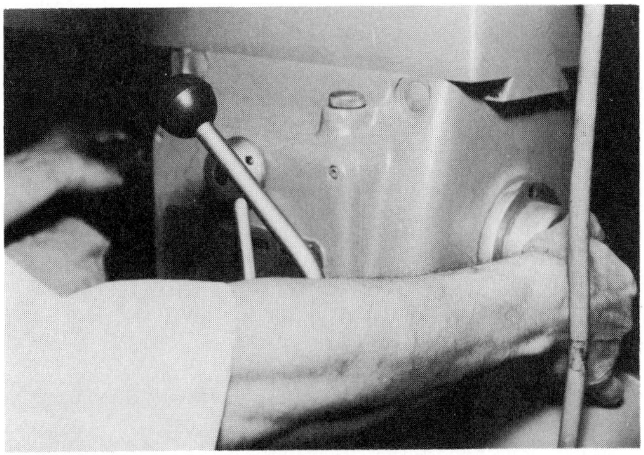

FIGURE 49–21
Hand tighten the jam nut before using a wrench on it.

Removing an Arbor

With the arbor support removed from the overarm and the milling machine still set in a low spindle speed, the arbor can be removed in the following manner:

1. Make sure the machine has been shut off.
2. Loosen the jam nut on the draw-in bar counterclockwise about two turns.
3. Loosen the draw-in bar not more than one full turn.
4. Holding the arbor with one hand, strike the end of the draw-in bar head with a soft-faced hammer to release the grip of the taper hold (Figure 49–22).
5. Still holding the arbor with one hand, unscrew the draw-in bar with the other hand to remove the arbor.

FIGURE 49–22
Strike the end of the draw-in bar with a soft-faced hammer to release the taper hold grip.

Mounting a Cutter

Mounting cutters has to be done with care. Cutters are sharp and should not be handled with bare hands. A table protector in the form of thin plywood or masonite should be placed on the table top to prevent damage when placing the various milling machine equipment on it.

With the milling machine placed in the lowest spindle speed, proceed as follows:

FIGURE 49-23
Remove the arbor support and place it on the table.

1. Make sure the machine has been shut off.
2. Remove the arbor nut and place it on the table protector.
3. Remove the arbor support and place it on the table protector (Figure 49-23).
4. Slide out the number of space collars required to locate the cutter.
5. Thoroughly clean the arbor and the cutter faces with a cloth.
6. With the cutter teeth positioned to rotate in the proper direction and held by a cloth or folded paper toweling, slide the cutter on the keyed arbor (Figure 49-24).
7. Clean and slide the spacer collars required to allow the bearing bushing to be as close as possible to the cutter (Figure 49-25).
8. Slide the bearing bushing on the arbor.
9. Slide the arbor support on the overarm and slip it over the bearing bushing (Figure 49-26).
10. Slide the number of spacer collars on the arbor to allow the arbor nut to tighten the cutter and spacers without damaging the arbor.
11. Tighten the arbor by hand.
12. Tighten the arbor support nut to lock it over the bearing bushing.
13. Tighten the arbor nut securely with a good-fitting wrench.

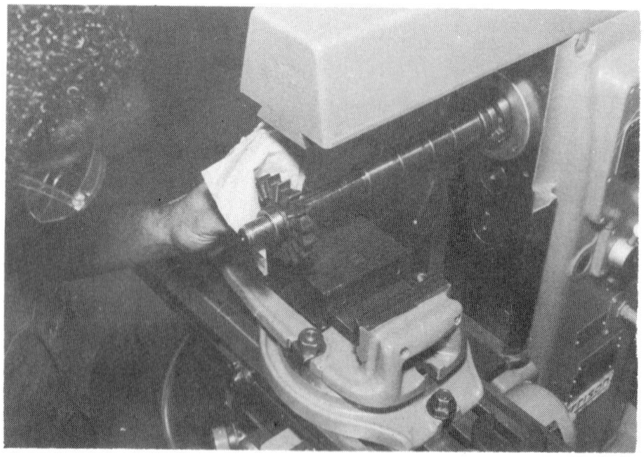

FIGURE 49-24
Hold the cutter safely with a cloth or folded piece of toweling.

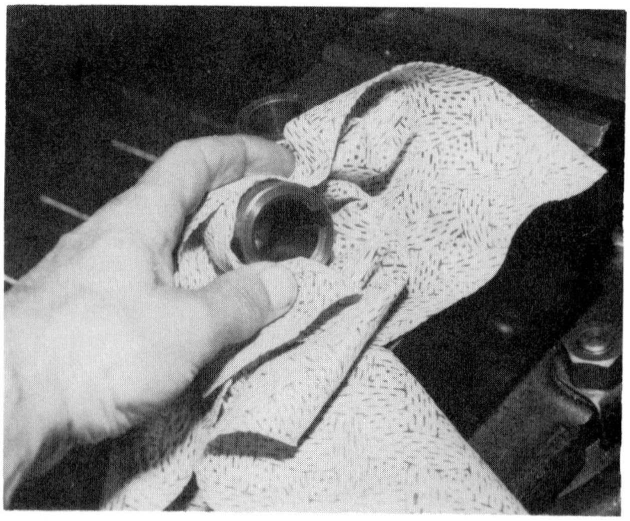

FIGURE 49-25
The edges of the spacers as well as the holes should be cleaned before mounting.

Removing a Cutter

In order to remove a cutter, the directions for mounting a cutter are generally reversed. Proceed as follows:

1. Shut off the machine.
2. Set the machine in the lowest spindle speed.
3. Loosen the arbor nut with a good-fitting wrench (Figure 49–27).
4. Loosen the arbor support nut.
5. Place a protector board on the table top (Figure 49–28).
6. Unscrew the arbor nut by hand and place it on the table protector.
7. Slide the arbor support out and place it on the table protector.
8. Slide out the spacing collars and place them on the table protector (Figure 49–29).

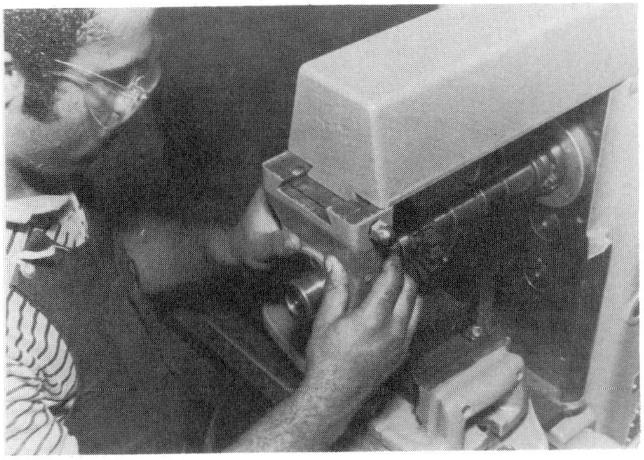

FIGURE 49–26
Slide the arbor support over the bearing bushing.

FIGURE 49–28
Place a protector board on the table top.

FIGURE 49–27
Loosen the arbor nut with a good fitting wrench.

FIGURE 49–29
Slide out the spacing collars and place them on the table protector.

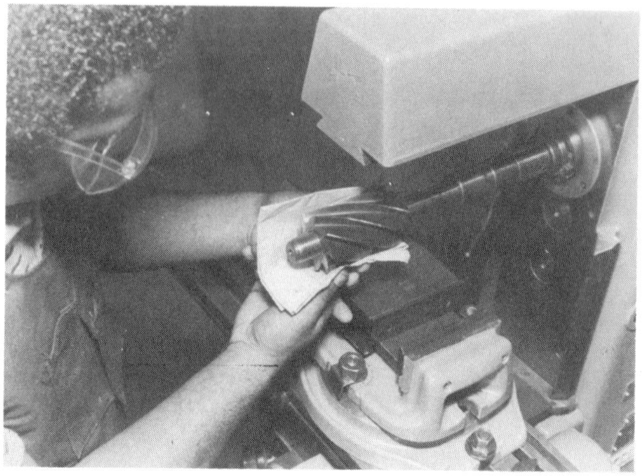

FIGURE 49-30
Slide the cutter out carefully, protecting the hands in the process.

FIGURE 50-1
Greater holding power can be had by milling against the vise jaws.

9. Hold the milling cutter with a cloth or folded paper toweling (see Figure 49-30) and slide the cutter out.

If the arbor nut is not to be removed, the spacing collars, bearing bushing, arbor support, and nut should all be cleaned and repositioned for future use.

UNIT 50 HORIZONTAL MILLING A FLAT SURFACE

In order to efficiently mill a flat surface on a horizontal milling machine, several factors will have to be taken into account. These include the work holding method, the type of cutter used, and the speed, feed, and depth of cut taken by the milling cutter. All these conditions will be examined in this unit.

Objectives
After completing this unit you should be able to

1. know how to safely set up a workpiece for a surface milling operation.
2. select the proper milling cutter.
3. select and calculate the proper speed, feed, and depth of cut to be taken by the cutter.

Work Holding Methods

Machining a flat surface in a horizontal milling machine requires a safe, yet rigid setup. The work must be held in such a way that the surface to be cut is free of any obstruction. The devices frequently used in the machine shop to hold work include the milling vise, angle plates, strap clamps, and fixturing.

The milling machine vise can hold a variety of shapes including round, square, and rectangular stock. Generally, work is held on parallels to elevate the work surface above the vise jaws and to rest it on a solid seat. (For holding work on a set of parallels in a vise, see Unit 53.) Work holding in a milling vise can be improved if the force of the cutter's rotation can be applied against the vise jaw (Figure 50-1).

Angle plates are often used by clamping the workpiece against an upright surface when the machined surface requires it to be at right angles to the clamped workpiece surface (Figure 50-2).

When work is large and bulky, it should be strapped or bolted to the table directly or by means of a set of T-bolts, straps, and step blocks (Figure 50-3).

A fixture is a specially made work holding device that allows a particular milling operation to be performed. This type of tooling is generally made for repetitive production operations or where the conventional work holding methods are not suitable.

FIGURE 50-2
By holding one work surface against the upright of an angle plate, the top surface will be milled square with it.

Selection of a Cutter

As with any milling operation, the type of cutter used will depend on the material being cut and the operation being performed. For wide surface cutting in school shops, the cutter is generally made of high-speed steel. The style of cutter would either be a face milling cutter or the shell end mill. Both are mounted in the spindle and are used for machining flat work surfaces near the edge of the table or extending from one end of the vise. In machining flat surfaces held in a vise, a variety of plain milling cutters are available, as indicated in Unit 49.

FIGURE 50-3
Large or bulky work should be bolted or strapped directly to the table, such as this special vise mounting.

Speeds and Feeds

Milling machine speeds are based on the same formula as used on lathes and drill presses (see Units 31 and 35). The principal question raised for effective milling is how fast should the cutter be rotating. Since the cutter used on the milling machine may range in diameter from a drill-like size used in a vertical milling machine to a wheel-type cylinder used on the horizontal milling machine, a quick solution would be to consult Table 50-1.

Feed is the rate at which the milling cutter is advanced into the workpiece. While the factors governing cutting are varied, depending on the depth and width of the cut, the material and sharpness of the cutter as well as the direction of the feed are also important.

TABLE 50–1
Milling machine cutting speeds

Material	*High-speed steel cutter* *sf/min*	*m/min*	*Carbide cutter* *sf/min*	*m/min*
Aluminum	600+	180+	1000+	150+
Brass (free cutting)	250–300	75–90	1000+	150+
Bronze (hard)	150–200	45–60	400	120
Cast iron (soft)	75–100	23–30	250–350	75–90
Machine steel	70–100	21–30	150–250	45–60
Magnesium	600+	150+	1000+	150+
Tool steel	60–70	18–20	125–200	37–60
Stainless steel	35–70	10–21	250–400	75–120

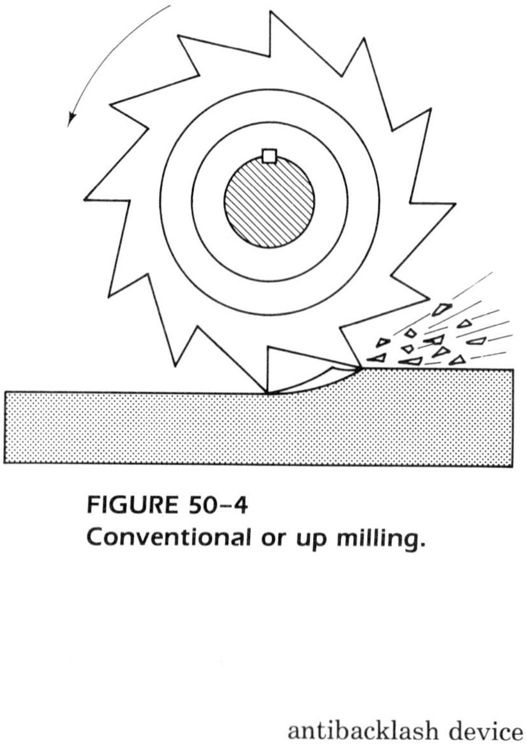

FIGURE 50–4
Conventional or up milling.

Direction of Cut

There are two methods of feeding work: with the rotation of the cutting edge or against the rotation of the cutting edge (Figure 50–4). When the workpiece is fed with the cutter rotation, there is a tendency for the workpiece to skid through. In appearance, the cutter appears to be climbing over the workpiece. Climb or down milling should never be undertaken when the machine is not equipped with an antibacklash device (Figure 50–5). In down milling, the work is pulled through. The operator loses accurate control of the feed, creating a lost motion (backlash) in the feed screw being turned (Figure 50–6).

In conventional or up milling, the work is fed against the rotation of the cutting edges. It is somewhat similar to swimming against the tide.

In production milling, feed rates are calculated according to the number of teeth the cutter contains, taking into account also the chip advanced and the number of rpm. Suggested feeds per tooth for various types of high-speed cutters under average conditions are listed in Table 50-2. The feed per tooth affects the chip thickness. It is also very important to the life of the cutter. On smaller toolroom machines that are hand fed, judgment is exercised in passing the workpiece past a cutter. Excessive feed will cause cutter damage, wear, or breakage (Figure 50-7).

FIGURE 50-7
Excessive feed will cause damage, wear, or breakage in a cutter.

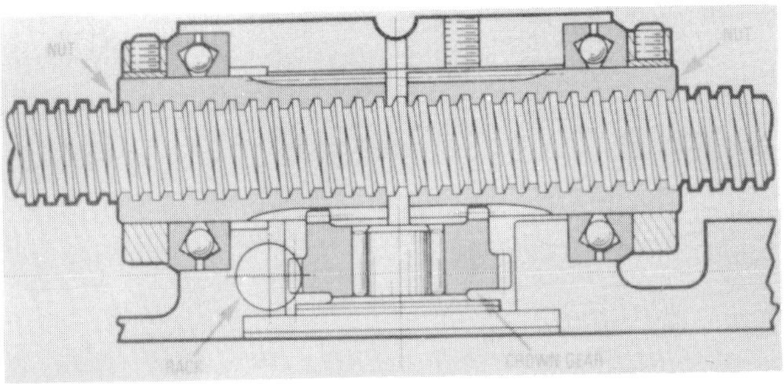

FIGURE 50-5
An antibacklash device such as this prevents the loss of an accurate feed control. (Credit: Cincinnati Milacron Inc.)

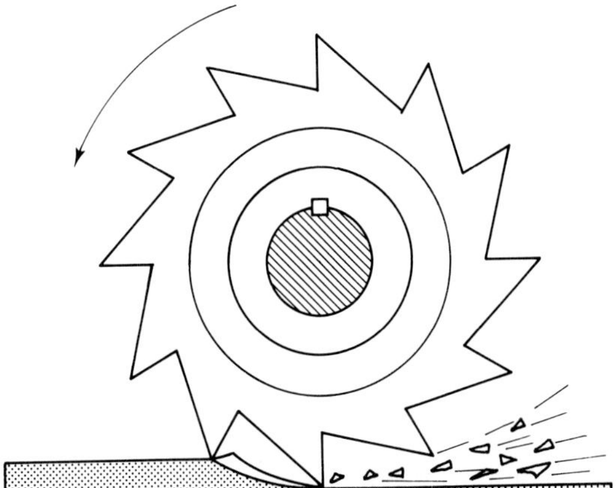

FIGURE 50-6
Climb or down milling.

TABLE 50–2
Suggested feed per tooth for high-speed steel milling cutters

	Face mills		*Helical mills*		*Slotting and side mills*		*End mills*		*Form-relieved cutters*		*Circular saws*	
Material	*in.*	*mm*	*in.*	*mm*	*in.*	*mm*	*in.*	*mm*	*in.*	*mm*	*in.*	*mm*
Aluminum	0.022	0.558	0.018	0.457	0.013	0.330	0.011	0.279	0.007	0.177	0.005	0.127
Brass (soft)	0.022	0.558	0.018	0.457	0.013	0.330	0.011	0.279	0.007	0.177	0.005	0.127
Bronze (hard)	0.009	0.228	0.007	0.177	0.006	0.152	0.005	0.127	0.003	0.076	0.002	0.050
Cast iron (soft)	0.016	0.406	0.013	0.330	0.009	0.228	0.008	0.203	0.005	0.127	0.004	0.101
Machine steel	0.012	0.304	0.010	0.254	0.007	0.177	0.006	0.152	0.004	0.101	0.003	0.076
Magnesium	0.022	0.558	0.018	0.457	0.013	0.330	0.011	0.279	0.007	0.177	0.005	0.127
Tool steel	0.010	0.254	0.008	0.203	0.006	0.152	0.005	0.127	0.003	0.076	0.003	0.076
Stainless steel	0.006	0.152	0.005	0.127	0.004	0.101	0.003	0.076	0.002	0.050	0.002	0.050

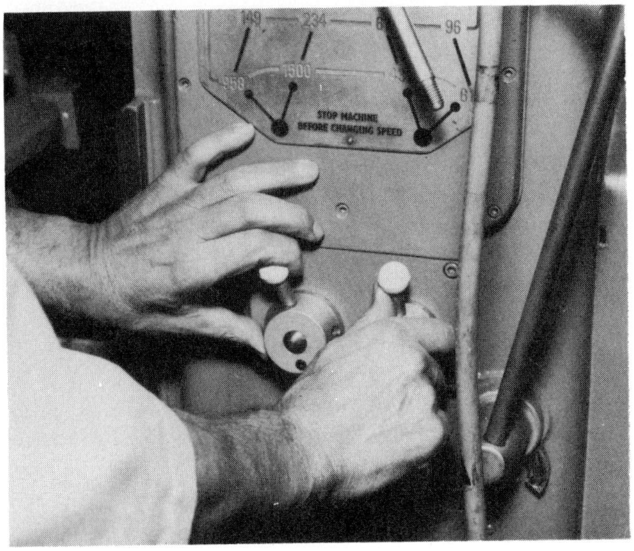

FIGURE 50-8
On this model, the machine must be stopped before changing speeds.

Setting Speed and Feed

Milling speeds can be determined by referring to the cutting speeds given in Table 50-1 and can be set by moving the machine's speed change levers to their proper positions (Figure 50-8).

In milling, feed is the distance the work moves past the cutter in inches or millimeters per minute. In operation, it is looked upon as feed per tooth or chip load. Recommended feeds per tooth are provided in Table 50-2. Since each tooth advances into the work with equal removal, the rate of feed can be incorporated in the formula

Feed = number of cutter teeth × recommended feed per tooth × rpm.

While the values in Table 50-2 are ideal and can be maintained with good production equipment, it should not be forgotten that feed is dependent on:

Material
Rigidity of setup
Type of cutter
Conditions of the machine
Shape of the workpiece
Cutter sharpness

Students would be advised to set the feed rate at approximately one-half the values shown in Table 50-2. This will prevent overfeeding, which can lead to cutter damage and premature wear.

Setting the Depth of Cut

There are two basic cuts to be considered in cutting work to size: roughing and finishing cuts. This is not to say that all work will require just two cuts. The roughing cut may require several cuts before completing the operation with a finishing cut. When this happens, it is advisable to take as heavy a roughing cut as the machine, setup, and cutter will accept. In general, a coarse tooth cutter will take a heavier chip.

Finishing cuts should be completed with as sharp a cutter as possible. Material removal should be at least 0.015 in. (0.38 mm) to allow the tool to cut under rather than rub on the surface, and a fine feed should be used. To prevent any tool marks or digging in, the feed should not be stopped with the cutter in contact with the work. The machine should also be stopped when returning the work past the cutter to its original feed start. This will prevent skipping tool marks.

Milling a Flat Surface

Milling the surface flat of a workpiece held in a vise is one of the common operations performed in a horizontal milling machine. It does not really matter whether the vise is perfectly square or parallel to the table travel. What does matter is:

- The workpiece is free of burrs before mounting
- The workpiece is held securely in the vise
- The cutter width is greater than the surface width being cut
- Setting the cutter safely to the work surface
- Locking the table height when taking depth of cuts

With these guidelines in mind, proceed as follows:

1. Remove all burrs from the workpiece with a file.
2. Clean the milling vise and workpiece.
3. Select a set of parallels that will elevate the workpiece surface slightly higher than the vise jaws.

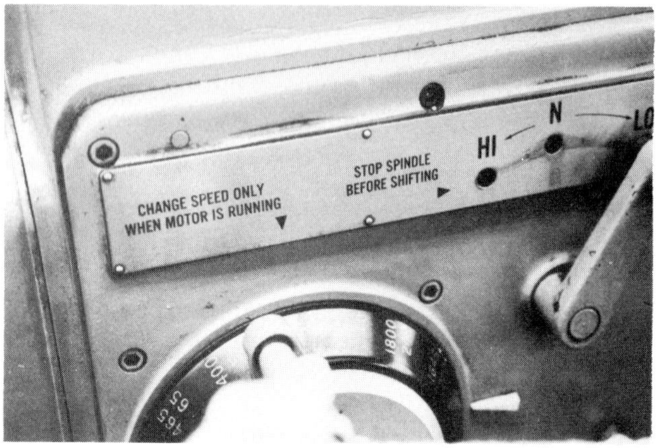

FIGURE 50-9
The motor must be running when changing speeds on this milling machine.

FIGURE 50-10
Use a long strip of paper to set the cutter to the work surface.

4. Mount the workpiece so that it rests on the parallels with paper feelers extending beneath its four corners (see Unit 53).

5. Tighten the milling vise securely by hand. *Note: Never use any hammer on the vise wrench for this purpose.*

6. Tap the workpiece corners down with a weighted mallet until the paper feelers have been pinched against the parallels.

7. Select a helical milling cutter wider than the workpiece surface to be cut.

8. Mount the cutter on the arbor to be used for conventional or up milling.

9. Consult Table 50-1 and set the calculated speed for the size of the cutter and workpiece material (Figure 50-9).

10. Tear a strip of paper about $\frac{1}{2}$ in. (13 mm) wide by at least 12 in. (300 mm) long.

11. Raise the table so that the workpiece is about $\frac{1}{16}$ in. (1.60 mm) below the cutter.

12. Holding one end of it, place the strip between the workpiece and the cutter.

13. Start the milling machine spindle.

14. Slowly raise the table until the cutter just begins to cut the paper (Figure 50-10).

15. Stop the milling machine spindle.

16. Move the table so that the workpiece clears the cutter.

17. Move the table to center the workpiece with the width of the cutter.

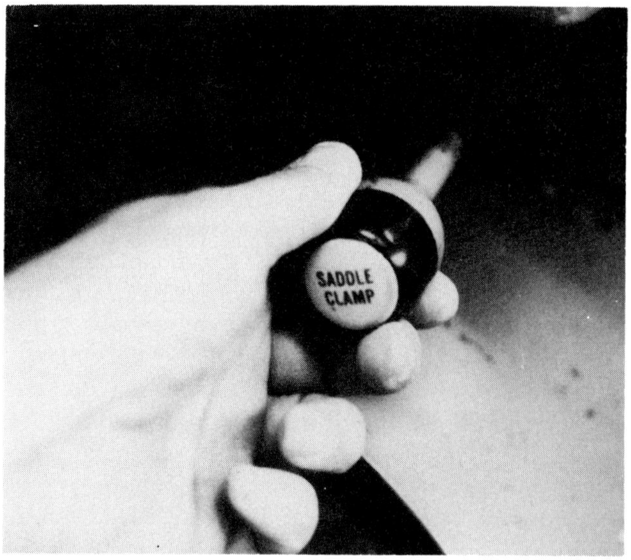

FIGURE 50-11
Lock the milling machine saddle to prevent the table from moving cross ways.

18. Lock the table to prevent it from moving cross ways (Figure 50-11).

19. Raise the table 0.003 in. (0.07 mm) to allow for the paper thickness and set the micrometer collar to zero (Figure 50-12).

20. Turn the vertical feed handle to raise the table to the desired depth of cut.

FIGURE 50–12
Once the initial cut has been established, set the micrometer collar on the vertical feed to a zero reading.

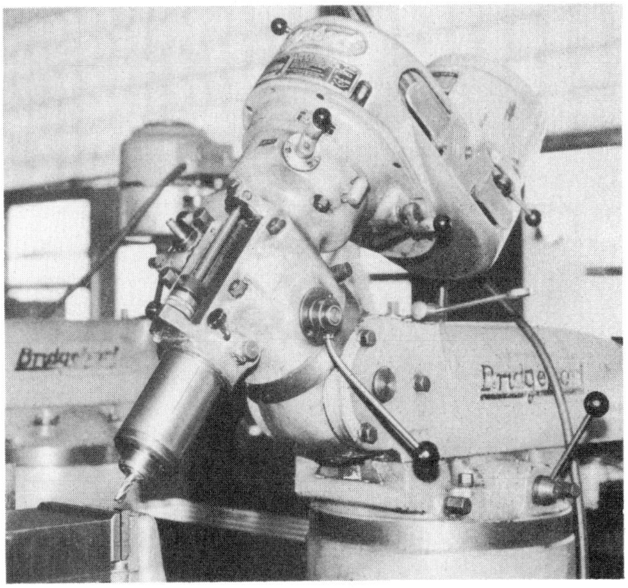

FIGURE 51–1
The spindle head on a ram-type vertical milling machine can be tilted forward or backward.

21. Lock the knee.

22. Start the milling machine spindle.

23. Advance the workpiece by hand so it can be trial cut about $\frac{1}{4}$ in. (6 mm) wide.

24. Back off the workpiece and stop the milling machine spindle.

25. Measure the workpiece to determine if the depth of cut is correct or needs to be adjusted.

The cut may then be completed using a cutting fluid along with automatic feed, if any, or hand feed.

UNIT 51 VERTICAL MILLING MACHINE

FIGURE 51–2
The spindle head assembly of a ram-type vertical milling machine can be rotated in a 180° semicircle. (Credit: Bridgeport Machines, Inc.)

The vertical knee and column milling machine has a spindle that can hold a cutter at 90° to the work table. On some vertical milling machines, the head in which the cutter is mounted can be set to lean forward or tilt backward at different angles (Figure 51–1) as well as rotate in a 180° semicircle (Figure 51–2). The vertical milling machine is capable of many milling operations such as end milling, fly cutting, slotting, keyway cutting, T-slotting, face cutting, as well as drilling, reaming, and jig boring. The flexibility of the vertical milling machine has been greatly increased by being fitted or having digital readout systems added (retrofitted) to their operational capabilities. Because of its versatility, it is widely accepted in the machine shop and is rated just behind the lathe for versatility.

Objectives
After completing this unit you should be able to

1. recognize and know the difference between the two vertical milling machines used in the machine shop.
2. recognize and know the purpose of the main operational parts of a vertical milling machine.
3. recognize and know the purpose of four standard and four special end mills.
4. recognize and know the purpose of four vertical milling machine accessories.

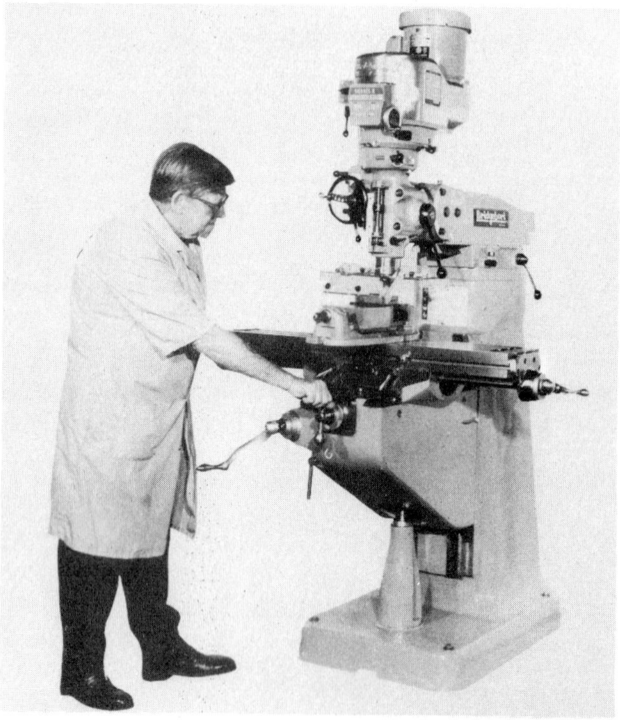

FIGURE 51-3
The main operating parts of a ram-type vertical milling machine. (Credit: Bridgeport Machines, Inc.)

Vertical Milling Machines

There are two types of vertical milling machines generally used in school shops. The spindle head assembly on the standard vertical milling machine cannot be swiveled or tilted and is designed to take heavy side milling loads. It is often used in face cutting and shell end mill operations. Due to the spindle's vertical milling location and short travel, the standard vertical milling machine is well suited to perform as a drilling and boring machine where hole machining locations require very close tolerances.

The ram-type vertical milling machine (Figure 51-3) is lighter in construction than the standard vertical milling machine but is more flexible and in more wide use as a general-purpose machine tool. The ram-type vertical milling machine will be discussed in this unit.

Vertical Milling Machine Parts

The *base* and *column* are cast as a one-piece construction. The base is cast iron ribbed and may act as a reservoir for coolant. The machined face of the column contains dovetail ways that permit vertical movement of the knee. The ways are scraped for accuracy and may also be chrome plated. Another set of dovetail ways are provided on top of the column to receive a *turret* on which a *ram* or *overarm* is mounted.

The turret may be turned in any position in a horizontal plane (Figure 51-4). The ram may be moved toward or away from the column, greatly increasing machining capabilities over the table.

The *head* is attached to the end of the ram or overarm. The head may be swiveled in two planes. The *spindle* is mounted on top of the head and is attached to the motor. Spindle speed adjustment is provided by the *motor* by means of variable-speed pulleys or, on some models, by means of V-belts and gears. The spindle is housed within a *quill.* The quill is raised or lowered vertically by means of a *quill feed hand lever* or the *quill fine-feed handwheel.* It can also be set in motion by automatic power feed.

The *knee* supports the *saddle* and *table* and is raised or lowered by means of a vertical traverse lever. Knee locks are provided to lock the knee at a given position. The saddle, which can be moved toward or away from the column by turning the crossfeed handwheel, supports the table. Saddle locks are also provided to secure the saddle at a given position. The table may be moved to the right or left of the column by turning the table traverse

FIGURE 51-4
The turret allows the ram to be swiveled so that either end can be used for machining. (Credit: Bridgeport Machines, Inc.)

handwheel. This, too, may be locked in position by means of table locks. While power knee and table feeds are available as options, general-purpose vertical milling machines are manually operated.

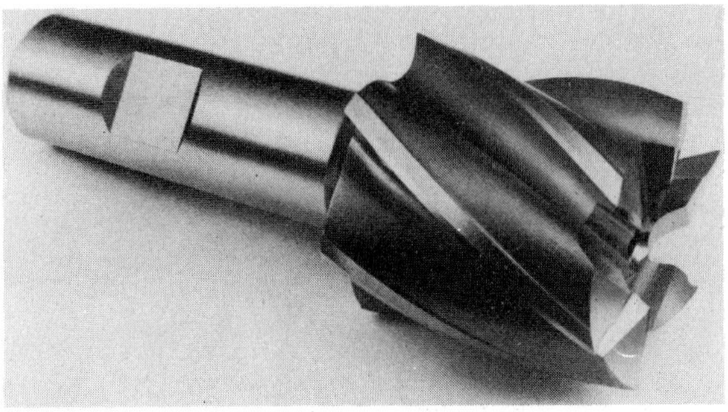

FIGURE 51-5
A straight-shank, large-diameter end mill. (Credit: Union/Butterfield)

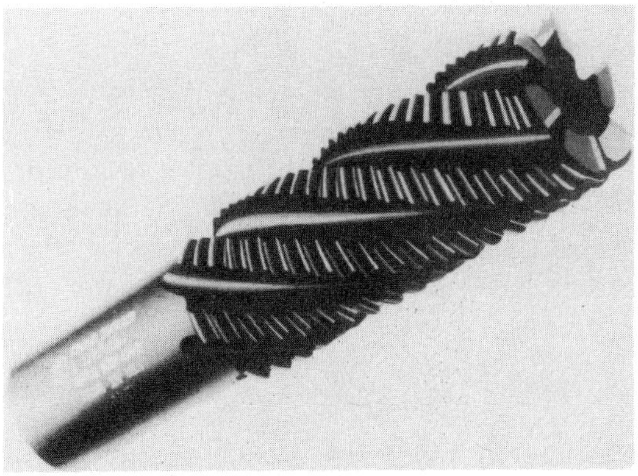

FIGURE 51-6
A roughing end mill will remove more metal in less time due to a "nibbling" cutting action. (Credit: Regal-Beloit Corporation)

End Milling Cutters

There are many types of end mills used in the machine shop; only some of the more commonly used end mills will be discussed.

Standard end mills used on vertical milling machines generally come with a straight shank, are made of high-speed steel, and are equipped with two to eight cutting edges that may be straight or helical (Figure 51-5). Cutters are also available for use as roughing end mills. These end mills are designed with chip breaking cutting edges (Figure 51-6) and will produce faster metal removal, run cooler during heavy cuts, require lower horsepower consumption, and leave a better work surface finish.

Two-fluted end mills are available as single or double-ended cutting tools (Figure 51-7). The two-fluted end mill is used for slotting, plunge cutting, profiling, and drilling. Its deep gullet construction provides room for chips to escape, making it ideal to be used when the tool is used in a deep cut. It is frequently used for milling aluminum and nonferrous metals.

Four-fluted end mills (Figure 51-8) are capable of greater metal removal than two fluted end mills for a given feed rate. Due to their ability to distribute the work load over more cutting edges, a four-fluted end mill will produce a finer finish and will last longer than a two-fluted end mill in the process.

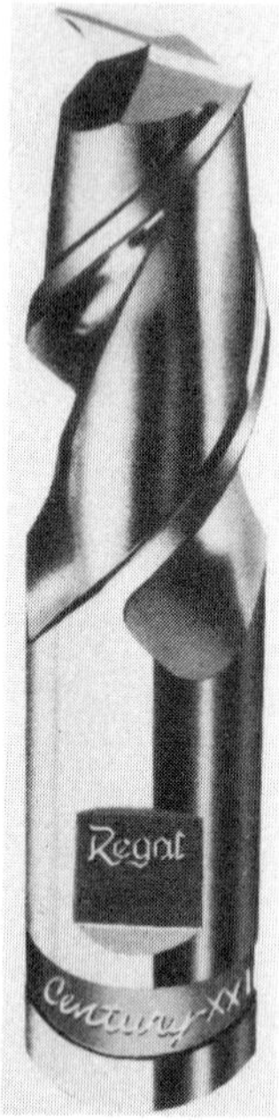

FIGURE 51-7
The deep gullets of a two-fluted end mill provide room for chips to escape. (Credit: Regal–Beloit Corporation)

FIGURE 51-8
For a given feed rate, four-fluted end mills can remove more metal than two-fluted end mills. (Credit: Regal–Beloit Corporation)

Since metal removal is based on the feed rate per revolution of the end mill, it will bear a direct relationship to the number of teeth in the cutter. For a given feed rate per tooth, a four-fluted end mill can be fed at twice the rate of a two-fluted end mill. Four-fluted end mills are used on easy and difficult to cut alloys. They are used most often after roughing cuts to finish off slots, grooves, and recesses. It is one of the most widely used high-volume production cutters.

Multifluted end mills (Figure 51-9) are available with five, six, or eight flutes and are generally made $\frac{3}{4}$ in. (19 mm) in diameter and larger to be used on an assortment of metals. Due to the many cutting edges, these cutters can be used at a higher feed rate, resulting in higher production levels and/or improved tool life.

Indexable insert-type end mill (Figure 51-10) uses replaceable inserts that are fastened on a tool steel cutter body. These inserts are made of high-speed steel or cemented carbide and contain several cutting edges. Once a cutting edge has become dull or worn, the insert is loosened from the cutter body, indexed to expose another cutting edge, and retightened. The original cutting diameter is thus restored without adjustment to the tool or machining program. Once all the cutting edges of the inserts have been used, the insert is thrown away or reconditioned.

Carbide end mills are available as carbide-tipped and replaceable blade end mills (Figure 51-11) and are considered production-type cutters. Not as easily affected by cutting temperatures, these end mills can be used at cutting speeds three to ten times faster than high-speed steel cutters. Due to the brittle nature of carbide, the successful use of these tools requires a solid, rigid setup of the workpiece and the choice of a tool that will minimize deflection.

Special-Purpose End Mills

There are a number of end mills that at one time were designed for a specific application. Due to frequent use and commonplace cuts, these cutters are generally regarded as "special standards."

The *Woodruff keyseat cutter* (Figure 51-12) is designed to cut keyseats in shafts and is available in straight and staggered tooth style. Undercut at the neck behind the cutting edges, as many as 16 teeth can circle the narrow face of the cutter. It is available in 41 standard sizes. Each size is designed to produce a slot to match a keyseat number desig-

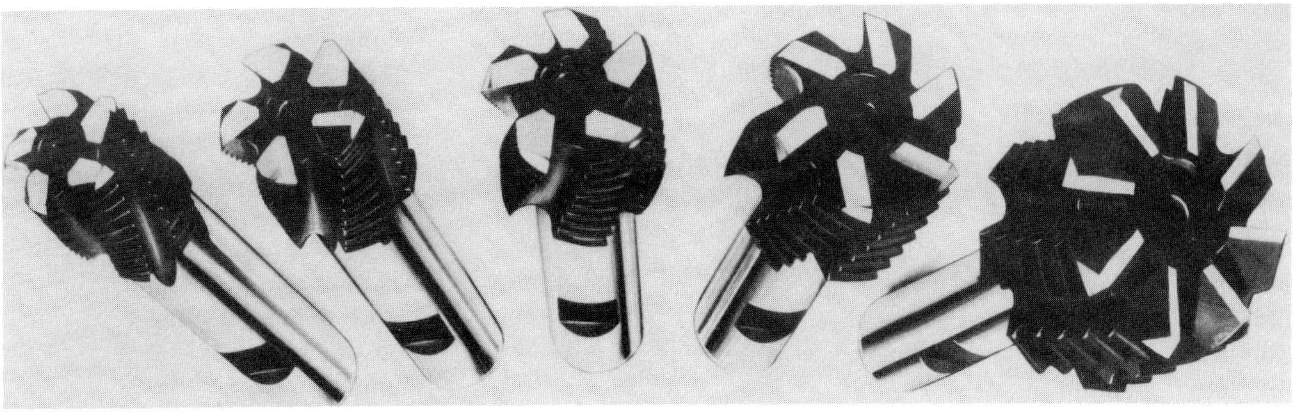

FIGURE 51-9
Multifluted end mills can have five, six, or eight cutting edges. (Credit: Union/Butterfield)

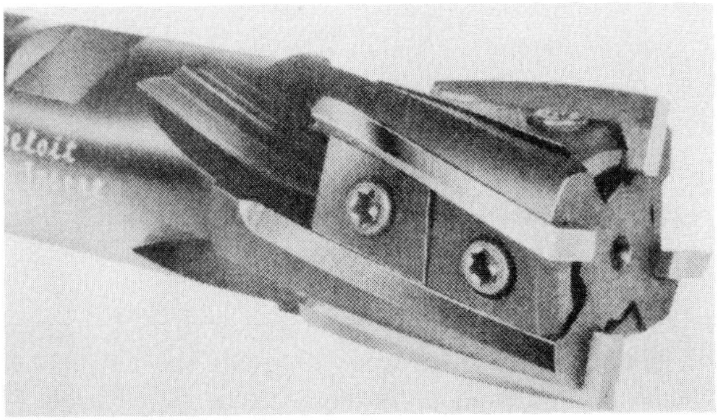

FIGURE 51-10
An indexable insert-type end mill. (Credit: Regal–Beloit Corporation)

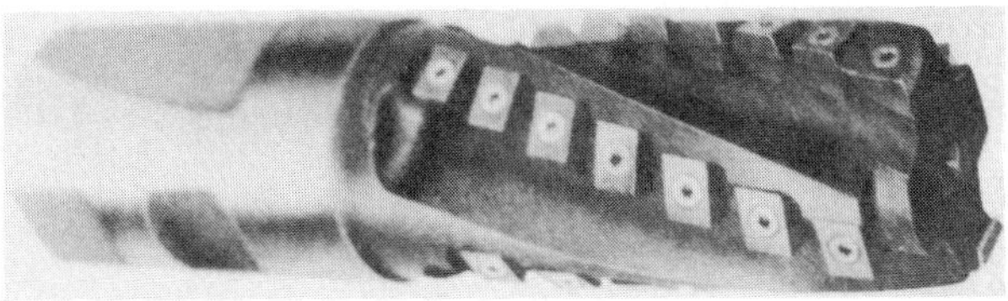

FIGURE 51-11
Carbide end mills can be used at cutting speeds three to ten times faster than high-speed steel end mills. (Credit: Regal–Beloit Corporation)

nated by the American National Standard Association. While the staggered tooth cutter has alternating teeth for freer cutting action, both styles of cutters are provided with clearance ground into the sides of the teeth to prevent clogging and binding during milling.

The *T-slot cutter* is similar in design to the keyseat cutter. The T-slot cutter (Figure 51-13) is available in a number of standard sizes to cut a recessed slot at right angles to a previously milled slot to provide room for boltheads. It is used for machining T-slots found in face plates, machine tool tables, rotary tables, and other work holding devices.

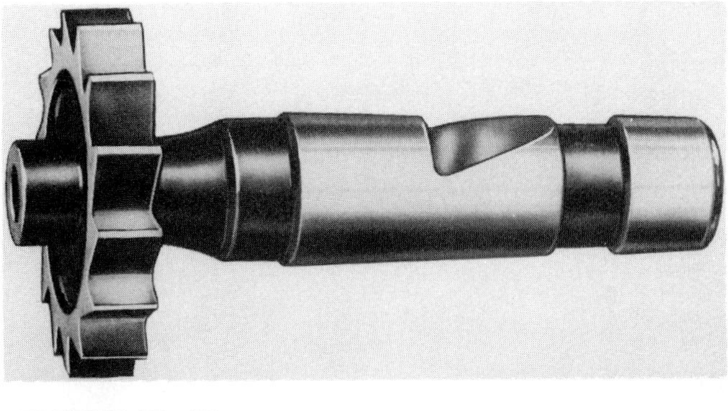

FIGURE 51-12
The Woodruff keyseat cutter is designed to mill keyseats in shafts. (Credit: Regal–Beloit Corporation)

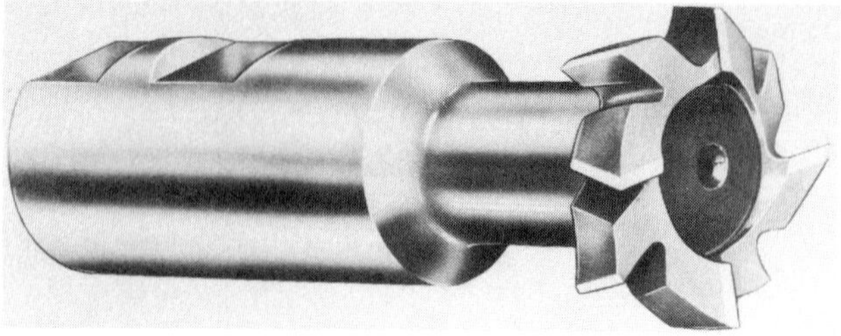

FIGURE 51-13
T-slot cutters are used to cut grooves in machine tool tables, face plates, and similar work holding devices. (Credit: Regal–Beloit Corporation)

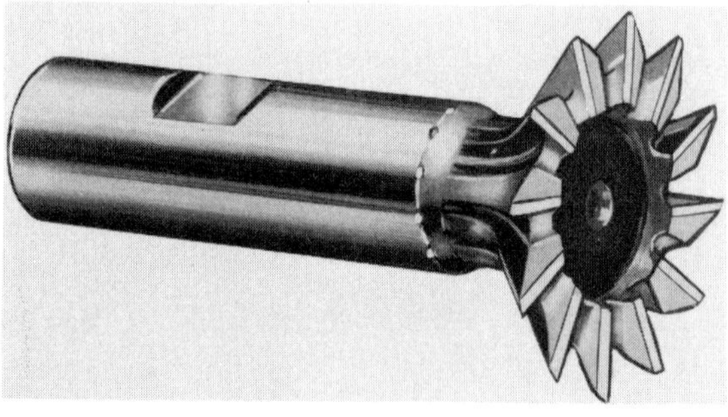

FIGURE 51-14
The dovetail cutter is used to machine slots that are required for parts to mate with a sliding fit. (Credit: Regal–Beloit Corporation)

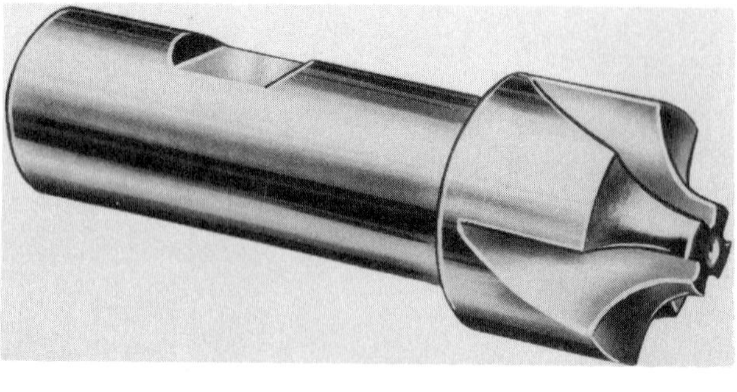

FIGURE 51-15
The corner-rounding (radius) end mill is used for smoothing sharp external corners. (Credit: Regal–Beloit Corporation)

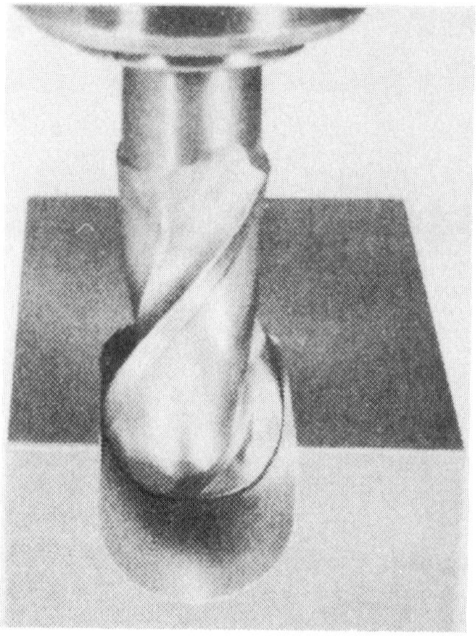

FIGURE 51-16
A ball end mill is used for milling contoured surfaces. (Credit: Regal–Beloit Corporation)

The *dovetail cutter* (Figure 51–14) is undercut at the neck and is generally made with a 60° included angle. Dovetail cutters are used to machine slots that are required for parts to mate with a sliding fit.

The *corner-rounding (radius) end mill* (Figure 51–15) is available for smoothing out sharp external corners. The tool is manufactured in a range of radii from $\frac{1}{16}$ to $\frac{1}{2}$ in. (1.6 to 13 mm).

The *ball end mill* (Figure 51–16) is designed for cutting contour surfaces such as rounded bottoms on holes and slots. It is an ideal tool for finishing fillets and other rounded corners on tool dies where sharp corners have to be avoided to prevent stress cracks during hardening. Filleting also increases the strength of the joint.

Accessories

As with the horizontal milling machine, a plain or swivel base vise can be used as the chief work holding device on a vertical milling machine table. The swivel vise, with its base marked with degree graduations to indicate the angled position of the vise, is generally favored for machine shop work. A set of keys attached to the bottom of the vise provides a quick and accurate alignment in the T-slots of the milling machine table (Figure 51–17).

A number of rotary dividing heads are avail-

FIGURE 51-17
A set of keys at the bottom of the base will quickly align and accurately locate a milling vise in the table's T-slots.

FIGURE 51-18
A small indexing device.

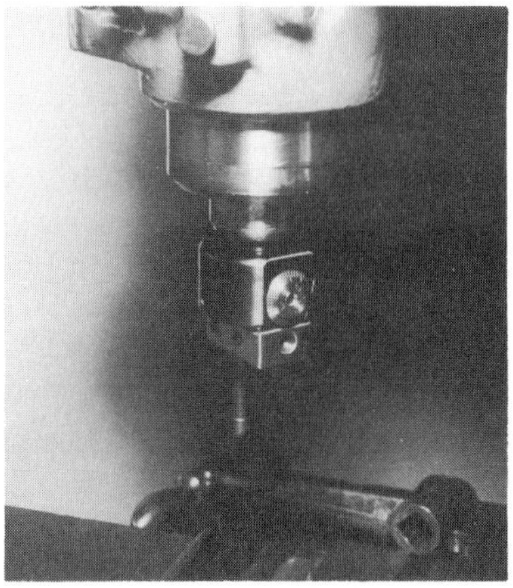

FIGURE 51-19
A boring head can accurately enlarge and straighten a previously drilled hole.

FIGURE 51-20
A right-angle attachment is a self-contained unit that can simplify right-angle work. (Credit: Bridgeport Machines, Inc.)

able for use on a vertical milling machine. Generally favored for light milling work is the small *indexing head* (Figure 51-18), which uses spring collets for securing the work and can be indexed to promote hole or surface locations.

A *boring head* (Figure 51-19) is used to enlarge holes at exact locations. The boring tool cutting edge is set on the centerline of the boring head and is adjusted by turning a micrometer adjustment screw. Readings in thousandths of an inch or hundredths of a millimeter are available. The machining operation will produce a hole that is round, is accurate to size and location, and has a better surface finish. The hole will also be straight, unlike the hole that drifts when drilled.

A *right-angle milling attachment* (Figure 51-20) permits cutting tools to be positioned in a horizontal plane. With it, many operations can be performed that would not be possible by any other method, such as milling racks. Drilling a hole that is difficult to reach or positioned at right angles to a work surface that cannot be removed from the milling vise is but one example. When used with an auxiliary support, they can be used with an arbor that will convert the vertical milling machine to a horizontal milling machine.

UNIT 52 MOUNTING AND REMOVING AN END MILL

Vertical milling machines require different toolholding devices to make machining operations as convenient as possible while still allowing ease of operations. There are five main methods of holding cutters in vertical milling machines: collets, special holders, adapters, arbors, and quick-changing toolholding systems. Each can play an important role for specific machining operations.

Objectives:
After completing this unit you should be able to

1. identify and explain how to use the different toolholding devices.
2. mount an end mill safely.
3. remove an end mill safely.

Toolholding Devices

The standard vertical milling machine may use a collet and adaptor system for its large spindle opening. Once the adaptor has been fitted into the spindle and held by a draw bar, a collet and end mill assembly is inserted and held in the adaptor by means of a retainer nut.

The principal means of holding a cutter in a ram-type vertical milling machine is by use of a tapered spring collet mounted in the matching taper spindle of the machine. Once positioned by aligning the keyway of the collet with the key inside the spindle, the spring collet is pulled up into the spindle by means of a threaded draw bar (Figure 52-1), closing the collet against any cutter shank as it is drawn against the spindle's tapered opening. Caution must be used in tightening the spring collet. If the cutter's shank is not held securely, the end mill will slip under the cutting pressure. This may cause the cutter to pull out of the collet, ruining the workpiece or the end mill or perhaps both.

End mill holders (Figure 52-2) are available with a solid, tapered designed shank to fit the spindle and a set screw to lock the straight shank of an

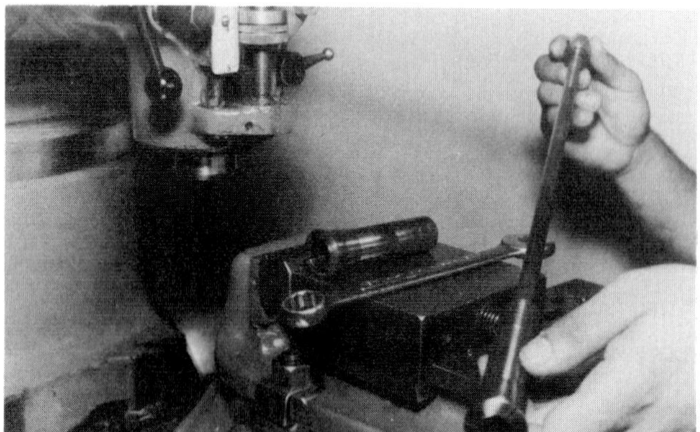

FIGURE 52-1
A spring collet being threaded by a draw bar.

FIGURE 52-2
A solid end mill holder with set screw for tightening.

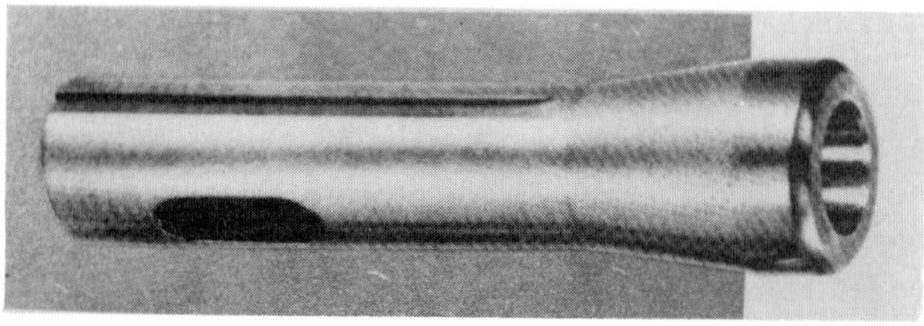

FIGURE 52-3
An adaptor for use with tapered shank end mills and drills. (Credit: Bridgeport Machines, Inc.)

end mill in place. Once the cutter has been inserted in the holder, the set screw is tightened against a flat surface on the cutter's shank to prevent slippage.

An adaptor is available for use with tapered shank end mills and drills (Figure 52-3). A reducing sleeve is also available for use on cutting tools smaller than the adaptor hole.

FIGURE 52-4
A shell mill with special holder extends the tooling capability of a vertical milling machine.

Shell mill holders (Figure 52-4) extend the tooling capabilities of a vertical milling machine. A shell end mill is mounted on this stub arbor and tightened in place with an arbor screw.

A flycutter is a single-point cutting tool whose end has been ground to a desired shape for machining flat surfaces, boring holes, or cutting grooves. The cutting tool is generally a high-speed steel, stellite, or carbide-tipped toolbit. Flycutters may consist of one or more single-point tools and are mounted in a special flycutter holder or adaptor (Figure 52-5).

There are a number of quick-change toolholding systems available that are generally used in production. They increase machining output by reducing tool changing time. These changes can take as little as 6 sec to make. A range of adaptors are preloaded with drill chucks, drills, reamers, end mills, or other cutting tools. Adaptors are then locked in place using a ball bearing locking system and released by twisting open the adapter locking fixture (Figure 52-6).

Precautions

The precautions involved in mounting and removing cutters from vertical milling machines are basically those of safety. No changing should be undertaken without the machine being shut off. Not only should all the parts be clean and free of nicks and burrs, but all sharp-edged tools should be handled with a cloth

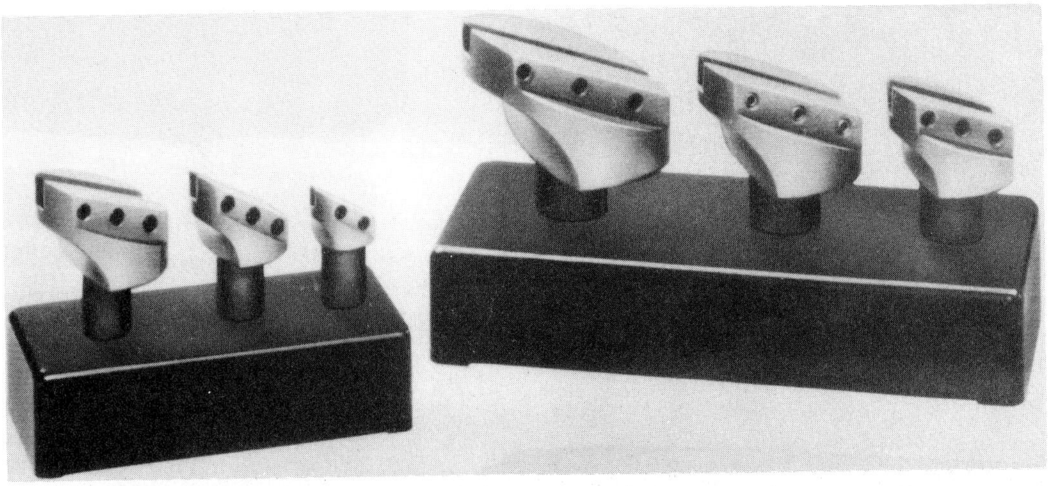

FIGURE 52-5
Holders for single-point flycutting toolbits. (Credit: DoAll Company)

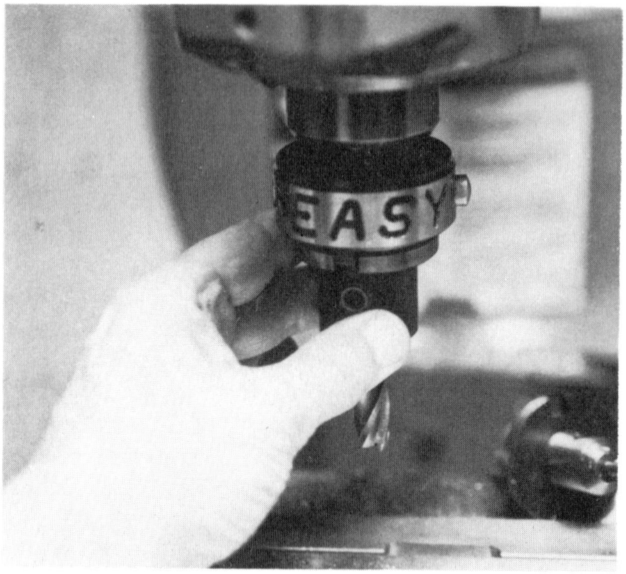

FIGURE 52-6
A quick-changing tool holding system may complete a change in about 6 sec.

FIGURE 52-7
Never hold and pull down against an end mill bare handed.

or paper toweling to prevent serious cuts. This is especially true when attempting to remove a cutter from a collet with a downward pull (Figure 52-7). The sharp edges can slice deep gashes into the hand that may require stitches or cause serious injury. Just as important, the table directly below the spindle opening should be covered with a protector such as a board of masonite. This can prevent nicks or other damage to the table due to falling objects. If a cutter is to be mounted or removed directly near a workpiece, this too should be protected with a cloth.

Mounting an End Mill

Since most commonly used vertical milling machines have an R8 spindle, the explanation used for mounting and removing an end mill will involve the use of an R8 spring collet.

Procedure

1. Clean the spindle taper.
2. Clean the outside of the spring collet.
3. Insert the collet inside the spindle opening so that it aligns with the key inside and push it up.
4. Turn the draw bar clockwise to thread into the back of the collet.
5. Holding the end mill by a cloth or folded paper toweling, insert the cutter into the collet and hold it in place.
6. With the other hand, reach up and tighten the draw bar finger tight.
7. Lock the spindle by pulling against the brake with one hand and securely hand-tighten the draw bar with a suitable wrench with the other (Figure 52-8).

Removing an End Mill

Removing an end mill from a vertical milling machine will involve the general reversal of the steps taken to mount the cutter. Removal should be undertaken with the utmost of safety.

1. Cover the workpiece or area below the end mill with a piece of masonite or cloth.
2. Lock the quill in its uppermost vertical position.
3. Prevent the spindle from turning by pulling against the spindle brake lever.

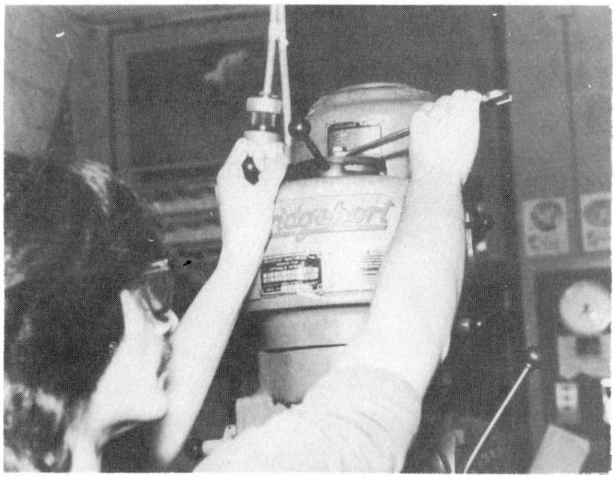

FIGURE 52-8
Lock the spindle by pulling against the brake with one hand and tighten the draw bar with a suitable wrench with the other.

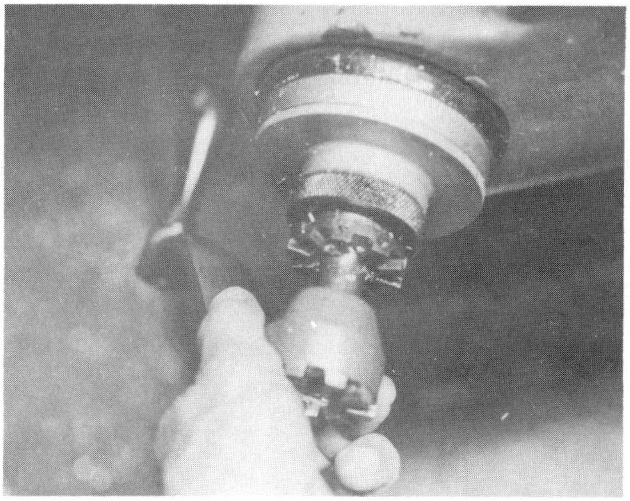

FIGURE 53-1
Remove the position of the vertical feed lever when not in use.

4. Loosen the draw bar counterclockwise with a suitable wrench so it can be turned by hand.

5. Unscrew the draw bar knob by hand no more than one or two complete turns.

6. Hold the end mill with a cloth and hit the top of the draw bar knob sharply with a lead hammer to remove the end mill.

UNIT 53 VERTICAL MILLING A FLAT SURFACE

One of the advantages of using a vertical milling machine is the flexibility it affords the operator. Milling a flat surface is an operation an operator is called upon to do time and time again. This operation can be carried out by using either a shell end mill, a face mill, or a flycutter.

Milling a flat surface may also be required on different operations, such as machining work square and parallel. For such operations, the toolhead will have to be positioned at 90° to the surface of the table and the vise will have to be positioned parallel to the table travel. All these operations will be examined in this unit.

Objectives
After completing this unit you should be able to

1. align the toolhead at right angles to the surface of the table.
2. align the vise parallel with the table travel.
3. mount work in a milling vise securely.
4. mill a flat surface.
5. machine work square and parallel.

Safety

1. Always wear appropriate safety glasses when in a machine shop.
2. Remove watches, rings, and all forms of jewelry to prevent them from getting caught in moving parts.
3. Do not wear loose-fitting clothing; roll up sleeves above the elbow.
4. Stop the spindle before attempting to change or adjust belts, pulleys, or gears.
5. Stop the spindle before attempting to take any measurements.
6. Keep hands free of the spindle start switch when attempting to change cutting tools.
7. Always stop the spindle before mounting or removing work from the vise.
8. To prevent workpiece and cutter damage, never start the spindle when the cutter is in contact with the workpiece.
9. To prevent objects from falling, remove all loose items such as chuck keys or vise handles from the table.
10. To avoid hitting it with your leg, reverse the position of the vertical traverse lever when not in use (Figure 53-1).

FIGURE 53-2
Use a safety shield in front of the cutting area whenever possible.

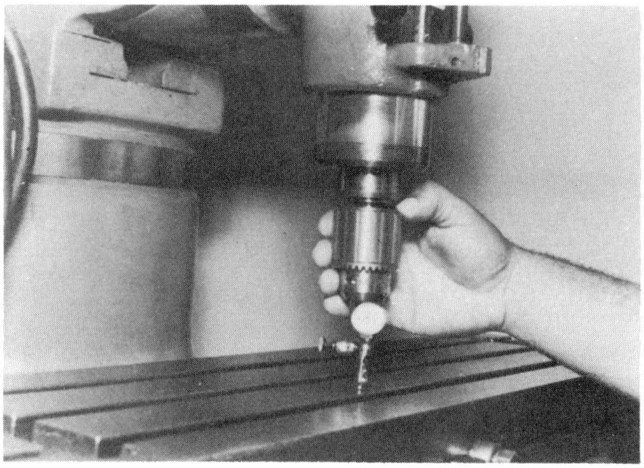

FIGURE 53-3
"Sweeping the table" will require indicating the table surface with a dial indicator.

11. Whenever possible, use a see-through safety shield in front of the cutting area (Figure 53-2).

12. Keep the work area clean and dry by removing all chips and oily substances.

Aligning the Toolhead

To assure accuracy in milling surfaces or drilling holes, the spindle of the toolhead must be square with the table. This condition can be checked out in an operation called "sweeping the table" or "tramming the table," where an indicator mounted in the spindle is turned in a small circle on the surface of the table. Since the toolhead of a vertical milling machine can be set at compound angles, that is, moving the toolhead clockwise or counterclockwise as well as tilting the toolhead down or away from the operator, four different points at 90° to each other will have to be checked and adjusted to get the head at right angles to the table.

Procedure

1. Tighten a drill chuck in the spindle.
2. Fasten a $\frac{1}{4} \times 2\frac{1}{2}$-in. (6.35 × 63-mm) L-shaped bent rod in the drill chuck.
3. Clamp a dial indicator near the end of the rod.
4. Position the table so that rotations of the indicator do not go beyond the width of the table.
5. Position the indicator either to the right or left side of the table and lock the quill.
6. Carefully raise the table to make light pressure (about one-quarter turn) against the indicator contact point.
7. Lock the table and set the indicator bezel reading to zero.
8. Rotate the chuck in a semicircle (180°) to measure the opposite side (Figure 53-3). *Note:* The indicator must be positioned as close as possible to the same relative table position for each reading.

If there is no difference in reading, the head is aligned between these two points. If there is a difference, turn the spindle to locate the indicator at the highest reading and do the following:

1. Loosen the head lock nuts but leave them snugged up.
2. Crank the head moving device until the indicator reading is half the difference of the two readings.
3. Turn the spindle to move the indicator slowly around in a circle to check the results.
4. Retighten the head locking nuts and recheck the indicator's reading.
5. Repeat steps 2 to 5 until both readings are the same.

While a different set of head locking screws control the tilting of the head in a rising or falling arc, the method for aligning the head is similar. Place the indicator near the front of the table with a low reading (approximately one-quarter turn) and lock the table height. Rotate the chuck by hand to register the point directly opposite this on the back side of the table and compare the readings. If there is no difference in reading, the head is square with the table. Any difference should be adjusted as previously described.

Aligning the Vise

The operation of aligning or squaring the vise is an important part of setting up work. Unless the vise is set at 90° to the column, parallel cuts to the table travel are not possible. The vise's accuracy cannot be gaged by eye and must be indicated. If the surface of the solid jaw is straight and smooth, an indicator can be used directly against it. Generally, a heavy parallel is used in the following manner:

1. Loosen the vise's lock nuts, leaving them semitight (Figure 53-4).

2. Select a thick parallel at least as long as the vise jaws (Figure 53-5).

3. Extend the parallel edge height about $\frac{1}{2}$ in. (13 mm) above the vise jaws and tighten the vise.

4. Extend the machine's indicator mounting rod or mount a drill chuck in the spindle and tighten a rod in it.

5. Fasten a dial test indicator to the mounting rod.

6. Raise the table so that the indicator's contact point is midway between the jaws and the top of the parallel.

7. Turn the crossfeed handle to adjust the right side of the parallel to make contact against the indicator's point with light pressure (approximately one-quarter turn) (Figure 53-6).

8. Set the indicator dial bezel to zero.

9. Turn the table feed handle to move the parallel past the indicator contact point, stopping about $\frac{1}{4}$ in. (6 mm) from the other end.

FIGURE 53-5
Select a thick parallel as long as the vise jaws.

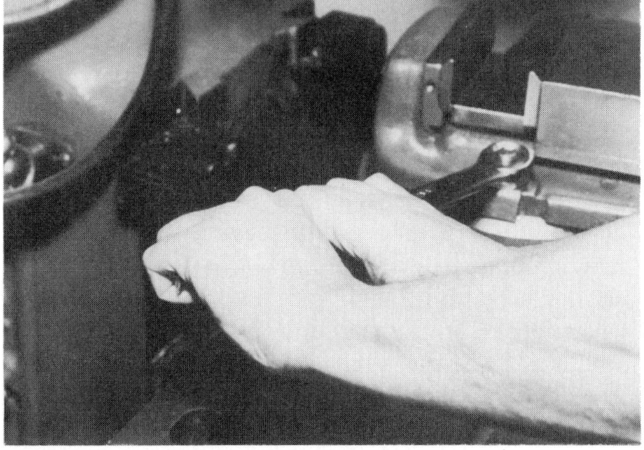

FIGURE 53-4
Prepare to move the vise by leaving the locking nuts semitight.

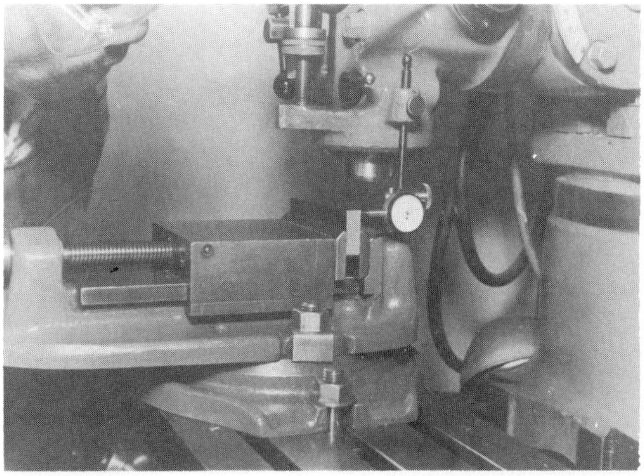

FIGURE 53-6
Start indication from one end of the vise jaws.

10. Leave or return the vise to the high-side reading.

11. Leave one nut loose and snug up one clamping nut as a pivot point.

12. With a lead hammer, lightly tap the end of the vise on the same side as the indicator to reduce the difference in reading by half. *Note:* Never tap the vise toward the indicator's contact point when moving it.

13. Turn the table feed handle to move the parallel past the indicator contact point to see that the difference in reading has been removed. If it has not, repeat steps 10 to 12.

14. Snug up the lock nuts a little at a time, first one side and then the other, until both nuts are secure. *Note:* Never tighten one side completely before tightening the other side.

FIGURE 53-7
Paper feelers should be placed on the parallels.

Mounting Work in a Vise

In order to machine work accurately, it must be held firmly and securely in a milling vise. All the surfaces and edges of the workpiece should be free of nicks and burrs, as should be the resting surfaces of the parallels and the gripping surfaces of the vise jaws. The height of the workpiece above the vise jaws is also important. Always grip as much of the workpiece in the vise as possible; only extend a little more than is required to take the cut.

Procedure

1. Select a flat file and remove and nicks and burrs from the workpiece.

2. Remove any burrs from the vise jaws with an oilstone.

3. Select a set of parallels that will raise the workpiece about $\frac{1}{4}$ in. (6 mm) above the vise jaws.

4. Examine the parallels and remove any burrs with an oilstone.

5. Thoroughly clean the workpiece, parallels, and vise jaws.

6. Cut four narrow strips of paper and place them on the ends of the parallels so that they will extend beneath the four corners of the workpiece (Figure 53-7).

7. Place the workpiece on the four strips of paper in the middle of the vise (Figure 53-8).

8. Tighten the vise securely.

9. Seat the workpiece down on the paper feelers with lead hammer blows (Figure 53-9).

FIGUIRE 53-8
Place the workpiece in the middle of the vise.

10. Alternate tapping one corner of the workpiece down.

11. Tug at the paper feelers at each corner. Any slippage of paper indicates that particular corner is not resting flat (Figure 53-10). *Note:* Once the workpiece is firmly seated, *do not retighten the vise* as it will force the work up and off the parallels.

To Mill a Flat Surface

Once the technique of mounting work in a vise has been learned and the vertical head is at right angles

FIGURE 53–9
Seat the workpiece down with lead hammer blows.

FIGURE 53–10
Test for workpiece seating by tugging at the paper feelers; they should be tight.

to the table, it remains to select a proper size cutter to machine a flat surface. Since it provides an economical means and can cut a wide area, a flycutter will be used in the following outline.

Procedure

1. Approved safety glasses must be worn before attempting any machining.

2. Mount the workpiece in the milling vise. *Note:* If the workpiece is big, it can be seated on paper feelers placed on the bottom of the vise (Figure 53–11).

FIGURE 53–11
Workpieces can be set directly on paper feelers on the bottom of the vise.

3. Mount a suitable flycutter toolholder in the machine spindle.

4. Extend and tighten the flycutter to produce a cutting circle slightly larger than the area to be machined.

5. Set the spindle speed for the size of the flycutter (cutting circle produced) and the material being machined.

6. Raise the spindle to its uppermost position and lock the quill clamp.

7. Start the machine spindle.

8. Raise the table until the flycutter just contacts the surface of the workpiece.

9. Move the table to clear the workpiece from the cutter.

10. Set the micrometer collar on the vertical feed to zero.

11. Raise the table about 0.030 in. (0.76 mm) and lock knee clamps.

12. Turn the table handwheel to take a trial cut a distance of about $\frac{1}{4}$ in. (6 mm) long.

13. Stop the machine spindle.

14. Turn the table handwheel to clear the workpiece from the flycutter.

15. Measure the workpiece and determine the amount to be removed.

16. Unlock the knee clamps and raise the table to take the depth of cut desired.

17. Lock the knee clamps and mill the surface to size.

Machining Square and Parallel

In Figure 53-12 an irregular surfaced block where two sides are parallel and two sides are not is shown to demonstrate the technique of machining square and parallel. The irregular shape will require the use of a round bar for holding purposes.

Procedure

1. Align the toolhead at 90° to the table (squaring the head).
2. Align the vise parallel with the table travel (squaring the vise).
3. Remove all burrs from the workpiece with a file and clean the vise.
4. Place four strips of paper on a set of parallels in the middle of a clean vise.

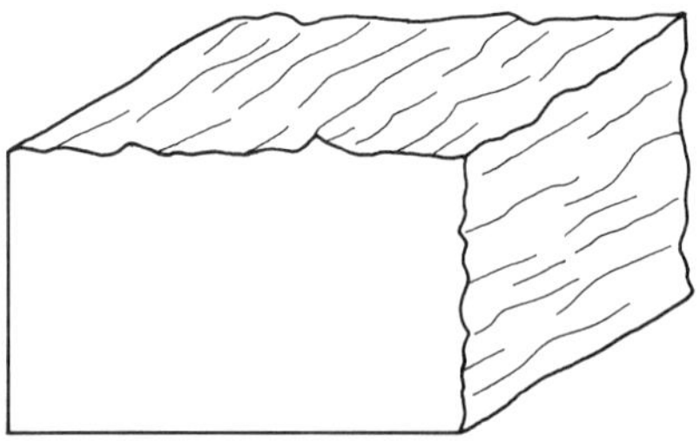

FIGURE 53-12
An irregular faced workpiece.

Machining Side 1:

5. Place side 4 of the workpiece against the solid jaw, side 3 to rest on the paper feelers (Figure 53-13).
6. Place a round bar against side 2 in the center of the amount of work held in the vise and tighten the vise securely.
7. Tap the top surface down with a lead hammer and test the paper feelers for slippage.
8. Mount a suitable end mill or flycutter.
9. Set the spindle speed for the size of the cutter and workpiece material.
10. Start the machine spindle and mill the top surface flat.
11. Remove and deburr the machined edges.

Machining Side 2:

12. Clean the vise and parallels, replacing the paper feelers if necessary.
13. Remount the workpiece, keeping side 4 against the solid jaw with side 1 resting on the paper feelers (Figure 53-14).
14. Tighten the vise securely and tap the workpiece down with a lead hammer until the paper feelers are tight.
15. Start the machine spindle and mill side 2 flat.
16. Take the workpiece out of the vise and remove all burrs with a file.

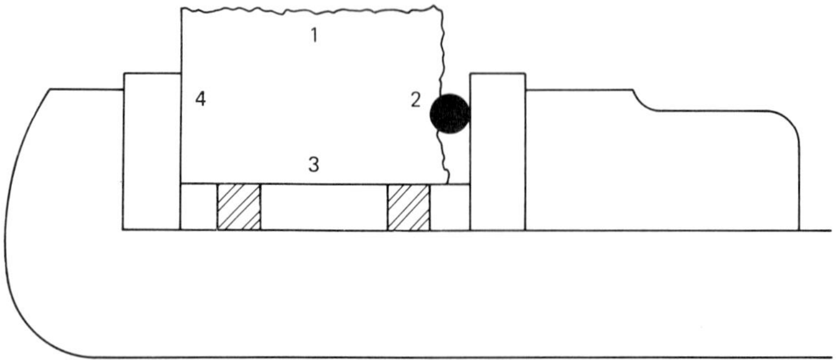

FIGURE 53-13
Setup for machining side 1.

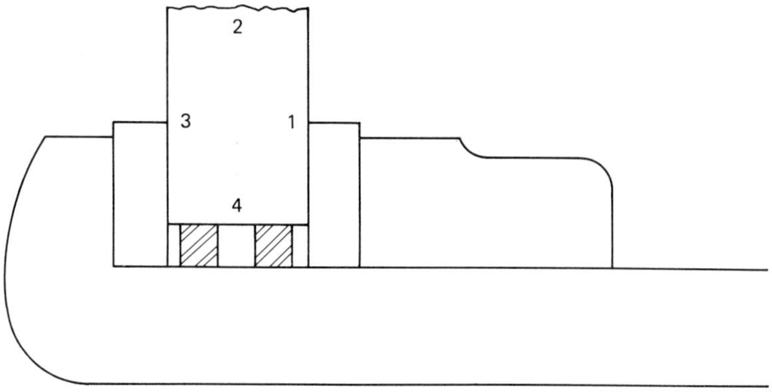

FIGURE 53-14
Setup for machining side 2.

FIGURE 53-15
Rechecking side for squareness.

At this point, all four sides should be straight with opposite sides parallel. If the ends of the block have to be machined square with these sides, proceed as follows.

Machining the Ends

1. Clean the vise.
2. Set a square on the base of the vise.
3. Place the work in the vise so that one of the machined surfaces can be butted against the blade of the square.
4. Tighten the vise just enough to hold the workpiece at a slight angle to the square.
5. With a soft-faced mallet, tap the side of the block to align the opposite edge with the blade of the square.
6. Tighten the vise securely.
7. Recheck the accuracy of the setup (Figure 53-15).
8. Mill the end of the workpiece flat and to size. This end should be square with its sides.
9. Place this end on a set of parallels and mill the opposite side parallel with it to complete milling all the sides square and parallel.

UNIT 54 THE EDGE FINDER

An edge finder is a tool that is used by the operator of a vertical milling machine for establishing the edge of a workpiece as a reference point. The work surface may be quickly and easily located with an edge finder and will permit the milling machine operator to establish this surface geometrically when used with the micrometer collar readings on the table traverse and crossfeed handwheels. It is this capability that makes it possible to accurately locate the centers for drilling and boring operations as well as for use against shoulders, grooves, and round work. The use of an edge finder becomes a commonplace operation whenever accurate machining is required on a vertical milling machine.

Objectives
After completing this unit you should be able to

1. explain how an edge finder works.
2. accurately use an edge finder to position the axis of the milling machine spindle in relation to the edges of a workpiece.

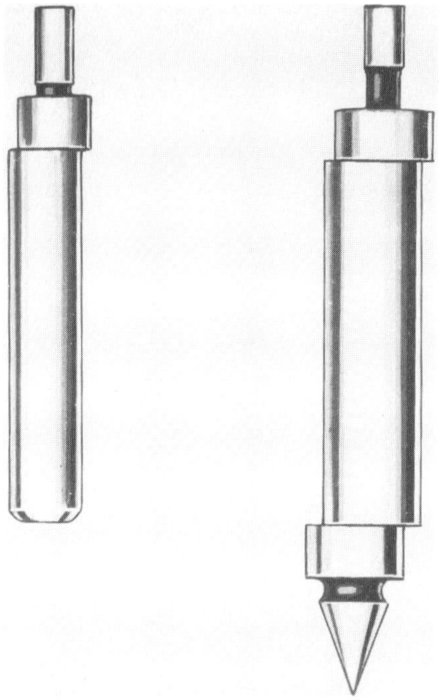

FIGURE 54–1
Edge finders are available with single, double, or pointed ends. (Credit: L. S. Starrett Company)

The Edge Finder

Edge finders are available single or double ended (Figure 54–1) and may be held in the spindle of a vertical milling machine by means of a drill chuck or a spring collet. Each edge finder is constructed so that the end contact diameter, which is retained in the main body by an internal spring, is able to be offset as it rotates to touch against the work surface. The contact ends of edge finders may measure 0.200 in. (6 mm) in diameter. At the moment that the contact diameter suddenly shifts or darts from its steadied rotation as it nears the work surface, the center of the spindle will be exactly one-half the measured diameter of the contact, or 0.100 in. (3 mm) from the work surface edge.

The Use of an Edge Finder

In order to achieve accuracy with an edge finder, certain conditions must be met. The tool head of the vertical milling machine should be square with the table, and the milling vise or squared workpiece, if clamped to the table, must be set parallel to the table travel. Of equal importance is the removal of any backlash that may exist in the feed screws when positioning the table. The crossfeed or table traverse handwheels should be backed off one complete turn and then reversed to advance the work surface toward the edge finder when establishing the surface location. For ease of layout calculations, the micrometer collar should be set so that when the work edge has been positioned directly under the machine spindle center, it reads zero. Due to the fact that one complete turn of either the crossfeed or table traverse handwheels on most vertical milling machines measures 0.200 in., the micrometer collar will have to be positioned on the 0.100-in. graduation. Once the micrometer collar has been set to indicate the surface edge location, it should be rechecked by repeating the process. All location movements using the crossfeed or table traverse handwheels should be made in one direction only. It will be the same direction that was used to locate the work edge.

Procedure

1. Check that the tool head is set square with the table.
2. Check that the milling vise is set parallel to the table travel.
3. Select a work block whose edges are square and parallel.
4. Mount the work block in the milling vise.
5. Mount an edge finder in a spring collet or drill chuck in the machine spindle (Figure 54–2) using 0.200 in. contact diameter.
6. Set the machine spindle speed between 600 and 700 rpm.
7. Loosen the micrometer collar nut on the crossfeed handwheel (Figure 54–3).
8. Start the machine spindle.
9. Lower the quill or raise the table so that the contact diameter of the edge finder may be brought against the work surface.
10. Use the end of a pencil against the contact diameter to make it run eccentrically (Figure 54–4).
11. Back off the crossfeed handle one complete turn and then reverse the movement to advance the work surface toward the rotating edge finder (Figure 54–5).

FIGURE 54-2
An edge finder can be used held in a drill chuck. (Credit: Philip Rapisarda)

FIGURE 54-4
Make the contact diameter run eccentrically. (Credit: Philip Rapisarda)

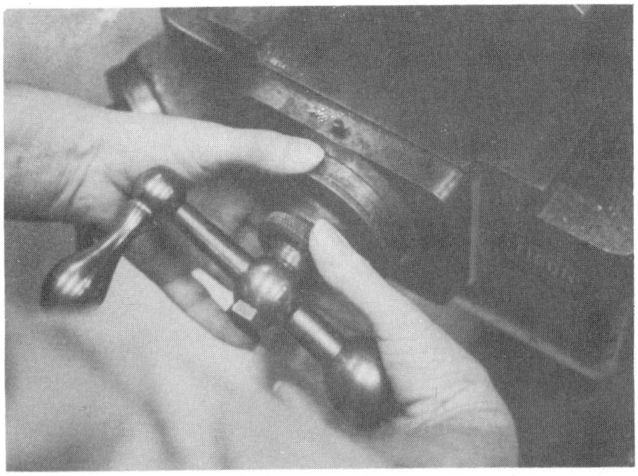

FIGURE 54-3
Loosen the micrometer collar nut. (Credit: Philip Rapisarda)

FIGURE 54-5
Back off and then reverse the crossfeed turning to eliminate the backlash in the feed screw. (Credit: Philip Rapisarda)

FIGURE 54-6
Lightly tap the crossfeed handle as the work surface comes in contact with the edge finder. (Credit: Philip Rapisarda)

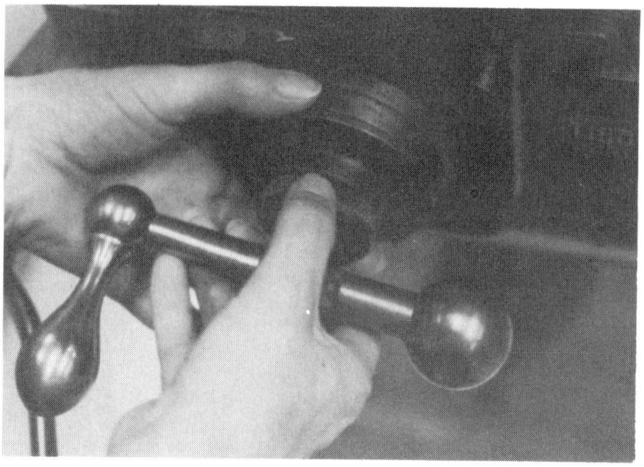

FIGURE 54-8
Set the micrometer collar to 0.100 in. and lock it. (Credit: Philip Rapisarda)

FIGURE 54-7
Edge location is obtained the moment the edge finder's contact diameter shifts to one side. (Credit: Philip Rapisarda)

12. Slowly advance the work to make contact with the edge finder, lightly tapping the crossfeed handwheel as the edge finder begins to run steady and concentrically (Figure 54-6).

13. By sensitively tapping the crossfeed handwheel with the hand, the contact diameter will suddenly shift or dart to one side (Figure 54-7). The machine spindle center is now one-half the contact diameter from the work surface edge.

14. Set the crossfeed micrometer collar to a 0.100-in. reading and lock it (Figure 54-8).

15. Back off the crossfeed handwheel one complete turn (Figure 54-9) and recheck the accuracy of the setup by repeating steps 9 to 13. If the positioning is correct, the micrometer collar should read 0.100 in. again.

16. Stop the machine spindle.

Using the Traverse Table Handwheel

Once the crossfeed position is set, the table traverse position must also be established. *Do not move the workpiece at any time during the setup.*

Procedure

1. Raise the quill or lower the table to position the edge finder above the workpiece.

2. Turn the table traverse handwheel to clear the workpiece from the edge finder.

3. Start the machine spindle.

4. Lower the quill or raise the table so that the contact diameter of the edge finder may be brought against the work surface.

5. Use the end of a pencil against the contact diameter to make it run eccentrically (Figure 54–10).

6. Loosen the micrometer collar nut on the table traverse handwheel (Figure 54–11).

7. Repeat steps 11 to 16, applying the guidelines to the table traverse handwheel.

Using the Pointed Edge Finder

The pointed edge finder can be used for locating scribed lines or center points (Figure 54–12), but it is not as accurate as the solid contact diameter. It depends more on visual alignment than physical alignment.

Procedure

1. Mount a pointed edge finder in a spring collet or drill chuck in the machine spindle.

2. Start the machine spindle.

3. Make the point run true by using the end of a pencil against the eccentrically running edge finder.

4. Lower the quill to bring the rotating point slightly above the layout lines or center points.

5. Move the table to position the layout lines or center points as close as possible to the spinning point.

6. Use a magnifying glass to determine the final positioning of the workpiece.

FIGURE 54–9
Back off the crossfeed handwheel and repeat the operational steps to check the accuracy of the edge finder setting. (Credit: Philip Rapisarda)

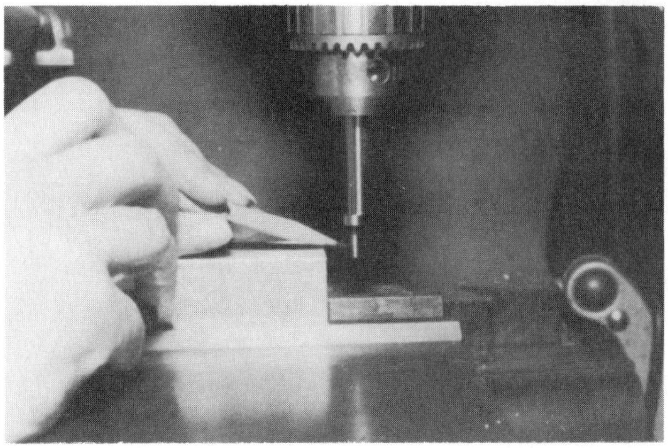

FIGURE 54–10
The edge finder is repositioned to locate the work surface edge affected by the table traverse. (Credit: Philip Rapisarda)

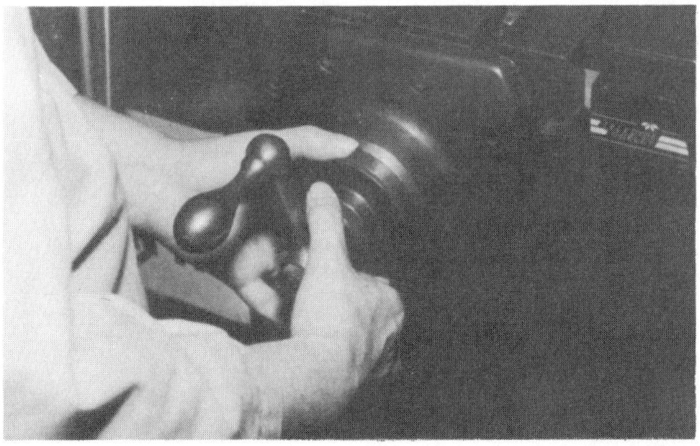

FIGURE 54–11
Loosen the micrometer collar nut on the table traverse dial. (Credit: Philip Rapisarda)

FIGURE 54–12
The use of a pointed edge finder is more visual than physical in practice. (Credit: Philip Rapisarda)

FIGURE 55-1
A DRO is generally located near the toolhead of a vertical milling machine. (Credit: Bridgeport Machines, Inc.)

UNIT 55 DIGITAL READOUT

A digital readout (DRO) is an electronic device that is used to monitor the table movements of machine tools. These systems are widely used in industry on lathes, grinders, jig borers, milling machines, measuring inspection devices, and monitoring systems for process control. These devices are available as original equipment or may be retrofitted to improve accuracy and increase productive machining on older, manually operated machines.

On a vertical milling machine, the DRO is generally mounted next to the toolhead (Figure 55-1) and will save the operator a great deal of time and eye strain in counting the number of handwheel turnings and peering at the micrometer collar graduations when making location measurements.

Objectives
After completing this unit you will be able to

1. explain the principle affecting the use of a DRO system.
2. name the three basic components in a DRO.
3. use a DRO system.

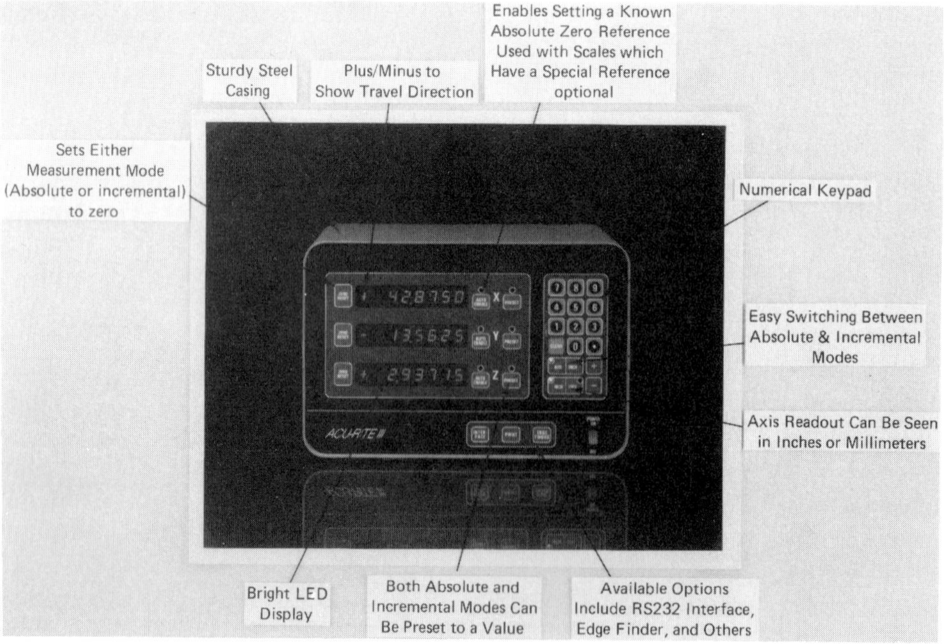

FIGURE 55-2
DROs are available with three-axis movement. (Credit: Acu-Rite Incorporated)

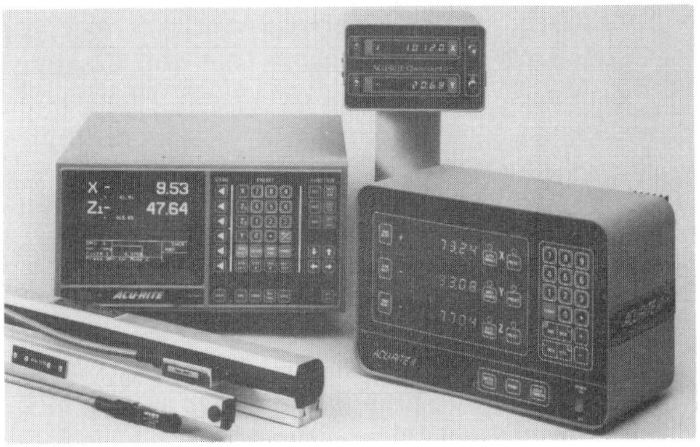

FIGURE 55–3
The components of a DRO system include an engraved glass scale, an electro-optical reader, and a console. (Credit: Acu-Rite Incorporated)

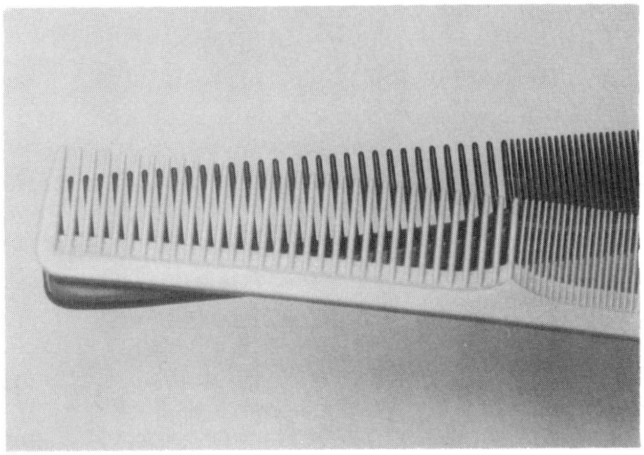

FIGURE 55–4
The principle of the Moire fringe pattern.

The Moire Fringe Principle

The DRO system acts as a coordinating measuring device and indicates direction, distance, and location. Direction is displayed as a plus (+) or a minus (−) from a starting point to indicate a left or a right movement along the *X* axis or a back and forth movement along the *Y* axis. On some devices, a third movement, that of the cutting tool (the *Z* axis) can also be measured in a down or up movement (Figure 55–2). Distance relates to the print dimensions, and on a vertical milling machine the two axes or table movements (*X* and *Y*) may be registered in relation to the cutting tool. These movements are accurately measured and read on the DRO console. Location refers to the actual point at which the machining takes place. These movements are constantly being monitored and displayed in the form of a numerical readout.

While there are a few DRO systems available, such as a rack-and-pinion-driven encoder and magna scales, the more widely used system uses a precision engraved glass scale and an electro-optical reader (Figure 55–3). This system operates on the Moire fringe principle, which can be illustrated by using a set of combs with equally spaced teeth. When one comb is placed above the other at a slight angle (Figure 55–4), a series of light and dark bands appears. As the upper comb moves, the bands will change direction up or down relative to the comb but at a faster rate of movement than the upper comb. Thus it can be seen that lateral movements of the upper comb is translated into vertical movements of dark bands. It is this principle that is applied to DRO systems to count the Moire fringes and convert them to a high-resolution accuracy of measurement in electronic pulses.

The Basic Components of a DRO System

The core of a DRO is the accuracy delivered by the workings of three basic components: a glass scale with a calibrated grating etched on a fine layer of chromium, a reading head, and a console with a DRO display.

The glass scale (linear encoder) of the system has lines etched along the face of its entire length 0.001 in. (0.02 mm) apart, rendering resolutions as small as 0.0001 in. (0.002 mm). Greater accuracy is available with systems that render resolutions as fine as 0.00005 in. (0.001 mm).

The linear encoder is attached together with the reading head to the moving table of the machine tool. The reading head also has a transparent index grating equal to the spacing on the etched glass linear encoder and is placed in a fixed position at a slight angle to the linear encoder. As these grated lines pass each other in movement, a fringe pattern is created that is picked up by a light and photoelectric cells within the reading head. These fringe patterns are sensed as electrical pulses. These pulses, in turn, are relayed and interpreted by the DRO console as numerical displays indicating location. The DRO readings in inch or millimeters are available at the touch of a key.

Due to the extreme accuracy and reliability of the DRO system, there is no longer any need for an operator to depend on the accuracy of the feed screws. There is no backlash to interfere with the DRO system.

Types of DROs

There are two types of DRO systems available: programmable and nonprogrammable. Each mode is intended to help the operator position the table in relation to the tool location or move the table in relation to a part reference point.

With the programmable type (also known as the incremental mode), table movement is measured using a point-to-point or hole-to-hole method. Once the length of travel has been made between dimensions, the display is reset to zero and the operator is ready for the next point-to-point move.

The nonprogrammable type (also known as absolute dimensioning) keeps track of the total distance the table travels from the reference point, which is generally set as a zero starting point, regardless of the number of stops made. The distance is displayed on the console when the ABS key is pushed (lit up) and will correspond with the total print dimension of the part. This type of programming eliminates much of the paper-and-pencil calculations involved when adding or subtracting measured distances between moves.

Key Functions (Figure 55-5)

ON-OFF POWER SWITCH: When pushed on, the console is activated to light up the ABS and INCR keys. The axis windows will be blank except for a vertical mark on each end. These are fault indicators that will disappear during normal system operation. Should they reappear, it will indicate a false measurement has taken place and the operational procedure will have to be restarted.

INCR: This key switches and displays from the absolute mode or preset value and displays a point-to-point table positioning when lit. After the table has been moved from one point to another, the display is reset to zero and the operator can continue to the next stop.

ABS: This key switches the display from the incremental mode or preset value and displays the absolute measurement when lit. The system keeps track of the distance the table travels or has traveled along an axis from a zero starting point regardless of the number of point-to-point stops made.

PRESET: This key permits the operator to preset a travel dimension into the counter and display, such as the radius or center of a cutting tool when milling a slot. The table can then be moved until the display reads zero to permit the machining of the location.

ZERO RESET: The opposite of the preset procedure, this key can zero the counter and display when a dimension is needed or at any time, for that matter, to permit travel to the next measurement.

FIGURE 55-5
An explanation of the key functions of a two-axis DRO console.

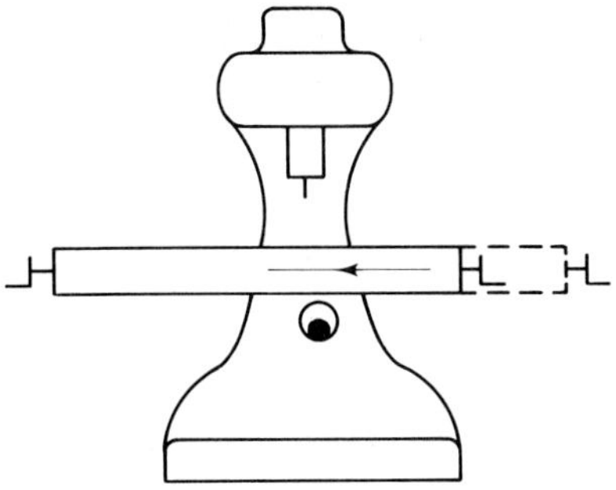

Table movement from right to left of the column should display a negative (−) directional count on the *X* axis.

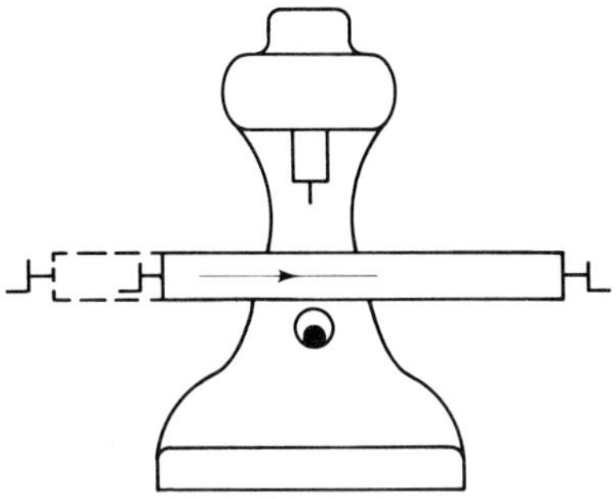

Table movement from left to right of the column should display a positive (+) directional count on the *X* axis.

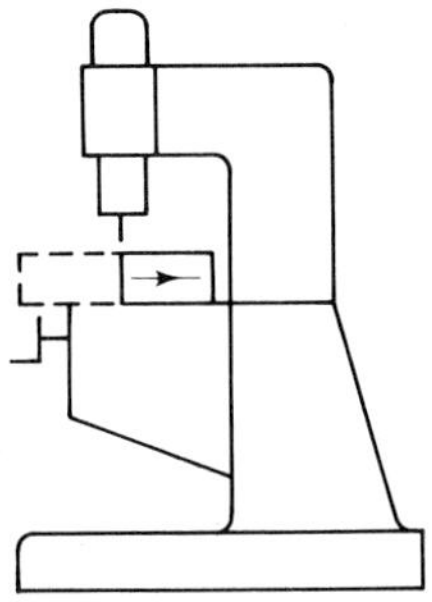

Cross feed table movement toward the column should display a negative (−) directional count on the *Y* axis.

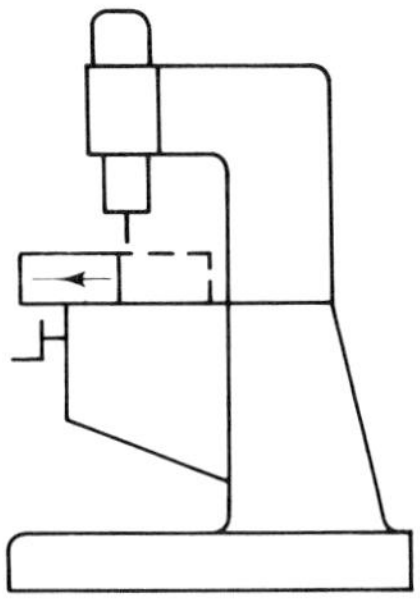

Cross feed table movement away from the column should display a positive (+) directional count on the *Y* axis.

FIGURE 55-6
Table travel direction.

INCH-MM: Axis displayed dimensions can be seen in inches or millimeters at the push of a key.

PLUS-MINUS: This key indicates travel direction when pushed following a digit keyboard entry or immediately following a preset key entry, as illustrated in Figure 55-6.

CLEAR: Will only function when the preset key is lit and will clear the value in print memory to blank the display. This will allow a new value entry into the preset memory.

DIGIT KEYBOARD: Will enter new values into preset memory when preset key is lit and clear key has been pushed.

AUTO ENABLE: Sets a permanent zero reference for machine tool travel; may also be used with an edge finder when applicable.

Using the DRO System

In order to effectively machine work, an operator will not only have to properly mount it on the machine tool table or vise but will also have to establish a starting zero point from which to take coordinate measurements. Mounting work at locations that have not been previously established (stop) may require the use of an edge finder.

By using Figure 55-7, the drilling of holes in a rectangular workpiece whose sides are square and parallel, one operational use of the DRO system will be explained.

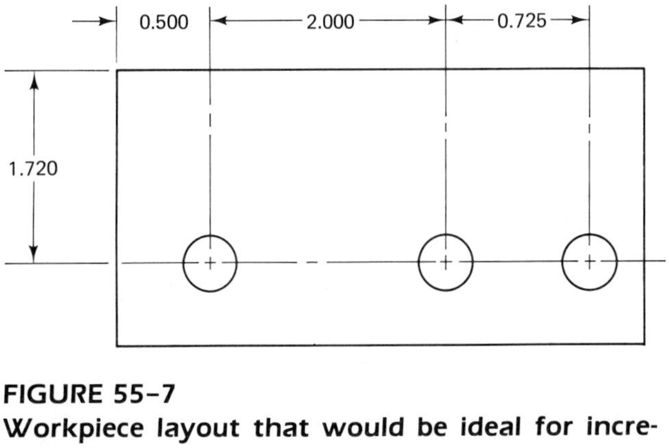

FIGURE 55-7
Workpiece layout that would be ideal for incremental mode programming.

FIGURE 55-8
To activate the console, push the power switch to the on position.

FIGURE 55-9
Push the ZERO RESET key for the *X* axis.

FIGURE 55-10
Push the ZERO RESET key for the *Y* axis.

FIGURE 55-11
Push the INCR key if using a point-to-point dimensioning program.

Procedure

1. Mount the workpiece on the table so that its long edge is parallel to the table travel.
2. Push the DRO power switch to the on position (Figure 55-8).
3. Push the ZERO RESET key for the *X* axis (Figure 55-9).
4. Push the ZERO RESET key for the *Y* axis (Figure 55-10).
5. As the drawing indicates a hole-to-hole layout, push the INCR key (Figure 55-11).
6. Push the ZERO RESET key for each axis to display the plus zeros (Figure 55-12).
7. Push the INCH or MM key for the measurement required (Figure 55-13).

The referencing of the workpiece will now have to be done. A 0.200-in.-contact-diameter edge finder

FIGURE 55-12
Push the ZERO RESET for each axis to display the plus zeros.

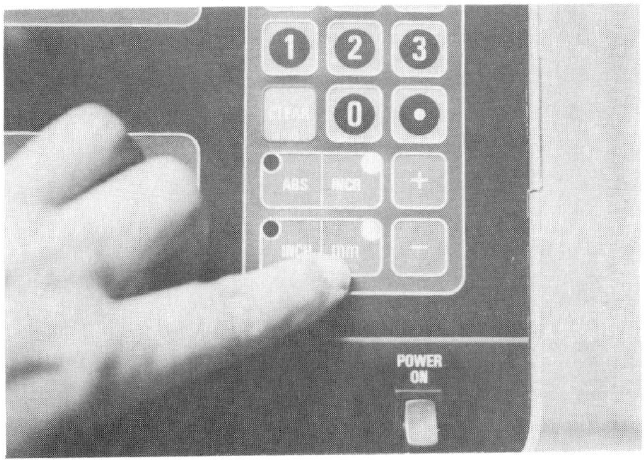

FIGURE 55-13
Push the proper measuring value key.

FIGURE 55-14
Push the PRESET key on the *X* axis.

FIGURE 55-15
Push the CLEAR key.

will be used to establish the starting *X* and *Y* coordinate points.

1. Mount a 0.200-in.-contact-diameter edge finder in the machine tool spindle.
2. Set the spindle speed between 600 and 700 rpm.
3. Start the machine tool spindle.
4. Move the table so that the edge finder will be placed to the left of the workpiece.
5. Slowly move the workpiece against the edge finder's wobbly rotation to make the contact diameter run concentrically. The moment the contact shifts direction as explained in Unit 54, stop the table movement.
6. Push the ZERO RESET key on the *X* axis.
7. Push the PRESET key on the *X* axis (Figure 55-14).
8. Push the CLEAR key (Figure 55-15).

FIGURE 55–16
Push the point and 1 keys to establish one-half of the 0.200 in. edge finder diameter.

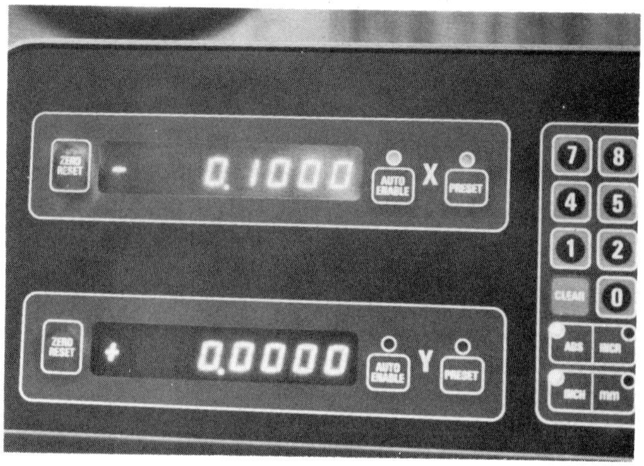

FIGURE 55–17
Push the minus key to indicate the edge finder's offset of 0.100 in.

9. Compensating for one-half the edge finder diameter (0.100 in.) will require two steps:
   (a) Push the digit keys, point and 1 (Figure 55–16).
   (b) Push the minus key to indicate that the centerline of the machine spindle is to the left of the reference surface (Figure 55–17).

10. The X axis display should now read −0.10000. The longitudinal reference edge has now been established so that when the light emitting diode (LED) display reads 0.00000, the centerline of the spindle will be directly over the reference edge of the workpiece.

11. Move the table along the X axis for the dimension required to locate each hole operation and reset to zero at the end of each stop.

The Y axis will also have to be established. Using the crossfeed table movement from the column face toward the operator (back to front), repeat steps 5 to 11 as they apply to the Y axis.

Advantages of Using a DRO System

Some of the advantages of using a DRO system are as follows:

1. Less operator skill is required to use it.
2. The digital display is clear and easy to read.
3. Positioning time is greatly speeded up.
4. It eliminates the use of gage blocks and measuring rods for precise table locations.
5. It greatly reduces or eliminates scrap or rework.

UNIT 56 THE DIVIDING HEAD AND ROTARY TABLE

The ability to divide the periphery of a circle into small, equal or many divisions is an important function in the metalworking industry. Not only does it play a role in the development of many mechanical devices, but it also is a factor in locating positions of holes, slots, and machining feed mechanisms such as teeth on a gear. This capability can be found in two groups of accessories used on the milling machine—the rotary table and the dividing head. The devices operate using different methods of measurement and are available manually or power fed.

The rotary table can be revolved continuously in a 360° circle and will divide and measure the divisions by a direct angular method down to a fractional part of a degree. Indexing tables are also available. These devices are rotary tables with very fine, precise means of measurement, indexing a quarter of a degree (15′) to an accuracy of 0.1″ (Figure 56–1).

FIGURE 56-1
A precision rotary jig borer. (Credit: Producto Machine Co.)

Dividing or indexing heads also divide a circle or circumference of a workpiece into any number of equal parts or angles and hold the work in the position required so it can be machined. A number of precision spacing devices are available ranging from special indexing arrangements to index computers where practically any number of divisions are possible. This unit will concern itself with the manually operated dividing head and rotary table.

Objectives
After completing this unit you will be able to

1. calculate and use a dividing head by simple indexing.
2. calculate and use a dividing head by direct indexing.
3. calculate and use a dividing head by angular indexing.
4. calculate and use a rotary table.

The Dividing Head

The dividing head (headstock) is often used in connection with the footstock, also referred to as the tailstock, and a center rest for work support (Figure 56-2). The main parts of the dividing head are the worm and worm wheel, an index plate, an index crank, and sector arms.

A 40-tooth worm wheel is attached to the index spindle and meshes with a single thread worm on a worm shaft connected to the index crank. Movement of the spindle is made by turning the index crank (Figure 56–3). Since there are 40 teeth in the worm wheel on conventional dividing heads, the dividing head has a ratio of 40:1. This means that it will require 40 complete turns of the index crank to rotate the spindle one complete revolution (360°). A circular indexing plate, which is provided with a series of equally spaced holes, is designed to accept a plunger pin at the end of the index crank to register a partial circle movement.

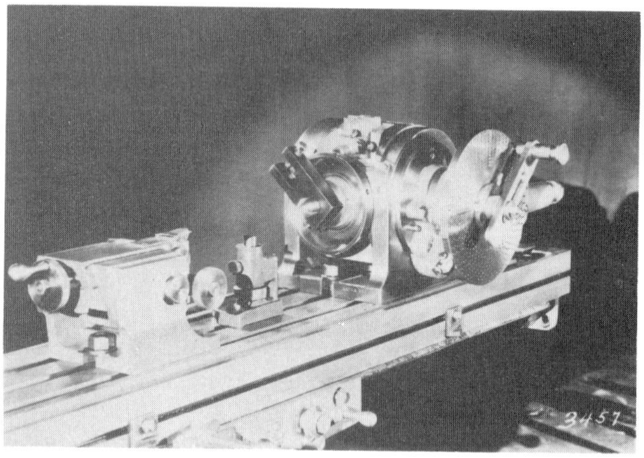

FIGURE 56–2
A universal dividing head set. (Credit: Cincinnati Milacron Inc.)

The spindle, which is constructed to accept a threaded chuck on its end and a tapered hole to suit a 60° center as well as tapered shank tools, is mounted on a swivel block positioned between housings cast with the base plate. By loosening or tightening a clamping bolt, the spindle may be set on an arc as low as 5° below the horizontal to 10° beyond the vertical. Graduations on the side of the head indicate the degrees of elevation within this 115° arc.

The footstock is used together with the headstock to support work that is held between centers or extends beyond the jaws of the chuck. The footstock may be adjusted longitudinally with a handwheel and is also adjustable to be raised or lowered vertically. To compensate for tapered work that is elevated to an angle, the footstock can also be elevated at an angle to maintain alignment with the headstock. An adjustable center rest is also used and can be locked in position to support long or slender work being machined.

Indexing Methods

There are several methods of indexing a dividing head for milling operations such as direct, simple angular and differential indexing. The most common methods used for dividing the circumference of a workpiece into accurate divisions are the direct and simple methods.

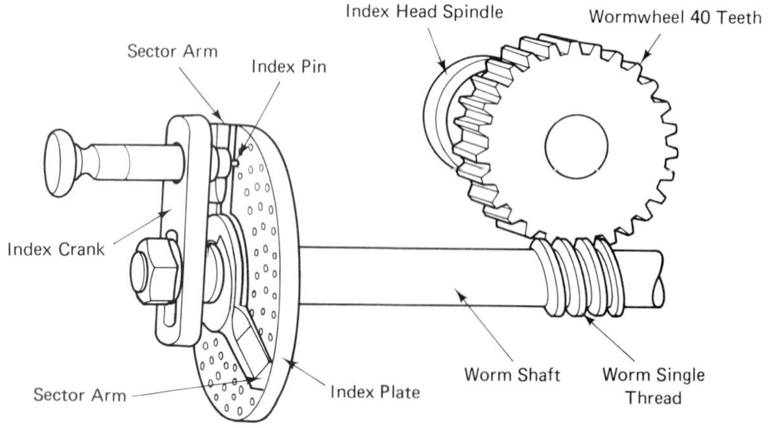

FIGURE 56–3
Illustration showing the construction and parts of an indexing mechanism. (Credit: Cincinnati Milacron Inc.)

TABLE 56–1
Direct Indexing divisions

Index plate hole circles														
24	2	3	4	—	6	8	—	—	12	—	—	24	—	—
30	2	3	—	5	6	—	—	10	—	15	—	—	30	—
36	2	3	4	—	6	—	9	—	12	—	18	—	—	36

Direct Indexing

In this operation (often called rapid indexing) the plunger pin in back of the direct indexing plate attached to the spindle is pulled back and removed from a hole in the direct indexing plate. By moving an eccentric device through a partial movement of the index crank (on some models), the worm and worm wheel are disengaged by lowering the worm. This permits the spindle to be turned freely by hand, and the movement is measured by the number of holes on the direct index plate that pass by the plunger pin.

A direct indexing plate generally contains three series of circled holes or slots evenly divided in 24, 30, and 36 locations. Spindle movement (divisions) will be limited by any number that is a factor of 24, 30, or 36. The whole numbers obtainable by direct indexing are listed in Table 56–1. Once the direct indexing plate has been moved, the plunger pin is inserted in the proper hole or slot to lock the spindle.

Example

Determine the indexing movement and setup required to mill a five-sided work blank 0.125 in. (3.18 mm) deep.

Solution:

Since a 30-hole circle is the series that can be evenly divided by 5, indexing would be as follows:

$$\text{Indexing} = \frac{\text{Number of index plate holes}}{\text{Required number of divisions}}$$

$$= \frac{30}{5}$$

$$= 6 \text{ holes.}$$

Note: Counting should begin with the first hole or slot next to the inserted plunger pin.

Procedure

1. Mount the work blank in the dividing head chuck or between centers.
2. Disengage the worm and worm wheel according to the dividing head being used.
3. Adjust the plunger pin behind the index plate attached to the spindle into the 30-hole circle or slot (Figure 56–4).
4. Mount a proper-size milling cutter.
5. Adjust the depth of cut and mill the first side.
6. Count 6 holes or slots starting with the hole or slot next to the plunger pin and mark this with a piece of masking tape or marking pen.
7. Remove the plunger pin from the direct indexing plate.
8. Turn the direct indexing plate so that the plunger pin can be inserted into the marked sixth hole or slot.
9. Complete the second cut.
10. Repeat steps 6 to 8 until all the sides have been milled.

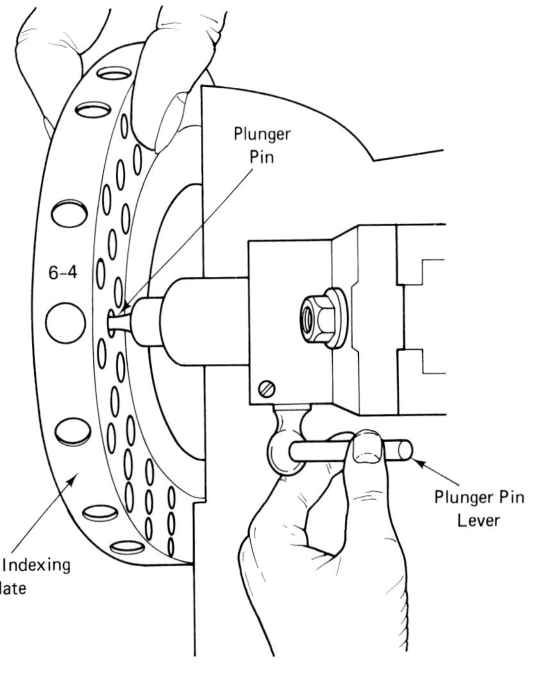

FIGURE 56–4
Adjusting the plunger pin for direct indexing. (Credit: Cincinnati Milacron Inc.)

FIGURE 56–5
A 36-division indexing whirl gig. (Credit: Suburban Tool, Inc.)

FIGURE 56–7
A built-in 3-in. sine bar permits accurate plate locations. (Credit: Suburban Tool, Inc.)

FIGURE 56–6
Positioning the indexing whirl gig. (Credit: Suburban Tool, Inc.)

Direct Indexing Devices

Simplified indexing devices (Figure 56–5) are also available that can index a circle into precise 36 divisions. The indexing is positioned by inserting a plunger pin in one of the 36 holes or slots that circle the dividing plate (Figure 56–6). By turning a handcrank, work fastened to the surface of the indexing plate can be made to turn continuously. This rotary motion makes it ideal to be used when grinding the radius on the neck of a hardened punch, for example. Generally, divisions with this device are limited to increments as small as 10° or numbers that are factors of 36. With the use of a built-in 3-in. sine bar (Figure 56–7), it is possible to locate workpieces at increments between the 10° divisions using standard gage blocks (Figure 56–8). Angular dimensions can be set to within 15 seconds of an arc.

Simple Indexing

Simple indexing, which is also referred to as plain indexing, includes the use of sector arms together with the index plate and crank to obtain many divisions not possible with the direct indexing method. The dividing head spindle can only be rotated by turning the index crank when the worm and worm

FIGURE 56–8
Setting the 3-in. sine bar with gage blocks. (Credit: Suburban Tool, Inc.)

wheel are in mesh. Since there are 40 teeth on the worm wheel of standard index heads, it will require 40 complete turns of the index crank to complete one full revolution of the spindle (360°). Thus, a ratio of 40 : 1 exists between the turns of the index crank and the dividing head spindle.

To determine the number of turns required for indexing most divisions, it is calculated by dividing 40 by the number of divisions required:

$$\text{Indexing} = \frac{40}{N}$$

where N is the number of divisions required.

Example

Determine the indexing required to mill a four-fluted cutter.

Solution:

$$\begin{aligned}\text{Indexing} &= \frac{40}{N}\\ &= \frac{40}{4}\\ &= 10 \text{ complete turns.}\end{aligned}$$

In this example, the number of turns required was a whole number. But what if a whole number is not possible and a mixed-fraction number has to be indexed? Suppose it were required to index for a six-fluted cutter? How many turns of the index crank would be involved?

$$\begin{aligned}\text{Indexing} &= \frac{40}{N}\\ &= \frac{40}{6}\\ &= 6\tfrac{2}{3} \text{ turns.}\end{aligned}$$

While the whole number would be easy enough to make, how would two-thirds of a turn be made? This would involve the use of a particular index plate and sector arms.

Index Plate and Selector Arms

The index plate has six or more circles of equally spaced holes into which the index plunger pin is inserted to lock the dividing head spindle. Once the plunger pin is pulled back, the index crank can be turned to rotate the dividing head spindle. A sector is formed by two radial arms placed in front of the index plate and is positioned to span the distance equal to any fractional part of a turn.

To be able to get two-thirds of a turn, select any hole circle from Table 56–2 that is a number of holes that is evenly divisible by the denominator 3, such as 24, 30, 39, and 42, found on one plate. Using circle 24 as an example, two-thirds of 24 is 16 holes. To obtain the six divisions, the index crank would have to be turned six full turns plus the 16 holes on the 24-hole circle.

TABLE 56–2
Index plate hole circles

Brown and sharpe	*Cincinnati standard plate*	
Plate 1: 15-16-17-18-19-20	One side:	24-25-28-30-34-37-38-39-41-42-43
Plate 2: 21-23-27-29-31-33	Other side:	46-47-49-51-53-54-57-58-59-62-66
Plate 3: 37-39-41-43-47-49		

Procedure

1. Mount the workpiece in the dividing head chuck or between centers.

2. Mount the proper-size milling cutter.

3. Mount an index plate with a 24-hole circle on the dividing head.

4. Loosen the index crank nut and position the index plunger pin into hole 24 in the 24-hole circle.

5. Tighten the index crank nut and check that the index plunger pin can easily enter and be withdrawn from the hole.

6. Turn the index crank half of a turn counterclockwise and then clockwise so that the pin enters hole 24 to remove any backlash.

7. Loosen the binding screws on the sector arms.

8. Slide the tapered edge of the left radial arm against the index plunger pin.

9. Count 16 holes in a clockwise direction on the 24-hole circle beginning with the hole next to the plunger pin. *Do not count the hole that has the index pin.*

10. Tighten the right sector arm with its tapered edge just beyond hole 16.

11. Align the milling cutter with the workpiece.

12. Start the machine spindle.

13. Set the cutter to the top of the workpiece by using a paper feeler.

14. Move the table so that the cutter clears the end of the workpiece.

15. Set and lock the depth of the cut to mill the first cut.

16. Return the table to the original starting position.

17. Pull back the index plunger pin and turn the index crank clockwise six full revolutions using the left sector arm as a reference point, pausing after each complete revolution.

18. Continue slowly moving in an arc, releasing the plunger pin to contact the index plate surface between holes 15 and 16 and gently tap the side of the index crank to permit the plunger pin to snap into hole 16, adjacent to the right sector arm.

19. Slide the sector arm furthest from the index plunger pin in a clockwise direction until the tapered edge butts against the index pin.

20. With the sector arms repositioned and the dividing head spindle locked, continue machining and indexing for the remaining flutes as described in steps 16 to 19.

In the event that the index crank is moved too far in a clockwise direction beyond the required hole, it must be corrected by revolving the crank counterclockwise at least one-half turn to remove any backlash. Then the index crank is carefully turned clockwise until the plunger pin snaps into place.

Angular Indexing

Simple indexing may also be used for angular indexing. Since it will require 40 turns on standard dividing heads to revolve the dividing head spindle 360°, one complete revolution of the index crank will turn the spindle one-fortieth of 360°.

The formula for determining angular dimensions in degrees is

$$\text{Indexing} = \frac{\text{number of degrees given}}{9}$$

Example

Determine the indexing for 36°.

$$\text{Indexing} = \frac{36}{9}$$

$$= 4 \text{ complete turns.}$$

Example

Determine the indexing for 30°.

$$\text{Indexing} = \frac{30}{9}$$

$$= 3\tfrac{1}{3} \text{ turns}$$

$$= 3 \text{ complete turns plus 8 holes on a 24-hole circle.}$$

When determining the indexing for a given dimension in degrees and minutes, it is necessary to first change the degrees to minutes in terms of one full turn ($9° \times 60' = 540'$), which is then added to the minutes required.

The formula for determining the indexing in minutes is

$$\text{Indexing in minutes} = \frac{\text{number of minutes required}}{540}$$

Example

Determine the indexing required for 54′.

$$\text{Indexing} = \frac{54}{540}$$

$$\text{Indexing} = \frac{1}{10}$$

The indexing for 54′ would be two holes ($\frac{1}{10}$) on a 20-hole circle plate.

Example

Determine the indexing for 30° 15′.

Solution:

Convert the 30° to minutes:

$$30 \times 60 = 1800'$$
$$+ \quad 15'$$
$$1815'.$$

$$\text{Indexing} = \frac{\text{number of minutes required}}{540}$$

$$= \frac{1815}{540}$$

$$= 3\tfrac{1}{15}$$

= 3 complete turns plus 1 hole on a 15-hole circle.

Rotary Table

The hand-operated rotary table (Figure 56–9) consists of a cast-iron base that can be bolted to a milling machine or jig borer table. Fitted within the base and located below the table is a precise worm wheel that meshes with and is driven by the worm shaft connected to the manual handwheel. When rapid table movement is required, the worm shaft may be disengaged. The table itself is slotted to accept T-bolts for clamping work to the surface and is constructed with a precise hole to accept a test plug. The hole is concentrically located and is designed to

FIGURE 56–9
A hand-operated rotary table.

speed up the alignment of the table with the machine spindle. Rotary motion is made by turning the handwheel and measured by the degree graduations lined along the lower outside edge of the table. A Vernier scale located on the handwheel collar enables measurements to be set and read to an accuracy within 2′ of a degree.

Unlike the standard dividing head that normally uses a ratio of 40 : 1 to complete a 360° turn, rotary tables may use ratios of 72 : 1, 80 : 1, 90 : 1, or as much as 120 : 1 for the larger tables. Calculations for spacing will be the same as indexing using the dividing head only the number of teeth in the rotary worm wheel is used.

Example

Calculate the number of turns required to locate six equally spaced holes on a circular workpiece using a rotary table with a 90 : 1 ratio.

Solution

$$\text{Indexing} = \frac{\text{number of worm wheel teeth}}{\text{number of divisions required}}$$

$$= \frac{90}{6}$$

= 15 turns.

FIGURE 56–10
Insert a close fitting test plug to indicate when aligning the rotary table with the machine spindle.

Aligning a Rotary Table with the Machine Spindle

Aligning the rotary table with the center axis of the machine spindle requires that the tool head be square with the machine table and that a small dial test indicator be mounted to measure the circumference of a test plug.

Procedure

1. Square the tool head with the milling machine table if necessary.
2. Mount the rotary table on the milling machine.
3. Insert a close-fitting test plug in the center of the rotary table (Figure 56–10).
4. Insert a drill chuck in the machine spindle.
5. Close the jaws of the drill chuck.
6. Lower the drill chuck so that the jaws are slightly above the test plug and lock it in this position.
7. Move the milling machine table to center the test plug by eye with the jaws of the drill chuck.
8. Raise the drill chuck.
9. Select a small dial test indicator with a swivel friction holder attachment.
10. Mount the dial test indicator in the drill chuck.
11. Mount the indicator to form a small angle with the center of the test plug.
12. Lower the drill chuck to bring the indicator contact point in line with the circumference of the test plug.
13. Move the indicator by hand until it contacts the surface of the test plug and registers about one-fourth of a turn on the dial.
14. Set the bezel to zero (0).
15. Slowly rotate the drill chuck by hand exactly one-half turn (180°) and note the differences in the indicator readings.
16. Adjust the longitudinal or crossfeed handle so that the indicator registers one-half the difference between the two readings.
17. Once both readings are the same, repeat step 16 on the other half of the circumference.
18. Lock the longitudinal and crossfeeds and recheck the adjustment.
19. Readjust if necessary.

SECTION REVIEW

Unit 48

1. Name two types of horizontal style milling machines to be found in the toolroom.
2. Describe the purpose of the following parts: (a) base, (b) table, (c) spindle, (d) elevating screw.
3. What safety precautions must be taken in the following situations: (a) changing belts or gears, (b) removing chips by hand.
4. How does the table handwheel differ from the crossfeed handwheel?

5. Describe the purpose of the following accessories: (a) arbor, (b) collet adaptor, (c) swivel base vise, (d) universal vise.

Unit 49

1. Briefly describe the cutting action of peripheral milling.
2. Briefly describe the cutting action of face milling.
3. Define the following milling cutter terms: (a) gullet, (b) land, (c) heel.
4. Describe five types of special-purpose cutters and explain their uses.
5. Explain the cutter holding method used on a horizontal milling machine.
6. How should arbor runout be treated?
7. When a milling cutter is mounted or removed, how should it be handled?

Unit 50

1. In addition to material, list five factors that will affect the efficiency of a milling operation.
2. Name five common work holding methods used when milling a flat surface.
3. What factors affect the selection of a cutter?
4. How is feed defined in a milling operation?
5. What is meant by up milling?
6. What is meant by down milling?
7. How is backlash defined?
8. Upon what factors are feed dependent?
9. In general, how much material removal should be considered when taking a roughing cut?
10. How deep a cut should be considered when taking a finishing cut?
11. In production milling, how is feed defined?
12. What is the purpose of placing a long strip of paper between the cutter and the workpiece when establishing a depth of cut?

Unit 51

1. How is the vertical mill rated for usage in the machine shop?
2. Name eight operations that can be performed in a vertical mill.
3. How does the ram-type vertical milling machine differ from the standard vertical milling machine?
4. State the purpose of the following parts: (a) column, (b) turret, (c) quill.
5. Describe an end mill.
6. What type of end mill is recommended for milling aluminum and nonferrous materials?
7. What advantage do indexable insert-type end mills have over standard production end mills?
8. What is an advantage and a disadvantage of carbide end mills?
9. What kind of end mill is recommended for finishing fillets and other contoured surfaces?
10. What is the purpose of the boring head accessory?

Unit 52

1. List the five main methods of holding cutters in a vertical milling machine.
2. What is the principal means of holding an end mill in a standard vertical milling machine?
3. What is the principal means of holding a cutter in a ram-type vertical milling machine?
4. List four precautions to be taken when mounting or removing an end mill from a vertical milling machine.
5. Briefly describe how to mount an end mill in a ram-type vertical mill.
6. How should an end mill be removed?

Unit 53

1. Name two economical types of cutters used on a vertical milling machine to produce a flat surface.
2. What is the purpose of "sweeping the table"?
3. What is another term for "aligning the vise"?
4. Why is it important to align the milling vise?
5. What tools are required to align the milling vise?

6. What is the purpose of laying strips of paper under the corners of a workpiece when mounting it in a vise?

7. How can a workpiece bearing opposite sides, one of which is flat and the other uneven, be fastened in a milling machine vise?

8. Draw a sketch indicating the order in which all four sides of a block will be machined. The block has two adjacent sides flat and at 90° to each other and two uneven adjacent sides.

Unit 54

1. What is the purpose of an edge finder?

2. Briefly describe how an edge finder works.

3. Name five conditions that should be met when using an edge finder on a milling machine without a DRO.

Unit 55

1. Define a digital readout (DRO).

2. Why have DROs been so widely accepted in the machine tool industry?

3. What are the three basic components in a DRO system?

4. Describe the Moire fringe principle.

5. How does the DRO system operate using the Moire fringe principle?

6. What are the two DRO types available?

7. List five advantages of using a DRO system.

Unit 56

1. Name the main parts to a dividing head.

2. How is work that is extended beyond the indexing head supported?

3. What is another term for direct indexing?

4. What is another term for simple indexing?

5. List two methods of indexing in addition to direct and simple indexing that may be performed using the dividing head.

6. How many turns of the crank does it take to make one complete revolution of the standard indexing head?

7. What series of three circled holes does a direct indexing plate generally contain?

8. Using a direct indexing method, calculate the indexing movement to machine an eight-sided work blank.

9. How are fractional parts of a turn handled when using a simple indexing dividing head?

10. Explain how angular indexing is calculated using a standard dividing head.

11. How does a rotary table differ from a standard dividing head when dividing angles?

12. How large a ratio can be found on a rotary table?

13. What is the purpose of inserting a test plug in the center of a rotary table?

Section 11

GRINDING MACHINES

Grinding is a process for removing and reshaping ferrous, nonferrous, or even nonmetallic substances. The grinding process differs from other metal removal processes in that grinding uses grainlike abrasives that are tough and hard. To keep it sharp and cutting well, the grinding wheel can be reconditioned to expose new cutting edges. Thus, the grinding wheel can be looked upon as an abrasive cutting tool where greater machining accuracy may be achieved due to the control that can be exercised in grinding.

UNIT 57 GRINDING MACHINES

Grinding machines play an important role in precision metal technology. Grinding machines make possible the grinding processes that are able to produce a variety of forms to extreme accuracy essential to modern manufacturing requirements. Many types of grinding machines are available to industry. In many cases, they provide the only means by which production can be carried out economically.

Objectives
After completing this unit you should be able to

1. explain the composition of a grinding wheel.
2. explain the difference between a hard and soft grinding wheel.
3. analyze the markings found on a grinding wheel blotter.
4. identify four different types of grinders used in industry and explain their functions.

Safety

The principles of safety that have been cited in other units also apply to the field of grinding. Improper use of grinding wheels, storage, and mounting are but a few of the considerations that grinding machine operators must weigh. The dangers involved in grinding can be seen and unseen. The fiery red or white sparks that accompany any grinding are actually tiny chips of metal that have been heated up. In addition to protecting the eyes, a vacuum system should be used to protect against breathing in these particles during dry grinding. What follows are a few of the general safety recommendations to be observed in grinding.

1. Always wear a protective face shield or safety glasses when grinding.
2. Never use a grinder without a proper exhaust or vacuum system to collect the dust and grinding chips (Figure 57–1).
3. Keep fingers and hands away from any moving wheel or workpiece.

FIGURE 57–1
Never use a dry-type grinder without an exhaust or vacuum system.

FIGURE 57-2
Always check to see that machine attachments will clear the workpiece. (Credit: DoAll Company)

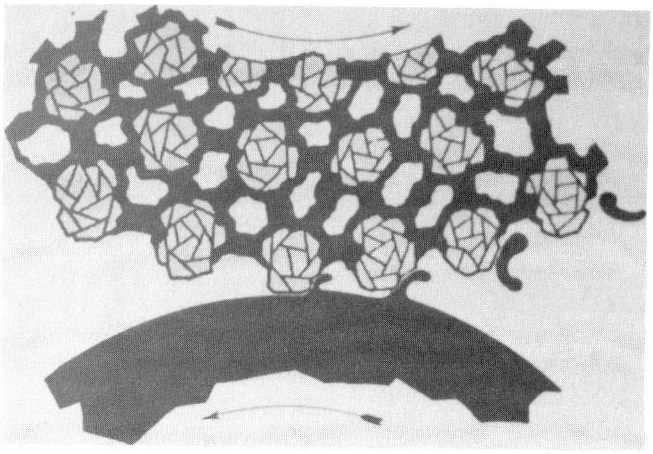

FIGURE 57-3
Illustration of how voids in a wheel structure serve to promote chip clearance. (Credit: Bay State Abrasives)

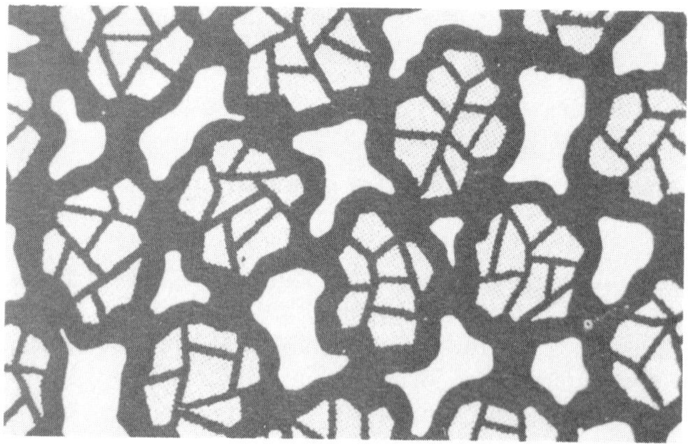

FIGURE 57-4
Bond "posts" holding abrasive grain particles. The dark areas around the grain are the bond "posts". (Credit: Bay State Abrasives)

4. Remove any jewelry or rings before attempting any grinding operation.

5. Roll up sleeves and tuck in loose clothing.

6. Always check the workpiece to see if it is held securely by the magnetic chuck before attempting any surface grinding.

7. Make sure all work is securely clamped in a vise or grinding accessory before placing it on the magnetic chuck.

8. Always check any attachments to see that they will clear the workpiece during grinding (Figure 57-2).

9. To prevent a wheel out-of-balance condition, always shut off the coolant before stopping the spindle.

10. Do not leave stationary wheels immersed in fluids, as it will create an out-of-balance condition.

11. Do not jam a wheel against the workpiece; instead, let it coast to a stop.

12. Do not plunge or gouge into a spot or corner of a workpiece when establishing grinding contact.

13. Always replace any safety guards before using any grinding machine.

14. Never stand in front of the direct rotation path of a grinding wheel, but rather to one side, when turning on a grinding machine.

15. Never mount a grinding wheel unless it has been ring tested first.

The Elements of a Grinding Wheel

A grinding wheel is composed of two basic elements, abrasive grain and bond. The abrasive grain is the element that does the cutting. The bond is the "cement" that holds and supports the abrasive grain in the form of the wheel.

The *structure* of the wheel makeup is designed by the percentage of the abrasive grains and bond. It is the spacing between these two elements that provides the pores, the necessary chip gullets that create the chip clearance and contributes to a cooler cutting action (Figure 57-3). In a manner of speaking, the abrasive grains in a wheel can be considered individual cutting tools; the bond is considered the toolholder. The bond should hold the abrasive grit against the workpiece until the abrasive becomes dull and worn. Then, in a process called fracturing, the bond should break down to tear away or eliminate the worn abrasive. A new series of sharp particles should appear on the wheel's grinding surface. This fracturing process is considered a self-dressing operation.

A wheel is classified hard or soft depending on how easily fracturing of abrasives occurs as they dull during cutting. If grains are eliminated too quickly during the grinding operation, it indicates that the bond has a weak holding power and the wheel will be classified soft. If the abrasive grains are retained during the grinding operation, it indicates the bond has a strong holding power and the wheel is classified hard. The wheel will exhibit a slow cutting action due to glazing or loading—the filling of pores with material (Figure 57-4).

Types of Grinding Wheels

There is no one ideal grinding wheel. Many factors will affect the performance of a grinding wheel. The nature of the work material, its hardness, and its sensitivity to heat are all important factors and will determine the abrasive selection and hardness of a wheel (Table 57-1).

TABLE 57-1
Basic grinding wheel shapes and applications

Standard type	Applications	Standard type	Applications
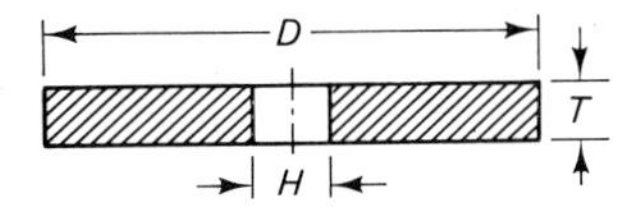 Type 1 Straight	Peripheral grinding wheel used for centerless, cylindrical, internal, surface, off-hand, snagging, and tool grinding operations.	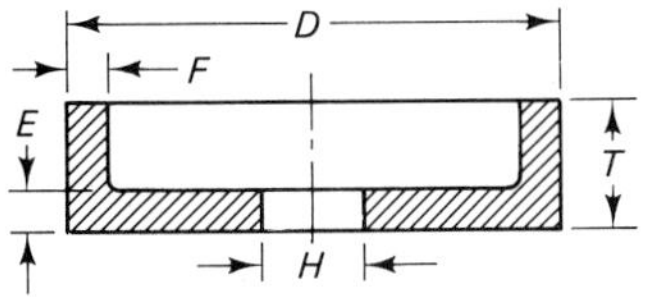 Type 6 Straight cup	Side grinding wheel used for snagging in portable machines and surface grinding operations.
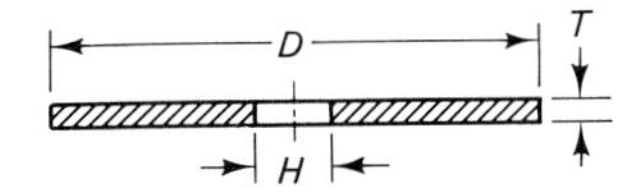 Type 1 Cut-off	Peripheral grinding wheel used for cutting or severing long sections to length on bars, pipe, tubing, and other structural shapes.	 Type 11 Bevel face Flaring cup	Recommended as a heavy-duty portable snagging wheel to be used on right-angle or vertical-type machines.
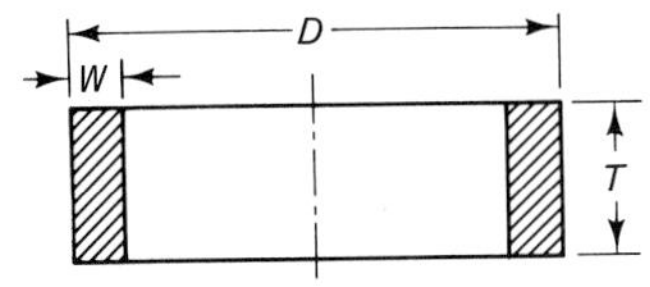 Type 2 Cylinder	Side grinding wheel used on horizontal and vertical spindle grinders.	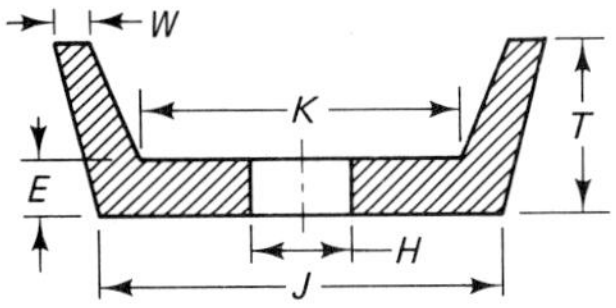 Type 11 Flaring cup	Side grinding wheel used for tool grinding cutters, mills, broaches, reamers, and taps as well as for portable snagging operations.
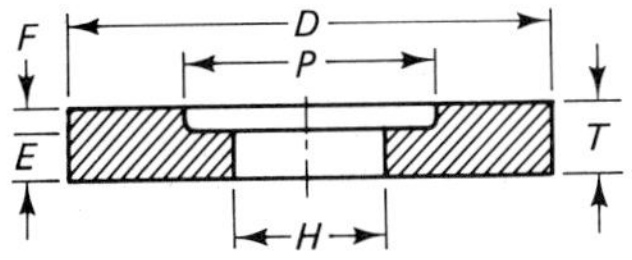 Type 5 Recessed 1 side	Peripheral grinding wheel used for centerless, surface, and cylindrical grinding operations. Recess accommodates mounting flanges and nuts.	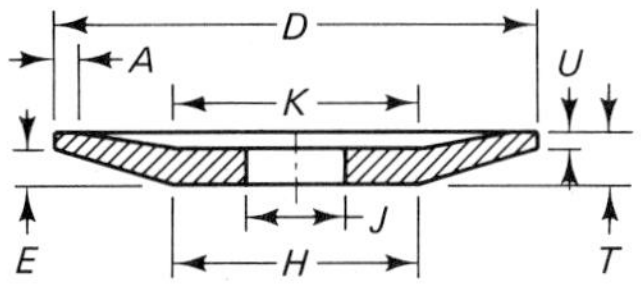 Type 12 Dish	Recommended for side or U-face grinding on cutting tools such as reamers, taps, and broaches.
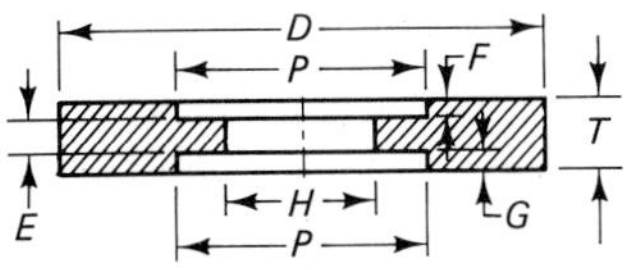 Type 7 Recessed 2 sides	Peripheral grinding wheel used for surface, centerless, and cylindrical grinding. Twin recesses accommodate mounting hardware, allowing more grinding clearance.	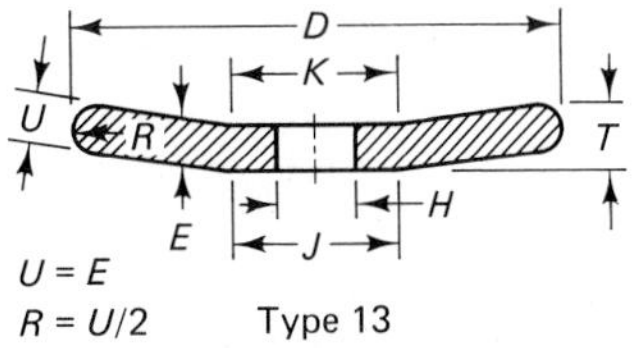 Type 13 Saucer	Peripheral grinding wheel recommended for reaching into narrow recesses to shape and sharpen saw teeth.

There are two general classifications of wheels. Both are classified by the abrasive grains used. One classification involves the use of conventional electric furnaces in the manufacture of natural abrasives. The other classification involves the use of manufactured artificial abrasives such as cubic boron nitride and manufactured diamond.

The electric furnace abrasive grinding wheel includes those composed of aluminum oxide, zirconia alumina abrasive alloys, and silicon carbide. Although the manufacturing process may vary, the main operations in making a grinding wheel conform to the following method.

1. The abrasive grains are precisely weighed and electronically controlled, as is the bonding material (Figure 57-5).

2. Both abrasive and bond are thoroughly mixed and screened until a uniform batch has been produced (Figure 57-6).

3. The mixture is then carefully distributed in steel molds of definite size, shape, and volume.

4. A hydraulic pressing within the steel mold produces a wheel slightly larger than the finished product. It is here that the wheel's density or grain spacing (structure) is formed (Figure 57-7).

5. The fusing of bond to abrasive is achieved through heat application using kilns (high temperature) or ovens (low temperature). Vitrified bonded wheels are produced in kilns in an operation that lasts almost a week. The temperature reaches nearly 2300°F (1260°C) and results in melting the bond to form a hard, shiny coating around the abrasive grain.

6. The grade or exact hardness of each wheel is checked by a number of scientific means such as sonic testing.

7. The arbor hole of the grinding wheel is fitted with bushing, steel back, or other mounting devices that are then turned to the thickness of the wheel.

FIGURE 57-6
Uniformity is accomplished by thoroughly mixing the grain and bond. (Credit: Bay State Abrasives)

FIGURE 57-5
Bond and abrasive grains are carefully weighed prior to mixture. (Credit: Bay State Abrasives)

FIGURE 57-7
Abrasives and bond are pressed in a steel mold by means of a hydraulic press. (Credit: Bay State Abrasives)

8. The wheel is next mounted on a special lathe and turned to the exact size and shape using a diamond tool or special grinding wheel.

9. Next the wheel is subjected to balancing (even-weight distribution) and a truing operation (concentric turning).

10. In a special protected case, the wheel is now speed tested in accordance with ANSI Safety Code B7.1, where rotational speeds reach at least 50% above the recommended operational speed. This margin of safety is added to assure the wheel will not break during recommended operating speeds and conditions.

11. The last operation is the application of stenciled markings on wheels or blotters (Figure 57-8).

Manufacturers' recommendations suggest that wheels composed of aluminum oxide stay sharper longer on hardened steels. Silicon carbide works best on low-tensile materials such as cast iron, while zirconia alumina abrasive works well under high pressure on high-speed grinding of cast and fabricated steels.

Analyzing Wheel Markings

As outlined in Figure 57-9, the makeup and structure of any grinding wheel (with the exception of diamond wheels, which have their own system) can be analyzed once six blocks of information are at hand. The coded information appears on the blotters of small and medium-sized wheels and is stenciled on the side of larger wheels.

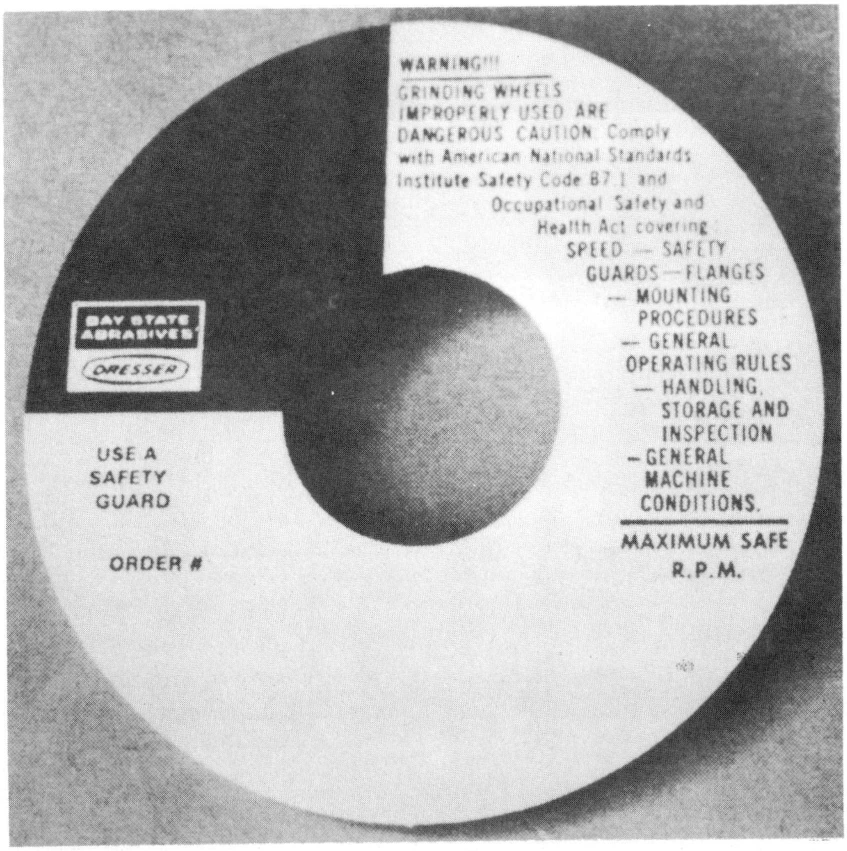

FIGURE 57-8
Stencil markings are one of the last operations performed when making a wheel. (Credit: Bay State Abrasives)

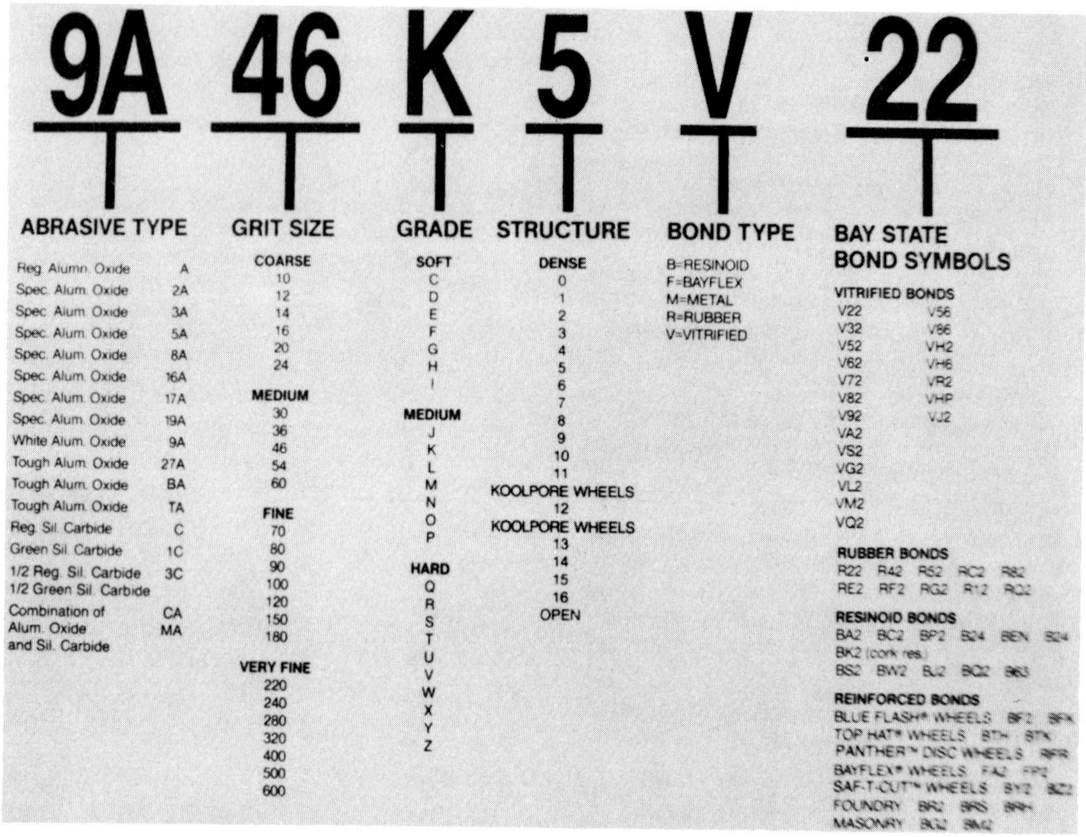

FIGURE 57–9
Abrasive standard marking system. (Credit: Bay State Abrasives)

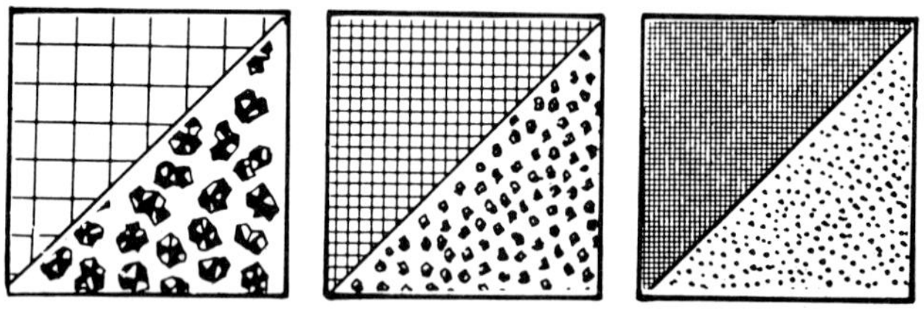

FIGURE 57–10
The grit size can range from 4 to 1000. Shown are sifted grain sizes 10, 24, and 60. (Credit: Bay State Abrasives)

Example:

9A	46	K	5	V	22

9A The letter designates the abrasive type. The letter "A" designates aluminum oxide. The letter "C" designates silicon carbide. The number identifies the specific abrasive type.

46 Designates the grit size. The size of grain abrasive ranges from 4 to 1000 (Figure 57–10).

K Designates the grade or bonding strength and ranges in hardness from A to Z. Letter "A" indicates more bond is used to hold the abrasive grain; letter "Z" provides more resistance to grinding forces (Figure 57–11).

5 Indicates the structure or grain spacing of the wheel, ranging in numbers from 1 to 16. The higher the number, the greater the spacing (Figure 57–12).

V22 The letter designates bond type, such as V for vitrified. The number is a particular company's modification denoting bond symbols.

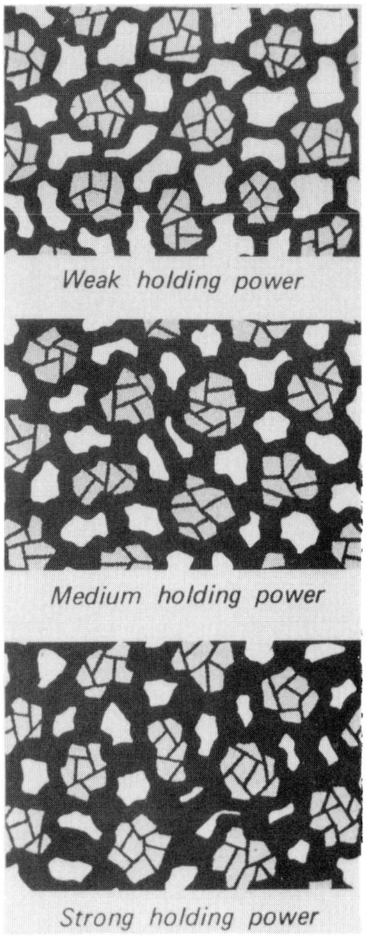

FIGURE 57–11
An illustration of the holding power of different types of bonds. (Credit: Bay State Abrasives)

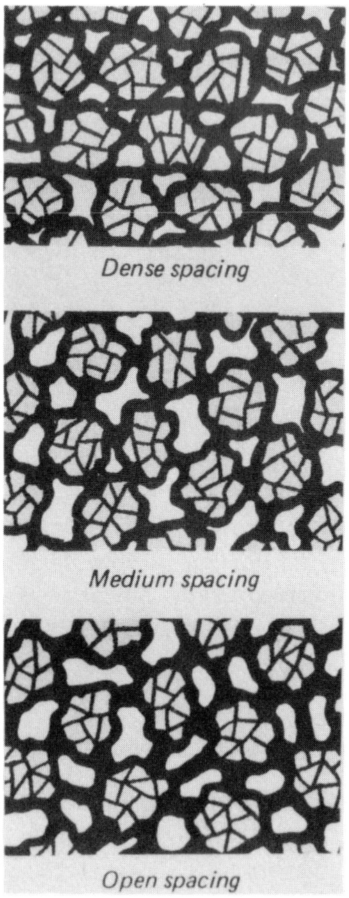

FIGURE 57–12
Grain spacing will vary from dense to open, affecting wheel performance. (Credit: Bay State Abrasives)

Bond Types

The purpose of a bond in a grinding wheel is to hold the abrasive grains together. There are five main types of bonds used for the manufacture of grinding wheels. These include vitrified, resinoid, rubber, shellac, and metal.

Vitrified (V). A ceramic- or glass-type bonding material, vitrified bond is used in most grinding wheels. The bond provides a rigid and brittle support for the abrasive grit, which breaks down easily on the grinding surface of the wheel to expose sharp abrasives during the grinding operation. Vitrified bond is best suited for grinding operations that require high surface finishes and close tolerance dimensions. Vitrified wheels are generally run at relatively low speeds on standard applications, between 6300 and 8500 sf/min (1920 and 2590 m/min).

Resinoid (B). Phenolic resins are bonded in resinoid wheels. They provide a tough, strong bond when fused at a relatively low temperature and can withstand severe grinding stresses. Able to operate at higher speeds than vitrified bonded wheels, resinoid wheels are cool cutting and remove stock rapidly. Resinoid wheels are ideally suited for rough grinding, cutoff operations, snagging, as well as roll grinding and superfinishes.

Rubber (R). Rubber bonded wheels have excellent polishing characteristics and are used to produce a

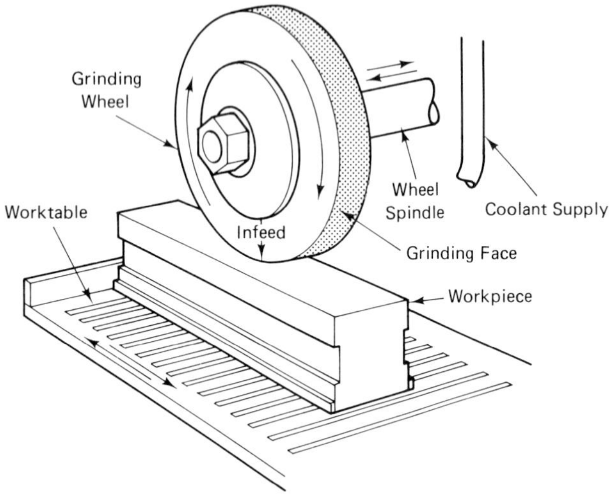

FIGURE 57-13
Illustration of a horizontal spindle grinder with reciprocating table. (Credit: Bay State Abrasives)

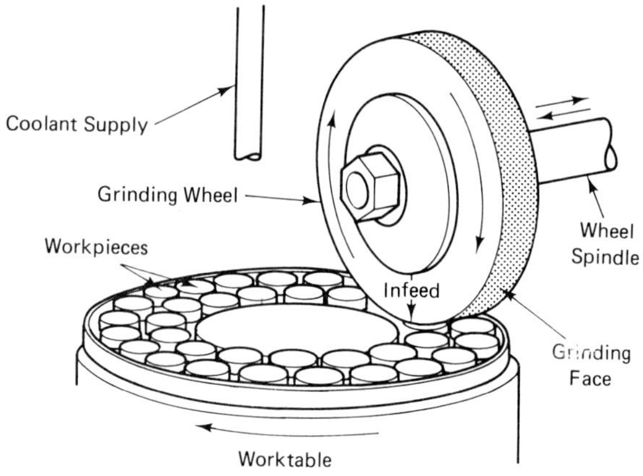

FIGURE 57-14
A horizontal spindle grinder with rotary table. (Credit: Bay State Abrasives)

range of products from knife grinding to ball bearing races. Rubber bonded wheels are also used for cutting-off operations and regulating wheels on centerless grinders.

Shellac (E). Shellac bonded wheels combine qualities of free cutting with high finishes. While limited in use, they are used mainly for cutlery grinding and in roll grinding wheels.

Metal (M). Metal bonds (generally nonferrous) are used in electrolytic grinding operations where an electric current must pass through the wheel. Metal bonds are also used in diamond grinding wheels where the metal's toughness and strength prevents the costly diamond abrasive from falling out.

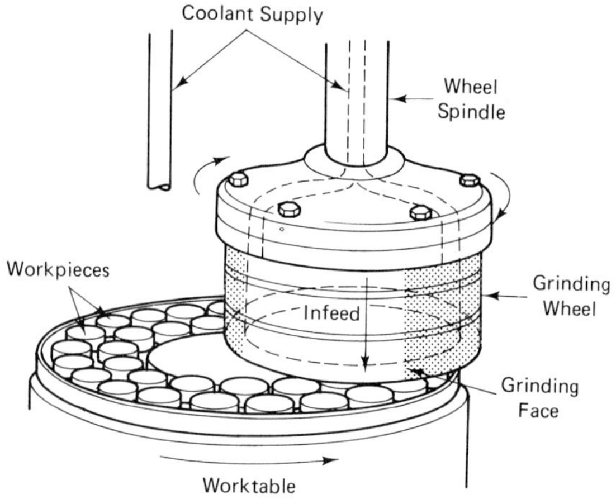

FIGURE 57-15
A vertical spindle grinder with a rotary table. (Credit: Bay State Abrasives)

Types of Grinders

There are many types of grinding operations, some of which require special grinding machines in their operations. The most common grinder found in the machine shop is the surface grinder and there are four common types available. While some of the machines and their operations will be outlined in other units, this unit will cover some of the most common grinding machines and their functions.

The *horizontal spindle grinder with a reciprocating table* (Figure 57-13) is the grinding machine usually found in a toolroom and is called a surface grinder. The surface grinder has a magnetic chuck fastened to a table that can be cross moved in or out as well as traversed either left or right past a grinding wheel. The grinding wheel can be fed in up or down movements.

The *horizontal spindle grinder with a rotary table* is considered one of the most precise means of production surface grinding (Figure 57-14). Workpieces are mounted on a flat magnetic chuck and then rotated beneath the periphery of a straight

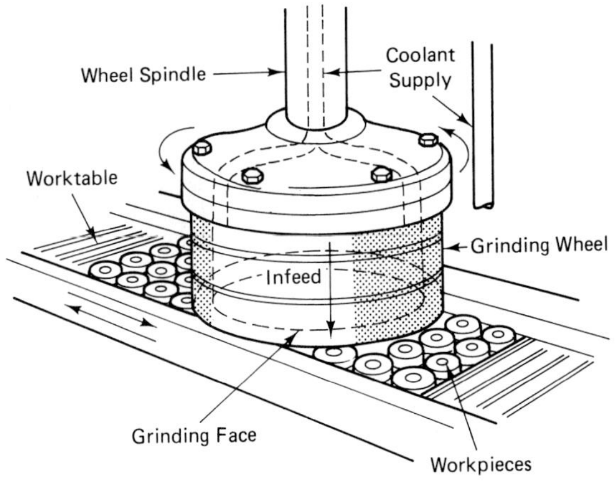

FIGURE 57–16
A vertical spindle grinder with a reciprocating table. (Credit: Bay State Abrasives)

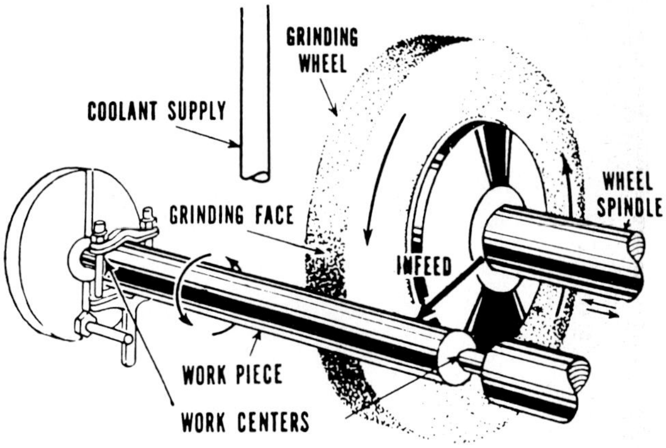

FIGUIRE 57–17
An illustration of the cylindrical grinding machine showing the directions of the workpiece and wheel motions. (Credit: Bay State Abrasives)

grinding wheel that is able to move back and forth across the table. This grinder is capable of grinding parallel faces on single pieces while producing a concentric finish.

The *vertical spindle grinder with a rotary table* is suited to high-production needs (Figure 57–15). The machine has its spindle positioned 90° to the magnetic chuck. The setting creates a greater contact between the wheel and the work surface, permitting heavier cuts to be taken. Grinding is accomplished on the face of the wheel as it moves toward and away from the center of the rotating magnetic chuck. The results are a pattern of overlapping arcs resembling cross-hatching.

The *vertical spindle grinder with a reciprocating table* (Figure 57–16) uses the face of the grinding wheel as the work on the magnetic chuck is slid past it from side to side. By tilting the spindle slightly and forcing the wheel to cut on its rim, a pattern of overlapping curves can be produced that makes an excellent finish for bearing surfaces.

The grinding of outside cylindrical surfaces such as external tapers and angles is usually done in a *cylindrical grinder* (Figure 57–17). In this operation, the workpiece is mounted in a chuck or between centers to be rotated while the periphery of a wheel grinds its outside surface.

The grinding of straight or tapered holes is accomplished in industry with a special *internal grinder* (Figure 57-18). On these machines, the workpiece is made to rotate while its inside is finished by an internal grinding wheel. The reconditioning of milling cutters, end mills, reamers, counterbores, and many other types of cutters to bring back their sharp edges is important in any machine shop. Using various attachments, these tools are set up and resharpened on a *tool and cutter grinder* (Figure 57–19).

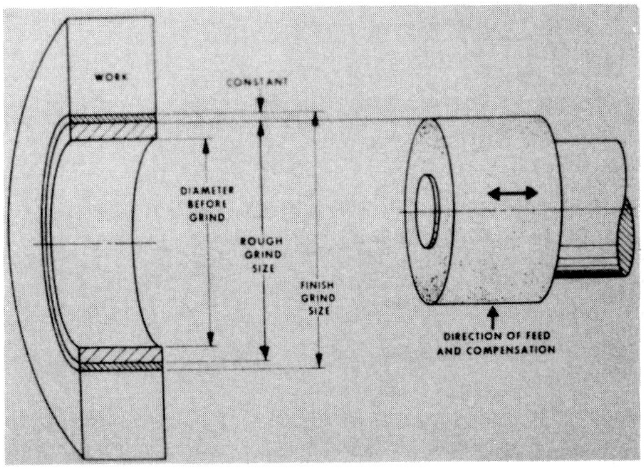

FIGURE 57–18
An internal grinder is used for grinding straight or tapered holes. (Credit: Bay State Abrasives)

FIGURE 57–19
A CBN wheel is used on a tool and cutter grinder for resharpening an end mill. (Credit: DoAll Company)

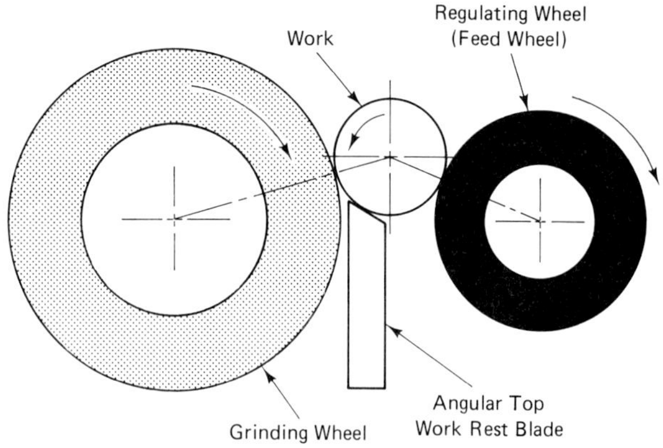

FIGURE 57–20
The workpiece on a centerless grinder rests on a steadyrest and is fed between two wheels, one of which grinds while the other regulates the feed. (Credit: Bay State Abrasives)

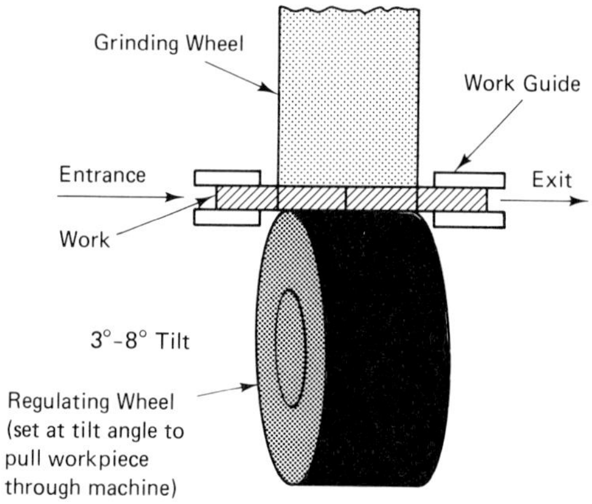

FIGURE 57–21
An illustration of through-feed centerless grinding. (Credit: Bay State Abrasives)

The workpiece on a *centerless grinder* rests on suspended blades and is fed between two wheels, one of which grinds while the other is used as a regulating feed wheel (Figure 57–20). This three-point system is able to produce parts to exact limits of size, shape, and finish.

Methods of Centerless Grinding

There are three primary methods of centerless grinding: through-feed, in-feed, and end-feed grinding.

Through-Feed Centerless Grinding. In through-feed centerless grinding (Figure 57–21), work is passed between the grinding and regulating wheels as it enters through one opening and exits through another. This method is widely used in the grinding industry, but it can only be used on straight cylindrical parts. The workpiece is pulled through the machine, and its movement is controlled by the speed and tilt of the regulating wheel. The tilt is generally 3° but can vary from zero to 8°.

In-Feed Centerless Grinding. In in-feed grinding, work is not traversed during the operation. It is similar in method to plunge cutting done on center-type grinders and will accommodate work having more than one diameter or work that is shouldered or contoured (Figure 57–22). The grinding wheel may be dressed to form for a particular shape in this opera-

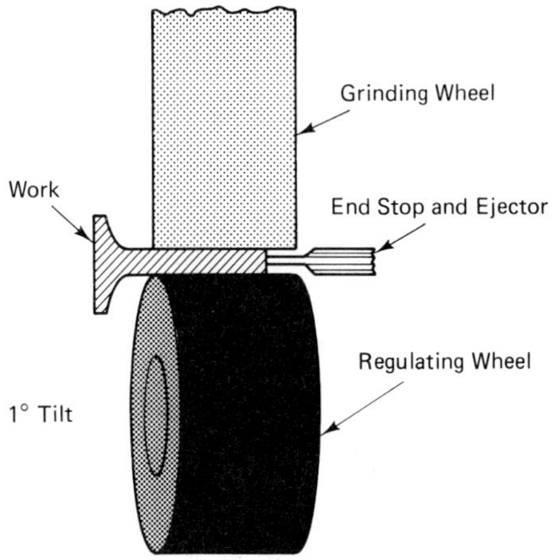

FIGURE 57-22
An illustration of in-feed centerless grinding.
(Credit: Bay State Abrasives)

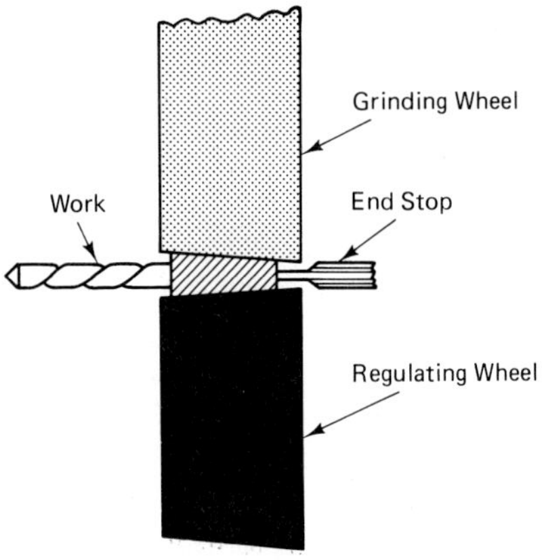

FIGURE 57-23
An illustration of end-feed centerless grinding.
(Credit: Bay State Abrasives)

tion, and generally the length of the grind is not as wide as the grinding wheel. When the grind length is wider than the wheel width, the work is supported on the rest blade on one end, while the other end rests on a roller support mounted on the machine.

The work is entered between the grinding and regulating wheels where it is fed up to a predetermined stop and steadied on the rest blade. At the conclusion of the operation, the work is ejected.

End-Feed Centerless Grinding. The end-feed method (Figure 57-23) is used primarily to grind tapered work. The grinding wheel, the regulating wheel, and the work rest blade all remain in fixed relationship to each other. The work is fed from the front, either manually or mechanically, to a fixed end stop. In all cases, the grinding or the regulating wheels or both must be dressed to the proper taper.

Grinding Fluids

A variety of work is ground dry on smaller types of grinding machines such as cutter grinders and the toolroom surface grinder. On the larger surface grinders used in production, grinding fluids are used in a recirculating manner to provide the following:

- Coolant for reducing the temperature of the work to reduce distortion and increase the surface finish and wheel wear.
- Lubrication between the work surface and the wheel to keep the wheel clean of sticking chips and maintain free cutting.
- A flushing system to wash away grinding debris composed of work chips, particles of bond, and loosened abrasive grains (swarf) and prevent corrosion from forming in the systen.
- An elimination of grinding dust to maintain a healthier environment.

Types of Grinding Fluids

A soluble oil and water mix, better known as *water-soluble oil,* is prepared by adding emulsifiable oil to water. A ratio of at least 30 : 1 (30 parts water to 1 part oil) is mixed to form a solution that is cloudy or milky white in appearance. These solutions have better cooling properties than cutting oil as well as excellent lubricating and corrosion-resistant properties.

Synthetic cutting fluids are a mixture of water and water-soluble chemical compounds. Properly mixed in deionized water, these solutions have excellent cooling properties and can be more economical to use than oil base cutting fluids. Additives are also used to inhibit corrosion and increase its resistance to skin irritation.

Straight oil grinding fluids are superior to other grinding fluids for lubricity properties. While not as effective as other grinding fluids for heat transfer qualities, straight oil grinding fluids are widely used as coolants for thread grinding, where high accuracy and fine finishes are required. In this operation, a steady flow of oil is directed toward the work and wheel at the point of contact.

Seeded-Gel Abrasives

Grinding technologies have developed an advanced abrasive that meets the requirements of precision production grinding operations needed for high-temperature alloys and other hard-to-grind materials. A non-superabrasive product, "SG" (seeded-gel) abrasives feature faster metal removal rates than conventional aluminum oxide.

Aluminum oxide is made in a fused process where it is heated in a furnace, allowed to cool, and then crushed to a desired particle size. Composed of ceramic aluminum oxide, SG abrasives are made in a sintered process and produce a grain that has a fine, keen-cutting edge. Each abrasive has a submicrometer crystal structure with billions of particles in each grain. This results in a durable, self-sharpening abrasive during grinding where the grains stay sharp as they fracture to expose new cutting edges. This self-sharpening feature also minimizes burning as well as metallurgical damage. The uniformity of grain size helps the wheel resist loading and wear, maintaining cutter efficiency and surface finishing. Longer wheel life has resulted in fewer wheel changes, lowering labor costs while increasing more parts per hour.

An SG abrasive is suitable for both ferrous and nonferrous metals with a hardness range up to R_c 67. Seeded-gel wheels greatly outperform the conventional aluminum oxide wheels, outwearing them as much as 10 : 1 when used in oil, as much as 5 : 1 when used dry, and as much as 3 : 1 in wet applications.

UNIT 58 THE SURFACE GRINDER

In the toolroom, workpieces that have been previously machined in a milling machine generally are finished to size on a horizontal spindle reciprocating table surface grinder, more commonly referred to as a surface grinder. The surface grinder can machine

FIGURE 58–1
Hydraulic-operated surface grinder. (Credit: DoAll Company)

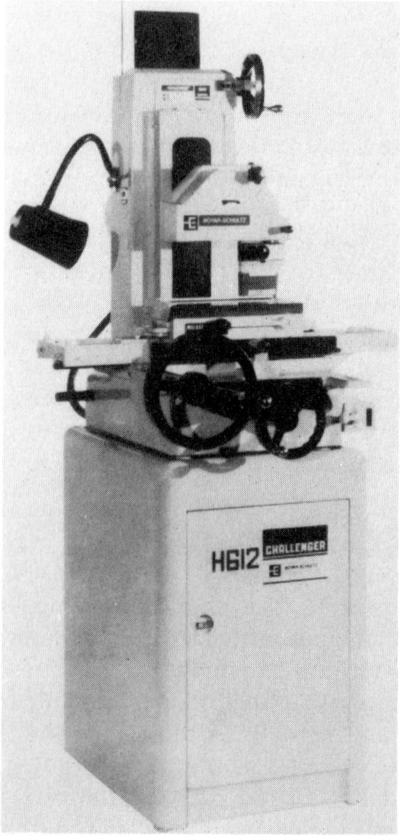

FIGURE 58–2
Manually operated surface grinder. (Credit: Boyar–Schultz Corporation)

FIGURE 58-3
Short workpieces should be blocked on the magnetic chuck as a safety measure.

FIGURE 58-4
A grinding wheel should be ring tested for cracks by tapping four different places, 90° apart, about $2\frac{1}{2}$ in. (63 mm) from the circumference.

parts to very close tolerances and produce a high-quality finish at the same time. The surface grinder is the most commonly used machine for precision grinding to be found in a machine shop. Surface grinders are available hydraulically operated (Figure 58-1) or manually operated (Figure 58-2). This unit will concentrate on the hand-operated surface grinder.

Objectives
After completing this unit you should be able to

1. properly mount a surface grinding wheel.
2. true and dress a grinding wheel.

Safety

The general safety rules outlined in the last unit also apply to any surface grinding operation. There are other general safety rules that should also be observed on a surface grinder.

1. Whenever possible, block short workpieces with parallels as extra protection against slippage on a magnetic chuck (Figure 58-3).
2. Always make sure the right type of wheel is being used for the job.
3. Make certain that the wheel is properly ring tested for fracture before mounting (Figure 58-4).
4. Make certain that the wheel is properly mounted before attempting grinding.

FIGURE 58-5
Any raised edges or burrs on the magnetic chuck surface should be smoothed out with an oilstone.

5. Check the magnetic chuck surface for raised edges or burrs and stone if necessary (Figure 58-5).
6. Do not stand in line with the rotation of the wheel when dressing it.
7. Always stop the machine spindle and move the workpiece clear of the wheel before attempting to remove the workpiece from the magnetic chuck.

Parts of a Manually Operated Surface Grinder

The design of the *base* is a boxlike cast-iron construction, the front of which is covered by a door. Behind the door may be found shelved storage areas.

The *saddle* is found to operate on hand-scrapped or machined ways mounted on top of the base. The saddle supports the *table,* upon which the *magnetic chuck* is mounted. The magnetic chuck is the principal means of holding work for grinding. The saddle may be moved toward or away from the column.

The ways of the table are situated at 90° to those on the base. This enables the table to move across the upper ways of the saddle in a reciprocating motion (side to side) past the grinding wheel, while the table and saddle assembly moves toward or away from the column (in or out) on the ways of the base.

The *spindle housing* and *wheelhead* operate on ways mounted on the *column,* which is located on the back of the frame. The wheelhead is raised or lowered vertically by means of a *wheel feed handwheel.* By turning the wheel feed handwheel, the grinding wheel can be set to a depth of cut.

Stop dogs are mounted on the front side of the table to limit the length of table travel during grinding. On power-fed surface grinders, the stop dogs are positioned to strike the table reverse lever to change the direction of the table travel at the end of each cut.

By manually turning a crossfeed handwheel, the table may be fed toward or away from the grinding wheel. On power-fed surface grinders this too can be set to move automatically at the end of each cutting stroke.

Magnetic Chucks

In grinding material on a surface grinder, work may be fastened to the table or clamped in a vise or fixture and bolted to the table. Generally, ferrous metals are held directly by means of a magnetic chuck that is clamped to the table. Magnetic chucks are available in different types, sizes, and shapes. The two types are the electromagnet chuck and the permanent magnetic chuck. They both work on the same principle of operation, by exerting magnetic force on the ferrous material.

The permanent magnetic chuck (Figure 58-6) houses a grid or magnetic pack that contains a series

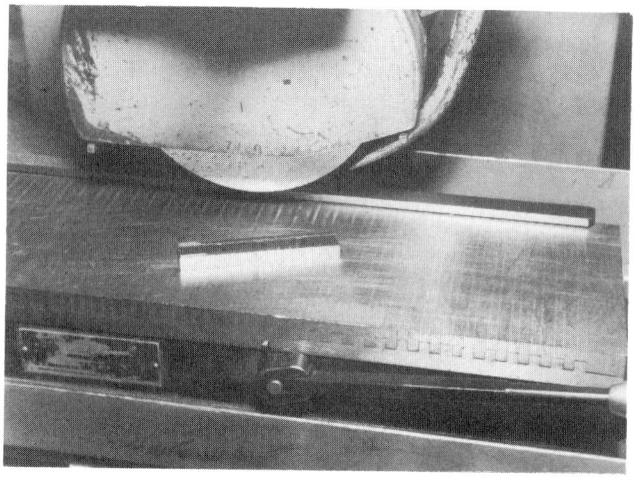

FIGURE 58-6
A permanent magnetic chuck.

of permanent magnets that are separated by grid conductor bars. This grid is located in a holding case that can be moved longitudinally by turning a handle on the outside of the chuck in an "on" or "off" position. The top surface of the magnetic chuck contains a series of oblong loops called inserts or pole pieces, which are separated from the rest of the plate by nonmagnetic metal. This separation provides a circuit to conduct a magnetic line of force. When ferrous work is placed on the magnetic chuck and the handle is turned clockwise to the on position, the grid conductor bars and inserts are placed in alignment. This permits the magnetic line of force to pass through the work, holding it down against the surface of the magnetic chuck. When the handle is turned counterclockwise to the off position, the grid conductor bars and inserts remain out of alignment, releasing any magnetic line of force. While the work may be removed from the magnetic chuck at this time, there will still be some residual magnetism left in the workpiece.

The electromagnetic chuck (Figure 58-7) uses electricity to provide a magnetic line of force. The electromagnetic chuck can use current from any direct-current circuit, not alternating current, and retain its holding power as long as the current is applied. Once the power is removed, the workpiece can be removed easily. There is a tendency for the workpieces to retain little of their residual magnetism when using an electromagnetic chuck.

A third chuck, the vacuum chuck (Figure 58-8), has one big advantage not enjoyed by either the permanent magnetic chuck or the electromag-

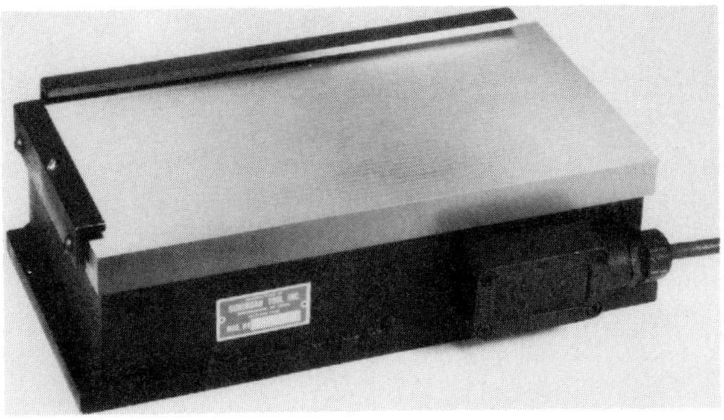

FIGURE 58-7
The electromagnetic chuck uses direct-current electricity to provide a magnetic line of force. (Credit: Suburban Tool, Inc.)

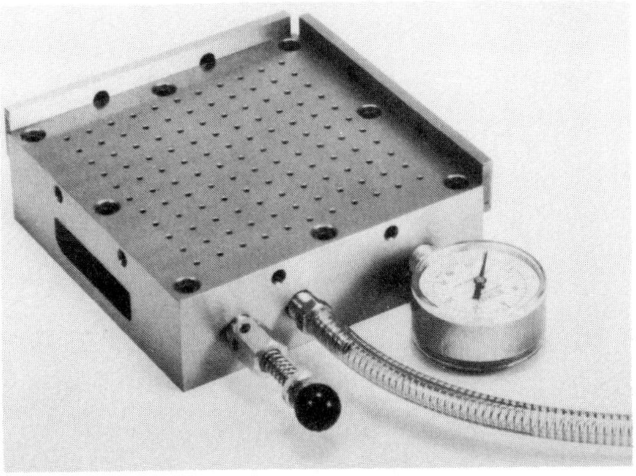

FIGURE 58-8
A vacuum chuck can hold ferrous, nonferrous, or nonmetallic materials directly to its surface. (Credit: Suburban Tool, Inc.)

netic chuck. It can hold ferrous, nonferrous, or nonmetallic materials directly to its surface without the aid of accessories. By blocking off the unused holes in its top plate with masking tape and pumping out the air directly under the workpiece, a strong suction is created with a holding power directly proportional to the area of the workpiece. Using the average vacuum of 12 psi (pounds per square inch), the total holding force for a work surface 2 × 3 in. would be 2 × 3 × 12 psi, or 72 lb. Work surface areas as small as 2 in. (50 mm) square can be held in this manner.

Cutoff (rubber, shellac, resinoid)	*9000–16000 sf/min
Cutter grinding	4500–6000 sf/min
Cylindrical grinding	*5000–12000 sf/min
Disc grinding	4000–5500 sf/min
Internal grinding	*4000–12000 sf/min
Surface grinding	4000–6500 sf/min
Tool grinding	5000–6000 sf/min

*Higher speeds may be permitted where safety devices, wheel bearings, and machine rigidity are adequate. In no case should wheels be operated faster than the maximum operating speed marked on the wheel.

FIGURE 58-9
Efficient grinding wheel operating speeds.

The vacuum chuck can be clamped to a machine tool table as well as to a fixture or sine plate device.

Grinding Wheel Speeds

As with any cutting tool, there is a factor of speed consideration. How fast should a grinding wheel be turning? Too slow a speed may result in abrasive grain wastage without getting the full work use of the wheel. Conversely, a hard cutting action can result from excessive speed, leaving the work surface burned in its wake. There is danger of the wheel breaking or being damaged. In general, vitrified bonded grinding wheels are well suited to general-purpose and precision grinding up to 12,000 sf/min. This speed, however, will depend on the machine's rigidity and operating conditions. For efficient cutting, vitrified wheels are recommended to operate at speeds between 6300 and 6500 sf/min.

Resinoid bonded grinding wheels use a thermosetting resin in its bonding material that gives them a greater toughness. Resinoid grinding wheels can be operated at maximum speeds up to 16,000 sf/min. The efficient cutting speed recommended for resinoid grinding wheels is around 9500 sf/min. In any event, grinding wheels should not be operated faster than the speed marked on the wheel by the manufacturer (Figure 58-9).

Grinding Wheel Feeds

There are two basic types of feeds in surface grinding: table feed and crossfeed. On surface grinders equipped with an automatic feed, table travel can be adjusted for the cut required. On dry-type surface grinders found in most toolrooms where the table reciprocates by turning a table handwheel, the table should move fast enough to prevent burning the workpiece surface being ground. Better finishes can be produced by taking a light depth of cut and increasing the rate of longitudinal table travel.

The crossfeed works in conjunction with the table feed. This too can be set to automatically feed on machines so equipped and should be set to move at the end of each cutting stroke. The amount of movement will depend on the type of finish to be produced. Generally, a crossfeed is set to move not more than 0.100 in. (2.54 mm) for rough cutting. The amount can be reduced as much as 75% for finish cuts. A rule of thumb would be 0.020 to 0.030 in. (0.50 to 0.76 mm) cross-cut feed.

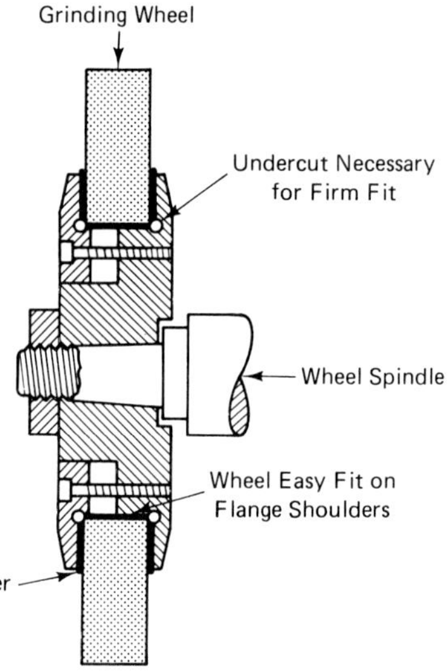

FIGURE 58–10
Blotters slightly larger than the flanges should be used on either side of the wheel. (Credit: Bay State Abrasives)

Mounting a Grinding Wheel

When attempting to mount any grinding wheel, protective safety glasses should be worn and the following basic rules of safety should be observed:

1. Use a wooden handle; always tap the selected wheel for cracks before mounting. If it does not ring true, the wheel is cracked and should not be used.
2. Never exceed the maximum speed printed on the wheel's blotter.
3. The wheel's center hole should fit snugly on the grinder's spindle without alteration or forcing.
4. The flanges should be at least one-third the wheel's diameter and free of burrs to fit at 90° to the axis used on mounting.
5. Use blotters that are slightly larger than the flanges on either side of the wheel (Figure 58–10).
6. Tighten the spindle nut only tight enough to hold the wheel firmly without over tightening.
7. Wheel guards should cover at least one-half of the grinding wheel and be closed before operating the machine.

FIGURE 58–11
A single-point diamond may be set at a 15° angle with the base and positioned pointing in the same direction as the wheel's rotation when dressing.

Truing and Dressing the Wheel

Truing and dressing the wheel are two different terms but are often used interchangeably. Truing is the removal of material from an uneven rotation, identified as "wheel bounce" or high spots, so as to make the wheel run concentric with the grinder spindle. Truing should always be done to a new wheel before it is used for grinding. A remounted wheel should also be trued before it is used.

Dressing is the restoration of a wheel's grinding face to expose a new, sharp cutting surface, removing any loading or dullness in or between the abrasive grains. Single-point diamond tools are the

FIGURE 58–12
A multipoint diamond dresser.

most widely used dressing tools. For greater efficiency, these tools may be mounted in a holder at a 15° angle with the base and should be pointed in the direction of the wheel's rotation when being used (Figure 58–11). Where once a single-point diamond was used, multipoint diamond dressers are now in common use (Figure 58–12). Manufacturers also produce a very strong single-point man-made diamond for wheel dressing. The diamond is thermally stabile and more resistant to wearing, dulling, or fracturing.

Procedure to True a Wheel

1. If using a single-point diamond dresser, examine the point for wear and turn it in the holder, if necessary, to expose a sharp cutting edge.
2. Thoroughly clean the magnetic chuck with a cloth and wipe away any remaining grime with your hand.
3. To prevent scratching the chuck in removal, place a piece of paper slightly larger than the dresser base on the magnetic chuck.
4. To prevent scratching and wear to the table caused by a stream of abrasive grains during dressing, place the single-point or multipoint diamond dresser on the left end of the magnetic chuck to span as many magnetic inserts as possible (Figure 58–13).
5. Energize the chuck.
6. With the dresser in front of the wheel, move the table so that the diamond is to the left of the centerline of the grinding wheel by about $\frac{1}{4}$ in. (6 mm).
7. Turning the wheel feed handwheel, raise the grinding wheel to a height of about $\frac{1}{32}$ in. (0.79 mm) above the diamond.

FIGURE 58–13
Place the diamond dresser as far left as possible on the chuck for dressing purposes.

8. Start the grinder spindle.
9. Using the hand crossfeed handle, move the diamond in and out beyond the revolving wheel.
10. Carefully lower the wheel in small increments as the diamond is passed underneath it.
11. At the first sign of dust particles, stop lowering the wheel and advance the crossfeed to clear the diamond dresser from under the wheel.
12. Lower the wheel 0.0005 in. (0.01 mm) per pass. For rough grinding, a fast rate of crossfeed should be used. To obtain a fine finish, use a slow rate of crossfeed without setting a downfeed on the last pass.
13. To obtain a sharper wheel face, a slightly heavier downfeed may be necessary.

UNIT 59 GRINDING A FLAT SURFACE

There are many types of operations carried out on a surface grinder such as form, angular, and vertical grinding. The most common operation performed on a surface grinder is grinding a flat surface. Achieving good results when grinding a flat surface requires a safe setup on the surface grinder as well as a properly selected and sharp cutting wheel to do the job.

Objectives

After completing this unit you should be able to

1. mount a workpiece on a magnetic chuck safely.
2. grind a surface flat on a surface grinder.
3. demagnetize any residual magnetism from work or accessories used on the magnetic chuck.

Work Holding Devices

Not all work to be ground can be held directly on the magnetic chuck. There are times when the work is fastened in a fixture or device that is clamped directly to the table. Other situations dictate some type of work holding accessories be used, either singularly or in conjunction with other tools, on the magnetic chuck.

Some accessories are used for a variety of setups as well as holding nonmagnetic materials such as brass, bronze, copper, stainless steel, aluminum, and other nonferrous metals. The use of hardened *V-blocks and clamps* (Figure 59–1) is a practical example. The *toolmaker's grinding vise* (Figure 59–2) is frequently used for holding a number of shapes and contours. The vise has its sides perfectly square with the holding surface jaws, which allows the vise to be turned on its side. Any workpiece that is held and extended beyond the edge of the jaws can have two sides ground square with each other without disturbing the hold.

The *magnetic V-block* (Figure 59–3), with its alternating sections of brass and soft iron, is highly magnetic. The construction may also be made of alternating layers of soft iron and fiber, producing a holding effect to nearly 80% of the magnetic chuck's holding power. Magnetic V-blocks lend themselves

FIGURE 59–1
Using V-blocks and clamps to hold nonferrous material for grinding.
(Credit: L. S. Starrett Company)

for holding square, round, and octagonal shaped bar stock for light surface grinding.

Magnetic parallels (Figure 59-4) also are of a laminated construction that makes them useful for holding surfaces that might have extensions or other projections on the side to be held. Magnetic parallels are generally used in pairs. Both the parallels and V-blocks should be stored in a wooden box and covered to prevent nicking and burring when not in use. Their flatness or holding surfaces can be restored by taking a few light finishing cuts under coolant. Overheating these surfaces to any degree while grinding should be avoided if accuracy in flatness is to be achieved.

The grinding of angles presents problems that are solved by various holding devices. One of the most basic setups is the use of the angle plate with work clamped to its surface (Figure 59-5), either for grinding one surface square with its side or for creating an angular slope. Another setup for grinding an-

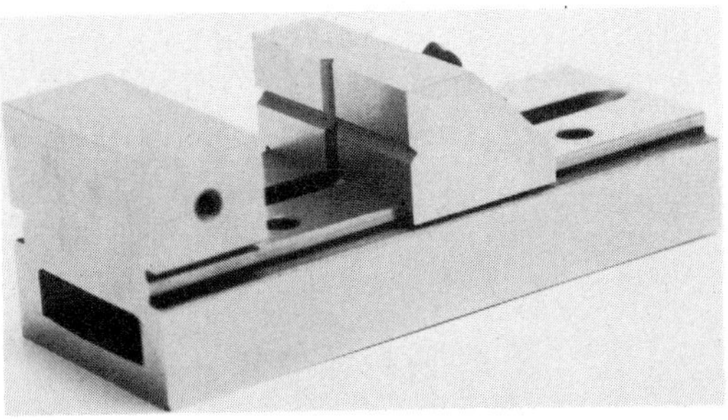

FIGURE 59-2
The toolmaker's grinding vise has its sides perfectly square with the holding surface of the jaws. (Credit: Suburban Tool, Inc.)

FIGURE 59-4
Magnetic parallels require perfectly flat surfaces in order to achieve flatness during grinding. (Credit: MTI Corporation)

FIGURE 59-3
Magnetic V-blocks produce a holding effect to nearly 80% of the magnetic chuck.

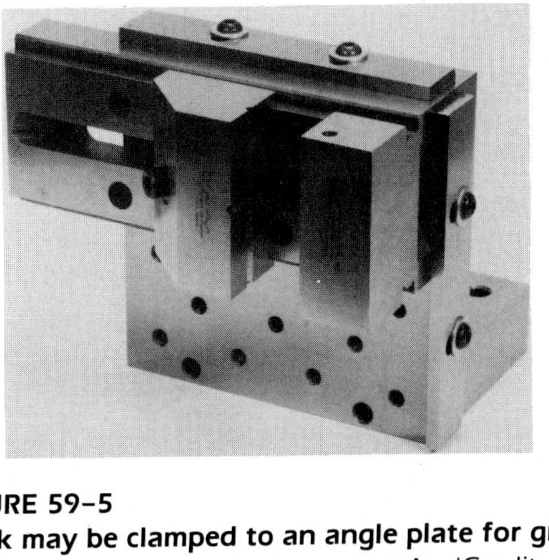

FIGURE 59-5
Work may be clamped to an angle plate for grinding the edges square or on an angle. (Credit: Suburban Tool, Inc.)

FIGURE 59–6
Angular flats on the end of short rods or shafts can be produced by using a toolmaker's V-block clamped to an angle plate. (Credit: L. S. Starrett Company)

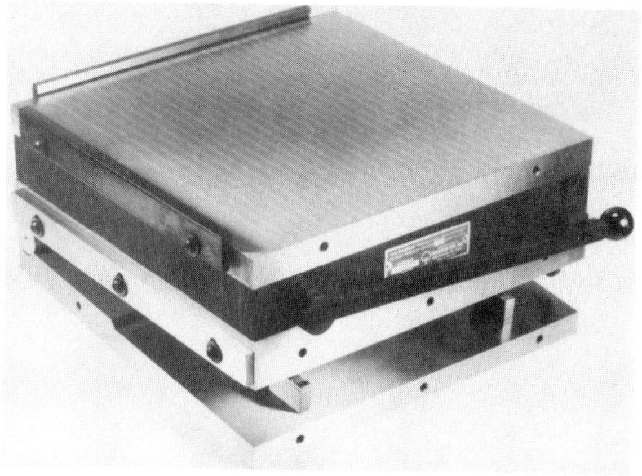

FIGURE 59–8
A magnetic chuck can be bolted to a sine plate to form a magnetic sine plate. (Credit: Suburban Tool, Inc.)

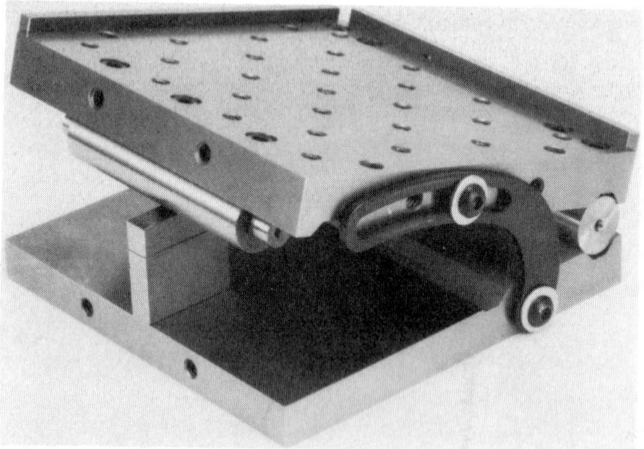

FIGURE 59–7
A wide-area sine plate. (Credit: Suburban Tool, Inc.)

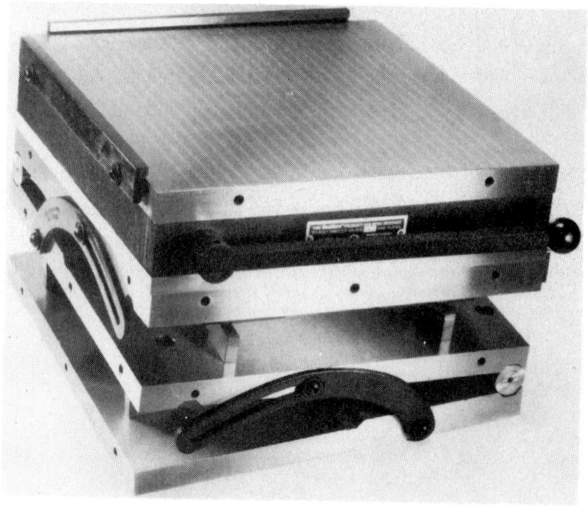

FIGURE 59–9
A magnetic compound sine plate allows an operator to grind compound angles. (Credit: Suburban Tool, Inc.)

gular surfaces involves the clamping of a multipurpose *toolmaker's V-block* to an angle place (Figure 59–6). This is especially useful for grinding flat surfaces on the ends of short rods and shafts.

A more accurate setup for grinding angles can be found in a *sine plate,* which has a wider surface than the sine bar (Figure 59–7) and is set up in the same manner. The work surface is elevated to its proper height by means of precise gage blocks. A one-unit-construction magnetic sine plate is available, as is a magnetic chuck that can be bolted directly to the sine plate to form a *magnetic sine plate*

FIGURE 59–10
A sine vise can be set on gage blocks to allow grinding of accurate angles. (Credit: Suburban Tool, Inc.)

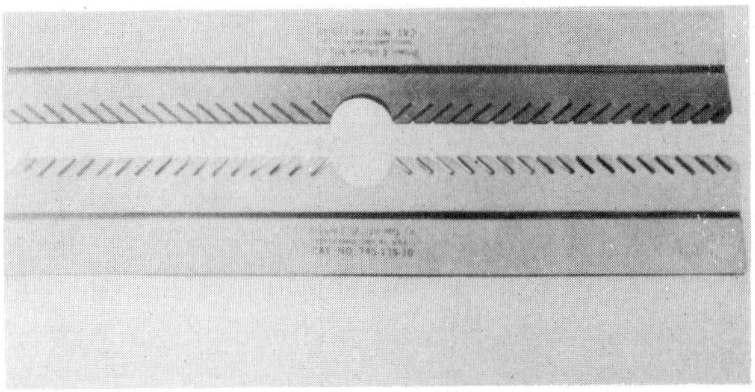

FIGURE 59–11
Tooth clamps make it possible to hold nonmagnetic metals on a magnetic chuck. (Credit: Brown & Sharpe Manufacturing Co.)

(Figure 59–8). By bolting a magnetic chuck to a compound sine plate, it is possible to form a *magnetic compound sine plate* (Figure 59–9), permitting an operator to grind compound angles. For holding and grinding smaller workpieces on an angle, sine vises (Figure 59–10) are available. Built with the precision of a toolmaker's vise, very small angles can be set and ground using standard gage blocks.

The availability of *tooth clamps* (Figure 59–11) makes possible the holding of nonmagnetic work or workpieces with small holding areas directly on a magnetic chuck. The workpiece is prevented from sliding by a series of comblike steel teeth. The construction of each tooth clamp consists of two bars of steel, one solid and one bearing a row of teeth along its length. Both bars are connected with a piece of spring steel, which raises the row of teeth slightly higher than the level of the solid bar. Holding is started by positioning the solid edge of one of the clamps along the back rail and above the face of the magnetic chuck. The workpiece is then placed on the chuck against the row of teeth and nested by placing the second row of teeth against the opposite edge. When the magnetic chuck is energized, the jaws of the clamps are pulled down against the magnetic chuck, trapping the workpiece in position.

One of the simplest ways of holding flat, thin nonmagnetic workpieces for light grinding on a magnetic chuck is by means of *double-faced tape* strips placed on the magnetic chuck. When the protective layer is peeled off, it allows the workpiece to be pressed on (Figure 59–12).

FIGURE 59–12
Holding thin flat work down with double-faced tape.

Mounting Work on a Magnetic Chuck

When work is mounted on a magnetic chuck, it should be clean of dirt and oil and free of burrs. Once mounted and the chuck energized, the operator should not assume that the work is being held securely. It should be tested by attempting to take it off.

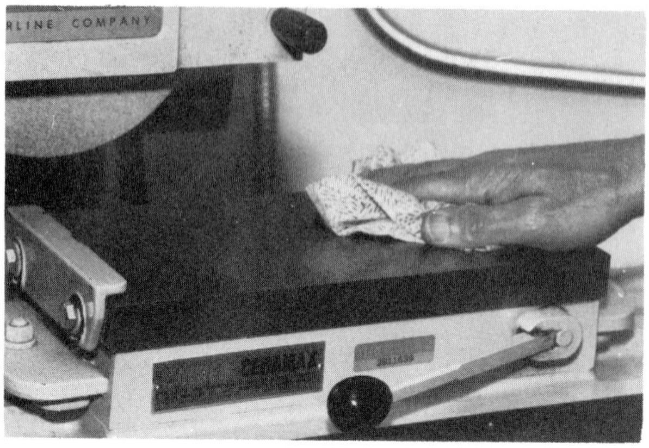

FIGURE 59-13
Wipe the magnetic chuck clean after dressing the wheel.

FIGURE 59-14
Place the work on a piece of paper slightly larger than its base.

Procedure

1. Wear safety glasses before doing any grinding.
2. Dress the wheel as explained in Unit 58.
3. Wipe the magnetic chuck surface clean with a soft cloth to remove any abrasive grains (Figure 59-13).
4. With a fine honing stone, remove any raised edges or burrs from the magnetic chuck surface.
5. Reclean the surface of the chuck.
6. Remove any burrs from the work surface and clean it free of dirt and grease.
7. Place a piece of paper slightly larger than the workpiece in the center of the table to straddle as many magnetic inserts as possible.
8. Place the workpiece on the paper (Figure 59-14).
9. Press the edges of the work to test if it is resting flat. If not, shim it with paper to prevent rocking. *Note:* If the workpiece is not large enough to span three magnetic inserts, block the work on both sides with a set of parallels less than the thickness of the work.
10. Move the handle to the on position to energize the chuck.
11. Move the surface grinder table to clear the work from the stationary grinding wheel.
12. With both hands, try to remove the workpiece from the magnetic chuck.

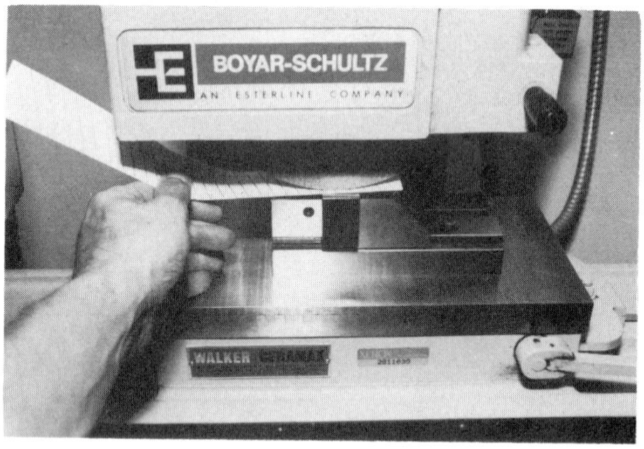

FIGURE 59-15
Lower the stationary wheel until it just touches the paper.

Setting the Wheel to the Work

Once the workpiece has been securely mounted on the magnetic chuck, the next operation is setting the wheel to the work surface.

Procedure

1. Establish the length of table travel by setting the table reverse dogs so that the grinding wheel can clear each end of the workpiece by about 1 in. (25 mm).

2. Take a sheet of paper and tear it in half lengthwise so it can be folded at least four times to produce a narrow strip about 1 in. (25 mm) wide and place it on the work surface.

3. Lower the stationary grinding wheel until it begins to touch the folded paper (Figure 59–15).

4. Move the table using the crossfeed handle until the wheel just overlaps the workpiece edge by about $\frac{1}{8}$ in. (3 mm).

5. Start the grinding wheel spindle.

6. As the table is moved by hand from side to side under the spinning wheel, carefully lower the grinding wheel in small increments until it sparks a high spot.

7. Move the table to the end of the cutting stroke and prepare to take a trial cut.

Trial Cut

In surface grinding, conditions may be such that just sparking the work may not be the height the wheel should be set. For example, a surface may be rough with hills and valleys, in which case sparking on one end may only represent a high spot in one area. An average wheel setting should be established by mentally dividing the workpiece in three general areas, with each area being tested for sparking. The readings of all three areas should be noted, for it will approximate the height of the wheel to start the trial cut.

Procedure

1. Raise the grinding wheel about 0.005 in. (0.12 mm) above the initial sparking.

2. Start the grinder spindle.

3. Move the table to locate area 1 under the wheel.

4. Carefully lower the grinding wheel as the table is moved from side to side until it sparks and compare this reading with the initial setting.

5. Raise the grinding wheel 0.005 in. (0.12 mm) to test areas 2 and 3 in the manner described in steps 3 and 4.

The highest setting of the grinding wheel is where the trial cut should be started to grind the surface.

Rough and Finish Grinding the Surface

Once the highest setting of the grinding wheel has been established, rough grinding the work surface can begin. In general, the crossfeed handwheel of the surface grinder should never be moved more than one-half the width of the grinding wheel face. The higher the crossfeed, the greater the wheel pressure is against the work surface, and unless coolants are used, it will result in higher heat. Decreasing the crossfeed will permit a higher surface finish. It can be seen that the amount of stock to be removed, the finish desired, and the capacity of the machine (power, bearings, looseness in the ways) will control the amount of crossfeed movement.

Procedure

1. With the grinding wheel in motion and set to its highest setting, move the crossfeed handwheel to take a 0.50-to-0.080-in. (1.27-to-2.03-mm) cut (Figure 59–16).

2. Set the table to travel automatically, or if using a manual model, turn the table handwheel just fast enough to prevent burning the work surface as the table is made to travel from side to side under the grinding wheel.

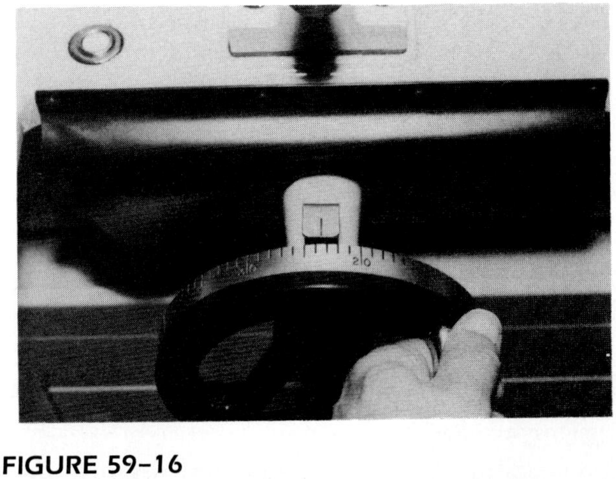

FIGURE 59–16
Advance the crossfeed handwheel to take a cut of about 0.050 to 0.080 in. (1.27 to 2.03 mm).

3. Rough grind the work surface, moving the crossfeed handwheel 0.050 to 0.080 in. (1.27 to 2.03 mm) with each pass of the wheel until the work surface has been completely ground.

4. Examine the work surface for finish. If the work surface has not been completely cleaned up or there remains unwanted hollows, lower the grinding wheel about 0.002 to 0.003 in. (0.05 to 0.07 mm) deeper if using a dry grinder or 0.005 to 0.015 in. (0.12 to 0.38 mm) if using coolants and reverse the crossfeed handwheel movement to complete grinding the surface.

5. Dress the wheel if necessary to improve the grinding finish.

6. Finish grinding the work taking cuts 0.0005 to 0.001 in. (0.01 to 0.02 mm) deep and crosscut feeds ranging from 0.005 to 0.015 in. (0.12 to 0.38 mm) per pass.

7. Upon completion of grinding, move the work clear of the grinding wheel.

8. Shut off the machine.

9. Remove the workpiece.

10. Clean the magnetic chuck.

11. Prepare to demagnetize the work.

FIGURE 59–17
A demagnetizer should remove any residual magnetism from the work articles held on a magnetic chuck.

Demagnetizing

Once the workpiece has been ground and removed from the magnetic chuck, all the parts used during the grinding operation should be demagnetized. By subjecting a workpiece to what first sounds like a steady humming when the demagnetizer is turned on (Figure 59–17), the article will vibrate in movement as it passes over, in and out of contact with iron plates at the top of the unit. While it may require several passes, nearly all of the residual magnetism can be eliminated in this manner. To avoid scratching, a sheet of paper should be placed on the demagnetizer before attempting to remove any magnetism.

UNIT 60 GRINDING A BLOCK SQUARE AND PARALLEL

Producing sides square and parallel on a workpiece is just as common to a grinding operation as it is to any other machine tool operation. Once the operations of mounting, truing, and dressing a wheel safely can be carried out, grinding a workpiece square and parallel can be accomplished using proper holding devices. Regardless of the holding device used, a procedure should be followed.

Objectives
After completing this unit you should be able to

1. grind a workpiece square and parallel using a grinding vise.

2. grind a workpiece square and parallel using an angle plate.

The Use of a Grinding Vise

The grinding vise is one of the most flexible means of grinding a workpiece square and parallel. Not only does it hold work firmly between its hardened and lapped vise jaws, but also its sides are perfectly ground and lapped square with the jaws. This will allow two surfaces that extend above and laterally beyond the vise jaws to be ground square without disturbing the setup. Once the work surface extending above the jaws is ground flat, the vise is turned 90° on its side to permit the second side to be ground.

While bar stock with four uneven, sawcut sides will be used to demonstrate the method of grinding square and parallel (Figure 60–1), one of the biggest problems to overcome will be to establish the initial

FIGURE 60-1
The demonstration will take place on bar stock with four sawcut sides.

FIGURE 60-2
In addition to the grinding vise, a square head with a spirit level and a soft-faced mallet may be used.

surface to be ground flat. For this purpose, additional tools used will include a square head with a spirit level, a soft-faced mallet, and an indicator mounted on a surface gage (Figure 60-2). Once the surface grinder has had its grinding wheel trued and dressed, proceed as follows.

Grinding Sides 1 and 2

Procedure

1. Clean the workpiece of any grease and dirt and deburr any sharp edges (Figure 60-3).
2. Place the workpiece in the grinding vise with one of its sides extending beyond the edge of the vise about $\frac{1}{4}$ in. (6 mm) and snug up the handle.
3. Set the square head on the workpiece surface and tap the block gently with the mallet, if need be, to position the work surface level.
4. Tighten the vise jaws (Figure 60-4).
5. Turn the surface grinder table handwheel to move the magnetic chuck so that any work placed on it will clear the grinding wheel.
6. Place a piece of paper that is slightly larger than the grinding vise on the magnetic chuck.
7. Set the grinding vise on the paper and turn on the magnetic chuck handle to energize the chuck (Figure 60-5).

FIGURE 60-3
Clean and deburr the workpiece of any sharp edges.

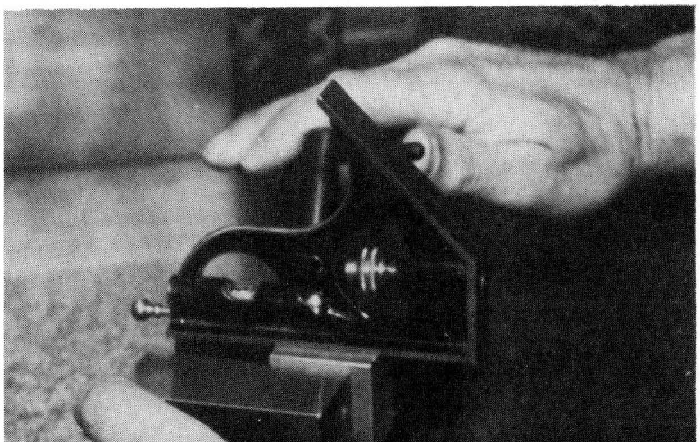

FIGURE 60-4
Tighten the vise jaws.

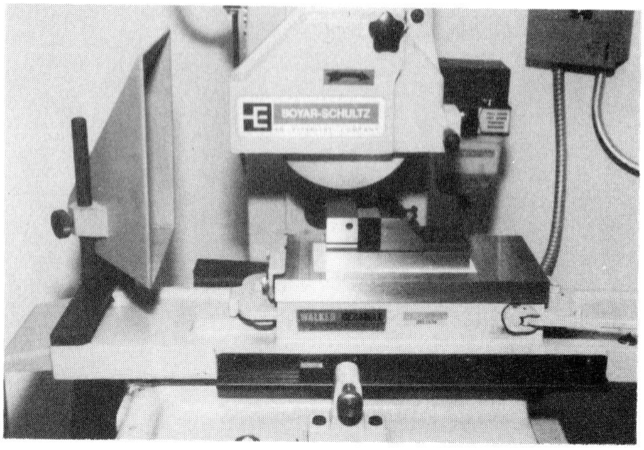

FIGURE 60–5
With the grinding vise set on paper, turn on the magnetic chuck handle.

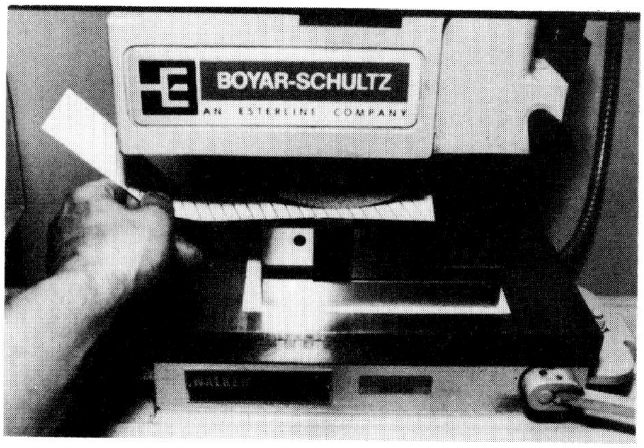

FIGURE 60–7
Lower the grinding wheel until it touches the folded strip of paper.

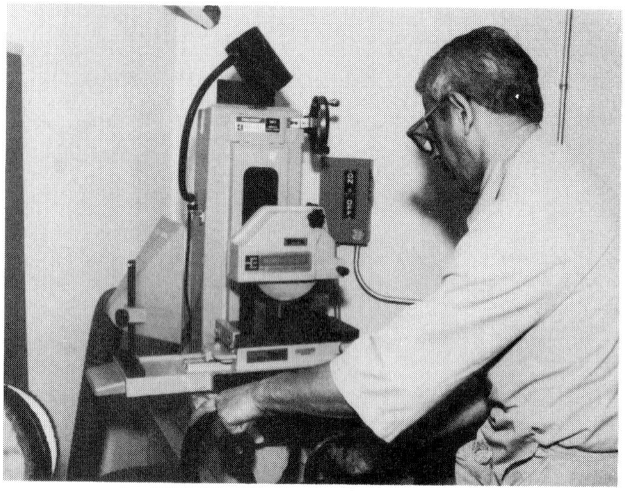

FIGURE 60–6
Position the workpiece under the grinding wheel.

8. Turning the table handwheel and crossfeed handle, position the workpiece under the grinding wheel (Figure 60–6).

9. Take a strip of paper that has been torn lengthwise and folded at least four times and place it on the work surface.

10. Lower the stationary grinding wheel until it just touches the folded paper (Figure 60–7).

11. Turn the crossfeed handle until the wheel clears the workpiece surface.

12. Start the grinding wheel spindle.

13. Turn the table handwheel to move the workpiece from side to side while turning the crossfeed handle until the grinding wheel overlaps the workpiece surface by about $\frac{1}{4}$ in. (6 mm).

14. As the table is moved from side to side under the spinning wheel, carefully lower the wheel in small increments until it sparks a high spot.

15. Continue feeding the table by hand until the entire surface has passed under the grinding wheel.

16. Use a 0.003-to-0.005-in. (0.07-to-0.12-mm) depth of cut for roughing cuts until the surface has been cleaned up. Take a 0.0005-to-0.001-in. (0.01-to-0.02-mm) depth of cut for finishing passes using the following methods:

(a) If the work surface is long, set the table reverse dogs so that the grinding wheel can clear each end by about 1 in. (25 mm) and start the table traveling automatically if this option is available.

(b) If a short work surface is involved, grind the surface by moving the crossfeed and table handwheel manually.

17. Stop the machine spindle.

18. Move the surface grinder table until the grinding vise clears the wheel.

19. Turn off the magnetic chuck.

20. Remove the grinding vise, clean the magnetic chuck, and replace the paper.

21. Turn the grinding vise on its side, work surface up, and place it on the paper (Figure 60–8).

FIGURE 60-8
Turn the grinding vise on its side, work surface up, and place it on the paper.

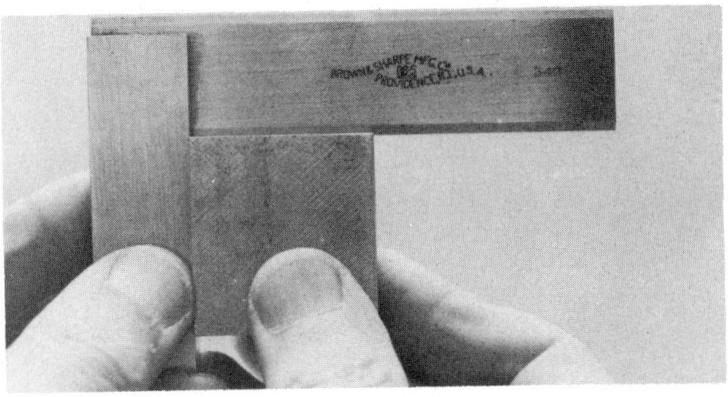

FIGURE 60-9
Check the ground sides to see if they are square.

FIGURE 60-10
Remove any grinding burrs.

22. Turn on the magnetic chuck handle to energize the chuck.

23. Raise the grinding wheel so that it is at least $\frac{1}{4}$ in. (6 mm) higher than the work surface.

24. Repeat steps 8 to 16 until side 2 has been completely finish ground.

25. Remove the vise from the magnetic chuck and demagnetize it.

26. Remove the workpiece from the vise and check the ground surfaces to see if they are square (Figure 60-9).

Grinding Sides 3 and 4

Once sides 1 and 2 have been ground square, one of the sides can be used as a reference surface by holding it in the grinding vise on paper feelers and then grinding sides 3 and 4 in the manner previously described.

Procedure

1. Clean the workpiece and remove any sharp edges (Figure 60-10).

FIGURE 60-11
Place two paper feelers on the bottom of the vise.

FIGURE 60-13
Tap the workpiece down tight against the paper feelers.

FIGURE 60-12
Tighten the vise with the workpiece set on the paper feelers.

FIGUIRE 60-14
Tear off any excess paper ends.

2. Place two strips of paper on the bottom of the grinding vise (Figure 60-11).

3. Set the workpiece on the paper feelers, extending an unground surface beyond the edge of the vise, about $\frac{1}{4}$ in. (6 mm), and tighten the vise (Figure 60-12).

4. Tap the workpiece down tight against the paper feelers with the soft-faced mallet (Figure 60-13).

5. Tear off the end of any paper that extends beyond the edge of the vise (Figure 60-14).

6. Repeat steps 5 to 16 as previously explained for grinding sides 1 and 2 until side 3 has been completely ground.

7. Repeat steps 17 to 26 as previously explained until side 4 has been completely ground and all sides have been checked for squareness.

8. Clean off the magnetic chuck (Figure 60-15).

9. Thoroughly clean the grinding vise and protect it from rusting with a film of oil before storing it (Figure 60-16).

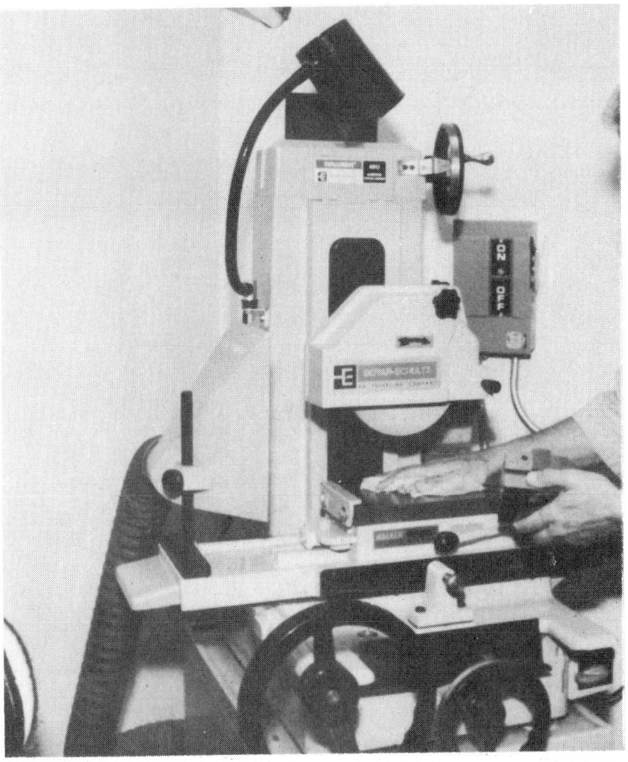

FIGURE 60–15
Clean off the magnetic chuck.

Using an Angle Plate

An angle plate with side rail can be used to grind a block square and parallel using the same tools that were used with the grinding vise as well as one or more C clamps for holding purposes (Figure 60–17).

Procedure for Grinding Sides 1 and 2

1. Remove or slide the side rail away from the main surface of the angle plate.

2. Clean the workpiece and deburr any sharp edges.

3. Lightly clamp the workpiece on one corner of the angle plate with one edge extended about $\frac{1}{4}$ in. (6 mm) beyond the edge of the angle plate.

4. Level the work surface using the square head and tighten the C clamp, keeping the clamp frame well below the surface to be ground (Figure 60–18).

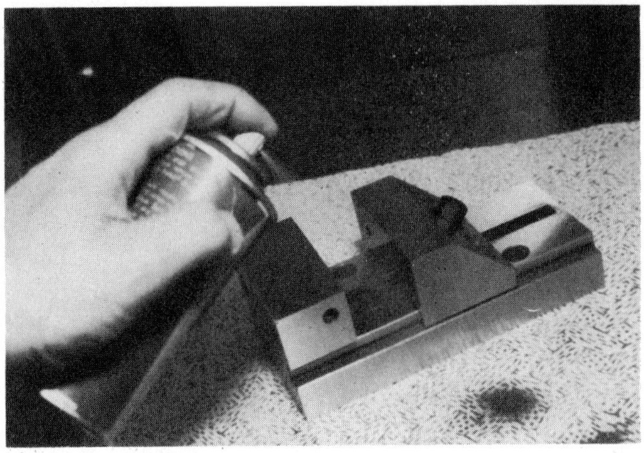

FIGURE 60–16
Protect the grinding vise against rusting with a film of oil before storing.

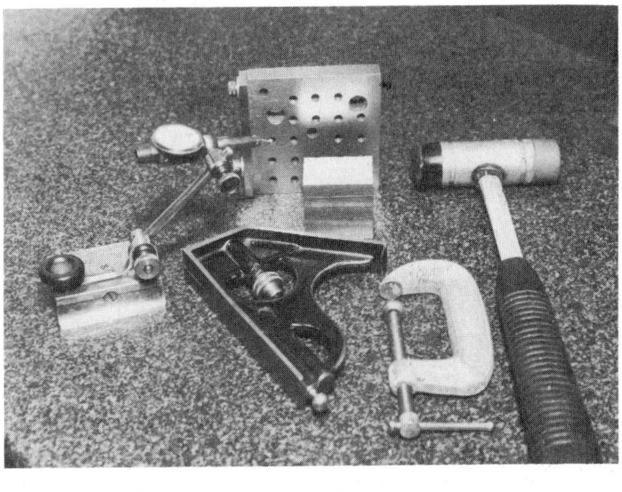

FIGURE 60–17
Some of the tools that may be used with an angle plate for grinding.

5. Turn the surface grinder table handwheel to move the magnetic chuck so that any work placed on it will clear the grinding wheel.

6. Place a piece of paper slightly larger than the angle plate base on the magnetic chuck.

7. Set the angle plate on the paper and turn on the magnetic chuck handle to energize the chuck (Figure 60–19).

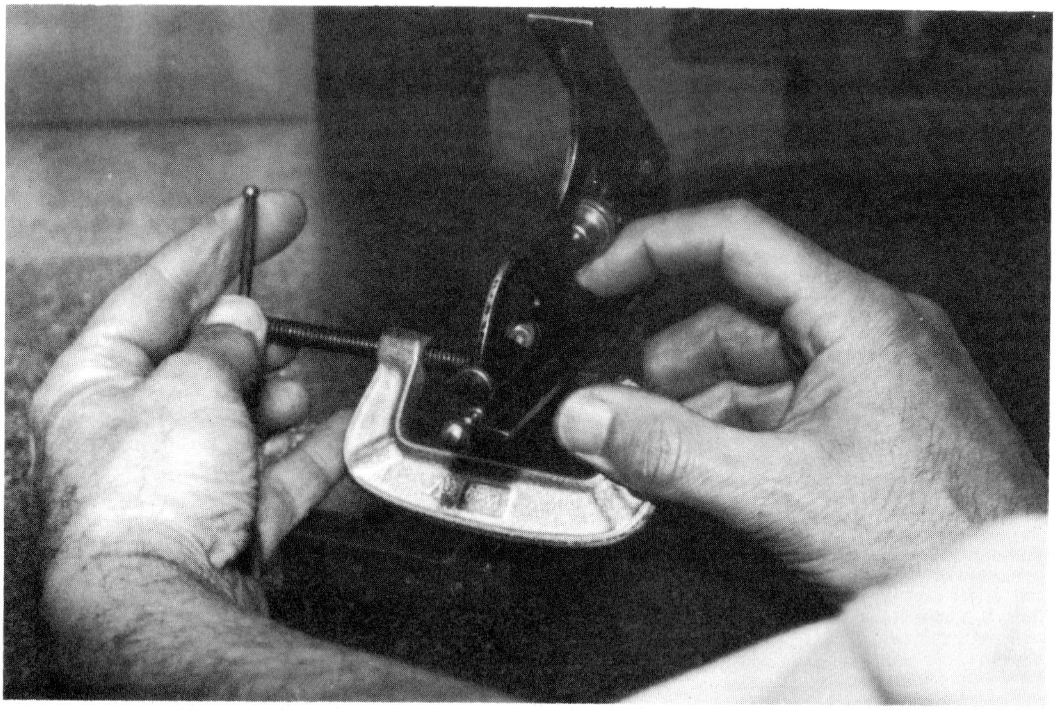

FIGURE 60–18
Level the work surface quickly using a square head with spirit level.

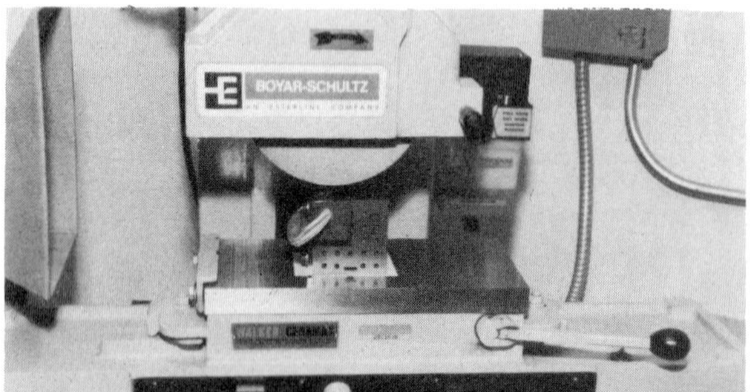

FIGURE 60–19
Set the angle plate on paper and energize the magnetic chuck.

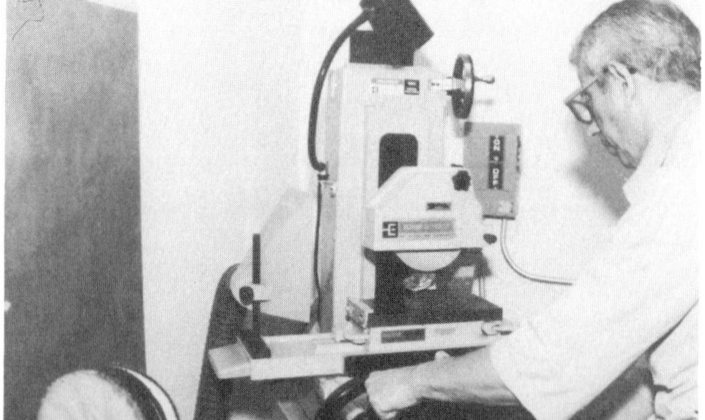

FIGURE 60–20
Position the angle plate under the grinding wheel.

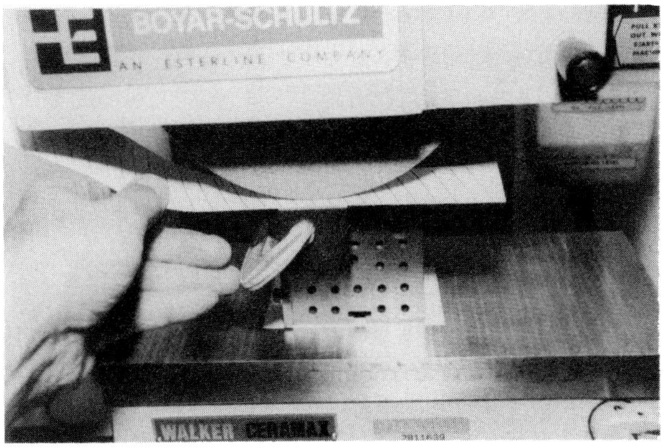

FIGURE 60-21
Lower the stationary grinding wheel until it begins to touch the folded paper.

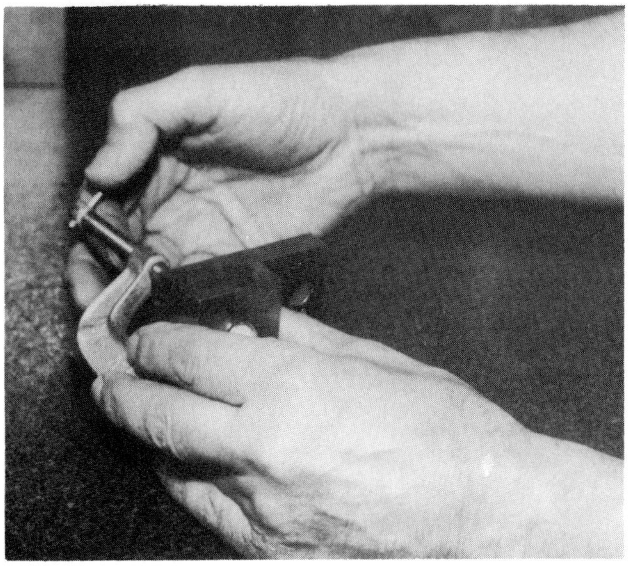

FIGURE 60-22
Loosen the C clamp so that the frame can be pivoted.

8. Turn the table handwheel and crossfeed handle to position the workpiece under the grinding wheel (Figure 60-20).

9. Take a long folded strip of paper and place it on the work surface.

10. Lower the stationary grinding wheel until it just touches the folded paper (Figure 60-21).

11. Repeat steps 11 to 17 of grinding sides 1 and 2 with a grinding vise.

12. Move the surface grinder table until the angle plate clears the grinding wheel.

13. Turn off the magnetic chuck.

14. Remove the angle plate and demagnetize it.

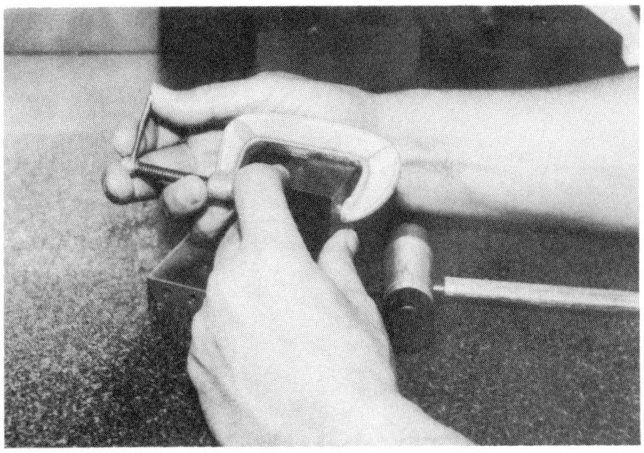

FIGURE 60-23
With the C clamp turned 90° half tighten the clamp screw.

Grinding Side 2

Once the first side has been ground, the C clamp will have to be turned so that the angle plate can rest on its side. To make sure that the workpiece has not lost the accuracy of the grinding setup, an indicator is used to measure the level of the surface.

Procedure

1. Loosen the C clamp so the frame can be pivoted (Figure 60-22).

2. Turn the C clamp 90° carefully and half tighten the clamp screw (Figure 60-23).

3. Indicate the ground surface to see it has not been disturbed, tapping any high spot level with a soft-faced mallet if necessary (Figure 60-24), and complete tightening the C clamp.

4. Complete grinding side 2 by repeating steps 5 to 14 as previously described for using an angle plate to grind square and parallel.

5. Check the ground surfaces to see that they are square.

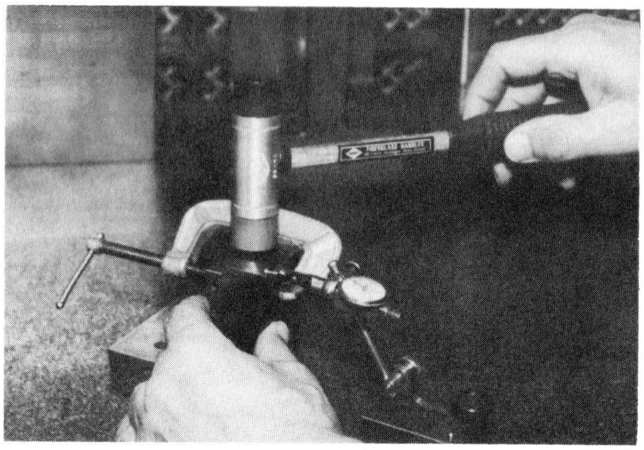

FIGURE 60-24
Tap the high spots level and completely tighten the C clamp.

Grinding Sides 3 and 4

Grinding sides 3 and 4 will require the use of a side rail, against which a ground surface may be placed. This will permit any surface that is extended to have its adjacent side ground at 90° to it. The ground surface will also be parallel to the opposite side.

Procedure

1. Extend and tighten a side rail to the angle plate to form a corner.
2. Place a ground surface of the workpiece against the side rail, extending the unfinished third side above the angle plate by about $\frac{1}{4}$ in. (6 mm) and clamp it in place.
3. Grind the top surface in the manner described in steps 5 to 19 when using a grinding vise.
4. Remove the angle plate and demagnetize it.
5. Remove the workpiece and deburr the edges.
6. Repeat steps 2 to 5 in the preceding procedure to complete grinding the unfinished fourth surface.

UNIT 61 SPECIAL PROCESSES

Many processes introduced in industry replace or complement conventional methods of machining, grinding, or metal finishing. These special processes are more efficient and may require less handling to complete the fabrication of metals that may vary in characteristics as they do in thickness and size.

But even as this is being written, different methods are being developed that will reduce hard-to-machine parts to standard production processes. These methods promise to produce parts more economically and with precision never before attainable with difficult-to-form alloying metals.

Objectives

After completing this unit you should be able to

1. name and explain the principles of four different electroprocesses.
2. state the advantages of these different processes.
3. name two different types of high-energy forming processes.
4. state the advantages and disadvantages of these processes.

Electrochemical Machining

Electrochemical machining (ECM), as it is properly called, involves chemical changes utilizing electricity and basic mechanical components as shown in Figure 61-1. The ECM shapes workpieces by removing deposits of metal in a bath of electrolyte, an electrical conductive solution of special chemicals, mirror imaging the tool contour in the cavity it machines.

In the process, a shaped tool (a negative electrode called the cathode) is placed near the workpiece (the positive electrode called the anode) at a controlled gap, with both members connected to a low-voltage direct-current power source. As the electrolyte is pumped through the system under high pressure, current is passed between the workpiece and tool to cause a chemical action to take place. It is this chemical action that dissolves metal from the workpiece surface. The waste is flushed away by the continuous flow of the electrolyte, cooling and preventing build-up and wear on the tool at the same time.

The ECM tools come in different sizes and shapes with controls for regulating the measured gap between the tool and the workpiece. Rate of removal is determined by the gap, rate of tool feed, and amperage flowing between the tool and the workpiece. Properly balanced, these factors will cause a mirror image tool cavity. Excessive amperage will cause overcut, an enlarged cavity, and tools are generally machined a little undersize to compen-

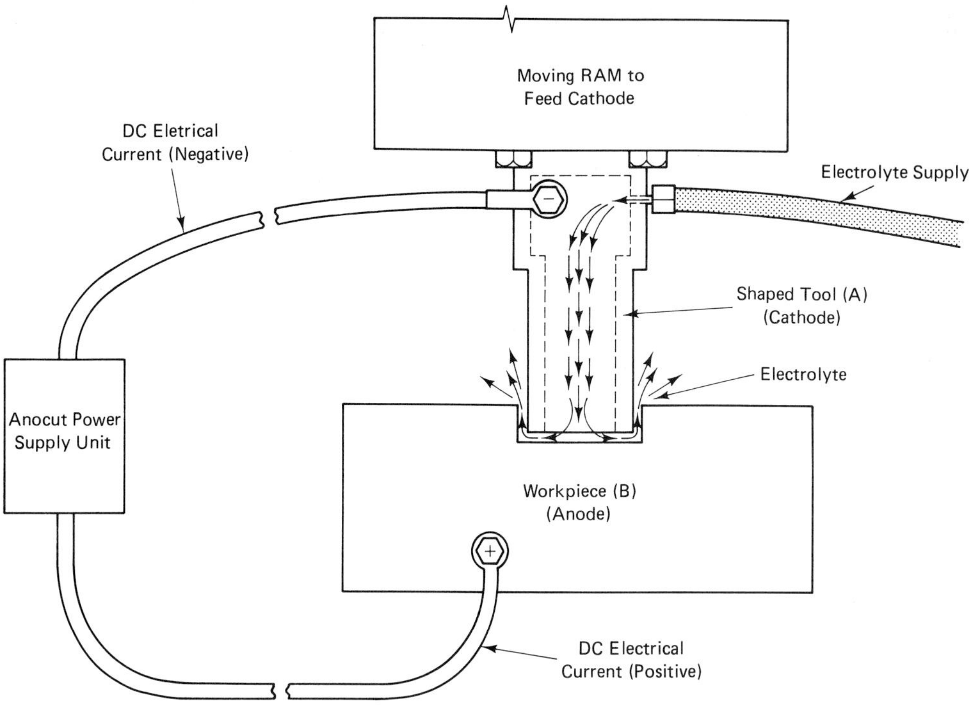

FIGURE 61-1
The elecrochemical machining (ECM) process. (Credit: Anocut Inc. and ASME)

sate for overcut. Hardened metal shapes or intricate, thin, and fragile parts can be cut with this process as long as they can conduct electricity. Electrochemical machining can produce parts that are burn free, burr free, stress free, and warpage free (Figure 61-2).

Electrical Discharge Machining

Electrical discharge machining (EDM) is a spark erosion process. With both the workpiece and tool connected to a power supply, an electric current sends rapid sparks through a nonconductive (dielectric) oil. The energy of the discharge melts away or vaporizes metal in a controlled manner. The process is initiated by submerging the workpiece in the dielectric while a shaped cutting tool (electrode) is placed near it to create a gap, causing an extremely hot spark to vaporize the metal from the workpiece (Figure 61-3). An EDM machine incorporates the means by which the preselected gap is maintained by either holding one and moving the other or by moving both the tool and the workpiece. The depth

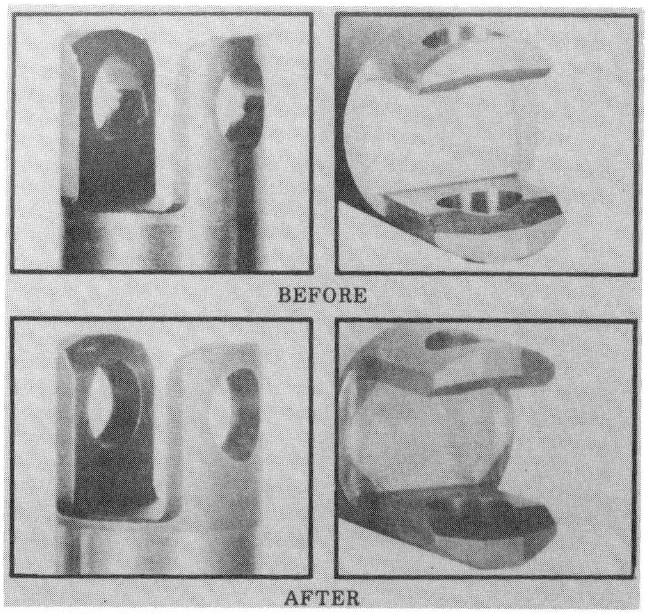

FIGURE 61-2
Parts produced by ECM machining are burn free, burr free, stress free, and warpage free. (Credit: Anocut Inc.)

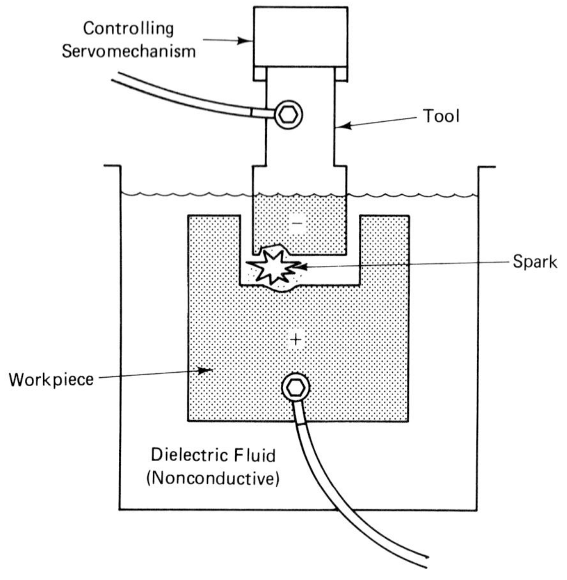

FIGURE 61–3
The electrical discharge machining (EDM) process.
(Credit: Anocut Inc. and ASME)

of the cutting action is determined by the intensity of discharge, producing minute chips in the form of sludge, which is flushed away from the gap by the dielectric. Metal removal is dependent on the energy level, gap width, and time duration of the current flow. Finer finishes are possible with increased sparking frequencies, while a higher current discharge level will lead to overcutting.

Electrical discharge machining is accurate and has been used to eliminate broken taps and drills in workpieces and for sculpting deep impressions such as letters and figures in steel. With a small wire electrode it can also be used in the diemaking trade to cut the punch and die blank from a slab of steel in one operation, eliminating all of the filing and fitting that was formally necessary. The principal disadvantage in EDM is that the sparks erode the work as well as the tool.

Electrolytic Grinding

Electrolytic grinding (ELG), also called electrochemical grinding, offers many advantages to industries seeking to grind difficult alloys. The grinding pressure is low in this process, there is little or no grinding heat, and delicate work can be ground free of distortion or warpage. The process offers an increase in grinding wheel life, an increase in grinding rate, and the elimination of grinding cracks. Other benefits can be found in sharpening carbide tools. Electrolytic grinding leaves the rake surface finish smoother and free of stress and thermal damage to produce a sharper cutting edge.

In electrolytic grinding, a metal-bonded abrasive wheel (cathode) and the workpiece are connected to a low-voltage power source (Figure 61–4). The electrolyte, a solution of water and special chemical, is directed on the wheel very much like a coolant. An electrochemical action takes place as electric current passes from the workpiece (+) through the electrolyte to the wheel (−), which results in a decomposition of material on the workpiece surface. The soft, nonconducting deplating film that forms is removed by the wheel's rotation and washed away by the steady stream of electrolyte. At the same time, the wheel's abrasives form tiny little spaces that carry the fluid across the work surface (Figure 61–5).

The current flow is directly responsible for the rate of electrochemical action. It is estimated that 90 to 95% of removal is the result of electrochemical action. The remaining percentage is caused by the wheel's abrasive action by wiping away the deplating film.

Photochemical Machining

Photochemical machining combines photography with chemical etching. With this process, a variety of shapes and thicknesses can be produced. The design may be flat or complex, solid or peppered with holes, prototype or production. Material thickness can range from 0.0005 to 0.090 in. (0.01 to 2.28 mm), and the parts will be left without burrs or internal stresses. Furthermore, there is no elaborate tooling or ensuing wear and storage problem.

In photochemical machining, the operation is carried out through a series of photographic steps that eliminate expensive tooling without sacrificing accuracy. An enlarged master drawing of the part is made. This drawing may be as large as 50 times the original size. The master is then reduced photographically, bringing the part down to working size, which is then produced on a glass master or film to precise detail. A photorepeating machine then produces multiple negatives containing several images. These negatives are then exposed on metal coated with a photosensitive surface. The metal is later dipped in etching acid where all surface not protected by the coating is eaten away by the chemicals.

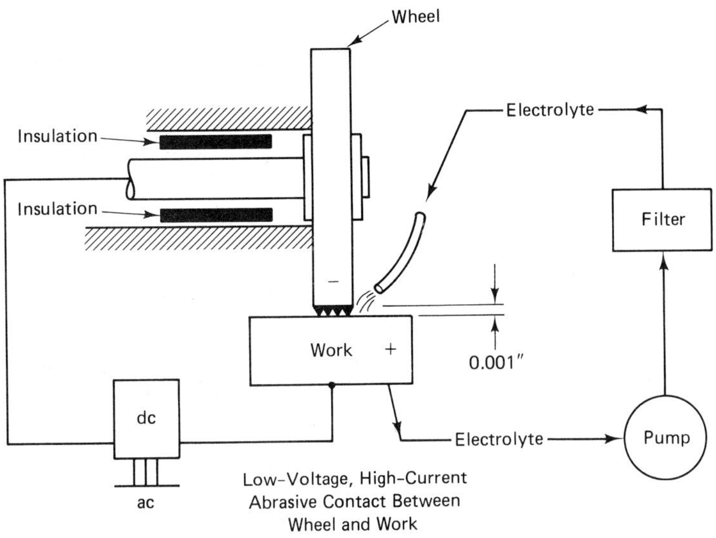

FIGURE 61-4
Electrolytic grinding (ELG) principle. (Credit: Everite Machine Products Co. and ASME)

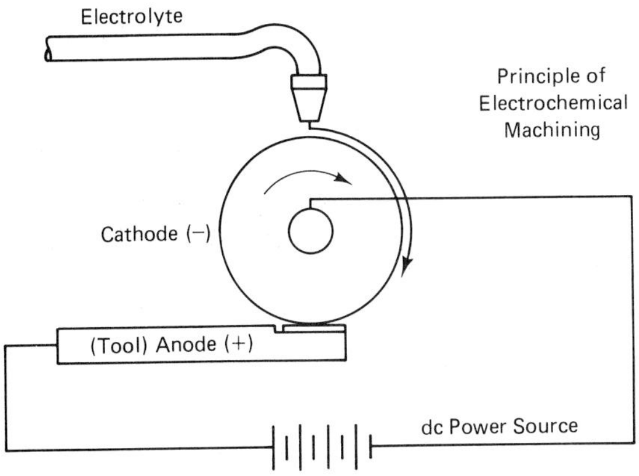

FIGURE 61-5
Electrolytic grinding can produce very smooth, very sharp cutting edges on carbide cutting tools. (Credit: Everite Machine Products Co. and ASME)

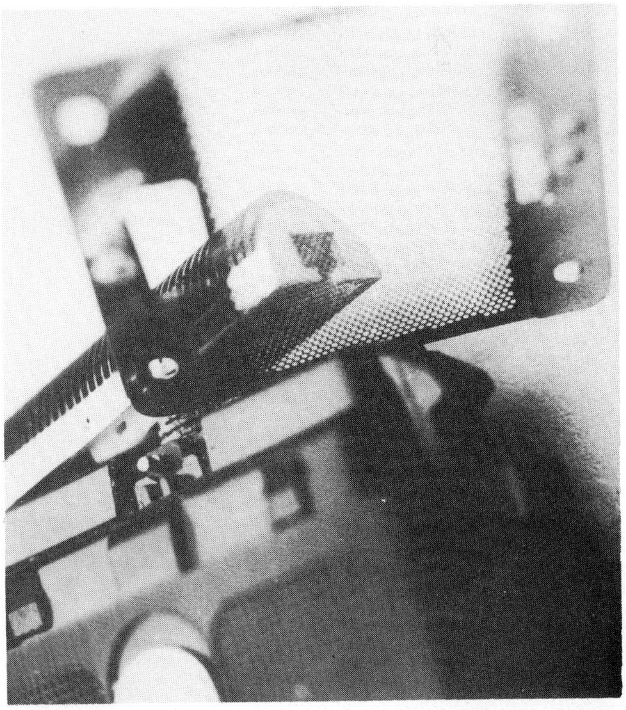

FIGURE 61-6
A shaving head screen is but one example of photochemical machining.

Products made by photochemical machining may vary from jewelry to shaving head screens (Figure 61-6) to radial and axial springs. Materials include a broad range of metals such as aluminum, copper and silver to tungsten, vanadium, and rhenium. Other nonmetallic substances such as glass and plastics can also be machined photochemically.

Ultrasonic Machining

Impact grinding is another name for ultrasonic machining (USM). In this process, a circulating fluid called a slurry is directed in the space between the tool and the workpiece. The slurry is composed of fine abrasive particles such as silicon carbide, boron carbide, or aluminum oxide suspended in liquid, generally water, containing 30 to 60% abrasives by volume. A power unit moves the tool up and down ultrasonically at a rate that can be adjusted from 15,000 to 30,000 cycles per second. While the up and down stroke of the tool may be equal to the thickness of newspaper, the rapid movement within the slurry is successful to cause an abrasive cutting action, impacting the abrasive particles in the work surface. It is this combination of erosion and chemical action that is responsible for low stock removal to take place.

There are a few disadvantages connected with the process. Tool wear is high, which requires having several roughing and finishing tools on hand. The tools themselves are made of soft brass or steel and must be made to the exact shape of the cavity they will produce. Another disadvantage is the relatively high cost of this machining process.

One of the principal advantages of USM is that it can machine extremely hard materials such as carbide and hardened steels as easily as it can nonelectrically conductive materials. The process also lends itself to machining holes of different contours, depths, and sizes that would otherwise be difficult to machine as long as a tool can be made and slurry can be fed to its end. Enlarged contours and even threading can be achieved by moving the workpiece during cutting.

High-Energy Forming

High-energy rate forming (HERF) is a high-energy transfer process that forms metal at a velocity greater than 50 ft/sec (15 m/sec). The HERF processes are used when large parts have to be formed that are too difficult or impossible to accomplish by conventional methods. They are often determined by the size and properties of the material involved, such as aluminum, magnesium, berrylium, titanium, zirconium, stainless steel, super alloys, as well as refractory metals.

While there are a number of factors to be considered in selecting a HERF process, such as equipment size, improvement of dimensional quality, and expedition of fabrication, the chief underlying factors are the cost per unit and favorable production techniques when compared to conventional methods. The HERF techniques are based on three methods of releasing energy: (1) chemical, (2) electrical, and (3) mechanical. Only a few techniques will be discussed in the text that follows.

High-Explosive Forming

Deep drawing of large complex parts in industry is often completed in stages, utilizing annealing between stages and expensive equipment. When one considers that the same operation can be completed in one operation using the pressure shock wave in explosive forming, one can appreciate the value of explosive forming. The process is widely used in missile and aerospace technologies and makes use of the violent pressure that takes place when an explosive is detonated under water. Only half a die is used, the cavity on which the metal blank rests. An inexpensive bulk charge is placed at a predetermined distance above the blank. While the explosive package varies with the metal being used, bulk charges of TNT are generally used in forming parts. At the moment of detonation, a pressure shock wave traveling nearly 25,000 ft/sec (7620 m/sec) collapses the surface of the blank to fill the walls of the die as if fitted by a punch (Figure 61–7).

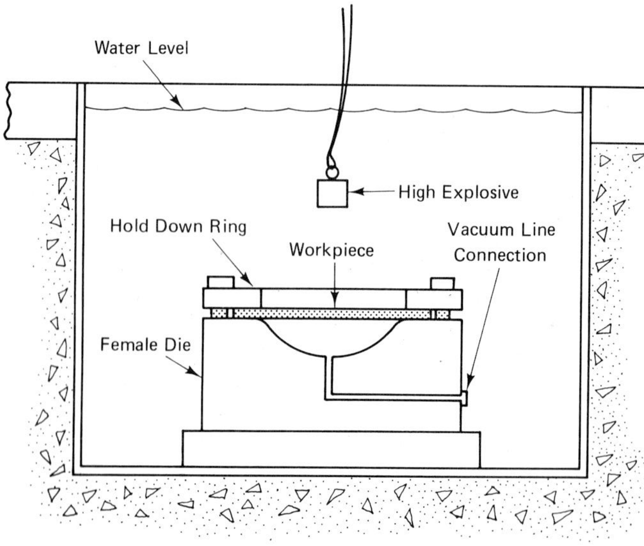

FIGURE 61–7
High-explosive forming can create a pressure shock wave that can complete a deep draw in one operation.

In addition to the advantages just stated, explosive forming can reduce equipment bulk, minimize tool costs, maintain quality finishes, eliminate or minimize springback, and promote other metallurgical properties such as hardening and strengthening through shock hardening. Some of the disadvantages at this writing are the low production rate and water temperatures have to be maintained to stabilize fabrication and working safely with explosives.

Spark Forming

Flat blanks of metal can be formed against die walls in a manner similar to explosive forming only the shock wave is triggered by electrical current passing through two electrodes. The process is called electrohydraulic forming (EHF), whereby the two electrodes are placed at a predetermined distance apart and above the metal blank fastened to the top of a die submerged in a medium of water or oil. When high voltage is passed through the electrodes, a spark is generated. The energy released between the electrodes is similar to an explosive shock wave, causing the metal blank to deform and filling the cavity of the die, which has been pumped out of air by a vacuum line (Figure 61–8). Accuracy of dimension is dependent on the quality and precision of the die as well as on the location and intensity of the energy discharge.

Spark forming is also conducted by bridging the two electrodes with special wire. When voltage is sent surging through, it vaporizes the wire, rapidly heating and expanding the water to create a shock wave. In this process, the wire must be replaced for every part produced.

Electrohydraulic forming is only used when conventional methods will not do. Capable of forming most materials such as aluminum alloys, stainless steel, nickel alloys, titanium, and a host of conventional metals, materials that are brittle or impact at velocities below 100 ft/sec (30 m/sec) should be avoided.

Gas Forming

In pneumatic-mechanical metal forming, compressed gas supplies the energy to drive a ram down a ram cylinder. Newer developments in forging equipment enables the ram to travel at velocities up to 200ft/sec (60 m/sec) by using special design.

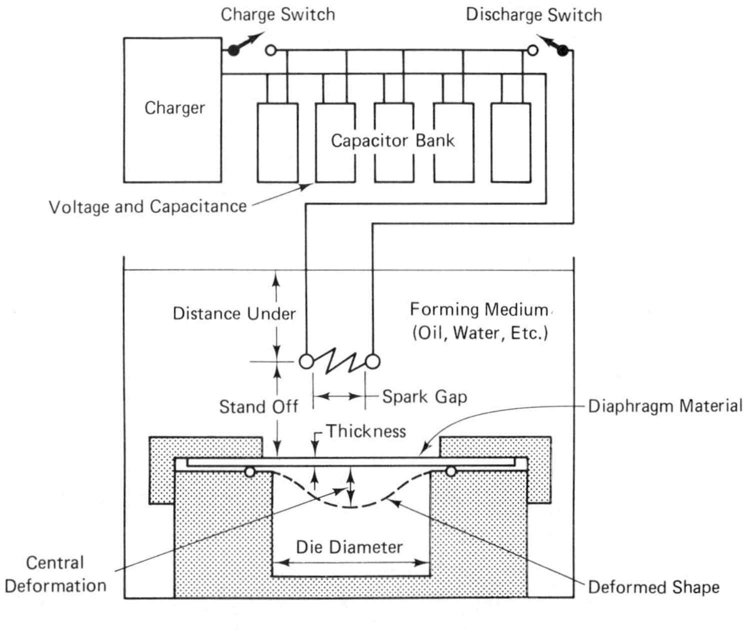

FIGURE 61–8
Spark forming is capable of forming most conventional materials such as aluminum alloys. (Credit: Cincinnati Milacron Inc.)

FIGURE 61–9
Pneumatic-mechanical forging presses compare favorably with drop hammer forge presses for accuracy. (Credit: American Iron & Steel Institute)

Utilizing either air or nitrogen at a predetermined pressure, pneumatic-mechanical presses are housed within a rugged frame, operating on a cycle that is similar to an internal combustion engine: intake (charge), ignition, compression, and exhaust. The forging punch and die are mounted on opposing rams or a ram and a bolster, which, upon release of the compressed gas, close with accelerated velocity at the midpoint. The kinetic energy is absorbed by the workpiece at the end of the stroke, resulting in the transference of a minimum of shock to the press.

While not applicable to all types of commercial forgings, pneumatic-mechanical machines are ideal for forging thin sections, intricate details, and symmetrical shapes such as gears, gear blanks, cones, disks, and similar designs. The accuracy achieved by pneumatic-mechanical forging presses compare favorably with hydraulic, mechanical, and drop hammer forge presses (Figure 61–9); however, they lack the versatility and production rate of the others.

SECTION REVIEW

Unit 57

1. Define grinding.
2. How does the cutting action of a grinding wheel compare to other machine tools?
3. How does grinding play an important role in industry?
4. What are the two basic elements that form the composition of a grinding wheel?
5. Briefly explain the process of fracturing.
6. Explain the cutting action of a soft wheel.
7. Explain the cutting action of a hard wheel.
8. What are the two general classifications of wheels?
9. What information does the coded wheel markings listed on the wheel's blotter impart?
10. What is the purpose of a bond? List five main types of bonds used in the manufacture of grinding wheels.
11. Describe four common types of surface grinders that are used in production.
12. On what type of grinders are milling cutters, end mills, and similar cutting tools sharpened?
13. List three primary methods of centerless grinding.
14. List four benefits to be found in the use of recirculating grinding fluids.
15. Define swarf.

Unit 58

1. Give four safety rules to follow when using the surface grinder.
2. Briefly explain the purpose of these parts of a surface grinder: (a) stop dogs, (b) saddle, (c) column.
3. Name three types of chucks used for holding work on a surface grinder.
4. Of the three, which can hold ferrous as well as nonferrous materials directly on its surface?
5. In general, how should short ferrous workpieces be guarded against for slippage on a magnetic chuck?
6. What efficient cutting speed is recommended for vitrified wheels? For resinoid grinding wheels?
7. List five precautions to be observed in mounting a grinding wheel.
8. Define the term "truing."
9. Define the term "dressing."
10. What devices are generally used for dressing standard aluminum oxide or silicon carbide grinding wheels?
11. What is the purpose of placing a piece of paper slightly larger than the dresser base on the magnetic chuck?
12. Where should the dressing device be placed on the magnetic chuck when dressing a wheel?
13. In what relative position to the wheel should the dressing device be placed?

Unit 59

1. How are nonferrous or nonmagnetic materials held on a magnetic chuck?
2. How are magnetic V-blocks or magnetic parallels constructed?
3. Why should magnetic V-blocks and magnetic parallels be stored in a wooden box when not in use?
4. How can flat, thin nonmagnetic workpieces be held for light grinding on a magnetic chuck?
5. Name four accessories used for grinding angles.
6. Briefly describe how to mount work on a magnetic chuck.
7. Describe how to test the magnetic chuck for holding power.
8. How long should the table travel be when grinding a flat surface?

9. Why should cutting fluids be used whenever possible during rough cutting?
10. How are workpieces neutralized of any magnetic force once removed from the grinder?

Unit 60

1. Explain how to grind a block square and parallel using a grinding vise

or

2. Explain how to grind a block square and parallel using an angle plate.

Unit 61

1. What processes do the following letters represent: ECM, EDM, ELG, and USM.
2. How does the EDM process operate?
3. What is the principal disadvantage of using the EDM process?
4. Name three advantages to photochemical machining.
5. What is another name for ultrasonic machining?
6. Describe the ultrasonic machining process.
7. What are the advantages and disadvantages of ultrasonic machining?
8. What do the letters HERF represent?
9. Name two processes that depend on the use of a shock wave to spread a flat metal blank against a die wall cavity.
10. What do the letters EHF represent and what is the common name for it?
11. What two gases are used for compression in the pneumatic-mechanical metal forming process?

SECTION 12

PROJECT PLANNING

The planning involved in machining a part according to print dimensions will require a good background of knowledge and skills. It is important to understand what machining steps to use and which steps should be completed first. The mental process for this is called job planning and is extended even further in fitting and assembling parts together. Undertaking any fabrication without looking ahead will only increase the chances of wasting time, money, and materials.

UNIT 62 MACHINING PROCEDURES

Machining procedures will vary according to the shape of the part involved. Is it round, square, octagonal, or flat? It will also vary with the type of material. Is it mild steel, tool steel, nonferrous, or alloy steel? Is it to be heat treated? What kind of machining accuracy is required? Will it have reamed holes, close fitting slots, threads, or angular grooves to a fractional dimension? As the list grows, so do the machining choices and procedures. There is no one method or procedure that will cover every situation. It is known that much time and effort will be saved by following basic machining practices.

Objectives
After completing this unit you should be able to

1. plan the sequence of operation and machine the tool shown in Figure 62-1.

Offset Tap Wrench

An offset tap wrench consists of two independent arms that are notched and clamped together against a tap's square to act as its turning device. The material involved will be two lengths of mild steel square stock and two $\frac{1}{4}$-in. cap screws 1 in. long.

The making of an offset tap wrench presents one major problem: the alignment of the square opening when the locking screws are tightened. This problem can be overcome if the project is undertaken by locating the lock screws in relation to an established square opening. For this reason, the project is laid out using one arm as the prototype, making a V-notch in both arms, drilling the holes in the laid-out arm, and then transferring the hole locations to the second arm while the V-notches are positioned against a tap's square. This method will eliminate any error that would occur if both arms were laid out, center punched, and drilled only to find that the V-notches would not form a square. Essentially, perfect alignment is assured by transferring both locations with a tap's square held between the V notches.

While the project is outlined in small visual details, it does not move along without reference to the print. No outline should be so complete without checking with the drawing. There has to be a coordinated meaning between outline and drawing. It reinforces the act of reading a print and interpretation of how to achieve the dimensions involved.

Layout and Machining Preparation

An examination of the print reveals that the tap wrench arms will be made to a fractional accuracy. Laying out can be made using a surface gage and rule or with a height gage. How will the V-notches be machined? The V-notch on both arms will be milled in one operation, with both arms set at a 45° angle. For this reason, the V formation need not be laid out on the workpiece. This 90° V will be formed by the shape of the milling cutter. In order to mill

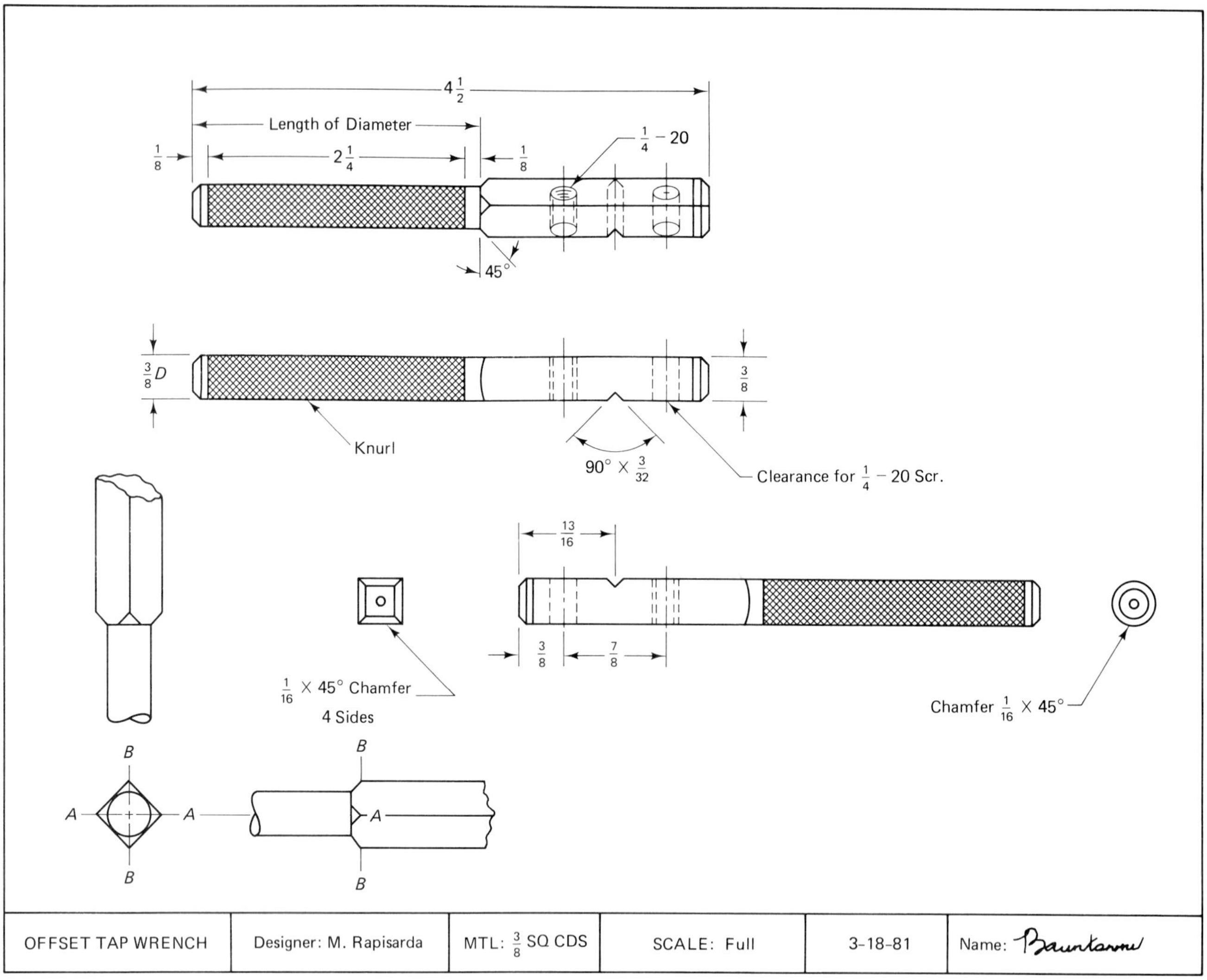

FIGURE 62–1
Offset tap wrench.

the notch to its proper opening, the width of the notch must be laid out. This can be calculated by using trigonometry as follows:

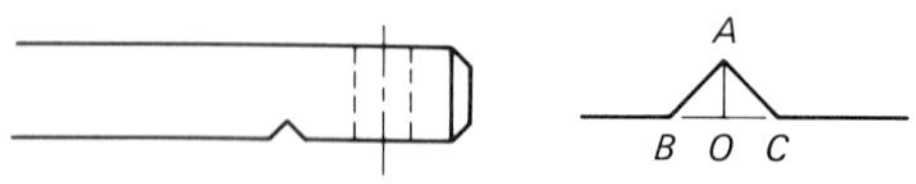

In triangle ABC, $AB = AC$. By constructing line AO, 45° angles are formed at BAO and OBA. The line BO will equal OC. Since AO equals $\frac{3}{32}$ in., what does BC equal?

Solution:

$$BO = AO \times \tan BAO,$$
$$BO = \tfrac{3}{32}\text{ in.} \times \tan 45°,$$
$$BO = \tfrac{3}{32}\text{ in.} \times 1,$$
$$BC = BO + OC,$$
$$BC = \tfrac{3}{16}\text{ in. (4.8 mm)}$$

Procedure

1. Study the work print.
2. Cut off $\frac{3}{8}$ in. square × $4\frac{3}{4}$ in. long CDS stock (Figure 62–2).

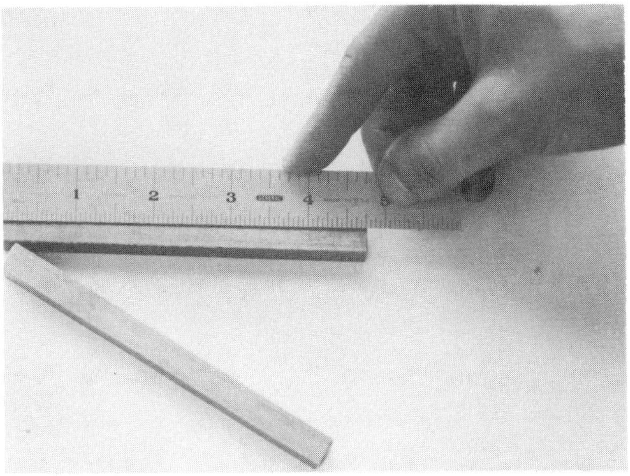

FIGURE 62-2
Cut off stock, allowing $\frac{1}{4}$ in. (6 mm) waste.

FIGURE 62-3
Square off one edge on both pieces.

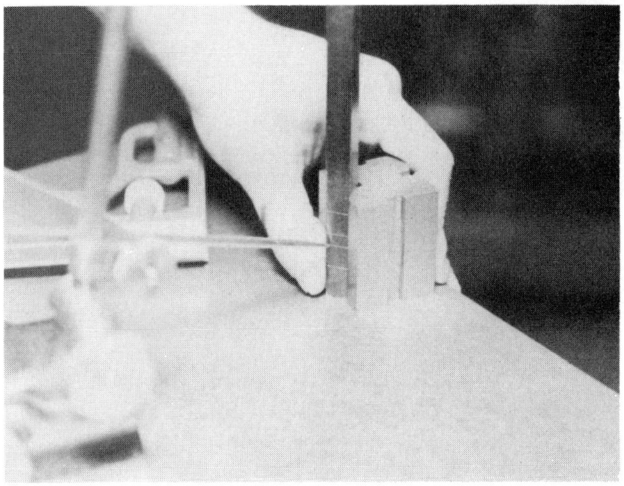

FIGURE 62-4
Layout the V-notch width, adding thickness of waste material to centerline location.

FIGURE 62-5
Align the edges of the handles against a rule on a magnetic chuck.

3. Square off one end of both pieces by machining in a milling machine (Figure 62-3).

4. Apply layout dye along one surface of one part.

5. Stand this coated workpiece on its machined end on a surface plate, holding it upright against a V-block.

6. Using a height gage or a surface gage for scribing, allow $\frac{1}{8}$ in. (3mm) waste and lay out the hole locations and length of the knurled handle (see print).

7. Lay out the width of the V-notch (Figure 62-4).

8. To assure uniformity in assembly, line up the machined ends of the handles, scribed surface up, on a magnetic chuck (Figure 62-5).

9. Energize the chuck and clamp both handles together above the intended V-notch (Figure 62-6).

10. Mount a sharp, multifluted end mill in a vertical milling machine.

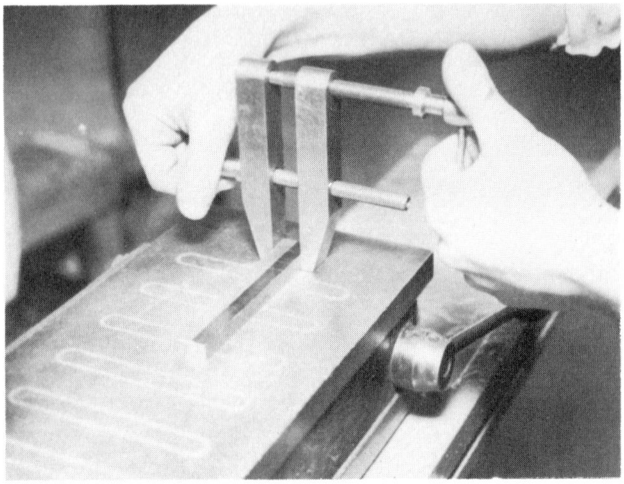

FIGURE 62–6
With the magnetic chuck energized, clamp both handles together above the V-notch.

FIGURE 62–8
Tighten the vise with the handles set against the 45° slope of the square head.

FIGURE 62–7
Place a square head in the milling machine vise.

FIGURE 62–9
When the bottom of the cutter has touched the laid-out line, lock the depth of cut.

11. To machine the 90° V-notch, place a square head in the milling vise (Figure 62–7).

12. Set the clamped handles against the 45° slope of the square head and tighten the vise (Figure 62–8).

13. Remove the clamp.

14. Set the machine to its proper cutting speed and start the machine spindle.

15. Turn the crossfeed handwheel to mill the middle section of the scribed notch opening, raising the knee to take small depth of cuts past both parts to form a step. Continue making these cuts until the bottom of the cutter touches the lowest scribed line and lock the depth of cut (Figure 62–9).

16. Move the longitudinal handwheel for each depth of cut, taking cuts past both parts using the

FIGURE 62–10
Continue cutting until the full width of the V-notch has been completed.

FIGURE 62–12
Insert the square of a short tap between opposing vees on a magnetic chuck.

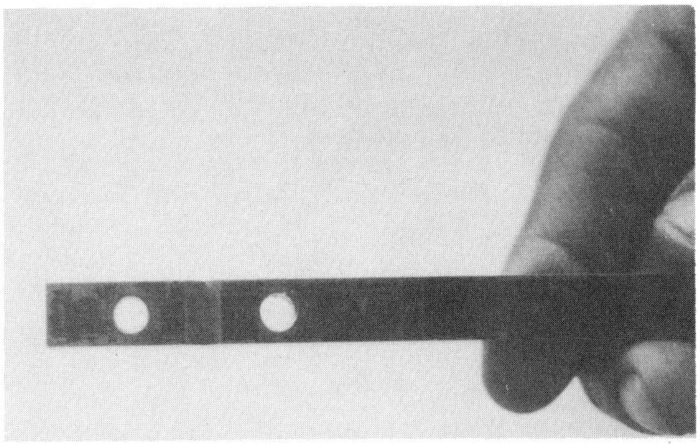

FIGURE 62–11
Drill both center-drilled holes through using the tap drill.

FIGURE 62–13
Energize the magnetic chuck and clamp both handles together.

crossfeed handwheel, until the side of the cutter barely touches the upper scribed line (Figure 62–10).

17. Prepare the hole location by center punching the layout lines.

18. Insert a suitable center drill in the drill press chuck and drill the locations to provide a guide for the tap drill to follow.

19. Replace the center drill with the proper-size tap drill (see print) and drill through (Figure 62–11).

20. Prepare to transfer the hole locations. Insert the square of a short tap between the opposing vees of the handles on a magnetic chuck (Figure 62–12).

21. Energize the magnetic chuck and clamp both handles together, making sure the holes are exposed for drilling purposes (Figure 62–13).

22. Transfer the hole locations into the second handle using the first handle as a guide, drilling as far as the full diameter of the tap drill (Figure 62–

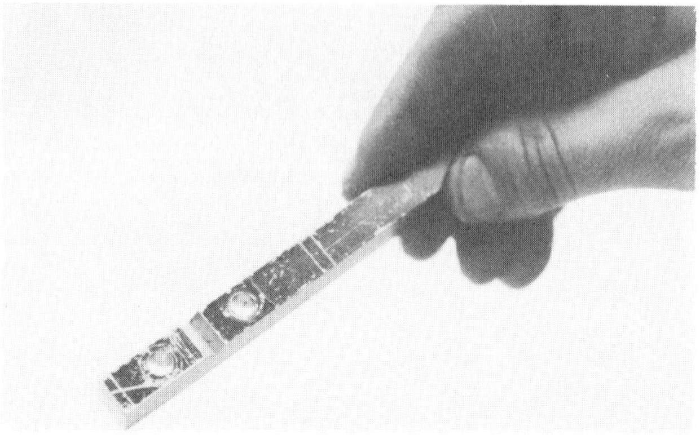

FIGURE 62-14
Transfer the hole locations, drilling as deep as the full diameter of the tap drill.

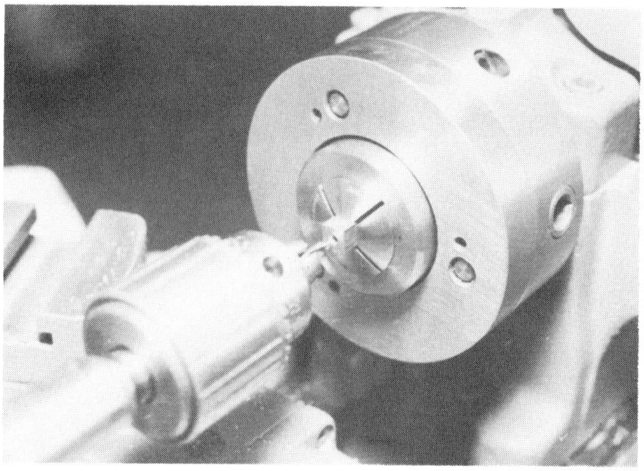

FIGURE 62-16
Center drill the end for mounting between lathe centers.

FIGURE 62-15
Prepare to mount the handle in the lathe using a square collet for holding.

FIGURE 62-17
Mount the handle between lathe centers.

14). (Prepare to turn and knurl the length of the handles to the dimension called for on the print. Since the operation will be performed between centers, the ends will have to be center drilled.)

23. Mount a square collet in the lathe's draw-in collet chuck attachment (Figure 62-15).

24. Tighten one handle within the collet, extending it not more than $\frac{1}{2}$ in. (13 mm) so that its end can be center drilled (Figure 62-16).

25. With all four ends center drilled to proper depth, mount one handle between lathe centers (Figure 62-17).

26. To obtain the 45° chamfer at the end of the knurled handle (see print), place the lathe toolbit on a 45° angle using the square head as a guide (Figure 62-18).

27. Machine the corners to obtain 0.365 in. (9.27 mm) diameter, stopping the cut on the layout line (see print).

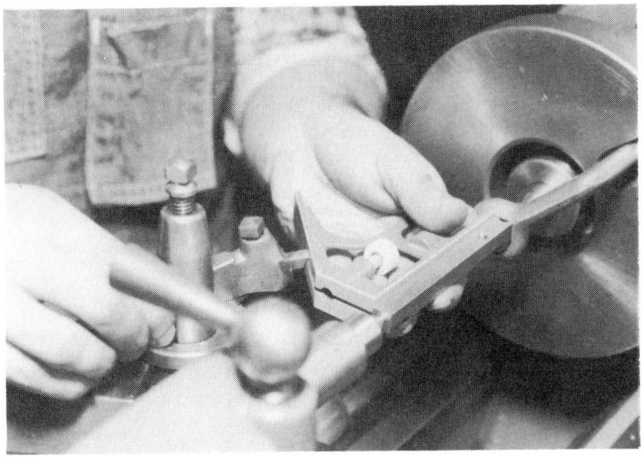

FIGURE 62–18
Set the toolbit to cut a 45° chamfer by using the square head as a guide.

FIGURE 62–20
Remove all sharp edges from both sides of the holes by countersinking.

FIGURE 62–19
Complete the knurling operation.

FIGUIRE 62–21
Complete the tapping operation.

28. Set up the lathe and quick-change gearbox for knurling.

29. Set up the knurling tool and complete the knurling (Figure 62–19). (Remove both handles from the lathe and prepare to complete the drilling operation. Examine the print to see which holes should be tapped and mark these with layout dye accordingly. The remaining two holes will have to be drilled as clearance holes for the screws to slip through.)

30. Complete drilling the tap holes through.

31. Complete drilling the clearance holes through.

32. Remove all sharp edges from both sides of the holes by countersinking (Figure 62–20).

33. Complete the tapping operation (Figure 62–21). (The remaining operations will be made to bring the handles down to length and chamfering the ends.)

34. Remount the square collet in the lathe's draw-in collet chuck attachment.

FIGURE 62-22
Face off the $\frac{1}{8}$ in. (3 mm) waste from the square end.

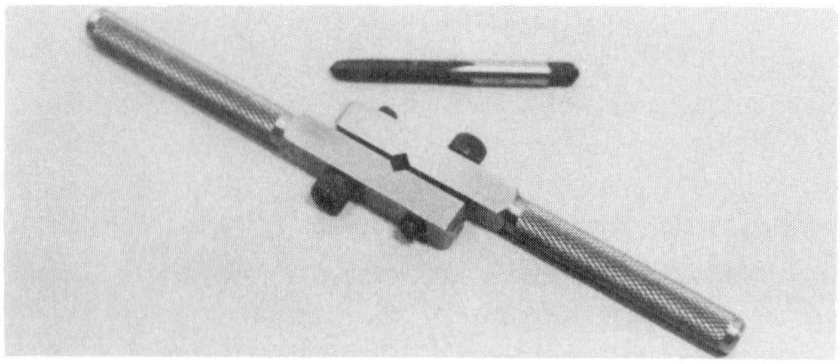

FIGURE 62-24
The offset tap wrench is assembled and ready for use.

FIGURE 62-23
Chamfer the round end using the compound rest set at a 45° angle.

35. Machine the $\frac{1}{8}$-in. (3-mm) waste from the square ends (Figure 62-22).

36. Remove the square collet.

37. Insert the knurled handle diameter through the back end of a round collet, extending it not more than $\frac{1}{2}$ in. (13 mm) from the front end and mount the collet in the lathe.

38. Face off the end of the handle until the length has been brought to size.

39. Set the compound rest to a 45° angle.

40. Advancing the compound rest handle, chamfer the end to size according to the print (Figure 62-23). (The only operation remaining is laying out the chamfer on the square end and milling it to depth.)

41. Stand the square end on a surface plate against a V-block and lay out the chamfer depth on all four sides (see print) for both handles.

42. Place the square head in the milling machine vise.

43. Set the square section against the 45° slope, extending the end of the handle about 1 in. (25 mm) beyond the jaws and tighten the vise.

44. Remove the square head and mill the end of the handle down to the laid-out line.

45. Repeat steps 42 to 44 until all the edges have been chamfered.

46. Assemble the offset arms using two $\frac{1}{4}$, 20 × 1 cap screws (Figure 62-24).

SECTION 13

HEAT TREATING OF STEEL

Heat treating deals with changing the internal structure or properties by means of heating and cooling in controlled degrees to adapt the metal to the rigors and stresses of the environment without changing their chemical compositions.

UNIT 63 HARDENING AND TEMPERING

The history of civilization has always been characterized by societies working with metals. Warriors returning from the Crusades during the eleventh century talked about the scimitar blade. It was light, flexible, and sharp. What they were talking about was Damascus steel and another level of metallurgy developed in the Middle East. Many metallurgists today believe that one of the reasons Damascus steel was so good was because it contained a high content of carbon, almost double that of our tool steels.

The Japanese Samurai sword is perhaps one of the finest cutting edges ever produced. It has been learned that the sword is a product of many, many layers of steel. The layers are a result of hammer blows that stretch the initial hot bar of steel to twice its length. It is folded over and over for better than thirty times. It is no wonder that the result of these layers of steel is sometimes referred to as "welded Damascus steel."

Thus it can be seen that a better understanding of metal can lead to sharper cutting tools, for example, or better products can be made by refining and dealing with the behavior of metal during the shaping and treating process. There is a relationship between metallurgy and machining. Many of the marvels of modern technology would not be possible without the science of heat treating metals.

Objectives
After completing this unit you will

1. understand the hardening process.
2. understand the quenching process.
3. understand the tempering process.
4. be able to evaluate a simple hardness test.

Grain Structure

The atomic structure of metal plays a key role in the heat treating process. This is due to the fact that hardness is dependent on the form or chemical composition of the carbon, such as iron carbide, for instance, as well as the carbon content. Without a change in the atomic structure through heating, however, there cannot be a chemical composition and distribution of carbon content for the hardening process to be completed.

The grain structure of any metal is composed of a network of three-dimensional atoms arranged in a geometric pattern. Two arrangements are illustrated and are called space lattices (Figures 63–1 and 63–2).

The composition of every metal contains its own particular pattern within its space lattice. The smallest characteristic group is called a cell or unit and will vary with the type of metal. The most common units to be found are classified in four categories and are briefly described as follows:

Body-centered cubic unit: Not as densely grouped as the others, this unit has an arrangement of eight atoms, one in each of the corners and one in the center (Figure 63–1). These units form and grow in strength in three different di-

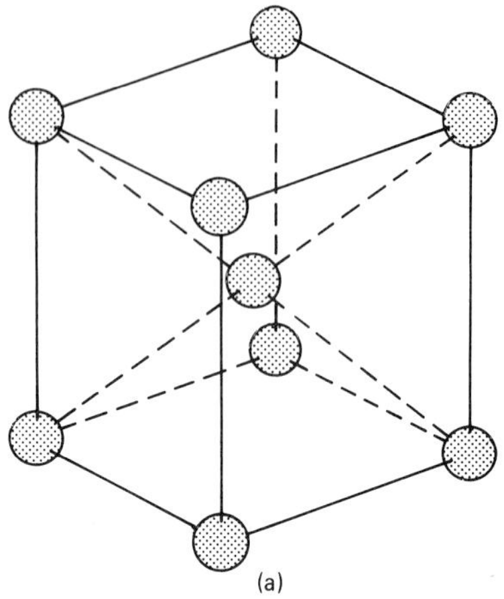

(a)

FIGURE 63–1
Atom arrangement in a body-centered-cubic unit.

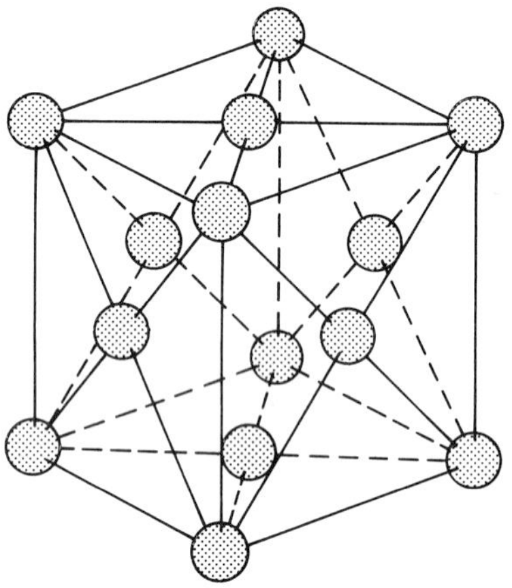

(b)

FIGUIRE 63–2
Atom arrangement in a face-centered-cubic unit.

rections when changing from a liquid molten mass into a solid.

Face-centered cubic unit: This unit has an arrangement of 14 atoms, one in each of the eight corners and one in the center of the six faces of the cube (Figure 63–2). Rearrangement from a body-centered cubic form takes place at elevated temperatures at which gamma iron and austenite form. At these temperatures, the iron carbide completely dissolves to form a solid solution.

FIGURE 63–3
Parts to enter a heat treating furnace.

Hexagonal close-packed unit: While this unit contains 17 atoms in its arrangement, it lacks plasticity and cannot be cold worked. Metals containing this atomic arrangement include magnesium, zinc, titanium, cadmium, and cobalt.

Body-centered tetragonal unit: This unit contains nine atoms arranged in a rectangular body in a manner somewhat similar to the body-centered cubic unit. It rearranges its atomic structure when quenched at hardening temperatures to become martensite.

Common Heat Treating Terms

To understand the general process of heat treating, a background must be provided and basic terms must be explained. Steel is an alloy of iron and carbon. There are other elements found in steel, such as silicon, phosphorus, and sulfur, and a few other impurities, but carbon is the key element needed for hardening. By itself, any tool made of steel would not necessarily stand up to prolonged use. It must have strength, wear resistance, and hardness. To have all these things, the grain structure is changed through heat treating (Figure 63–3).

Essentially, steel is heated to certain temperatures, known as the critical or transformation range, held there to saturate the heat (soaking), and then cooled off (quenching) to become hardened.

Heat treating includes many technical terms that should be understood. Many of these heat treating definitions follow:

FIGURE 63-4
Microstructure of low-carbon steel with large areas of ferrite scattered with small areas of pearlite. (Courtesy of Metcut Research Associates Inc., Cincinnati, Ohio)

FIGURE 63-5
Microstructure of austenite in a nickel-base material. (Courtesy of Metcut Research Associates Inc., Cincinnati, Ohio)

Alpha iron: Pure iron at room temperature. Its atomic structure is body-centered cubic.

Gamma iron: Iron above the transformation temperature. Its atomic structure is face-centered cubic.

Ferrite: Alpha iron, which contains carbon or any other element in solid solution (Figure 63-4).

Cementite: A combination of iron carbides and other carbides to add hardness and strenth to steel.

Pearlite: The microstructure of non-heat-treated steel that is composed of alternate layers of ferrite and cementite.

Austenite: The solid solution found in a mixture of carbon and iron by dissolving the pearlite at a temperature known as the decalescence point (Figure 63-5).

Martensite: The grain structure that results when transforming austenite by quenching and cooling rapidly. It is the carbon content that is instrumental in affecting and increasing this hardness (Figure 63-6).

Decalescence point: The temperature at which pearlite transforms into austenite when heat treating carbon steel.

Recalescence point: The temperature at which austenite transforms to pearlite, generally taking place during slow cooling.

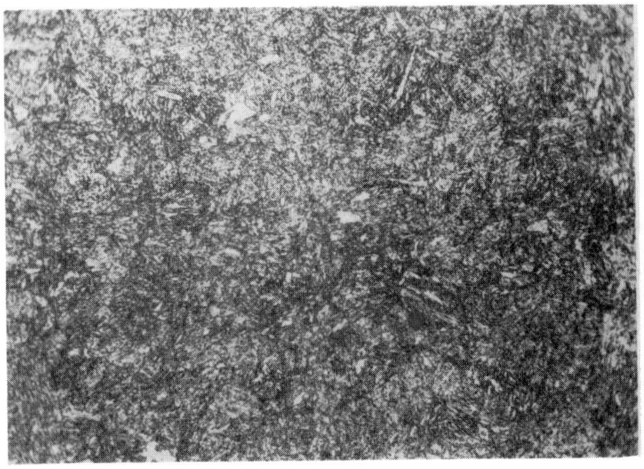

FIGURE 63-6
Microstructure of tempered martensite in medium-carbon alloy steel. (Courtesy of Metcut Research Associates Inc., Cincinnati, Ohio)

Lower critical temperature: The lowest temperature at which steel may be hardened. The same temperature as the decalescence point.

Upper critical temperature point: The maximum temperature to quench steel while achieving hardness and fine grain structure.

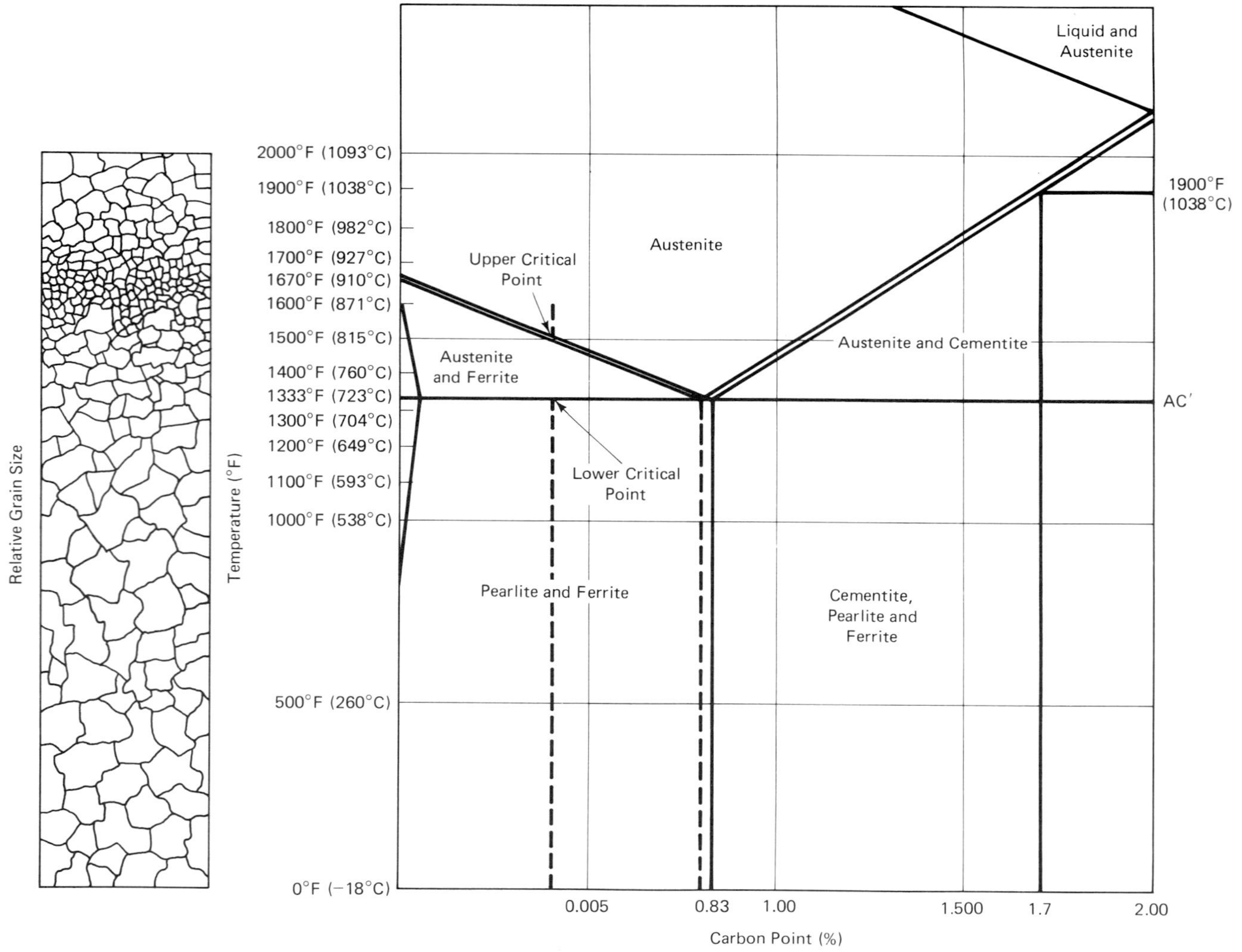

FIGURE 63-7
A simplified diagram illustrating the relationship between the upper critical temperature and the carbon content of the steel.

Critical Range: The temperature zone that falls between the upper and lower critical temperature points (Figure 63–7).

Temper: The slow heating process that removes some of the extreme hardness, reduces brittleness, while it increases toughness in hardened steel.

Drawing: Another term for tempering.

Annealing: To soften and bring back to a machining state. Annealing is accomplished by heating steel above the upper critical point and allowing it to cool off gradually in the furnace or buried in sand or lime.

Normalizing: A process whereby steel is returned to "normal" by heating it above the upper critical point and allowing it to cool gradually to room temperature in still air. This removes internal strains and stresses while improving grain structure.

Eutectoid steel: A steel whose carbon content ranges from 0.80 to 0.85% (80 to 85 points). Unhardened it has an all-pearlite grain structure. Unlike other compositions, eutectoid steel will not range in temperatures when changing from pearlite to austenite (or vice versa). It is noted for its transformation at a single temperature.

Hypereutectoid steel: A steel whose carbon content is above the 0.80 to 0.85% range. Unhardened it is surrounded by cementite boundaries. This steel must be held above the upper critical temperature for maximum hardness.

Hypoeutectoid steel: A steel whose carbon content is less than the 0.80 to 0.85% range. Due to an excess of ferrite, carbon and other alloying agents will not go into solution unless held above the upper critical range.

Spheroidizing: A modified annealing process of heating and cooling steel that produces a rounded or globular form of carbide. For best results, these carbides should be small and evenly distributed. This will provide better cold forming conditions in hypoeutectoid steels and increase machinability in hypereutectoid steels.

Isothermal annealing: A controlled temperature annealing process that produces a fairly soft ferrite carbide structure when transforming the austenite by heating above the critical temperature. This is followed by a cooling period in which the steel is held following a prearranged temperature range and cycle of time.

Quenching: A process of rapid cooling heated metal by immersion in various media, such as gases, liquids, or solids, to achieve hardness.

Soaking: The saturation of heat in metal at a fixed temperature to achieve a more complete and uniform penetration.

Transformation range: Heating temperature in which pearlite changes to austenite. In cooling, the temperature at which austenite changes to pearlite. These temperatures often overlap but seldom occur at the same reading (Figure 63–8).

Hardening Process

Steels to be hardened will fall into two classes: those with a carbon content high enough to be heat treated in the manner previously described and those whose carbon content is not high enough and must have a shell of hardness added to its surface. This is called case hardening and will be examined in Unit 65.

Steels with a minimum of 75 points carbon (0.75%) will harden but cannot be used for cutting tools. Steels used for cutting tools range from 90 to 150 points but seldom average more than 130 points. These steels are called high-carbon steels or simply tool steels (Figure 63–9). It is the hardening of tool steel that will be explained.

Furnaces

The heat treating of tool steel takes place in a furnace heated by gas, oil, electricity, or solid fuel. Its heating chamber atmosphere is read by a pyrometer (Figure 63–10). A sensing device consisting of two dissimilar metals, called a thermocouple, is mounted inside the furnace and connected by leads to the pyrometer. Changes in temperature can be read on the pyrometer, which can be set to shut off the heat to the furnace automatically when a predetermined heat temperature has been reached. The factors that will influence hardening are rate of heating, hardening temperature, soaking time, and method of cooling. To cover a variety of situations, steel producers publish handbooks that not only offer steel selections but also list such factors as the critical temperature of the steel supplied, furnace atmosphere, and tempering time (Figure 63–11).

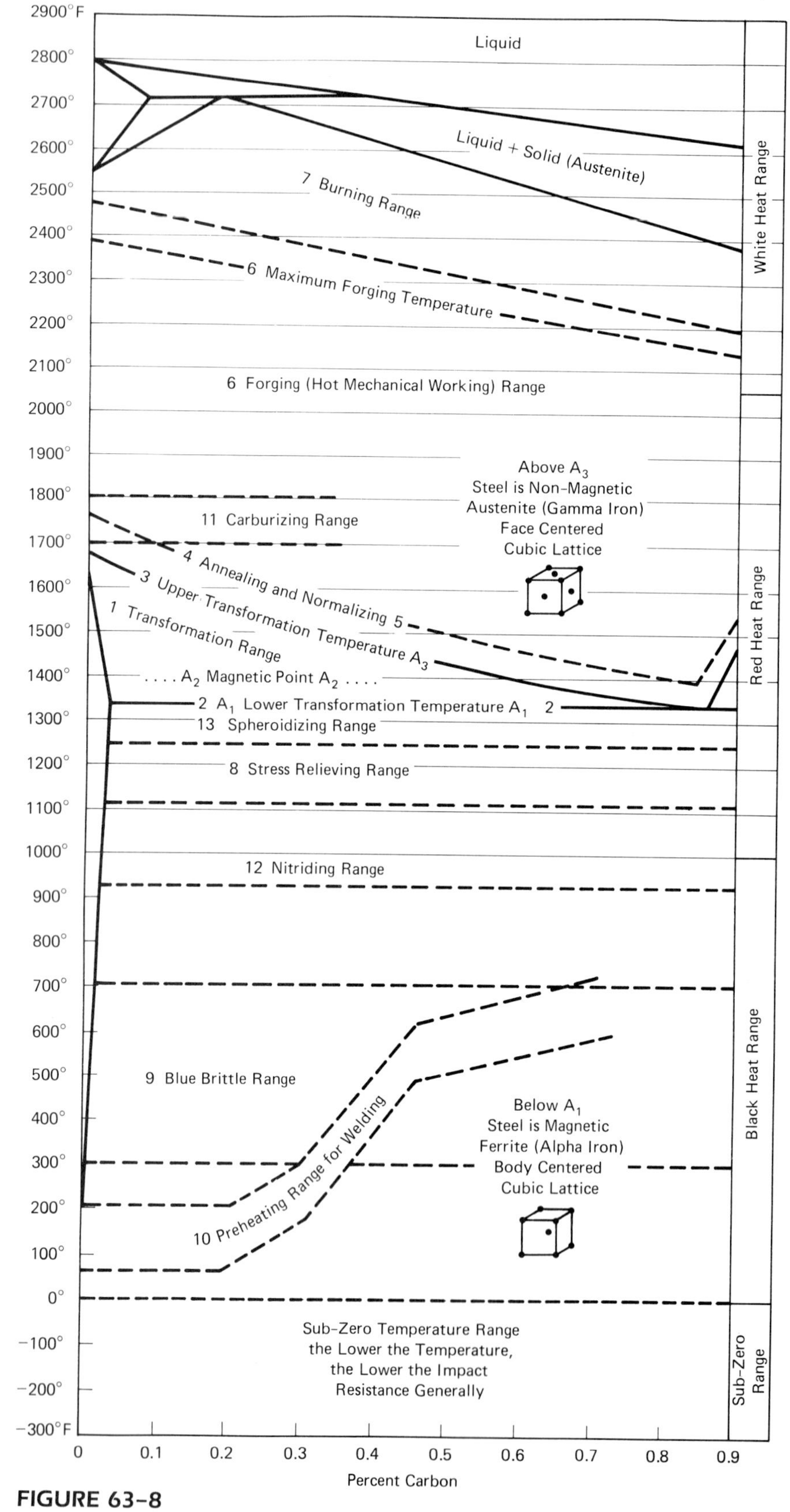

FIGURE 63–8

Basic guide to ferrous metallurgy. (Credit: Tempel, Big Tree Industries, Inc.)

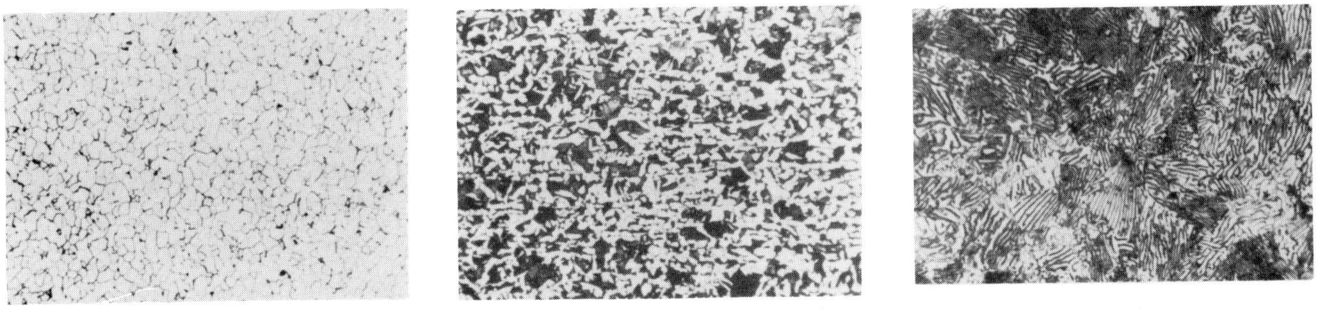

FIGURE 63-9
Microstructure of low-, medium-, and high-carbon steel. (Credit: American Iron & Steel Institute)

FIGURE 63-10
A heat treating furnace with indicating temperature control.

Quenching

The hardening of steel takes place by slowly bringing the metal up to the recommended critical temperature range and allowing it to soak to provide time for internal structural changes to occur in the metal. At the decalescence point, or slightly above, the metal is quenched in the recommended medium: oil, water, brine, or air cooled, depending upon the type of steel or hardness. Quenching at this point should produce the smallest grain size and greatest hardness.

FIGURE 63-11
Pot-type heat treating furnace.

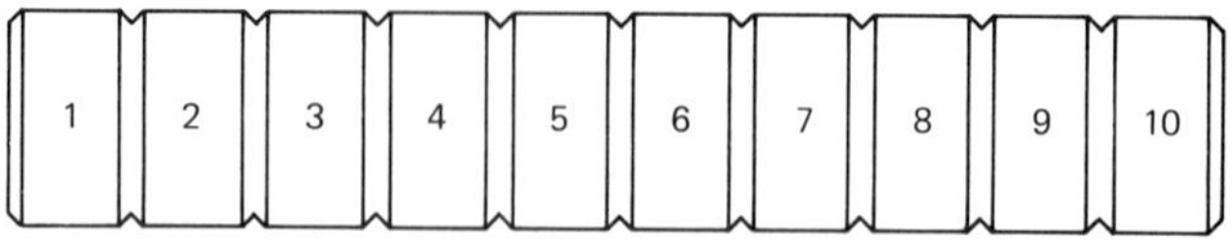

FIGURE 63-12
A typical test piece used for Metcalf's experiment.

Quenching will also involve important factors such as metal thickness and temperature of the medium. Cooling too rapidly will produce internal stress, which can lead to cracking. Cooling too slowly may result in uneven or incomplete hardening. It is best to follow the steelmaker's recommendations as to choice of quenching medium and conditions involved. Once immersed in the medium, the agitation generally followed is that of making figure eights or up and down movements within the medium to prevent warpage.

Metcalf's Experiment

A simple test can be made to show the effects of overheating and underheating tool steel. The test is conducted on a length of rod grooved and hardened to have its sections broken off (Figure 63-12).

Procedure

1. Select a length of $\frac{1}{2}$-in. (13 mm) drill rod that is 5 in. (125 mm) long.
2. Hold the test piece in the lathe chuck.
3. With a sharp pointed toolbit, make a series of shallow grooves along the length $\frac{1}{2}$ in. (13 mm) apart.
4. Stamp each solid section numbers 1 to 10.
5. Using an acetylene torch, concentrate the heat to the number 10 section to bring it to white heat, keeping sections 5 to 7 to a cherry red color but *do not heat sections* 1 to 4.
6. Quench in cold water.
7. Dry off the test piece.
8. Using the edge of a worn-out file, test each section for hardness.
9. Break off as many sections as possible and examine the inner core.

It will be seen that the overheated sections break off easily and that the grain structure is very coarse. The underheated sections 1 to 4 will be easily cut with the file. The inner core will reveal a pearlite condition—an uneven but fine grain structure. The middle sections 5 to 7 will display the greatest hardness and resistance to breaking off.

Tempering Process

When steel has been properly hardened, it will feel glassy hard when a file is used on it. A file is not the most accurate way of testing for hardness, but it is a practical way by which diemakers and skilled tradesmen quickly test for hardness. Steels cannot be used in this state. They must be tempered to remove the extreme hardness and brittleness that steels take on upon hardening. Tempering also increases toughness and relieves internal strains and stresses the steel undergoes in quenching. Not to do so will only result in cracking or fracturing in use.

In tempering, steels are heated slowly and brought to a temperature range and immersed in the same bath it was initially quenched to harden it. Generally, tempering should begin as soon as the cooled steel is comfortable enough to hold in the bare hand.

Tempering temperatures will range as low as 300 to 1200°F (149 to 649°C), depending on four factors:

1. Nature of the material (carbon steel or alloy steel).
2. Carbon content.
3. Hardness requirement.
4. Toughness requirement.

Although heat treating directions can be found in technical handbooks, particular tempering information is best supplied by the steelmaker.

Modern methods of tempering will find hardened steels immersed in baths of heated oil, molten lead, or other liquids that are temperature controlled. Upon soaking in these media for recommended lengths of time, which is based on the workpiece thickness, the steel is quenched. This type of tempering results in greater hardness uniformity.

Another method of tempering is called drawing. It depends on the use of changing colors to indicate heat temperature and acts as a quenching guide. Upon hardening the steel, the surface is cleaned of scale and polished. If a large section like a die is going to be tempered this way, it is placed on

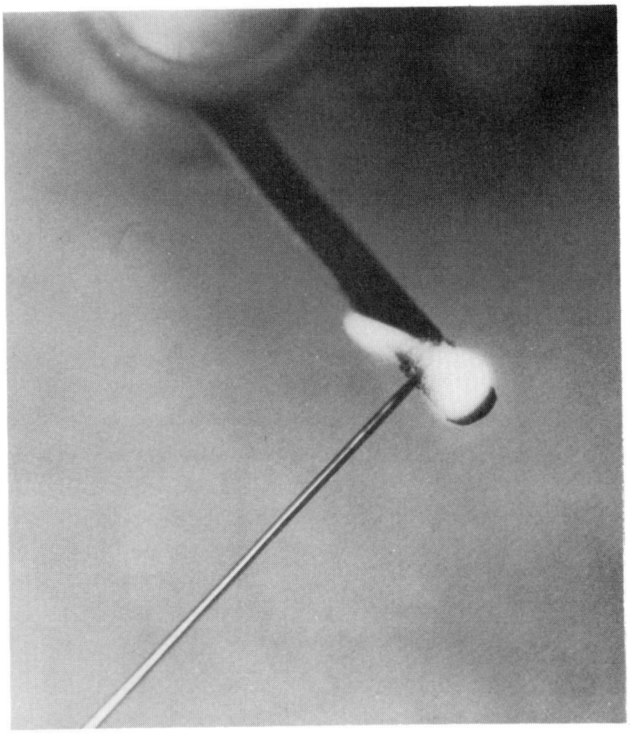

FIGURE 63-13
Color bands of oxide will travel as a lit match is held under the end of a needle.

alternating heated blocks so that the heat can slowly spread from the back side up toward the cutting edge.

If a tool is small enough, like a center punch, it can be held by tongs. Slowly, heat from a torch is played against the back end, never holding the torch steady on one spot. Rotating the center punch so it can be heated uniformly, it will be seen that the polished surface soon gives way to bands of colors—a pale yellow to straw to brown, dark brown, and dark blue. The colors are actually films of oxide and only indicate heat changes. It will do the same thing on other surfaces, hardened or not. Figure 63-13 demonstrates how the color band will form and travel by holding a lit match to the end of a needle. When the working surface of the hardened steel has achieved the desired color as noted in Figure 63-8, it is quenched.

Most press dies are now made with air hardening shock steel because of its stability in heat treating.

Careers

In the field of metalworking, there is no more exciting field than that of metallurgy—the study of the composition of metals. It is the metallurgist who helps develop and examine the methods and means for achieving better grades and combinations of metals. In the research area, metallurgists work with other professionals such as chemists to achieve high-performing metals and determine how they can best be heat treated. In industry, a metallurgist's job may consist of testing and controlling the quality of metal used in fabrication.

Working closely with metallurgists in industry are the technicians. Using hardness testing equipment or similar types of machines that examine the metal's properties—why metals break down (fatigue) or why they stand up (tensile strength)—technicians of this type have to be highly trained and are in demand.

Just as important to the metal trades are the heat treaters. In addition to the full-time heat treating technicians that work in various processes involved, there are a variety of trades that require heat treating skills. Diemakers, toolmakers, experimental machinists, modelmakers, and any trade that requires making tools or gets involved in machining has to have a background in the science of heat treating.

UNIT 64 ANNEALING

Annealing is the process of bringing hardened steel to a pliable state or softening so it can be machined. Hardened steel will, from time to time, be annealed so it can be reworked. Hardened machine parts is a good illustration of this sort of thing. When clutch parts have to be changed to accommodate a repair or product change, they must undergo an annealing process so they can be machined.

Another example of the use of annealing lies in reducing the brittleness and hardness as well as relieving internal stress caused by cold working, rolling, or drawing wire.

Objectives
After completing this unit you will be able to

1. understand the different levels of annealing.
2. soften a hardened workpiece by full annealing.

FIGURE 64–1
This vacuum furnace may be used for applications such as annealing, hardening, or carburizing, where uniform heating and quenching in a controlled atmosphere leaves work with a bright, clean finish. (Credit: Lindberg, a unit of General Signal)

FIGURE 64–2
The air–gas ratio is manually controlled in this toolroom furnace.

3. process anneal low-carbon steel workpieces.
4. undertake the normalizing process.

Annealing Terms and Processes

There are several levels of annealing that range from full annealing to normalizing. Each is done with a specific purpose in mind.

Full Annealing

Full annealing is done by completely softening the steel (Figure 64–1). It will also relieve internal stresses and restore machinability in the metal. A high-carbon tool steel die would have to be fully annealed before major repairs could be done to it. To minimize scaling and decarburization (loss of surface carbon) during heat treating, it is recommended using a controlled-atmosphere furnace where a specially manufactured gas composition is used to protect the work. For best results in full annealing, the recommendations of the leading steel producing companies should be followed.

Procedure

1. Place the die in a controlled-atmosphere furnace.
2. Set the pyrometer to 1330 to 1350°F (721 to 732°C).
3. Heat the die and allow it to soak at this temperature for approximately 1 h/in. (25 mm) of die thickness.
4. Cool very slowly at the controlled rate of not more than 20°F (7°C) per hour until the furnace is black.
5. Shut off the furnace and allow it to cool naturally.

In toolrooms where atmosphere furnaces may not exist, one where the atmosphere is controlled by manually adjusting the air–gas ratio (Figure 64–2), full annealing can take place by heating the object 100°F (38°C) above the upper critical temperature and removing the workpiece from the furnace after soaking and burying it in a bed of lime, sand, or dry ashes. There, under cover of these poor heat conductors, the workpiece is allowed to cool naturally.

Process Annealing

Materials that have become work hardened during cold working are frequently stress relieved (Figure 64–3). Process annealing is generally carried out on

FIGURE 64–3
Process annealing can be controlled in a continuous atmosphere-controlled annealing furnace. (Credit: Carpenter Technology Corporation)

low-carbon steels and is done in the following manner:

1. Place the workpiece in a furnace and set the pyrometer to read 30°F (-1°C) below the lower critical temperature and heat slowly.
2. Allow the workpiece to soak at this temperature for approximately 1 h/in. (25 mm) of thickness.
3. Remove the workpiece from the furnace and place it on firebricks to cool in the air.

Spheroidizing

When carbon steel parts require a special grain structure that is soft and easy to machine, the parts are spheroidized. The microstructure reveals that the cementite particles become rounded globules (spheroids) and are surrounded by ferrite (Figure 64–4). The result is a tougher steel that is easier to bend, twist, or draw than by using the full annealing method. While processes for spheroidizing vary due to heat treating conditions and steel composition, the following is one method used.

Procedure

1. Place the part in a furnace, set the pyrometer to read approximately 30°F (-1°C) below the lower critical temperature, and heat slowly.
2. Allow the part to soak at this temperature for several hours.
3. Shut off the furnace and allow the part to cool down to about 1000°F (537°C).
4. Remove the workpiece from the furnace and allow it to cool in the air.

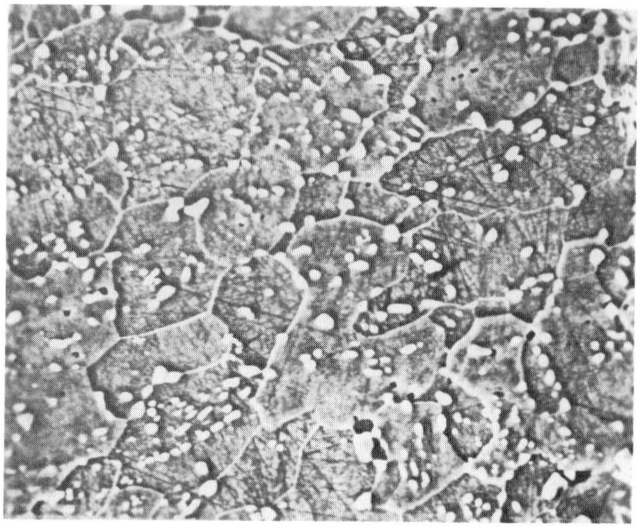

FIGURE 64-4
Microstructure of spherodites. (Credit: U.S. Steel)

Normalizing

While the normalizing process may seem to be the same as full annealing, there are several differences in results. Both relieve internal stresses and improve machining. But in removing any hardness to return the workpiece to "normal," there is not uniform removal; therefore, hard spots may be encountered during normalizing. The grain structure will be harder and the workpiece will not be able to be shaped as easily as it would if it were fully annealed (Figure 64-5).

Procedure

1. Place the workpiece in the furnace and set the pyrometer to 1500°F (816°C) temperature and slowly heat the furnace.
2. Upon reaching this temperature, allow the workpiece to soak 1 h/in. (25 mm) of thickness.

FIGURE 64-5
Large size normalizing is carried out in huge box furnaces used for annealing and tempering. (Credit: Carpenter Technology Corporation)

3. Remove the workpiece from the furnace and allow it to cool slowly in air to room temperature. *Note:* Thin workpiece may harden in air during the normalizing process. To prevent this, place the part in lime or ashes to slow the cooling process.

UNIT 65 CASE HARDENING

Early civilizations found that heating, hammering, and cooling iron as it was subjected to a traditional charcoal fire would result in a veneer of hardness. By continually working the iron in this fashion, the surface would absorb the carbon from the open fire. The resultant hardness was not understood at the time. Today, this process is called case hardening. Steel with a low carbon content is heated and carbon in the form of gas or hydrocarbon liquid is added and quenched. The process leaves the outer surface coated with a hardness while its interior remains structurally soft.

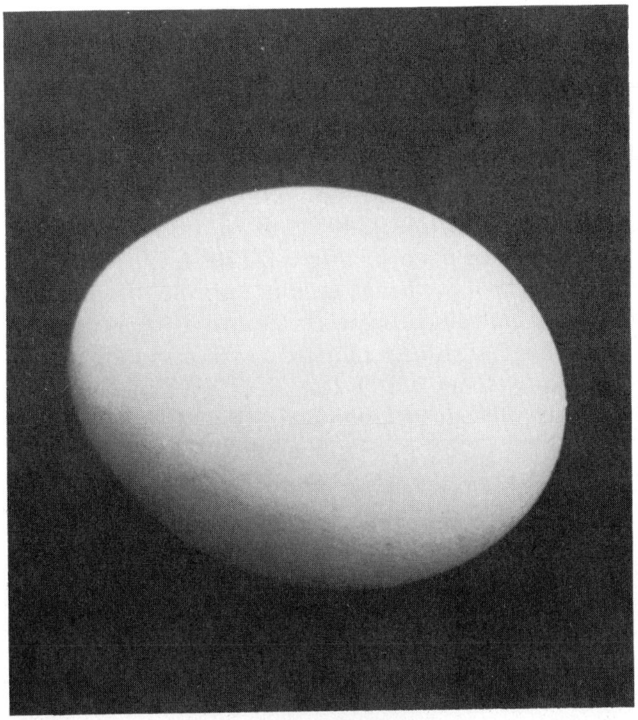

FIGURE 65–1
A raw egg illustrates the properties of case hardening: a shell of hardness that surrounds a soft interior. (Credit: Philip Rapisarda)

Case hardening has been compared to a raw egg: a shell of hardness that surrounds the soft interior (Figure 65–1). Commercially, case hardening is a cheap method of hardening steel.

Objectives
After completing this unit you will be able to

1. identify five different methods of case hardening steel.
2. case harden a piece of machine steel.

Case Hardening Materials

Low-carbon steels are widely used in industry and require careful heat control, double refining, and double quenching to minimize grain growth. The low-carbon, low-alloy steel contains higher levels of nickel, chromium, and molybdenum, which assures little grain growth and the steel can be heated to lower temperatures than the plain low-carbon steel.

Carburizing Process

There are several case hardening processes, with a subdivision of methods within each process. These include carburizing, carbonitriding, and nitriding.

Carburizing

Like a sponge, the surface of a low-carbon steel (containing below 0.3% carbon) is made to soak up carbon from carbon-rich (carbonaceous) materials when they are heated together. The depth of hardness is controlled by the carburizing material, temperature, and time exposure. Carburizing may be completed by any of three methods: pack carburizing, liquid carburizing, or gas carburizing.

Pack Carburizing Process

1. Place a 1-in. (25-mm) layer of carbonaceous material such as wood, bone, or charcoal within a metal container (Figure 65–2).

2. Place the parts on the layer (Figure 65–3).

3. Pack around and cover the parts with at least $1\frac{1}{2}$ in. (38 mm) carbonaceous materials.

4. Shake the container to settle the contents.

5. Close the container with a metal cover and seal it with fire clay (Figure 65–4).

6. Place the container in a furnace and heat it to about 1700°F (926°C).

FIGURE 65-2
Pack carburizing begins by forming a bed of carbonaceous material.

FIGURE 65-4
Place a lid on the container and seal it with fire clay.

FIGURE 65-3
Lay the workpiece on the carbonaceous material.

7. Soak at this temperature for depth-of-hardness penetration according to the steelmaker's recommendation.

8. Shut off the furnace and allow the contents to cool.

9. Remove the contents upon cooling.

10. Reheat the parts to its critical temperature and quench as recommended by the steelmaker.

Liquid Carburizing

Another means of introducing carbon and nitrogen into the surface of low-carbon steel parts is by carburizing them in a molten salt bath. The bath is a mixture of carbonates, chlorides, and cyanides. It is the mixture that controls the amount of carbon or nitrogen absorbed. Generally, carbon is absorbed the most.

As with any case hardening process, the depth of hardness is controlled by temperature and time exposure to the hardening environment. The temperature in liquid carburizing will range from 1650 to 1750°F (898.9 to 954.4°C), and a depth of penetration ranging from 0.020 to 0.120 in. (0.50 to 3.04 mm) is possible. Since carburizing may liquify any cyanide contained, which is explosive when it comes in contact with any moisture, parts should be preheated to about 800°F (426.6°C) before suspending them in salt baths. Upon being carburized to any desired depth, parts are immediately quenched.

Gas Carburizing

In this process, gas is used as the carbonaceous material as it is being heated and circulated in a special furnace. Natural gas or propane is pumped into a sealed gas container equipped with inlet and exhaust ports. It is this container that holds the parts to be treated. As the furnace is heated to its carburizing temperature, exhaust gases from the container are burned off to prevent them from seeping back into the container. It is during this process that carbon from the gas is absorbed by the parts. Upon

FIGURE 65-5
Carburizing chain components. A two-chamber furnace provides a higher carbon environment and direct radiation to the work loads to allow faster heat-up rates and accept heavier loads. (Credit: Lindberg, a unit of General Signal)

achieving the time-temperature relationship—usually about 4 h at a temperature of 1700°F (926.6°C)—the parts may be quenched. The depth of case produced will vary between 0.020 and 0.030 in. (0.50 and 0.76 mm). A deeper case can be produced with a prolonged time exposure (Figure 65-5).

It is preferred to quench directly from the carburizing temperature to eliminate scaling. If the parts are allowed to cool, they must be reheated to the critical temperature and then quenched.

Carbonitriding Processes

Carbonitriding will introduce carbon and nitrogen to the steel's outer surface to produce a hard, shallow case and can be done in any of two methods.

Cyaniding

This form of heat treating is a liquid carbonitriding process. Parts are processed in a salt bath composed of 30% cyanide, 40% carbonate, and 30% chloride. In some cases, greater amounts of cyanide may be added to maintain hardening levels for specific applications. Although the cyaniding is carried out in a pot-type furnace, there cannot be a strong enough warning issued on the dangers of working with cyanide salts. The fumes can be lethal if inhaled and are highly toxic if exposed to open wounds or scratches.

Parts are suspended in cyaniding temperatures that range just above the steel's critical temperature, usually from 1400 to 1600°F (760 to 871.1°C). Greater penetrations are possible when working with temperatures that range from 1550 to 1600°F

(843.3 to 871.1 °C). In this range, a depth of 0.003 to 0.005 in. (0.08 to 0.13 mm) is obtained in 30 min. Although seldom exceeding 0.020 in. (0.51 mm), a 0.015-in. (0.38-mm) depth can be obtained in about 2 h. Parts can be quenched in water, oil, or brine depending on the type of steel used. Cyanide-hardened parts are easy to wash due to the fact that all the components of the salt bath are water soluble.

The equipment used for cyanide hardening is similar to that used for liquid carburizing. Not only is there less distortion involved, but also the lower temperatures and shorter hardening cycles prove less costly as well.

Carbonitriding

This process is really gas cyaniding and is similar in operation to gas carburizing. A mixture of nitrogen and carburizing gas is allowed to enter and circulate through a special furnace that holds the workpieces. Ammonia is the source of nitrogen. A gaseous atmosphere is heated in the furnace whose elevated temperature is regulated to range from about 1400 to 1700 °F (787.7 to 926.6 °C). It is in this heat range that the workpiece absorbs both the nitrogen from the ammonia and the carbon from the carburizing gas. Quenching follows in a bath of oil, which results in less distortion of the parts being hardened.

The success of carbonitriding hardening depends on the temperature, time exposure, atmosphere, and type of steel. The higher the temperature, the less effective is the ammonia mixture. The lower the temperature, the more concentration of nitrogen there is on the steel. The ammonia content must be decreased as the content of the steel's alloy is increased.

Hardened case depths of 0.001 to 0.003 in. (0.02 to 0.07 mm) are commonly used on thin parts for light-duty wear resistance. A depth of 0.030 in. (0.76 mm) can be produced in 4 to 5 h at a temperature of 1600 °F (871.1 °C).

Nitriding Processes

When alloy steels have to have exceptionally hard surfaces on parts already machined and heat treated, they are subjected to nitriding. The nitrides combine with the steel's alloys, which are nitride forming elements, to produce a harder case than any other heat treatment. The alloys include chromium, vanadium, molybdenum, and nickel. Nitriding can be performed in either gas or a salt bath. In either case, no quenching is required.

Gas Nitriding

Ammonia gas in a protected atmosphere furnace is the environment generally used in nitriding. Steels immersed in this atmosphere must be hardened and tempered before being nitrided. The nitriding temperature for all steels is between 925 and 1050 °F (496.1 and 565.5 °C). Because of this, steels being nitrided should have a tempering temperature high enough to withstand any structure instability, at least 50 °F (28 °C) higher than the maximum nitriding temperature.

Gas nitriding is a slow process, requiring as much as two days to complete its hardening process to a depth of 0.020 in. (0.50 mm). As the heated ammonia is circulated within the furnace, a chemical action takes place. The ammonia breaks down to form hydrogen and nitrogen, with nitrogen uniting with the steel's alloy to form hard nitrides. Due to the comparatively low temperatures used, nitriding produces less deforming and distortion than conventional hardening and will not affect the inner core of the steel (Figure 65-6).

FIGURE 65-6
Large loads can be carburized in a gas-fired pot furnace such as is seen here. (Credit: Lindberg, a unit of General Signal)

Salt Nitriding

Nitriding can also be done using a molten bath of salts. Hardened parts are suspended in the bath whose temperature range is similar to that of gas nitriding. In this case, only nitrogen will be absorbed by the steel's surface. Cutting tools such as taps, dies, drills, and milling cutters are often nitrided to increase their tool life by two and three times (Figure 65-7).

Cyanide salts as well as other salt mixtures are available for nitriding, but they are deadly poisons. As with the use of other salts, extreme care and caution should be followed in using them (Figure 65-8).

Flame Hardening

Flame hardening is a process used to heat treat long parts, such as lathe ways and flat bearing surfaces of machine tools. It can also be used to harden crankshafts, gear teeth, cams, and similar items to varying depths of from 0.003 to 0.250 in. (0.76 to 6.35 mm).

A specially designed torch burning oxyacetylene gas is guided over a designated area where it rapidly brings the surface to its hardening temperature. This is immediately followed by a quenching spray. On smaller parts (Figure 65-9) a flame is aimed at them and the parts are spray quenched automatically.

The tempering of flame-hardened parts should follow immediately. Small parts should be tempered in a furnace, while large parts are tempered by a flame method followed by spray quenching.

FIGURE 65-8
The continuous furnace system shown here can be used for carburizing and carbonitriding. It features a quench tank in which two loads can be quenched at the same time. (Credit: Lindberg, a unit of General Signal)

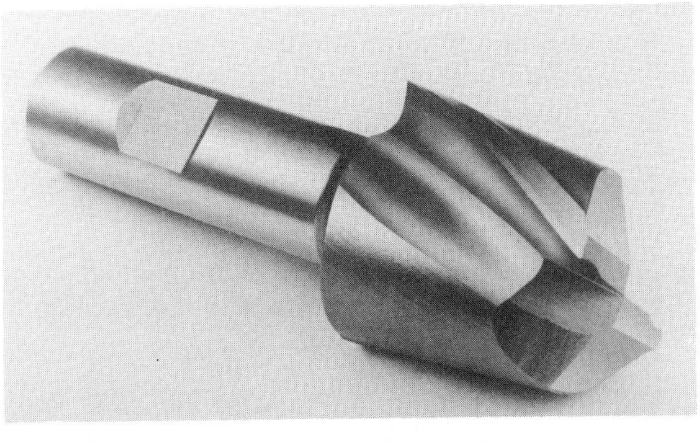

FIGURE 65-7
Niriding can increase cutting tool life two to three times. It can also minimize corrosion. (Credit: Union/Butterfield)

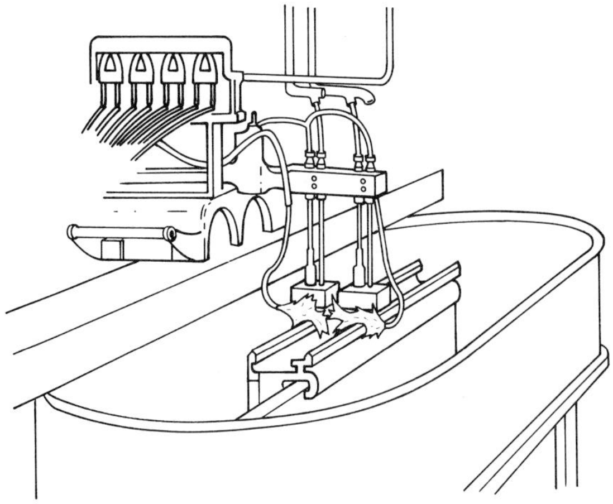

FIGURE 65-9
In flame hardening, the surface is quickly brought to hardening temperature and immediately quenched.

FIGURE 65–10
Hardening temperatures are developed by sending high-frequency alternating currents through induction coils to create a magnetic field, such as is used in this vertical tooth-by-tooth induction hardener. (Credit: Lindberg, a unit of General Signal)

Induction Hardening

Ferrous metals can be hardened by means of high-frequency alternating currents that are sent through induction coils surrounding the workpiece (Figure 65–10). The currents create a magnetic field around the coils and workpiece that induces eddy currents to heat up the surface of the workpiece to the required hardening temperature within seconds.

Quenching of induction-hardened parts may take place in water, oil, or compressed air and may be applied as a spray or bath, either manually or automatically, depending on the equipment involved.

Depth of hardness is controlled by the type of frequency used, which can vary from 1 kHz/sec to 2 MHz/sec. High frequencies are used for shallow hardness, ranging from 0.010 to 0.060 in. (0.25 to 1.52 mm) in depth. Lower frequencies can produce deeper case depths, ranging from $\frac{1}{16}$ to $\frac{1}{4}$ (1.6 to 6 mm). Very high frequencies tend to have a very shallow penetration.

Selective hardening can also be done. Splines, crankshafts, and gear teeth are but some of the parts selectively hardened. Through hardening is also possible on parts such as rods and shafts. For these applications, a lower frequency is used, ranging between 180 and 3000 Hz. By using low-temperature induction coils, parts can also be tempered using induction heating in a manner similar to the hardening process.

Laser Hardening

The technique of case hardening has been updated to include laser hardening. Due to the high concentration of focused heat and rapid cooling, hardenable materials self-quench, eliminating a liquid bath. This has the advantage of producing a hard surface for wear resistance while its inner core remains ductile. Since less heat is put into the surface than under conventional methods, surface stress is minimized.

The actual depth and width of hardness is directly related to the type of material used. Generally, areas $\frac{1}{2}$ in. (13 mm) wide and, on flat surfaces, depths of 0.003 in. (0.76 mm) are possible. On certain shapes, depths of 0.075 in. (1.90 mm) can be produced.

Unlike laser welding, laser beam hardening does not require any special atmosphere in which to operate. It can be used in any conventional shop, and the beam can be focused into rings, gears, cylinder bores, and other difficult areas. In addition, it can produce a hard substructure to support any hard surface coating such as ceramic, chrome, and tungsten carbide without chipping or flaking the coating.

UNIT 66 HARDNESS TESTING

In the metalworking industry, fabrication of parts exist because they are made to some plan, program, or print. The physical and mechanical properties of hardened metals are also measurable. Metallurgical tests have shown that the performance of metals is predictable when the physical and mechanical values are at certain levels. It goes beyond trying to cut into the hardened surface by passing a file over it.

To determine these measurable values, nondestructive hardness testing exists. The five principal means of measuring hardness are the Brinell, the Rockwell, the Vickers, the Knoop, and the Scleroscope testing devices. All but the Scleroscope will use a system measuring depth or width of indentation to determine hardness. The procedure for these testing standards are established by ASTM (American Society for Testing and Materials).

Objectives
After completing this unit you will

1. understand at least three different methods of hardness testing used in the machine shop.
2. be able to perform a Rockwell hardness test on a test specimen.
3. be able to perform a hardness test using a Scleroscope stand mounted device.

Hardness Terms

In order to understand how these nondestructive testing devices work, it is best to explain a few hardness terms in relation to the metal's physical and mechanical makeup:

Hardness: resisting permanent indentation or scratching.
Wear resistance: being able to resist surface removal through use.
Machinability: being able to be cut by machine tools.

It is only through hardness testing that these properties can best be evaluated.

Brinell Hardness Tester

The Brinell hardness tester (Figure 66-1) uses a 5- or 10-mm ball penetrator to indent the surface of a test specimen. The indent is formed under a steady load of weights ranging between 500 and 3000 kg using a penetrator that may be a hardened steel ball or one made of tungsten carbide. The indent is made over a span of time, varying between 15 and 30 sec, specific periods of which may be necessary to halt the plastic flow of material.

Once the load has been removed, the circular impression is viewed through a 20 × or 40 × microscope and measured. The millimeter measurement can be mathematically converted to reveal the Brinell hardness number (BHN) of the metal by dividing the applied load by the surface area of the impression (in square millimeters). Generally, conversion tables such as the E-10 document supplied by ASTM provide the BHN for given loads, ball diameters, and indent diameters. The higher the BHN, the harder the material.

Used for both laboratory and shop work (Figure 66-2), Brinell hardness testers can be used on ferrous and nonferrous materials. Softer materials or thin materials are indented using a 500-kg load through the penetrator. Case-hardened surfaces do not lend themselves to an adequate indentation and should not be used for Brinell hardness testing. Conversely, forged metals, cast metals, and powdered metal parts are effectively measured.

FIGURE 66-1
Brinell hardness tester. (Credit: Wilson Instruments)

FIGURE 66–2
Automatic Brinell testing on drill pipe casing eliminates manual intervention while providing automatic data. (Credit: Wilson Instruments)

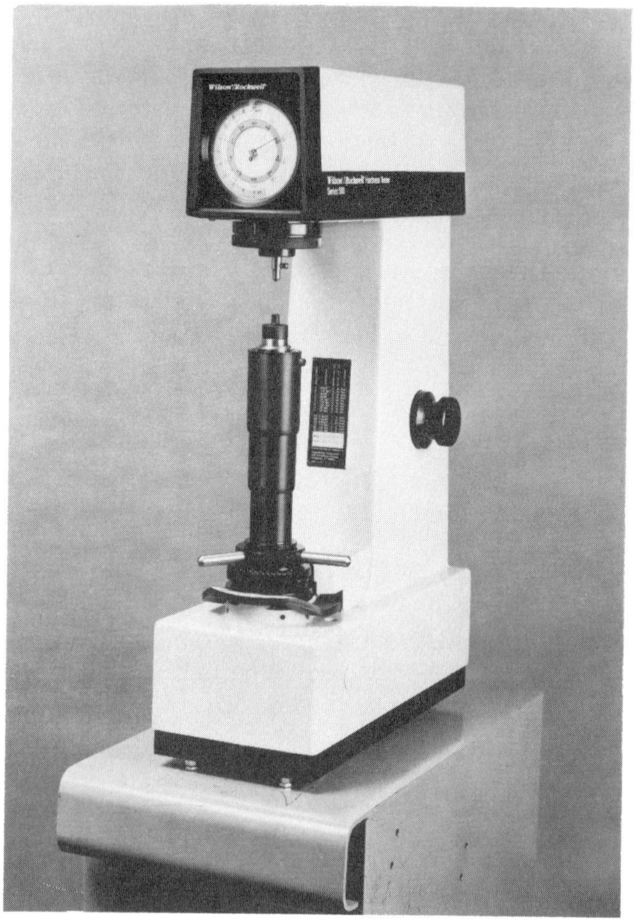

FIGURE 66–3
The Rockwell hardness tester is the most commonly used tester in the United States. (Credit: Wilson Instruments)

Rockwell Hardness Tester

The Rockwell hardness tester is the most commonly used tester in the United States. There are two types of indenters used in Rockwell hardness testers. In accordance with ASTM document E-18, a 120° diamond cone with a radius of 0.20 mm (Brale) is used for hardened steels and cemented carbides. A steel ball measuring $\frac{1}{16}$, $\frac{1}{8}$, $\frac{1}{4}$, or $\frac{1}{2}$ in. in diameter is used for the softer materials that range from brass to plastics. Unlike the Brinell tester, the Rockwell hardness tester (Figure 66–3) uses two loads for its indentation, a major load and a minor load. The minor load is applied first followed by the major load. The depth of indentation is read on one of the several scales, the most commonly used being the "B" and "C" scale. Rockwell hardness is read by combining letters and numbers to indicate the major load and type of penetrator.

For example, the outer black "C" scale is read in testing the hardness of steel. The diamond Brale is used under pressure exerted by a major load of either 60, 100, or 150 kgf (kilogram-force). The softer the steel, the lower the reading. Machinable steel has a reading of R_c 25. A hardened high-speed steel cutting tool like a tap, on the other hand, will register R_c 63 to R_c 66.

A number of anvils are available to accommodate the variety of shapes and sizes requiring hardness testing (Figure 66–4). Basically, the following principles are used for Rockwell hardness testing.

Procedure

1. Select and mount a proper-size anvil.
2. Wipe the specimen's surface clean of any oil, grit, or scale.

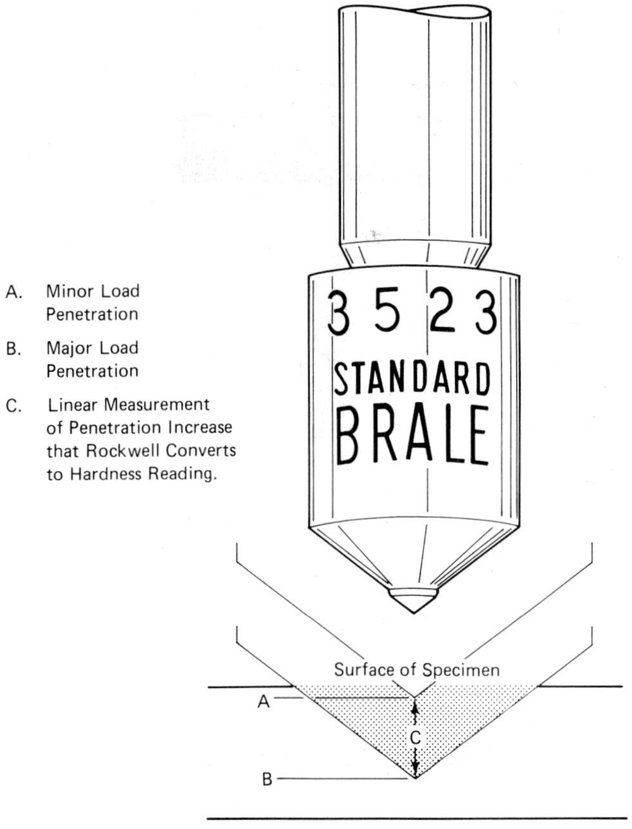

FIGURE 66-4
An illustration showing how the penetration difference between the minor and major load is converted to a Rockwell hardness reading. (Credit: Wilson Instruments)

3. Select and mount a proper-size indenter: Brale if using hardened steel, a $\frac{1}{16}$-in. ball for softer materials.

4. Place the specimen on the anvil.

5. Apply the minor load (10 kgf) by elevating the anvil until the specimen touches the indenter; then slowly raise the specimen until a small pointer is opposite the red dot on the dial face.

6. Adjust the outer dial (bezel) to zero.

7. Apply the major load: 150 kgf if using the Brale "C" scale, 100 kgf if using the "B" scale ball penetrator.

8. When the dial hand has come to rest, read the black figures on the outer "C" scale if using the Brale; read the red figures on the inner "B" scale if using a ball penetrator. The reading is the depth of indentation—the difference between the minor and major loads—and represents the Rockwell hardness number (Figure 66-5).

9. Release the minor load and remove the specimen.

More advanced testing models have electronic features such as digital readout, maximum programming, statistical analysis, and an automated elevating screw (Figure 66-6). On these automated production systems, there is no interpretation of a dial gage reading. All hardness test numbers are displayed on a digital screen, and testing can be conducted in seconds (Figure 66-7).

FIGURE 66-5
Various standard anvils used for holding narrow, round, or flat objects for testing. (Credit: Wilson Instruments)

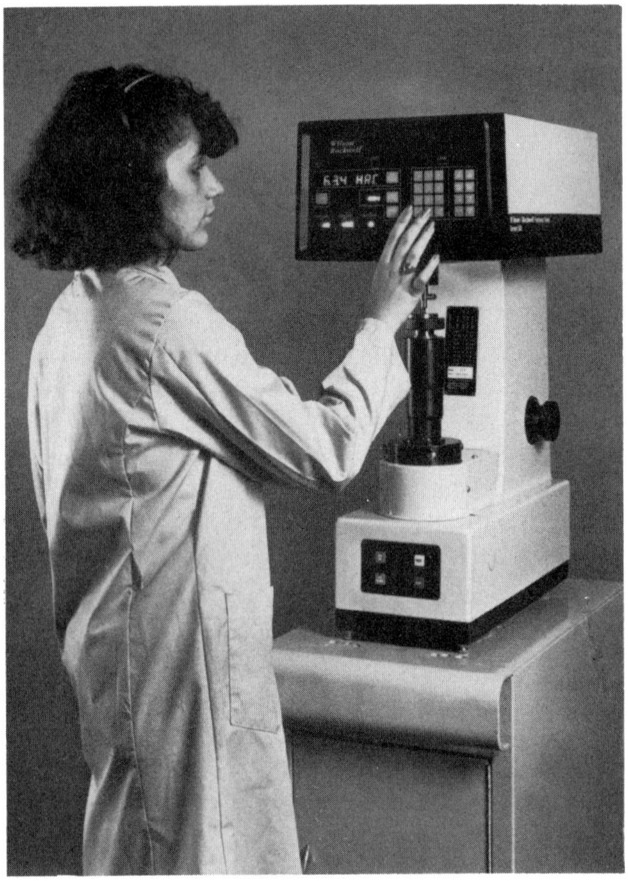

FIGURE 66–6
A digital readout Rockwell hardness tester can test parts within seconds. (Credit: Wilson Instruments)

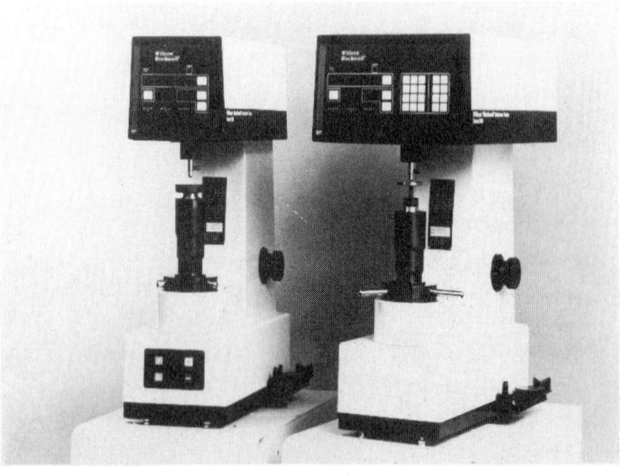

FIGURE 66–7
High-production-rate Rockwell hardness testers. (Credit: Wilson Instruments)

FIGURE 66–8
Mobile cylindrical hardness tester. (Credit: Wilson Instruments)

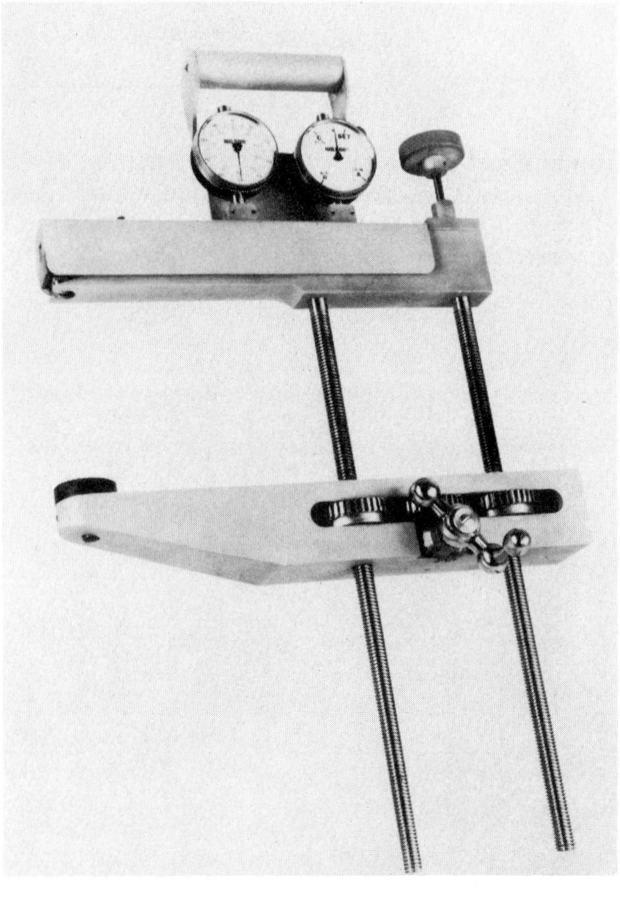

FIGURE 66–9
A portable hardness tester. (Credit: Wilson Instruments)

FIGURE 66–10
Rockwell hardness test blocks and calibration set. (Credit: Wilson Instruments)

As with any Rockwell, such as the portable model shown in Figures 66–8 and 66–9, the testing devices should be checked out for accuracy with test blocks provided with known hardness values (Figure 66–10). For best results, readings in five different locations on the test block should be taken and averaged out (Figure 66–11). The tester is calibrated if the average falls within the tolerance indicated on the side of the block.

Microhardness Testing

In the field of microhardness testing, there are two scales used for measuring hardness: the Vickers and the Knoop scale. An extremely small indenter is used, generally under a load of up to 1000 kgf. The indenter has to be perfectly shaped, just as the specimen's surface should be polished and free of rough edges.

The Vickers testing results are shown in Figure 66–12. These indents are made by a 136° diamond pyramid (Figure 66–13). The results of the Vickers tester are similar to the Brinell hardness testing results in that the indent is read with a microscope. Since the loads used make the penetration very light, the microscope used to read the minute indent must be able to magnify at least 500 times.

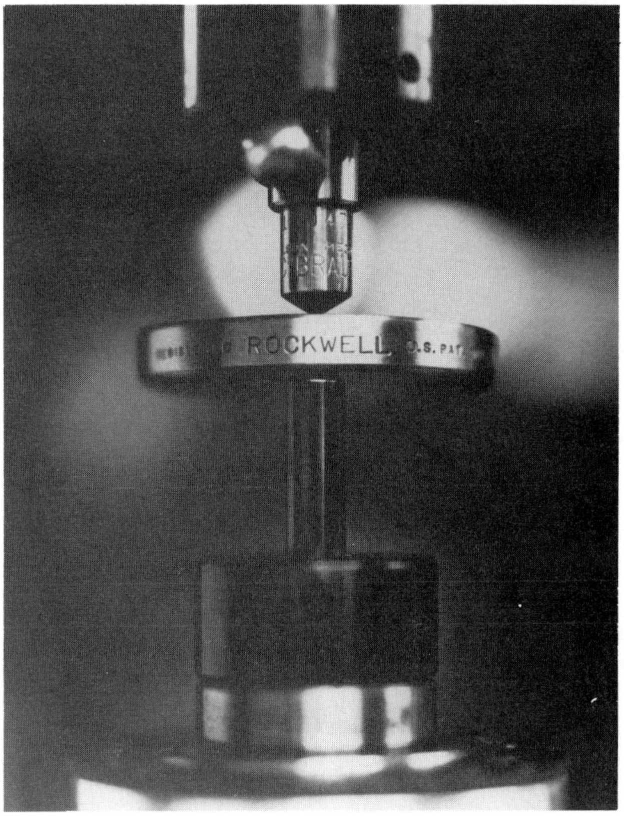

FIGURE 66–11
Test readings should be conducted on five different locations and averaged out. (Credit: Wilson Instruments)

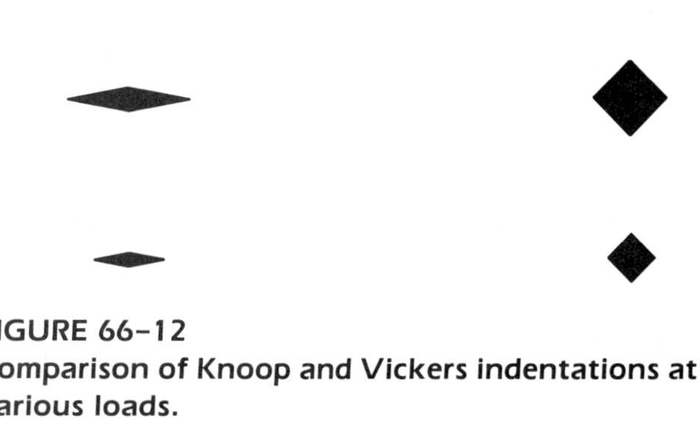

FIGURE 66–12
Comparison of Knoop and Vickers indentations at various loads.

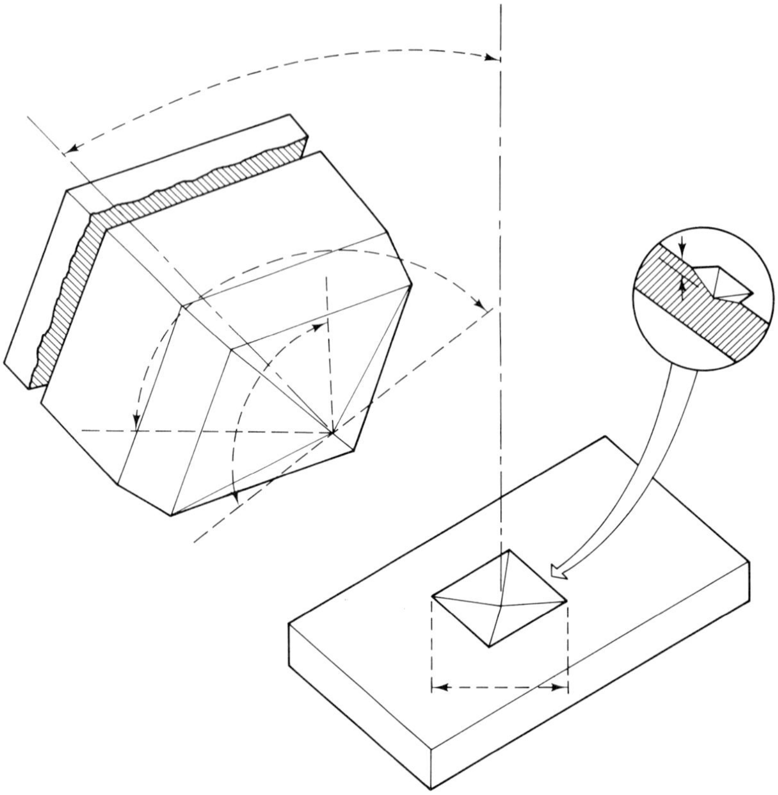

FIGURE 66–13
Vickers indenter.

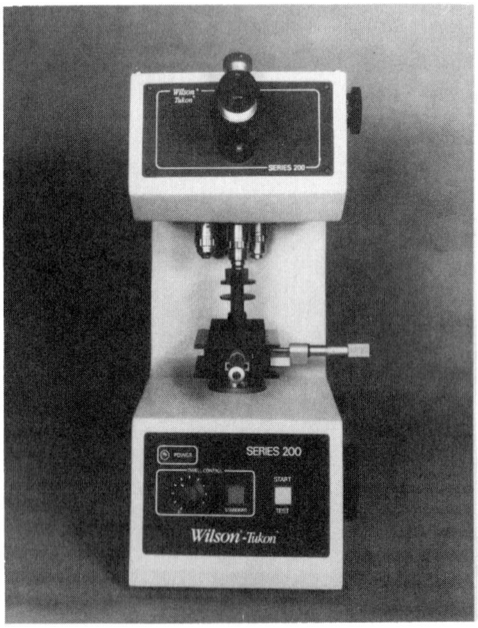

FIGURE 66–14
Microhardness tester is used for measuring steel's microstructure in laboratories. (Credit: Wilson Instruments)

While a microhardness tester is used to test the steel's microstructure in laboratories, it can also examine small delicate parts such as razor blades, very thin tubing, precision instruments, and wafer-thin hardened parts.

The Knoop indenter (Figure 66–14) was developed by the U.S. Bureau of Standards to produce the imprint shown in Figure 66–15. Under equal loads as the Knoop, it will leave an indent three times longer than the Vickers and can be used in testing brittle materials such as glass.

Scleroscope Testing

The Shore Scleroscope uses a diamond-tipped hammer (indenter). The system measures the height of rebound made by the hammer that has dropped from a fixed height through a guiding glass tube. The hammer is elevated pneumatically by means of a rubber bulb that is pumped by squeezing and releasing. When it hits, the diamond hammer makes a

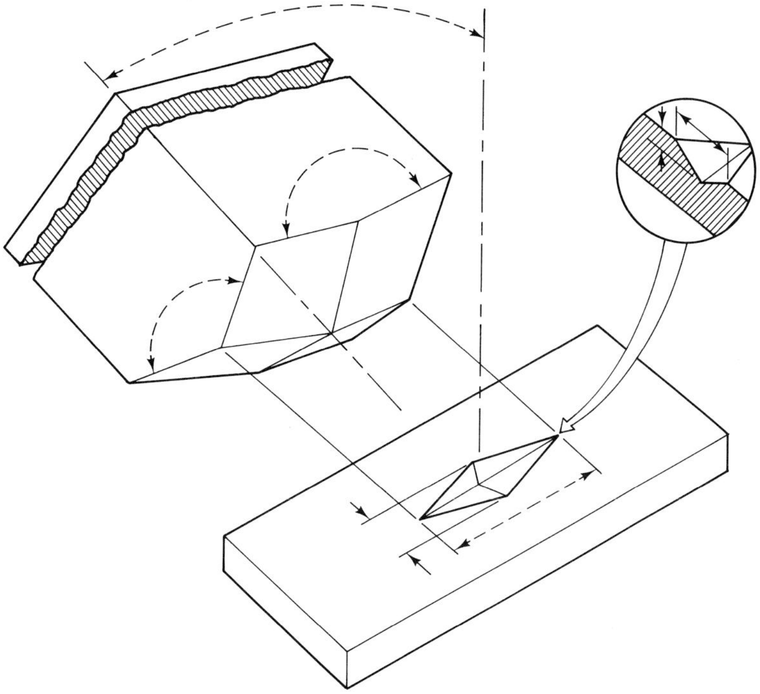

FIGURE 66–15
Knoop indenter.

small indent, the size of which will depend on the hardness of the specimen being tested. The hammer then rebounds; the higher the bounce, the harder the metal. The Scleroscope is a dynamic rather than a static load-type hardness tester.

Hardness is measured by a scale inside the glass tube. The scale is graduated to higher than 100 equal parts to cover superhard metals. Generally, quenched tool steels will register at levels between 95 and 105.

One of the first commercially available hardness testers produced, the Scleroscope may be used to test the hardness of all metals, ferrous or nonferrous, as well as polished or unpolished surfaces. There are many styles of Scleroscope testing devices from which to choose. The Scleroscope model shown in Figure 66–16 may be used on a number of surfaces. Its versatility can be demonstrated held free hand to test objects of unlimited size, mounted on a flat horizontal base, testing irregular surfaces, or mounted on any cylindrical base.

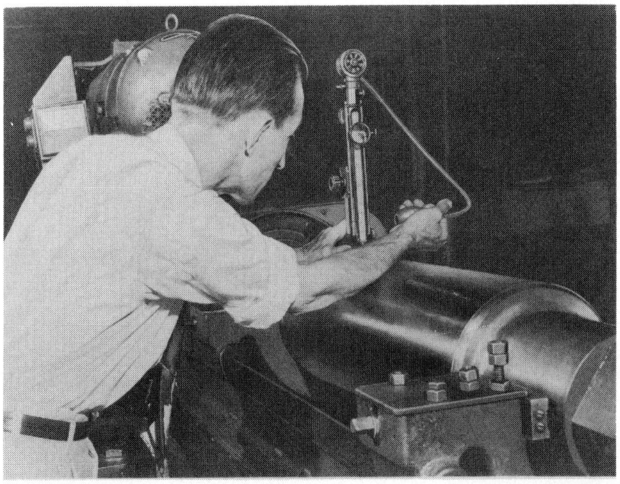

FIGURE 66–16
A portable-style Scleroscope hardness tester.
(Credit: Shore Instrument & Mfg. Co. Inc.)

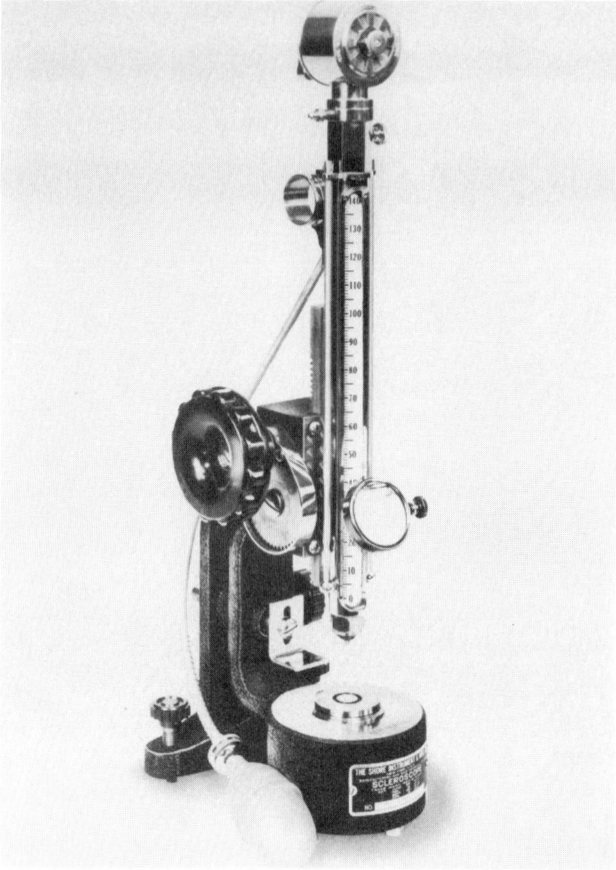

FIGURE 66–17
A hardness testing Scleroscope mounted on a clamping stand. (Credit: Shore Instrument & Mfg. Co. Inc.)

FIGURE 66–18
Nondestructive testing: a necessary tool in heat treating.

A variety of parts not exceeding the height limitation of the model mounted in the clamping stand (Figure 66–17) can be tested. Hardened thicknesses as little as 0.005 to 0.006 in. (0.13 to 0.16 mm) can be measured accurately as well as case-hardened materials. Metals bearing a soft inner core will require a minimum case hardness 0.015 in. (0.38 mm) thick. To be effective, parts must be free of rust, scale, and decarburized material, they must be firmly held, and the diamond-tipped hammer should hit on a 90° drop.

Procedure

1. Using the spirit level as a guide, level the clamping stand by means of the thumbscrews located on either side of the base.
2. Mount the specimen firmly on the anvil, directly below the hammer.
3. Turn the hand knob to lower the vertical barrel until the barrel cap contacts the specimen. (This contact must be maintained throughout the test.)
4. Rapidly squeeze and release the rubber bulb to raise the diamond-tipped hammer to engage a pneumatic latch.
5. While still maintaining barrel cap pressure against the specimen using the hand knob, again rapidly squeeze and release the rubber bulb to allow the hammer to drop.
6. Note and record the highest point to which the top of the hammer rebounds on the first bounce. This reading will indicate the specimen's hardness.
7. Turn the hand knob to raise the vertical barrel, move the specimen for another test location, and turn the hand knob to again contact the barrel cap with the specimen.
8. Again rapidly squeeze and release the rubber bulb to raise the diamond-tipped hammer to engage its latch.
9. Focusing several points below the first rebound reading, repeat steps 5 to 7 to complete two or three more tests. The correct hardness is indicated by the average of several tests made.

Careers

Many industries use hardness testing as a means of detecting the physical condition of the parts manufactured. Articles made for the aerospace industries are constantly checked for flaws, fatigue cracks, and other failures through nondestructive testing.

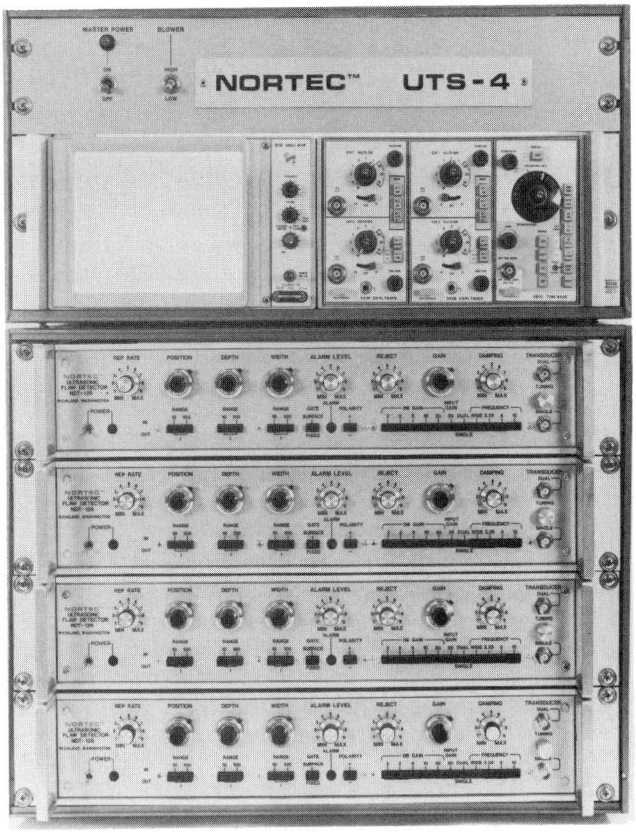

FIGURE 66-19
Ultrasonic testing device.

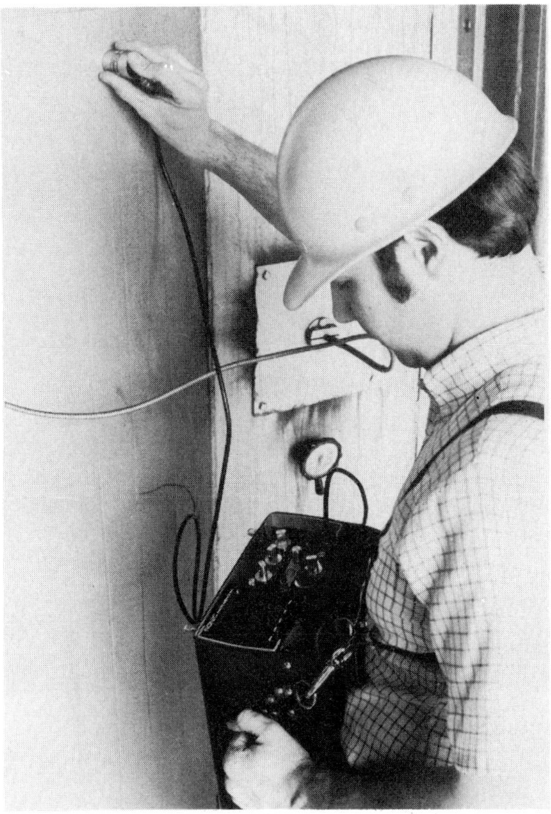

FIGURE 66-20
Portable ultrasonic flaw detector.

Heat treaters in job shops are another group that have wide use for hardness testing (Figure 66-18).

Not to be overlooked are the areas of research and development. In its pursuit for better steel and steel mill products, the steel industry is constantly testing and researching new methods of recording and analyzing information given by nondestructive testing. Technicians, engineers, and metallurgists also have to involve themselves with other nondestructive inspection devices such as ultrasonic, radiographic (x-rays), liquid penetrant, magnetic particle, and eddy current (circulating electrical currents) devices.

Ultrasonic devices (Figure 66-19) have the ability to produce an electrical output that is directly related to the size and shape of the reflector. The versatility of these instruments enable them to analyze stress in bulk as well as surface designs. These nondestructive testing devices can detect fractures and flaws in metal 30 ft. (9.15 m) thick. A portable flaw detector (Figure 66-20) can accurately gage wall thicknesses, erosion, corrosion, and wall thinning in tanks, vessels, pipes, and other metal containment structures measuring as much as 10 in. (254 mm) thick.

Applications of ultrasonic devices make it an important precision search unit for industry, scientific, and medical testing.

UNIT 67 METAL TESTING PROCESSES

In designing machinery or structural projects, engineers base their estimates of material requirements on the strength of the material. Testing is conducted on materials to see how well they will behave under stress (a force that resists change in a body). These tests reveal strength as well as whether they can change shape without fracturing. Obviously, when the right type of material is not used that can with-

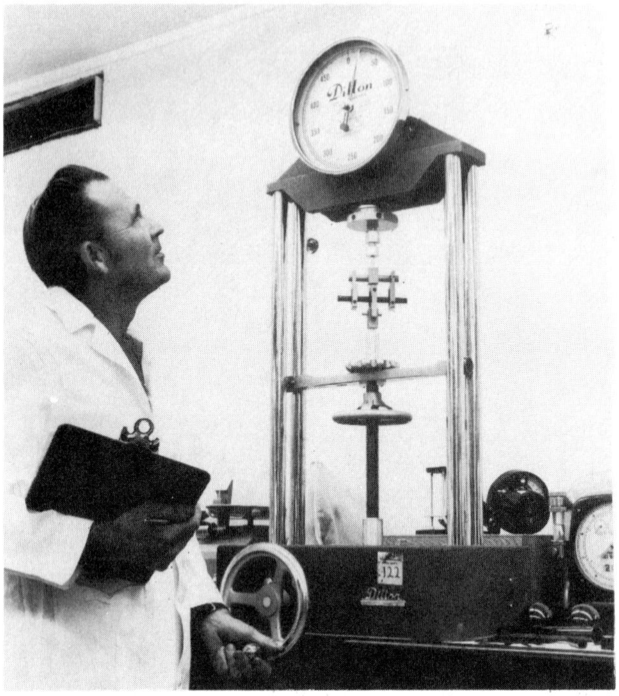

FIGURE 67-1
Tensile testing machine. (Credit: W. C. Dillon and Co. Inc.)

FIGURE 67-2
In tensile testing, a mounted specimen is pulled apart until it ruptures. (Credit: W. C. Dillon and Co. Inc.)

stand many types of stresses, the part or project will fail. Imagine what it would be like to build a bridge and the material used could not handle the everyday wind, weather, or traffic conditions. Would you cross it? Have you ever seen a bridge that has collapsed or felt a bridge sway with the wind?

In destructive testing, metals are subjected to loads (pressure or weights) in machines where their life span is determined as they undergo stresses. These stresses can either stretch the metal (tensile stress), shorten the metal (compression stress), or divide it into layers (shear stress).

Objectives
After completing this unit you will be able to

1. calculate different tensile strength tests.
2. identify two different types of impact testing devices.
3. identify three other metal testing processes.

Tensile Strength

The tensile strength of metal is the maximum pull that material can take without breaking. Tensile strength testing is conducted in a machine such as is shown in Figure 67-1. Here material is placed between two viselike grippers and made to fracture under load. It is like two teams of horses pulling an object in different directions at the same time.

Other data collected during tensile strength testing will include yield, percentage of elongation, and percentage of area reduction of the material.

Yield is a stretch of metal that occurs as the testing is carried out (Figure 67-2). The yield point is reached when the metal continues to stretch without increasing the load. Some materials such as aluminum and copper will not yield under tensile testing but will suddenly fracture under an increasing load. Some ferrous materials, such as cast iron, will stretch for a shorter period before fracturing.

Tensile Strength Testing

Tensile strength testing is conducted in machines of varying capacities (Figures 67-3A and 67-3B). While larger shapes and sizes may be tested, tensile testing specimen measuring 0.505 in. (12.82 mm) in diameter and bearing a cross-sectional area of 0.2 in. (129.03 mm) are generally used as standards for this test. Some smaller machines may be used to conduct

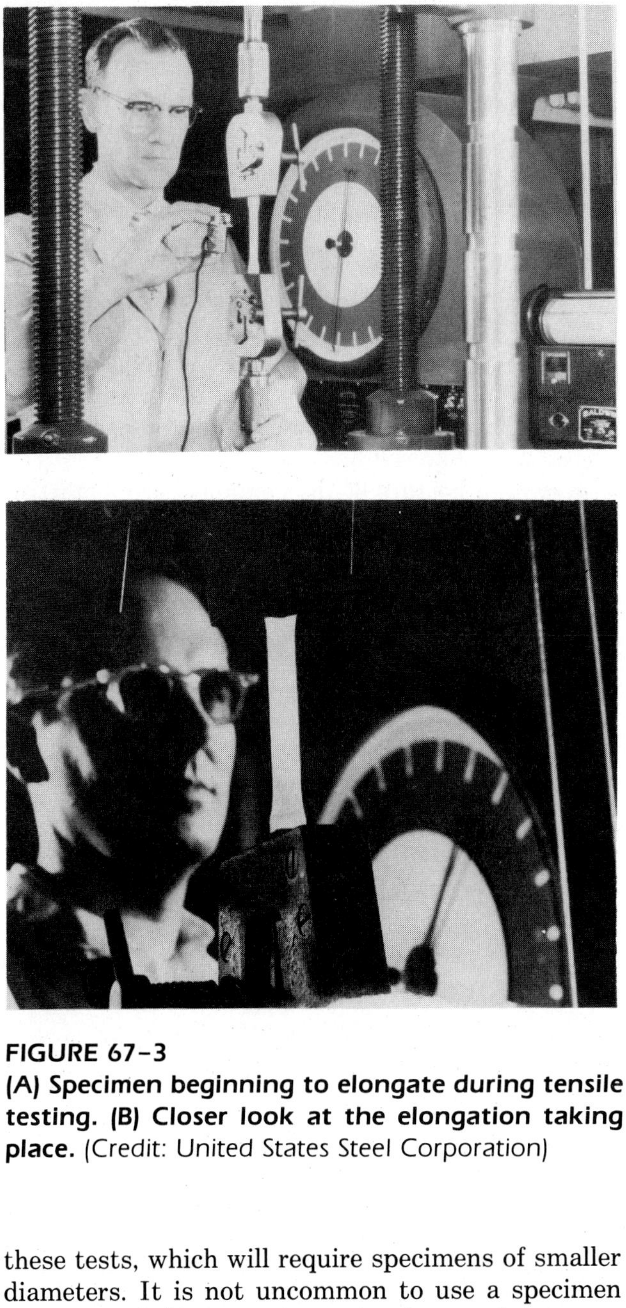

FIGURE 67-3
(A) Specimen beginning to elongate during tensile testing. (B) Closer look at the elongation taking place. (Credit: United States Steel Corporation)

these tests, which will require specimens of smaller diameters. It is not uncommon to use a specimen measuring half the cross-sectional area of the standard specimen. A test sample on a smaller machine might measure 0.252 in. (6.40 mm) in diameter.

The tensile strength of material in the inch system can be calculated using the following formula:

$$\text{Tensile strength} = \frac{\text{maximum load}}{\text{specimen cross-sectional area}}$$

where Tensile strength = psi
Maximum load = lb
Area = πR^2

Example

Using a test load of 5000 lb, determine the tensile strength of a specimen measuring 0.505 in. in diameter.

Solution:

$$\text{Tensile strength} = \frac{5000 \text{ lb}}{0.2 \text{ in.}^2} = 25{,}000 \text{ psi.}$$

Metric Calculations

The tensile strength of material in a metric tensile testing machine is calculated in kilograms per square centimeters. The cross-sectional area of a specimen is calculated in square centimeters or square millimeters. The tensile testing machines are graduated in millimeters.

As with the inch system of measurement, the basic tensile strength formula is the same except that metric values have been substituted:

Tensile strength = kg/cm
Maximum Load = kg
Area = πR^2

Example

Determine the tensile strength of a specimen 2 cm in diameter under a maximum load force of 3720 kg.

Solution:

$$\text{Tensile strength} = \frac{\text{maximum load}}{\text{specimen cross-sectional area, cm}^2} = \frac{3720}{3.1416} = 1184 \text{ kg/cm}^2.$$

Elongation

Elongation is a permanent stretch in length undergone by the test rod near the fracture point. It is expressed as a percentage of the original specimen's length and can be calculated using the following formula:

$$\text{Percentage of elongation} = \frac{\text{amount of elongation}}{\text{original length}} \times 100$$

Example: Identifying marks placed exactly 50 mm apart before the start of the test were found to measure 55 mm upon completing the stretch testing. What was the percentage of elongation?

Solution:

$$\text{Percentage of elongation} = \frac{\text{amount of elongation}}{\text{original length}} \times 100$$

$$= \frac{5}{50} \times 100$$

$$= 10\%.$$

Area Reduction

Area reduction is also expressed as a percentage of the original area and represents the difference between the original specimen diameter and the final specimen diameter measure at the point of fracture. Ductile materials will have a greater reduction than those that are brittle.

Example:

Suppose the original specimen diameter measured 0.505 (12.82 mm) and was reduced to 0.379 in. (9.62 mm) after fracturing. What was the percentage of area reduction?

Solution:

$$\text{Amount of reduction} = \text{original diameter} - \text{reduced diameter}$$

$$= 0.505 - 0.379$$

$$= 0.126 \text{ in.},$$

$$\text{Percentage of area reduction} = \frac{\text{amount of reduction}}{\text{original diameter}} \times 100$$

$$= \frac{0.126}{0.505} \times 100$$

$$= 25\%$$

Impact Test

As excessive hardness (brittleness) in steel is reduced by tempering after hardening, the toughness increases. Impact testing is conducted on metal to see how much sudden shock and impact it can withstand. It is really a test of the metal's toughness by dynamic load.

The two common tests used are the Izod and Charpy impact tests. In either test, force is applied by a swinging pendulum against a specimen 0.3937 in. (10 mm) square. In both tests, the specimen is notched. A standard mass pendulum is raised to a fixed height that is based on the specimen involved, released, and allowed to strike the notched area. The difference in heights of the pendulum from start to end of impact stroke is measured on a gage and indicates energy required to fracture the specimen.

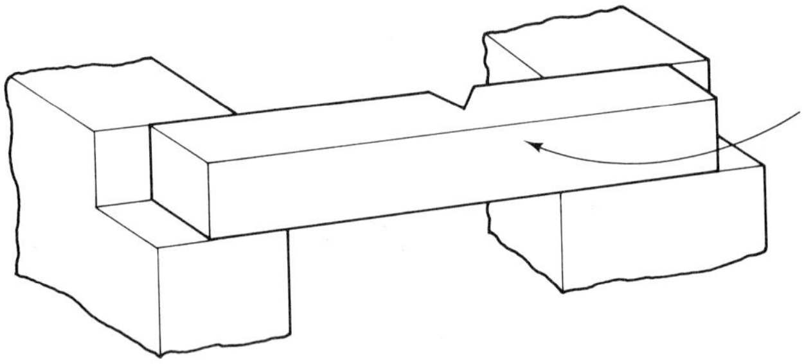

FIGURE 67–4
The principle of the Charpy impact test.

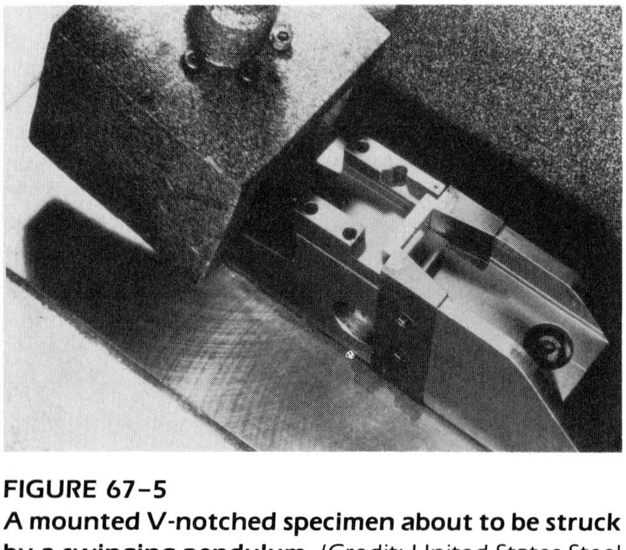

FIGURE 67–5
A mounted V-notched specimen about to be struck by a swinging pendulum. (Credit: United States Steel Corporation)

In the Charpy impact test, a knife edge pendulum strikes the center and just behind the specimen's notch (Figures 67–4 to 67–6).

In the Izod test, the specimen is clamped so that the pendulum can swing against the exposed V-notch (Figure 67–7).

Compression Testing

A compression is performed to see how well a specimen will stand up as compression forces tend to squeeze and shorten the specimen. While this test is done on a tensile machine, its force is applied by a pushing load rather than a pulling load. Specimens that fail the compression test will buckle or shear. The strength involved will also be measured in pounds per square inch (kilograms per square centimeter). It should also be noted that the specimens tested are generally used in situations that may require strength and stiffness in design. Cast-iron power presses would be a good illustration. There is little stretching of this material before it ruptures.

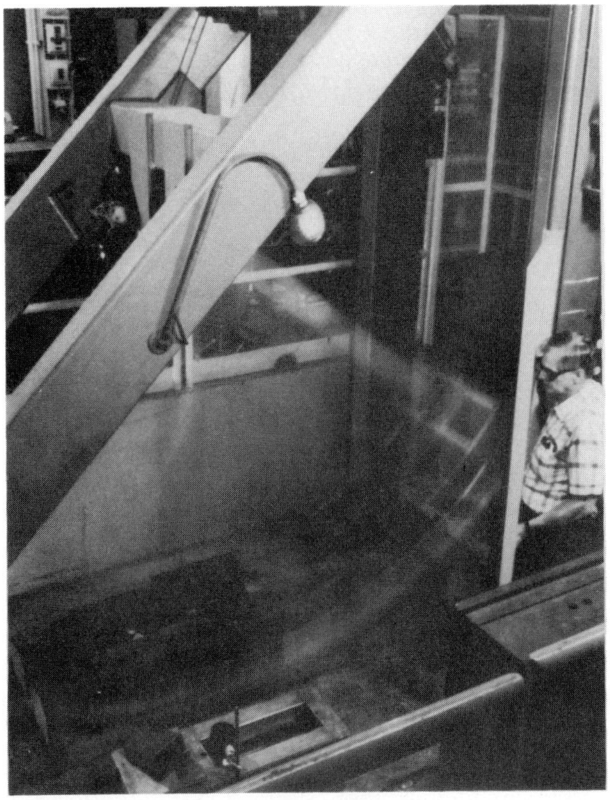

FIGURE 67–6
The height of the pendulum arc and the broken specimen indicate the completion of the Charpy impact test. (Credit: United States Steel Corporation)

Shear Strength

Shear strength is the resistance of material to being cut (sheared) by two cutting edges or loads working in opposite directions. Cutting tools such as the ones used by sheetmetal workers are good examples. Hand shears are used to overcome the material's shearing strength in order to cut it.

Another illustration used is that of testing the shear strength of rivets and bolts (Figure 67–8). This shear strength test demonstrates a sliding failure. Parts are slid or pulled in opposite directions to each other.

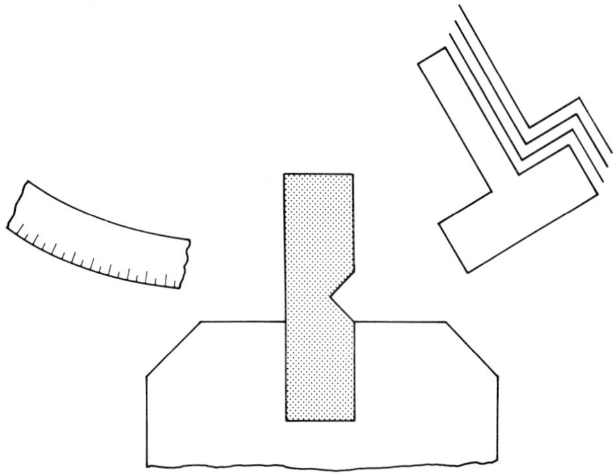

FIGURE 67–7
In the Izod impact test, the amount of energy directed by a swinging pendulum is registered on an indicating scale.

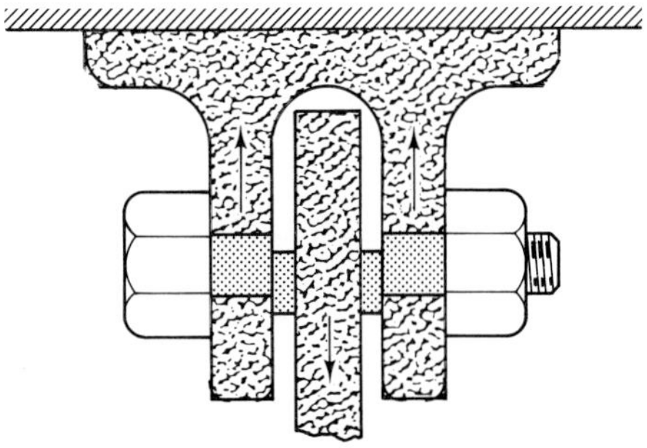

FIGURE 67–8
The principle of a shear strength test—parts being slid or pulled in opposite directions to each other.

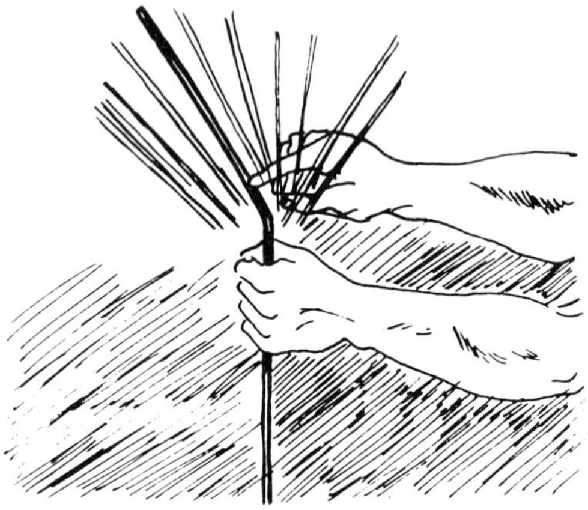

FIGURE 67–9
The principle of fatigue testing. (Credit: Paul Rapisarda)

Shearing angles have to be provided on the cutting edge of dies in order to reduce the shearing stress of the material. If a shear is not provided, there is danger of a violent shock to the press frame and equipment when the pressure load is released after blanking.

Fatigue Testing

Fatigue is another name for endurance. A simple illustration of this would be the continuous bending and unbending of a piece of metal. Eventually, the metal will fracture due to fatigue, its broken ends brittle.

Before being used in industry, many parts have to have certain fatigue requirements. Parts used for aircraft and helicopters, for instance, are constantly subjected to vibrations. Accessories for automobiles and trucks also have similar problems. Items such as signal lights and side-view mirrors take a jarring up and down vibration on the road.

In fatigue testing, metal is subjected to repeated cycle loads or pulsating loads (Figure 67–9). At first, the strain caused by every cycle is very minute. As the cycles increase, so do the strains. In time, the strained areas become brittle and tiny cracks appear, a small amount at first, increasing with the cycles until suddenly—snap! Fatigue failure is the result.

There is more than one testing machine for doing fatigue tests. The cycles are recorded to establish the limit of endurance of the metal. One of the most common types is a rotating-cam device that subjects a fastened specimen to undergo alternating compression and tension with each cycle.

SECTION REVIEW

Unit 63

1. Define heat treating.
2. Define alpha iron.
3. What results when iron and carbon are united to produce iron carbides?
4. What is the minimum temperature for hardening steel called?
5. Name the factors that will influence hardening.
6. For what purpose is a pyrometer used?
7. Define quenching.
8. Name four media used for quenching in heat treating.
9. What factors affect quenching?
10. What is the purpose for conducting Metcalf's experiment?
11. Name a practical but not very accurate method of testing for hardness.
12. What is another term for tempering?
13. Why is it necessary to temper steel after hardening?

14. What range of temperatures are used for tempering steels?
15. What four factors affect the process of tempering?

Unit 64

1. Define annealing.
2. How is annealing accomplished?
3. For what purpose is process annealing done and on what kind of steels?
4. How do normalizing and full annealing compare? Differ?
5. Define spherodizing.

Unit 65

1. Define case hardening.
2. Explain what type of steel is widely used in industry to be case hardened.
3. Name three methods of case hardening steel.
4. How is depth of hardness controlled in case hardening?
5. Why should parts be preheated when using any form of cyaniding in heat treating?
6. Describe carbonitriding.
7. What is nitriding?
8. Name two nitriding processes and the quenching medium required.
9. On what type of surface is flame hardening used and how are the surfaces quenched?
10. List five advantages to laser hardening.

Unit 66

1. Define hardness.
2. Describe how a Brinell hardness test is carried out.
3. How does a Rockwell tester operate?
4. Describe how nonferrous materials are hardness tested on a Rockwell.
5. Describe how the Scleroscope hardness tester operates.

Unit 67

1. Define stress.
2. Name three principal stresses involved in destructive testing of metals.
3. Describe how tensile testing is carried out.
4. Briefly explain what yield is.
5. How is the yield point gaged?
6. Describe how the impact test is carried out.
7. What are the two principal methods of conducting an impact test?
8. Describe the principle of shear strength testing.
9. What are the two characteristics of a compression test in fracturing the specimen?
10. How is metal tested for limits of endurance?
11. Under what type of load conditions are metals tested for limits of endurance?

Section 14

MASS PRODUCTION

The concept of accuracy is the basis of mass production, the means by which articles are manufactured in abundance. The end result of manufacturing are goods and products that are made available to a nation's economy. It is this availability together with the ability to purchase these items that raises the standard of living of any nation. To this end there are a number of processes, systems, workers, and management teams organized to maintain and improve the quality of manufactured products, both in value costs and control.

UNIT 68 DEVELOPMENT OF MASS PRODUCTION

Manufacturing and the concept of division of labor—the assignment of one task to a worker—might very well have originated in Egypt almost at the same time that commerce began to flourish. The earliest evidence of this appears in a painting on the wall of an Egyptian tomb circa 1600 B.C. that illustrates the use of specialists involved in one task.

Prior to the Industrial Age, there was no such thing as mass production. Articles were still made by eye or by hand skill accuracy. No factory-made goods were yet available.

Objectives
After completing this unit you will be aware of

1. the development of the early foundation of mass production.
2. the effect that mass production has on the standard of living.
3. the development of twentieth-century mass production methods.

The Early Practitioners

The factory system in America began on two unrelated levels. As the Industrial Revolution was beginning in England and shortly before the American Revolutionary War took place, the manufacture of cotton thread and cloth began to flourish in factory towns located at the source of power, generally water power. When the war cut off English imports of cloth goods, the water mills, which were also used for manufacturing iron and slitting machinery, were utilized to take the place of hand-operated tools.

Oliver Evans, an inventor and mechanical engineer, who later constructed the first high-pressure steam engine in America, was one of the first to develop techniques of mass production. Evans put his concept to work increasing flour production by tieing in a variety of machines to the same water wheel of a flour mill. By using endless horizontal and vertical conveyor belts with buckets attached, material was lifted, moved back and forth, raked, and sifted by machinery. The operation required two workers: one to empty bags of wheat at the start, another to close and roll away barrels of flour at the finish of the process. The year was 1787, and Oliver Evans had succeeded in automating the flour industry.

Samuel Colt, inventor of the Colt revolver, was one of the first to use Whitney's industrial mass production method, preceding Henry Ford by some 50 years. Colt had precision forging dies made in his forge shop that enabled his factory to produce interchangeable parts at high speed.

Another shrewd industrialist was the man who put America on wheels—Henry Ford. It was written of Ford that he had never invented anything but merely adapted everything for mass production. He, too, borrowed from Whitney's system of interchangeability and added a few wrinkles of his own.

From the beginning of the Ford Motor Company, there was a subdivision of labor. Borrowing the idea of a moving conveyor from the meatpackers, Ford tackled labor waste and lost motion. He felt that no worker should take more than one step or bend down to do the job. No small detail escaped his attention for picking the right tool for the right job. Subdivision of labor was also incorporated to speed up completed operations.

Ford's labor saving devices and methods were responsible for raising wages and lowering the workers' hours while cutting the cost of the automobile. As a result, consumer savings increased as did profits of mass production due to the assembly line. Ford had revolutionized the automobile industry.

New Technology

Other studies for cutting waste were also presented to industry near the end of the nineteenth century. The scientific management of production was an outgrowth of studies made by an efficiency engineer, Frederick Winslow Taylor, who, with Manuel White, developed a special process of heat treating high-speed steel to enable cutting tool edges to increase their cutting capacity 200 to 300%. Taylor based his "scientific" findings on the specialization of men and machines. If tools could cut faster, operators could also run machines faster. Applying the same scientific yardstick to management as he had to cutting tools, where he discarded the poorer cutting tool for the longer lasting cutting tool, Taylor reasoned that the weaker worker would be replaced by the stronger worker. It was an attitude he later softened.

While totally rejected by organized labor in the beginning, industry pursued the "scientific study" of workers' motions. Inefficiency in management was costing millions of dollars. The result of cutting inefficiency lowered consumer prices as well as increased the workers' wages to improve the standard of living.

Along with the studies came the construction of laboratories, such as the one Thomas Edison pioneered, for the development of new products and tools. Shortly before World War I, over 50 labs were in existence, including Eastman Kodak, DuPont, General Electric, and Bell Telephone. Mass production was the result of these new methods and machines. High on the list of mass production advancement was the invention of the transistor, which helped usher in the electronic age. Many of the electronic marvels that followed, from word processors to the guidance systems used in rocketry, contained the technical descendants of this device.

The Computer Age

The success of the automobile system was instrumental in spawning other labor saving devices and systems that mechanized work and working parts to a high degree of sophistication. The evolution was slow and did not take hold until World War II. The period that followed brought a greater demand for products that could be mass produced with quality and high reliability. While social changes were at the root of new product standards such as improved performance and longer use, safety, financial and social responsibility, profits, etc., planning for mass production and quality was largely made possible by the computer. It was the computer that made a large array of systems work.

Although there were a few contributions made to the computer technology during the early part of the twentieth century, World War II intervened to channel all technological developments to the war effort. British intelligence built a computerlike device called Ultra that was instrumental in unlocking German military codes being sent by their cipher machine called Enigma. Ultra stood about 8 ft (2.44 m) high and measured 8 ft (2.44 m) at the base, with its insides wired with tons of circuits, rotors, and vacuum tubes.

The first electronic digital computer was built in 1946 by J. Presper Eckert and John W. Mauchly of the University of Pennsylvania and could calculate at a speed 1000 times faster than the then-current electromechanical machine. It accomplished in 2 h what it would have taken 100 engineers to calculate in one year. This device was the forerunner of Univac I, which used self-checking devices and could easily handle both alphabetical and numerical information.

By 1947, Bell Laboratory introduced an innovation that made the vacuum tube obsolete. Developed by William Shockley, John Bardeen, and Walter H. Brittain, the transistor did not work in a vacuum, nor was there a glass shield to cover it or a filament to burn out. Neither did it have to warm up like a vacuum tube. It came on full strength the moment it was switched on, consumed small amounts of power, and generated almost no heat. It could also be made as small as a pea, or smaller, to take up very little space. Used as a switch to control the flow of electronics, the transistor was only one of the building blocks of the computer age.

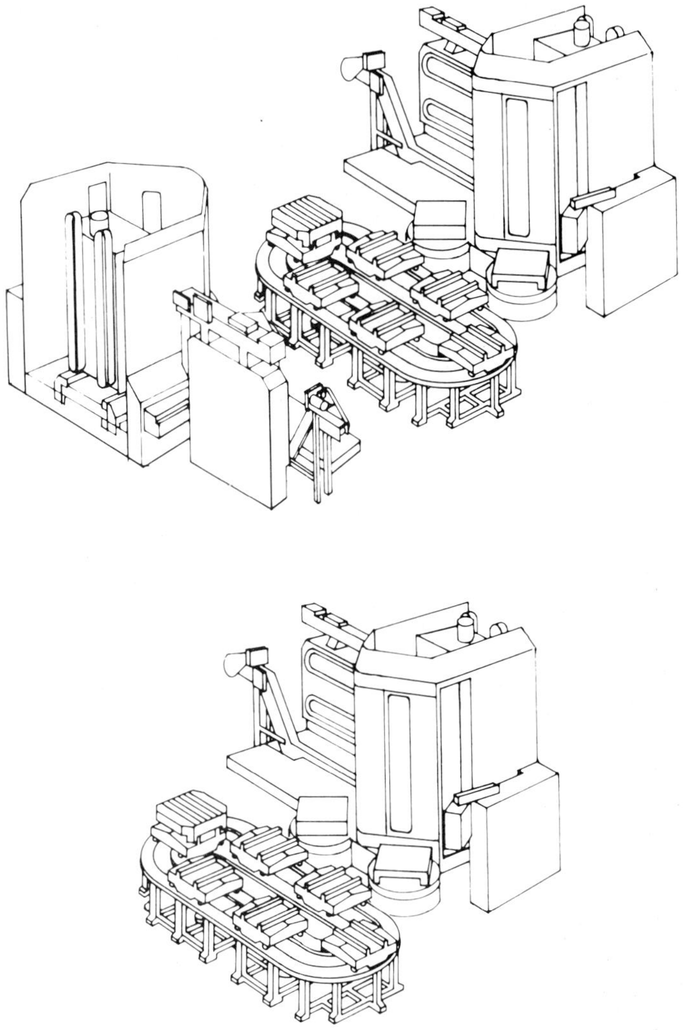

FIGURE 68–1
A family of parts can be produced in this manufacturing center without leaving the group of integrated machine tools. (Credit: Kearney & Trecker Corp., Milwaukee, Wis.)

The integrated circuit was made possible by Kilby of Texas Instruments and Noyce of Fairchild in the late 1950s and eliminated the extensive hand wiring. It consisted of a series of transistors that were connected and etched on a strip of silicon. These circuits served as one function of a computer's operation such as a memory register. Thousands of pieces of information can be stored on this chip, but their use was limited when first developed due to the rigid placement of the circuits and the narrow duties for which they were designed. By 1971, Intel Corporation had developed the microprocessor, a chip that could be programmed to do any number of commands.

With its ability to collect, process, record, and recall data rapidly, computer-assisted equipment play an important role in manufacturing systems (Figure 68–1). Computers are now being used for monitoring, controlling, material handling (Figure 68–2), numerical control machining, designing, and assisting in other manufacturing applications. It is within the realm of possibility that total factory automation will be integrated and systematized using the computer (Figure 68–3).

FIGURE 68–2
The traditional material handling equipment has been modified to include automatic guided vehicles. (Credit: Kearney & Trecker Corp., Milwaukee, Wis.)

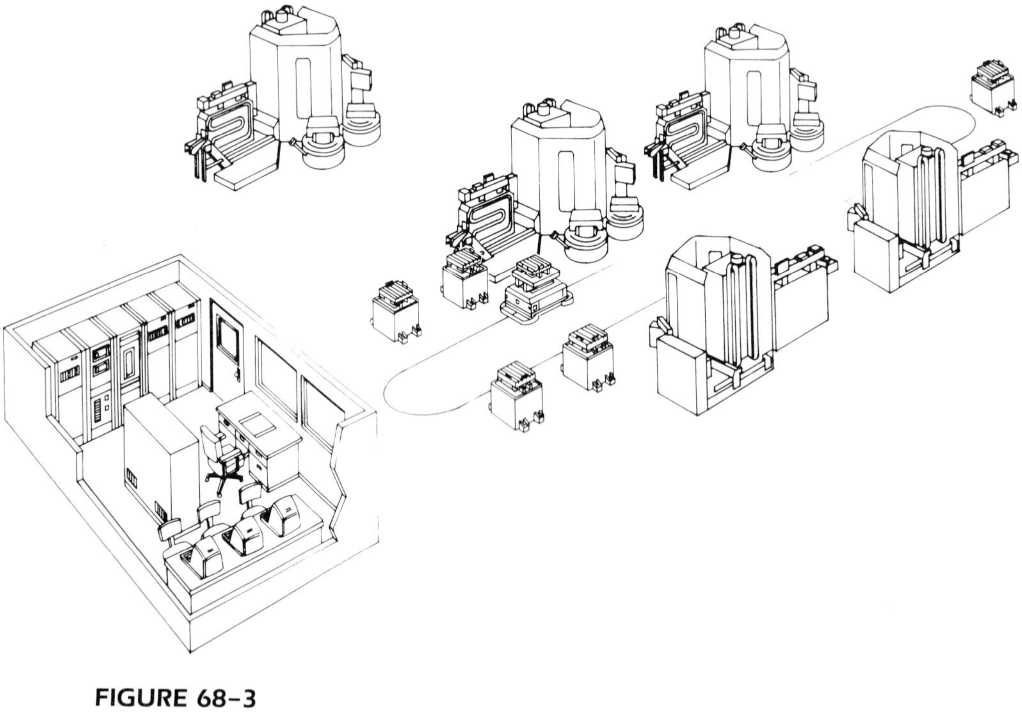

FIGURE 68–3
A series of machining operations can be performed without human intervention and controlled with a computer system. (Credit: Kearney & Trecker Corp., Milwaukee, Wis.)

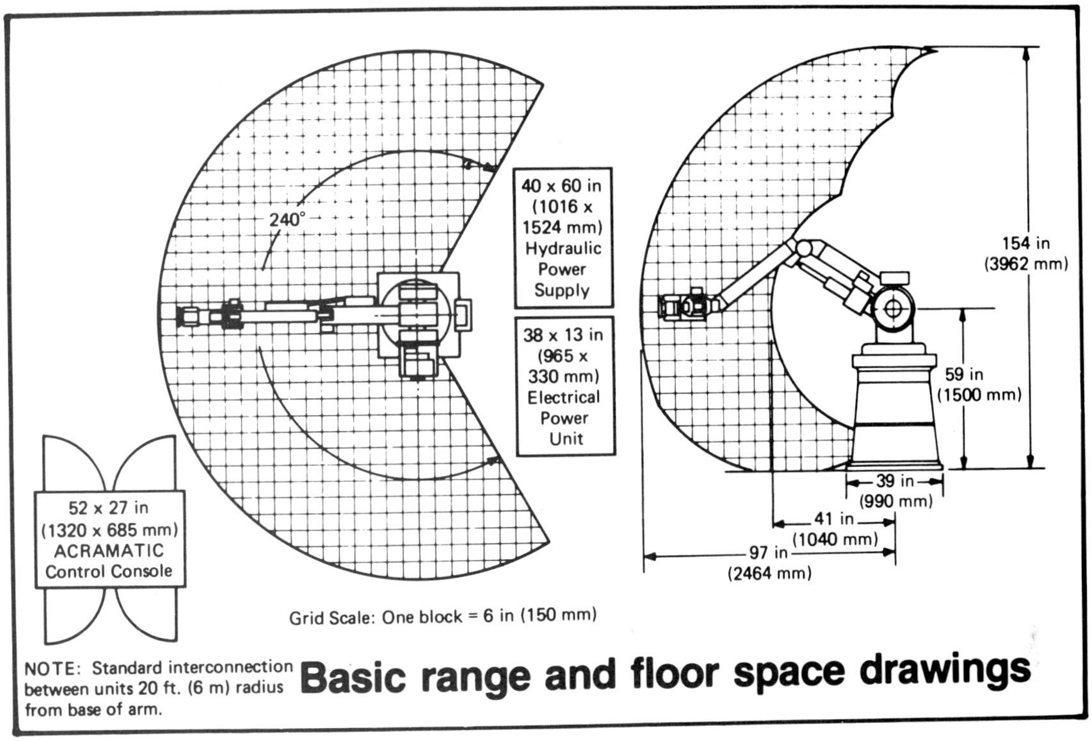

FIGURE 68-4
Industrial robots are capable of handling a variety of jobs from material handling to assembly tools. (Credit: Cincinnati Milacron Inc.)

Industrial Robots

The other side of automation deals with the process of feeding and handling materials. Industrial robots come in all shapes and sizes and are used for many purposes (Figure 68-4). They have been used in a variety of work situations to date: inspection, assembly, painting, material handling, and stacking and are widely used in the automobile industry for part transfer and welding.

The first generation of industrial robots were fixed-sequence devices. They were programmed by an operator who recorded the sequence of operations in the robot's control unit for use in production. Each part had to be precisely located for the operation to be successful. Once a new sequence of operation took place, it had to be "taught" all over again.

The second generation of industrial robots were controlled by minicomputers. This meant more flexibility. Using a special computer language, these robots were "taught" and programmed. It was during this stage that robots were developed that could move six joints (Figure 68-5). Not only could they track moving parts on an assembly line, they could also calculate how to move depending on which part arrived at the work station first and how to position it. But they lacked capabilities to adapt to a clutter of different shapes and sizes.

The third-generation robot is also controlled by a minicomputer and may be equipped with various devices such as television cameras. Serving as the "eyes of the robot," television cameras help make selections as parts move on a conveyor belt or an automated assembly line, for example see Figure 68-6.

The use of control computers offers a range of applications to these single-arm devices. Because the computer permits the robot to make logical decisions, the device can perform one operation and yet remain flexible enough to be "taught" to do another.

Operating Concepts

The Tomorrow Tool (T3), manufactured by Cincinnati Milacron, relies on two operating concepts: the controlled path and tool center point (TCP) concepts.

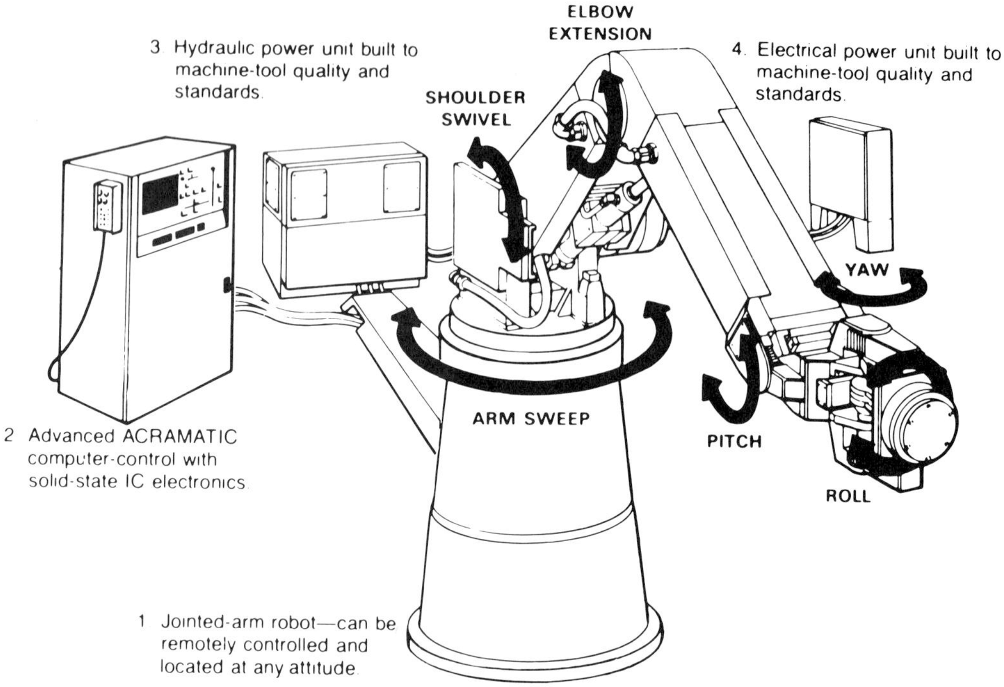

FIGURE 68–5
The robot's arm has a wide range of flexibility, being capable of six rotary axes of motion. (Credit: Cincinnati Milacron Inc.)

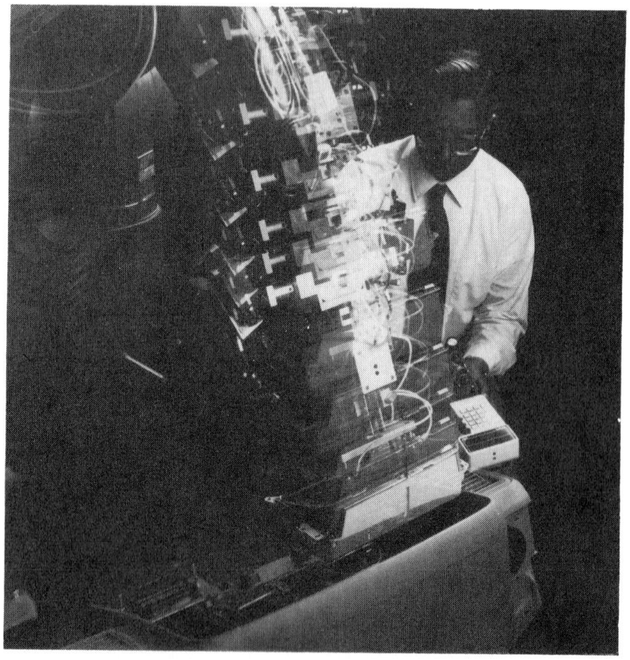

FIGURE 68–6
Stereo cameras mounted on the end of the robot's arm provides three-dimensional positioning and orientation of an automated assembly. (Credit: Ford Motor Company)

The controlled path operation (Figure 68–7) refers to the system's ability to provide coordinated control of all axes during both teaching and automatic operations. During teaching, the operator can position the robot to a desired location by direct commands for tool motion, as opposed to directing individual axis motion. Furthermore, the operator need not generate the required path, but only identify path end points.

The tool center point (TCP) is a point that lies along the roll axis, whose distance from the roll face plate is designated by the user (Figure 68–8). This point follows the straight-line path as the arm moves from one programmed location to another. In addition, all movements of the wrist axes take place about the tool center point.

The Control Systems

The robot's control system is designed to direct the robot through complex jobs and also to provide the means for an operator to "train" the robot to perform the moves and sequence of motions required for an operation. Through the use of sensors and adaptive controls, robots are able to function and

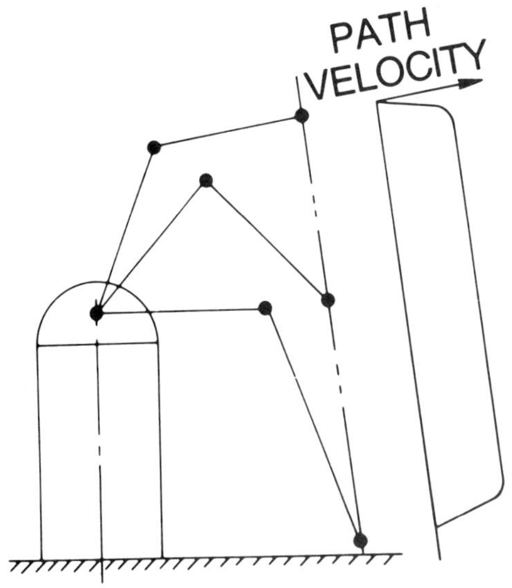

FIGURE 68–7
The controlled path operation provides coordinated straight-line movements of all axes during teaching and automatic operations. (Credit: Cincinnati Milacron Inc.)

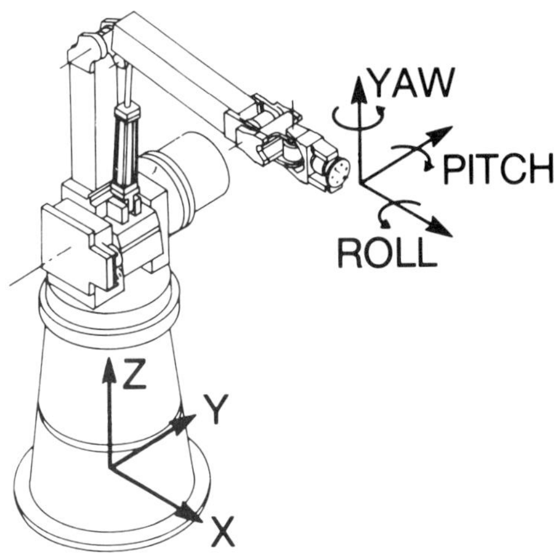

FIGURE 68–8
The tool center point provides rectangular and angular coordinates. (Credit: Cincinnati Milacron Inc.)

withstand changing or hostile environments. The robot's control unit (Figure 68–9) consists of a minicomputer, a cathode ray tube (CRT) and a keyboard, a control panel, a teach pendant, axis servo-control electronics, a servo test panel, and terminals for communication with other computers (Figure 68–10).

In order to function effectively, the robot's control system should:

1. Enable the robot to analyze and make logical decisions concerning its movements and to test for completion of operation based on information supplied by two-way communication with the working environment.
2. Enable a robot to analyze problems that occur to minimize downtime and loss of production.
3. Coordinate all axes when moving in a straight line from position to position so that all axes start and finish movements at the same time.
4. Provide the human teacher with the capacity to teach the arm to move or perform along the axes in a coordinated system that can be a rectangular, cylindrical, or hand system.
5. Provide communications between human intervention (teacher) and the control that permits the teacher to edit or delete information or direction of movement previously stored, if necessary.

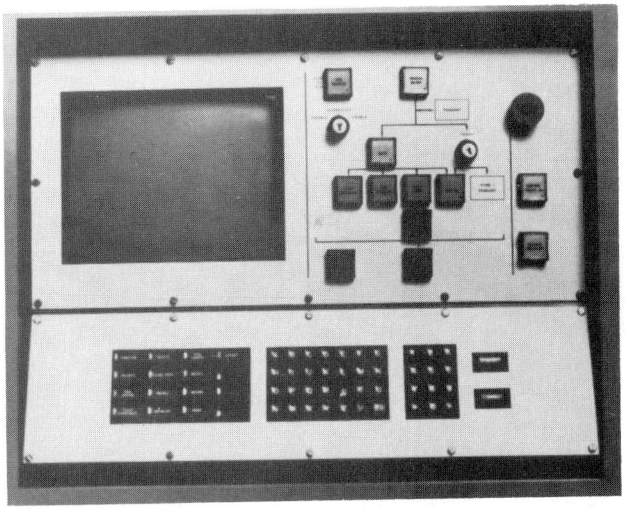

FIGURE 68–9
The robot's control unit is designed to direct the robot through complex as well as simple jobs. (Credit: Cincinnati Milacron Inc.)

Modes of Operation for Robots

The T3 has three modes of operation: the teach mode, the manual mode, and the automatic mode. Each mode of operation is selected on the control console.

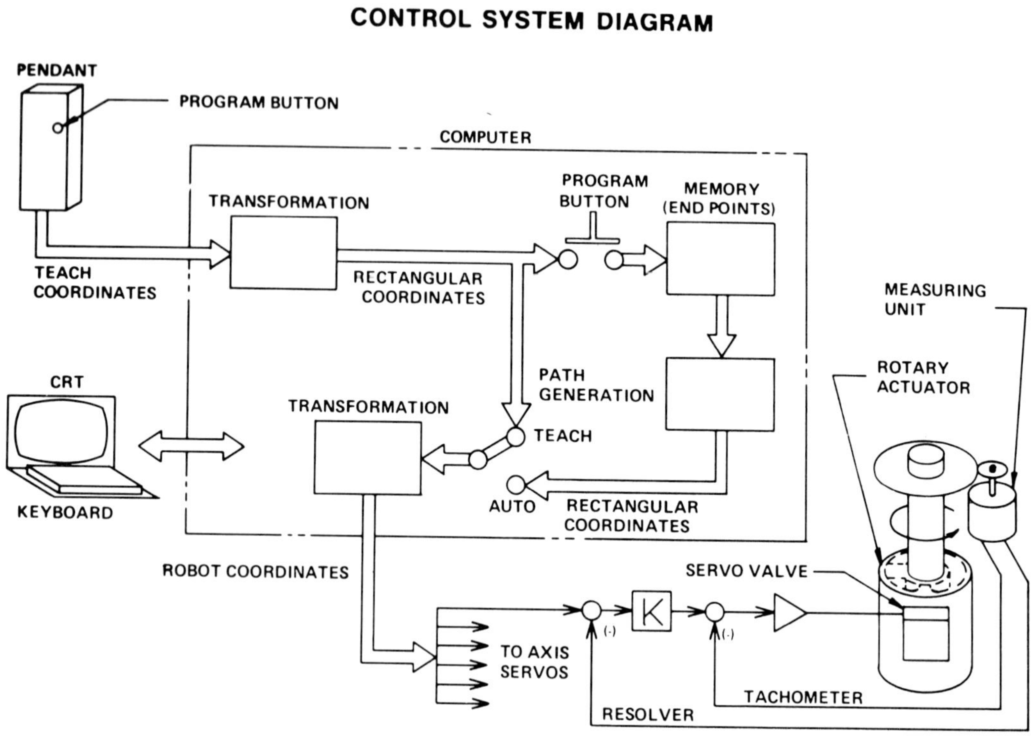

FIGURE 68–10
The T3 computer control system. (Credit: Cincinnati Milacron Inc.)

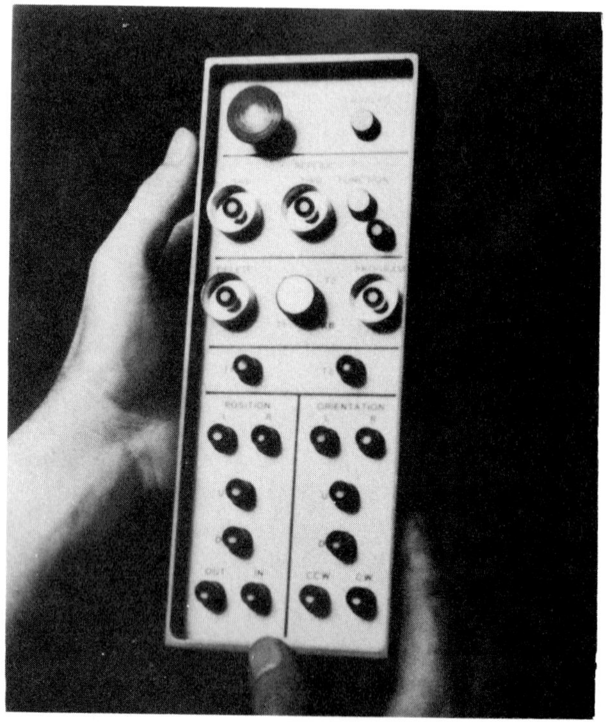

FIGURE 68–11
The robot is "taught" through the use of a hand-held pendant. (Credit: Cincinnati Milacron Inc.)

1. *The Teach Mode:* It is in this mode of operation that the teacher can direct the arm to perform a designated operation, communicating the axes to move in a coordinated fashion. The three-coordinate-system options available are rectangular, cylindrical, or hand, with each coordination taking place within the coordinate system chosen. Movement is accomplished through the use of position buttons on the hand-held pendant (Figure 68–11). The two upper position buttons will cause the TCP to move left or right on the rectangular system, the next two center buttons cause the TCP to move up or down, while the bottom two buttons cause the TCP to move in or out. It is the use of these buttons that permits the teacher to "train" the robot to perform within any of the coordinate systems.

2. *Manual Mode:* The coordinate axis control does not apply to this mode. What can be accomplished is a movement of each arm axis that is independent of all the others with the use of the lower buttons on the teach pendant. The primary purpose of the manual mode is to permit the operator to align the arm to its "home" position. Home is a unique location where the six coordinates of the system are known. The arm must be aligned to home before the human operator can program the robot through the

teach mode or enter the automatic mode. To assist in the alignment, a special console CRT display and arrows on the robot arm are used.

3. *Automatic Mode:* In this mode, the robot will continue to move through the taught cycle until either the "end of cycle stop" or "interrupt" buttons on the computer console are depressed. The "end of cycle stop" halts the movement of the arm the next time it reaches the last point in its cycle. The operator can then depress the "return" button to send the arm back to the "home" position. The "interrupt" button freezes the arm motion immediately, which can be resumed by depressing the "cycle start" button. An error code will appear in the control console display whenever something goes wrong.

Application Flexibility

Industrial robots come equipped with special devices such as grippers for holding or moving parts securely. Double grippers are also available if the part shape changes radically—one for holding the part before machining and one for holding the part after machining. They may also be used to load or unload machine tools, packaging, and palletizing, a system of loading parts or material on pallets for transportation to another work station (Figure 68–12). Industrial robots may be powered with more than one type of power source. They may be hydraulically operated, pneumatically operated, or electrically operated. Their servo system allows the arm movement to be controlled with accuracy and repeatability.

Today's industrial robots are tough and tireless workers that are helping to increase productivity while reducing manufacturing costs. It is claimed that the Egyptians had only three labor saving devices at the time the Pyramid of Gizeh was built: the lever, the wedge, and the inclined plane. With modern technology developing robots that will be able to "see," "feel," and "hear," the best is yet to come.

UNIT 69 METAL STAMPING PROCESSES

The ability to pierce, form, and bend sheet metal in quantity is an integral part of our industrialized society. It is unfortunate that sheet metal fabrication

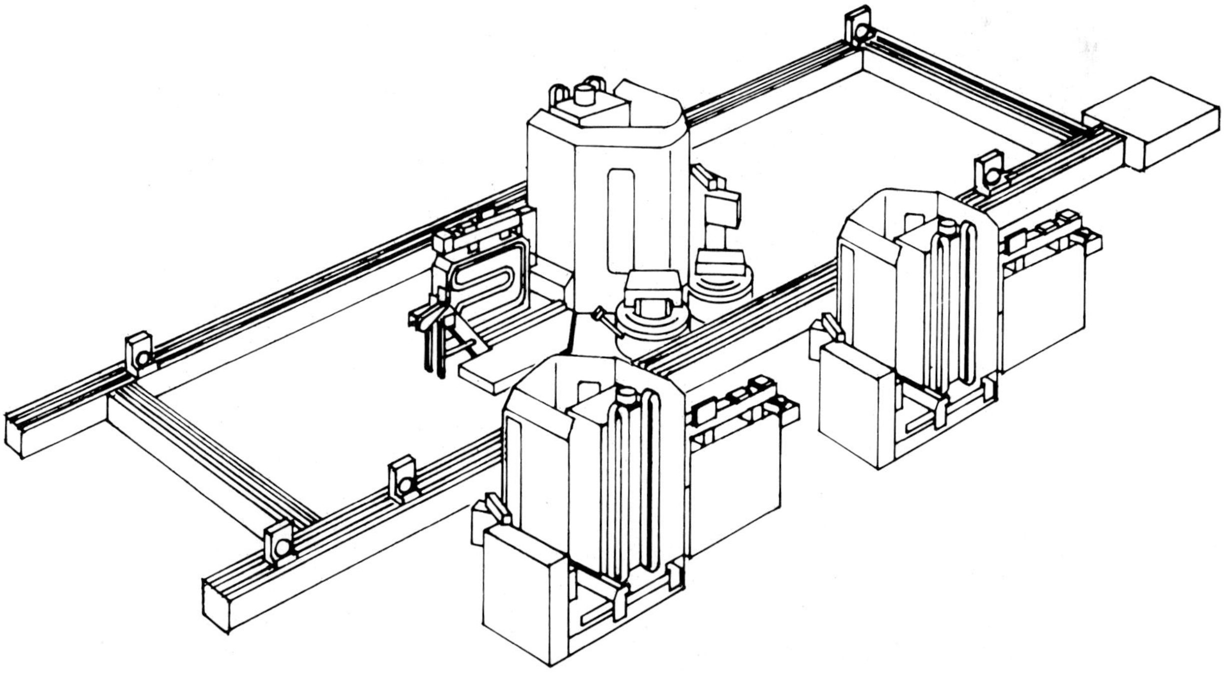

FIGURE 68–12
The pelletized cell can be applied to high-variety, low-volume production problems. (Courtesy of Kearney & Trecker Corp., Milwaukee, Wis.)

is thought of as a hammer-and-stake operation. In reality, our society would find it difficult to do without the fabrication of sheet metal parts that go into items such as automobiles, home appliances, and the aerospace industry. It is with several methods of metal forming in a power press that this unit will concern itself.

Objectives
After completing this unit you should be able to

1. explain what a blanking operation is and how it works.
2. explain what a drawing operation is and how it works.
3. explain what a trimming operation is and how it works.

Blanking Dies

The trade of tool and diemaking consists of making many of the tools and developing processes for mass production. High-volume output is made possible through the fabrication of a die.

A paper punch (Figure 69-1) can best illustrate the concept of a die. A steel punch is made to be accurately pushed through a sharp opening, which is considered the die. The paper disc that is pushed through in piercing the paper is considered a blank if it is useful; if it has no purpose, it is considered scrap. The difference between a blank or a scrap piercing is the use involved. Essentially, a die works by placing material between a punch and die, and when the punch is pushed through, it results in some sort of change in the material: piercing, blanking, trimming, slitting, or some other reshaping operation.

While a paper punch does not require much force to operate, power presses are used in industry (Figure 69-2).

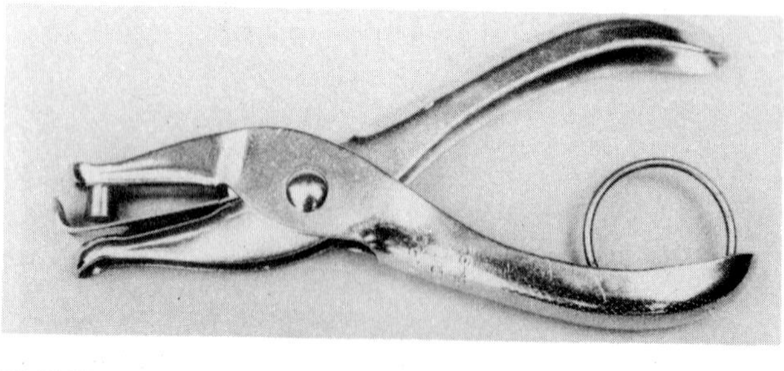

FIGURE 69-1
A paper punch: a simple example of a punch and die.

To assure accuracy in alignment, the punch and die are mounted in a die set (Figure 69-3), which is provided with a set of accurate guide posts and bushings for the many times the punch and die have to slide open and close against each other.

The making of a simple steel washer (Figure 69-4) can illustrate the concept of how piercing and blanking work in production. Sheet metal of the proper thickness is passed through a stripper between the piercing and blanking punches and the die below them. Material is first positioned and the hole is pierced and moved into position so that a blanking punch can pick up the pierced hole. As the material is pushed along with each opening, the scrap hole made by the blanking punch is automatically stopped by a trip finger that will always position the material so that with each closing the material is pierced and blanked and the trip finger is raised slightly, allowing the new scrap hole to be repositioned, producing washers in an on-going process.

Besides the multistage progressive dies, industry is now using numerical control machines that

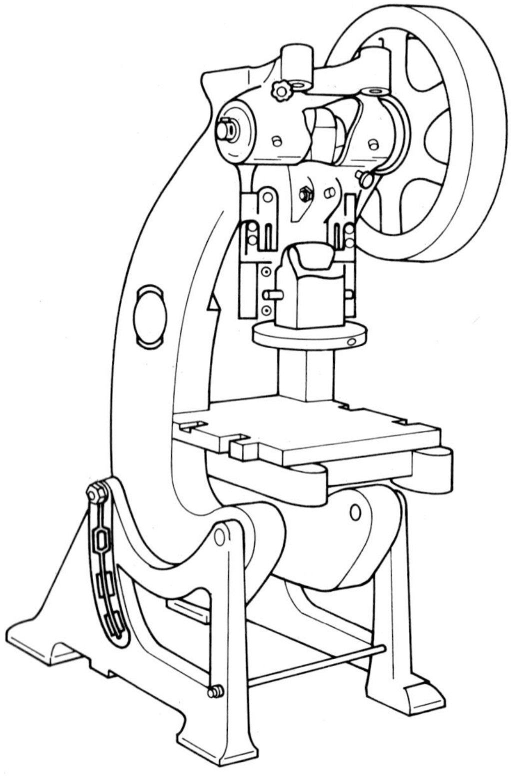

FIGURE 69-2
Punch presses in industry are supplied with either mechanical or hydraulic power. (Credit: Paul Rapisarda)

can take a whole sheet and notch out blanks very fast and efficiently, eliminating the cost of fabricating and maintaining expensive dies. While more will be said about automated machinery in a later unit, Figure 69-5 illustrates an example of tape control piercing and notching and the resultant blanks.

Forming Dies

In industry, many stampings are shaped in forming dies. Quite often the terms "bending" and "forming" are used interchangeably. Bending changes the plane of a blank into a different position, such as

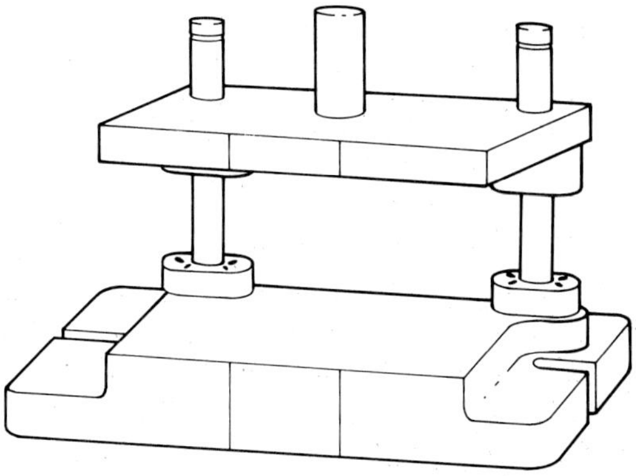

FIGURE 69-3
For production operations, a punch and die is mounted in a die set such as is shown here.

FIGURE 69-5
Numerical control machines not only eliminate the expense of building and maintaining a die but can also do the job efficiently.

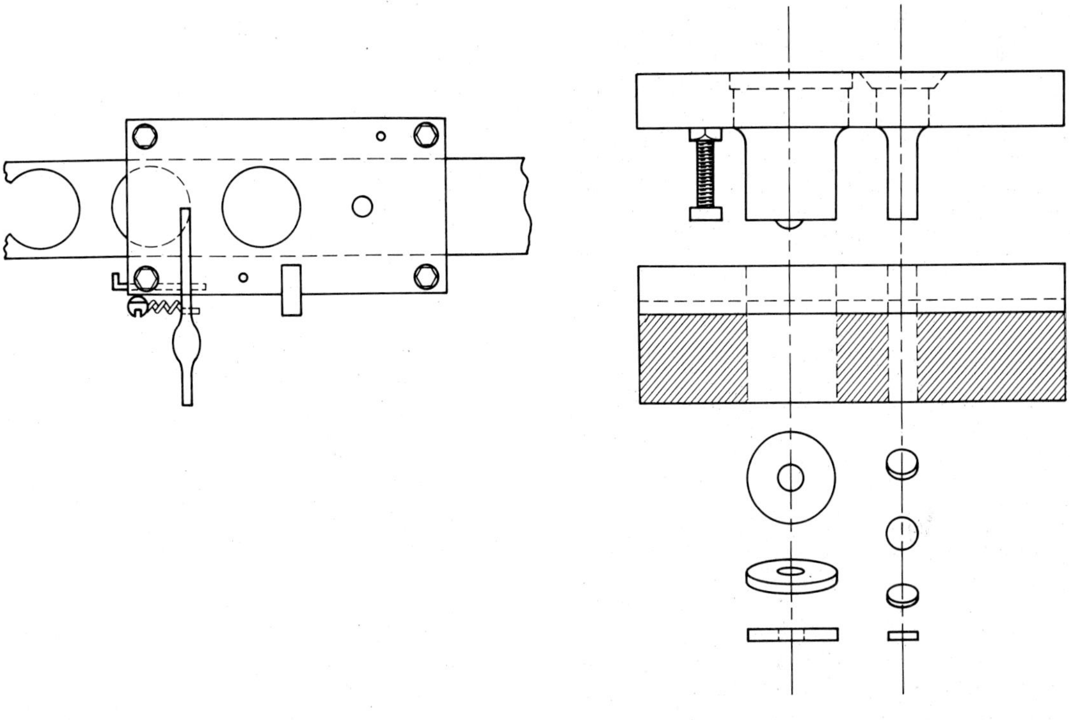

FIGURE 69-4
An example of a two-stage progressive die for making steel washers.

FIGURE 69–6
This aluminum door panel was partially made in a form die. (Credit: Reynolds Metals Company)

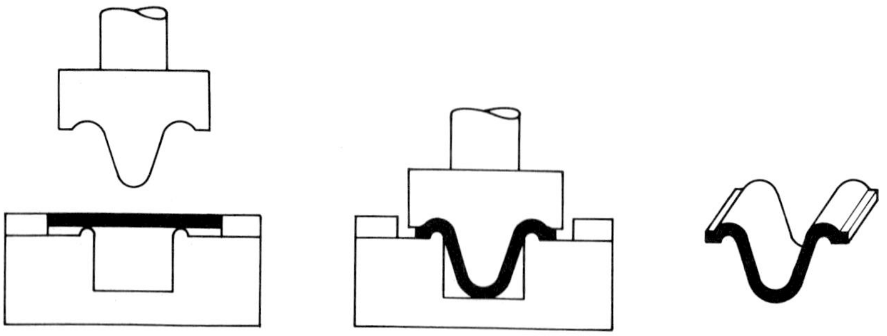

FIGURE 69–7
This bracket was shaped and curled in a form die. (Credit: Paul Rapisarda)

FIGURE 69–8
Wrinkling around the edges suggests an imperfect draw.

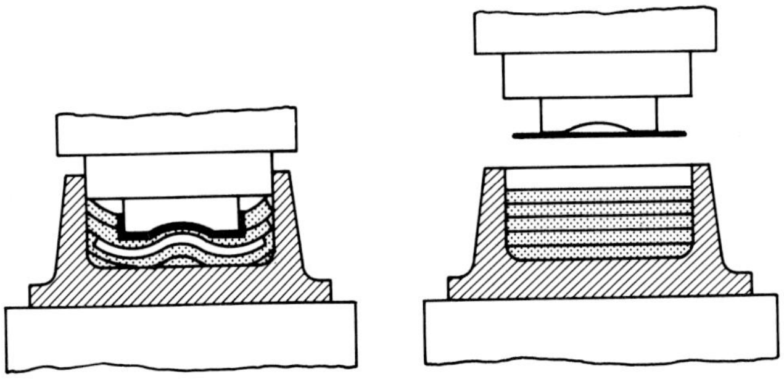

FIGURE 69–9
The Guerin process: forming with a rubber die, a method that is widely used in forming shallow metal parts for aircraft.

making angles up to 180°. Forming, on the other hand, changes the contour of the metal to conform to the shape of the punch whether it be round, curved, or irregular. One only has to look at the automobile to realize that forming dies are responsible for the contours that range from fenders to door panels (Figure 69–6). While forming dies can be simple in design, as shown in Figure 69–7, other forms may be deeper.

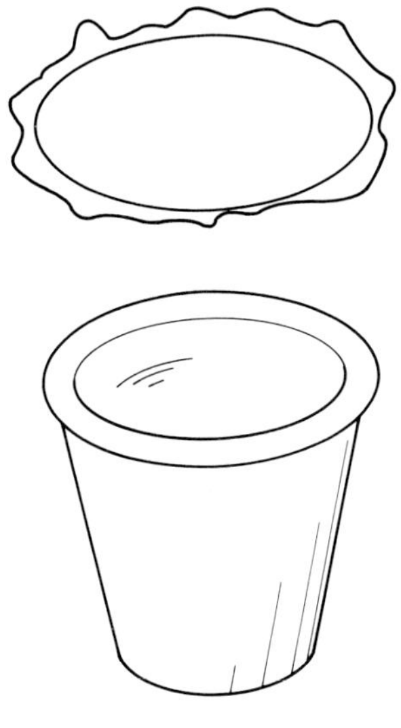

FIGURE 69–10
Trimming is a necessary operation to eliminate the wavy edges that surround the opening and bring it to size.

Draw dies can be made to produce long hollow shells, referred to as deep drawings, or can be made to have a slight depth. In either case, the sheet metal blank has to be held firmly in place by the blank holder (pressure pad) or the metal will not slide evenly as the cup is formed, wrinkling the edges, as shown in Figure 69–8.

A third method for forming sheet metal parts is that of using rubber forming, commonly called the "Guerin process." Used in short-run production, only the form die is made of metal; the upper half is composed of a slab of hard rubber that is surrounded by a heavy steel frame. A blank is placed over the die (Figure 69–9), and when the press is closed, the rubber exerts pressure and forces the metal down and around the exact shape of the form.

Trimming Dies

Another operation performed by dies is that of trimming. In draw dies, for example, a blank has to be used that will be large enough to hold onto as the shell is being drawn. The wavy edges around the opening must later be trimmed to complete the shell to a finished size (Figure 69–10). Trimming is also done in drop forging operations where excessive metal is eliminated from the main drop-forged parts by trimming. Nor is trimming restricted to drawn objects and drop-forged parts. Surplus material that forms around the contour of zinc, brass, or aluminum castings are also trimmed to a final dimension.

FIGURE 69–11
Sheet metal is reduced in thickness as it passes at high speed through these cold rolling mill stands. The metal is then coiled for annealing or shipping. (Credit: American Iron & Steel Institute)

Cold Working

The cold working of metal involves working it in a cold state, usually about room temperature. There can be property changes brought about in cold-working metal by hammering, rolling, twisting, bending, or vibrating it. These changes are classified as work hardening. If a narrow strip of soft copper is bent, the copper will become work hardened. It cannot be straightened again by hand, nor can it be bent again in the same place. It would have to be annealed (softened) to be reworked.

Coiled sheet steel (Figure 69–11) is also an example of how rolling causes work hardness, after which the metal is annealed before it can be used commercially. A deep-drawn shell (see Figure 69–10) is yet another example of how cold working affects metal. Tearing and splitting of the shell can occur when extensive cold working is performed. Many deep-drawn shells cannot be made without annealing.

Cold-working materials are used extensively in industry for machining and sheet metal stampings. Cold-rolled steel bars, for example, are used for a variety of purposes after they have been prepared by the steel mills.

UNIT 70 MANUFACTURING PROCESSES

The fabrication of raw materials such as sheet metal, aluminum bars, and steel rods into useful component parts and assemblies is called manufacturing. When one considers that the manufacturing of products utilizes one-quarter of the American workforce and is responsible for over 25% of every dollar earned, one can begin to understand how manufacturing plays such an important role in an industrialized society. In global competition, manufacturing plays an even greater role as newer and innovative methods are introduced to improve manufacturing and lower the costs of productivity.

Objectives
After completing this unit you should be able to

1. know the purpose and function of a CNC system.
2. know the purpose and function of a DNC system.
3. know the purpose and function of the CAD/CAM system.

Managing People

A network of leadership has to manage a pool of human resources in order to make any manufacturing operation successful. In industry, the personnel department is responsible for hiring the skilled, unskilled, and professional workers that will fill the various job slots available.

It is in the personnel office that plans for recruiting, placement, training, rating, lay-offs, and discharging of workers are carried out. Records are maintained on workers, job standardizations, classification, and wage setting.

In recruiting people for different positions, the personnel department will have to do so within the framework of equal opportunity. Not only is an applicant's work history and previous experience examined, but so is a person's appearance.

Does the applicant with the excessively long hair present an insurance risk when working around machinery?

Is an applicant really trying to sell himself or herself or is the person just going through the motions of an interview?

In many situations technical applicants are referred to department heads who are better qualified to make the decision on the applicant's worthiness. In addition to hiring and firing, personnel looks after the fringe benefits (medical insurance, life insurance, sick pay, vacation, etc.) that management packages to attract workers. All in all, personnel plays an important role in the quality of workers an organization hires. The personnel department provides one of the first impressions an applicant will get of a company.

Processing Materials

The manufacture and completion of a product involves processing and handling of materials from its raw material stage to finishing and assembly and final shipment to the marketplace (Figure 70-1). No plant can operate without the coordinated steps of process planning that assure a smooth flow of handling at different stages. While there is no one successful plan of action to fit every plant's needs, every company must adopt some planning form in order to be successful and compete economically. One such plan is labeled just-in-time manufacturing.

Just-in-Time (JIT) Manufacturing

There are many things that JIT manufacturing represents and many things it does not represent. It does not represent:

- A zero-defect program
- A zero inventory program
- A lead-time reduction program
- A shop floor control replacement

It does represent:

- A goal and philosophy to achieve overall business excellence
- A strategy to make and ship parts when needed and control productivity to meet demand

FIGURE 70-1
Orders in the process of being shipped to the marketplace. (Credit: DoAll Company)

Born of necessity, JIT manufacturing was the brain child of Taiichi Ohno, who started his work career as a thread production engineer and wound up in the automobile business when his company was absorbed by Toyota in 1943. By 1948, he was managing an engine plant. In the intervening years, he had managed an automobile assembly shop where general-purpose machinery was equipped with special tooling to overcome the lack of special or updated machinery.

Not really knowing the automobile business, Ohno drew on his past experiences with automatic looms. Little by little, scheduling was rearranged or changed so that parts were produced only as required, with stock not built for the next operation. Every detail in the manufacturing process was worked out to eliminate waste. By rearranging machinery, product moves were minimized; setup time and downtime were reduced. Quick-changing dies were put into use in presses. Noticeably absent to any observer was the lack of storage space. What space there was could be found at machines that were not part of an assembly line. They were reserved for raw and finished parts. So dramatic were the results that completed assemblies were timed to come off the lines and went directly into shipment trucks. There was no room provided for storage, and if trucks failed to show up for loading, the assembly line would have to shut down.

The traditional practice of maintaining reserve stock for "just-in-case" emergencies was replaced by a new philosophy that called for first-time machine capability, teamwork, as well as minimizing inventory.

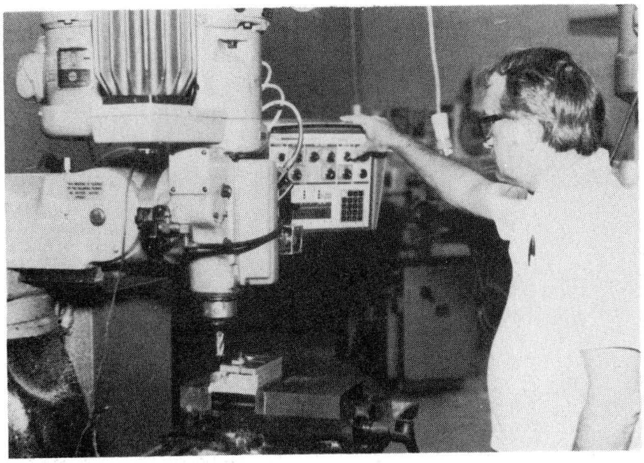

FIGURE 70-2
Special tooling provided by jigs and fixturing do much to simplify the machining of castings or odd-shaped work.

Production Tooling

In metalworking plants, extensive tooling is used to properly produce the varied shaped components. The use of stamping dies and manufacturing devices requires the services of a highly skilled labor force: tool designers and tool and diemakers. In addition to the diemakers building the permanent tools that will transform and shape metal by piercing, slotting, and forming, toolmakers also have to make jigs and fixtures (holding devices), such as are seen in Figure 70-2, by which castings and similar bulky articles can be machined for interchangeability. The key to processing is the simplicity by which these work holding devices operate. They are designed for easy loading and unloading of the part, combining as many machining operations with each use and eliminating as many tool changes as possible.

While the trend in the future is toward unmanned machine tools and other computer-controlled developments to maintain efficiency, quality, and cost, many jobs will fall in the low-volume category. These jobs will continue to be more cost effective when managed by skilled and trained people.

Manufacturing Systems

Manufacturing processes are going heavily into systems that are controlled or managed automatically through the use of numerical control (NC) and/or computers. A number of machining systems are used for large-scale, high-quality production runs or jobs that require many tool changes and handling (Figure 70-3).

FIGURE 70-3
A robot "trained" to load tools in a tool magazine.
(Credit: Kearney & Trecker Corp., Milwaukee, Wis.)

Early NC machining produced the means by which the positioning of tools and workpieces could be accomplished by tape programming. These machines were considered hard wired. This meant that the directing of coded instructions was limited by the fixed connection of circuits and storage devices. Changes could only be made by changing the connections. This system was improved with the introduction of computer-assisted programming and would become known as the *CNC* (computer numerical control) system. The CNC operations would give machine tools such as lathes (Figure 70-4) and production grinders (Figure 70-5) a greater machining flexibility than before.

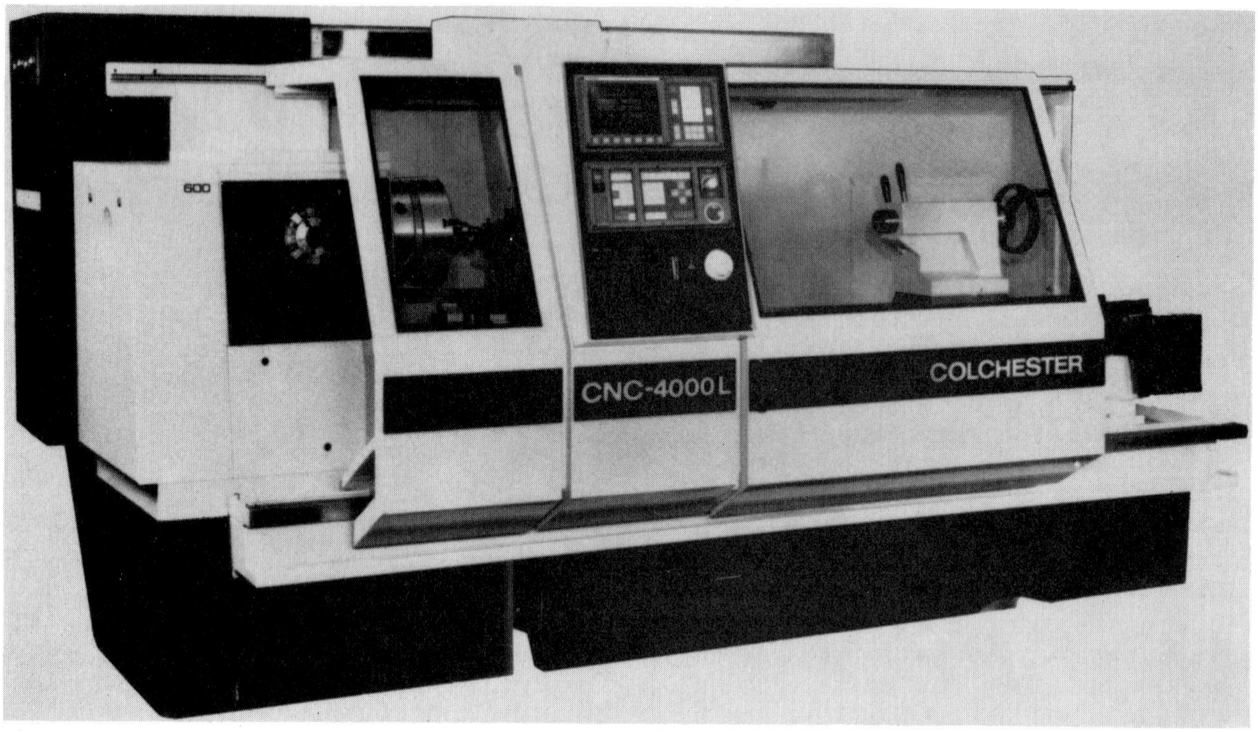

FIGURE 70–4
A CNC production lathe. (Credit: Clausing Industrial Inc.)

FIGURE 70–5
A CNC production grinder. (Credit: Litton Industrial Automation, Landis Grinding Machines)

FIGURE 70–6
DNC systems act upon instructional commands directly. (Credit: Kearney & Trecker Corp., Milwaukee, Wis.)

The next development to become operational was the direct numerical control (DNC) system. Connected to a series of minicomputers that were linked together or by combining a large computer with a minicomputer to form a common computer memory, the DNC system made possible the numerical control of machining or processing by a computer. In both the NC and CNC operations, instructional commands that had to be read by the machine control unit were then translated into specific machine operations. Instructional commands were acted upon directly in a DNC system (Figure 70–6). Machining was made possible through time sharing the computer. Resulting benefits would be similar to those of the CNC, including absolute and incremental dimensioning in either the inch or metric system, simple reprogramming and editing at any time, program storing and memory recall, cutting tool compensation for wear and length, and lowered program costs by using interchangeable software containing NC requirements.

FIGURE 70-7
Developing computer modeling of a sheetmetal forming process.
(Credit: Ford Motor Company)

CAD/CAM Systems

The DNC system also became the lifeline of the CAD/CAM system of manufacturing. Computer-aided design/computer-aided manufacturing (CAD/CAM) are twin technological advances that use the computer to speed up design calculations for the development of a product (CAD) and to communicate the instructions to automatic equipment to speed up the handling and processing of materials to completion (CAM) (Figure 70-7).

Of great importance to a designer is the CAD system. Computer drawings can quickly produce hard copies of component parts with speed, skill, and precision unmatched by the capabilities of a draftsperson who still uses conventional drafting tools and instruments. The computer is instrumental in calculating workpiece size and will eliminate the repeated tasks of drawing dimension lines, extension lines, and arrowheads. The computer is just as capable of erasing lines and dimensions as it is of drawing them, eliminating unwanted symbols, characters, arcs, circles, or other special designs. The CAD system can also be programmed to draw sectional or auxiliary views. All information can be stored in memory to be recalled at the push of a key on the keyboard as a three-dimensional drawing or a multiview shop print.

The CAD system can also generate graphic pictures that are displayed on a televisionlike screen, the cathode ray tube (CRT), which receives electronic signals sent by the main computer. By adding a light pen or an electronic tablet and plotter to the desk-top computer keyboard and CRT screen to the main computer, a designer can produce any view or drawing. The CRT screen permits a very close examination of any part drawn or recalled. Not only can the print be examined as a whole, but also sections can be separated from the whole and enlarged on screen for analysis and workability. The part may be manipulated and viewed from another angle. Mirror images of mating parts can be quickly drawn and

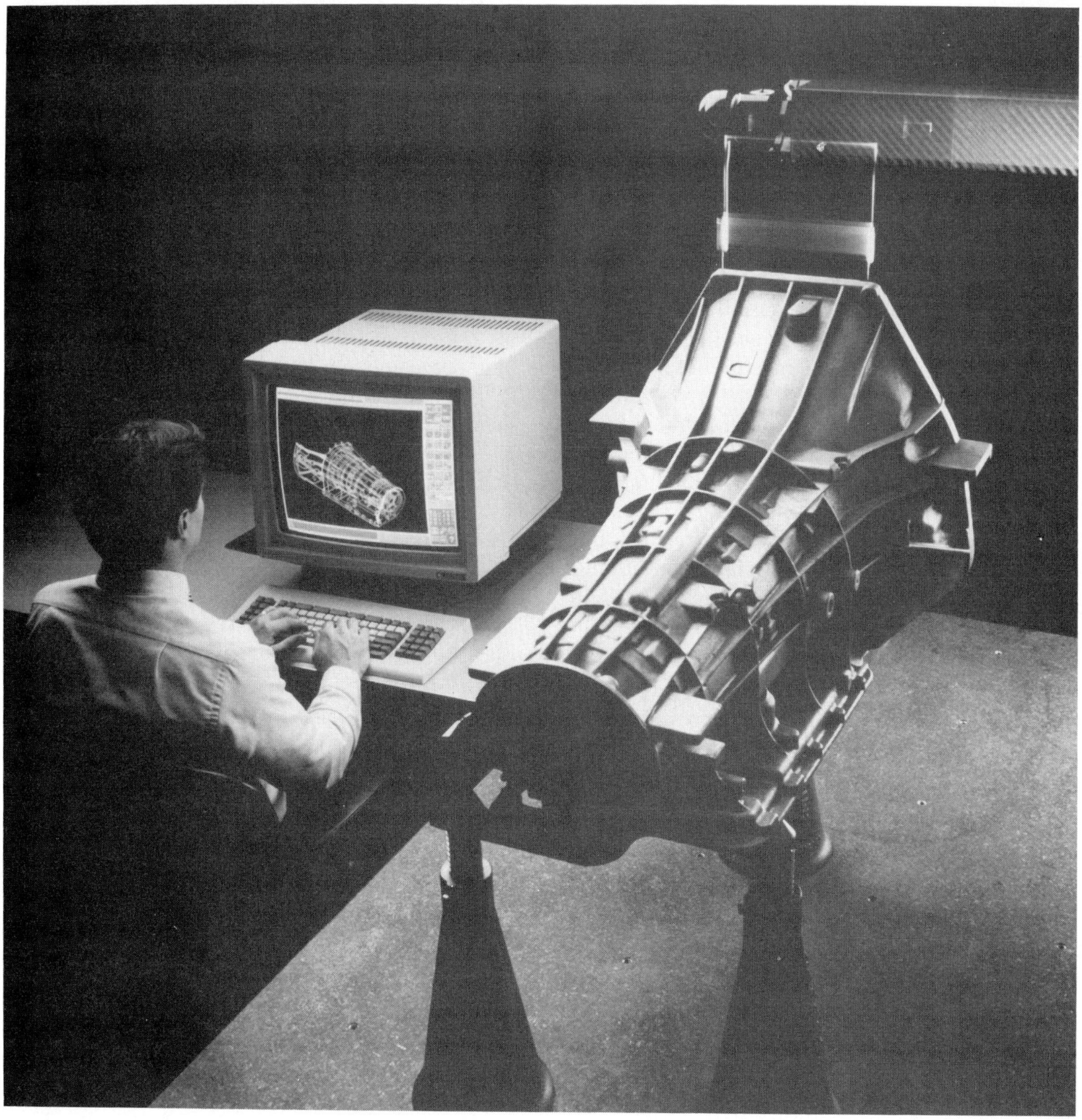

FIGURE 70–8
Using a CAD/CAM work station to develop the coordinate measuring locations for a machining inspection system. (Credit: Ford Motor Company)

"assembled" for possible design modification. Changes may be made on the screen by using the light pen, and these changes are recorded automatically in the computer memory.

Using a typewriterlike keyboard, the CAD system can also generate a numerical control tape for the manufacturing of the part. This feature both saves time and greatly minimizes production-related costs.

Total manufacturing processing can be coordinated by the scheduling of production and inventory control; inspection, gaging, and testing; machine tool control and monitoring; and management systems. This has resulted in higher productivity, reduced cost per item, and improved product quality (Figure 70–8).

The combined use of the CAD/CAM system has made the sharing of timely product information required by manufacturing available at electronic speed, linking management, design engineering, manufacturing support, and factory personnel in an information network. It has played an important role in moving production closer to total factory automation.

Materials Handling

The problem of storing and moving packaged materials by means of a fork lift truck and simple conveyor belts and rollers has been surpassed by the most technologically advanced means that increases the accuracy of the picking and sorting system.

In large warehouses where the automobile parts are stored, high-volume direction and package shipping has been made by means of manual and automated equipment. Stockpilers are sent to the storage areas in rail-mounted cars where they pick and transfer cases from storage shelves onto a moving conveyor belt. Here a worker places color bar labels on the top and sides of the cases (Figure 70-9). The labels are computer printed and identify the product, customer, and bill of lading (package contents) (Figure 70-10). The labels are "read" as they move past laser scanners, where a sorting land number will lead them into the proper trailer truck line (Figure 70-11). The scanner's input controls and divides the flow of cases. Since many of these lines can accumulate and merge, the sorting or separation of these lines can be manually controlled to override automated controls or a computer's program to unlock traffic, very much like controlling merging train traffic entering a terminal.

A recirculating loop of ten sorting lines is formed with this system that enables packages to be loaded into trailer trucks while continuously accepting, dividing, and feeding new orders.

Career Opportunities

Technology in the metalworking industry is constantly changing. Systems of automation will require people with a great deal of machining capabilities to enable these systems to work more effectively and faster. Programmers with a basic understanding

FIGURE 70-9
Placing a color bar label on packaging that will be processed for shipping. (Credit: General Motors Corporation)

FIGURE 70-10
Labels are computer printed to identify the product customer, direction, and sorting control. (Credit: General Motors Corporation)

FIGURE 70–11
The scanner's input controls and divides the flow of cases to waiting shipping trucks. (Credit: General Motors Corporation)

of machining, metals, and processes will be in great demand. In turn, numerical control devices will enable those who are in machining to work more effectively and quickly and will tend to extend and prolong their careers as skilled tradespeople rather than eliminate them. When one considers that a numerical control machine with a single cutting tool can generate holes, contours, angle blends, and fillets, thus eliminating a number of tools and tool changes, it becomes evident that such a device can be an aid never before possible. Rather than calculate by hand the mathematical moves to achieve tool positioning, it can be done simply by the press of a key. A numerical control machine will do anything it is asked to do if it is asked properly. Essentially, the machine will only be as intelligent as the person programming it.

Technology updating will play a key role in the career of any individual working in the manufacturing field. Technology is a tool that cannot go to "rust." It can greatly increase the effectiveness of time and work. In the machine tool industry, for example, technology seems to double every two or three years. This makes the manufacturing processes of four or seven years ago obsolete. No industrialized nation can afford to rely on obsolete factories, obsolete technology, or obsolete manufacturing processes.

FIGURE 70–12
Engineers and technicians will have to stay knowledgeable or be updated with the latest technology to be successful in industry.
(Credit: DoAll Company)

It is necessary for the engineer/technician of today to stay knowledgeable of the latest manufacturing technologies if the industrialized nation that he or she works in is to survive (Figure 70–12).

UNIT 71 TURNING AND CHUCKING CENTERS

One of the most widely used machine tools in the metalworking shop is the toolroom lathe. For all its flexibility, however, it was never designed for production work. A host of technological advances such as design features, electromechanical devices, hydraulics, computers, and numerical control operations have led machine builders to the development of the computer numerical control (CNC) turning and chucking centers. With these technological advances, machining is carried out accurately, efficiently, and economically. In the toolroom, the lathe has been updated to now incorporate vision systems of operations, machining work faster and to accuracies never before possible.

Objectives
After completing this unit you will be familiar with

1. the purpose and function of chucking and turning centers.
2. the purpose and function of vision systems for lathes.

CNC Chucking Centers

Chucking and turning centers are designed to meet high manufacturing needs with a minimum of handling. Chucking centers (Figure 71-1) are versatile and productive turning systems that feature chucks as the chief work holding devices. These gripping devices include self-centering, collet, and countercentrifugal chucks among others. Ruggedly built, these chucks range in sizes from 8 to 36 in. (200 to 900 mm) in diameter. Chucking centers are made by a number of manufacturers with machine standard features that are similar in functions. Consequently, only the chucking lathe designed as a four-axes model will be described.

This CNC chucking center features a set of dual-cutting turrets seated and held firmly by 16 tons of force. These turrets can operate independently or together to machine a workpiece at the same time. Provided with 14 tool stations, the inside and outside diameters of a workpiece may be machined simultaneously (Figure 71-2). In situations that may require mostly internal machining, both turrets can concentrate their tooling efforts in this area (Figure 71-3). Total outside-diameter (OD) machining can also be achieved by equipping the upper turret with turning tools. This is indicated in Figure

FIGURE 71-2
Inside and outside diameters may be machined at the same time. (Credit: Cincinnati Milacron Inc.)

FIGURE 71-1
CNC chucking centers feature chucks as the chief work holding devices. (Credit: Cincinnati Milacron Inc.)

FIGURE 71-3
Both turrets of a chucking center can be used for internal machining. (Credit: Cincinnati Milacron Inc.)

FIGURE 71-4
External machining can also be carried out using both turrets. (Credit: Cincinnati Milacron Inc.)

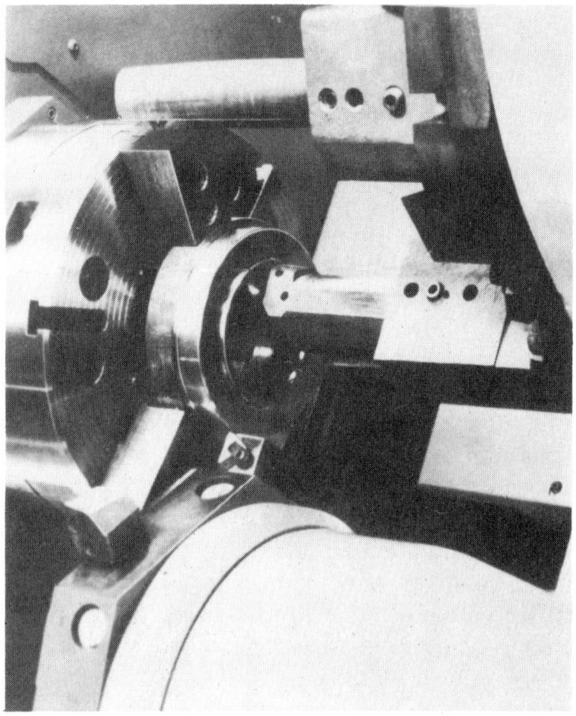

FIGURE 71-6
Facing and turning operations may be carried out at the same time by using the twin turrets. (Credit: Cincinnati Milacron Inc.)

FIGURE 71-5
Long shafts may be rigidly supported on one end for machining by a center mounted in the upper turret. (Credit: Cincinnati Milacron Inc.)

71-4, which shows the simultaneous machining of a chamfer and the diameter of a workpiece. Free machining adaptability is maintained by the design feature that positions the OD tooling interference free in a clearance area during inside-diameter (ID) machining operations. The capability of the OD turret operation is flexible enough to machine the largest specified workpiece diameter without interference from the carriage and cross slide. In addition, boring bars as large as $2\frac{1}{2}$ in. (75 mm) in diameter may be used for ID machining.

The upper turret may also be loaded with a center to support the end of long work shafts (Figure 71-5), enabling the lower turret to complete any OD machining operation that may be required.

Other advantages provided:

- Simultaneous facing and turning (Figure 71-6)
- Internal and external threading (Figure 71-7)
- Chamfering, internal radii, or retaining grooving

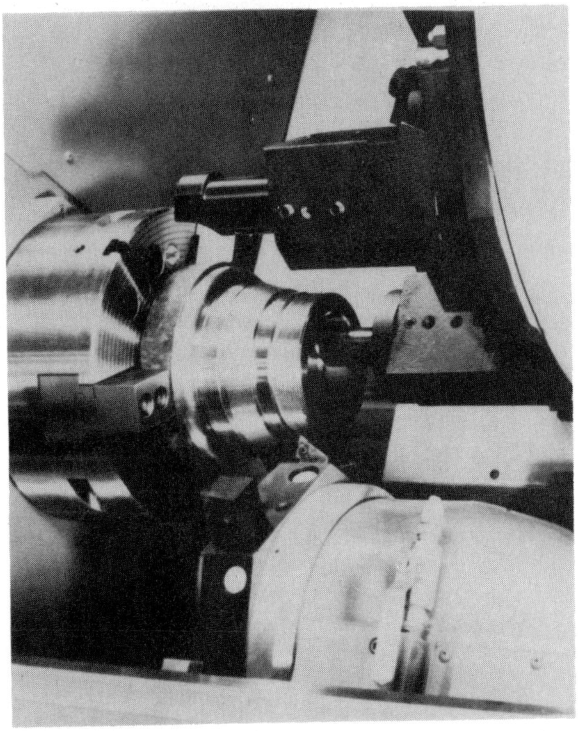

FIGURE 71-7
Threading poses no problem as internal and external threads may be cut at the same time. (Credit: Cincinnati Milacron Inc.)

FIGURE 71-8
Cutting tools are preset by using a tool setting gage to assure location accuracy for programming purposes. (Credit: Cincinnati Milacron Inc.)

- No adjacent tool interference
- Turnings right up to the chuck jaws with minimum turret withdrawal for indexing

Construction

Rigidity is the key to the machining success of the chucking center. Its foundation is constructed of a one-piece cast-iron bed that eliminates bolted joints and provides excellent damping (vibration absorbability). The machine frame provides additional strength and rigidity with its heavy walls, angular support braces, and ribbing throughout the length of the bed. Loading and unloading of parts is more easily achieved due to the 30° slant from the vertical plane bed design, which allows chips and coolant to be collected at the bottom of the bed. A heavy steel sliding shield is also included as a protective device. Mounted on a heavy-duty front rail, the shield completely encloses the chip cutting area. Failure to close the sliding shield completely causes the machine to stop.

Accurate positioning and travel of the cutting tools is provided by direct-current (DC) servo motor drives. Dependable and repeatable turret cross-slide movement is made possible by high-precision ball screws that assure repeatability to an accuracy of ± 0.0002 in. The servo motor can produce fast machining cycles, generating turret slide speeds as fast as 200 in./min (5080 mm/min). Rapid traverse speeds are rated twice as fast as the turret slide speeds.

Tooling

Chucking centers can accommodate seven types of tools in both the upper and lower turrets. A gage is used for presetting cutting tools (Figure 71-8), which can then be mounted in the lower turret for machining ODs. These tools will only need replaceable inserts when worn. Dovetail tooling is used and mounted in the upper turret for machining IDs. As with the lower turret tools, these tools are qualified (preset) by using a tool setting gage to assure common tool point location for programming.

FIGURE 71-9
The console is the heart of the computer control system. (Credit: Cincinnati Milacron Inc.)

CNC Operation

A wide range of standard and optional programming features are possible with the control console shown in Figure 71-9. The console is the heart of the computer control system that features a minicomputer, a CRT, a tape reader, part program storage and edit, buffer storage, and a maintenance diagnostic display.

The minicomputer controls instructional logic, mechanical operations, and input-output commands. The buffer storage enables a control system to act quickly on stored information without waiting for the tape reader to react. The CRT provides a video display of a number of operations and information, such as spindle surface speed, slide positions, sequence numbers, diagnostics (analysis of system fault conditions), preparatory functions, keyboard data, and operator interface (instructions).

The keyboard is the means by which input with the CNC system is made. Information such as setup and tooling data on new parts or programs are entered; revisions or tool correction data may also be entered. Input data concerning diagnostics or troubleshooting the system as to machining conditions, tool wear, and thermal gaging may also be entered on the keyboard. The results of these questions are revealed on the CRT video display.

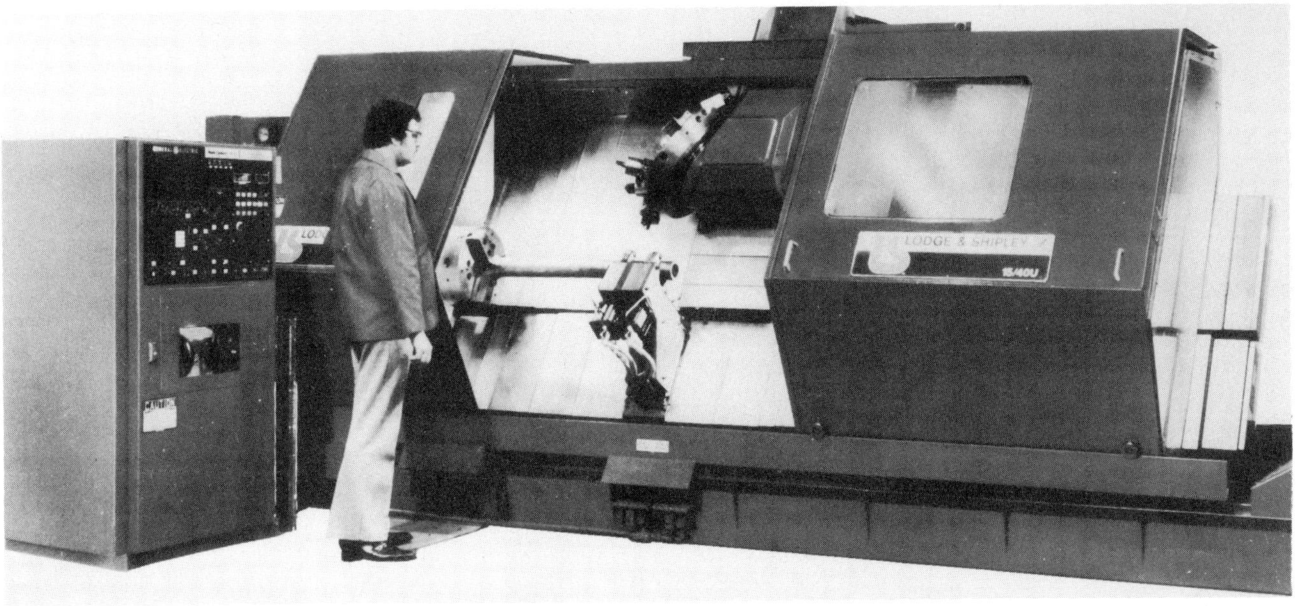

FIGURE 71-10
A CNC turning center is designed for efficiency and productivity.
(Credit: Lodge & Shipley)

CNC Turning Centers

Computerized numerical control turning centers (Figure 71-10) are very similar to the chucking centers in that they are also equipped with multiple tooling and flexible automation equipment. One of the benefits of turning centers is that they are modular in design. This means that additions or expansion to different levels of automation can be made.

Turning centers can stand alone or become integrated as components of flexible manufacturing cells (FMCs). These manufacturing cells consist of two or more machines, a material handler system, and a cell controller (Figure 71-11).

Machine Construction

Design and construction of turning centers has improved to increase rigidity and reduce vibration. Stiffness has also been increased for minimizing deflection. Where once flat beds were predominant, slant beds are now produced to permit greater rigidity, reduce floor space requirements, and improve access to loading and unloading parts.

Like the chucking center, the turning center base and bed is mainly cast in one piece, although some are weldments. The materials and methods used to produce this construction will vary with the manufacturer. Some manufacturers are bypassing any weldments, making their beds and/or bases of granite or reinforced cement. Beds made of reinforced cement or an epoxy and granite powdered base have been found to have six to eight times the damping qualities of cast-iron beds. In addition, heat transfer may be about 20 times lower, contributing to highly accurate, consistent machining.

Automatic Tool Changing

Automatic tool changes for turning centers are similar to those featured on a chucking center four-axis machine: two opposed turrets, each capable of holding seven different tools, are mounted on separate cross slides. Due to the fact that one turret is placed above while the other is placed below the centerline of the workpiece, the cutting forces applied by the turrets are equalized. This permits extremely heavy cuts to be taken when the work is supported by the

FIGURE 71-11
A cell of four CNC lathes and turning centers involves automatic part loading, gaging and tool changing. (Credit: Lodge & Shipley)

tailstock. The cutting units of the twin turrets provide many advantages to using turning centers, such as:

- Quality workpieces due to improved accuracy and reliability
- Increased productivity due to faster setups
- Finish turning and threading at the same time
- Roughing and finishing cuts in one pass
- Greater flexibility due to reduced manual tool changing
- Machining two different sections of a shaft at the same time
- Machining different diameters on a shaft at the same time (Figure 71-12)
- Cost savings due to shortened machining cycles

Quality Control

Quality control inspection is an operation that can be carried out both in process and postprocess. In-process gaging (workpiece measurement takes place

FIGURE 71-12
Different diameters on shafts can be machined by both turrets on a turning center. (Credit: Cincinnati Milacron Inc.)

during the machining operation) uses a surface sensing probe, an electronic triggering device that contacts the workpiece. The probe is loaded into the tool storage holder and is automatically selected and changed by the tool changer in exactly the same way as any cutting tool. This has advantages and disadvantages. The advantages include low cost, flexibility, and programmability. Disadvantages include false measurements due to a hostile environment and increased cycle time.

Postprocess gaging (measuring work removed from the work area after machining) may require the use of electronic gages, optical gages, and in some cases laser gages.

Lathes with Vision

One of the more advanced systems to be introduced in the lathe technology is a vision readout system (Figure 71-13). Evolved from the digital readout system, lathes equipped with vision readout improve accuracy and repeatability and reduce scrap as well as operator fatigue to make lathes more productive.

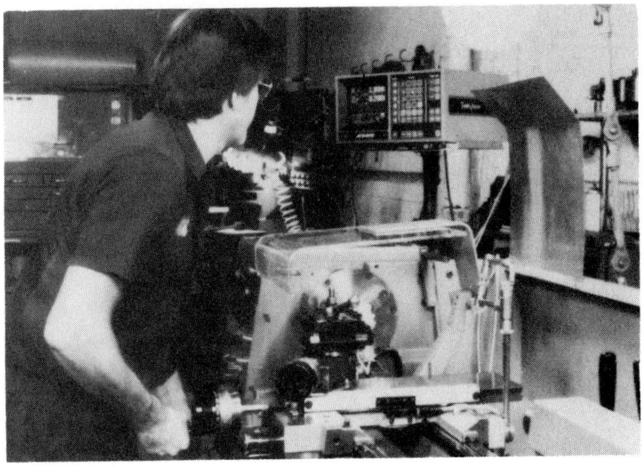

FIGURE 71-13
A vision readout system designed for lathe operations. (Credit: Acu-Rite Incorporated)

FIGURE 71-14
Both *X* and *Z* axes are prominently displayed. (Credit: Acu-Rite Incorporated)

The vision readout system permits the setting of all operations at the keyboard and can provide a number of features for production lathes never before possible, such as:

- A readout counter that is either graphic or digital
- Calculations and measurements in either the inch or metric system
- Eight-digit axis display
- Axes display on screen (Figure 71-14)

FIGURE 71–15
This vision readout system features a user-friendly software for lathe operations. (Credit: Acu-Rite Incorporated)

- Automatically adjusts and displays the *X* and *Z* axes in relation to the movement of the compound crossfeed (*Y* axis)
- Built-in tapering calculations for figuring angles
- Display in degrees, minutes, and seconds
- Capability of entering speeds and feed rate
- Tool offset can be entered manually or automatically
- Twenty-four programmable tool offsets
- A user-friendly software that is written specifically for lathes (Figure 71–15)

Vision readout systems can be retrofitted to lathes in the same manner that digital readout systems have been added to milling machines, jig borers, and other manual mills to greatly increase their efficiency and accuracy. Vision readout systems can reduce lathe positioning and greatly eliminate the manual machining operations and skills associated with lathe turnings.

UNIT 72 SUPERABRASIVES

The natural extension of manufacturing processes that characterized the Industrial Revolution has created a second industrial revolution. Spurred by competition, today's manufacturing materials are a result of automated machine tools, robots, and even factories controlled by some form of numerical control. The development of superabrasive cutting tools that outwear, outlast, and outperform conventional cutting tools has resulted in manufacturing processes that are faster, efficient, and more reliable. When used in connection with computers and robotic systems, superabrasive cutting tools improve the user's competitive position in the world market.

Objectives
After completing this unit you will be able to

1. explain how manufactured diamond and cubic boron nitride abrasives are produced.
2. identify the properties of a manufactured diamond and its comparison to other abrasives.
3. identify the properties of cubic boron nitride and how it compares with other abrasives.
4. condition (true and dress) a superabrasive grinding wheel.

Production Cutting Tools

Along with the development of machine tools came the development of cutting tools with which to increase productivity. Historically, cutting tools fell in two broad categories: those made from hard durable metal and materials and those that were abrasive based.

Carbon tool steels were the principal source of cutting tools for many years. Cutting tools made of carbon tool steel featured poor hardenability characteristics and lost their cutting edges at relatively low temperatures. Their use in cutting tools is almost nonexistent in industry today.

High-speed steel was introduced about 1900 and is alloyed with tungsten, molybdenum, vanadium, chromium, and cobalt to increase its wear resistance and retain hardness at high temperatures. High-speed steel can be operated at speeds twice as fast as carbon tool steel and still maintain a sharp cutting edge. Cutting tools made of high-speed steel can withstand shock and will perform well on ferrous and nonferrous metals.

In addition to carbon, *cast alloys* are composed of three main elements: tungsten, cobalt, and chromium. They are cast in molds that are moderately chilled during cooling, which gives their outer edges extra hardness and wear resistance qualities. While they may be operated at cutting speeds more than twice that of high-speed steel, their cutting edges may soften slightly in their red hardness range but will return to original hardness upon cooling.

Although tungsten carbide was discovered in 1890, it was not until 1927 that *tungsten cemented carbide* cutting tools were available commercially. Made with cobalt to increase high hardness and with varying amounts of titanium or tantalum carbide, tungsten cemented carbides have excellent red hardness, have high compressive strength, and can be operated at cutting speeds three or four times faster than high-speed cutting tools.

Coated carbides have a thin vapor-deposited coating of titanium carbide and titanium nitride to increase productivity by greatly prolonging tool life and making possible a greater metal removal rate.

Ceramic cutting tools were first introduced in 1956 and are manufactured primarily from aluminum oxide and varying amounts of titanium oxide or titanium carbide. The cutting speed and tool life of ceramic cutting tools are four times greater than carbide cutting tools on certain applications and can produce chips five times faster than high-speed steel cutting tools. Machining at increased speeds will improve surface finish and maintain dimensional accuracy due to less heat distortion generated in the part being cut.

The History of Manufactured Diamond

By 1954, General Electric scientists had unlocked the secrets to producing diamond. This was the end of a long road that originated when the scientific community conclusively proved that diamond was a form of carbon. While most of the claims for manufacturing diamond over the years proved to be false, there were some breakthroughs and allied developments that resulted during these experiments. In 1880, the Scot chemist James Hannay nearly lost his life when 77 of 80 iron cannon tubes stuffed with bone oil, lithium, and hydrocarbon exploded in a heated furnace. His laboratory was shattered; left behind to be uncovered in the debris were 12 crude, little crystals acknowledged to be true diamond chips.

The manufacture of the third hardest known substance, silicon carbide, by E. A. Acheson in 1891 was the result of a failed attempt to manufacture diamond. Previously an assistant to Thomas Edison, he was also actively involved in graphite and lubricants containing graphite.

A five-year program to study a process for manufacturing diamond was undertaken in 1941 in a joint effort by Norton, Carborundum, and General Electric. Due to World War II, the program was put aside after a few years but not before experiments had been conducted using high-pressure/high-temperature methods on graphite contained in a press. Although studies by the Norton Company continued after the war on a limited basis, all efforts to continue the joint development program with General Electric were rejected. The General Electric Company would work independently on the project. It was free of any commitment or joint obligations with any other company to pursue and develop a method of producing diamond.

The years of team research and investigation, fruitless exploration, failures, and frustration would be many before the announcement was made on February 15, 1955. General Electric had successfully manufactured a diamond! Viewed under a microscope, a slide of the manufactured crystals looked grayish green and yellow in color. They were also unimpressive in appearance. Man-Made* Diamonds, as they were called, proved to be superior to other industrial abrasives. General Electric's production methods of using high-pressure/high-temperature technology spawned a long list of products labeled "superabrasives."

A few years later, General Electric introduced the cubic boron nitride (CBN) crystal that had been produced by Dr. Robert H. Wentorf Jr.

Properties of Manufactured Diamond

Manufactured diamond contains physical properties almost the same as that of natural diamond. The Knoop hardness test is generally used to describe the *hardness of diamond* (Figure 72-1). Manufactured diamond is very close to the hardness of a single-crystal diamond.

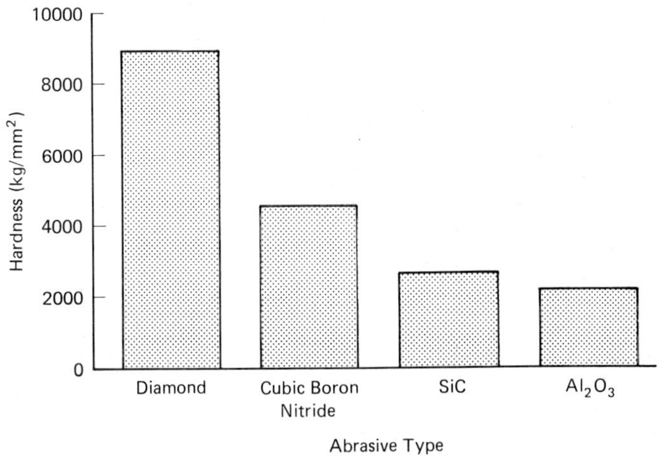

FIGURE 72-1
Knoop hardness of abrasives. (Credit: GE Superabrasives)

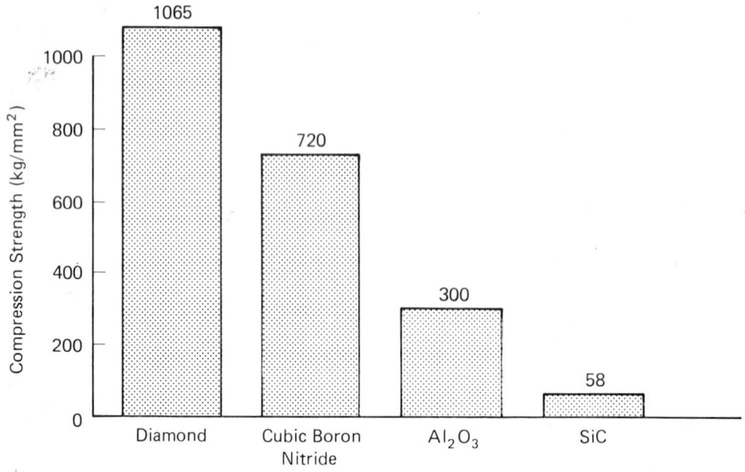

FIGURE 72–2
Compressive strength of abrasives. (Credit: GE Superabrasives)

Compressive strength may be defined as the maximum stress in compression a material will take before rupturing. As the bar graph in Figure 72–2 reveals, the compressive strength of diamond is the highest of all abrasives.

Diamond is also noted for its *high thermal conductivity*, that is, the ability to carry off heat rapidly. The point of a diamond cutting tool will not become hot due to this property and adds greatly to its abrasion or wear resistance in cutting. Very high temperatures pose a problem with a diamond cutting tool. When the temperature ranges close to 700°C (1292°F) in an inert environment, diamond turns to graphite. In air, a diamond begins to oxidize and burn at temperatures of 600 to 700°C (1112 to 1292°F).

A Background to Abrasive Grinding

It is not known when the system of hieroglyphic writing originated in Egypt, but it is known that Egyptians were using natural abrasives from vast deposits of sandstones as early as 8000 years ago. Not only did the Egyptians develop an early form of grinding wheel, but many of their architectural splendors could not have been possible without the use of abrasive technology. Their temples and pyramids were constructed with fairly close fitted stones. Their edges had been cut and abrasively polished to a dimensional finish. Despite the early introduction of natural abrasives, it was not until the 1890s that silicon carbide and aluminum oxide were mass produced in electric furnaces.

In the areas of grinding, natural diamond abrasives had been used for over a century to manufacture saw blades used in the natural stone industry. For nearly half of that time, diamonds were bonded on grinding wheels for use in cutting ceramics, glass, and cemented carbides. The need for an abrasive tool with extraordinary hardness and strength was apparent but did not materialize until the mid-twentieth century.

Superabrasive Grinding Wheels

Superabrasives, such as Man-Made* Diamond and CBN, the hardest abrasive materials available, were first used in grinding wheels. They are both made by the high-temperature/high-pressure technology to far exceed aluminum oxide and silicon carbide abrasives in wear and efficiency. Superabrasives provide the metalworking industry with high-performance abrasives for grinding abrasive ferrous and nonferrous materials. Diamond wheels are used primarily to grind nonferrous metals, nonmetallic materials, and cemented carbides. Cubic boron nitride wheels are mostly used in metalworking for grinding hardened ferrous metals, cast iron, and nickel-base and cobalt-base high-temperature superalloys. This section will deal primarily with the use of a CBN grinding wheel in the metalworking industry.

*A trademark of General Electric Company, USA.

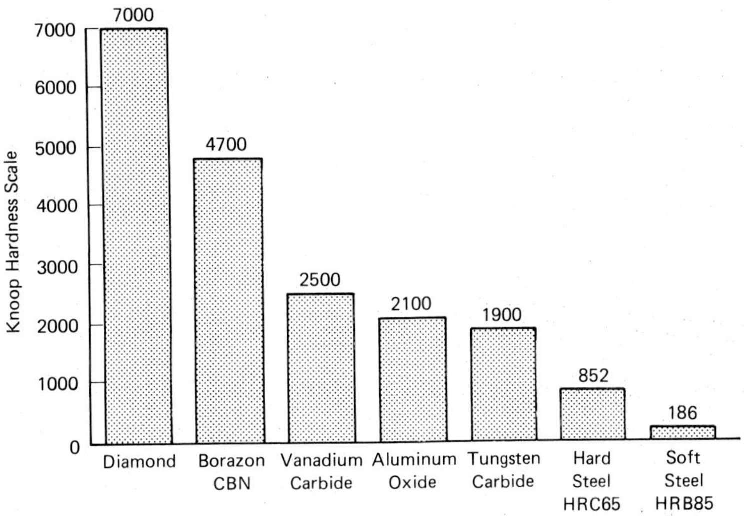

FIGURE 72-3
A hardness comparison of various metals and abrasives. (Credit: GE Superabrasives)

Due to the extraordinary hardness and wear resistance of the CBN crystal, grinding wheels made from this abrasive have dramatically proven to be the superior cutting tool for grinding superalloys and difficult-to-machine metals. From their initial application, when they were used in the toolroom and for resharpening high-speed steel cutting tools, CBN wheels have grown in use worldwide and have been found to cut manufacturing costs while increasing part quality and productivity, even when grinding ultrahigh strength alloy steels.

CBN Crystal Properties

The benefit to be had from using CBN grinding wheels can be traced back to the properties of the abrasive crystal.

Hardness. The typical aluminum oxide grain registers 2100 on the Knoop hardness scale, silicon carbide registers around 2480, while the CBN crystal registers 4700. This makes the CBN crystal almost twice as hard as the conventional aluminum oxide and silicon carbide abrasives. If one were to grind high-speed steel that contains amounts of vanadium, tungsten carbide, titanium carbide—giving this steel a hardness that ranges between 700 and 1000 on the Knoop scale (Figure 72-3)—one would understand why it is virtually impossible to keep the conventional abrasive wheel sharp and free cutting. Conventional abrasives dull rapidly and generate high frictional heat, while CBN crystals remain sharp longer and have a low coefficient of friction. This reduces the chances of wheel glazing and prevents thermal damage to the workpiece due to heat buildup (Figure 72-4).

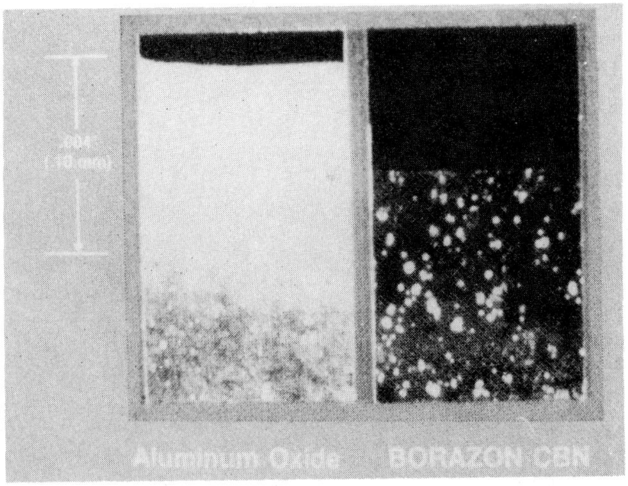

FIGURE 72-4
Comparison of workpiece microstructure: part on left shows thermal damage left by aluminum oxide wheel; part on right shows no thermal damage left by a CBN wheel. (Credit: GE Superabrasives)

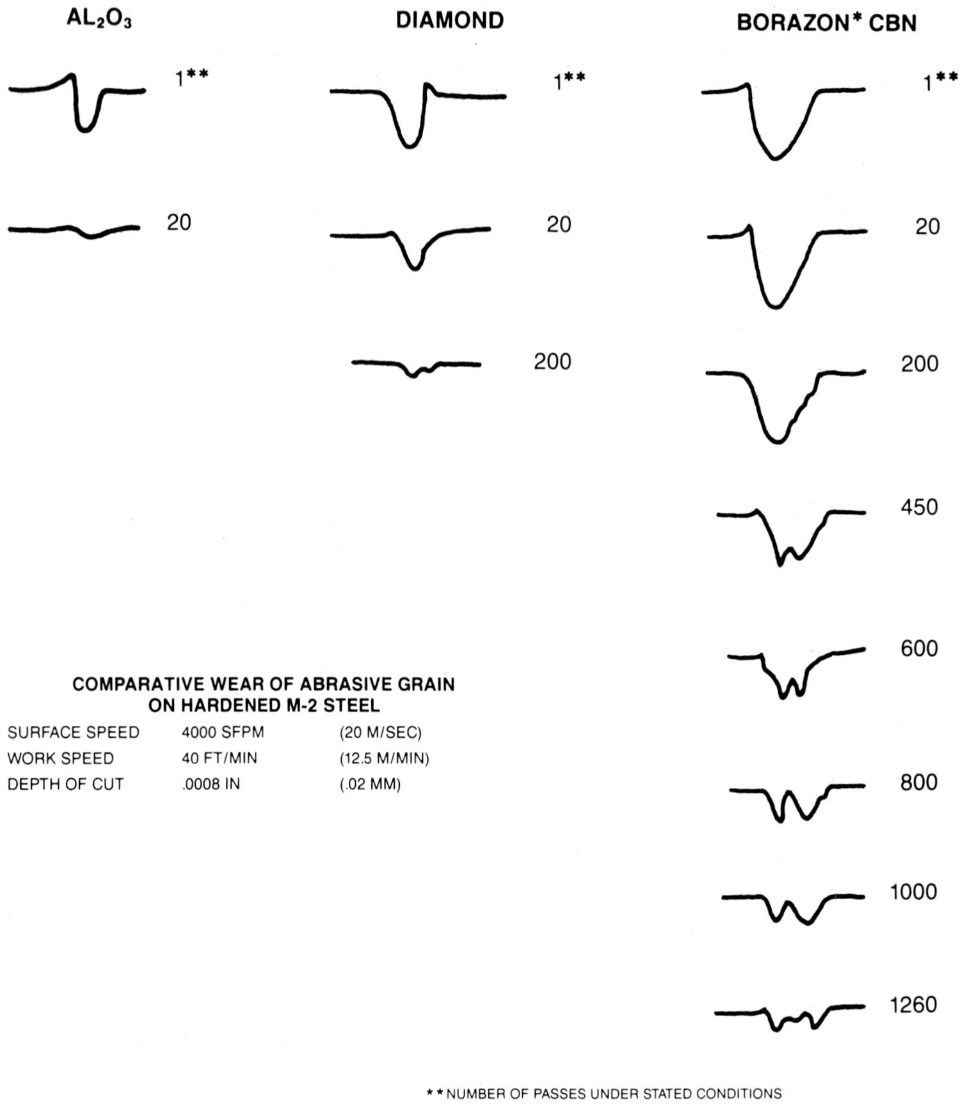

FIGURE 72–5
Comparative wear of abrasive grain on hardened M-2 steel. (Credit: GE Superabrasives)

Abrasion Resistance. The CBN crystals have a much longer wear resistance than either diamond or aluminum oxide grains. This is dramatically shown in Figure 72–5, where it can be seen that hardened M-2 steel was subjected to the grinding wheels composed of a single-grain CBN, diamond, and aluminum oxide. After 20 passes, the aluminum oxide grain was completely worn out. The CBN grain was still useful after completing more than 1260 passes. The ease with which material removal is obtained when grinding with CBN wheels makes their use a routine application. The results are consistent in quality.

Chemical Stability. The CBN crystals are chemically inert and do not react with ferrous metals as do silicon carbide and diamond. Most steels are low in carbon and seek a steady source of carbon to satisfy this need. Both silicon carbide and diamond contain carbon, which supplies the carbon sought by the carbon-hungry steels. This chemical reaction leads directly to their abrasive wear. While aluminum ox-

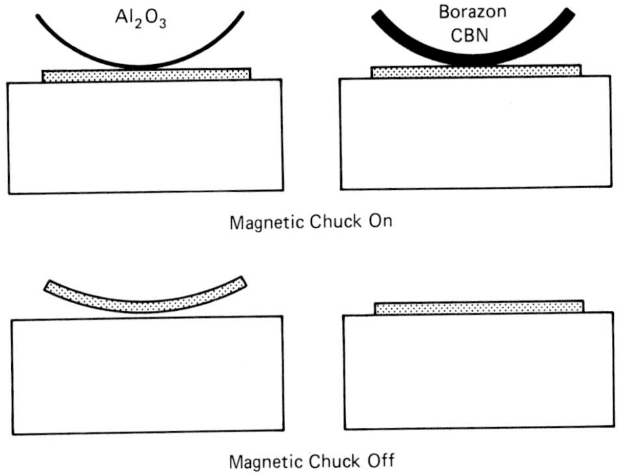

FIGURE 72-6
The cool grinding characteristics of CBN wheels make it easier to control residual stresses, preventing thin parts from warping or bowing when released from the magnetic chuck. (Credit: GE Superabrasives)

ide does not react with ferrous metal, CBN is the superior grinding abrasive because it is harder and lasts longer.

High Thermal Conductivity. Unlike diamond, which begins to oxidize and burn at temperatures ranging from 600 to 700°C (1112 to 1292°F), CBN crystals develop a thermal barrier of solid oxide (B_2O_3) at this temperature and can withstand grinding temperatures as high as 1371°C (2500°F). When properly used, there is almost no danger of burning the workpiece with CBN grinding wheels. The cool grinding characteristics of CBN wheels make it easier to control residual stresses when grinding thin parts. This is often not the case when grinding with aluminum oxide wheels, in which case thin parts may warp or bow when released from the magnetic chuck (Figure 72-6).

Wheel Bond Type

There are four common types of bonds used for the manufacture of CBN grinding wheels: resin, metal, vitrified, and electroplated.

Resin bond CBN wheels represent the most widely used CBN wheels worldwide.

- They perform with long life and a high material removal rate.
- Grinding ratios (ratio of work material removal to the volume of wheel material consumed during grinding) may be 100 times greater than aluminum oxide wheels grinding the same material under the same conditions.
- They can be successfully used on practically all types of grinding—ID, OD, jig grinding, and centerless grinding.

Metal bond CBN wheels are the most durable and longest lasting of the bond-type wheels because the metal bond grips the superabrasive particles more firmly.

- They are excellent for deep grinding of hardened tool and die steels.
- They are used for ID grinding as well as creep feed grinding.
- They have wide use in verticle spindle rotating grinding.

Vitrified bond CBN wheels are in wide use in grinding operations where excellent finishes are required.

- They perform well in ID grinding and centerless grinding.
- They promote long grinding life at a controlled rate.

Electroplated bond CBN wheels have metal cores upon which a layer of CBN superabrasive particles are electroplated to the grinding surface. As such, they provide many economic benefits.

- They cost less than other CBN grinding wheels of the same size.
- A new layer of superabrasive can be electroplated to a wheel once the worn layer has been removed.
- They can be operated at high speeds and feeds required for high material removal rates.
- They hold their exact machined profile, which is of great benefit for grinding purposes.

Surface Grinding Speeds and Feeds

Wheel Surface Speeds. The CBN wheels perform best at speeds of at least 5500 to 6500 sf/min (28 to 33 m/sec). While CBN wheels will grind very effectively at the conventional speeds just noted, their effectiveness in many materials may be improved as

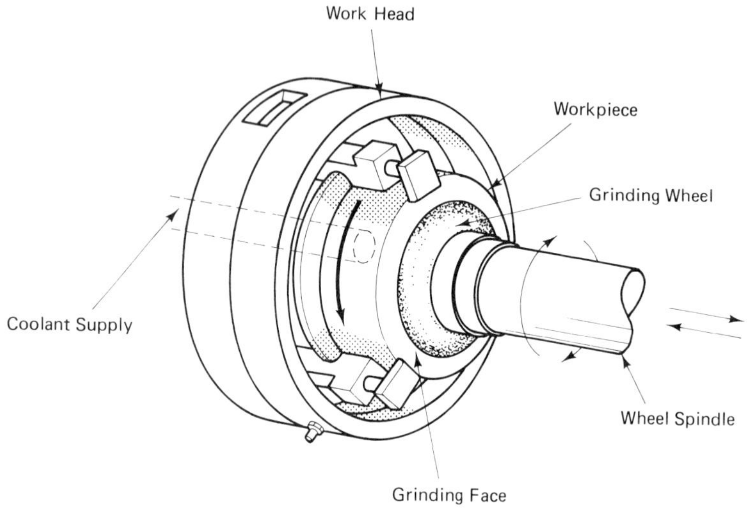

FIGURE 72–7
CBN wheels run at high speeds for small internal grinding aided by a coolant supply. (Credit: Bay State Abrasives)

speeds are increased. For example, CBN wheels run at standard speeds of 14,000 to 16,000 sf/min (71 to 81 m/sec) or higher for small ID grinding (Figure 72–7).

Table Speeds. High table speeds, up to the maximum speed of the grinder, are recommended for best grinding performance. Since material removal rates are high, the economic benefits are higher. A greater production output will reduce labor and overhead cost per piece.

Crossfeed. Large crossfeed movements are recommended for general-purpose grinding, ranging between one-quarter and one-half of the wheel width. Smaller crossfeed increments for finishing passes can be used.

Downfeed. The correct downfeed for CBN wheels will depend on the material being ground. On easy-to-grind materials (such as M-2 and T-1 steels), a downfeed of 0.001 to 0.002 in. (0.02 to 0.5 mm) per pass is recommended. For difficult to grind material (such as M-4 and T-15), a downfeed of 0.0005 to 0.001 in. (0.01 to 0.02 mm) per pass is recommended.

Finishing Feeds. In order to improve surface finish, a reduction of feeds should be used. It is recommended that the downfeed be reduced 0.0002 to 0.0005 in. (0.005 to 0.013 mm) at the same time that a reduction of crossfeed is being used. The crossfeed should measure about one-eighth of the wheel width. If the desired results are not possible using these reductions, it is recommended that the wheel be changed and replaced with a CBN wheel containing a finer abrasive mesh size.

Grinder Conditions

Grinding is just another means of making chips. Like any other machine tool used in the chip making process, any grinder using a CBN wheel must operate within a set of conditions in order to produce parts effectively (Figure 72–8).

Spindle Bearings. High rigidity is required to obtain high material removal rates. The CBN wheels are made to hold their size and form. Loose spindle bearings will result in excessive vibrations that are passed on the workpiece in the form of chatter marks, poor finish, and inaccurate dimensions and result in short wheel life.

Machine Slides. Loose slides also contribute to excessive vibrations and reduce the effectiveness of any CBN grinding wheel.

Spindle Speeds. The CBN wheels run at higher speeds than conventional grinding wheels, which results in a longer wheel life and greater efficiency. "Faster is better" is not an absolute rule, however; the wheel speed must be consistent. Safety requirements dictate to *never run any grinding wheel faster than the maximum safe speed listed.*

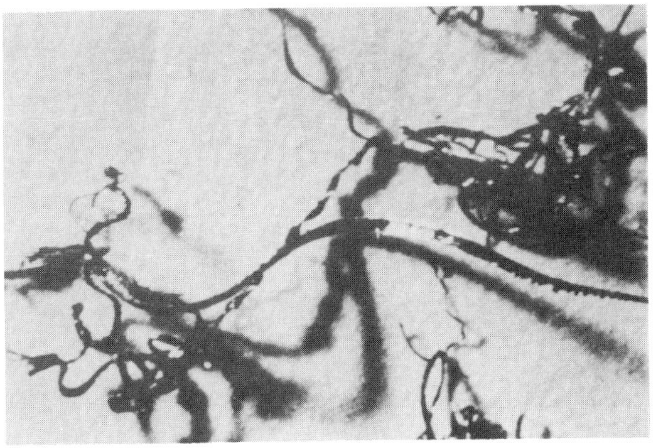

FIGURE 72-8
Metal chips produced by a grinding wheel on an abrasive machining operation. (Credit: Bay State Abrasives)

Feeds. Play or wear in a grinder's feed system will lead to poorly ground parts and must be eliminated in order to produce accurate parts.

Coolant System. While speed is an attempt to make smaller chips, one of the advantages in using a CBN wheel is that the cutting tool stays sharper longer and generates chips for a longer period of time. The purpose of a coolant system is to keep the surfaces of the wheel and workpiece cool while flushing away the chips to keep the wheel sharp and free cutting. A steady supply of cutting fluid, when properly applied, will reduce excessive wear of the wheel's bond, which will cause the abrasive crystals to fall out and maintain surface quality (Figure 72-9). A good coolant system can increase the life of a superabrasive grinding tool by as much as 10 times (Figures 72-10 and 72-11).

Horsepower Recommendation. Constant spindle speed is required for high production grinding of hard abrasion-resistant material. The power required to remove a given amount of material will vary widely depending on the wheel speed, wheel size, part material, and cutting fluid. As a general rule, the recommended horsepower for a 1-in.- (25-mm-) thick wheel should be one-half the CBN wheel diameter, i.e., a 20 × 1-in. (508 × 25-mm) CBN wheel requires 10 horsepower.

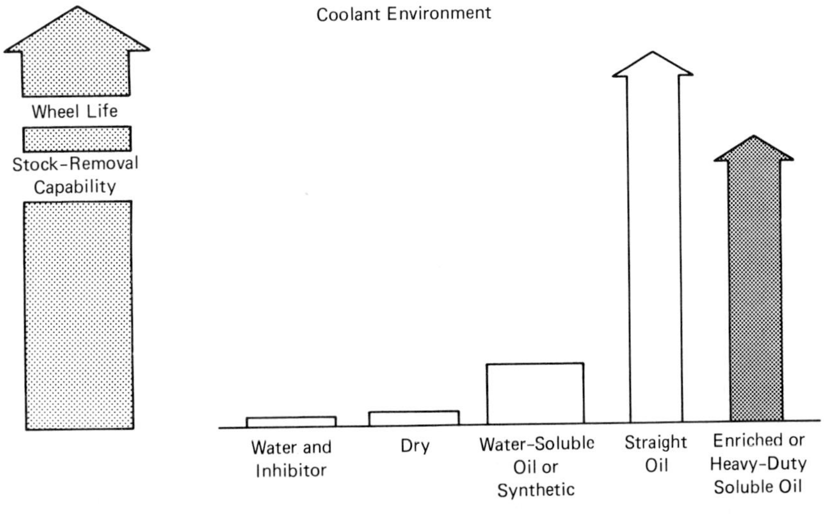

FIGURE 72-9
A steady supply of coolant will prolong wheel life. (Credit: GE Superabrasives)

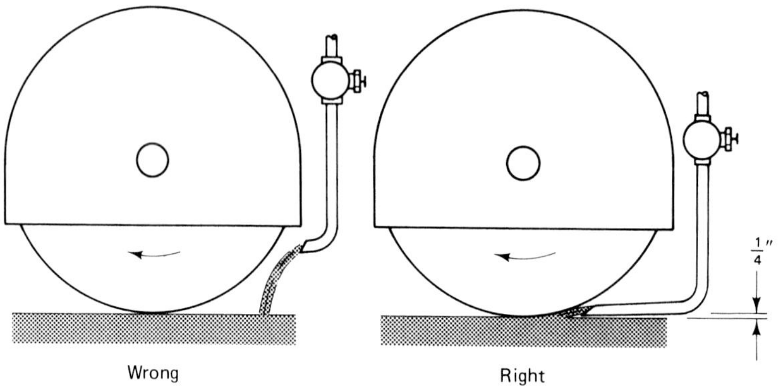

FIGURE 72-10
The coolant nozzle should be placed as close as possible to the work and wheel interface. (Credit: GE Superabrasives)

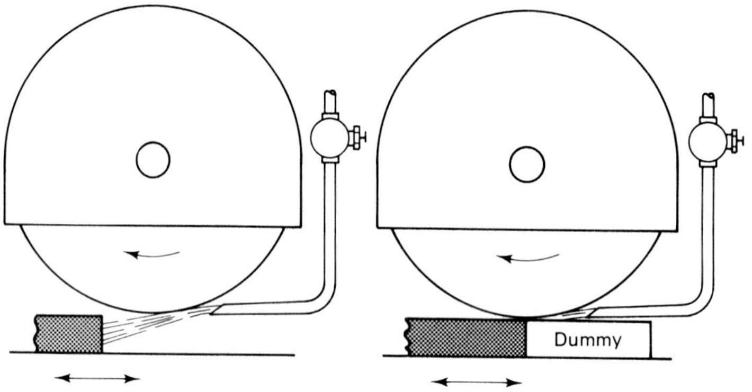

FIGURE 72-11
To assure a steady supply of coolant, a dummy block should be placed at the right end of the workpiece. (Credit: GE Superabrasives)

Preparing the CBN Wheel for Grinding

When setting up a CBN wheel for grinding, two things must be done if it is to work effectively. The wheel must be properly conditioned (trued and dressed) (Figure 72-12). Truing is the process of cutting or abraiding the wheel's contour with a tool to change its profile to make it round and turn concentrically. Dressing is the process of removing bond material from between the abrasive grains to expose the crystals that form the cutting edges. It is absolutely essential that a CBN wheel be properly conditioned. Truing and dressing is generally performed in one operation on conventional grinding wheels. The wheel truing devices used for conventional wheels will not be effective on CBN wheels. A single-point diamond or cluster-type diamond dresser will soon develop a flat spot due to the hardness of the CBN abrasive. It is also important that the grinding machine have the necessary rigidity and is capable of a 0.0001-in. (0.002-mm) incremental downfeed.

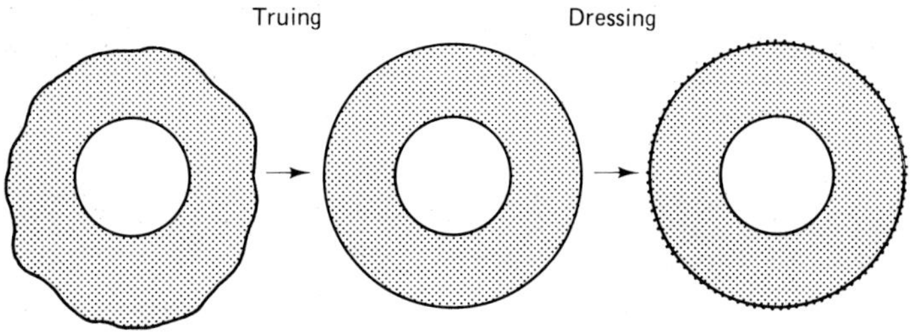

FIGURE 72–12
It is important that superabrasive wheels be properly trued and dressed. Without proper conditioning, these wheels cannot work efficiently. (Credit: GE Superabrasives)

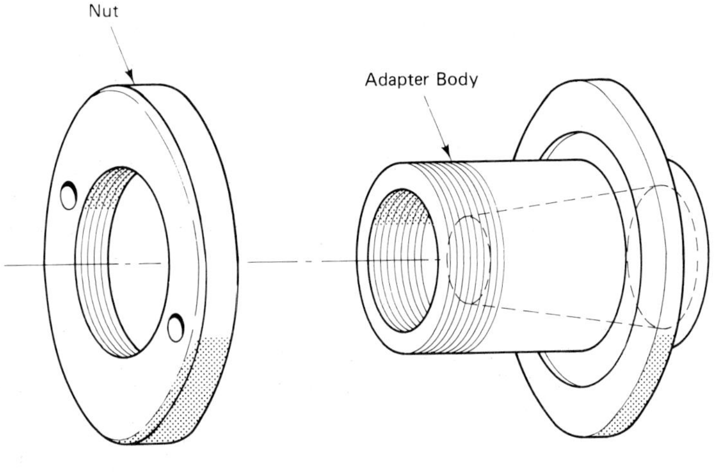

FIGURE 72–13
The internal tapered diameter of a standard spindle adapter is made to match the grinder's spindle taper. (Credit: DoAll Company)

Truing

CBN grinding wheels fall into two categories: impregnated bond and nonimpregnated bond.

The impregnated bond classification fits those wheels that have the abrasive grain impregnated (evenly spaced) in the bond system. These include metal, vitrified, and resin wheels. These particular wheels require truing and in most cases dressing.

The electroplated wheel falls into the nonimpregnated wheel category due to the fact that only a single layer of abrasive covers the wheel. These wheels do not require truing and dressing. They will, however, require that they run concentrically and should be trued with an indicator before using.

Different types of CBN wheels exhibit different wheel surfaces when trued. The vitrified wheel does not generally require dressing due to the fact that the vitrified (glass) bond breaks down by fracturing during truing. This fracturing continues once the grinding operation begins to expose sharp abrasive crystals and becomes a self-sharpening operation.

In contrast, metal and resin bond wheels will be smooth after truing. They cannot cut in this condition and will require dressing to lower the wheel bond and expose the abrasive crystals.

Mounting the Wheel

Special emphasis is placed on correctly mounting a CBN wheel. Unlike the conventional abrasive wheel that is slipped on and tightened against the surface grinder spindle, a CBN wheel is mounted on a standard adapter (Figure 72–13). This adapter has an in-

FIGURE 72-14
A wheel loading assembly fixture. (Credit: M. Rapisarda and GE Superabrasives)

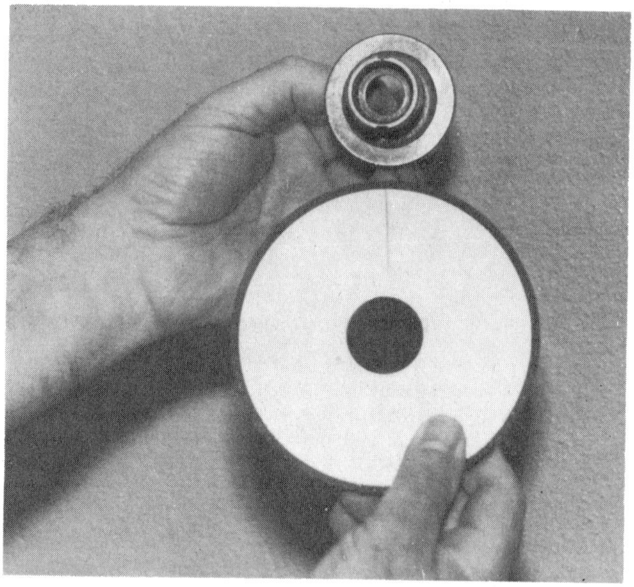

FIGURE 72-15
Mark the wheel and adapter for alignment so they may be remounted without runout if separated. (Credit: M. Rapisarda and GE Superabrasives)

ternal tapered diameter to match the grinder's spindle taper. The outside diameter, the front section of which is threaded, is concentric and parallel to match the hole in the wheel. A solid shoulder on the adapter serves as the inner flange against which the wheel is placed, while a threaded nut, which is used to tighten the wheel against the adapter's shoulder, serves as the outer flange (Figure 72-14).

The wheel must be mounted squarely and securely to a quality adapter. It is also recommended that the wheel and adapter remain together as a unit for the life of the wheel. In this manner, should the wheel and adapter ever be removed from the grinder's spindle, the unit can be replaced without taking the time to retrue the wheel.

Procedure

1. Before attempting to mount the wheel on a close-fitting adapter, both the wheel and adapter should be clean and free of burrs.

2. For ease of insertion, spray the adapter internal taper with a light film of oil.

3. To eliminate runout in the event of separation and remounting, file or scribe a groove on the adapter end and mark a line with a felt pen on the side of the wheel for alignment purposes (Figure 72-15).

4. Inspect the grinding wheel sides and hole for dirt and burrs and clean them thoroughly.

FIGURE 72-16
Carefully slide the wheel but do not force it on the adapter. (Credit: M. Rapisarda and GE Superabrasives)

5. Carefully slide the wheel on the adapter. *Never force it on* (Figure 72-16).

6. Place this unit on a wheel assembly fixture.

7. Align the marking on the wheel with the line or groove on the adapter.

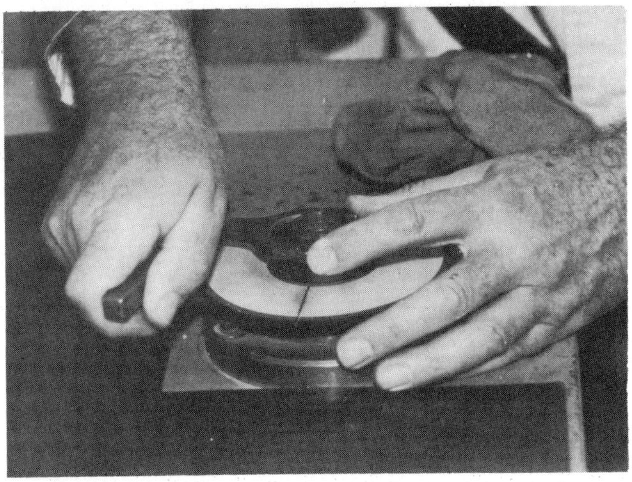

FIGURE 72-17
Hand tighten the nut securely. (Credit: M. Rapisarda and GE Superabrasives)

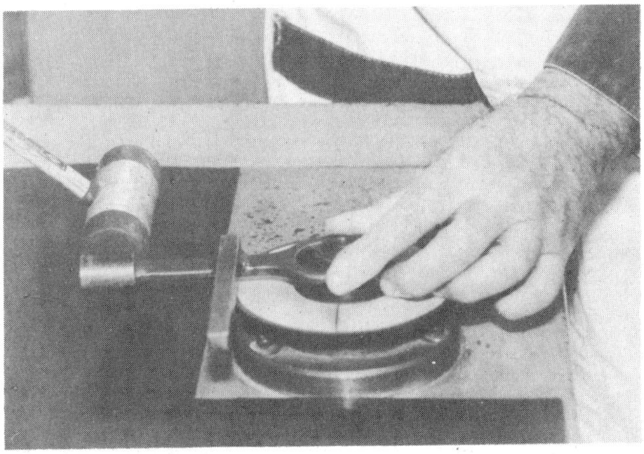

FIGURE 72-18
Secure the wheel against the adapter by applying light force against the wrench. (Credit: M. Rapisarda and GE Superabrasives)

FIGURE 72-19
Test the wheel for slippage by hand turning it against the adapter. (Credit: M. Rapisarda and GE Superabrasives)

FIGURE 72-20
Remove any burrs from the spindle and flange before mounting the wheel. (Credit: M. Rapisarda and GE Superabrasives)

8. Hand tighten the adapter nut to keep the wheel and adapter aligned.

9. Place an adapter wrench in the holes of the nut and hand tighten the nut securely (Figure 72-17).

10. Hold one end of the adapter wrench with one hand and secure the wheel against the adapter shoulder by tapping against the end of the wrench with a soft-faced hammer (Figure 72-18).

11. Test the assembly unit by trying to turn the wheel (Figure 72-19). If successful, the wheel cannot turn. It indicates that the nut has been tightened against the wheel body and not on the adapter threads.

12. Prepare to mount the assembly. Clean the grinder spindle and flange and remove any burrs (Figure 72-20).

FIGURE 72-21
Align the wheel and spindle markings. (Credit: M. Rapisarda and GE Superabrasives)

FIGURE 72-23
Tighten the grinder spindle nut securely by applying light force on the end of the wrench. (Credit: M. Rapisarda and GE Superabrasives)

FIGURE 72-22
Seat the wheel-adapter assembly on the grinder spindle. (Credit: M. Rapisarda and GE Superabrasives)

13. Align the wheel marking with a corresponding mark on the grinder spindle (Figure 72-21).

14. With a sharp snap, seat the wheel-adapter assembly on the grinder spindle (Figure 72-22).

15. Tighten the grinder spindle nut securely by applying force on the end of the wrench with a soft-faced hammer (Figure 72-23).

16. Mount a dial indicator and test the wheel for runout. Runout should not be more than 0.001 in. (0.02 mm). If the runout is greater, recheck the assembly procedure.

[On wheels where the circumference is used for grinding, the runout should never be more than 0.0005 in. (0.01 mm). Cup wheels may have to be shimmed between the adapter shoulder and the back of the wheel core, but the runout on the wheel face should not be more than 0.001 in. (0.02 mm).]

17. Adjust the coolant nozzle as close as possible to the point of the workpiece-wheel contact (Figure 72-24).

Truing and Dressing the CBN Wheel

There are a wide variety of truing devices used for CBN grinding wheels. Generally, each device works best on a specific type of bond system. Three of the more commonly used devices include:

A diamond-impregnated nib: This device provides a large number of cutting edges to distribute heat over a wide area that will reduce

FIGURE 72-24
Adjust the coolant nozzle close to the workpiece–wheel contact. (Credit: M. Rapisarda and GE Superabrasives)

FIGURE 72-25
The use of a brake-controlled truing device is considered a gentler truing process. (Credit: M. Rapisarda and GE Superabrasives)

excessive wear. It may be used to true CBN resin and vitreous bond wheels not larger than 8 in. (200 mm) in diameter.

A metal dressing block: A block made of mild steel or molybdenum may be used to be ground by an untrued, undressed CBN wheel until the grinding surface is worn "true" and will also expose sharp abrasive crystals.

A brake-controlled truing device: This device uses a silicon carbide braking device that makes it revolve at a slower surface speed than the CBN wheel being trued. Commonly used for truing straight-faced metal and resin bond grinding wheels, this process involves a fast wear rate of the truing wheel but is considered a gentler truing process (Figure 72–25).

Once the wheel has been trued, it must be dressed. A number of dressing methods and devices are used for this purpose. Only a few of the more commonly used devices will be described.

A water-soluble wax stick: This device is a wax stick consisting of loose aluminum oxide grains that is applied to and carried by the revolving wheel face of the grinding wheel positioned near the workpiece. The abrasive grains are squeezed between both surfaces to erode the bond matrix.

Wire brush dressing: By using this method, the grinding wheel's bond matrix is eroded by the stiffness of a wire brush held at a 45° angle to that of the grinding wheel's rotation. Despite its stiffness, the wires are flexible enough not to be ground away.

Aluminum oxide dressing stick or block: A soft, fine-grade aluminum oxide stick is brought into contact with a rotating wheel surface that is dipped in a coolant to create a slurry between the dressing stick and the grinding wheel to erode the bond matrix and expose the sharp CBN crystals.

In the process that follows, a brake-controlled truing device, which generally uses a 60-grit, "L"-hardness silicon carbide wheel, will be used to true a resin CBN wheel. The device used for dressing will be an aluminum oxide dressing stick.

Procedure

1. Mount and indicate the CBN wheel on the grinder spindle to within 0.001 in. (0.02 mm) or less maximum runout.

2. Tighten the wheel securely on the grinder spindle.

3. Thoroughly clean the magnetic chuck with a cloth and smooth over it with the palm of the hand.

FIGURE 72–26
Position the center of the truing wheel slightly to the left of the grinding wheel centerline. (Credit: M. Rapisarda and GE Superabrasives)

FIGURE 72–27
Lower the stationary CBN wheel until it just contacts the paper. (Credit: M. Rapisarda and GE Superabrasives)

4. Place a sheet of paper that is slightly larger than the base of the truing device on the left end of the magnetic chuck.

5. Place the brake-controlled device on the paper so that the spindle is in line with the spindle of the grinder and energize the magnetic chuck.

6. Lower the CBN wheel to a level just above the height of the truing abrasive wheel.

7. Move the surface grinder table longitudinally to position the center of the truing wheel slightly to the left of the grinding wheel centerline by about $\frac{1}{2}$ in. (13 mm) (Figure 72–26).

8. Lock the grinding wheel table in this position.

9. Place a long strip of paper between the stationary CBN wheel and the truing wheel.

10. Lower the stationary CBN wheel until it just contacts the paper (Figure 72–27).

11. Remove the paper and turn the CBN wheel by hand to see that it does not contact the truing wheel. If it does, raise the CBN wheel about 0.003 in. (0.07 mm) (Figure 72–28).

12. Set the grinder spindle to run at 1000 to 2000 sf/min (5 to 20 m/sec).

13. Using a wax marking crayon, coat the surface of the rotating CBN wheel (Figure 72–29).

FIGURE 72–28
Remove the paper and raise the CBN wheel about 0.003 in. (0.07 mm). (Credit: M. Rapisarda and GE Superabrasives)

14. Rotate the truing wheel with a long pencil and slowly bring the CBN wheel down in increments of about 0.001 in. (0.02 mm) downfeed until both wheels are in contact.

15. Turn the grinder's crossfeed handle until the truing wheel clears the edge of the CBN wheel.

FIGURE 72–29
Coat the rotating CBN wheel with a wax marking crayon. (Credit: M. Rapisarda and GE Superabrasives)

FIGURE 72–30
Select a 200-grit, G-grade aluminum oxide dressing stick. (Credit: M. Rapisarda and GE Superabrasives)

16. Turn on the grinder's coolant and adjust the stream so that it is carried directly between the contact of the two wheels.

17. Prepare to traverse the truing wheel under the CBN wheel. Set the traverse rate of the truing wheel between 3 and 12 in. (75 and 300 mm) per minute. The truing wheel must clear the edge of the CBN wheel with each pass.

18. Lower the CBN wheel in small increments until the crayon has been removed from the cutting surface of the CBN wheel. The following incremental downfeeds are recommended: coarse-mesh wheels, 0.001 to 0.002 in. (0.02 to 0.05 mm); fine-mesh wheels, 0.0002 to 0.0005 in. (0.002 to 0.005 mm); and finish passes, 0.0002 in. (0.0005 mm). *Note:* If chatter develops during truing, reduce the downfeed rate.

19. Inspect the CBN wheel occasionally to see if all the hollows and crayon marks have been removed. *Note: Do not overtrue.* A harsh sound while truing will indicate the wheel is not completely true. A smooth sound generally indicates the wheel has been trued.

20. Remove the brake-controlled truing device from the grinder.

21. Select a 200-grit, G-grade aluminum oxide dressing stick (Figure 72–30).

22. Dip the dressing stick in coolant so that a slurry is created when the stick is brought into contact with the CBN wheel.

FIGURE 72–31
Feed the dressing stick into the wheel to open up the wheel face by eroding the bond. (Credit: M. Rapisarda and GE Superabrasives)

23. Hold the dressing stick firmly in both hands and bring it into contact with the rotating wheel.

24. Keep feeding the dressing stick into the wheel to open up the wheel face (Figure 72–31).

25. Once the dressing stick starts to wear rapidly, the CBN wheel is dressed.

To assure that the wheel has received the full benefit from dressing, several passes across the workpiece should be taken, with each depth of cut measuring 0.0005 in. (0.005 mm) per pass.

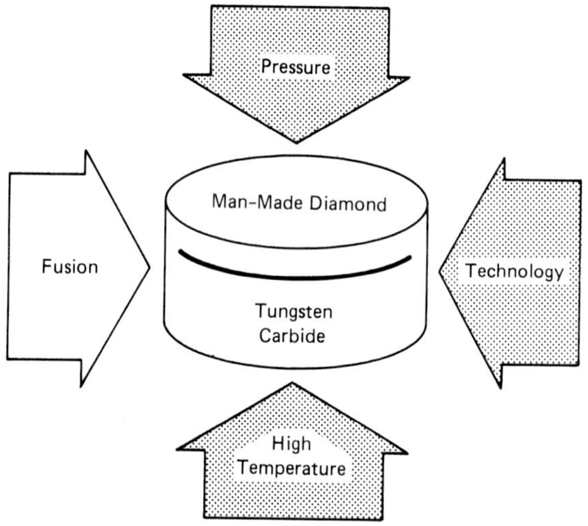

FIGURE 72–32
The manufacture of superabrasive materials is made with an advanced high-pressure/high-temperature process. (Credit: GE Superabrasives)

Polycrystalline Diamond (PCD) Tools

A natural diamond is crystalline in structure with the bond between atoms arranged in a definite pattern. As a consequence, the hardness and abrasion of the diamond will vary according to the concentration of the bond across a crystal surface.

The manufacture of a PCD tool blank consists of fusing together a layer of diamond about 0.020 in. (0.50 mm) thick on a cemented tungsten carbide substrate (base) by means of a high-pressure/high-temperature process (Figure 72–32). The temperature is somewhere between 1700 and 1800°C (3092 and 3272°F) and the pressure measures around 1,000,000 psi. The process produces a cutting tool blank that combines the hardness and abrasion resistance of a diamond in all directions and the toughness of a cemented tungsten carbide base. Their ability to resist the shock of impact as well as their increased wear resistance make PCD cutting tools superior to natural diamond cutting tools. In many milling and turning operations on nonferrous and nonmetallic materials, PCD machining tools (Figure 72–33) have replaced cemented carbide cutting tools, offering the following benefits:

- Better work surface finish.
- Long cutting tool life.
- Control of part dimensional accuracy.
- High productivity.
- Less machine downtime.
- Cost-effective machining.

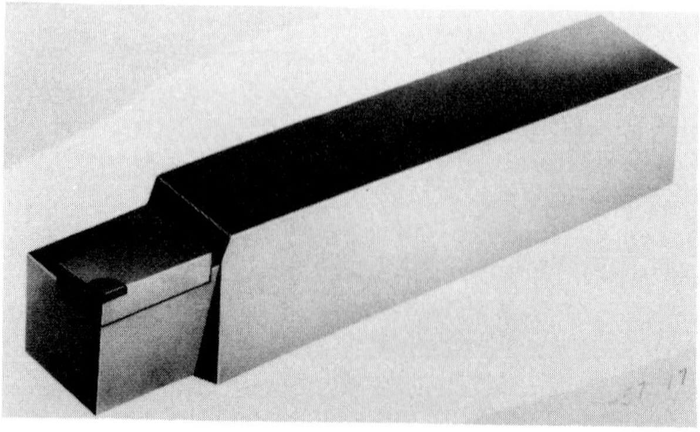

FIGURE 72–33
Polycrystalline diamond (PCD) machining tools have replaced cemented carbide cutting tools on nonferrous and nonmetallic materials. (Credit: DoAll Company)

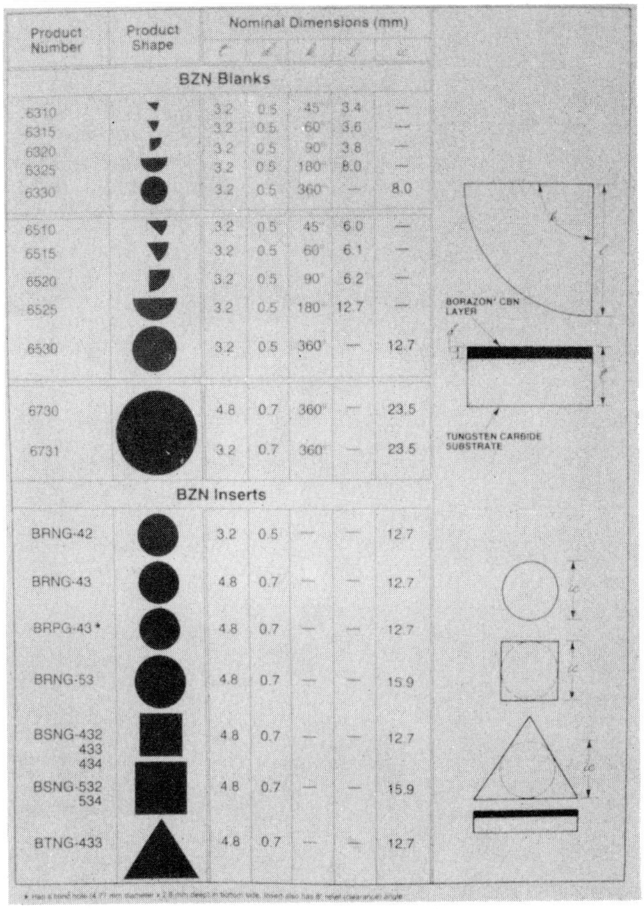

Product Number	Product Shape	Nominal Dimensions (mm)				
BZN Blanks						
6310		3.2	0.5	45°	3.4	—
6315		3.2	0.5	60°	3.6	—
6320		3.2	0.5	90°	3.8	—
6325		3.2	0.5	180°	8.0	—
6330		3.2	0.5	360°	—	8.0
6510		3.2	0.5	45°	6.0	—
6515		3.2	0.5	60°	6.1	—
6520		3.2	0.5	90°	6.2	—
6525		3.2	0.5	180°	12.7	—
6530		3.2	0.5	360°	—	12.7
6730		4.8	0.7	360°	—	23.5
6731		3.2	0.7	360°	—	23.5
BZN Inserts						
BRNG-42		3.2	0.5	—	—	12.7
BRNG-43		4.8	0.7	—	—	12.7
BRPG-43*		4.8	0.7	—	—	12.7
BRNG-53		4.8	0.7	—	—	15.9
BSNG-432 433 434		4.8	0.7	—	—	12.7
BSNG-532 534		4.8	0.7	—	—	15.9
BTNG-433		4.8	0.7	—	—	12.7

FIGURE 72–34
The process of manufacturing PCBN blanks produces the same type of beneficial characteristics as in the case of PCD blanks. (Credit: GE Superabrasives)

Polycrystalline Cubic Boron Nitride (PCBN) Tools

Cubic boron nitride (CBN) is not found in nature. It is manufactured from hexagonal boron nitride, sometimes known as white graphite. To make cubic boron nitride, equal numbers of boron and nitrogen atoms are subjected to superpressure sintering technology.

In a manner similar to the production of PCD, polycrystalline cubic boron nitride (PCBN) cutting tools and inserts are a combination of a layer of cubic boron nitride approximately 0.020 in. (0.50 mm) thick that has been fused together on a cemented tungsten carbide substrate using high-pressure/high-temperature technology. The result is a tough, strong, hard, three-dimensional bonded crystalline structure with sharp edges (Figure 72–34).

PCBN cutting tool blanks and inserts contain certain superabrasive properties, such as:

Hardness. Cubic boron nitride cutting tools have high hardness and strength in all directions due to the random orientation of tiny CBN crystals. Cubic boron nitride rates next to diamond in hardness, registering around 4700 on the Knoop hardness scale, easily surpassing the conventional abrasives of silicon carbide and aluminum oxide. High impact resistance and the ability to withstand shock of interrupted cuts is supplied by the toughness and strength of the cemented tungsten carbide base.

Abrasion resistance. PCBN cutting tools have almost the same abrasive resistance of diamond and three times that of silicon carbide (Figure 72–35). The abrasion resistance is uniform in all directions,

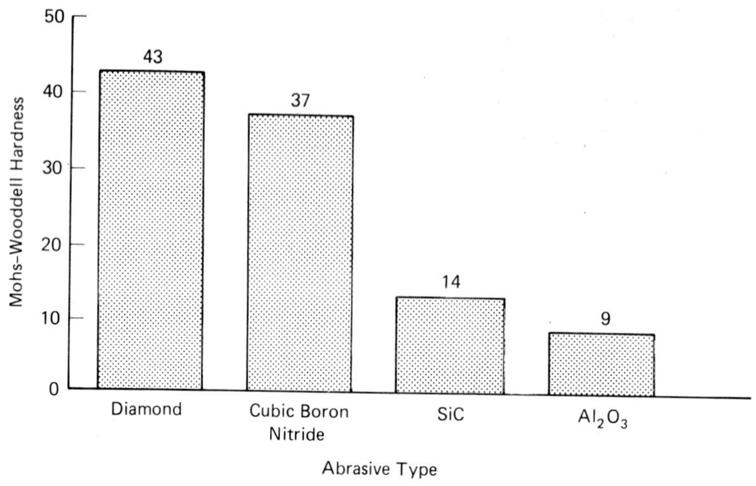

FIGURE 72-35
Relative abrasion resistance of different materials. (Credit: GE Superabrasives)

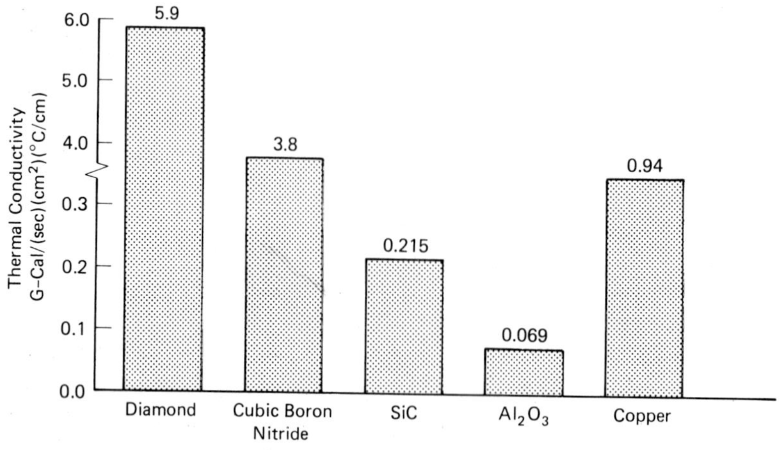

FIGURE 72-36
Thermal conductivity of different abrasive materials. (Credit: GE Superabrasives)

making them ideal for machining hard, tough material. It also means that PCBN cutting tools maintain their sharp cutting edges longer, increasing productivity while producing dimensional accuracy.

Compression Strength. Cubic boron nitride is fairly close to that of diamond in compressive strength. This enables PCBN tools to withstand the forces created during hard metal machining and shock of severe interrupted cuts.

Thermal Conductivity. Cubic boron nitride abrasives are next to diamond in their ability to conduct or transfer heat (Figure 72-36). PCBN cutting tools retain their hardness and strength at the cutting edge at temperatures up to 1000°C (1830°F) and also resist oxidation at this temperature.

Chemical Stability. While PCD and PCBN have similar properties that make them superabrasives, there is one property that separates the two: chemical stability. It has been found that one of the reasons natural abrasives wear so quickly, despite their hardness, is that they are chemically reactive with steels. Very much like the human body that may lack water, steels are deficient in carbon. They "thirst" for a continuous stream of carbon. Natural abrasives such as silicon carbide and diamond are rich in carbon and supply this drink; hence, they wear faster (Figure 72-37). Cubic boron nitride contains no carbon and is therefore chemically inert with ferrous materials.

Advantages of Using PCBN Cutting Tools

The advantages of using PCBN cutting tools, which should only be used on ferrous and ferrous alloys, are listed in Figure 72-38. A summary of some of their efficiency is as follows:

- Superior cutting edge will outperform ceramic and cemented carbide cutting tools 10 to 700%.
- Can withstand heat and excessive wear at the high temperatures created by increased speeds and feeds.
- Higher material removal rates—as much as three times that of carbide cutting tools—to reduce total machining cost.
- Reduced tool wear results in better workpiece quality and less machine downtime.
- Can machine hard materials that at one time was done by grinding ferrous metals with a hardness of Rockwell C 45 and higher and high-temperature alloys (cobalt base, nickel base) with a hardness of Rockwell C 35 and higher.
- Better surface quality finishes while eliminating slow finishing operations: 20 to 30 μin. finish while rough cutting; single-digit microinch number for finish cuts.
- Reduced machine downtime results in a lower cost per piece and increased productivity.

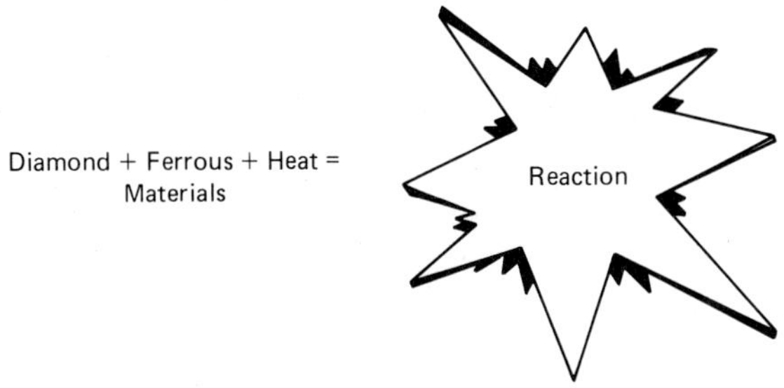

FIGURE 72-37
Ferrous materials react chemically with diamond cutting tools and will wear out these tools faster. (Credit: GE Superabrasives)

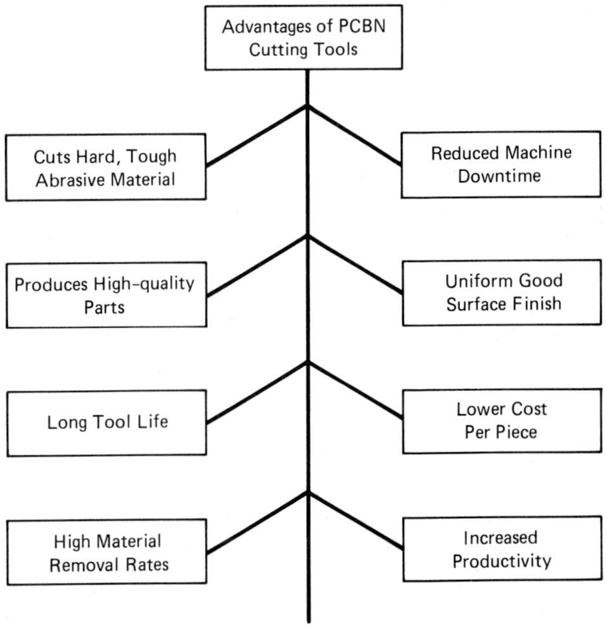

FIGURE 72-38
Advantages of using PCBN cutting tools. (Credit: Kostel Enterprises)

SECTION REVIEW

Unit 68

1. Who is credited with having automated the flour industry in 1787?
2. Name two early American practitioners to use Whitney's methods of industrial interchangeability.
3. What innovation is credited with having ushered in the electronic age and who were its developers?
4. Describe the integrated circuit developed in the late 1950s and what was its purpose.
5. Name five different ways that computers are used in manufacturing systems.
6. Name at least five different types of industrial operations in which robots are used.

Unit 69

1. Name a simple example that illustrates the concept of a punch and die.
2. How is the distinction made between a blank and a scrap piercing?
3. In industrial application, how are dies mounted and where are they used?
4. Explain the difference between bending and forming operations.
5. Trimming, the removal of excessive metal, is an important operation in the metal stamping process. Name three different operations that would use trimming dies.
6. Name three ways materials can change their physical state while cold worked.
7. How can work-hardened metals be brought back to a pliable state for further machining?

Unit 70

1. Describe the functions of a personnel department.
2. What is JIT manufacturing and what does it represent?
3. What is CAD and what is its purpose?
4. What is CAM and what is its purpose?
5. What early high-technology machining system played an important role in the development of the CAD/CAM system?
6. In addition to manual handling, name two other systems that are used for identifying and sorting packaged materials.

Unit 71

1. For what purpose were chucking and turning centers designed?
2. What factors led to the development of chucking and turning centers?
3. What is the purpose of slant-bed designing in the construction of chucking and turning centers?
4. Briefly describe the construction of the dual-cutting turrets on chucking and turning centers.
5. How are heavy cuts taken on long shaft turnings on chucking and turning centers?
6. Name three types of chucks used as gripping devices on chucking centers.
7. How does a turning center differ from a chucking center?
8. Describe the in-process quality control method used on turning centers.
9. What are the advantages of using in-process quality control? The disadvantages?
10. Name six advantages to using vision readout systems on lathes.

Unit 72

1. List the properties that make a manufactured diamond a "superabrasive."
2. At what temperature in an inert environment will diamond turn to graphite?
3. For what purpose are diamond wheels used in the metalworking industry?
4. For what purpose are cubic boron nitride wheels used?

5. For what purpose was the CBN grinding wheel first used?

6. State four reasons why CBN crystals are superior to conventional grinding abrasives.

7. Name four common types of bonds used for the manufacture of CBN grinding wheels and which is the most widely used.

8. List four benefits to be found in a metal-bonded CBN wheel.

9. How are electroplated CBN wheels produced?

10. List four benefits in using an electroplated CBN wheel.

11. List any five conditions that a grinder must meet in order to use a CBN wheel effectively.

12. Explain which CBN wheel does not require truing and dressing before using.

13. What is the advantage of keeping a wheel and adapter together as a unit for the life of the wheel?

14. What are the most important factors in using a CBN wheel effectively?

15. For what purpose is a diamond-impregnated nib used?

16. How should the brake-controlled truing device be positioned on the magnetic chuck?

17. Briefly explain how a brake-controlled truing device works.

18. Name three of the more commonly used dressing methods.

19. How are polycrystalline diamond and cubic boron nitride tools manufactured?

20. List any five benefits to be had using polycrystalline superabrasives.

21. Why should you not use diamond superabrasive cutting tools on ferrous metals?

Section 15

HIGH-TECH SYSTEMS

UNIT 73 NUMERICAL CONTROL

Perhaps no other machining system has done more for twentieth-century manufacturing processes than the use of computer numerical control (CNC) machines. These machines offer to industry a capability never before available, permitting less restrictions in design and manufacturing. Where once human effort and skill were utilized in the completion of a manufacturing process, machines are now capable of being instructed to work on command by coded instructions. It would give the user all the planning and skill required for any metal cutting operation, as well as the means for storing this information for future use.

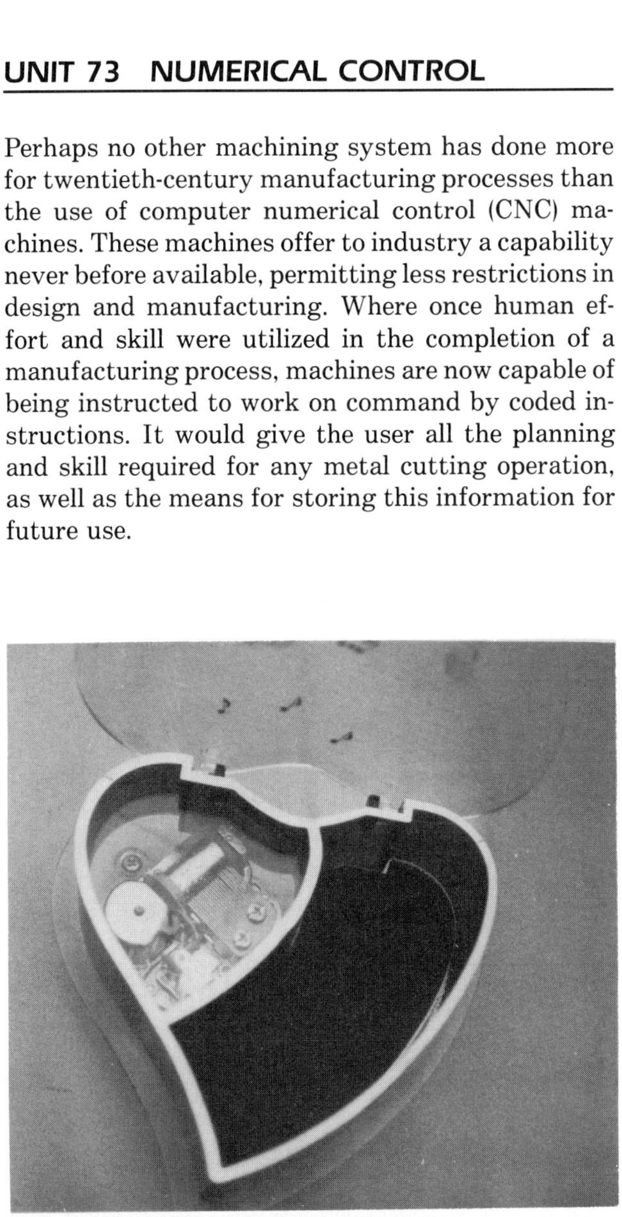

FIGURE 73-1
The music box operates by using a rotating drum with prepositioned pins.

Objectives
After completing this unit you should be able to

1. name an early pioneer in the use of numerical data to control machine tool operations.
2. name three types of input program media in addition to the computer.
3. define servo controls.
4. describe the function of the machine control unit.

History

The evolution of numerical control equipment probably started during the eighteenth century when chimes in European cathedrals and some American churches were controlled by rotating drums bearing prepositioned pins. The same principle was later used in music boxes (Figure 73-1). In 1863 the automatic player piano was introduced by M. Fourneaux who used perforated paper rolls. The holes allowed air to pass through to move the piano keys in proper sequence, as well as control the tone, sound intensity, and roll speed.

Modern numerical control machine came into being shortly after the introduction of the first electronic digital calculator. This development was to be found with John Parsons of the Parsons Manufacturing Company of Traverse City, Michigan, a defense contractor who at the time was confronted with a very difficult and time-consuming problem. As a producer of helicopter blades, how could the inspection of the complex contour of a rotor blade airfoil be speeded up to meet production schedules? The problem was solved by gathering incremental measurements with inspection probes placed along the blade and feeding these measurements into the early model computer. Electronically, the data was digested and compared to a mathematical formula representing the basic curve of the airfoil. It worked. Any variation from the formula could be singled out. Since this system had worked on parts already made, could it not also work on parts that were to be manufactured?

In 1948 the U.S. Air Force entered the picture when it contracted Parsons to continue research work on the use of numerical data to control machine tool operations. In turn, Parsons subcontracted a study project to the Servomechanism Laboratory of Massachusetts Institute of Technology (MIT). By 1952 MIT has successfully demonstrated three axes cutting tool movements on a vertical spindle milling machine, a Cincinnati Hydrotel, which had been retrofitted to make parts. In 1957 many of these automated machine tools began appearing in factories contracted by the air force. While these early numerical control machines were capable of only point-to-point movement, they greatly increased both the accuracy and productivity in manufacturing. Improvements from the electronic industry in cost, controls, and reliability were to come later. Time would prove that there would be very little metalworking applications that would not be affected by the advent of computer numerical control machining.

Numerical Control

Numerical control is a versatile method of automatically operating machine tools by means of a series of coded instructions that can be received and acted upon by the machine control unit. These instructions are largely numbers and letters of the alphabet but may also contain other symbols that may be found on the keyboard of a typewriter, such as the asterisk, the comma, the plus sign, the negative sign, the period, the dash, the dollar sign, etc.

A milling machine, where the workpiece is mounted on the table to be moved past a cutter, is a good example of how a machine can be set in motion by an operator. When one handle is turned, the table is moved in a parallel direction to the cutter (longitudinal feed); when another handle is turned, the table is moved at right angles to the cutter (crossfeed). When both handles are turned together, an angular motion of the table is created. In numerical control operations the handles of the machine are replaced by electric servo motors that can control table movements and act as a brake to stop on exact locations with extreme accuracy.

As these numerical control (NC) machines became more and more popular in machine shops, it was found that increasingly faster feed rates were needed to keep up with the demands of industry. The early NC machines were nothing more than automated versions of old standby manual machine tools. It was quickly discovered that while these machine designs were capable of very accurate and reliable machining, they could not handle the high speeds required. This led to many advances in machine tool design in the areas of lubrication systems, tool turrets, slideway geometry, and spindle heads. One of the most notable developments to come out of this effort was the recirculating ballscrew.

On most machine tools the slides that supported the cutting tools were moved along the slideways by means of an Acme screw. This consisted of a long shaft with a ground bearing surface at each end and a continuous screw thread along its length. A block containing the screw nut would be mounted on the movable slide and the screw would be attached at each end of the nonmoving slide base. By rotating the screw a precalculated number of revolutions, the nut would travel along the length of the screw and thus the slide could be moved a precise distance. This system worked fine for manual machines, but the frictional forces between the screw and nut limited the speed at which the slide could be moved.

The *recirculating ballscrew* looks very similar to the Acme screw, with the exception that the continuous thread along the length consists of a groove that is rounded at the bottom instead of the V-shaped threads of the Acme screw. The nut also has the same kind of rounded grooves, precisely matched to those of the screw. When the screw and nut are assembled, ball bearings that exactly fit between the grooves of the screw and nut are inserted into a recirculating track within the nut. This provides for the nearly frictionless movement of the nut along the length of the screw. By using this superior screw design along with the electric servo motors, the higher feed rates needed by industry were now obtainable.

To control the movements of the machine tool, the coded instructions of the numerical control system are converted into electric pulses of current that activate the motors and controls of the machine tool. Additionally, the up and down movements of the cutter can be controlled, giving the machine flexibility of movement in six directions; lateral, cross, and cutter feeds can take place at the same time if desired.

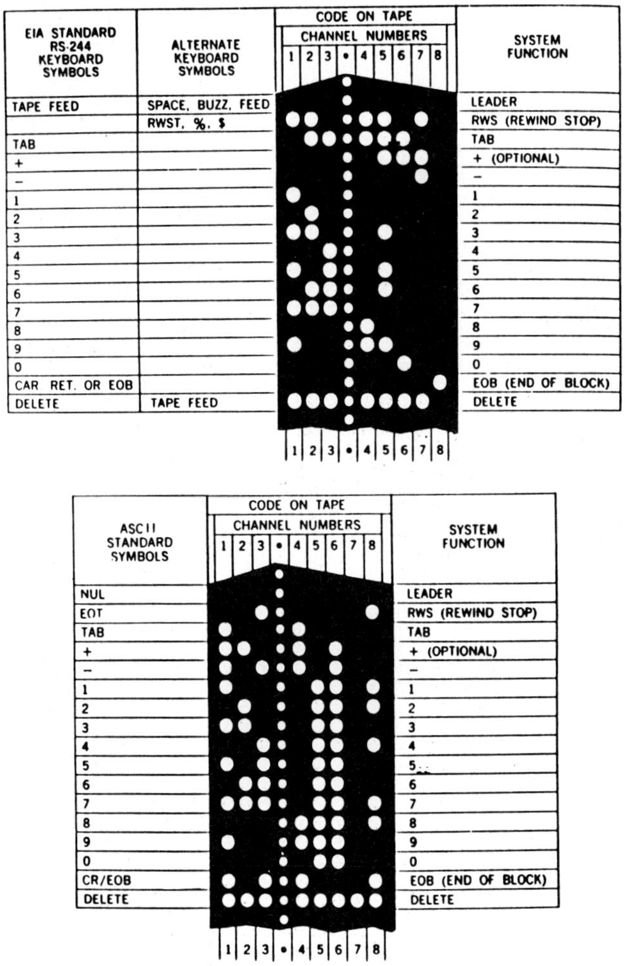

FIGURE 73–2
Note the similarities and differences between the EIA and ASCII standard coding systems. (Credit: Superior Electric Co.)

Types of Input Program Media

Commands to a numerical control system are communicated to the machine's control unit by means of coded instructions. These instructions are largely an established number language that has been standardized through the efforts of the Electronics Industries Association (EIA) and the American Standard Code for Information Interchange (ASCII) (Figure 73–2). These coded instructions are prepared in advance to control the sequence and performance of the machine tool operations. In the early development of numerical control, a number of different media was used to communicate the instructions. These instructions may be input to the control unit directly when connected to a computer [direct numerical control (DNC)] or by a reader supplied by the control. With a reader, data for machining can be input by punched tape, punched cards, magnetic tape, and diskette (floppy disk) (Figure 73–3).

Today, punched tape is the most widely used of the input systems used for NC programs, with a slow movement toward the use of diskettes and DNC.

The coded instructions for machining operations are contained in several rows called a *block* (Figure 73–4) and are based on the binary number concept. Binary numbers are based on the combinations of 1 and 0. In the punched tape "language" the 1 and 0 correspond to a hole/no-hole combination. Because a tape has a hole or no-hole positioning, the machine control unit quickly decodes this informa-

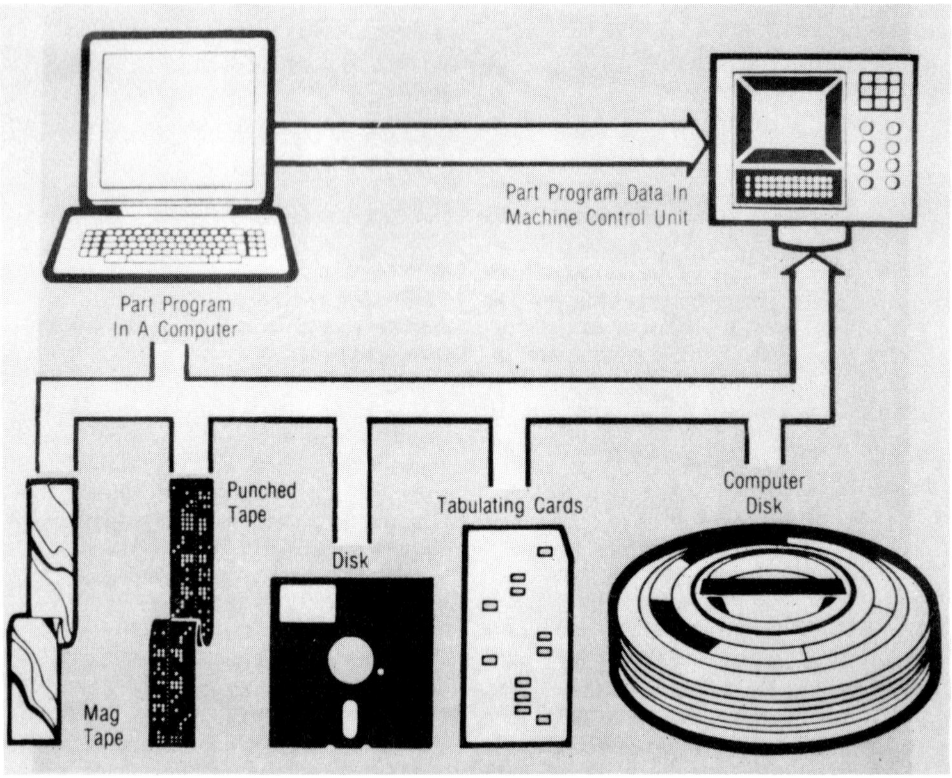

FIGURE 73-3
Some of the more common methods used to put information into a computer. (Credit: Modern Machine Shop, NC/CIM Guidebook)

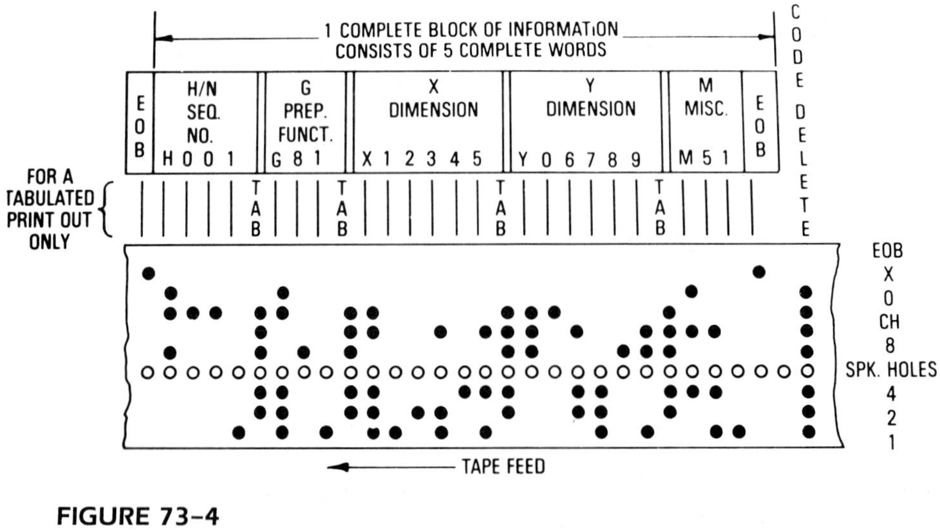

FIGURE 73-4
An example of a block of information punched on tape. (Credit: Cincinnati Milacron Inc.)

tion and converts it into electrical pulses. Special tape punching machines are used to prepare NC or computer-aided NC programming. The combination of punched holes represents a different number of symbols by EIA and ASCII standards.

Punch cards or tabulating cards were at one time used in machine tool programming (Figure 73-5). Programmed instructions were recorded on these cards by means of a card punching machine using a perforation format for each given numerical digit or symbol. Today these have been replaced by punched tape or computer programs. Indications are the future will see the use of no tape.

Operational Outline

There are many steps involved in NC machining. The beginning of NC machining starts at the design stage where the print is dimensioned for ease of fabrication and handling. Once the part has been "digested" by a person who can visualize the machining steps before they are physically performed, the program is written on a workpiece programming manuscript. If satisfied that the machining steps will work, the programmer will then convert the manuscript to punch tape or a computer program that is then fed into the machine control unit. The NC machine tool is then put into operation, adjusted for any tool error, studied for wasted motion, to finally make the first part. The first piece will then go to inspection where the part is or is not approved for a production run. The successful conclusion to any NC machining programming is the storing of the tapes or computer program for future use.

The flowchart in Figure 73-6 is a summary of the steps necessary for the completion of a good NC program tape.

Feedback Systems

The machine control unit (MCU) receives instructions from the tape reader or the computer program and will convert this coded data into command signals that are transmitted to the machine as needed. These command signals are in the form of electrical pulses and can produce a movement in the machine table slide position as small as 0.00001 in. (0.00025 mm).

Computer numerical control systems are designed to direct a machine tool through a variety of operations as indicated by the instructions given to the punched tape. In order to assure the machine tool's accuracy, a system of checks has been devised. Has the machine tool table slide moved as much as required by the input data on the punched tape? *Feedback devices* send back data to the MCU. If there is any variation between the machine tool's table slide position and the required dimension of the input data, the drive motors will automatically

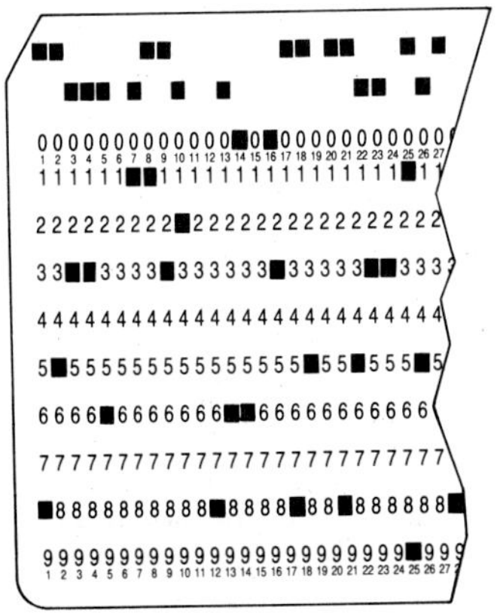

FIGURE 73-5
In the early days of numerical control programming, instructions were recorded on tabulating cards.

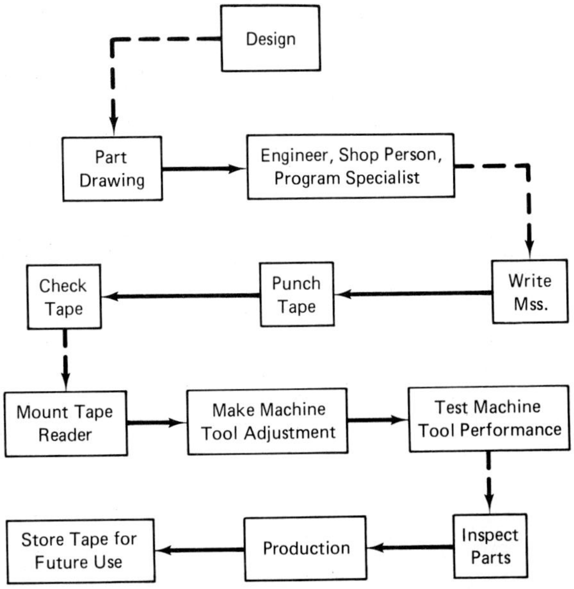

FIGURE 73-6
An NC programming flowchart.

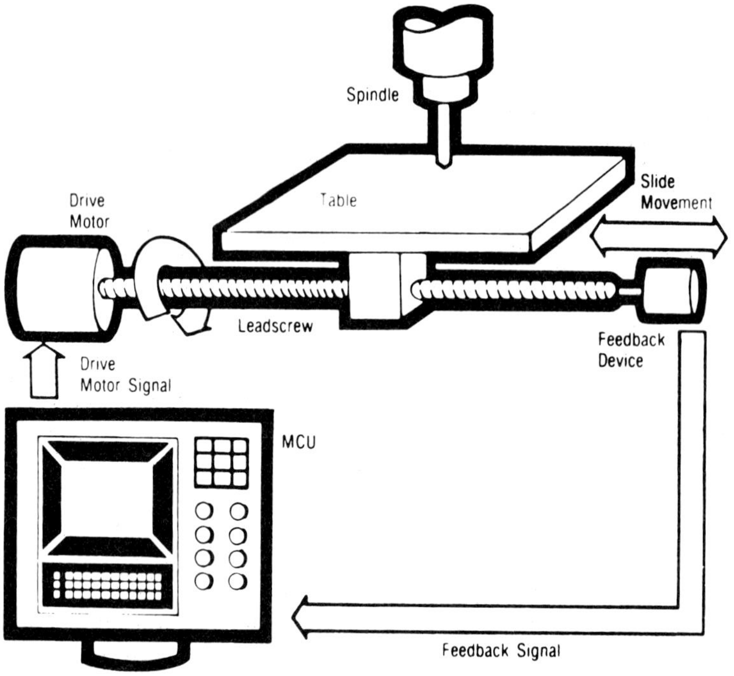

FIGURE 73–7
Feedback devices are mounted on the machine tool slide or lead screw to indicate machine table position. (Credit: Modern Machine Shop, NC/CIM Guidebook)

make corrections through the transmission of command signals (electrical pulses) by the MCU.

Feedback devices are mounted on the machine tool slide or lead screws to indicate the position of the machine table (Figure 73–7). The more common feedback devices are pulse coders, linear scales, and tach generators.

Pulse coders are optical devices composed of a glass disk coated with silver and mounted on a shaft supported on each end by precision bearings. Very precisely placed, evenly spaced lines are cut into the surface of the disk from the center to the outer edge. These lines provide "slots" through which light from a light emitting diode (LED) mounted below the disk shines through to a photodiode mounted above the disk. As the ballscrew to which the pulse coder is attached rotates, the resulting flashes of light are counted by the MCU, giving the control a precise indication of actual machine movement.

Linear scales work in much the same way as do pulse coders, except rather than indicate machine positioning by rotary movement, it uses straight line or linear movement. The main body of the scale is usually mounted to the nonmoving portion of the machine slideway and the sensor head is mounted on the movable slide. As the slide moves along the slideway, the sensor head traveling along the linear scale sends pulses back to the machine control, this indicating the position of the slide.

Feedback devices that use pulse-type signals are referred to as *digital signal devices.* This is due to the fact that pulsing on or off signals mimic the 0 or 1 binary states of a digital computer.

The *tach generator,* generally found used in conjunction with servo and spindle motor systems, is attached directly to the shaft of a motor. As the motor rotates, the generator produces a voltage proportional to the speed of the motor. A common system uses a scale of 0 to 10 V, with 0 V being equal to zero speed and 10 V being equal to top speed. Thus, when the machine control unit receives a signal of 5 V, it knows that the motor is running at exactly half speed and calculates positioning or feed rate information accordingly.

Devices, such as the tach generator, that utilize a variation in voltage or current as a means of relaying information are referred to as *analog signal devices.*

Servo Systems

A servo system generally consists of three components: a servo controller, a servo motor, and a feedback device.

The *servo controller* receives positioning signals from the machine controller, interprets those signals in terms of the type of servo system, and then sends the appropriate signals out to the servo motor.

The *servo motor* provides the driving force to move a machine slide along the slideway by rotating a ballscrew by means of either a direct coupling to the screw shaft or by means of a gearbox. Whether electrical, pneumatic, or hydraulic in design, the servo motor rotates an exact number of revolutions based on signals it receives from the servo controller to position the machine slide.

Servo systems of this type are referred to as *open-loop* systems. This is due to the fact that while the MCU calculates the required movement, and the servo controller sends out the proper signals to the servo motor to produce the required movement, the MCU really does not know whether or not the machine has even moved at all. Since the open-loop type of servo system leaves so much room for improvement in checking the accuracy of movement, it is generally found on many of today's low-cost CNC equipment (Figure 73–8).

The improvement came with the introduction of a *feedback device* to the servo system. Now when the servo controller sends the positioning signals to the servo motor and the slide is moved along the slideways, the feedback device in turn sends signals back to the MCU indicating the exact position of the slide. If the MCU determines that there is a difference between the movement called for the movement made, it sends out the appropriate signals to the servo controller to correct the positioning difference.

Servo systems that use a feedback device are referred to as *closed-loop systems* (Figure 73–9).

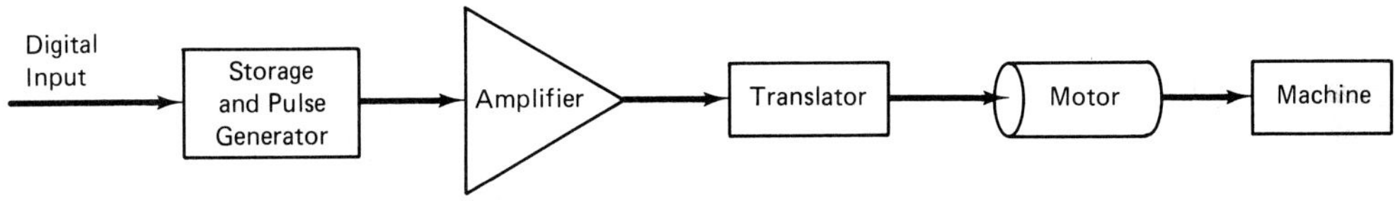

FIGURE 73–8
An open-loop servo control is generally found on low-cost NC equipment.

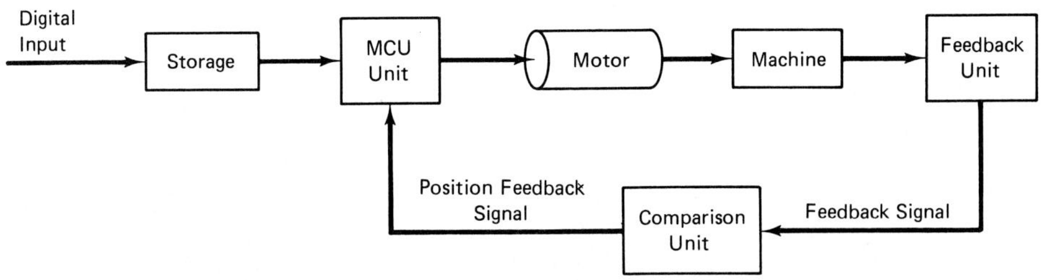

FIGURE 73–9
A closed-loop system is very accurately adjusted for error.

UNIT 74 SIMPLE PROGRAMMING

Numerical control of tools, production equipment, and machine tools represents the most flexible form of automation. To be effective, instructions for machine control operations are placed on the control tape or computer for the machine to follow. These instructions will be detailed steps where all the sequence of operations are organized to produce machine parts efficiently, accurately, and faster to lower production costs. To aid the programmer, a number of positioning systems are available with which the programmer should become familiar and understand in order to do any simple programming.

Objectives
After completing this unit you should be able to

1. explain the Cartesian coordinate system.
2. explain the purpose of preparing a manuscript.
3. explain absolute positioning.
4. briefly explain how an incremental positioning program works.

Coordinate System

The Cartesian coordinate system was named after the seventeenth-century French philosopher and mathematician, René Descartes, who developed the mathematical method of determining the location of any scientific point along two or three perpendicular axes. Referred to as rectangular coordinates, this method is used by a machinist to layout measured points on a rectangular workpiece, with a height gage. Two straight lines are drawn from a measured edge, one vertical and one horizontal, to form right angles with each other. Where they meet forms the location of a measured point on the workpiece.

This is the same concept an NC programmer uses to move the workpiece under a spindle location of a machine tool. The difference between the two methods is that the lines drawn with the height gage are now called axes and the intersection point is called the origin or zero point (Figure 74–1). To illustrate the direction of movement using a vertical milling machine, the X and Y axes represent table movements, with the X axis representing the horizontal table feed (right or left movement) and the Y axis representing the table crossfeed (toward or away from the column).

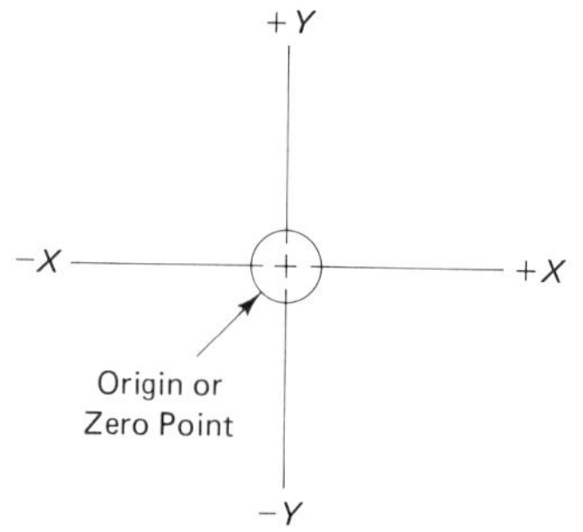

FIGURE 74–1
The origin or zero point is created with lines that intersect at right angles.

When the X and Y axes are drawn and charted on graph paper, they form quadrants and are lettered in a counterclockwise direction (Figure 74–2). All positions located in the upper right-hand corner or quadrant 1, would be positive X $(+X)$ and positive Y $(+Y)$. In quadrant 2 all positions would be negative X $(-X)$ and positive Y $(+Y)$. In the lower left-hand corner of the graph or quadrant 3, all positions would be negative X $(-X)$ and negative Y $(-Y)$. In quadrant 4 all positions would be positive X $(+X)$ and negative Y $(-Y)$.

As an example of how this system would work, assume that every square or unit on the graph equals 1 in. Starting from the point of origin, the location of point A would be two units above the X axis and three units to the right of Y axis. Thus the location of position A would be $X = +3.000$, $Y = +2.000$. Point B would be located two units to the right of the Y axis and two units below the X axis. For point B the location would read $X = +2.000$, $Y = -2.000$. Since positive signs are understood and need not be indicated, the above locations would be indicated as follows:

$A = X3.000 \quad Y2.000$

$B = X2.000 \quad Y-2.000$

A two-axes dimensioning system is widely used by numerical control programmers. A third axis, the Z axis, which is perpendicular to the X and Y axes, represents the up and down movement of the spindle or knee and is added to establish a three-dimension coordinate system (Figure 74–3).

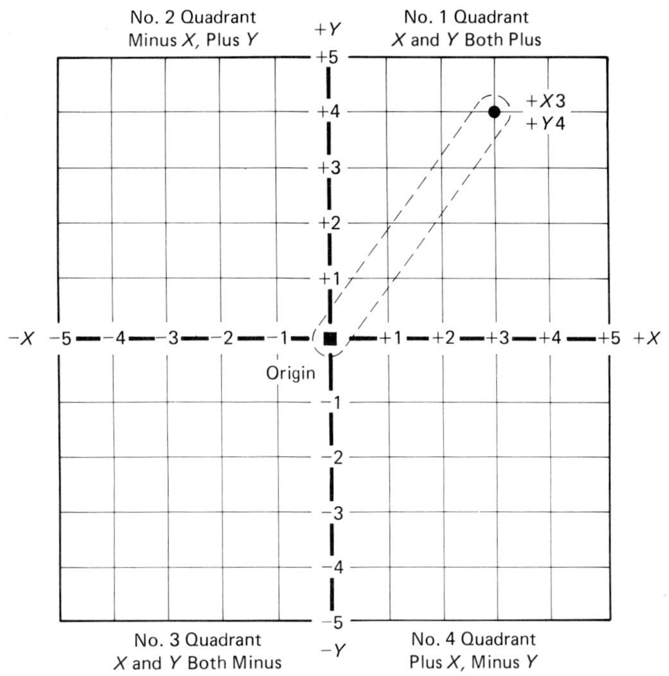

FIGURE 74–2
The quadrants of a Cartesian coordinate system are lettered and numbered in a counterclockwise direction. (Credit: Modern Machine Shop, NC/CIM Guidebook)

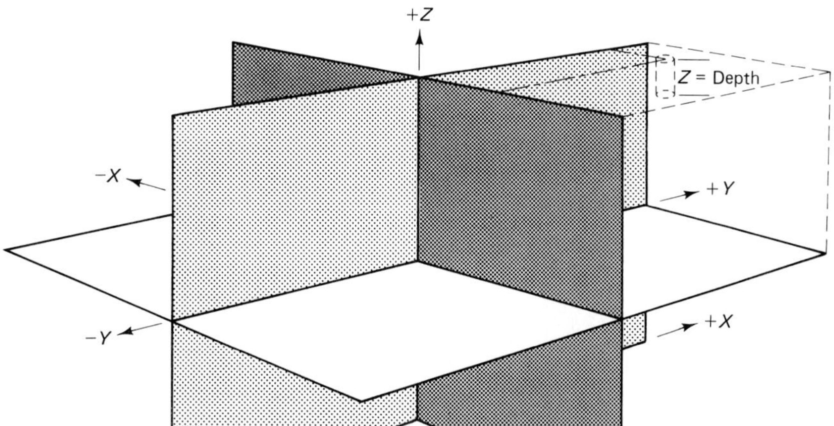

FIGURE 74–3
A three-axes movement system is used on NC machine tools. (Credit: Modern Machine Shop, NC/CIM Guidebook)

Guidelines

Since numerical control programming is highly dependent on the system of rectangular coordinates, some established guidelines should be used by all concerned. This will put the engineer, draftsperson, programmer, or machine operator on the same level of understanding as to what steps will be required.

1. Use the Cartesian coordinates *X*, *Y*, and *Z* to define all the part surfaces.
2. Whenever possible, use reference points on the part itself. This will make it easier for quality control to check the parts later.
3. Establish reference planes along part surfaces that are parallel to the machine axes.
4. Describe the part in terms that will be acceptable to the control system being used.
5. Dimension the part clearly and simply so that the cutter path may be easily determined and programmed.
6. Work within the allowable tolerance of the design throughout the cycle.
7. Establish all dimension points firmly to eliminate any false assumption or calculation as to the physical shape of the part.

Manuscript

In order to communicate to the machine tool what a programmer wants it to do, a programming language has to be used. The language is based on spelling, format, and the proper sequence of words. To record the programming language, an NC programmer uses a manuscript or a process planning sheet. Once the information is taken from the part drawing, the programmer will list all the details necessary to make the part. The instructions may be a line written in long hand or by a typewriter. Each line on the manuscript is called a block.

A block may be composed of words, characters, or digits and may contain several commands, as long as these commands do not conflict or oppose each other at the same time (see Figure 73-3). For example, commanding the spindle to turn clockwise and counterclockwise in the same block would be in conflict. The number of blocks in a manuscript will reflect the number of specific tool movements. The greater the number of blocks, the longer will be the tape or the computer program to contain all the information. The manuscript layout will be the same as the format to be used.

An NC machine cannot think for itself and will produce parts according to the instructions or commands directed by the programmer. In order to do any programming, a good programmer should have a good background in machining procedures and machinist skills. Not only should the programmer know and understand tools, work holding devices, and setups, the person should also know speeds and feeds and understand the NC tape format. When a machinist makes a part, the machinist uses previous skills and experience as the machining steps are undertaken; seldom if ever does a skilled machinist write the steps on paper. A good programmer will have to put all the program steps in a language that the machine control unit will understand.

There are several recommendations a programmer should follow in preparing a program. They are as follows.

1. Prepare a dimensioned sketch of the part.
2. Determine the best possible means of fixturing to locate and hold the workpiece in position.
3. Select the origin or zero position on the machine or workpiece, which will be used as a reference. All machined dimensions will be measured and directed from this point.
4. Rough out the sequence of operations and edit before entering in the manuscript. Be sure to include:
   (a) tools and availability of quick-change toolholders
   (b) the first tool change where the part may be loaded or unloaded in the fixture easily

Data Description

Data for each block in the manuscript is usually recorded left to right. Information describing characteristics are generally published in company manuals that suit the NC system being used, based on the Electronic Industries Association (EIA) and the American Standard Code of Information Interchange (ASCII) standards.

Each command in the block must be written in the letter address system that is listed in the NC machining manual. Note that each letter follows a publication format and is followed by a number. The letter identifies the type of command. The number identifies the specific information. For example,

N0100 (sequence number)

The letter N is the address letter used for the sequence letter and is the first item entered on the manuscript. The letter is followed by up to a six-digit number and is used for the purposes of research, recall, and editing.

G71 (preparatory function)

The letter G is the address for the preparatory function and is used to set up the controls to receive machining data such as direction of spindle rotation, program mode, etc. As many as four G commands can be used on each block data as long as they do not conflict. G70 and G71 would be in conflict in the same block. G70 is the inch input mode, G71 is the metric input mode.

Some G commands are modal and some are nonmodal. These commands are identified as such in published manuals. Modal means in memory, that a command will not change once entered unless it has been changed or canceled. Nonmodal means that the command is only effective for the one block of program data.

$X-3.0000$ (X increment)

The table movement along the X axis is indicated in this location. The letter address is followed by as many as seven digits that can be either positive or negative. In the inch mode this would be written three digits, decimal point, followed by four digits. In the metric mode it would be four digits, decimal point, followed by three digits. In either case the leading and trailing zero digits need not be written.

$Y-12.0000$ (Y increment)

The table movement along the Y axis is indicated in this location. A movement up or away from the zero point would be a positive (+) movement. A movement down from or toward the zero point would be a negative (−) movement. In this example the table must be moved 12.0000 in. below the X axis.

$Z-1.0000$ (Z increment)

The spindle movement along the Z axis is indicated in this location. A movement down (below the Z zero point or gage height) would be a negative (−) movement and a movement up (above the zero point) would be a positive (+) move. In this case the drill or end mill held in the spindle of the machine tool would move 1.0000 in. below the zero point.

M03 (miscellaneous function)

The letter M is the address for the miscellaneous function and is followed by a one- or two-digit number. Miscellaneous functions include on/off type operations. This would include turning on or off spindles, coolants, tool transfer, etc. M03 indicates turning on the spindle in a clockwise direction.

EOB (end-of-block character)

An end of block character indicates the end of a block of tape information and is punched before the first block. It is used to stop the tape reader after the tape has been read. It is also punched at the end of every block to separate the blocks in a program.

Reference Point Systems

There are two modes of programming used in describing the coordinate points in NC machining: absolute and incremental systems (Figure 74-4). Both systems are in wide use with neither system being superior to the other. Many controls found on machine tools today are capable of using either absolute or incremental programming. With the use of the G code, this will allow the programmer to switch from one operating mode to the other whenever it is found to be easier to describe any coordinate point.

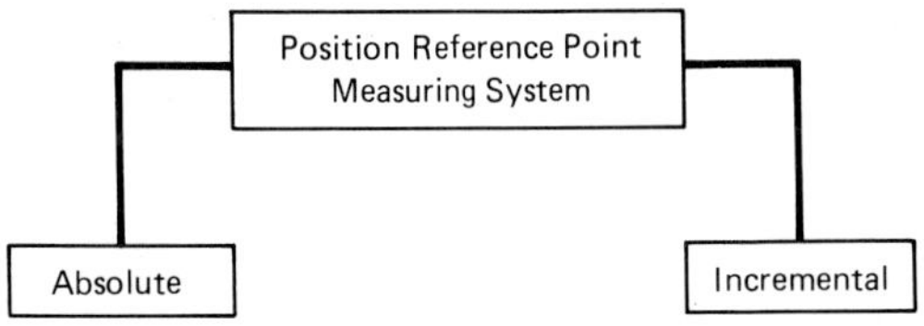

FIGURE 74-4
Reference point systems.

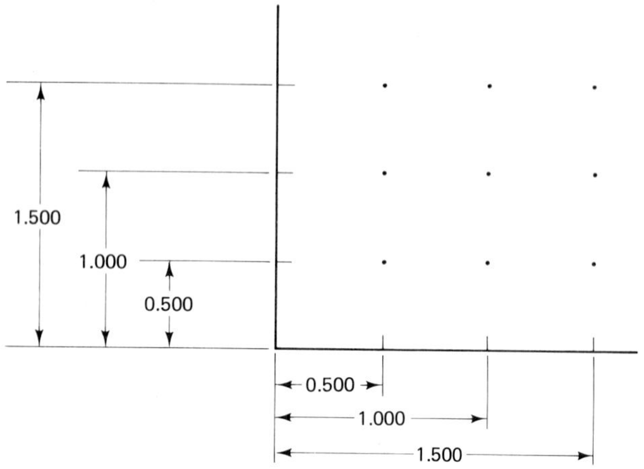

FIGURE 74–5
An example of absolute positioning. Notice how all the dimensions have a common starting point.

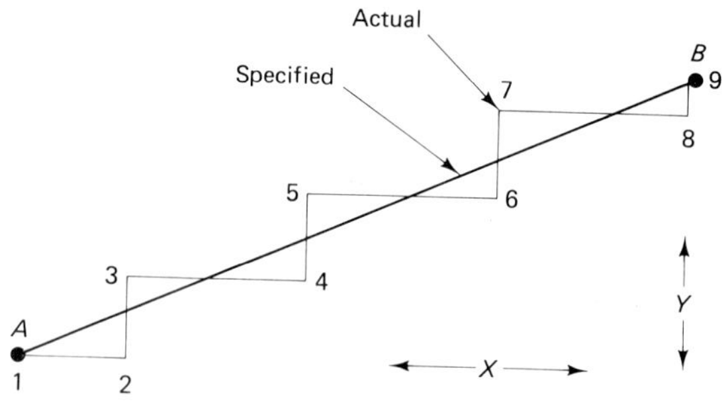

FIGURE 74–6
An example of incremental positioning, also known as point-to-point dimensioning, and how it compares with straight cut control. (Courtesy Cincinnati Milacrom Inc.)

Absolute System

In the absolute system all the coordinated points are located from a fixed zero reference point. All distances and directions used in locations are based on the relationship to the fixed point. There is an advantage to this, however. Errors that are made in such movements are not carried on to other locations. The mistake will apply to one particular error each time. It is helpful to use sketches or drawings that use a baseline dimensioning system when using the absolute programming system. If the drawing bears conventional dimensioning, they should be resketched to indicate a baseline dimensioning system.

Absolute Programming
The G90 command will instruct the computer and machine control unit that programming is to be carried out in the absolute mode. The command codes that instruct the machine tool to move the table, knee, or spindle will be explained as if using a vertical milling machine to carry out the absolute programming as illustrated in Figure 74–5, where each point or location on the workpiece is based on the relationship to a fixed zero reference point.

A plus X $(+X)$ commands the spindle to be positioned to a location by moving the machine table to the left of the zero reference point.

A minus X $(-X)$ commands the spindle to be positioned to a location by moving the machine table to the right of the zero reference point.

A plus Y $(+Y)$ commands the spindle to be positioned to a location by moving the machine table toward the column from the zero reference point.

A minus Y $(-Y)$ commands the spindle to be positioned to a location by moving the machine table away from the zero reference point.

Incremental System

In incremental positioning, or point-to-point positioning, a dimension or movement is made to a new point using the last previous known point as a reference.

In point-to-point programming the movement or location of the workpiece or the cutting tool is directed on a straight line, with each position or location treated as a new starting point. For example, if several holes in a workpiece had to be countersunk (Figure 74–6), the programmer would instruct the machine tool to move directly to a specific location, countersink, retract the countersink, move a measured distance from the operation to the next location, countersink, retract the countersink, move a measured distance from this operation to the next location, countersink, retract the countersink, and so on until all the holes were countersunk.

If the travel of the machine tool positioning is to remain high—rated at about 150 to 400 in./min

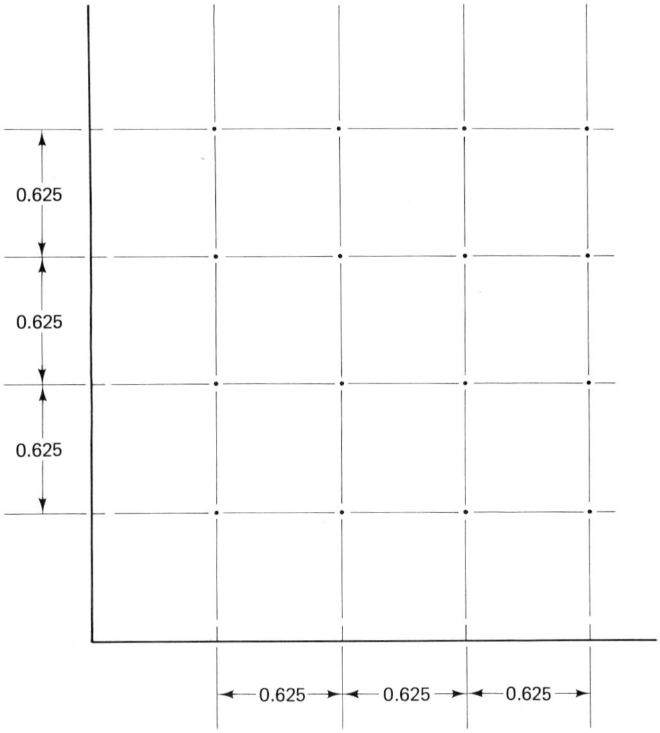

FIGURE 74-7
Incremental programming.

(38 to 100 m/min)—contact between the workpiece and the cutting tool must be avoided. That means that the workpiece would be located under a fixed spindle during movement, or that the spindle would be above a fixed workpiece when any table movements are made. It does not matter whether movements are made along the X and Y axes first. Clamps must be avoided. Some CNC machines will generate motions along both axes at the same time as well as develop a 45° angle between both axes with no change in cutting tool depth along the line of travel. A good programmer eliminates as much wasted motions or nonmachining travel as possible.

Incremental Programming

The G91 command instructs the computer and machine control unit that programming is to be carried out in the incremental mode. The command codes that instruct the machine tool to move the table, knee, or spindle will be explained as if using a vertical milling machine to carry out the incremental programming as illustrated in Figure 74-7 where distances and direction are taken from the preceding point.

A plus X ($+X$) command positions the spindle to a location by moving the machine table 0.625 in. to the left of the last point.

A minus X ($-X$) command positions the spindle to a location by moving the machine table 0.625 in. to the right of the last point.

A plus Y ($+Y$) command positions the spindle to a location by moving the machine table 0.625 in. above the X axis or away from the column.

A minus Y ($-Y$) command positions the spindle to a location by moving the machine table 0.625 in. toward the column.

UNIT 75 MISCELLANEOUS PROGRAMMING

Contour programming shares a working relationship with linear programming. They both use information taken from a drawing that is then converted to layout the points on a rectangular field of coordinates that will affect the axes of movement. The relationship ends there, however. Contour programming will also require a path of tool travel, the means by which the tool is kept in contact with the changing workpiece outline at a controlled rate of feed.

Objectives
After completing this unit you should be able to

1. define continuous-path machining.
2. briefly explain interpolation and how it is used in programming.
3. name five different methods of interpolation.

Continuous Path

Continuous-path machining is another term for contouring. In continuous-path programming machining is planned so that a coordinated movement takes place that permits the cutting tool to remain in constant contact with the workpiece. Contoured-machined workpieces, such as those produced by the vertical milling machine (Figure 75-1) or the lathe, are examples of continuous-path programming.

FIGURE 75-1
An example of continuous-path machining.

In continuous-path programming two or more axes of the machine tool or the cutting tool have to be controlled for simultaneous movements. This permits the cutting tool to work while the workpiece is moving continuously around it. For example, if grooved cuts have to be made on a workpiece, one with a 30° and one with a 60° cut from the horizontal (Figure 75-2), it can be seen that the X-axis movement will be greater than the Y-axis movements creating the 30° groove. In order to accomplish this, the X-axis slide has to move a greater distance and at a faster feed rate than the Y-axis slide. The opposite is true when developing a 60° groove.

Programming by Interpolation

Interpolation is a process of making a mathematical adjustment between two given values to establish a new value. In contour programming the means by which the machine cutting tool flows from one point to another is by use of an interpolator. The interpolator may be considered a small, fixed-program computer and is built in most of today's machine control unit (MCU). The blending of the axis points with a predetermined tool path is accomplished through the interpolator, which is constantly "looking ahead" to tiny line segments to do its calculating. Thus, each movement change is made and transmitted as a signal to the servo controls to assure continuity between the work and tool relationship.

There are currently five different methods of interpolation: linear, circular, parabolic, helical, and cubic. Some linear interpolation can be made as fine as 0.0001 in. (0.0025 mm) and may be programmed to approximate slopes, arcs, and parabolic curves. All contouring controls provide linear interpolation, with many controls capable of both linear and circular interpolation. Parabolic, helical, and cubic interpolation are considered more complex or a higher-order interpolation and are used in industries that produce complicated contours or complex shapes. These would include the aerospace and automobile industries.

Linear Interpolation

Many CNC systems are capable of linear interpolation, that is, the moving of locations in a straight line as well as angles and tapers. If the span length or movement from one location to another is extremely short, however, the interpolator does not do any calculations to move the slides. The amount of movement that can be managed will depend on the

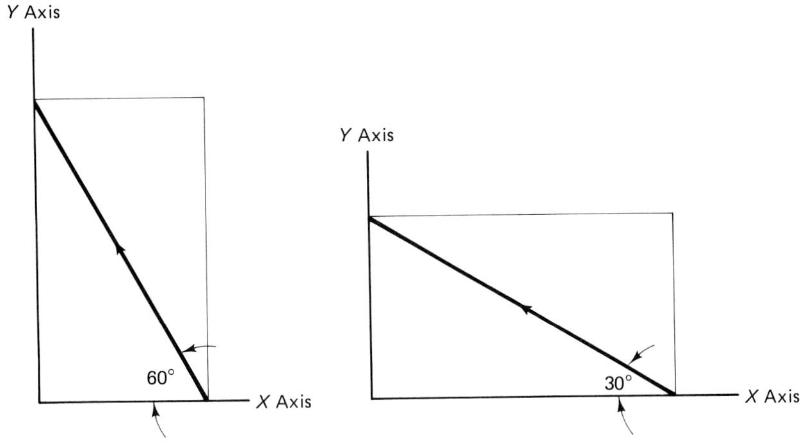

FIGURE 75–2
An angular groove could not be machined if both the *X* and *Y* slides operated with the same feed rate.

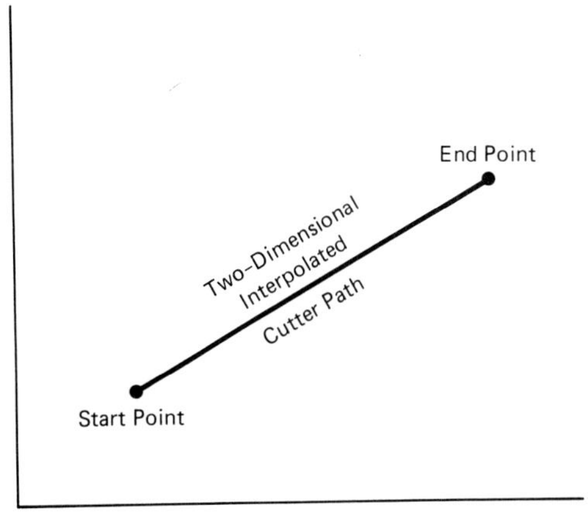

FIGURE 75–3
Two-axes linear interpolation requires programming the beginning and ending point of a line. (Credit: Modern Machine Shop, NC/CIM Guidebook)

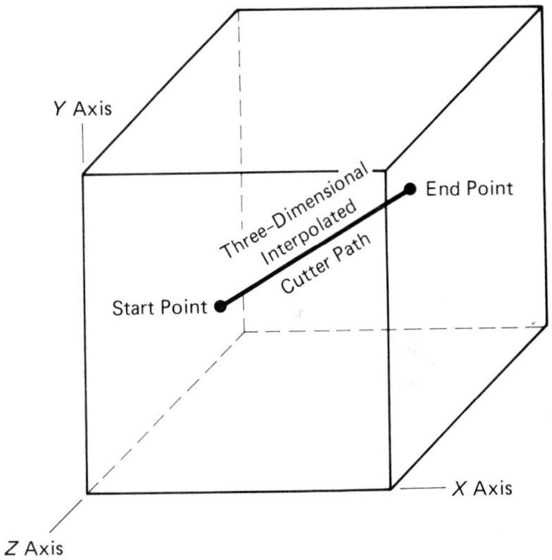

FIGURE 75–4
Three-axes linear interpolation uses the three axes of the machine at the same time. (Credit: Modern Machine Shop, NC/CIM Guidebook)

output resolution of the system. Some linear interpolation can be made as fine as 0.0001 in. (0.0025 mm).

The principal function of a linear interpolator is to create straight-line moves. This would involve the programming of coordinates (the *X* and *Y* coordinates in a two-axes work) that indicate the beginning and ending of each line (Figure 75–3). If many lines or segments have to be connected, the end locations of one line becomes the start point for the next line. This process continues on throughout the cycle. A linear interpolator would also have an application in these axes movements, such as in a milling machine, where the *X*, *Y*, and *Z* movements are straight. The work can be moved to the right or left as well as toward or away from the operator, while the cutting tool, such as a drill, is moved up or down toward the workpiece to meet at a common point (Figure 75–4).

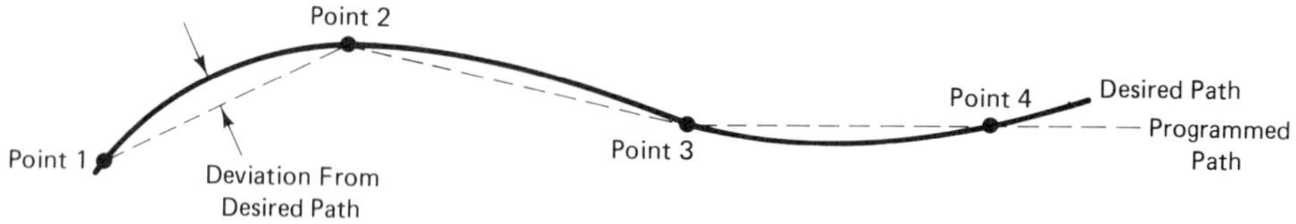

FIGURE 75-5
Linear interpolation requires many program points to approximate a curve that cannot be defined mathematically. (Credit: Modern Machine Shop, NC/CIM Guidebook)

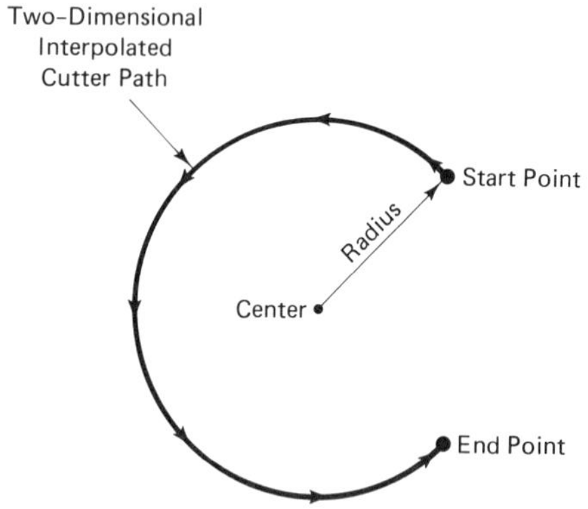

FIGURE 75-6
A two-dimensional circular interpolation uses any two linear axes of the machine at the same time.
(Credit: Modern Machine Shop, NC/CIM Guidebook)

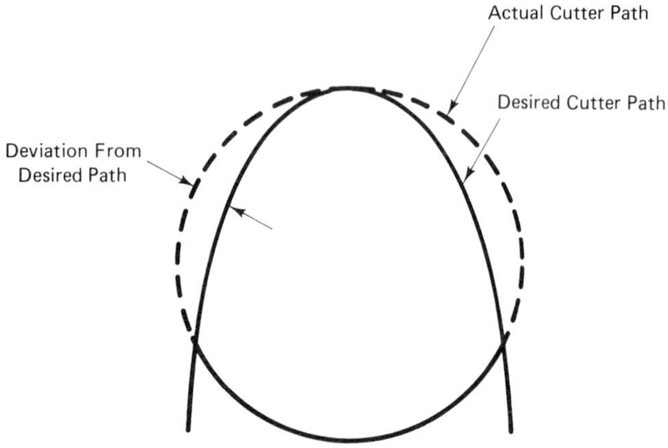

FIGURE 75-7
Circular interpolation can closely approximate a curve on a contour not constructed of true arcs.
(Credit: Modern Machine Shop, NC/CIM Guidebook)

However, many machined surfaces will require moves other than straight-line moves, such as the development of arcs, circles, or curved slopes, for example. Due to the fact that the contours of these surfaces are constantly changing, the coordinates are many and would have to be programmed in many, many very small line segments. The finer the line segments, the truer will be the contour. The process would be difficult and involved as each increment would require a new block of tape or computer information. The program would result in a very long tape or informational input. For contours of this type a linear interpolator would be limited and would not be very practical (Figure 75-5).

Circular Interpolation

The use of a circular interpolator has greatly reduced the task of developing higher-order contouring such as arcs and circles. Where many thousands of programmed line segments were calculated to form a circle or arc using a linear interpolator, all the programmed input that is required to generate an arc up to 90° can be compacted into five pieces of information. What are the *X* and *Y* coordinate positions of the center or radius of the circle? What are the start and end points of the arc being cut? And what is the direction in which the arc is to be cut: clockwise or counterclockwise? (Figure 75-6)

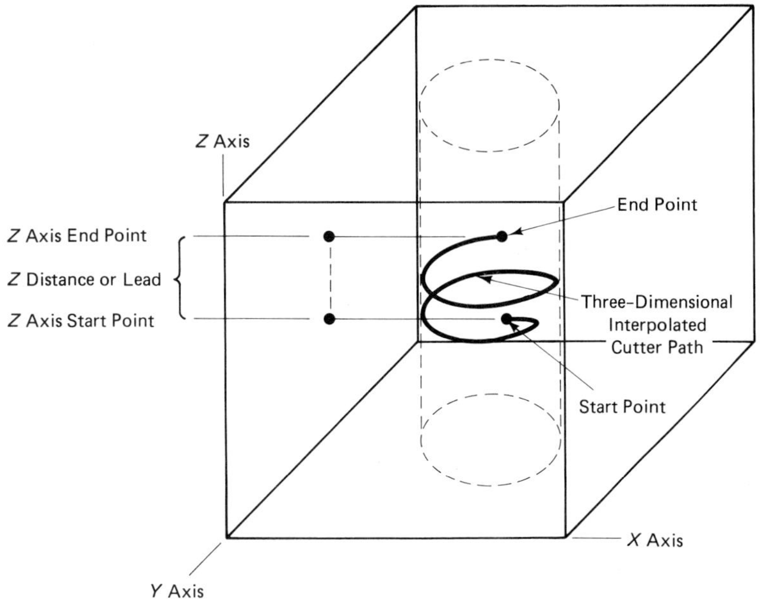

FIGURE 75–8
Helical interpolation combines a two-axes circular interpolation with the linear movements of the third axis. (Credit: Modern Machine Shop, NC/CIM Guidebook)

While the circle is the most frequently encountered contour, the circular interpolator in the MCU is limited to one quadrant or a 90° arc. If a full circle is to be programmed, it must be calculated in successive coordinates. This is done by reducing the measurement of each circular span (chord line) into very small measurements created by a single pulse output, which can be as fine as 0.0001 in. (0.0025 mm). Enough output pulses are automatically calculated to describe the desired circle, providing enough information to control the cutter centerline as well as the feed rate for each axis.

The circular interpolator can only produce a circle in the same plane as the *X, Y,* and *Z* coordinates. It cannot produce a tilted circle. This form becomes an elliptical circle and requires a special form of interpolator.

Free-form shapes as well as second- and third-degree curves can be closely approximated with the use of an arc and a circular interpolator (Figure 75–7). This method will provide a truer circle and smoother finish than by using linear interpolation.

Helical Interpolation

Helical interpolation is the third most common form of calculating contours. In this method a two-axes circular interpolation is combined with the linear movements of the third axis. These three movements are created to move at the same time to develop a helical spiral cutter path (Figure 75–8). The most common application of helical interpolation lies in milling large internal diameter threads and similar helical forms.

UNIT 76 CNC MILLING MACHINE

There is no segment of machine tool operations that has not been affected by CNC programming. Advanced computer software has greatly increased the use and acceptance of personal computers in business, medicine, and the home. Advanced software development has also been responsible for simplifying part programming for CNC machine tool operations. Gone is the time-consuming and nerve-wracking computations involved to machine a part such as a cam or other irregular outlines, for example. Color graphics may display the part and tool path that is automatically developed by the CNC system. And while programming has been made easier and faster with less chance of error, it has not completely replaced the background and skills required of a good programmer.

Objectives

After completing this unit you should be able to understand

1. the definition of software.
2. the application of a program for a CNC milling machine operation.

Software

Software may be defined as stored information, the collection of ingredients composed of programs, documents, and routines essential for computer operations. Language is the arrangement in which software is coded, used, or represented to relate information or meaning. In order to do any programming, the programmer must be familiar with the particular machine's language. Not all CNC systems program the same way. The information is contained in the programming manual published by the machine tool builder. Without this language information, the operator may not be able to tell the machine tool what to do.

CNC Milling Machine

A CNC milling machine offers flexibility and accuracy to toolroom productivity. It can drill, tap, bore, and mill workpieces up to 1000 lb (454 kg). Made to perform as a stand-alone machine or tie in with a computer-aided design/computer-aided manufacturing (CAD/CAM) system or be part of a machining cell, computerized numerical control milling machines are made of a heavy-duty cast-alloy construction to withstand deflection and repetitive machining demands (Figure 76–1).

Although there may be as many as 100 different programming languages available, machine tool manufacturers generally use one or two numerical control formats. It is not a specific language like BASIC, FORTRAN, or COBOL, but rather a language format. The variations are dependent on specific functions of that particular manufacturer's machine. This machine tool may be capable of features other machines are not. Extra G or M codes may be added to do functions within the program. For example:

Canned cycle: A single block of data that is used to control the duties of arithmetic and trigonometry to allow preset tool lengths and diameters to force the machine to produce the correct finished contour.

Macro or subroutines: A command identifying a group of text statements that may be generated in the same program by a special character, from an asterisk sign to a dollar sign, and terminated by another special character.

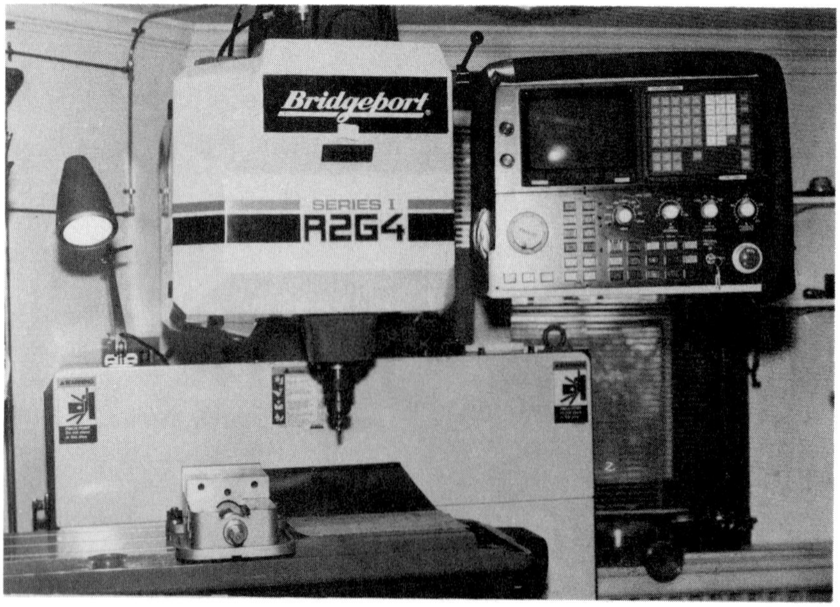

FIGURE 76–1
A CNC milling machine offers flexibility and accuracy to toolroom machining.

Special mill cycles: Face milling cycles (wide-area cutting), pocket milling cycles (internal milling of steps or contours), and internal bore cycles (the milling of a hole to size) may be added features to a CNC milling machine.

Preparatory Codes for a CNC Milling Machine

These G, or preparatory function codes, "prepare" the CNC milling machine to treat the information supplied in a particular way. While they all vary, the following are a listing of the G codes to be found on some CNC milling machines.

G Code	Function
G00	Rapid positioning
G01	Linear interpolation
G02	Circular interpolation clockwise (cw)
G03	Circular interpolation counterclockwise (ccw)
G04	Dwell (seconds)
G20	Inch data input (foreign-made machines)
G21	Metric data input (foreign-made machines)
G27	Zero return check
G50	Coordinate position setting
G70	Inch data input (American-made machines)
G71	Metric data input (American-made machines)
G77	Face mill cycle
G78	Pocket mill cycle
G79	Internal hole mill cycle
G80	Fixed cycle cancel
G81	Drill cycle
G82	Spot-facing cycle
G83	Deep hole drill
G85	Boring cycle (feed in, feed out)
G86	Boring cycle (feed, stop, and rapid out)
G87	Chip breaking cycle
G89	Boring cycle
G90	Absolute input
G91	Incremental input
G92	Establishing program zero point
G98	Feed in in./min
G99	Feed in mm/min

Miscellaneous Functions

Miscellaneous functions (M codes) are designed to control machine tool operations and will vary from turning on the coolant, turning off the coolant, to starting or stopping the spindle. Only nonconflicting commands can be used in one block of information. An example of conflicting commands would be M03 and M04. M03 is spindle on clockwise, M04 is spindle on counterclockwise. Again, it should be emphasized that not all machine tools use exactly the same miscellaneous function codes. Some of the more commonly used codes are as follow:

M Code	Function
M00	Program stop
M01	Optional stop (controlled by operator)
M02	End of program
M03	Spindle on clockwise
M04	Spindle on counterclockwise
M05	Spindle off
M06	Tool change (manual)
M07	Mist coolant on
M08	Flood coolant on
M09	Coolant off
M25	*Z*-axis home
M30	End of tape (rewind)
M98	Main program to subroutine
M99	Subroutine to main program

Sample Machining Program

The machining of an aluminum plate 0.300 in. thick, using a $\frac{1}{4}$ in. diameter end mill, and bearing milled cutouts and contours has been undertaken by writing the operations in a CNC format.

A programmer would see first the dimensioned print shown in Figure 76–2.

The next print, Figure 76–3, indicates the proposed tooling moves to machine the outside surface.

On the print shown in Figure 76–4, the programming codes have been entered in the appropriate places. These moves can be followed in the Sample Program for a CNC Milling Machine, which follows, where a line-by-line explanation of the coded instructions has been entered for ease of understanding.

Once the outside machining has been programmed, the planning shifts to milling the rectangular and circular slots and bores (Figure 76–5). The rectangular slots are machined first, starting with the N105 G00 X2 Y1 Z.05 macro command listed in the program instructions.

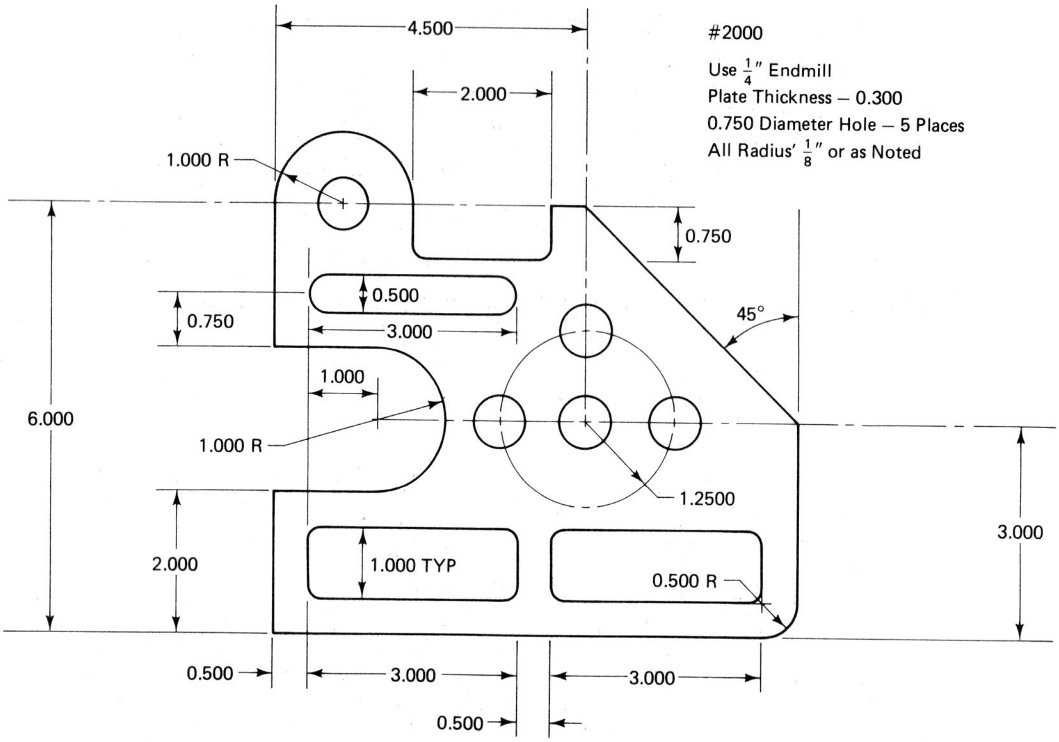

FIGURE 76-2
The challenge—to program the part to be machined to dimension. (Credit: Jim Washburn and Tom Burmeister)

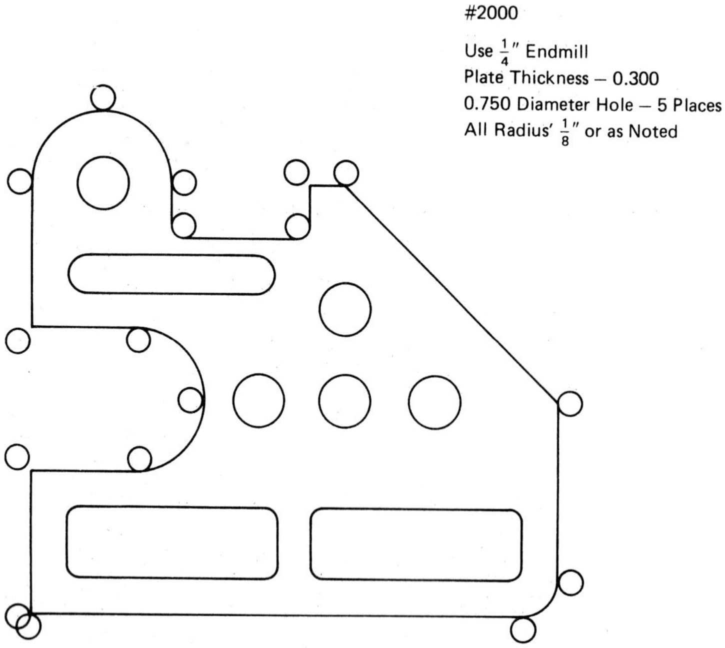

FIGURE 76-3
The proposed machining moves. (Credit: Jim Washburn and Tom Burmeister)

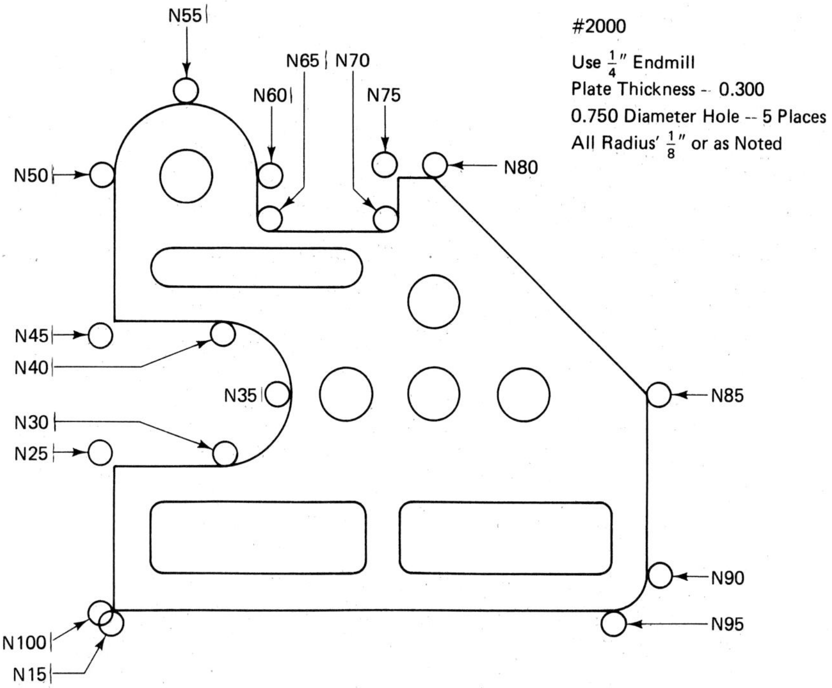

FIGURE 76-4
The programmed codes entered in the appropriate places. (Credit: Jim Washburn and Tom Burmeister)

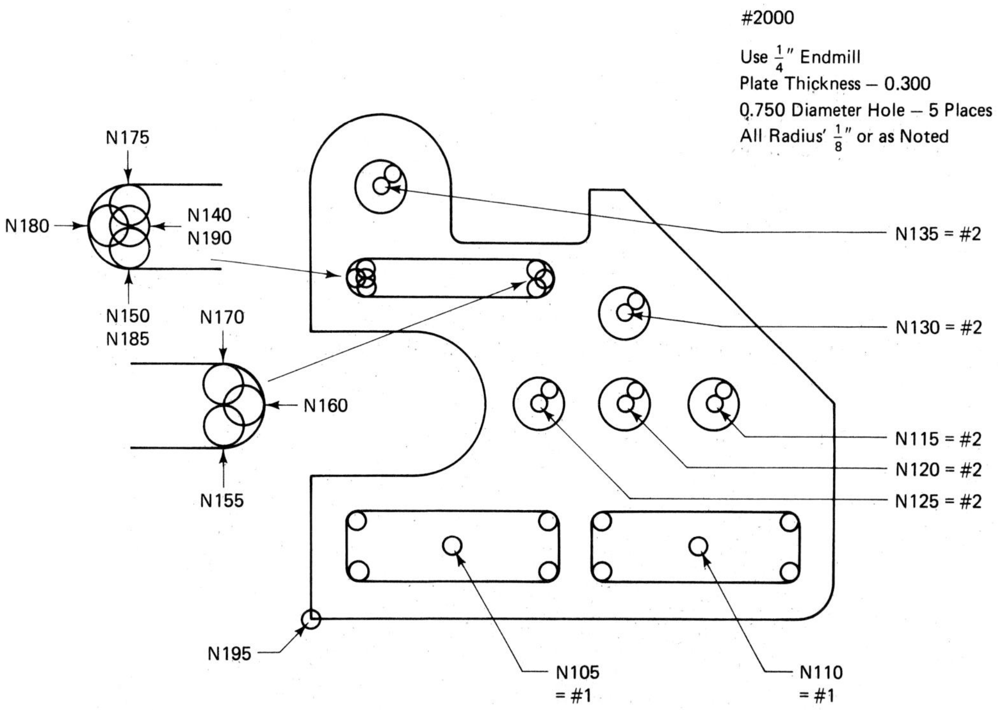

FIGURE 76-5
Programming to mill the rectangular and circular slots and bores.
(Credit: Jim Washburn and Tom Burmeister)

Sample Program for a CNC Milling Machine

Programming for the Outside Surface

N10 G00 X0 T1 M6	(TOOL CHANGE)
N15 X-.125 Y-.2 Z.05	(RAPID MOVE XY FIRST, Z LAST)
N20 G01 Z-.25 F100	(LINEAR MOVE, FEED RATE 10 in./min)
N25 Y2.125 F150	(LINEAR MOVE, FEED RATE 15 in./min)
N30 X1.5	(LINEAR MOVE)
N35 G03 X2.375 Y3. J.875	(CCW CIRCULAR MOVE)
N40 X1.5 Y3.875 I.875	(CCW CIRCULAR MOVE)
N45 G01 X-.125	(LINEAR MOVE)
N50 Y6.	(LINEAR MOVE)
N55 G02 X1. Y7.125 I1.125	(CW CIRCULAR MOVE)
N60 X2.125 Y6. J1.125	(CW CIRCULAR MOVE)
N65 G1 Y5.375	(LINEAR MOVE)
N70 X3.875	(LINEAR MOVE)
N75 Y6.125	(LINEAR MOVE)
N80 X4.548	(LINEAR MOVE)
N85 X7.625 Y3.048	(LINEAR MOVE 45 DEG.)
N90 Y.5	(LINEAR MOVE)
N95 G02 X7. Y-.125 I.625	(CW CIRCULAR MOVE)
N100 G01 X-.2	(LINEAR MOVE)

Programming for the Rectangular and Circular Slots

N105 G00 X2. Y1. Z.05	(RAPID MOVE Z FIRST, XY LAST)
#1	(MACRO START WILL STORE INFO.)
N1 G01 Z-.3 F100	(LINEAR MOVE, FEED RATE 10 in./min)
N2 G78 X1.375 X.125 Y.875 Y.079 F120	(MILL POCKET, 1 LINE INFO.)
$	(END OF MACRO)
=#1	(MACRO CALL, WILL ACT ON STORED INFO #1)
N110 G00 X5.5 Z.05	(RAPID MOVE Z FIRST, X LAST)
=#1	(MACRO CALL, WILL ACT ON STORED INFO #1)
N115 G00 X5.75 Y3. Z.05	(RAPID MOVE Z FIRST, XY LAST)
#2	(MACRO, STORED INFO TO BE USED LATER)
N3 G01 Z-.3 F100	(LINEAR MOVE, FEED RATE 10 in./min
N4 G79 J.25 F80	(MILL CIR. BORE, 1 LINE INFO, FEED RATE 8 in./min)
$	(END OF MACRO)
=#2	(MACRO CALL, WILL ACT ON STORED INFO #2)
N120 G00 X4.5 Z.05	(RAPID MOVE Z FIRST, X LAST)
=#2	(MACRO CALL, WILL ACT ON STORED INFO #2)
N125 G00 X3.25 Z.05	(RAPID MOVE Z FIRST, X LAST)
=#2	(MACRO CALL, WILL ACT ON STORED INFO #2)
N130 G00 X4.5 Y4.25 Z.05	(RAPID MOVE Z FIRST, XY LAST)

=#2	(MACRO CALL, WILL ACT ON STORED INFO #2)
N135 G00 X1. Y6. Z.05	(RAPID MOVE Z FIRST, XY LAST)
=#2	(MACRO CALL, WILL ACT ON STORED INFO #2)
N140 G00 X.75 Y4.75 Z.05	(RAPID MOVE Z FIRST, XY LAST)
N145 G01 Z-.3 F100	(LINEAR MOVE, FEED RATE 10 in./min)
N150 Y4.625 F120	(LINEAR MOVE, feed rate 12 in./min)
N155 X3.25	(LINEAR MOVE)
N160 G03 X3.75 Y4.75 J.125	(CCW CIRCULAR MOVE)
N170 X3.25 Y4.875 I.125	(CCW CIRCULAR MOVE)
N175 G01 X.75	(LINEAR MOVE)
N180 G03 X.625 Y4.75 J.125	(CCW CIRCULAR MOVE)
N185 X.75 Y4.625 I.125	(CCW CIRCULAR MOVE)
N190 G01 Y4.75	(LINEAR MOVE)
N195 G00 XO YO M2	(RAPID MOVE Z HOME FIRST, XY LAST)
	(M2 GO BACK TO TOP OF PROGRAM)

UNIT 77 QUALITY CONTROL

For any company to successfully compete, it must produce high-quality products at low cost and also provide quality service to satisfy its customers. The high-quality production and service cannot be maintained unless engineering and manufacturing problems are solved and eliminated. The process to monitor the standard of quality productivity is done through examination and interpretation of inspection data that is taken from random samplings, the key to quality control. Standard analysis methods reduce product defects, which in turn increases productivity.

However, good quality control does not mean perfection. It means maintaining a level of uniformity and dependability, at low cost, with quality that will be supported in the marketplace by the customer.

Objectives
After completing this unit you will be able to

1. identify the founder of Japan's quality control program.
2. identify the purpose of SQC and how it is applied.
3. identify SPC and how it is used in a quality control program.
4. name two advanced systems of quality control inspection and how they operate.

The Third Wave

If the Industrial Revolution that began in England with the mechanization of the textile industry was considered the first wave in mass production, and Henry Ford's labor saving methods was the second wave, then Dr. W. Edwards Deming would be considered the founder of the third wave. Deming, who at one time was an adviser in statistical sampling to the Bureau of Census in America, developed the method of analyzing and interpreting engineering and manufacturing statistics to determine what was wrong with a manufacturing system.

By invitation in 1950, Deming was advising 50 leading Japanese industrialists on how to transform their output from what was considered poor-quality products into products of quality. The Japanese, who were having trouble breaking into foreign markets at the time, were more than skeptical at Deming's statements that his methods of statistical analysis could end their manufacturing problems in five years. His prediction was wrong; it would only take four years, and Deming would go on to be acclaimed as the father of Japan's quality control program.

Among the benefits that could be compiled with strong quality control methods, Deming declared, would be:

- Uniformity of product
- Reduced cost
- A decrease in reject rate
- Increased productivity
- A happier work environment

- Decrease in manufacturing defects
- Reduction in manufacturing costs
- Inspection costs would be reduced
- A smoother working company organization
- An improvement in company credibility

Quality Control

The term "control" consists of four important steps that are connected and endless in operation (Figure 77-1):

- Planning
- Examination
- Measurement
- Correction

As an example, the performance of a production department is evaluated by the quality, cost, volume of production, and delivery. All of these steps are repeated until the desired product can be made.

The quality control system is a means of producing products reliably, economically, and at a low cost. In order to implement a quality control (QC) program, the following conditions must be met.

- The management of a company establishes its policy and enforces its rules
- Cooperation and understanding of allied department heads must be had in order to have a successful QC program
- The QC department must be staffed with competent people (Figure 77-2).
- The cooperation and good will of all company personnel must be obtained.

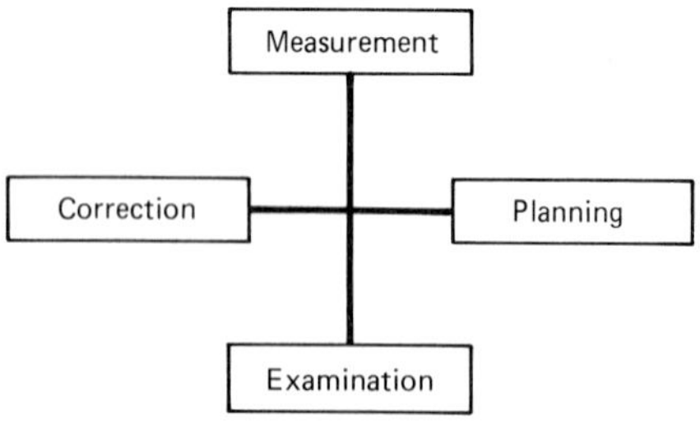

FIGURE 77-1
Quality control is an endless operation.

- A QC program must be carefully planned and supported by all concerned

Thus, it can be seen that in order to produce a quality product, a number of requirements must be met that go beyond the department of quality control. There is a combined effort, a dependence on other departments to cooperate to achieve a common goal. The activity can best be described as a total quality control. There are no shortcuts to good quality control. The principles of statistical quality control will have to be learned, adopted, and applied to the manufacturing conditions that exist.

Statistical Quality Control

The essence of statistical quality control (SQC) goes beyond establishing standard gage control procedures, calibration tests, or using varied measuring tools. Company standards have to take into account customer needs, production volume, and defective production costs. Records have to be maintained on efficiency. Reports that analyze defective operations have to be filed. Due to the nature of different procedures and their characteristics, the quality control engineer must work on the same level as a designer or manufacturing engineer in order to keep abreast of technology and product development.

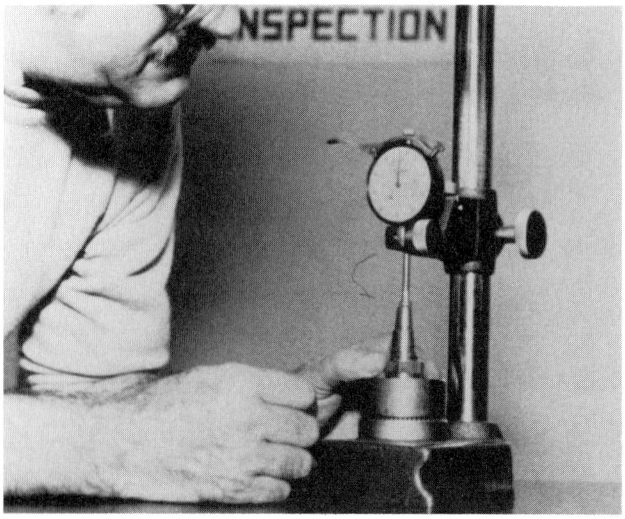

FIGURE 77-2
A quality control department must be staffed with competent people. (Credit: Photocom Productions)

FIGURE 77–3
The findings of measured random samplings are fed into mathematical tables by SPC in order to project acceptance or rejection of the lot in question. (Credit: DoAll Company)

Statistical quality control is designed to intercept flaws and defects in high-volume production. By taking random samplings of products at key stations located throughout the manufacturing cycle and checking them against the print specs for dimensions, finish, and/or hardness, for example, quality control is able to project the acceptance or rejection of the lot in question (Figure 77–3). Statistical quality control is done using mathematical tables establishing the laws of probability. The size of the sampling is generally based on the number of defective parts a company is willing to tolerate. Their in-process system is just as important as their final inspection and testing system.

Quality control will also act as liaison (go between) in reporting customer complaints to management and troubleshoot the area that a company's product fails to live up to. Their management also extends into other areas such as a company's supplier. Purchased items also have to live up to a predictable level of uniformity and dependability, at low cost, that is suitable in quality to a company's needs.

Statistical Process Control (SPC)

Quality control is based on deed. It is the control of

a product *after* it is made. It is also the quality of the element that goes into the assembly of a product. In general steps the part is tooled for production, machined, heat treated, and just before it is assembled it is tested for dimensional accuracy. It is there, at testing, that the part is placed in one of three piles: "acceptable," "reject," and "rework."

Statistical process control, on the other hand, is involved in quality assurance. It measures the process and not the part. The whole idea of statistical process control is found in the words, abbreviated SPC.

The letter *S* stands for statistical, the science of statistics (numerical facts). Through accepted formulas, random measurements taken can be evaluated. The measurement may be taken in a certain order or use a variation that combines random or patterned measurements. While each process is being made, an individual can determine ahead of time when a machine is going to start making bad parts. Since no machine or cutting tool will last indefinitely, what SPC can do is to tell the operator far in advance when the replacement has to be done. It is possible to know at what part, at what time of day it has to be done, so that a bad part is never processed.

The letter *P* stands for process, the system of manufacturing. Statistical process control is one of the ways used to find out the capability of the machine that is going to be worked on. For example, an operator may have a need for a machine to hold a 0.001-in. (0.03-mm) tolerance on a hole. Through SPC, the operator can discover without making a part or any scrap whether that machine can handle making that workpiece.

The letter *C* stands for control, either in the form of human intervention or through automation. Devices of control are used in manufacturing, which is part of the process. The control can be a sensor that advises the operator that the machine requires an adjustment or it can be in the form of a signal, a digital signal. An electronic impulse may be sent to the controller of the NC machine and automatically adjusts or tells it to change its tool or shut the machine down, speed it up, etc.

Since most parts are measured after they have been made during the quality control process, a certain amount of scrap will be made. On a bad day a large amount of scrap will be had. The theory behind SPC is that the process is being continually checked so that a bad part will never be made (Figure 77–4).

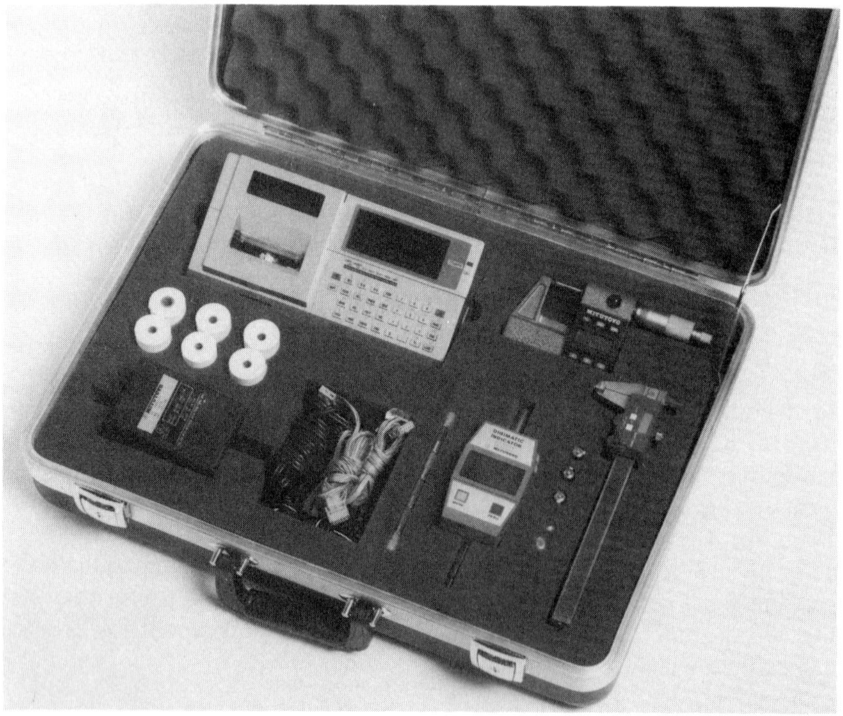

FIGURE 77–4
This kit includes a micrometer, caliper, and indicator that will accommodate most SPC measurement operations. (Credit: MTI Corporation)

Statistical Quality Assurance

Data is collected and recorded in over a dozen different diagrams and charts as well as different systems in statistical quality assurance (SQA) programs. Only a few of the more widely used methods will be discussed at this time. X-bar R ($\overline{X}$-R) charts and histograms are two of the most commonly used methods of recording quality and production conditions because they are easily recognizable by machinists and are the easiest to use.

X-bar R

The $\overline{X}$-bar R control diagram is a mean average chart. Early information was collected and marked in pencil on the chart in the form of a dot by the operator as quality data. As an example, a machinist or machine operator would take a measurement of a part after it had been machined in a certain work station, say, a precise hole boring operation. Once bored, every fifth part would be measured, with the data marked on a printed chart as a dot between two lines. An upper line drawn on the chart would indicate the upper tolerance of the part, while a lower line would represent the lower tolerance. The difference between the two lines would represent the R range, with the series of dots connected between these two lines. As long as these little dots stayed within the lines, everything was fine. But a pattern would be indicated if the dots started moving their way up or down from the center of the ideal measurement. Once a pattern began, which was usually four dots in a row climbing or four dots in a row dropping, a warning was indicated to the operator that the system was either tending to go out of process or was already out of process. The operator could tell when to anticipate the first bad part and when it was out of tolerance, just by the way the dots were positioned on the chart (Figure 77-5).

Industry found this method to be cumbersome. The machinist was spending more time drawing dots and connecting them than manufacturing. In order to speed up the process, automated data gathering is done at various points in the system. There is a gage or a sensor that can mark the data itself by sending a signal to a data gathering box or a data processor (Figure 77-6). That signal can go directly

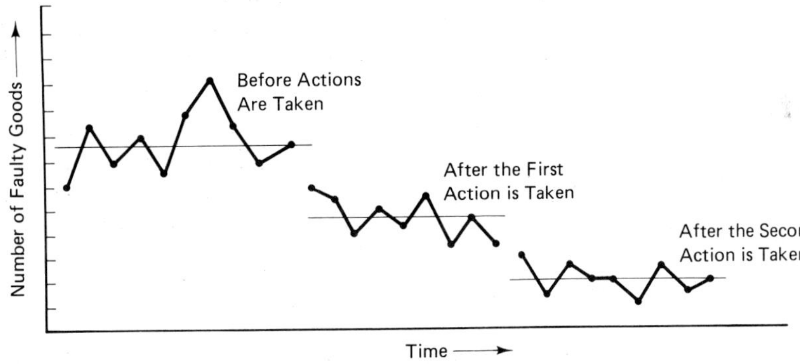

FIGURE 77-5
An early dot-to-dot chart that was penciled in and connected by the operator. (Credit: MTI Corporation)

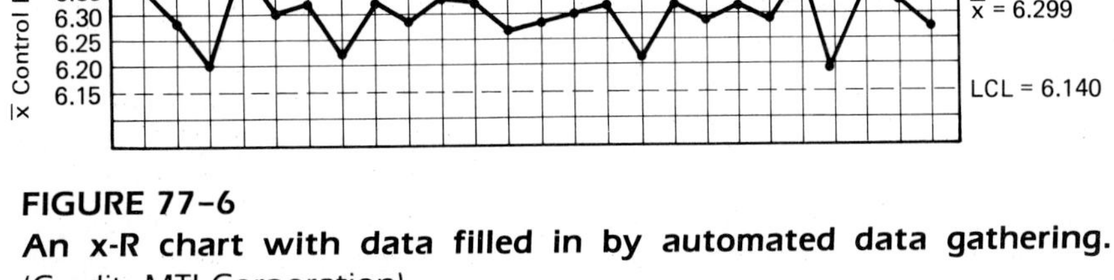

FIGURE 77-6
An x-R chart with data filled in by automated data gathering. (Credit: MTI Corporation)

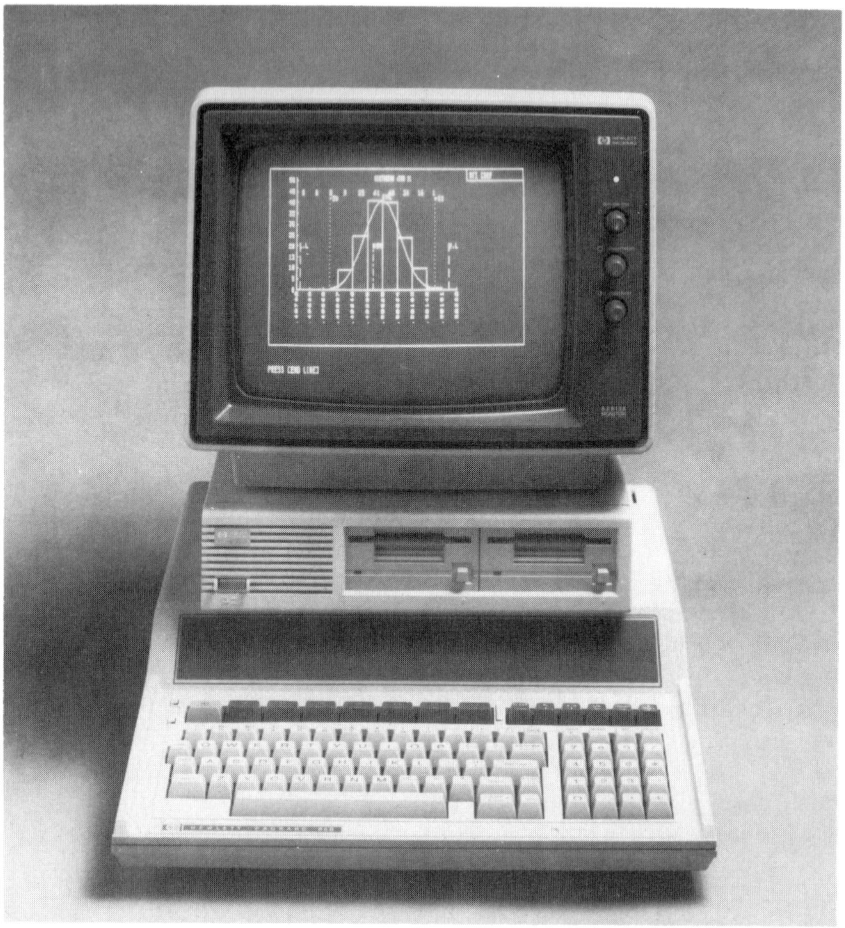

FIGURE 77–7
A histogram that has been superimposed over a bell curve using an SPC system linked to a computer. (Credit: MTI Corporation)

to a computer, like a personal computer, and the information can also be received by a series of computers that form a large network of computers.

Histograms

A histogram is a chart that puts the readings in a bar chart form as opposed to dots on a long chart. The histogram gives the individual a distribution based on the bell curve. In this case the bell curve represents the statistical average in manufacturing and creates a visual presentation of the allowable measurement range. If the data is inside the bell curve drawing, the operator is within tolerance or a certain degree of tolerance. The further out the data appears near the outside of the bell curve, the bottom part of the bell, the further away the operator is from the desired center of dimensions. The histogram is superimposed over this bell curve (Figure 77–7). Ideally, the operator wants this histogram to fit within that bell curve. When the operator starts straying outside the bell curve, one way or another, it tells the operator what is happening in the process. A lot of information is gathered from the histogram. Basically, the operator needs to know that the process is in or out of tolerance, or when the operation is starting to drift in or out of tolerance so that corrective actions can be taken.

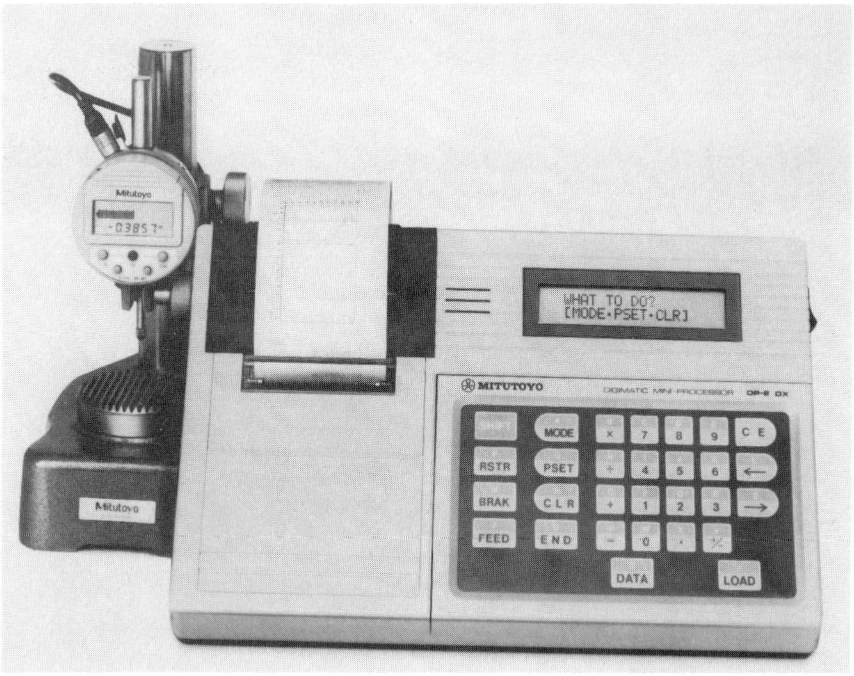

FIGURE 77–8
An electronic indicator with SPC output. (Credit: MTI Corporation)

Like the $\overline{X}$-R chart, the drawings of histograms have been connected to advanced electronic equipment such as a personal computer, which has a software program that involves statistical manufacturing data. An operator simply takes a measurement reading, pushes a little button on the electronic gage, and sends a signal to the computer. Once that signal is sent out, the computer digests the information and automatically gives the machinist the basic interpretation of the histogram. The purpose of the electronics is to speed up a system, make it more precise, more accurate, and be able to read performances in processes faster. This is very necessary and ideal in processes that are making parts at a speed of one every second or one every few seconds. More and more companies in industry are demanding electronic SPC output because human beings using a pencil on a chart can make mistakes (Figure 77–8).

Statistical process control also allows its user to obtain the greatest efficiency from the machine's performance. As an example, say a machine's ideal rpm rate is thought to be 1000 rpm. A machinist may find out through SPC that the machine may actually run better at 1100 rpm, speeding up the process and still make acceptable parts. Conversely, the machinist may also find the machine's rpm to be too high. The required speed should actually be 960 rpm and running the machine at 1000 rpm will dramatically increase scrap.

The purpose of SPC is (1) to eliminate scrap, (2) control the process by determining the machine's capability, and (3) reflect the operator's performance.

The Inspection Process

A support for the NC machine tools has been made possible by the use of the coordinate measuring machine (CMM). Coordinate measuring machines cannot only inspect first-run components more rapidly but can inspect them 100% in one setup without human intervention, reducing manual inspection by some 80%. They can also provide permanent records of inspection findings when equipped with a small line printer. Available in either manual or automatic (computer-assisted) models, CMMs (Figure 77–9)

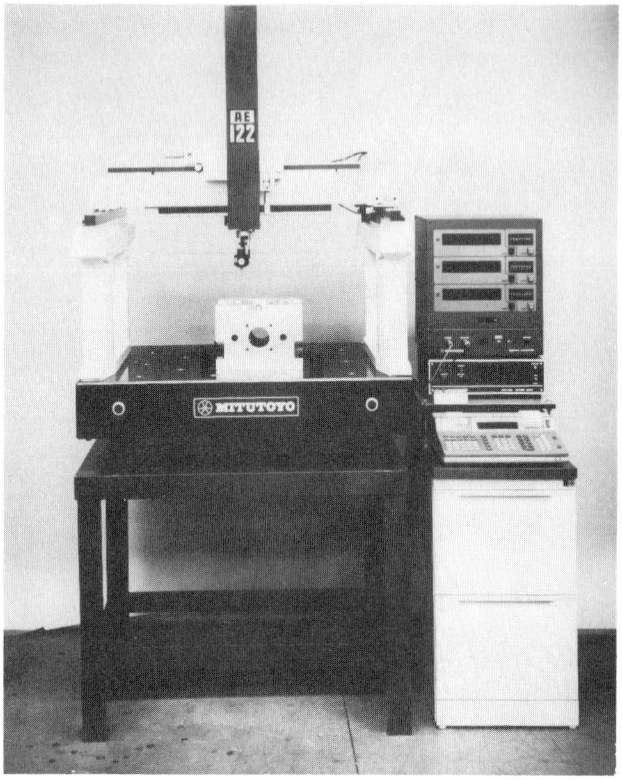

FIGURE 77-9
CMMs are available in either manual or computer-assisted models. (Credit: MTI Corporation)

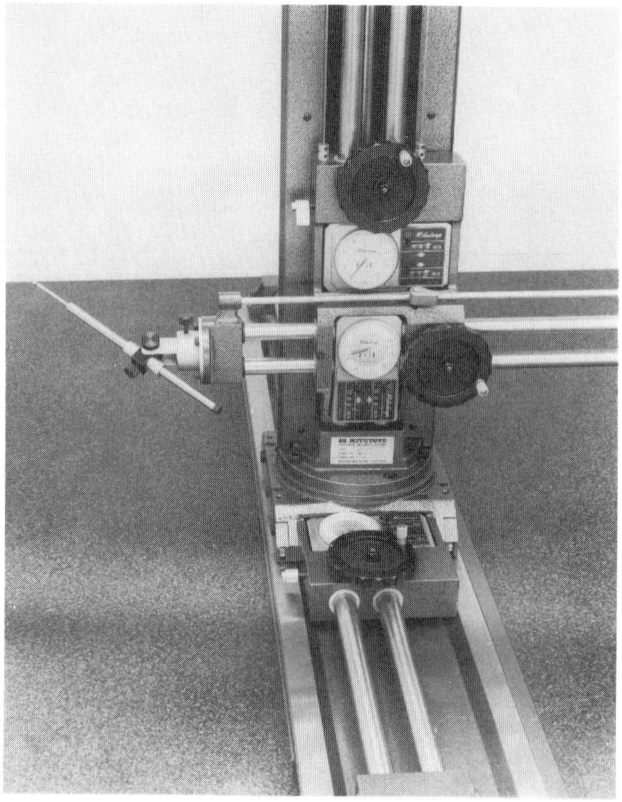

FIGURE 77-10
CMMs can also be used as very precise layout machines. (Credit: MTI Corporation)

and can also be used as a precise layout machine (Figure 77-10.)

Several manufacturers make these measuring centers, which come in a wide range of sizes and represent a very sensitive and accurate measuring system. The surface upon which all measuring takes place is a very large granite table that is below a moving gantry that accurately locates a multisensing probe head anywhere in the work area without any interference. It is this freedom of movement that allows the workpiece surface to be measured on three axes by the probe's head electronic sensors. The sensors may be of different shaped tips to accommodate a variety of surfaces. The resultant *X*, *Y*, and *Z* measurements are recorded to establish a common mathematical point that is set to a digital readout display of zero. This information is stored for use with a high-performance computer system whose video displays numeric results (Figure 77-11), functions, as well as other data and a high-speed hard copy printer.

The coordinate locations (*X*, *Y*, and *Z*) of any workpiece is begun by placing it in any position on the granite table. By using two or three electronic surveying points each have an extremely high and accurate resolution (clear picture), computer-assisted probing is capable of producing a three-dimensional crossing of line drawings on a video screen whose total error is less than the thickness of human hair at 10 ft (3.05 m). Testing can be done in a manual mode or by computer-assisted movements with the sensor contacting the workpiece surface to produce an impulse (signal) that locks in the common starting point. Once the workpiece has been tested on a number of surface points and the information has been recorded, the workpiece is mathematically aligned and all points are then referred to the established workpiece coordinate system.

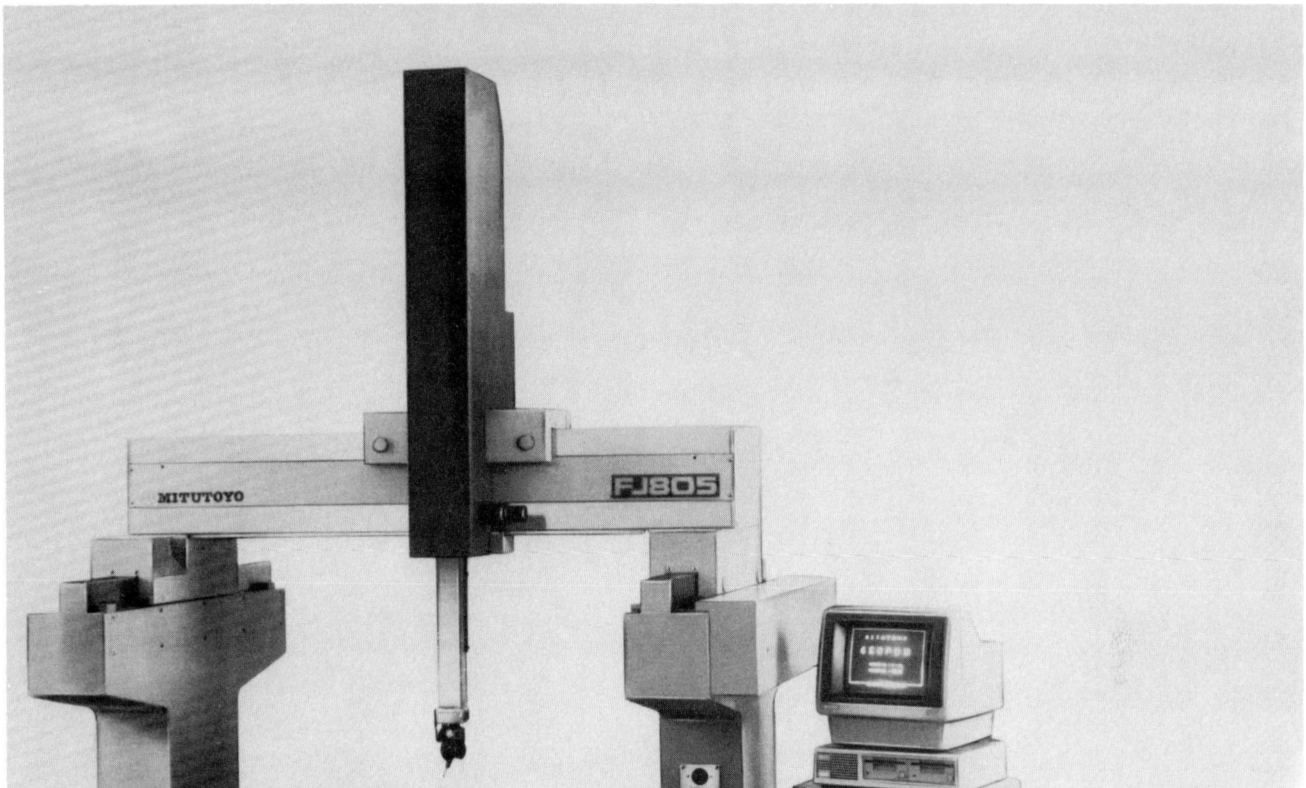

FIGURE 77–11
Computer systems provide video numeric displays for the *X, Y,* and *Z* axes. (Credit: MTI Corporation)

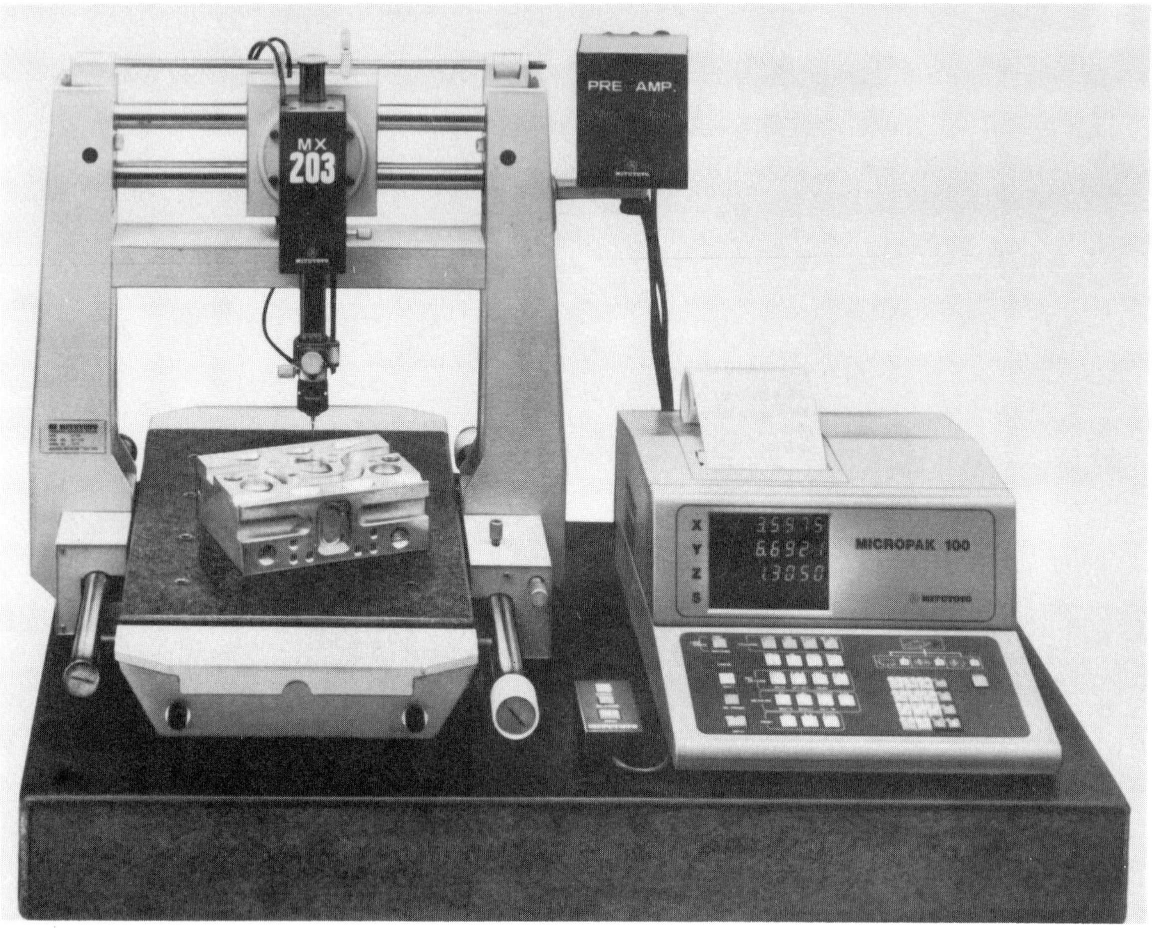

FIGURE 77–12
Reliable measurements can be taken, regardless of whether the part is very simple or highly contoured. (Credit: MTI Corporation)

The coordinate measuring machine has filled a need that existed between production technology and measuring technology to greatly reduce NC downtime. Highly accurate and reliable measurement data can now be made whether the part is very simple, very complicated, or spiral shaped (Figure 77–12). Advanced CMMs equipped with state-of-the-art electronics, optics, and lasers have revolutionized the inspection process.

Vision Machines

A product of the missile guidance system program, vision machines offer a high-speed, automated, and noncontact measurement, inspection, and alignment system that is capable of on-line (active production) or off-line usage. Much like the CMM, which features three axes movements, vision machines use high-resolution video cameras mounted on an inspection stage that enable them to digitize the video picture or signal on screen (Figure 77–13). Vision machines move with speeds and accuracy never before possible. Imagine an off-line system where an inspection table is able to move in an X or Y direction at a rate of up to 8 in. (200 mm) per second; furthermore, its X-, Y-, and Z-axes movements are guaranteed to an accuracy of 1 μm, $\pm$ 0.000040 in. ($\pm$ 0.001016 mm).

Vision machines operate by means of either edge finding or correlators, a system of template matching. Edge finders are used to detect both strong and weak edges by means of their varying shades of black to white or gray levels. Computer

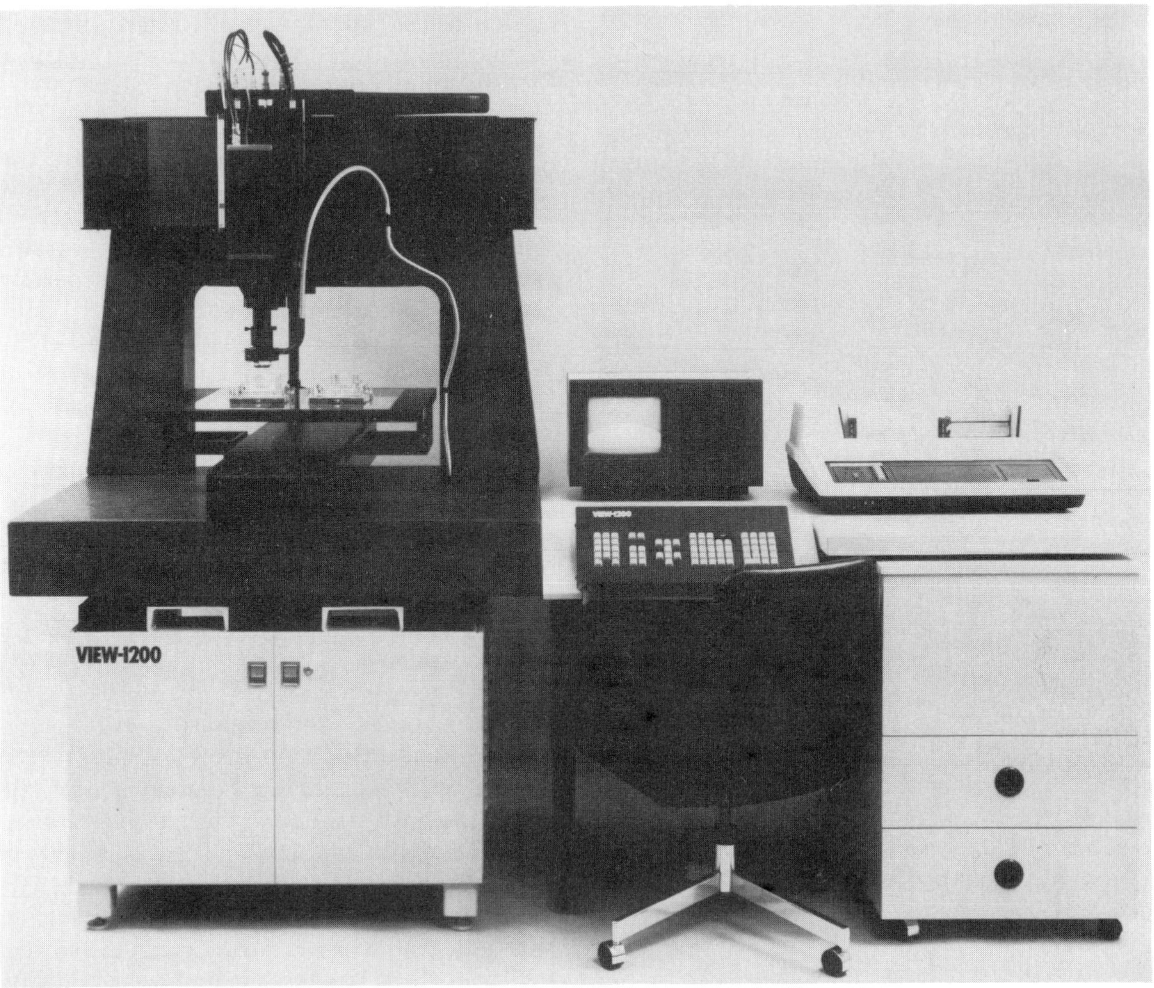

FIGURE 77-13
A vision machine can offer a high-speed, automated, and noncontact measurement system for either moving or stationary parts.
(Credit: View Engineering)

programming is used to store the reference picture. Once stored, the system will search and compare whatever view is present until the best match is accomplished. It is the correlator system that is used extensively to select, inspect, align, or position integrated circuits, for example.

The metrology of either technique is based on the video screen being divided into 512 × 512 lines that create a grid pattern of 262,144 squares. Each square is called a pixel (a picture element or resolution) and represents a mathematical value that is a function of the field of view; so if the field of view represents an area of say 0.050 in. (1.27 mm), each pixel would have a value equal to 1/262,144 or 0.000004 in. (0.000102 mm) of this area.

Inspection on off-line vision machines is begun by positioning the workpiece under the autofocusing, moving height video camera by command from a keyboard (Figure 77-14). Illumination and threshold levels (color shading) are controlled by the operator who may use through-the-lens top lighting or back lighting to eliminate false edge identification. The computer processor picks up the picture or edge signal. This information is then computed by the camera's location dimension, which takes into account the number of pixels from the edge of the digitized image to the edge of the object. The inspection pattern that follows can be entered manually on the keyboard (Figure 77-15) or can be recalled from a previously written program stored on a floppy disk.

FIGURE 77-14
Vision machines operate by means of an autofocusing, moving height video camera.

Ease of programming is assured by the listing of measurement processes in a menu that appears on screen. Menu programmable language (MPL) allows the operator to select the program entirely by menu simply by moving the system's joystick selector (Figure 77-16). Programs may also be written and edited, by line or character, if the operator chooses. In addition to color graphics, the field of view's measurement can be aided by a series of inspection tools such as radius, simple edge, alignment, line fit, as well as a min/max tool to establish the field of measurement. These tools can be positioned, rotated, or moved in combination by joystick to increase multiple inspections in a field of view.

Vision machines will make a measurement in about a second and while complex parts with many dimensions will take longer, they will dramatically reduce the time required by conventional methods. Equipped to register in either inch or metric, vision machines can also be equipped with a temperature compensation package to eliminate errors of differences in temperatures.

Reliability, repeatability, accuracy, speed, time, efficiency—vision machines play a complete role in an ever-changing manufacturing technology from the first production piece to final inspection.

There is a relationship between quality control and measurement control. Data for quality control can only be obtained through measurement. It is important that measurement be rapid and accurate. Without those two factors, data would only be a collection of worthless statistics.

FIGURE 77-15
Inspection patterns picked up by the camera can be entered manually on the keyboard.

FIGURE 77-16
Measurement processes are listed and appear on the video screen. These programs are selected by the operator by moving the system's joystick.

SECTION REVIEW

Unit 73

1. Name three early beginnings of numerical control.
2. Who was the first to use numerical information as a means of operating a machine tool?
3. Define numerical control.
4. What two developments led to the higher feed rates of numerical control machine tools required by industry?
5. What do the letters EIA and ASCII represent and why are they important?
6. Name five types of input media that has been used in NC machines.
7. What is the coded information contained in one block based on?
8. In punched tape "language," what do the numbers 1 and 0 represent?
9. Explain the role of a programmer.
10. To what accuracy can movement in the machine table slide position be?
11. What is the function of the feedback devices?
12. Name three common feedback devices.
13. Name the three principal components of a servo system.
14. What type of servo system is generally found on low-cost NC equipment?
15. How are servo systems that use feedback devices referred to?

Unit 74

1. What role did René Descartes play in the development of NC machining?
2. What is the intersection point of a rectangular coordinate system called?
3. Describe the X axis as it relates to a vertical NC milling machine.
4. Describe the Y axis as it relates to a vertical NC milling machine.
5. Describe the Z axis as it relates to a vertical NC milling machine.
6. Illustrate and letter the four quadrants of a two-dimensional coordinate system.
7. List four qualifications that a good programmer should have.
8. List five guidelines to NC programming.
9. Explain the letter address system.
10. On what is programming language based?
11. Explain the difference between modal and non-modal commands.
12. What relative position does the following command indicate: Z-0.500?
13. Explain how dimension is carried out in the absolute system.
14. Explain how dimensioning is carried out in the incremental system.

Unit 75

1. Define continuous-path machining.
2. Explain interpolation.
3. Name five different methods of CNC interpolation.
4. How accurate can linear interpolation be made?
5. Explain why a linear interpolator should not be used for curves, arcs, or slopes.
6. For what purpose is a circular interpolator used?
7. What five pieces of information are necessary to generate an arc when programming for the circular interpolator?
8. What three movements are created when programming a helical interpolator?

Unit 76

1. Define software.
2. Define canned cycle.
3. Define macro or subroutine.

4. What are G codes?

5. What are miscellaneous functions?

6. Name any three auxiliary commands that miscellaneous functions perform.

Unit 77

1. Who is said to be the founder of the third wave, the era of modern quality control?

2. List eight benefits that could be had from strong quality control methods.

3. What do the letters SQC represent?

4. What is the purpose of SQC? Explain how it functions.

5. What do the letters SPC represent?

6. Explain how SPC differs from quality control.

7. What do the letters SQA represent?

8. Explain how records are maintained using x-bar charts in SQA programs.

9. Explain how records are maintained in a SQA program using histograms.

10. What are the benefits of a SPC program?

11. What do the letters CMM represent?

12. Explain how measurements are taken using a CMM.

13. Briefly explain how vision machines make non-contact measurements.

14. To what accuracy can a vision machine's *X*, *Y*, and *Z* axes move?

15. What two methods are used for measurement detection when using vision machines?

16. Briefly explain the metrology of a vision machine.

17. What is the relationship between quality control and measurement control?

GLOSSARY

***A* axis** axis of rotary motion about the *X* axis

abrasive any material or substance used for grinding, polishing, lapping, smoothing, etc.

absolute positioning measurement taken from a fixed zero point on the machine or in reference to a point on the workpiece

acute angle an angle measuring less than 90°

allowance the intentional difference in dimensions between mating parts with regards to intended minimum clearance or maximum interference

alloy a combination of two or more metals that are melted together to form a new metal

amorphous carbon a carbon in which there exists no particular kind or uniform carbon arrangement

anneal a process to bring hardened steel to a pliable state or softening so it can be machined

ANSI abbreviation for American National Standards Institute

ASCII abbreviation for American Standard Code for Information Interchange

ASME abbreviation for American Society of Mechanical Engineers

ASTM abbreviation for American Society for Testing and Materials

austenite the solid solution of carbon in iron that exists between the upper and lower critical temperature

automation the technique of automatically controlling a system, an apparatus, or a process

axis the center or centerline about which a body or object revolves or rotates

backlash a condition that is caused by space, looseness, or play between interacting mechanical parts

bauxite a white to red earthy aluminum hydroxide; the ore from which aluminum is made

***B* axis** the axis of rotary motion about the *Y* axis

binary numbers a number system based on two digits, zero and one

Brinell measure of relative hardness of an alloy, normally used with soft material

brittleness tending to break or fracture easily without deformation

CAD an abbreviation for computer-aided design; an electronic-graphic system that is applied to the function involved in developing a product idea

calibration to determine, check, or adjust the amount of variation from absolute accuracy in a measuring instrument

CAM an abbreviation for computer-aided manufacturing; an electronic-graphic system that is applied to the functions involved in producing a product, including NC and DNC programming

carburizing a process where the carbon content of low-carbon steel is increased by heating the metal to a level below the melting point while in contact with carbon-rich materials

Cartesian coordinates a mathematical system that enables the location of any point along two or three perpendicular axes

C axis the axis of rotary motion about the *Z* axis

CBN an abbreviation for cubic boron nitride, a super hard abrasive.

CIM an abbreviation for computer-integrated manufacturing; a system designed for improving and maintaining overall efficiency in manufacturing

CNC an abbreviation for computer numerical control wherein a computer is used to store or perform some or all of the numerical control functions

cemented carbide a cutting tool material that is produced by a powdered metallurgy process where tungsten, titanium, or vanadium powders and cobalt are fused together under heat and pressure

cementite a combination of iron carbides and other carbides to add hardness and strength to steel

chatter noisy irregularities or vibration caused by a lack of rigidity in cutting tools, machine bearings, or parts

chip breakers grooves, notches, or similar features that are ground into and along cutting edges to produce small and uniform chips

cleavage a principal plane of division along which fracturing of a diamond or an abrasive grain takes place

climb milling see *down milling*

complementary angles two angles whose sum is equal to 90°

computer an electronic device capable of storing and retrieving information

concentric having a common center

contouring machining accomplished by control of one or more axes at the same time

conventional milling see *up milling*

crater a depression in the face of a cutting tool caused by the heat and pressure of the chip sliding over the cutting tool face

creep-feed grinding a full depth of cut grinding process that is completed by using very low table speeds in one or two passes

critical temperature the temperature at which certain changes take place in the chemical composition of steel during heating and cooling

cubic boron nitride a superabrasive manufactured by a high-temperature/high-pressure process from hexagonal boron nitride and nitrogen. It is the second hardest substance known, rated behind diamond in hardness

cutting speed the distance a given point on a drill or end mill cutting edge moves through the work, generally expressed in surface feet per minute (sf/min) or meters per minute (m/min)

datum a theoretical exact point, axis, or plane of reference forming the origin from which locations and measurements are taken

deflection the wavering or deviation of a cutting tool from its normal path of cut

DNC direct numerical control, whereby NC programming can be developed by means of direct inputs

down milling also called climb milling, where the milling cutter rotates in the direction of the feed and tends to pull itself into the work

downtime the waiting period or interval during which a machine tool, apparatus, or device is not working

drawing another term for tempering

dressing restoring a wheel's grinding face to expose new, sharp cutting abrasives.

DRO an abbreviation for digital readout; a digital measuring system that improves and maintains accuracy of machine tools and other *X-Y* positioning devices

ductility the property of a metal that allows it to be permanently deformed without breaking, such as drawing wire

dwell a momentary pause or delay in tool motion, such as in spot facing or counterboring, where a smooth surface is required

EDM an abbreviation for electrical discharge machining, a spark erosion process

EIA an abbreviation for the Electronics Industry Association

ELG an abbreviation for electrolytic grinding; an electrochemical process of removing metal in association with a metal-bonded abrasive wheel

flank the end of a single-point cutting tool that is located adjacent to and below the cutting edge when the turning tool is positioned horizontally

flute a helical groove or channel cut around the body of a drill that helps form the cutting edge, admits cutting fluid, and permits chips to escape from the hole

FMS an abbreviation for flexible machining system

friability a term describing the capability of an abrasive grain to break or fracture during the grinding operation, thereby exposing sharp abrasives

gage height a term used in NC programming to establish a tool's measured height above the workpiece

grindability the relative ease or difficulty with which material can be ground

hardness resisting permanent indentation or scratching

heat treating changing the internal structure or properties by means of heating and cooling metal

incremental positioning dimensional positioning that is made to a new point using the last previous point as a reference

interrupted cut cut in which the cutting edge(s) exit and reenter the work as in moving past a slot or groove

isotropic having one or more properties that are uniform in all directions

jig a device that holds and locates a workpiece and guides the tools that operate upon it

LCD an abbreviation for liquid crystal display

loading the filling or clogging of a grinding wheel's pores with chips from the material removed

machinability the relative ease or difficulty with which a given material can be cut

martensite the grain structure that results when transforming austenite by quenching and cooling rapidly

mesh size an internationally approved sieving method of measuring abrasive grain size

metallurgy the art or science of separating metals from their ores by smelting or alloying; the study of metals

metrology the science of precision measurement

modal an NC command that has been entered in the control and will remain in effect until cancelled or changed

nonmodal an NC command that has been entered in the control and will remain in effect for one block data only

NC an abbreviation for numerical control, a method used for operating machine tools and similar equipment by means of numerically coded commands

normalizing a process whereby steel is returned to "normal" by heating it above the upper critical point and allowing it to cool gradually in still air to room temperature

pearlite the microstructure of non heat-treated steel that is composed of alternate layers of ferrite and cementite

peck drilling a fixed cycle, during NC machining, of up and down movements along the *Z* axis to allow waste to escape when deep hole drilling

pitch the distance from the center of one thread or gear tooth to a corresponding point on the next thread or gear tooth

pitch diameter an invisible diameter formed in the thread's profile between the major and the minor diameter as it passes through a point where the width of the thread and groove are equal

polycrystalline a crystalline body whose composition consists of many tiny crystal particles that are positioned randomly

profilometer a stylus-type recorder that measures "hills and valleys" or surface roughness in terms of microinches

quenching a process of rapid cooling heated metal by immersion in various media

residual stresses internal forces found in a free metal state that may be caused by excessive heating during grinding, severe temperature changes as in quenching, or by chemical differences brought on by heat-treating processes

SAE an abbreviation for the Society of Automotive Engineers

self-sharpening the capability of a grinding wheel to continually expose sharp cutting abrasives through a process of micro- or macrofracturing during grinding

software a bank of information; stored intelligence

sparkout a condition where a grinding pass is taken over a surface without further downfeed to eliminate any previous material, grinder spindle, or wheel deflection

substrate the cemented carbide base that serves as support and strengthens the bonding of a polycrystalline superabrasive to its surface

superabrasive a general term to describe diamond and cubic boron nitride abrasives

superalloys a term defining alloys that have been designed specifically for high-temperature, highly stressed performances

temper the slow heating process that removes some of the extreme hardness, reduces brittleness, while it increases toughness in hardened steel; another term for drawing

tensile strength the maximum lengthwise pull that a metal can take without rupturing

thermal conductivity the ability of a material to dissipate or transfer heat

tool work interface the area of contact formed by the edge of the cutting tool and the workpiece

up milling also called conventional milling, it is a condition where the cutter is rotated in a direction opposite to the feed

wear resistance being able to resist surface removal through use

work harden a condition of increased strength and hardness in certain metals due to machining or forming operations

***X* axis** an axis of motion whose direction is either left (+) or right (−) on a horizontal plane and parallel to the table travel

***Y* axis** the crossfeed movement of a machine table perpendicular to the *X* axis, whose direction is either away from the machine tool column (+) or toward the machine tool column (−) and horizontal to the floor

***Z* axis** the plane of movement that is always parallel to the principle spindle of the machine tool

ANSWERS TO UNIT REVIEWS

Unit 1

1. Toolmaking would mark the beginning of human culture and the start of group behavior.
2. The Stone Age, the Bronze Age, and the Iron Age.
3. Thomas Newcomen and James Watt.
4. Henry Maudslay.
5. Eli Whitney.
6. Lightness, corrosion resistance, performance with high strength, reliability, and can withstand high temperatures.
7. They are able to develop the engineer's drawing into a physical part to see if it will work.
8. Technicians.
9. Metallurgists develop methods for extracting metal from the ore, refine it, and develop new alloys for better performance under varying conditions.
10. Industrial engineering.

Unit 2

1. Tensile strength.
2. Compression strength.
3. Elasticity.
4. Fatigue.
5. Toughness.
6. Brittleness.
7. Hardness.
8. Ductility.
9. Machinability.
10. Fusibility.
11. Basic oxygen furnace and the electric furnace.
12. Limestone acts as a cleansing agent and combines with the impurities in iron to form slag.
13. The basic oxygen furnace will take 45 to 50 min to make about 300 tons of steel.
14. The electric furnace.
15. The continuous-casting process.

Unit 3

1. Low-carbon steel, medium-carbon steel, and high-carbon steel. Low-carbon steel ranges from 0.15 to 0.30% in carbon content, medium-carbon steel ranges from 0.30 to 0.60%, while high-carbon steel ranges from 0.60 to 1.65%.
2. The carbon content of steel is generally referred to as points or hundredths of 1%.
3. Gray cast iron has a carbon content ranging between 2.00 and 4.50%. Its machinability is due to an excess of graphite, which acts as a lubricant.
4. One system is used by the SAE and one system is used by the AISI. The chief difference between the systems is that the AISI uses a prefix letter to indicate the steel making process.
5. Color coding and the spark test.

Unit 4

1. Nonferrous metals contain little or no iron.
2. Gold, silver, platinum, copper, aluminum, magnesium, lead, tin, and zinc.
3. *Copper* is ductile, malleable, noncorrosive, nonmagnetic, as well as containing high heat and electrical conductivity properties.

   Aluminum is light and strong, naturally high in heat and electrical conductivity, nonsparking, nonpoisoning, and reflects light and heat.

   Magnesium is too soft to be used in a pure state but is one-third lighter than aluminum, noncorrosive, and will burn with a fierce, intense light.

   Lead has little tensile strength; it also provides a protective shield against radiation.

   Tin is nontoxic, does not rust, and is highly malleable in its natural state. It is widely used as a plating (coating) material for other metals.

   Zinc is corrosion resistant and is widely used as a plating agent, such as in the galvanizing process.

Unit 5

1. An alloy is a combination of two or more metals that are melted to form a new metal.
2. Manganese, chromium, nickel, tungsten, molybdenum, vanadium, and silicon.
3. Stainless steels are made by adding large quantities of chrome and nickel. The two general grades include the 300 and the 400 series.
4. To alter and strengthen the base metal that might be too soft or subject to rusting or corrosion conditions.
5. Brass and bronze.
6. Aluminum–bronze does not contain tin.
7. Monel and inconel.
8. Babbitt, solder, and art metal.
9. Tin-base babbitt reacts well with materials that are used to make parts that move and create friction, such as cranks, axles, and shafts. Lead-base babbitt make bearing parts that will be subjected to strong mineral acids.
10. Magnesium, silver, zinc, manganese, copper, and silicon.

Unit 6

1. The inch system and the metric system.
2. The wavelength of monochromatic light from krypton 88 gas.
3. SI or *Système International d'Unités.*
4. The meter.
5. One ten-millionth of the distance measured from the North Pole to the equator as it traveled through Paris.
6. There is no relationship between units in the inch system. In the metric system units are related to each other by multiples of 10.
7. The inch system uses Fahrenheit; the metric system uses Celsius.
8. 0.000,001 in. is the value of the microinch.
9. Milli = one thousandth, centi = one hundredth, deci = one tenth, kilo = one thousand, mega = one million.
10. The liter.

Unit 7

1. The edges are divided into eighths, sixteenths, thirty-secondths, or sixty-fourths.
2. Metric rules are graduated in one millimeter and one-half millimeters.
3. Divisions of $\frac{1}{64}$ in.
4. The rule should be held on a straight edge, about 90° to the measured surface.
5. The rule should be placed so that the 1-in., or 1-cm, line is used as a reference point.
6. The spring joint caliper with adjustment nut and the hand-controlled firm joint caliper.
7. Outside calipers are usually set against a rule. In setting the calipers, one curved leg touches the end of the rule while the other leg is adjusted so that its edge splits the intended measurement graduation.
8. Inside calipers are used for measuring slots, holes, and similar openings.
9. Place the calipers over a rule, with one leg butted against a common stop. The other leg is adjusted on a parallel plane so that its edge splits the graduation line indicated.

10. The vernier caliper, dial caliper, and the electronic caliper.

Unit 8

1. There are 25 threads per inch on a standard inch micrometer.
2. Each line on the sleeve represents 0.025 in.
3. Each numbered line on the sleeve represents 0.100 in.
4. Each line on the thimble represents 0.001 in.
5. When a line on the thimble matches a line on the vernier scale, this number is added as the fourth place to the regular reading.
6. There are 50 measuring threads on a standard metric micrometer.
7. The value of each division on the sleeve above the index line is 1 mm.
8. The value of each division on the sleeve below the index line is 0.5 mm.
9. The value of each division on the thimble of a metric micrometer is 0.01 mm.
10. (a) Back off the thimble to expose the adjustment ring. (b) Insert the spanner wrench into the notched opening. (c) Turn the adjustment ring clockwise to remove all play between the threads. (d) Close the micrometer to check that the friction thimble or ratchet stop can advance the spindle freely.

Unit 9

1. Carl Johansson.
2. Alloy steels are seasoned in repeated cycles of deep freezing and extreme heat to prevent warpage and growing.
3. Class 1 set (laboratory or master set) is used for checking the accuracy of inspection and measuring gages. Class 2 set (inspection gage blocks) is used for setting and measuring inspection tools and manufactured parts. Class 3 (working set) is used for a variety of common machine shop uses, such as layout and surface plate measurement.
4. (a) Wipe the contact surfaces very clean. (b) Hold the bottom gage block firmly and slide the top block forward with slight inward pressure. (c) Continue wringing blocks together until the total buildup has been accomplished.
5. They should be rested at room temperature before using.
6. (a) Testing and calibrating measuring tools. (b) Set the accuracy of height for tools. (c) Inspect fixed gages to determine wear or warpage. (d) Establish machine tool setups. (e) Aid in precision layouts. (f) Aid in inspection of fixed workpiece locations. (g) Set the accuracy of sine bar or sine plates.
7. (a) Keep gage blocks in a closed box. (b) Avoid moisture between contact surfaces. (c) Avoid nicks or damaging by careless handling. (d) Eliminate dust or fingerprints on contact surfaces. (e) Wipe gage blocks with a solvent to clean oily surfaces. (f) Store in a room free from moisture and direct sunlight. (g) Protect blocks with a film of antirust oil before storage.
8. (a) They do not rust or stain when handled. (b) They will not be affected by temperature changes. (c) They are nonmagnetic. (d) They will not distort. (e) They require no special maintenance. (f) They do not require special storage consideration. (g) They will not absorb moisture, dust, or abrasive particles.

Unit 10

1. Fixed gages are simplified precision tools and supply quick, accurate, and easy-to-understand devices.
2. Fixed gages are made to an accuracy of 10% of the product's total tolerance applied in the proper direction to the part tolerance.
3. The inch dimensions are always marked to at least three decimal places, four places when called for. Millimeter markings are expressed to at least two decimal places; three decimal places are stamped as is.
4. The go gage is made longer than the not-go gage. Holders are often color coded.
5. A plain ring gage.
6. The major diameter, the minor diameter, the thread angle, the lead, and the pitch diameter of a thread are all checked with a thread gage.
7. Pin gages.

Unit 11

1. It can measure an angle to 5 min of an arc (one-twelfth of one degree).

2. Laboratory master angle gage blocks have a ± one-fourth-second accuracy. Inspection grade angle gage blocks have a ± one-half-second accuracy. Toolroom grade had a ± 1-second accuracy.

3. A sine bar rests on two accurately ground and lapped steel rollers of exact diameters. The center distance of the rollers may be 5 or 10 in. and are in precise alignment and parallel with the top surface of the sine bar.

4. The right triangle and the sine trigonometric function of one of its angles.

5. $$\text{Sine } C = \frac{c}{a}$$

$$\text{Sine } 25° = \frac{c \text{ (gage block height)}}{a \text{ (sine bar)}}$$

$$0.42262 = \frac{c}{5 \text{ in.}}$$

$$c = 5 \times 0.42262$$

$$c = 2.1131 \text{ in.}$$

6. $$\text{sine } C = \frac{c}{a}$$

$$\text{sine } C = \frac{c \text{ (gage block height)}}{a \text{ (sine bar)}}$$

$$\text{sine } C = \frac{2.4240}{5}$$

$$\frac{2.4240}{5} = \text{sine } 0.4848$$

$$= 29°$$

7. The compound sine plate consists of two sine plates, one above the other in order to create angles in two directions at the same time.

8. The solid square and the beveled edge square are commonly used in the machine shop; the cylindrical square is generally used in inspection work.

Unit 12

1. Comparison gaging entails the measurement, transference, and comparison of the measurement with a known standard or graduated measuring device.

2. An adjustable snap gage resembles a large C frame that contains one stationary anvil and two adjustable pins or button anvils whose limits may be preset by gage blocks as standards, locking the adjustment screw with sealing wax.

3. A feeler gage.

4. Adjustable parallels can be used for measuring purposes, layout purposes, as well as parallels in machine tool applications.

5. The precision height gage provides a quick and accurate method of setting any comparative height within the range of the many ground and lapped measuring steps or disks, which are spaced at measured intervals.

Unit 13

1. Adjustable gaging is another form of comparison gaging, one which visibly registers any change or variation from a set standard.

2. Indicating instruments, comparators, air gages, and electronic gages.

3. A standard micrometer has a fixed anvil and a movable spindle. An indicator micrometer has a movable anvil and a fixed spindle and is commonly used for inspection work.

4. The system's optical projector throws off a concentrated light beam that illuminates any object placed before it. The projector lens transmits and enlarges this image onto a projector screen as a highly accurate, dark shadow of the object being checked.

5. Optical comparators are ideal for testing alignment, squareness, flatness, as well as measuring small irregular shaped workpieces. Objects such as springs, plastic, and rubber are also easily checked.

6. Pneumatic gage.

7. Dimensional measurement is indicated by the escape of air through a series of holes in the periphery of a gage that registers against the workpiece surface. This air movement is compared to a setting or reading indicated by a master or known standard gage.

8. Free flow, back pressure, differential, or a combination of these principles.

9. (a) Adaptable to tool positioning. (b) Will not scratch the inside of a bore. (c) Multiple jet tooling can check a number of dimensions. (d) Adaptable for machine or bench setup. (e) Does not require operator skill for accuracy. (f) Long lasting due to reduced measurement friction. (g) Can machine gage from a distance. (h) Bores can be more easily checked than with mechanical hardware.

10. The lever type, the cartridge-type gaging head, and the reed-type electronic gage.

Unit 14

1. The purpose of surface measurement is to assure that parts will be made to work better as well as last longer. There is no economic advantage to having a part machined to a high-quality finish when it is not necessary.

2. It is a surface texture gage. By placing it alongside the workpiece surface, a comparison of the surfaces is made by the operator by drawing the tip of the fingernail across each surface at right angles to the tool marks.

3. 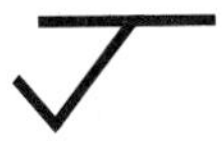

4. Lay: the direction of the predominant surface pattern. Flaw: irregularities such as cracks, blow holes, scratches, etc., which occur infrequently. Waviness: widely spaced irregularities found in the form of waves. Nominal surface: the theoretical accurate design that is shown and dimensioned on a drawing. Roughness: the relatively finely spaced irregularities that are narrower than the waviness pattern.

5. X: lay angular in both directions to the line. M: lay multidirectional. C: lay approximately circular relative to the center of the surface. R: lay approximately radial relative to the center of the surface. P: lay particulate, nondirectional, or protuberant.

6. Ra: roughness averaging. It is used to average the profile height deviation from a mean line. RMS: root-mean-square averaging. RMS is considered obsolete and is no longer used.

Unit 15

1. Interferometry is the scientific measurement of distance using the interference of two light waves as standards.

2. The science of optical flats.

3. The wavelength of a band of helium light is 23.2-millionths of an inch (589-millionths of a millimeter).

4. Optical flats vary from inexpensive glass to very expensive saphire to high-quality quartz.

5. (a) A straight and equally spaced fringe pattern will reveal that the measured face is perfectly flat. (b) Concentric rings that appear to move outwardly indicate the surface is dome shaped or convex. (c) Concentric rings that appear to move inwardly indicate the surface is hollow or concave.

Unit 16

1. The process of drawing lines, angles, circles, or patterns on metal and other materials for the purpose of guidance for material removal.

2. (a) Made of cast iron or granite, surface plates are extremely accurate in flatness and are used for layouts and inspection work. (b) Layout dye is a coating applied to metal or other material to produce a background that will eliminate eyestrain when drawn upon.

3. (a) An angle plate is a holding device for fastening work in an upright position. (b) Parallels are rectangular bars of steel that are uniform in height and thickness and are used for elevating work to different heights when laying out. (c) A V-block is a tool for holding round work. (d) A C-clamp is a simple C-frame, screw-type holding device. (e) A ball-peen hammer uses its flat face for striking center punches or other such work; its rounded, ball-shaped end is used for riveting or peening. (f) A surface gage is a tool with a base and adjustable scriber construction used for drawing semiprecise lines.

4. Holes and circles are scribed with dividers; larger arcs and radii can be drawn by the use of trammels. Irregular contours are drawn by the use of hermaphrodite calipers.

Unit 17

1. (a) A thorough understanding and transference of drawing dimensions, (b) a thorough under-

standing of the use of precision layout tools, and (c) a thorough understanding of mathematics and trignometric functions.

2. A vernier height gage, a dial height gage, and a digital readout height gage.

3. A vernier height gage consists of a base, a beam, a vernier plate, and a scriber.

4. By the use of an offset scriber.

5. The vernier principle of measurement uses calibrated lines on adjoining surfaces that can be read when two lines match to form one long line. To read, first examine the zero line on the vernier plate, record the beam value below it, and add to it the value of the vernier plate determined by matching lines.

6. The workpiece must have two adjacent sides that are finished at 90° to each other, from which the X and Y coordinates will be drawn to form the intersection of lines. The X represents the horizontal plane, the Y represents the vertical plane set at 90° to the X plane.

7. Sine bar and gage blocks.

Unit 18

1. The American Society of Mechanical Engineers are responsible for establishing engineering standards and codes.

2. ANSI is an abbreviation for the American National Standards Institute, which serves as a clearinghouse that publishes some 9000 approved standards and codes.

3. Lines that extend at 90° from a surface in two or more planes so as to represent the exact shape of the object.

4. A drawing that contains the correct shape, exact dimensions, and related information such as heat treatment, finish, and material specification necessary for the purpose of fabricating a part or object.

5. A sectional view.

6. (a) Isometric view: object is drawn around three major lines. Base angles to the right or to the left form 30° angles, the vertical axis forms a 90° angle. (b) Pictorial view: object appears in one view while revealing two or three of its surfaces.

7. (a) A visible line or object line. (b) A hidden line or an invisible edge line. (c) A cutting plane line or a viewing plane line. (d) A section line or a cross-hatch line.

8. (a) Basic dimension: the theoretical size; (b) limits of size: permissible maximum and minimum sizes; (c) tolerance: amount of variation between the upper and lower limits; (d) dual dimensioning: a dimensioning system where both the inch and metric units of measurement are shown; (e) bilateral tolerance: the total tolerance in two directions, including plus and minus dimensions; and (f) datum: a theoretical exact point, axis, or plane of reference forming the origin.

9. The title of the part, date drawn, and by whom, choice of material, stock size, heat treatment, scale of drawing, and tolerance listing.

10. BHN, Brinell hardness number; CDS, cold-drawn steel; CBORE, counterbore; CSK, countersink; HDN, harden; IN, inch; QTY, quantity; RH, right hand; THD, thread; TYP, typical.

Unit 19

1. Safety is a matter of mutual concern and respect for yourself, your fellow worker, and the equipment you will be using.

2. Eye protection will either be (a) safety glasses with side shields, (b) a face shield, or (c) goggles.

3. (a) Long sleeves or loose clothing should be avoided. (b) Shop aprons should be properly strung and neatly tied. (c) Shop coats should not be baggy and must not have buttons or sleeves that can get caught easily. (d) Protective clothing must be worn in the welding, forging, or foundry areas.

4. Gloves may be worn to handle or transport sharp-edged materials. Gloves are never worn when operating machinery.

5. (a) Remove all forms of jewelry. (b) Long hair should be tucked under a cap or hair net. (c) Beards should be neatly trimmed. (d) Sneakers or sandals should never be worn; safety shoes are recommended. (e) Hearing should be protected against high noise levels with ear protectors.

6. Frayed or loose wires should be reported to your shop supervisor at once.

7. Oily rags and combustible waste materials should be stored in a metal container. Paints, oils, etc. should be returned to the proper metal storage area.

8. (a) Know where every fire extinguisher is located and how to use it. (b) Know the location of the nearest fire alarm box and how to use it. (c) Know the location of the nearest fire exit and the procedures to follow in case of a fire.

9. (a) Do not attempt to operate any machinery unless supervised or allowed to by your instructor. (b) Never attempt to operate or machine work without safety guards or shields in place. (c) No machine should be operated by two people. (d) Avoid touching surfaces of moving parts or moving machinery.

10. Never use a file without a snug-fitting handle.

11. Compressed air can drive and wedge particles between and damage moving parts, as well as pollute the air with oil, lubricants, and other contaminants.

12. When lifting, keep the back straight, bend the knees, and lift with the legs. When an object is too heavy, get someone to help you or use a hoist.

13. Report every injury to your shop supervisor no matter how slight the injury might be.

Unit 20

1. The hacksaw, the abrasive saw, the cold saw, the bandsaw, and the friction saw.

2. In the forward direction.

3. The pitch and set.

4. (a) The 14-pitch blade is used for cutting thick sections. (b) The 18-pitch blade is used for general-purpose work. (c) The 24-pitch blade is used for tubing and thin stock. (d) The 32-pitch blade is used for very thin walled sections.

5. Power hacksaws are either wet or dry cutting.

6. An abrasive cutoff wheel as well as carbide-tipped or high-speed steel blades.

7. Circular in structure, cold-saw blades are made of chrome–vanadium steel or bear carbide-tipped cutting edges.

8. They support the saw blade and prevent it from cutting through the material on a slant.

9. The base, column, and the head are considered the main parts of a contour bandsaw.

10. Selection of the correct saw and saw speed for the material being cut.

11. Friction cutting is faster than the conventional teeth method and cutting is accomplished through heat, softening the metal just ahead of the blade.

12. Heavier metals can be cut using a rocker technique.

13. Raker set, wave set, and straight set.

Unit 21

1. A blade or body and a tang.

2. Under no conditions should a file ever be used without a file handle.

3. Single-cut, double-cut, rasp, and curved.

4. Square, triangle, circular, and rectangular.

5. Needle files.

6. The elbow height of a filer's bent arm.

7. Sandwiched between and raised slightly above two blocks of wood to prevent bending and chatter.

8. Pinning.

9. File card.

10. Draw filing.

Unit 22

1. Changing a flat surface into another direction.

2. Bench vise, ball-peen hammer, a wrench, a form block, and some buffer material to eliminate hammer marks.

3. One-half the material thickness for every bend.

4. Thinning and stretching the metal.

5. Acute angles.

6. Obtuse angles.

7. Distortion and buckling.

Unit 23

1. Archimedes.
2. The nut and bolt.
3. Setscrews are used for locking gears or pulleys against a shaft. The setscrew is identified by its point and head style.
4. They provide a bearing surface, distribute the load over a wide area, protect the adjoining surface, and may lock assembled parts.
5. To position and prevent shifting or lateral movements between assembled or mating parts.
6. A key is a metal insert fitted between the shaft and hub or pulley to provide alignment and a positive drive.
7. Bursting can occur.
8. Thermosetting and thermoplastic or a combination of both.
9. Structural adhesive bonds are examined by destruction testing.
10. Conditions include shear, tension, peel, and cleavage.
11. Welding and riveting.

Unit 24

1. The Whitworth thread.
2. 60° V-type, square, buttress, and acme threads.
3. Hand cutting external threads on rods, shafts, and similar round work.
4. Machines and hand tools.
5. Adjustable round split die, two-piece die, and a solid rethreading die.
6. Major diameter and the number of threads per inch.
7. Counterclockwise.
8. Clockwise.
9. (a) Solid rethreading die. (b) Any wrench that will fit.
10. Screw plate set.

Unit 25

1. Teeth or threads, shank, and square for holding.
2. Flute.
3. T-handle and straight handle tap wrenches.
4. Tap drill size; 75% standard thread depth.
5. Taper, plug, and bottoming taps.
6. Plug tap.
7. Gun tap.
8. Spiral fluted tap.
9. Screw pitch gage.
10. Tapping square.

Unit 26

1. Grinding that is done by holding the object by hand against the abrasive wheel.
2. Aluminum oxide and silicon carbide.
3. The coarse wheel is designed for rapid metal removal. The fine wheel is used for smoothing or touching up.
4. They will clog the wheels and leave them unusable.
5. By wedging a piece of wood to hold the wheel while the wrench turns the nut in the same direction as the wheel cuts.
6. No more than one-eighth in. (3 mm) away from the grinding wheel, to prevent any object or work from jamming between the rest and wheel.
7. A spark guard.
8. The wheel should be mounted on a wooden dowel or handle and struck lightly with wood. A ringing sound indicates a good wheel. A dull sound indicates it is cracked.
9. Always stand to one side when turning on a grinder. A shattering wheel can kill anyone in its path.
10. Mechanical, abrasive, and diamond.
11. This mechanical dresser is also called a star dresser.

Unit 27

1. Cutting tools severed by pushing metal apart.
2. Concentrated forces are applied to the cutting edge(s). The many tons of pressure compress

and deform work material in front of the tool face, generating frictional heat that will plasticize material. Chip will either flow or rupture.

3. The shear angle in the cutting zone.

4. Discontinuous, continuous, and continuous chip with built-up edge.

5. A depression or wear zone on the tool face that enlarges in size with cutting, eventually causing tool failure.

6. (a) Safety hazard to operator, (b) are difficult to handle, and (c) may twist around and clog machinery.

7. (a) Production of short broken chips, (b) easier disposal of broken chips, and (c) produces a better flow of cutting fluid to the cutting edge.

8. The result of the tool flank coming in contact with the fragments freed from the workpiece and chip.

Unit 28

1. The ease or difficulty with which metals may be machined.

2. The microstructures and chemical composition.

3. Adding one or more contaminants to the alloying process at the mill.

4. (a) Sulfur or lead, (b) lead, (c) lead or bismuth, and (d) sulfur or silenium.

5. Lead is insoluble and forms leaded globules that tend to lubricate and act as chip breakers.

6. Excessive heat at the cutting edge and gradual tool wear.

7. (a) Compression and deformation of metal in chip forming, (b) the thickness of the chip, (c) high friction on the tool face.

8. Coolants reduce or remove heat in the cutting zone. Lubricants reduce the frictional heat generated in the cutting zone.

9. Turning very hard material and on heavy interrupted work surface cuts.

10. (a) Dull tools, (b) speeds and feeds, (c) heat generated during machining, and (d) machine conditions.

11. Unpleasant odors or skin disorders may erupt if in contact with them.

Unit 29

1. Center drilling, countersinking, counterboring, reaming, tapping, and spot-facing.

2. Bench-type or floor model sensitive drill presses.

3. Any chuck key so left can be thrown out with great force once the drill press spindle is switched on.

4. The operator's hand can be punctured or maimed this way.

5. On a gang drill machine a group of independent drill presses are placed side by side to perform a variety of operations. The turret drill press has many tool holding devices that can be revolved and repositioned quickly.

6. The radial drill press is used for large heavy work and castings.

7. (a) The base is the foundation of the drill press; (b) the table is the surface on which the workpiece or drill vise rests; (c) the spindle is the tool holding shaft; and (d) the drill chuck is fitted into the spindle and is the gripping device for holding different size drills.

Unit 30

1. High-speed steel.

2. The shank, the body, and the point.

3. The straight shank and the tapered shank.

4. They aid in eliminating chips, allow lubrication to enter, and help form the cutting edges.

5. The land is the surface between the flutes and acts as a body clearance to reduce friction during cutting.

6. The web is the surface thickness between the flutes.

7. Numbers, letters, fractional, and metric systems.

8. Twist drills are measured across the cutting edge to the full diameter of the margin, using a micrometer or a drill plate.

9. A standard drill chuck, a keyless drill chuck, a tapered spindle, or by means of a sleeve or socket.

10. The drill chuck should be tightened using all three holes.

11. A drill drift is driven against the shank end of the drill or holding device. To prevent the cutting tool from striking the table upon release, a wooden block should first be placed below it.

Unit 31

1. Cutting speed is the distance an edge on the cutting diameter will travel on a straight line in one minute.
2. Cutting speed plays an important role in a tool's life. If it turns faster, the tool is burned out. If it turns slower, time is wasted.
3. Feed is the distance a drill travels into the workpiece with each revolution. Feeds are governed by the material, the diameter of the drill, and rigidity of the machine and setup.
4. Carbide drills require a rigid work setup and the use of a machine tool free of conditions that can lead to tool chatter.
5. Drill depths not more than six times the drill's diameter.
6. Carbides will crack if suddenly cooled after heating up (thermal shock).

Unit 32

1. The drill may break or may not cut the material cleanly, resulting in the lifting of the workpiece as it spirals up the drill or rotating in the same direction that the drill rotates.
2. The drill vise, the V-block, the bench block, the bolt and clamp system, and the angle plate.
3. On the workpiece not the step block.
4. (a) Coolant, (b) lubricant, (c) surface finish, (d) prevent rust, (e) chip removal, (f) harmless to health, and (g) harmless to work or machine.
5. The bell-type center drill has a secondary angle larger than the 60° slope and is designed to protect the edges of the center hole during production handling.
6. The hardened drill blank is placed and clamped exactly on the circumference of a hole that's been laid out and then used as a guide for drilling.
7. When drilling holes larger than one-half in. (13 mm) in diameter.
8. (a) The workpiece should be clamped to the table, (b) a pilot hole should be drilled, and (c) proper spindle speed and the proper cutting fluid should be used.
9. A pilot drill size should never be much more than the web of the final size drill to be used.

Unit 33

1. To enlarge and finish a previously drilled hole clean.
2. Hand reamers are designed to be turned by a suitable wrench and contain a square on the end of the shank for this purpose.
3. Solid and expansion-type hand reamers.
4. About 50% of the speed used for drilling.
5. This will ruin the cutting edge.
6. In separate containers and provided with a film of oil to prevent rust.
7. To enlarge the top of a previously drilled hole to a limited depth, generally to recess items like bolt heads.
8. To smooth and true the surface around the top of a hole to provide an area to seat washers, springs, nuts, etc.
9. To enlarge or produce cone-shaped holes, generally to accommodate flat head screws.
10. Maintaining tap and hole alignment. Once the tap drill has been used, the hole location should not be moved.

Unit 34

1. Wear, web thinning, and drill pointing.
2. (a) Greater pressure is required to penetrate the work surface; (b) greater heat is generated to shorten tool life; and (c) increased chisel edges can cause oversize diameters.
3. A 118° drill point angle with a lip clearance of 8° to 12°.
4. A 135° drill point angle and a lip clearance between 6° and 8°.
5. A 90° drill point angle with a lip clearance between 12° and 15°.
6. To produce holes to accommodate springs and other flat-bottomed hardware.

7. To prevent digging in and breaking.

8. (a) Control chip formation; (b) control size and shape of chip; (c) control of chip as it slides up flutes; (d) control size of drilled hole; (e) control amount of burr produced; (f) increases cutting lip strength; (g) reduces rate of wear; (h) reduces drill pressure; (i) reduces heat generated; and (j) permits a wider range of speeds and feeds.

Unit 35

1. The maximum diameter of work that can be turned or swung over the bed and the length of work turned between centers.

2. The bed, the headstock, the carriage, the quick-change gearbox, and the tailstock.

3. (a) A long tubular shaft that extends from the headstock, one end of which is mounted work holding devices; (b) part of the carriage, the saddle is machined and fitted to move along the ways of the bed; (c) responsible for moving the carriage; (d) provides cross movement for the cutting tool; and (e) support for the cutting tool, which can be set to an angle to permit taper cutting.

4. High-speed steel, carbide, cast alloys, ceramics, diamond, and cubic boron nitride.

5. The speed at which a lathe should be turning is referred to as the number of revolutions the spindle will turn in one minute (rpm). Cutting speed (CS) refers to the amount of material removal recommended for effective machining.

6. Feed is the distance a cutting tool will travel along the workpiece's length with each revolution.

7. To remove excessive stock rapidly.

8. A smoothing operation that brings the workpiece down to size and produces a good finish.

Unit 36

1. Three-jaw chuck, four-jaw chuck, six-jaw chuck, collet chuck, and Jacobs collet chuck.

2. Clean surfaces are required for mounting accuracy. The lathe should be disengaged from its electrical power to prevent any accidental turn on.

3. To prevent personal injury to the operator or damaging the lathe surfaces.

4. A key is used for alignment, while a threaded lock ring holds the mounted device.

5. Place a wooden block below the chuck, lock the headstock, and use a thick bar of steel or a heavy wrench for leverage to break the chuck loose.

6. Clean the jaws and jaw slides, insert the jaws sequentially, and close the jaws to test if they close on a common center.

7. With each jaw independent, the chuck can accommodate round, square, octagonal, and odd-shaped work to be set up and trued.

8. Using the concentric rings circling the chuck face, the chalk method, and surface gage method.

9. Dial test indicator method can true work to within 0.001 in. (0.02 mm).

Unit 37

1. Machining to resurface the end of a workpiece.

2. To establish a reference surface (face) for measurement, provide a surface 90° to the turned diameter, and to machine a workpiece to length.

3. Placing it slightly below center will produce a stub of metal; slightly above center will raise a bump on the middle of the surface.

4. The compound rest is set parallel to the lathe ways or it may be set 30° to the right.

5. It should not extend more than three times the diameter of the workpiece.

6. The tool can be fed from the center outward or may be fed from the circumference toward the center.

7. Alignment of centers, an undercut half center, proper work setup, lathe speed, and facing tool setup.

8. The end of the cutting tool should be slightly rounding and honed with an oilstone.

9. It will prevent the center hole from being burred in.

Unit 38

1. Micrometer collars have either a 1 : 1 or 2 : 1 relationship. The 1 : 1 is equal to the number of

graduations registered. For the 2 : 1 the total movement is twice the amount of graduations registered.

2. A straight or left-hand toolholder.

3. Not more than one-half in. (13 mm) beyond the toolholder and clamped tightly.

4. A finger's width between the holder's screw and the post.

5. It is an experimental cut to control the size of cut.

6. The toolbit should be set on center or slightly above and positioned with a lead angle so that the edge of toolbit contacts the workpiece first.

7. About one-thirty-second in. (0.8 mm) should be left for a finish cut.

8. If a finish cut is taken with the workpiece overheated, the diameter will change in size when the workpiece has cooled.

9. Shouldering involves the machining that begins as one diameter and suddenly increases in overall size.

10. It is generally added to increase strength to the part.

11. Angular shouldering is produced to eliminate weight, sharp edges, and also to improve appearances and reduce size.

Unit 39

1. It will not be possible to machine parallel turnings; the work will be tapered.

2. Visual alignment, trial cut method, and test bar and dial indicator method.

3. A parallel test bar is mounted between centers, allowing no end play. Once a dial test indicator has been tightened in the tool post, the indicator's contact point is positioned on the center of the test bar with a small amount of pressure against it. The indicator is then slowly moved along the test bar surface by turning the carriage handwheel. If the indicator's reading remains the same throughout the movement, the centers are aligned. If the indicator reading varies, the centers will have to be aligned.

Unit 40

1. Work is supported on both ends to permit heavier cuts to be taken, remounting a workpiece for machining once it has been removed becomes possible, and it becomes possible to complete a concentric machining operation by turning the shaft end for end.

2. The lathe should be disconnected from any electrical source to assure no accidental running while cleaning.

3. To maintain proper bearing contact between the hole and the center. Too shallow a hole will result in little or no support; too deep will result in a false bearing.

4. A small piece of brass or copper is placed under the hardened setscrew.

5. A live center is soft, requiring no lubrication as it rotates with the workpiece. The dead center is inserted in the tailstock spindle, is stationary, and requires a lubricant.

6. Hardened center, carbide-tipped center, half center, and the revolving center.

7. A dial test indicator is tightened in the tool post. Upon light contact with the contact point, the drive plate is turned by hand while observing the indicator needle for movement.

8. Soft centers are smoothed by filing, while hardened centers require an oilstone.

9. Excessive binding will cause friction during rotation. This heat will expand the workpiece between centers, preventing machining.

10. If using a dead center, the tension should be such that no end play exists between centers. When placed in a horizontal position and raised by hand, the lathe dog tail should drop from its own weight.

    If using a revolving center, there should be enough tension to rotate the center with the workpiece.

Unit 41

1. A gradual increase or decrease in diameter of a workpiece measured along its length.

2. Self-holding taper.

3. A steep or self-releasing taper that contains an included angle of more than 16°.

4. Self-releasing tapers used as lathe spindle noses for mounting a chuck or a drive plate.

5. The Jarno taper and tapered pins.

6. Length of taper, large diameter, small diameter, and degree of slope.

7. Form tool, compound rest, offsetting the tailstock, and the taper turning attachment.

8. $\text{tpf} = \dfrac{(D - d) \times 12}{\text{TL}} = \dfrac{1.5 - 1.375 \times 12}{3}$

$\text{tpf} = 0.5 \text{ in.}$

$\text{Tailstock offset} = \dfrac{\text{tpf} \times \text{OL}}{24}$

$= \dfrac{0.5 \times 3}{24} = 0.0625 \text{ in.}$

9. As a ratio of 1 mm per unit of length.

10. $D = 15 \text{ mm} + \dfrac{90}{30}$

$= 15 + 3$

$= 18 \text{ mm}$

11. (a) Center remains aligned, (b) any workpiece length can be machined, (c) work can be machined being held in a chuck or between centers, (d) external or internal tapers can be machined, (e) a wider range of tapers can be machined, (f) a power feed can be used, (g) it can be used with an offset tailstock.

12. (a) Standard micrometer, (b) a taper micrometer, (c) a sine bar, (d) a shadow graph, and (e) a ring gage.

13. The entire surface of the taper is covered with chalk or a light coating of Prussian blue. Once the ring gage has been slipped over the surface, it is rotated carefully about one-half turn. The gage is then removed. Rub marks will reveal the high spots.

Unit 42

1. Type of file, lathe speed, and safety.

2. Safety. There is a greater chance of damaging the hand against a spinning chuck or drive plate in a right-hand filing technique.

3. Aluminum oxide abrasive cloth for ferrous metal; silicon carbide abrasive cloth is recommended for nonferrous metal.

4. The lathe ways should be covered to protect them against falling abrasives.

5. About two times the speed for filing.

6. A machine operation that is used to change the finish and appearance for a workpiece and provide a better gripping surface.

7. The rollers of the knurling tool should be set against the workpiece so that the full width of the rollers are tracking, use plenty of oil, and the lathe made to turn at one-fourth the turning speed of the material.

8. Never disengage the automatic feed until the depth of the knurl has been completed.

9. With the use of a tool post grinder.

Unit 43

1. Three-jaw chuck, four-jaw chuck, spring collet chuck, and the Jacobs collet chuck.

2. Between 1200 and 1500 rpm.

3. The surface should be faced off or spotted by a center hole.

4. By use of a 60° shaped spotting tool and by use of a 60° countersink that contains one flat, single cutting edge.

Unit 44

1. They all require the cutting of a hole.

2. It is measured from the reference surface to the bottom of the hole at the full diameter.

3. Generally, a graduation on the tailstock spindle is used as a reference line to gage the depth of the hole being drilled.

4. An operation performed in a lathe in order to enlarge holes or to clean and true the surface of the hole itself.

5. (a) Boring bar should be as solid and heavy as possible; (b) it does not extend too far beyond the tool holder; (c) it enters the hole freely without obstruction.

6. To remove metal quickly to bring a hole close to a finish size.

7. About two-thirds the required drilling speed for the same material.

8. An oversize hole.

9. By supporting the end of a tap or tap holder with a lathe center in the tailstock spindle. As long as the workpiece hole has not been disturbed after machining, tapping can be accomplished with plenty of lubrication and the normal advancing and backing off the tap with the lathe center being used to keep the tap square in the hole.

Unit 45

1. A helical groove formed around the inside or outside of a cylinder or cone.

2. (a) The largest diameter to be found on an external or internal diameter, (b) the smallest diameter to be found on an external or internal diameter, (c) the top surface of the two sides of a thread, (d) an invisible diameter midway between the major and minor diameters.

3. Counterclockwise direction.

4. Both are measured in the same manner: the distance from a point on one thread to a corresponding point on the thread next to it.

5. Both thread forms have a 60° included angle. The ISO thread has a difference in thread depth that affects the root and crest. The Unified has uniform crest and root.

6. By the measurement of the pitch diameter.

7. Pitch $= \dfrac{1}{N} = \dfrac{1}{20} = 0.05$

$$\begin{aligned} \text{Depth} &= 0.64952 \times \text{P} \\ &= 0.64952 \times 0.05 \\ &= 0.0324 \end{aligned}$$

$$\begin{aligned} \text{Minor diameter} &= \text{major diameter} - (D + D) \\ &= 0.500 - (0.324 + 0.0324) \\ &= 0.500 - 0.0648 \\ &= 0.4352 \text{ in.} \end{aligned}$$

8. Pitch $= 2$ mm

$$\begin{aligned} \text{Depth} &= 0.54127 \times P \\ &= 0.54127 \times 2 \\ &= 1.082 \text{ mm} \end{aligned}$$

$$\begin{aligned} \text{Minor diameter} &= \text{major diameter} - (D + D) \\ &= 14 - (1.082 + 1.082) \\ &= 11.836 \text{ mm} \end{aligned}$$

9. A center gage bearing a 60° included angle.

10. For Unified, 0.125 × *P;* for metric, 0.250 × *P.*

11. It enables the operator to enter the threading tool in the same groove if engaged properly.

12. At 29° swung to the right.

13. To set the lathe for the required pitch of the thread to be cut.

14. Safety in machining. It allows the operator to withdraw the cutting tool without damaging the adjoining surface.

15. To avoid a sharp edge burr.

16. For 8, 10, 18, and 20, the split-nut lever should be engaged at any line on the dial face. For 11 and 13, engagement can be at any numbered line.

17. (a) Set lathe in thread cutting mode; (b) lock compound 29° to the right; (c) set threading toolbit even lathe center height; (d) place center gage against workpiece; (e) direct toolbit against V-notch of gage; and (f) tighten toolbit on lathe center height and against center gage.

18. Turned clockwise and lock 29° to the left.

19. Toolbit is placed square against axis of workpiece, not tapered section.

20. The pitch diameter.

Unit 46

1. They are designed to impart or transmit power.

2. Multiple-start threads are used where rapid advancement of travel along the threads is required, as in the construction of machinery.

3. (a) Using an accurate multislotted drive or face plate, (b) using the lathe's spindle gear for indexing, (c) using the thread chasing dial, (d) using the compound rest to index the threading tool.

4. The lead and diameter of the thread.

5. A 1° clearance is provided in addition to the slant angle.

6. Grinding a toolbit 0.002 in. (0.05 mm) under or using a roughing and finishing cutting tool.

7. With an Acme thread gage.

8. $$\text{Width of point} = \frac{0.3707}{N} - 0.0052$$
$$= \frac{0.3707}{6} - 0.0052$$
$$= 0.056$$

9. It is placed 14.5° to the right.

Unit 47

1. High-speed steel, cemented carbide, cast alloys, ceramics, and cermets.

2. Cubic boron nitride and diamond.

3. Rigidity of the workpiece, rigidity of the cutting tool, rigidity of the machine tool, and cutting fluids and coolants.

4. High hardness, wear resistance, shock resistance, and heat resistance.

5. Tools are made from pulverized tungsten carbide, or two or more carbides, fused together with a binder of cobalt to be molded into shape under high pressure.

6. Ground with a special grade of silicon carbide or diamond grinding wheel.

7. Brittleness.

8. (a) Bottom surface of the cutting tool shank, (b) surface upon which the cutting chip slides, (c) contoured tip of the cutting tool, and (d) main body portion of the toolbit.

9. (a) Slopes away from the top of the cutting edge to the back of the toolbit, designed to give a slicing action and (b) clearance angle between the cutting tool and the finished workpiece.

10. Recommended angle is 6° to 10°.

11. The angle that slopes away from the cutting edge to the opposite side that is designed to control the chip direction and free cutting.

12. Three.

13. Honing with an oilstone.

Unit 48

1. The plain knee and column and the universal knee and column horizontal milling machines.

2. (a) Foundation of the machine, (b) supports the vise and work, (c) provides the opening and drive for arbors and other attachments, (d) enables the knee and table to be raised or lowered vertically.

3. (a) The machine should be stopped before attempting any changing of gears or belts; (b) never remove any chips by hand, always use a brush.

4. The table handwheel moves the table in a side-to-side motion in front of the operator. The cross-feed handwheel moves the table toward or away from the operator.

5. (a) The keywayed shaft used with spacers for mounting and positioning milling cutters; (b) used for holding single- or double-end end mills; (c) main method of holding work in the milling machine, can be swiveled on any angle; (d) dual setting permits compound angles to be machined.

Unit 49

1. The workpiece surface is machined by the outer rotation of the cutting teeth of a wheel-like cutter.

2. In face milling, the cutting teeth are located on the face of the cutter, generally at right angles to the axis of the cutter.

3. (a) Space between the teeth of a multitooth cutter, (b) surface directly behind the cutting edge, (c) portion of cutting tooth behind the land.

4. (a) Convex or concave cutter, used for producing half circles or forms in workpieces; (b) corner rounding, used for radius milling; (c) angle cutter, used for angular as well as side cutting; (d) double angle cutter, used for cutting V-grooves; (e) hob, used for cutting gear teeth; (f) multiple-thread milling, available for internal or external thread cutting.

Unit 50

1. Speed, feed, depth of cut, cutter, and work holding method.

2. Milling machine vise, angle place, T-bolts, strap and step blocks, and fixturing.

3. The material being cut and the operation being performed.

4. The rate the workpiece is moved past the cutter in inches or millimeters per minute.

5. Formerly called conventional milling, the work is fed against the rotation of the cutter.

6. Formerly called climb milling, the work is fed with the rotation of the cutter.

7. Lost motion or play in the feed screw being turned.

8. Material, setup, cutter, machine condition, shape of workpiece, and cutter sharpness.

9. As much as the machine, setup, and cutter will accept.

10. At least 0.015 in. (0.38 mm) deep.

11. Feed per tooth or chip load.

12. Safety, which enables the operator to set the micrometer collar on the vertical feed to a zero reading.

Unit 51

1. It is widely accepted, rated just behind the lathe for versatility.

2. End milling, fly cutting, slotting, keyway cutting, T-slotting, face cutting, drilling, reaming, and jig boring.

3. The spindle head assembly on a standard vertical milling machine cannot be moved. The ram type is more flexible because of it.

4. (a) Cast with the base, the column contains dovetail ways, which permits vertical movement of the knee; (b) equipped with a ram, the turret can turn in any position on a horizontal plane; (c) housing for the spindle.

5. Generally, end mills come with a straight shank, are made of high-speed steel, and are equipped with two to eight cutting edges that may be straight or helical.

6. Two-fluted end mill.

7. Longer tool life. Once a cutting edge has become dull or worn, the insert is indexed to expose another cutting edge to restore the original cutting edge.

8. Advantage: can be used at speeds three to ten times faster than high-speed steel. Disadvantage: cutter is brittle and must be used within a solid setup.

9. Ball end mill.

10. To enlarge holes at locations that are round, accurate to size, and with a better finish.

Unit 52

1. Collets, special holders, adaptors, arbors, and quick-changing toolholding systems.

2. A collet and adaptor system.

3. A tapered spring collet.

4. (a) The machine should be shut off, (b) never grab a cutter bare handed, (c) cover the table surface to protect it against nicks or damage, (d) cover the workpiece to protect it against falling objects or cutters.

5. Clean the spindle opening and tapered collet. Insert and align the collet within the spindle. Thread the back of the collet by turning the draw bar. Hold the cutter with a cloth and insert it into the collet. Hand tighten the draw bar. Lock the spindle and securely tighten the draw bar with a wrench.

6. Generally, by reversing the steps taken in mounting the cutter.

Unit 53

1. A shell end mill and a flycutter.

2. To align the toolhead square or at right angles to the table.

3. Squaring the vise.

4. Unless the vise is set at 90° to the column, parallel cuts to the table travel are not possible.

5. A dial test indicator, a heavy long parallel, a suitable wrench, and a soft-faced hammer.

6. To indicate whether the workpiece is resting flat and firmly on the base. Any slippage of paper indicates a raised surface at that corner.

7. A round bar or rod is placed behind the uneven surface before tightening the vise.

8. 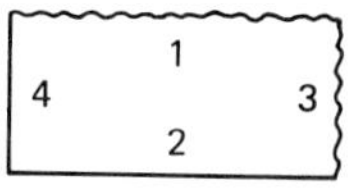

Unit 54

1. It enables an operator to establish the edge of a workpiece as a reference point.

2. The measuring diameter of the edge finder is made to rotate and is brought into a steady moving contact against the workpiece surface. At the moment the diameter suddenly darts or shifts, the center of the spindle will be exactly one-half the measured diameter from the work surface edge.

3. (a) The toolhead should be square with the table, (b) the vise or workpiece should be parallel with the table travel, (c) the spindle speed should be set between 600 and 700 rpm, (d) all back lash in the feed screw mechanisms should be removed, (e) the micrometer collar should be set to compensate for a zero reading when measuring contact has been made against the workpiece.

Unit 55

1. An electronic device that is used to monitor the table movements of machine tools.

2. There is no backlash to interfere with DROs, ensuring extreme reliability and accuracy.

3. A glass scale with calibrated etched gratings, a reading head, and a console with DRO display.

4. Lateral movements of light that can be interpreted as vertical movements of dark bands.

5. The DRO system counts the Moire fringes and converts them to a high-resolution accuracy of measurement.

6. Programmable and nonprogrammable.

7. (a) Less operator skill, (b) clear and easy-to-read display, (c) positioning is greatly speeded up, (d) eliminates gage blocks and measuring rods, (e) greatly reduces or eliminates scrap or rework.

Unit 56

1. Worm and worm wheel, an index plate, an index crank, and sector arms.

2. The footstock.

3. Rapid indexing.

4. Plain indexing.

5. Angular and differential.

6. 40

7. 24, 30, and 36.

8. $$\text{Indexing} = \frac{\text{Plate holes}}{\text{Hole divisions}} = \frac{24}{8} = 3 \text{ holes}$$

9. An indexing plate with holes that are evenly divisible by the denominator is used. A sector is formed by two arms placed in front of the plate and is positioned to span the distance equal to the fractional part.

10. Since a standard dividing head requires 40 turns to revolve 360°, one turn will rotate the spindle one-fortieth of this amount, or 9°. If finer divisions are required, each turn can be calculated in minutes.

11. A rotary table measures the division by a direct angular method right down to a fractional part of a degree.

12. As much as 120:1.

13. Aligning the rotary table with the center axis of the machine spindle.

Unit 57

1. Grinding is a process for removing and reshaping material by an abrasive wheel.

2. A grinding wheel is an abrasive cutting tool. It is a gradual cutting process rather than an abrupt cutting process represented by other machine tools.

3. It plays an important role in precision metal technology, making it possible to produce parts to extreme accuracy.

4. Abrasive grain and bond.

5. In fracturing, the wheel's bond should break down to tear away or eliminate the worn abrasive, exposing a new series of sharp particles.

6. The grains are eliminated too quickly during the grinding operation, indicating that the bond has a weak holding power.

7. Due to a strong bond, the wheel will exhibit a slow cutting action and will become loaded or glazed.

8. One involves the manufacture of natural abrasives; the other involves producing artificial abrasives such as cubic boron nitride and manufactured diamonds.

9. It imparts the wheel's characteristics, such as abrasive type, grit size, grain spacing, and bond type.

10. Bond is the "cement" that holds the abrasive grains together. Common bonds used are (a) vitrified, (b) resinoid, (c) rubber, (d) shellac, and (e) metal.

11. (a) Horizontal spindle grinder with reciprocating table, (b) horizontal spindle grinder with rotary table, (c) vertical spindle grinder with rotary table, (d) vertical spindle grinder with reciprocating table.

12. Tool and cutter grinder.

13. Through-feed, in-feed, and end-feed grinding.

14. (a) Acts as a coolant to reduce work temperature, increase surface finish; (b) reduces wheel wear; (c) lubricates wheel and work surfaces; (d) flushes away debris; (e) eliminates grinding dust.

15. Debris composed of work chips, particles of bond, and loosened abrasive grains.

Unit 58

1. (a) Use the right wheel for the job, (b) make sure the wheel is not cracked before mounting, (c) check the magnetic chuck for burrs or raised edges, (d) do not stand in line with the wheel's rotation when dressing it, (e) stop the machine spindle and move the workpiece clear of the wheel before attempting to remove it from the chuck.

2. (a) Limit the length of table travel during grinding, (b) supports the table upon which rests the magnetic chuck, (c) helps support the wheelhead and spindle housing.

3. Permanent magnetic chuck, electromagnetic chuck, and vacuum chuck.

4. The vacuum chuck.

5. They should be blocked off with parallels or flat pieces of steel.

6. Between 6300 and 6500 sf/min for vitrified wheels; around 9500 sf/min for resinoid wheels.

7. (a) Wheels should be tested for cracks, (b) wheel should fit snugly without altering, (c) recessed flanges should be one-third wheel diameter, (d) blotter should be larger than flange, (e) wheel should never be tightened too severely, (f) wheel guard cover should be at least one-half the grinding wheel, (g) the wheel's printed maximum speed should never be exceeded.

8. The removal of material from an uneven rotation to make the wheel run concentrically with the grinder spindle.

9. The restoration of a wheel's grinding face to expose a new sharp cutting surface.

10. Single-point or multipoint diamond dressers.

11. To prevent scratching the chuck in removal.

12. On the left end of the magnetic chuck, to prevent scratching and wear to the table caused by a stream of abrasive grains during dressing.

13. In front and to the left of the centerline of the grinding wheel by about $\frac{1}{4}$ in. (6 mm).

Unit 59

1. The work is generally fastened in a fixture or device that can be held directly on a magnetic chuck.

2. In alternating sections of brass and soft iron or soft iron and fiber.

3. To prevent nicking and burring.

4. By means of double-faced tape.

5. Angle plate, V-block, sine bar, and sine plate.

6. Once the magnetic chuck has been smoothed of burrs or raised edges, it is cleaned. A piece of paper slightly larger than the workpiece is placed on the chuck, the workpiece is positioned on it, and the chuck is energized.

7. With the chuck energized, move the surface grinder table to clear the work from the stationary grinding wheel. With both hands on it, the workpiece is attempted to be moved.

8. The wheel should clear each end of the workpiece by about 1 in. (25 mm).

9. Heavier grinding cuts will result in higher heat, which will result in loading or dulling of the grinding abrasives, leading to a shorter wheel life.

10. By a demagnetizer.

Unit 60

1. Once the workpiece has been cleaned and deburred, it is placed and tightened in a grinding vise with one work surface extending slightly beyond the edge of the vise. Placed on a piece of paper, the magnetic chuck is energized and the top surface is ground clean. The vise is turned over on its side, and the second surface is ground like the first one.

   Removed from the magnetic chuck, one of the ground sides is placed on the bottom of the vise, an unfinished side extended beyond the edge of the vise. Side 3 can now be ground flat and parallel to the opposite side. Once done, the vise can then be turned over on its side to grind the last side.

2. An angle plate with a side rail may be used, with the rail removed for grinding the first two sides.

   Upon cleaning and deburring, the workpiece is lightly clamped in one corner with one edge extended slightly beyond the side of the angle plate. The top surface is leveled, clamped tightly in place, and ground flat. Removed from the magnetic chuck, the C clamp is pivoted, the ground surface is indicated, and if found not to be disturbed, it is tightened in place. Turned over on its side, the second side is ground. With both sides ground, the side rail is mounted on the angle plate and used as a reference surface against which to clamp a finished side. Each of the unfinished sides is ground flat in this manner.

Unit 61

1. ECM: electromechanical machining, EDM: electrical discharge machining, ELG: electrolytic grinding, and USM: ultrasonic machining.

2. The EDM process uses an electrical current that sends small sparks through a nonconducting oil. The sparks are caused by a gap between a shaped electrode and the workpiece. The sparks are extremely hot and eat away or vaporize metal in a controlled manner.

3. The principal disadvantage of the EDM process is that the sparks erode the tool as well as the work.

4. No elaborate tooling is used, there is no tool wear or storage problem, parts are left without burrs or internal stresses.

5. Impact grinding.

6. A circulating abrasive fluid, called a slurry, is directed in the space between the tool and the workpiece. A power unit moves the tool up and down rapidly, ultrasonically, to cause an abrasive cutting action.

7. An advantage of USM is that it can machine extremely hard material such as carbide and hardened steel as easily as it can electrically conductive material. The principal disadvantage lies in high tool wear.

8. HERF is an acronym for high-energy rate forming.

9. Explosive forming and spark forming.

10. EHF, called electrohydraulic forming, is referred to as spark forming.

11. Air or nitrogen.

Unit 63

1. Changing the physical properties by means of heating and cooling to adapt the metal to environmental changes without changing its chemical composition.

2. Pure iron at room temperature.

3. Cementite.

4. Lower critical temperature point, also called decalescent point.

5. Rate of heating, hardening temperature, soaking time, and method of cooling.

6. To read the heating chamber atmosphere of a heat treating furnace.

7. A process of rapid cooling heated metal by immersion in various media to achieve hardness.

8. Oil, water, brine, and air.

9. Metal thickness and the temperature of the medium. Cooling too rapidly will produce internal stresses and lead to cracking. Cooling too slowly may result in uneven or incomplete hardening.

10. To show the effects of overheating and underheating tool steel.

11. Using a file on the hardened surface.

12. Drawing.

13. The steel will crack or fracture in use.

14. 300 to 1200°F (149 to 649°C).

15. (a) Nature of material, (b) carbon content, (c) hardness requirement, (d) toughness requirement.

Unit 64

1. The process of bringing hardened steel to a pliable state or softening it so it can be machined.

2. Steel is heated above the upper critical point and allowed to cool in the furnace or is buried in lime or sand.

3. To relieve stress for materials that have become work hardened during cold working in manufacture. This is generally done on low-carbon steels.

4. Both relieve internal stresses and improve machining. Normalizing differs in that there is no uniform hardness removal. Hard spots may be left behind.

5. Spherodizing is an annealing process that leaves metal with a special grain structure that is soft and easy to machine, bend, twist, or draw.

Unit 65

1. Adding a shell or case of hardness to the steel's outer surface while its inner core remains soft.

2. The low-carbon, low-alloy steels contain higher levels of nickel, chromium, and molybdenum. This assures little grain growth and can be heated to lower temperatures than plain low-carbon steels.

3. Carburizing, carbonitriding, and nitriding.

4. By the carburizing material, temperature, and time exposure.

5. Parts should be free of moisture. Liquified cyanide becomes explosive when in contact with moisture.

6. Carbon and nitride are introduced and absorbed by the steel's surface during carbonitriding to leave a hard, shallow case.

7. A case hardening process that is used to produce an exceptionally hard surface to parts that have already been machined and heat treated.

8. Gas and salt nitriding. Neither process uses quenching.

9. Flame hardening is used to heat treat long parts such as lathe ways and flat bearing surfaces of machine tools. Flame hardening is followed by spray quenching.

10. (a) Materials self-quench, (b) a hard wear surface with a ductile inner core can be produced, (c) surface stress is minimized, (d) varying depths of hardness can be produced, (e) can be done in any conventional shop, (f) a hard substructure can be produced without damaging a hard top surface.

Unit 66

1. Resisting permanent indentation or scratching.

2. A hard steel ball is driven under pressure into the surface of the workpiece. The indentation mark left is then measured with a microscope and read against a Brinell hardness chart. The deeper the penetration, the softer the metal, the lower the number.

3. A diamond cone (Brale) is driven into the surface of the workpiece using pressure from two loads, a major and a minor load. This measurement is then read on the outer "C" scale of a graduated dial.

4. A steel ball penetrator is driven into the surface using the major and minor loads. The measurement is then read on the inner "B" scale of a graduated dial.

5. A diamond hammer is dropped from a fixed height onto a hardened surface. The bounce of the hammer is then measured against a scale. The higher the bounce, the higher the reading, the harder the surface.

Unit 67

1. A force that resists change in a body.

2. Tensile stress, compression stress, and shear stress.

3. Metal is held and pulled in a straight line under load until it fractures.

4. Yield is a stretching that takes place as the metal is pulled apart during tensile testing.

5. The yield point is reached when metal continues to stretch without increasing the load.

6. A swinging weighted pendulum is made to strike a notched specimen from a fixed height.

The difference in the heights of the pendulum before and at the conclusion of the impact stroke is then registered. This will be the energy required to fracture the specimen.

7. The Izod and Charpy impact test.
8. Shear strength is the resistance of material to being cut (sheared) by two cutting edges or loads working in opposite directions.
9. The specimen is squeezed and shortened in the test.
10. Fatigue test.
11. Pulsating or cycle loads.

Unit 68

1. Oliver Evans.
2. Samuel Colt and Henry Ford.
3. The transistor, developed by William Shockley, John Bardeen, and Walter Brittain.
4. It consisted of a series of transistors that were connected and etched on a strip of silicon and was designed to eliminate hand wiring. Its purpose was to serve as a computer's memory record.
5. Monitoring, controlling, material handling, NC machining, designing, as well as collecting, processing, recording, and recalling data rapidly.
6. Inspection, assembly, painting, material handling, stocking, part transfer, and welding.

Unit 69

1. The paper punch.
2. The usefulness or intent of use for the piece involved.
3. Dies are mounted in die sets for use in power presses that may be powered mechanically or hydraulically.
4. Bending will change a blank into an angle. Forming changes the contour of the metal to conform to the shape of the punch.
5. Drawn shells, die castings, and forged parts.
6. Hammering, rolling, twisting, or vibrating.
7. By annealing.

Unit 70

1. Personnel is responsible for screening new job applicants. It is within its authority to enforce a company's policy in regards to hiring, firing, and retiring as well as informing applicants of their fringe benefits, which would include vacation, sick pay, medical, and life insurance.
2. Just-in-time (JIT) manufacturing represents a goal and a philosophy to achieve overall business excellence and a strategy to control productivity to meet demand.
3. CAD: computer-aided design. It is a system that uses the computer to speed up design calculations for the development of a product.
4. CAM: computer-aided manufacturing. It is a system designed to communicate instructions to automatic equipment to speed up the handling and processing of materials.
5. The direct numerical control (DNC) system enabled instructions to move one or a series of NC machines at the same time.
6. Computers and laser scanners.

Unit 71

1. To meet high manufacturing needs with accuracy and a minimum of handling.
2. Advances in design factors such as hydraulics, electromechanical devices, and computers.
3. There is greater rigidity, floor space requirement is reduced, loading and unloading becomes easier, and chips and coolants can collect at the bottom of the bed.
4. The turrets are seated with 16 tons of force and can operate together or individually. Seven different types of tooling can be used and are preset for programming purposes. They can machine internal or external surfaces, thread, groove, or undercut with no tool interference and minimum withdrawal for indexing.
5. The upper turret may be loaded with a center to support the end of a long shaft, while the lower turret completes the OD operation.
6. Self-centering chucks, collet chucks, and countercentrifugal chucks.
7. Turning centers are modular in construction and can stand alone or become integrated as compo-

nents of flexible manufacturing cells, whereas chucking centers are stand alone equipment.

8. An electronic trigger device contacts the workpiece during machining. This surface seeking probe is loaded into the tool storage holder and is automatically selected as the cutting tool.

9. The advantages include low cost, flexibility, and programmability. Disadvantages include false measurements due to a hostile environment and increased cycle time.

10. They can (a) reduce lathe positioning, (b) greatly eliminate manual machining skills, (c) improve accuracy and dependability, (d) reduce scrap, (e) reduce operator fatigue, and (f) make lathe more productive.

Unit 72

1. Hardness, abrasion resistance, compressive strength, and thermal conductivity.

2. Temperature ranges close to 700°C (1292°F).

3. Diamond wheels are primarily used to grind nonferrous metals, nonmetallic materials, and cemented carbides.

4. For grinding hardened ferrous metals, cast irons, nickel-base and cobalt-base high-temperature superalloys.

5. Sharpening high-speed steel cutting tools in the toolroom.

6. (a) Hardness: CBN crystals register around 4700 on the Knoop scale, almost twice as hard as the conventional abrasive. (b) Abrasion resistance: a longer wear resistance than either diamond or aluminum oxide. (c) Chemical stability: CBN crystals are chemically inert and do not react with ferrous metals. (d) High thermal conductivity: CBN crystals develop a thermal barrier and can withstand excessively high grinding temperatures.

7. Resin, metal, vitrified, and electroplated. Resin bond wheels are the most widely used CBN wheels.

8. It is (a) the most durable and longest lasting, (b) excellent for deep grinding hardened tool and die steels, (c) used for ID grinding and creep feed grinding, and (d) widely used in vertical spindle rotating grinding.

9. Layers of CBN particles are electroplated to the grinding surface of a metal core.

10. (a) Costs less than similar wheel size, (b) a new layer can be electroplated on, (c) can operate at high speeds and feeds, (d) wheel holds its exact machined profile.

11. (a) Tight wheel bearings, (b) tight machine slides, (c) fast but safe spindle speeds, (d) a feed system free of play or wear, (e) well working coolant system, (f) sufficient horsepower.

12. The electroplated wheel does not require truing and dressing as only one layer of abrasives covers the wheel. It does require running concentrically and should be trued with an indicator before using.

13. Should the wheel and adaptor ever be removed from the grinder's spindle, the unit can be replaced without taking the time to retrue the wheel.

14. Truing and dressing.

15. To true CBN resin and vitreous bond wheels not larger than 8 in. (200 mm).

16. On a sheet of paper toward the left end of the magnetic chuck so that the brake spindle is in line with the grinder spindle.

17. A silicon carbide braking device contacts and revolves at a slower speed than the CBN wheel being trued. The process involves a fast wear rate of the truing wheel.

18. A water-soluble wax stick, a wire brush dressing, and an aluminum oxide dressing stick or block.

19. A superabrasive tool blank consists of fusing together a layer of diamond or cubic boron nitride, about 0.020 in. (0.50 mm) thick, on a cemented tungsten base by means of a high-temperature/high-pressure process.

20. (a) Better work surface finish, (b) long cutting tool life, (c) control of part dimensional accuracy, (d) high productivity, (e) less machine downtime, (f) cost-effective machining.

21. Ferrous metals are carbon seekers. Due to the carbon nature of diamond, it answers this need and tends to wear out faster because of it.

Unit 73

1. Church chimes, music boxes, and automatic player pianos.
2. John Parsons.
3. A method of operating machine tools by means of coded information that can be received and acted upon.
4. The recirculating ballscrew and electric servo motors.
5. The Electronics Industry Association and the American Standard Code for Information Interchange have standardized the number language used by NC machines.
6. Punched tape, punched cards, magnetic tape, diskette, and direct input (DNC).
7. The binary number concept.
8. A hole/no-hole combination.
9. Once the part has been digested as to difficulty or ease of operation, a program is written outlining the machining steps of the part. The manuscript is converted to tape or computer program and then fed into the machine control unit. The NC machine is then put into motion and studied for correction.
10. As small as 0.00001 in. (0.00025 mm).
11. Feedback devices send back data to the machine control unit (MCU) to make corrections in machine measurements.
12. Pulse coders, linear scales, and tach generator.
13. A servo controller, a servo motor, and a feedback device.
14. Open-loop systems.
15. Close-loop system.

Unit 74

1. He developed the system of rectangular coordinates.
2. Origin or zero point.
3. The X axis represents the horizontal table feed (right or left movement).
4. The Y axis represents the table crossfeed motion (toward or away from the column).
5. The Z axis is perpendicular to the X and Y axes and represents the up and down movement of the spindle or knee.
6. 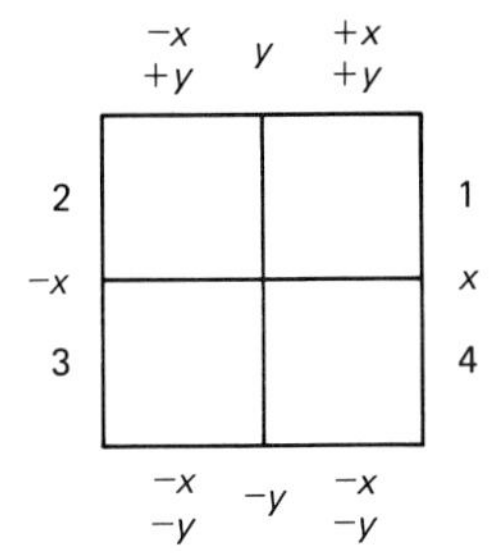

7. (a) A good background in machining procedures, (b) machinist skills, (c) an understanding of speeds and feeds, and (d) an understanding of NC tape format.
8. (a) Use X, Y, and Z coordinates to define all part surfaces, (b) use reference points on the part itself, (c) use reference planes parallel to machine axes, (d) describe part in NC language, (e) dimension part clearly, (f) work within allowable tolerance of design, and (g) remove any false assumption or calculations.
9. Each command in a block is written in a letter address system, which is listed in the NC machining manual. Each letter follows a publication format and is followed by a number.
10. Language is based on spelling, format, and the proper sequence of words.
11. Modal is a command that will not change once entered unless it has been changed or canceled. Nonmodal indicates that the command is only effective for one block of program data.
12. A spindle movement 0.500 in. below the zero point.
13. All coordinated points are located from a fixed zero reference point.
14. A dimension or movement is made to a new point using the last previous known point as a reference.

Unit 75

1. Continuous-path machining is another term for contouring.
2. Interpolation is a process of making a mathematical adjustment between two given values to establish a new value.

3. Linear, circular, parabolic, helical, and cubic.

4. 0.0001 in. (0.0025 mm).

5. The process would be very involved and the program would be too long, making the linear interpolator limited.

6. Contouring arcs and circles.

7. What are the *X* and *Y* coordinates for the radius of the circle? What are the start and end points of the arc? In which direction will the arc be cut: clockwise or counterclockwise?

8. A two-axes circular interpolator is combined with the linear movements of the third axis.

Unit 76

1. Stored information, the collection of ingredients composed of programs, documents, and routines essential for computer operations.

2. A single block of data used to control the arithmetical duties to allow preset tools to force the machine to produce the correct finished contour.

3. A command identifying a group of test statements that may be generated in the same program by a special character.

4. G or preparatory function codes prepare the CNC machine tool to treat the information supplied in a particular way.

5. Miscellaneous functions are designed to control machine tool operations.

6. Program stop, end of program, spindle on, spindle off, coolant on, coolant off, tool change (manual).

Unit 77

1. Dr. W. Edwards Deming.

2. (a) Product uniformity, (b) reduced cost, (c) rejection rate decrease, (d) increased productivity, (e) happier work environment, (f) defect decrease, (g) inspection costs reduced, (h) smoother work organization, (i) improved company credibility, and (j) reduced manufacturing costs.

3. Statistical quality control.

4. Statistical quality control is designed to interrupt flaws and defects in high-volume production. By taking random samplings and checking these against print specs, QC is able to project acceptance or rejection of the lot in question.

5. Statistical process control.

6. Quality control is based on deed. It is the control of a product after it is made. Statistical process control measures the process, not the part. It is involved in quality assurance.

7. Statistical quality assurance.

8. The control diagram is a mean average chart. Early information was recorded as penciled-in dots. Automated data gathering was later devised by using an electronic signal sent to the data gathering box.

9. A histogram is a chart that puts readings in a bar chart form as opposed to dots on a long chart.

10. (a) To eliminate scrap, (b) control the process, and (c) reflect the operator's performance.

11. Coordinate measuring machine.

12. Measurement takes place on a granite table that is below a movable gantry that accurately locates a multisensing probe head anywhere on the work without interference. This three-axes information is recorded and stored for use with a computer with a video display.

13. Vision machines use high-resolution cameras mounted on an inspection stage that enables them to digitize the video picture or signal on screen.

14. To $\pm$ 1 μm.

15. Edge finders and correlators, a system of template matching.

16. The video screen is divided into a grid pattern of 262,144 squares. Each square represents a mathematical value that is the function of the field of view, so that each square has a value of 1/262,144 of the area shown.

17. Data for quality control can only be obtained through measurement. The measurement has to be rapid and accurate to be effective.

INDEX